High Altitude

Erik R. Swenson • Peter Bärtsch
Editors

High Altitude

Human Adaptation to Hypoxia

Springer

Editors
Erik R. Swenson
VA Puget Sound Health Care System
University of Washington
Seattle, WA, USA

Peter Bärtsch
Medical University Clinic
Heidelberg, Germany

ISBN 978-1-4614-8771-5 ISBN 978-1-4614-8772-2 (eBook)
DOI 10.1007/978-1-4614-8772-2
Springer New York Heidelberg Dordrecht London

Library of Congress Control Number: 2013953322

Printed on acid-free paper

Springer is part of Springer Science+Business Media (www.springer.com)

Preface

Mountains stir our imagination and not solely for their beauty and magnificence. Perhaps also because they tax our physical limits as individuals and as a population. As scientists and nonscientists we find this an irresistible subject. The rarified air of the highest terrestrial altitudes poses one of the greatest biological challenges to functioning and survival that our species has faced in its radiation "out of Africa" over the globe. In moving into highlands above 2,500 m whether transiently or permanently, humans have relied upon an array of successful responses to hypoxia that range across the physiological continuum from organs to cells and genome. Unraveling what transpires with successful and unsuccessful adaptation to acute and chronic high altitude is inextricably linked to better understanding and care of clinical conditions involving ischemia, hypoxemia, and cellular hypoxia.

In this effort, which owes its lineage and draws great inspiration from its predecessor "High Altitude: An Exploration of Human Adaptation" edited by Thomas F. Hornbein and Robert B. Schoene and published in 2001, we have reassembled many of the same authors and brought on a number of new active younger researchers to update the field. Many of our contributors are mountaineers, who have taken the "bench to mountainside" and back approach in impressive transnational collaborations. Over the past decade there has been substantial work in better understanding the classic "oxygen cascade" of O_2 transport from inhaled air to the mitochondria. Equally exciting has been the extent to which genetics and molecular biology have now been brought to bear on hypoxia adaptation and maladaptation.

The 23 chapters comprising this book can be divided into several sections. The first chapters focus on how oxygen is sensed at the cellular level and how changes in O_2 tension evoke signaling cascades that mediate rapid nongenomic responses and set the stage for more sustained tolerance by altering gene expression. Particularly in this latter aspect, there has been a virtual explosion in our understanding of how hypoxia-inducible factors (HIFs) are regulated and how up- or down-regulation of many hundreds of genes accomplish tolerance for hypoxia. It is a very complicated story that only promises to become more interesting and relevant to high-altitude physiology and clinical medicine, since HIFs may be more than just gene transcription factors responsive to hypoxia. The next section takes up the role of many organ systems involved in oxygen delivery to the tissues, and the costs and limits of their compensatory capacity to contribute to adequate tissue oxygenation.

Of no less critical importance are other organs whose functioning in overall homeostasis and population survival (nutrition, infection control, waste elimination, and reproduction) succeed or fail in the face of hypoxia. The unique and differing patterns of hypoxic tolerance in highly adapted populations living at high altitude for many millennia are explored at the genetic and physiological levels. Evolution over many generations involves strategies entirely different (and likely more successful and economical) when compared to how lowlanders "partially adapt" over much shorter time durations. The last chapters are devoted to the unique acute diseases of sojourners at high altitude, chronic hypoxic maladaptation of high-altitude residents, and problems in people with preexisting medical conditions who wish to work or take enjoyment in the mountains.

We are grateful to many colleagues and friends over the years who have contributed to this volume. It is our hope that our combined work will stimulate more questions and research, ultimately to yield better care of hypoxic patients and safer sojourning and residence to those seeking the pleasure and excitement of life and livelihood at high altitudes.

Seattle, WA, USA — Erik R. Swenson
Heidelberg, Germany — Peter Bärtsch

Contents

Contributors

Philip N. Ainslie, Ph.D. Centre for Heart, Lung and Vascular Health, School of Health and Exercise Sciences, University of British Columbia, Kelowna, BC, Canada

Aaron L. Baggish, M.D. Division of Cardiology, Massachusetts General Hospital, Boston, MA, USA

Damian Miles Bailey, Ph.D. Neurovascular Research Laboratory, Faculty of Life Sciences and Education, University of South Wales, Mid-Glamorgan, UK

Peter Bärtsch, M.D. Division of Sports Medicine, Department of Internal Medicine, Medical University Clinic, University of Heidelberg, Heidelberg, Germany

Cynthia M. Beall, Ph.D. Department of Anthropology, Case Western Reserve University, Cleveland, OH, USA

Remco R. Berendsen, M.D. Department of Anesthesiology, Leiden University Medical Center, Leiden, The Netherlands

Konrad E. Bloch, M.D. Pulmonary Division, Sleep Disorders Center, University Hospital of Zürich, Zürich, Switzerland

George A. Brooks, Ph.D. Department of Integrative Biology, University of California, Berkeley, CA, USA

Victoria E. Claydon, Ph.D. Department of Biomedical Physiology and Kinesiology, Faculty of Science, Simon Fraser University, Burnaby, BC, Canada

Mark J. Drinkhill, Ph.D. Division of Cardiovascular and Neuronal Remodelling, Faculty of Medicine, University of Leeds, Leeds, UK

Joachim Fandrey, M.D. Institut für Physiologie, Universität Duisburg-Essen, Essen, Germany

Stilla Frede, Ph.D. Institut für Physiologie, Universität Duisburg-Essen, Essen, Germany

Peter H. Hackett, M.D. Clinical Professor of Surgery, Division of Emergency Medicine, Department of Surgery, Altitude Research Center, University of Colorado Denver School of Medicine, Denver, CO, USA

Institute for Altitude Medicine, Telluride, CO, USA

Roger Hainsworth, Ph.D. Division of Cardiovascular and Neuronal Remodelling, Faculty of Medicine, University of Leeds, Leeds, UK

Noor Hamad, B.M. B.Ch. (Oxon) M.A. (Cantab) Perranporth Surgery, Perranporth, Cornwall, UK

Susan R. Hopkins, M.D., Ph.D. Division of Physiology 0623A, Departments of Medicine and Radiology, University of California, San Diego, La Jolla, CA, USA

Hans Hoppeler, M.D. Department of Anatomy, University of Bern, Bern, Switzerland

Luis Huicho, D.Sc. Departamento Académico de Pediatría, Universidad Peruana Cayetano Heredia, Lima, Peru

Departamento Académico de Pediatría, Universidad Nacional Mayor de San Marcos, Lima, Peru

Departamento Académico de Pediatría, Instituto de Salud del Niño, Lima, Peru

Christopher H.E. Imray, Ph.D., F.R.C.S., F.R.C.P., M.B.B.S. Department of Vascular Surgery, Warwick Medical School, University Hospitals Coventry and Warwickshire NHS Trust, Warwick, UK

Benjamin D. Levine, M.D. University of Texas Southwestern Medical Center, Dallas, TX, USA

Institute for Exercise and Environmental Medicine, Texas Health Presbyterian Hospital Dallas, Dallas, TX, USA

Andrew M. Luks, M.D. Division of Pulmonary and Critical Care Medicine, Harborview Medical Center, University of Washington, Seattle, WA, USA

Carsten Lundby, Ph.D. Center for Integrative Human Physiology (ZIHP), University of Zürich, Zürich, Switzerland

Marco Maggiorini, M.D. Medical Intensive Care Unit, University Hospital, University of Zurich, Zurich, Switzerland

Robert S. Mazzeo, Ph.D. Department of Integrative Physiology, University of Colorado, Boulder, CO, USA

James S. Milledge, MBChB., M.D., F.R.C.P. Center for Altitude, Space and Extreme Environment Medicine UCL, University College London, London, UK

Matthias Mueller, Ph.D. Department of Anatomy, University of Bern, Bern, Switzerland

Susan Niermeyer, M.D., Ph.D. Pediatrics, Section of Neonatology, University of Colorado School of Medicine and Children's Hospital Colorado, Aurora, CO, USA

Yvonne Nussbaumer-Ochsner, M.D. Pulmonary Division, Sleep Disorders Center, University Hospital of Zürich, Zürich, Switzerland

Niels V. Olsen, M.D., D.M.Sc. Copenhagen University Hospital, Copenhagen, Denmark

Jean-Paul Richalet, M.D., Ph.D. University of Paris 13, Bobigny, France

María Rivera-Ch, D.Sc. Facultad de Ciencias y Filosofía, Departamento Académico de Ciencias Biológicas y Fisiológicas, Universidad Peruana Cayetano Heredia, Lima, Peru

Robert B. Schoene Clinical Professor of Medicine, Division of Pulmonary and Critical Care Medicine, Department of Medicine, University of Washington, Seattle, WA, USA

Bay Area Pulmonary/Critical Care Medical Associates, Berkeley, CA, USA

Paul T. Schumacker, Ph.D. Division of Neonatology, Department of Pediatrics, Feinberg School of Medicine, Northwestern University, Chicago, IL, USA

Fabiola León-Velarde, D.Sc. Facultad de Ciencias y Filosofía, Departamento Académico de Ciencias Biológicas y Fisiológicas, Universidad Peruana Cayetano Heredia, Lima, Peru

Erik R. Swenson, M.D. VA Puget Sound Health Care System, University of Washington, Seattle, WA, USA

Luc J. Teppema, Ph.D. Department of Anesthesiology, Leiden University Medical Center, Leiden, The Netherlands

Simon Travis, D.Phil. F.R.C.P. Translational Gastroenterology Unit, John Radcliffe Hospital, Oxford, UK

Francisco C. Villafuerte, D.Phil. Facultad de Ciencias y Filosofía, Departamento Académico de Ciencias Biológicas y Fisiológicas, Universidad Peruana Cayetano Heredia, Lima, Peru

Michael Vogt, Ph.D. Department of Anatomy, University of Bern, Bern, Switzerland

Mark H. Wilson, B.Sc., MB, BChir, F.R.C.S., F.I.M.C., F.R.G.S., M.R.C.A. Department of Neurosurgery, Imperial Hospitals NHS Trust, London, UK

Eugene E. Wolfel, M.D. University of Colorado Health Sciences, Denver, CO, USA

University of Texas Southwestern Medical School Center, Dallas, TX, USA

Cellular and Molecular Mechanisms of O_2 Sensing

1

Paul T. Schumacker

Abstract

Molecular oxygen is required for mitochondrial energy generation, and all eukaryotic organisms have the capacity to sense the abundance of cellular O_2. When oxygen levels decrease, these sensors activate a wide range of transcriptional, translational, and posttranslational responses that protect the cell and the organism from the consequences of severe oxygen deprivation and O_2-limited metabolism. A number of different mechanisms of oxygen sensing have been proposed, and it is likely that diverse cell types employ multiple means of sensing changes in molecular oxygen. However, growing evidence points to the mitochondria as an important site of O_2 sensing in diverse cell types. These organelles release reactive oxygen species (ROS) signals to the cytosol where they trigger transcriptional and posttranslational adaptive responses. Although the precise molecular mechanisms underlying the increase in ROS generation under low O_2 conditions are not fully understood, these signals appear to originate from complex III of the electron transport chain. The responses regulated by this mechanism include the stabilization of Hypoxia-Inducible Factor-1α and the acute vasoconstrictor response to alveolar hypoxia in the lung. Future studies of prolonged hypoxia are still needed to determine whether the responses to chronic hypoxia are also regulated by this pathway.

The Evolution of Oxygen Sensing Systems

A wide range of metazoan phyla evolved rapidly during the Cambrian Explosion period. The successful evolution of unicellular organisms into complex multicellular animals required the development of sophisticated tissue systems for delivering adequate oxygen to supply the needs of each and every cell. Moreover, these systems

P.T. Schumacker, Ph.D. (✉)
Division of Neonatology, Department of Pediatrics, Feinberg School of Medicine, Northwestern University, 310 E. Superior Street Morton Building 4-685, Chicago, IL, USA
e-mail: p-schumacker@northwestern.edu

E.R. Swenson and P. Bärtsch (eds.), *High Altitude: Human Adaptation to Hypoxia*,
DOI 10.1007/978-1-4614-8772-2_1, © Springer Science+Business Media New York 2014

needed to respond rapidly to changes in the metabolic needs of the organism, in order to meet the changing demands associated with feeding, reproduction, escape from predators, and other energy-intensive activities. Throughout evolution, the oxygen transport systems also adapted by acquiring the dynamic ability to regulate regional oxygen supply in accordance with local tissue oxygen demand. Today, mammalian cardiovascular and pulmonary systems have the ability to augment systemic oxygen transport above resting levels by 10- to 20-fold during exercise and to direct the majority of the associated increase in blood flow to working muscle and to tissues involved in heat dissipation such as skin. Similarly, organs such as the digestive system efficiently regulate oxygen supply in accordance with gastrointestinal activity. At the tissue level, the precise regulation of oxygen supply in accordance with oxygen demand requires a feedback signal, allowing the tissue being supplied to signal its oxygenation status to the system controlling the oxygen delivery. Although multiple factors participate in the regulation of local tissue blood flow, the tissue response to a fall in oxygenation below the normal level (i.e., tissue hypoxia) requires an ability to detect cellular oxygenation levels in the microenvironment and to transduce that information into a biological signal that is conveyed to the oxygen transport system. Clearly, without the ability to sense local and systemic tissue oxygenation, the evolution of complex oxygen transport systems in higher organisms would not have been possible.

Systemic and Specialized Mammalian Oxygen Sensing Systems

Systemic oxygen sensing systems at the organismal level complement the local tissue oxygen detection/signaling systems that regulate local blood supply and microvascular blood flow distribution. At the organismal level, increases in metabolic activity and blood flow during exercise must be matched with increases in lung ventilation in order to maintain blood oxygenation and excrete carbon dioxide. The ability to regulate lung ventilation in accordance with overall metabolic activity required the evolution of a complex arterial chemotransduction system capable of monitoring blood oxygenation and carbon dioxide levels. The carotid and aortic bodies are the sites where arterial blood oxygen levels are sensed; decreases in arterial oxygen tension (PO_2) elicit increases in afferent nerve firing by triggering neurotransmitter release from the chemosensitive glomus cells. These nerves convey a feedback signal to the ventilatory control systems in the medulla, which elicit a corresponding increase in activity of the muscles of respiration. Thus, the oxygen-sensitive cells of the arterial chemotransduction system play an essential role in regulating lung ventilation, thereby assuring adequate arterial oxygenation. That system has been the focus of extensive study for more than 50 years, but the underlying oxygen sensing mechanism still is not known.

Multiple specialized oxygen sensing systems also evolved in response to the unique stresses that accompany reproduction. For example, the oxygen requirements of the mammalian embryo are supported by placental gas exchange. During development in utero, the pulmonary vascular resistance is maintained at a high level in the lung, possibly to prevent the escape of fetal oxygen from pulmonary capillary blood into the hypoxic amniotic fluid. This response, termed "hypoxic pulmonary vasoconstriction," is rapidly reversed when lung oxygenation increases during the first breaths taken at the moment of birth. The rise in lung oxygenation and the corresponding increase in blood oxygen levels are detected by the vascular cells, where it triggers relaxation in the pulmonary artery and constriction in the ductus arteriosus. Simultaneously, the ductus arteriosus dilates under hypoxic conditions, facilitating the extrapulmonary shunting of blood from the pulmonary artery to the aorta and minimizing pulmonary capillary blood flow. Without the ability to adjust pulmonary vascular tone in accordance with the level of alveolar oxygenation, fetal development would be hindered and successful transition from placental to lung gas exchange at birth would be threatened. Indeed, the clinical syndrome of "persistent pulmonary hypertension of the newborn" is a life-threatening human condition that arises when the newborn lung circulation fails to vasodilate normally at birth. Interestingly, the ability to regulate pulmonary arterial smooth muscle tone in response to changes in oxygenation is retained throughout adulthood, and lung diseases that produce large areas of alveolar hypoxia can cause pulmonary arterial hypertension and may lead to right ventricular remodeling as a consequence. The oxygen sensing response associated with hypoxic pulmonary vasoconstriction is intrinsic to the smooth muscle cells of small precapillary arteries, as myocytes isolated from these vessels exhibit increases in cytosolic ionized calcium [Ca^{2+}] and they contract when exposed to a low oxygen environment. However, the oxygen sensing mechanism that controls this response is not fully understood.

The oxygen sensing systems described above are critical for assuring proper regulation of systemic and local tissue oxygen supply. However, the normal operation of those systems cannot always guarantee adequate cellular oxygenation in response to stressful stimuli, in response to growth, or in response to chronic illness. For example, intense aerobic exercise training creates a need for remodeling of the muscle microvascular system in terms of capillary volume density. The increase in capillary growth that occurs in exercise-trained muscles reflects a local response of the myocytes and other cells to exercise-induced hypoxia. Oxygen sensors in the muscle cells detect the low oxygen levels and activate the expression and secretion of vascular mitogens by the muscle. These vascular growth factors diffuse toward local blood vessels and trigger new capillary growth by binding to cognate receptors on vascular endothelial cells. Vascular Endothelial Growth Factor (VEGF) is one such mitogen [1], whose low basal secretion by parenchymal cells helps to sustain the existing capillary network, and whose increased expression in response to hypoxia stimulates new capillary growth that augments the ability of the tissue to extract oxygen from a limited supply (capillary angiogenesis). The importance of this system for maintaining capillary density has been demonstrated in diverse animal models. One elegant demonstration utilized transgenic mice carrying an inducible gene encoding a VEGF-binding protein in cardiac myocytes. Administration of the antibiotic doxycycline in that system activates the expression of the binding peptide, which is secreted into the extracellular space [2]. The secretion of that protein leads to a scavenging of secreted VEGF, preventing it from reaching nearby endothelial cells. The resulting depletion of local tissue VEGF levels deprives the endothelium of an essential survival factor, resulting in a significant decrease in capillary density. The resulting tissue hypoxia then induces hypoxia-mediated hibernation and a deficit in cardiac function due to the limited supply of O_2 to the ventricles. In other models it has been shown that genetic deletion of even a single allele of the VEGF gene during embryonic development evokes lethal consequences [3]. These examples illustrate that individual parenchymal cells are sensitive to the "normal" levels of tissue oxygenation and are capable of responding to a decrease in PO_2 by activating transcription of genes that help to augment tissue oxygen supply.

Characteristics of an Oxygen Sensor

Despite the existence of elaborate systemic and local systems that serve to regulate and sustain cellular oxygen supply, each cell responds to the onset of hypoxia by activating a host of protective responses, in anticipation that the oxygen supply could become critically limiting. These responses are initiated by even mild levels of hypoxia, and they become progressively more activated as the degree of hypoxia increases. The functional responses to hypoxia fall into two general categories. Posttranslational responses involve a suppression of metabolic energy-consuming activities, such as protein synthesis, combined with an upregulation of glycolysis through the translocation of glucose transport proteins to the plasma membrane. Transcriptional responses involve an increase in de novo expression of glycolytic enzymes, glucose transporters, and other genes that enhance the ability of the cell to maintain ATP production in the absence of oxygen. Both the posttranslational and the transcriptional responses act to suppress mitochondrial respiration, which lowers the demand for oxygen and thereby lessens the degree of hypoxia in the tissue microenvironment. It is important to note that the activation of gene expression requires the activation of transcription factors, the generation of mRNA from the target genes, the translation of that message into a peptide at cytosolic ribosomes, and the folding, posttranslational modification, and delivery of the new proteins to their sites of action. The time required for the completion of this process can be significant, so it is important for the cell to initiate these responses at the first sign of hypoxia. Clearly, a response that is not activated until the cell is already experiencing lethal hypoxia will not be useful in preventing that situation from developing in the first place. Many of the posttranslational responses to hypoxia are mediated by the activation of the Adenosine Monophosphate Kinase system (AMPK) [4], while the majority of the transcriptional responses are mediated by activation of the transcription factor Hypoxia-Inducible Factor (HIF) [5]. Cellular oxygen sensors are involved in the activation of both of these systems.

Definition of an Oxygen Sensor

By definition, an oxygen sensor is a molecular system capable of interacting with O_2 in a reversible manner. The binding affinity of this system for O_2 must correspond with the biological range of oxygen tensions that it seeks to measure. This is essential, as an oxygen sensor that becomes fully saturated with O_2 at a certain oxygen tension clearly would be incapable of detecting further increases in oxygenation. Another defining characteristic of an O_2 sensor is that it must be capable of transducing changes in its O_2 binding, allowing it to signal that information to other systems. Without that ability, it would be impossible to elicit a response to a change in oxygenation. It is also important to note that the terms "O_2 sensor" and "O_2-sensitive" are frequently used interchangeably, but they are not equivalent. Responses in a cell may be O_2-sensitive, in the sense that they change as the oxygenation state of the cell changes. But the fact that the activity of an enzyme or ion channel may change in response to hypoxia does not define it as an O_2 sensor, as it may instead represent a downstream effector that has been activated by an upstream sensor.

Biophysical Properties of an O_2 Sensor

To understand how oxygen sensing functions in a cell, it is useful to define hypoxia in terms of the relationship between cellular bioenergetic activity and extracellular oxygenation. To begin, consider the relationship between extracellular oxygen tension and mitochondrial respiration (Fig. 1.1). Such a relationship can be obtained experimentally using a Warburg respirometer and an aliquot of dissociated cells (Fig. 1.2). A suspension of cells in media is transferred to the respirometer, which consists of a sealed chamber containing an O_2 detector (typically a polarographic Clark oxygen electrode). The liquid completely fills the chamber and displaces air, so the cells must rely on dissolved O_2 for their metabolic needs. As the cells metabolize O_2, the oxygen tension in the vessel decreases steadily over time. The media is stirred continuously to keep the cells in suspension and to prevent the development of O_2 gradients. Initially, the dissolved O_2 tension is equilibrated with ambient air, corresponding to a dissolved O_2 tension of ~150 Torr. This level of oxygenation is more than enough to satisfy the mitochondrial demand for O_2, which is regulated by cellular metabolic activity and ATP consumption by the cells. As they consume oxygen, the PO_2 in the solution decreases while the rate of oxygen consumption (calculated from the rate of decrease in PO_2) remains constant. This process continues until the oxygen tension in the solution falls to a critically low level (5–7 Torr). At that point the rate of oxygen consumption slows because oxygen availability limits mitochondrial electron transport. Further decreases in PO_2 cause even greater decreases in oxygen consumption, as the supply of O_2 limits the activity of the mitochondrial electron transport chain.

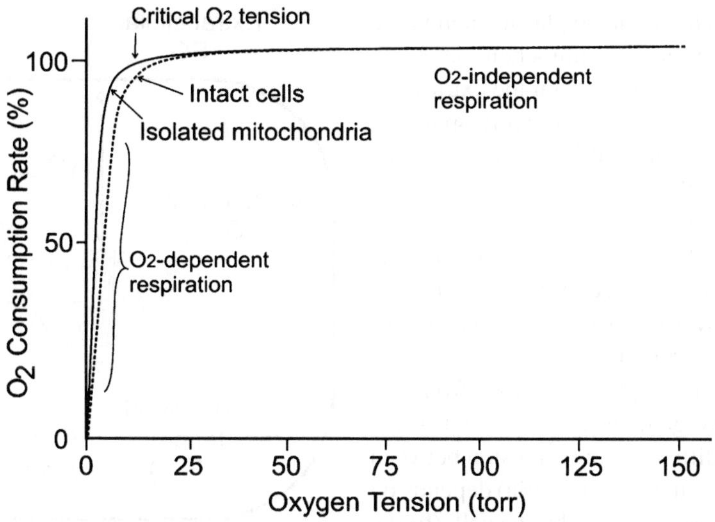

Fig. 1.1 Relationship between mitochondrial respiration and oxygen tension. For intact cells, respiration remains independent of extracellular PO_2 until a critical O_2 tension is reached, in the range of 5–7 Torr. Below that level, respiration becomes limited by oxygen availability at the mitochondria. Respiration by isolated mitochondria does not become O_2 limited until the oxygen tension reaches ~1 Torr, because the diffusive transport resistance encountered in the cytosol is absent

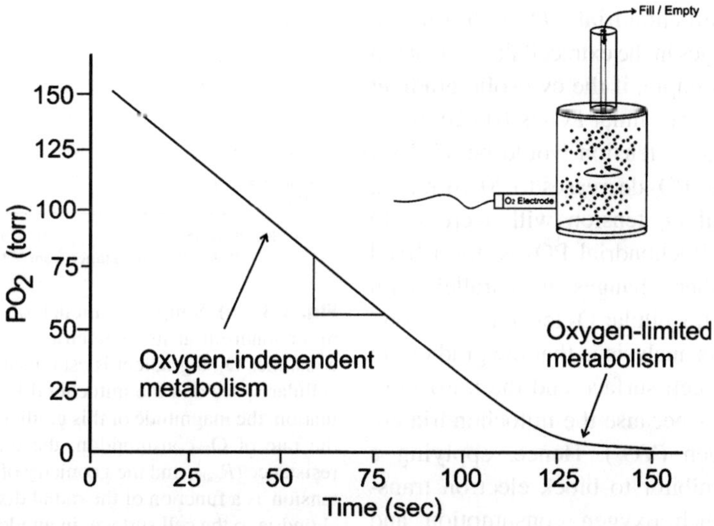

Fig. 1.2 The Warburg respirometer. Cells or mitochondria in solution are loaded into the reaction chamber via a long capillary tube. A magnetic stir bar keeps the cells in suspension, while the buffer PO_2 is assessed with an oxygen electrode. Above the critical O_2 tension, cellular respiration depends on metabolic activity, which determines mitochondrial respiration rate. This is manifested by a linear decrease in O_2 tension over time. Below the critical O_2 tension, the rate of oxygen consumption slows because O_2 availability limits respiration. Data from such an experiment is used to construct the relationship shown in Fig. 1.1

The decrease in cellular O_2 uptake below the critical threshold occurs because the diffusive O_2 transport resistance of the cytosol creates an O_2 gradient between the plasma membrane and the mitochondria. Mitochondrial cytochrome oxidase has a high apparent affinity for oxygen, allowing the organelles to function normally as long as the mitochondrial PO_2 exceeds 1–2 Torr. Because an

O_2 gradient exists between the plasma membrane and the mitochondria, the mitochondria reach their critical PO_2 (1–2 Torr) when the extracellular PO_2 reaches a level of 5–7 Torr (corresponding to an intracellular gradient of 3–5 Torr) (Fig. 1.1). The existence of this gradient can be demonstrated by repeating the above experiment using isolated mitochondria, which lack this cytosolic O_2 gradient. Under those conditions, the rate of O_2 consumption by isolated mitochondria remains constant down to a PO_2 of 1–2 Torr.

Several important implications arise from this analysis. First, the gradient in O_2 tension in the cytosol (i.e., the difference in O_2 tension between the cell surface and the mitochondria) depends on (a) the rate of O_2 consumption by the cell, (b) the diffusive transport resistance of the cytosol, and (c) the geometric distribution of mitochondria throughout the cytoplasm. For a given rate of oxygen consumption, if we assume that a single giant mitochondrion sits at the center of a spherical cell, then mitochondrial PO_2 will vary in parallel with changes in the extracellular O_2 tension (Fig. 1.3). For example, if the cytosolic gradient is 5 Torr, then if extracellular PO_2 is 100 Torr then the mitochondrial O_2 tension would be 95 Torr. If the extracellular PO_2 decreases to 20 Torr, then the mitochondrial O_2 tension will decrease to 15 Torr. Thus, mitochondrial PO_2 is not a fixed number, but rather changes in parallel with changes in the extracellular O_2 tension. A second implication of this analysis is that the gradient in PO_2 between the cell surface and the mitochondria (ΔPO_2) arises because the mitochondria are consuming oxygen (VO_2). Hence, applying a mitochondrial inhibitor to block electron transport would abolish oxygen consumption and cause the mitochondrial PO_2 to increase to the extracellular level. In the above example, this would result in a 5 Torr increase in mitochondrial PO_2. This point has important implications that will become clear further on.

How steeply does the PO_2 decrease as we move into a cell? The answer to this question is difficult to estimate analytically in a real cell, because (a) the mitochondria are distributed in a network, (b) the diffusion coefficient for O_2 is affected by the distribution of aqueous and lipid

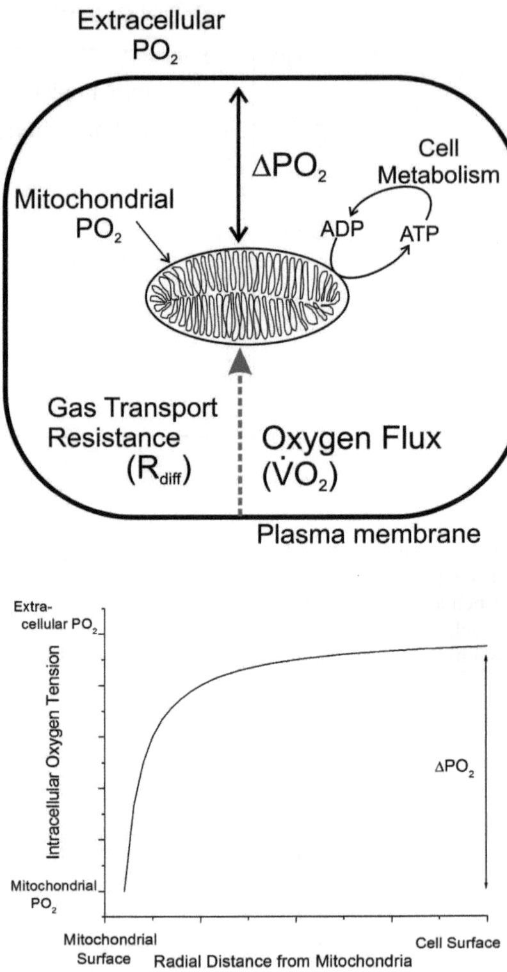

Fig. 1.3 (**a**) Simplistic model of a cell containing a mitochondrion at its geometric center. As the organelle consumes O_2, a gradient is established between the extracellular space and the mitochondrion. As a first approximation, the magnitude of this gradient (ΔPO_2) depends on the rate of O_2 consumption, the diffusive gas transport resistance (R_{diff}), and the geometry of the cell. (**b**) Oxygen tension as a function of the radial distance from the mitochondria to the cell surface, in an idealized spherical cell. Studies suggest that a normal ΔPO_2 is approximately 2–5 Torr; hence, mitochondrial oxygen tension is normally 2–5 Torr less than extracellular PO_2

components comprising the cell structures, (c) the geometry of most cells is far from spherical, and (d) the respiratory rates of individual mitochondria can vary depending on local differences in ATP demand throughout the cell. Nevertheless, a first approximation of the radial oxygen tension gradient can be achieved if we assume that the cell is spherical, that a single mitochondrion at its geometric center is responsible for all of the oxygen consumption, and that the cell interior has a uniformly distributed gas transport conductivity. Under those assumptions, the PO_2 decreases inversely between the cell surface and the mitochondria (Fig. 1.3).

In terms of defining an "ideal" cellular location for a molecular oxygen sensor, the above analysis suggests that the location is relatively unimportant. Clearly, the lowest oxygen tension will occur at the sites where oxygen is consumed, and the highest concentration will occur at the plasma membrane. However, except in cases where the cellular oxygen consumption rate is unusually high, the difference between the maximal and minimal O_2 tensions only amounts to a few torr. Empirical studies of intracellular O_2 gradients have yielded conflicting results regarding the magnitude of these gradients. In working muscle preparations, reflectance cryospectroscopy has been used [6] to estimate myoglobin saturation in rapidly frozen muscle samples [7–11]. Based on the relationship between myoglobin saturation and oxygen tension, the spatial distribution of intracellular PO_2 could be inferred from the oxymyoglobin dissociation curve. Those studies indicated that intracellular O_2 gradients in myoglobin-containing tissues were very small, on the order of 1 Torr [12]. As expected, this gradient was found to decrease when O_2 uptake was inhibited with cyanide in isolated cardiomyocytes [13, 14]. In hepatocytes, other studies have estimated the gradient based on a comparison of the Km for O_2-dependent oxidases and the observed O_2-dependence of respiration by intact cells [15]. These studies tend to confirm that the difference in PO_2 between the cell surface and the central core of the cell generally amounts to only a few torr. It is therefore reasonable to conclude that the intracellular O_2 tension is only a few torr less than the extracellular level.

Proposed Models of O_2 Sensing

A number of possible oxygen sensing systems have been proposed by different investigators over the past 4 decades. Some of these arose from investigators interested in the arterial chemotransduction system and the carotid body response to hypoxia, while others arose from the group of scientists interested in the mechanisms responsible for activating the expression of erythropoietin during hypoxia. Some of these models have subsequently been disproven, a few have evolved in response to new discoveries, and others have not been pursued. Each model meets the definition of an oxygen sensor by describing a putative interaction between a sensor and molecular oxygen with an affinity that corresponds to a physiological concentration of O_2. In some cases the models incorporate a signaling mechanism that would function to transmit information regarding oxygen abundance (or lack) to downstream targets.

O_2 Sensing Heme Proteins

More than 45 years ago it was recognized that cobalt salt administration to rats induced a severe polycythemia, which was triggered by excessive secretion of the hormone erythropoietin [16]. Treatment with cobalt or other transition metals was later shown to stimulate erythropoietin secretion from hepatoma cells in culture [17] through a mechanism involving transcriptional activation of the erythropoietin gene [18, 19]. This led Goldberg and colleagues to hypothesize that the O_2 sensor might be a heme protein that would undergo a structural change in response to the binding and release of molecular oxygen [18]. It was thought that cobalt ions could replace the iron in newly synthesized heme groups, thereby preventing coordination with O_2, locking the sensor into a "deoxy" configuration and activating the hypoxic response. However, subsequent studies showed that cobalt-induced activation of erythropoietin expression by hepatoma cells is not prevented by inhibitors of heme synthesis [20]. Those findings discount the idea that cobalt incorporation into heme is responsible for hypoxic activation of the sensor, but they do not rule out the possibility that the interaction of O_2 with a sensor involves a heme protein. While a protein–heme system could be still be involved, no such system has been identified to date.

Therefore, while the idea remains sound, the veracity of this model has not been established.

NAD(P)H (NOX) Oxidases

NAD(P)H oxidases are multisubunit membrane-associated complexes that remove electrons from the reduced dinucleotides nicotinamide adenine dinucleotide (NADH) or nicotinamide adenine dinucleotide phosphate (NADPH) and transfer them to molecular O_2, yielding superoxide anion. The first description of NADPH oxidase focused on the system in phagocytic cells involved in host defense [21, 22]. The phagocytic form of the enzyme is comprised of multiple components, including membrane-associated cytochrome b_{558}, the gp91^{phox} (NOX2), p22^{phox}, the cytosolic p47^{phox} and p67^{phox} proteins, and a number of associated proteins including rac-1 or rac-2, p40, and rap1A. The release of oxidants in phagocytic cells occurs across the plasma membrane, resulting in ROS secretion into an intracellular vesicle after phagocytosis. Other types of NAD(P)H-linked oxidase systems have been described in non-phagocytic cells, each of which contains a redox-active NOX subunit and at least one other subunit [23, 24]. These systems have been suggested to function as O_2 sensors in a variety of cell types, including blood vessels [23].

The original implication of NAD(P)H oxidase as an oxygen sensor was based on the expectation that the O_2 dependence of its activity should lead to a decrease in superoxide generation as the oxygen tension decreases. The decrease in oxidant generation would then create a "reductive stress" in the cell, which could affect the redox status of various target proteins mediating the cellular response. Based on immunohistochemistry, expression of various subunits of the NAD(P)H oxidase system has been shown to occur in a number of O_2-sensitive cells including type I cells of the carotid body [25], vascular cells [26], endothelium [27], and other cell types. However, demonstration of the involvement of these oxidases in the O_2 sensing pathway requires a genetic deletion, which should abolish the hypoxic response.

In that regard, Archer et al. examined the hypoxic pulmonary vasoconstrictor response in mice carrying a genetic deletion of the gp91^{phox} subunit of the NAD(P)H oxidase system [28]. Although the pulmonary response to hypoxia in the mouse is modest, the response was not abolished in the absence of gp91^{phox}. Those findings indicate that the oxygen sensing response in the lung does not require that protein. Subsequently, Weissmann and colleagues tested the hypoxic response in lungs from mice with a genetic deletion of the p47phox subunit and found that the acute response was significantly attenuated but not abolished [29]. Collectively, these results suggest that NAD(P)H oxidase systems may participate in the signaling pathways activated by hypoxia but may not be essential for triggering the response. Notably, the loss of p46^{phox} should have attenuated ROS generation by the oxidase, yet this attenuated the hypoxic response. This suggests that increases in ROS signaling contribute to the hypoxic response, which contradicts the intuitive expectation that ROS production should decrease when O_2 levels decrease.

O_2-Sensitive Ion Channels

An O_2-sensitive ion channel represents an attractive model to explain cellular oxygen-dependent responses. Such a channel could function as an O_2 sensor by interacting with molecular O_2, which would modify its conductance and thereby regulate ion transport and/or membrane potential. The glomus (type I) cells of the carotid body are electrically excitable cells that secrete neurotransmitters in response to hypoxia. The Na^+, Ca^{2+}, K^+, and other ion channels expressed in these cells have been the focus of extensive study in the search for an O_2 sensing mechanism. Current understanding of glomus cells suggests that membrane delayed rectifier potassium channels undergo a decrease in open channel probability without undergoing a change in single channel conductance [30, 31]. In rats, the hypoxia-sensitive channels appear to be comprised of maxi-K channels [32, 33], whereas in mice the voltage-dependent Kv 3.x channels fulfill this role [34]. In either case, inhibition of these K^+ channels induces membrane depolarization, leading to the opening of voltage-dependent calcium channels, Ca^{2+} entry, and neurotransmitter release [35].

Today there is little doubt about the existence of O_2-sensitive ion channels. A wide range of O_2-regulated ion channels have been described by a number of laboratories and in a number of different cell types. However, a central unanswered question is whether the membrane ion channels are directly responsive to O_2 or whether they are regulated by signals originating in a nearby O_2 sensor. One intriguing study discovered that Heme Oxygenase-2 (HO-2) co-precipitates with heterologously expressed maxi-K^+ channels. HO-2 utilizes O_2 as it degrades heme into biliverdin, iron, and carbon monoxide (CO); it is expressed in carotid body glomus cells; and siRNA knock-down of HO-2 abolishes the O_2 sensitivity of expressed channels. Interestingly, maxi-K^+ channels in membrane patches from glomus cells were activated by substrates that enhance HO-2 activity [36]. These findings raised the possibility that CO released from the O_2-dependent activity of HO-2 and activating nearby maxi-K^+ channels could mediate the O_2 sensitivity of the carotid body. Such a system would fully meet the definition of an O_2 sensor. However, subsequent studies of carotid body function in the HO-2 knockout mouse failed to detect any difference in sensitivity to graded levels of O_2 compared with wild-type controls [37]. These findings appear to rule out the possibility that HO-2 is functioning as a transducer of oxygen levels in the carotid body glomus cells.

Hypoxia-Inducible Factors and the Role of Prolyl Hydroxylases as O_2 Sensors

The activation or suppression of gene expression in response to hypoxia requires an oxygen sensor. The principal transcription factors involved in the hypoxic response are the Hypoxia-Inducible Factors, principally HIF-1 and HIF-2. The search for these proteins began in the early 1990s, among investigators interested in understanding how the erythropoietin gene was regulated by oxygen. It had long been known that oxygen deficiency triggers increased expression of erythropoietin by the kidney, and that certain transformed cell lines, notably Hep3B and Hep3G hepatoma cells, also respond to hypoxia by activating the expression and secretion of erythropoietin [17]. Analysis of DNA regions flanking the erythropoietin gene had shown that certain regions were required for the increase in expression during hypoxia. This observation suggested the existence of an O_2-regulated transcription factor, which led to a search for proteins capable of binding to oxygen-sensitive regions near the erythropoietin gene. The subsequent cloning of Hypoxia-Inducible Factor-1 (HIF-1) by Gregg Semenza's laboratory at Johns Hopkins University [38–40] represents an important milestone in our understanding of the mechanisms regulating the expression of genes under hypoxia. Although the initial interest in HIF was centered its role in regulating erythropoietin expression in specific cell types, it was later discovered that the 3′ enhancer of the EPO gene could trigger O_2-dependent gene activation in a wide range of other cell types genes [41]. That discovery led to the realization that O_2 sensing system regulating the EPO gene is broadly operative in many cell types. Subsequent studies have made enormous progress in understanding which genes are regulated by HIF, and how HIF itself is regulated in response to changes in the oxygen level [42].

Activation of HIF-1 triggers expression of glycolytic enzymes, membrane glucose transporters, vascular growth factors, proteins contributing to the regulation of vascular tone, cell death and survival genes, and other transcription factors that indirectly regulate expression of other genes. Hundreds of genes have been shown to be regulated by HIF, including a handful of other transcription factors. However, not all of these genes are expressed in every cell type. HIF-1 is essential for embryonic development, and genetic deletion of HIF-1α induces embryonic lethality [43–45]. HIF-1, and a related distinct gene HIF-2, contribute to oxygen homeostasis after birth and throughout adulthood by contributing to processes such as cell apoptosis, wound healing, erythropoiesis, and regulation of vascular tone. Increasing evidence reveals that HIFs also contribute importantly to the regulation of tumor angiogenesis and tumor progression. Hence, HIF plays important roles throughout life, both in health and in disease states.

As a transcription factor, HIF functions as a heterodimer comprised of α and β subunits [46]. Under normoxic conditions its activity is relatively low, but as the oxygen level decreases below 5 % it becomes progressively more activated [47]. After the discovery of HIF, investigators focused their attention toward understanding how its activity was regulated by oxygen. One early observation was that both subunits are transcribed and translated constitutively in cells, but only the α subunit is regulated by cellular oxygenation. Interestingly, the abundance of the α subunit was found to vary reciprocally with the [O_2], such that under "normoxic" conditions the α subunit levels are minimal whereas under "hypoxic" conditions the concentration of HIF-α increases. It was subsequently found that the α subunit is rapidly degraded by the proteasome system, and this O_2-dependent degradation depends on a region of the protein known as the oxygen-dependent degradation domain (ODD) [48]. Moreover, expressing a peptide that contains the ODD from HIF-1α confers oxygen-dependent stability to the protein. An important subsequent discovery was that the degradation of HIF-1α in a normoxic cell requires the activity of von Hippel–Lindau (vHL) protein [49]. The vHL comprises an E3 ubiquitin ligase, which targets the protein for proteasome degradation by ligating ubiquitin residues to lysine. Degradation of the α subunit begins with its hydroxylation at two conserved proline residues located in a region of the protein that has been shown to be required for O_2-dependent regulation. Three different 2-oxoglutarate-dependent enzymes, termed prolyl hydroxylases, carry out this modification. Prolyl hydroxylases require Fe^{2+}, 2-oxoglutarate, and molecular oxygen, so hydroxylation of HIF is prevented by complete anoxia, iron chelators, and competitive inhibitors that obstruct the binding of 2-oxoglutarate to the enzyme. Hydroxylation of the proline residues facilitates interaction of HIF-1α with vHL protein, an E3 ubiquitin ligase that labels the protein for destruction in the proteasome. Although HIF-1α is constitutively expressed, most of the transcribed protein is rapidly degraded by the cell. If cells are incubated under hypoxic conditions, the levels of the α subunit increase markedly within 1 h. However, if normoxic conditions are restored, the protein is rapidly degraded within a few minutes. Of the three prolyl hydroxylases involved in HIF regulation, PHD2 is the principal enzyme involved in the oxygen-dependent regulation of the α subunit [50]. Not surprisingly, genetic deletion of PHD2 causes embryonic lethality in mice due to uncontrolled activation of HIF-dependent transcription [51]. Peter Ratcliffe's group at Oxford and William Kaelin's group at the Dana Farber Cancer Institute in Boston simultaneously discovered that hydroxylation of these proline residues enables recognition of the protein by the vHL protein [52, 53]. The vHL is an E3 ubiquitin ligase that interacts with HIF after proline hydroxylation has occurred. The Ratcliffe lab had previously shown that vHL ligates ubiquitin to HIF, which labels the protein for immediate proteolytic degradation [49, 54]. This process is extremely rapid, such that newly transcribed HIF-1α has a half-life of less than 1–2 min in the normoxic cell. Von Hippel–Lindau syndrome is an autosomal dominant disease arising from mutations in the vHL gene. The loss of functional vHL is associated with the formation of highly vascular tumors such as renal clear-cell carcinomas that exhibit constitutive activation of HIF and associated excessive production of vascular growth factors.

In terms of oxygen sensing, the critical question then focuses on how prolyl hydroxylases are regulated by molecular oxygen. One simple and attractive idea is that prolyl hydroxylase itself is the oxygen sensor. As the enzyme uses O_2 as a substrate, hypoxia could limit PHD activity and attenuate hydroxylation of HIF-α, thereby suppressing the first step in its degradation. For this to work, PHD2 would need to demonstrate an affinity for O_2 that is compatible with the observed relationship between cellular [O_2] and

HIF-1α stabilization [55]. In vitro studies using recombinant PHD protein suggest that PHD has a relatively low affinity for oxygen, which is compatible with this idea. However, a recombinant protein from a bacterial expression system lacks the posttranslational modifications that may regulate its function within an intact cell, so further tests of this hypothesis will require an experiment where the O_2 dependence of PHD2 can be monitored in the cell. Thus, the specificity of PHD in terms of its ability to transduce the local oxygen concentration into the stability of HIF-α subunits is not yet clear. However, it is clear that this system provides a robust ability to detect anoxic conditions, as the enzymes split molecular oxygen in the hydroxylation process and cannot function in the absence of that substrate.

HIF-dependent gene expression is also modified by the *HIF asparaginyl hydroxylase*, known as FIH. This 2-oxoglutarate-dependent enzyme hydroxylates HIF-1 and HIF-2 at a conserved asparagine residue in the carboxy terminal regions of the proteins [56, 57]. Hydroxylation under normoxic conditions leads to an inhibition of the interaction between HIF and the p300 transcriptional co-activator. Under hypoxic conditions FIH activity declines, allowing the accumulating HIF-α protein to initiate gene transcriptional activation. This mechanism therefore provides a second line of defense against abnormal HIF activity – even if HIF-α is stabilized, its activity is restricted. This mechanism may be more important in some cell types than in others. In any case, its inhibition during hypoxia may be mediated by ROS signals, which are generated during hypoxia and can inhibit FIH [58].

Nitric Oxide Participation in the O_2 Sensing Pathway Regulating HIF

An interesting model to explain the regulation of hypoxic HIF stabilization by nitric oxide (NO) was proposed by Moncada and colleagues [59]. That model is based on the empirical observation that low concentrations of NO tend to inhibit HIF-1α stabilization in hypoxic cells. It was already known that NO can inhibit mitochondrial respiration by competing with O_2 at the binuclear center of cytochrome oxidase (mitochondrial complex IV), thereby displacing O_2 and causing oxygen consumption to fall [60]. According to the Moncada et al. model, low concentrations of NO prevent the stabilization of HIF-1α by restricting mitochondrial respiration. As described earlier, inhibition of mitochondrial respiration abolishes the gradient in [O_2] between the plasma membrane and the mitochondria. Thus, when NO limits mitochondrial function the cell experiences an increase in the "average [O_2]" in the cytosol. This would augment the oxygen available to PHD, whose function was already limited by the hypoxic conditions. The augmented PHD function would then accelerate HIF-1α hydroxylation and thus degradation.

To be clear, this model is founded on the idea that PHD functions as the O_2 sensor regulating HIF-α stabilization. The veracity of the model appears to hinge on the magnitude of the O_2 gradient between the plasma membrane and the mitochondria, in relationship to the affinity of PHD for oxygen. For example, if the "average cytosolic PO_2" in a cell is only a few torr less than the O_2 tension at the plasma membrane, then abolishing mitochondrial respiration will produce, at most, an increase of only a few torr in the cytosol. For the model to work, this small increase would have to be sufficient to restore PHD activity and HIF-1α degradation. Thus, while intriguing, some conceptual questions related to this model still need to be addressed.

The biological role of NO signaling in the response to hypoxia is complex. NO synthase requires molecular oxygen as a substrate, so the simplistic expectation is that NO production by NO synthases should decrease during hypoxia. However, studies in yeast led to the discovery that mitochondrial cytochrome oxidase can continue to function under anoxic conditions, indicating that an electron acceptor alternative to O_2 can be utilized [61]. This alternative is nitrite, which binds to the terminal oxidase in the mitochondrial electron transport chain and is reduced by the enzyme, generating NO. This process becomes active at low oxygen concentrations (below 20 μM) and also occurs in mammalian hepatocytes.

Low pH conditions also favor this process, raising the possibility that it may be important in regulating cellular responses to tissue ischemia, where very low O_2 levels co-exist with low pH conditions. This interesting discovery raises the possibility that cytochrome oxidase, by generating NO signals at low oxygen conditions, could function as an oxygen sensor in eukaryotic cells [62]. Further work is needed to establish the biological significance of this interesting response and to determine whether it operates at moderate levels of hypoxia.

Mitochondrial Oxygen Sensing

As the principal site of oxygen utilization in the cell, the mitochondria have long been considered to be attractive as a potential site of O_2 sensing. Cytochrome *c* oxidase, also referred to as cytochrome aa_3 because it carries the *a* and a_3 heme centers, comprises the terminal complex of the mitochondrial electron transport chain. This complex consists of approximately 13 subunits in mammalian cells [63] although tissue-specific expression of certain subunits may occur [64, 65]. The electrons used to reduce O_2 to H_2O are obtained from the oxidation of metabolic intermediates in glycolysis and in the tricarboxylic acid cycle. Sequential transfer of electrons to cytochrome c oxidase from cytochrome c drives the proton translocation process at that complex [66]. Changes in the availability affecting O_2 binding at the oxidase could therefore potentially affect the redox status of electron transport complexes at more proximal sites in the electron transport chain. Redox changes in those electron carriers might then activate a signaling pathway, thereby allowing the mitochondria to "sense" O_2 and to telegraph that information to the rest of the cell.

However, that model is confounded by the observation that mitochondrial respiration remains independent of the local $[O_2]$ until the oxygen tension at the oxidase falls to near-anoxic levels (1–3 Torr, Fig. 1.1). A low apparent Km of cytochrome oxidase for O_2 would therefore suggest that mitochondrial redox should remain unaffected by changes in $[O_2]$, as long as the oxygen levels remain above the critically low levels where electron transport becomes limited. If so, then the mitochondria would function well as sensors of anoxia, but would be incapable of detecting moderate levels of hypoxia in the cell.

One possible explanation of this paradox was suggested by Mills and Jobsis [67], who hypothesized that O_2-sensitive tissues such as the carotid body may contain a second "isoform" of cytochrome oxidase with significantly lower affinity for O_2. A low affinity cytochrome could potentially serve as an O_2 sensor because its redox state would change dramatically over the physiological range of oxygen tensions. In support of that idea, one group studying perfused carotid body preparations noted that the fall in tissue PO_2 in response to a cessation of blood flow became slower as tissue oxygen tension reached lower levels [68, 69]. The rate of O_2 disappearance (dPO_2/dt) was analyzed using a mathematical model that took into account the influence of O_2 release from hemoglobin trapped in the tissues. That analysis yielded results consistent with a two-cytochrome model in the carotid body [70]. However, the results were indirect and could conceivably be explained by alternative mechanisms [71]. For example, it is possible that the lower rate of O_2 consumption at lower tissue PO_2 could be caused by a suppression of metabolic demand for ATP, or a decrease in non-mitochondrial oxidase activity, rather than a limitation of electron transport by cytochrome oxidase. Hence, although the idea that cytochrome oxidase isoforms in O_2-sensitive tissues may allow the oxidase to function as an O_2 sensor is conceptually appealing, direct evidence in support of this hypothesis has not been reported.

Cytochrome oxidase can bind O_2 with high affinity at the a_3 site when the heme is in the ferrous state, and it can bind HCN, HN_3, or H_2S when heme a_3 is in the ferric state [72]. When reduced, heme a_3 can also bind CO, albeit with much lower affinity than O_2. Nevertheless, at high partial pressures, CO will form complexes with heme a_3 in cytochrome oxidase, thereby blocking electron transfer to O_2 and causing the proximal electron carriers in the chain to become fully reduced. Administration of CO had previously been found to mimic the effects of hypoxia [73, 74]. This led Lahiri et al. to hypothesize that the interaction of CO with cytochrome oxidase in the carotid body might explain its effects on chemotransduction [74, 75]. One difficulty in understanding this model is the issue of which heme(s) the CO may be interacting with to cause the change in chemosensory discharge. A second as-yet unresolved issue relates to the identity of the secondary signal transduction signals that link the O_2 sensor with the increase in type I cell Ca^{2+} elevation. The interpretation is further clouded in the intact carotid body, where carotid sinus nerve activity may be influenced by cell-specific differences in cytochrome reduction, CO binding, and sensitivity to metabolic inhibitors. At present, this remains an intriguing concept but a number of critical questions remain unanswered.

The Mitochondrial ROS Hypothesis

A growing body of evidence implicates mitochondria-derived ROS signals in triggering both transcriptional and posttranslational responses to hypoxia in a wide range of cells. The principal source of these hypoxia-induced oxidant signals appears to be complex III. In terms of oxygen sensing, this implicates the electron transport chain as an O_2 sensor, with oxidant signals released to the intermembrane space and the cytosol functioning as second messengers. The process of ROS generation by mitochondria, which have long been known to generate ROS [76, 77], is described next.

ROS generation by mitochondria can occur when a single electron is transferred from the electron transport chain to molecular oxygen, generating a superoxide radical. Hydrogen peroxide can subsequently be generated by the dismutation of two superoxide molecules by superoxide dismutase. Until recently, these oxidants were considered to be toxic by-products of the electron transport chain that contributed to cell injury in ischemia–reperfusion injury [78–83] and other disorders [84, 85]. However, more recent work suggests that some mitochondrial oxidants may function as signaling messengers in response to a variety of biochemical events [86] including the oxygen sensing transduction pathway [87, 88]. A potentially important mechanism by which mitochondria could signal changes in O_2 tension would be to alter the generation of ROS. Mitochondrial oxidative phosphorylation is regulated at multiple sites [89], and the redox state of the electron transport system is responsive to a wide range of metabolic signals. Changes in mitochondrial redox status could therefore provide a mechanism for altering the generation of ROS. The redox signal resulting from the change in ROS production would then be subject to further modification by antioxidant systems that exist in various compartments.

In terms of the site of production, mitochondrial superoxide production can occur at complex I [90–93], II [94], or III [95–101], principally through the escape of electrons from reduced iron–sulfur groups, flavin-containing proteins, or from the free radical ubisemiquinone in the Q cycle of complex III. Depending on the site of superoxide generation, the oxidants may be released to the matrix compartment or to the intermembrane space. Superoxide dismutase is expressed in each of these compartments leading to H_2O_2 generation. Most studies assessing the sites of mitochondrial ROS production have used pharmacological inhibitors to restrict electron flux to specific regions of the electron transport chain, while simultaneously measuring oxidant production. Depending on where electrons are fed into the chain, these inhibitors can either increase or decrease superoxide production from specific sites. For example, rotenone, an inhibitor that acts at the downstream end of complex I, increases superoxide production from that complex when NADH is used as a substrate [90], whereas ROS production from complex I is decreased by rotenone if electrons are supplied using succinate at complex II, as reverse electron flux into complex I is prevented [91].

In the electron transport chain, complexes I and II each deliver pairs of electrons to the mobile electron carrier ubiquinone, yielding ubiquinol. The ubiquinol then migrates to the Qo site in complex III, where the two electrons are removed sequentially. The first electron is passed to the Rieske iron–sulfur protein (RISP) in complex III, which subsequently passes the electron to cytochrome c_1, to cytochrome c, and finally to cytochrome oxidase. Removal of the first electron from ubiquinol by RISP generates the transient free radical, ubisemiquinone, at the Qo site. Normally, the second electron is rapidly removed by the b cytochromes, thereby returning ubiquinone to the membrane pool. However, if the removal of the unpaired electron from ubisemiquinone is delayed, superoxide may be generated if the electron is instead captured by molecular O_2. Thus, the lifetime of the ubisemiquinone at complex III represents a potential mechanism for controlling ROS generation at the Qo site. Interventions that prolong the lifetime of ubisemiquinone lead to a marked increase in superoxide formation at the Qo site. The toxin antimycin A, for example, prevents removal of the second electron and causes a large increase in superoxide generation [95].

Hypoxia-Induced ROS Signaling in the Mitochondrial Intermembrane Space

A growing body of data indicates that hypoxia increases ROS generation at the Qo site of complex III, possibly by increasing the lifetime of the ubisemiquinone at that site. As this pocket is near the outer surface of the inner mitochondrial membrane, superoxide generated there should enter the intermembrane space and could subsequently migrate to the cytosol. Genetic studies implicating complex III were reported by Guzy et al., who assessed ROS production during hypoxia. To assess ROS signals they used a novel FRET-based redox sensor, HSP-FRET, to measure thiol oxidation in the cytosol [102]. Interestingly, they detected increases in oxidation of the sensor during acute hypoxia, which were attenuated when expression of the RISP was suppressed using short hairpin RNA interference. Consistent with that work, Mansfield and colleagues studied ROS production in embryonic cells from mice with a genetic deletion of cytochrome c [103]. In the absence of cytochrome c, the RISP in complex III remains fully reduced, which blocks Q cycle operation and thereby prevents the generation of superoxide at the Qo site of complex III. Both the RISP knockdown and the cytochrome c-deficient cells failed to generate an oxidant signal in the cytosol during hypoxia, and they also failed to stabilize HIF-1α. However, the same cells retained the ability to stabilize HIF-α in response to prolyl hydroxylase inhibition by DMOG, indicating that the impaired response was specific to the detection of hypoxia. These findings implicate electron flux through complex III as a critical event in the detection of hypoxia in cells.

A fuller test of the mitochondrial ROS hypothesis required the ability to measure oxidant signals in subcellular compartments. This goal has been hindered by the lack of tools for assessing subcellular redox status. Recent studies by Waypa et al. began to address this by targeting redox-sensitive fluorescent sensors to various subcellular compartments [104]. These sensors (roGFPs) are mutants of green fluorescent protein (GFP) that contain two cysteine residues in the outer surface of the chromophore [105]. Oxidation of a cysteine thiol by H_2O_2 or superoxide leads to dithiol formation, which alters the fluorescence properties of the protein [106]. This behavior allows roGFP to function as a thiol redox sensor. In practice, the protein is expressed in cells and is then interrogated by excitation with light at two different wavelengths (typically 405 and 488 nm) during an experiment. The emission at 525 nm (when excited at 405 and 488 nm) provides a ratiometric measure of the redox status that is independent of protein expression levels, excitation intensity, or other factors. Moreover, oxidation or reduction of the protein is reversible, allowing the sensor to be calibrated at the end of the experiment by applying chemical reducing and oxidizing agents. To accomplish this, the emission ratios are measured during the experiment, and at the end of the study the cells are bathed in dithiothreitol (to reduce cellular thiols) followed by aldrethiol (to oxidize all cellular thiols). The ratios obtained in states where the protein is

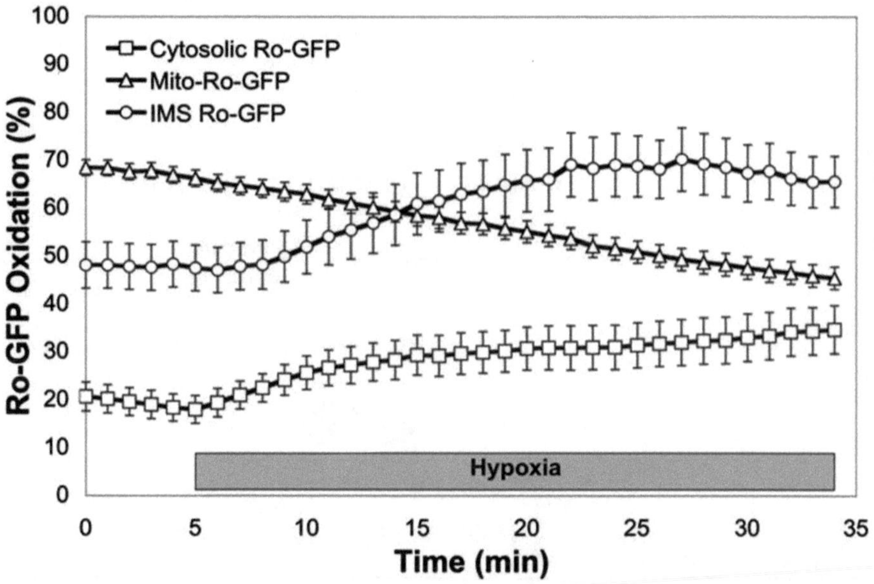

Fig. 1.4 Oxidant signaling response to hypoxia in cytosol, mitochondrial intermembrane space, and matrix compartment, in pulmonary artery smooth muscle cells. The redox-sensitive protein roGFP was targeted to each compartment, and cells were studied under controlled O_2/CO_2 conditions on a fluorescence imaging microscope. Differences in baseline levels of protein thiol redox state exist in these compartments, as does the direction and magnitude of change during 1.5 % O_2 challenge

fully reduced or oxidized are then used to calculate the redox status corresponding to the ratios obtained during the study.

Targeting expression of the protein to specific subcellular compartments is accomplished using genetic engineering. For example, roGFP expression can be directed to the mitochondrial matrix by appending the mitochondrial targeting sequence from cytochrome oxidase subunit IV (another matrix protein) to the amino terminus of roGFP [104]. To target the sensor to the intermembrane space, roGFP was appended to the carboxy terminus of glycerol phosphate dehydrogenase (GPD). The carboxyl tail of this protein normally resides in the intermembrane space, while the mid-region of the protein is embedded in the inner mitochondrial membrane. In both targeting schemes, immunogold localization studies confirm the expected localization of the sensor proteins [104]. Conceivably, a wide range of methods could be used for targeting expression of the sensor to specific cellular domains. Since the fluorescence from the cell originates exclusively from the compartment where the sensor is expressed, simple whole cell fluorescence measurements provide a measure of the redox conditions in the compartment of interest.

Pulmonary arterial smooth muscle cells respond to hypoxia by contracting via increases in cytosolic ionized calcium levels. To assess redox signaling in that response, Waypa et al. compared redox responses in the cytosol, intermembrane space, and mitochondrial matrix during acute hypoxia, using the targeted roGFP sensors described above (Fig. 1.4). Under baseline normoxic conditions, roGFP in the cytosol was approximately 20 % oxidized, whereas it was ~45 % oxidized in the intermembrane space and 70 % oxidized in the matrix compartment. During acute hypoxia, increases in oxidation were detected in the cytosol (35 %) and in the intermembrane space (65 %). By contrast, roGFP oxidation in the matrix decreased significantly during hypoxia. These findings reveal several important points. First, significant differences in basal redox conditions exist at different sites

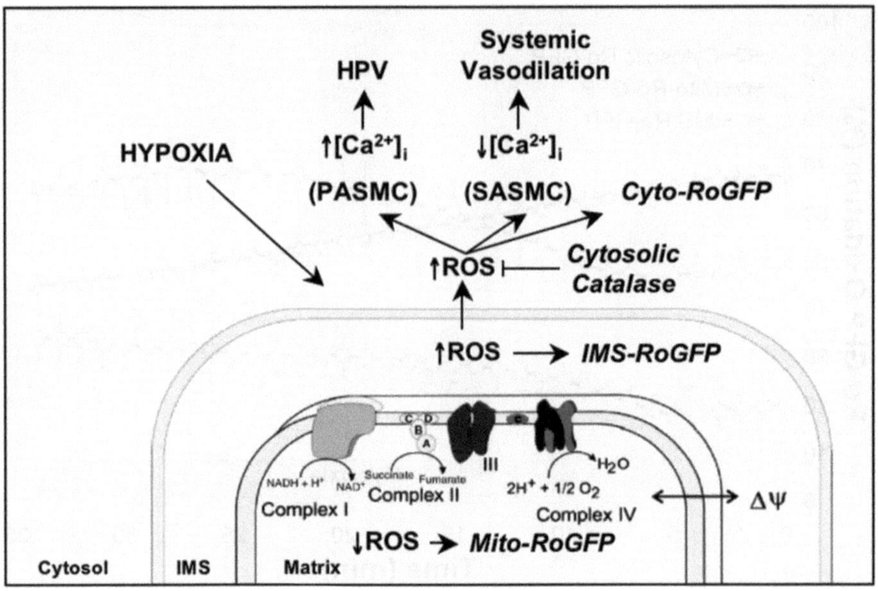

Fig. 1.5 Proposed relationship between the mitochondrial O_2 sensor and the vascular responses to hypoxia in pulmonary and systemic arterial smooth muscle cells. From [104], with permission

within the cell. Second, hypoxia causes an increase in oxidation within the intermembrane space and the cytosol, which is consistent with the hypothesized release of superoxide from the outer surface of the inner membrane. Third, important differences in the redox response to hypoxia can occur among different compartments, with some undergoing oxidation and another undergoing reductive stress. It is reasonable to propose that the generation of ROS in the matrix compartment is nonspecific and proportional to the O_2 concentration, whereas the ROS production in the intermembrane space is a regulated event controlled by an O_2 sensing system located in the inner membrane. Importantly, antioxidant scavenging of ROS during hypoxia abrogates the increase in cytosolic ionized calcium, indicating that increases in ROS are required for the signal transduction system and the functional response to hypoxia in these cells [107–109]. A summary description of the mitochondrial O_2 sensing system, ROS signaling, and the systemic (vasodilatory) and pulmonary vascular (vasoconstriction) responses to hypoxia is shown in Fig. 1.5.

Independent Confirmation of the Mitochondrial O_2 Sensing Hypothesis

Biological screening is a method frequently useful for identifying naturally occurring small molecules with interesting biological effects. A recent screen of a library of crude microbial extracts revealed that an extract of a fungal strain, *Embellisia chlamydospora*, blocked the ability of endothelial cells to respond to hypoxia, without causing cytotoxicity [110]. It also blocked angiogenesis in the chick chorionic membrane. Further studies to identify the responsible molecule yielded terpestacin, a small molecule with a bicyclo sesterterpene structure. When the purified molecule was administered to mice, it effectively blocked the angiogenic response in tumor xenografts. Further studies using a phage display amplification to identify the protein target interacting with this molecule identified UQCRB, a subunit of mitochondrial complex III, as the target. As this complex had previously been implicated as a potential oxygen sensor, further studies were carried out to determine whether terpestacin might affect the generation of ROS by the complex. Using dichlorofluorescein to assess oxidant production, it was found that terpestacin blocked the increase in ROS production during hypoxia. Moreover, the drug abolished the stabilization of HIF-1α during hypoxia, but not in response to direct inhibition of prolyl hydroxylase. These interesting findings, which involved an unbiased search for small molecules that block the hypoxic response, provide further support for the role of complex III in the oxygen sensing pathway and for hypoxia-induced ROS production in triggering hypoxic responses by the cell.

Initial evidence linking mitochondrial function to the O_2-dependent regulation of HIF-1α came from studies using mitochondria-deficient cell lines. These ρ^0 cells lack mitochondrial DNA, which encodes a set of protein components of complexes I, III, IV, and V. In the absence of these proteins, the electron transport chain is disabled and all cellular ATP must derive from glycolysis [111]. These ρ^0 cells fail to stabilize HIF-1α during hypoxia [87], but they can still respond to anoxic conditions [112] and to exogenous H_2O_2 administration [102]. Pharmacological mitochondrial inhibitors, such as myxothiazol or stigmatellin, prevent ubiquinol binding at the Qo site and thereby prevent the formation of superoxide at complex III. They also prevent the hypoxic stabilization of HIF-1α, essentially by preventing the generation of an ROS signal [113]. Unfortunately, the inhibition of electron transport also blocks oxidative phosphorylation, and many cells cannot survive the resulting bioenergetic crisis. Interestingly, terpestacin interaction with complex III interferes with the hypoxia-induced ROS production, yet it does not interfere with the main electron transport function in the complex. This ability to inhibit the O_2 sensing function of complex III without disrupting electron flux and ATP production underlies its ability to inhibit hypoxic HIF-1α stabilization without blocking cellular respiration.

In the regulation of HIF, ROS signals may act to inhibit PHD and FIH directly [58], or by triggering a set of posttranslational modifications that inhibit their functions allosterically. An important unanswered question relates to how the ROS signals trigger HIF stabilization in the cell.

Mitochondrial ROS Regulate AMPK Responses to Hypoxia

Oxidant signals arising from mitochondria in hypoxic cells also regulate some of the posttranslational responses of the cell, including the activation of AMPK. In a study of lung epithelial cells, Gusarova et al. found that catalase overexpression, or depletion of mitochondrial DNA (ρ^0 cells), blocked the ability of alveolar epithelial cells to internalize the apical membrane ATPase in response to hypoxia [114]. In another study, Emerling et al. found that hypoxia-induced AMPK phosphorylation was blocked in ρ^0 cells, that hypoxia augmented oxidant stress, that antioxidant compounds prevented AMPK activation in hypoxia, and that exogenous H_2O_2 administration was sufficient to activate the kinase [115]. To further investigate the role of mitochondrial ROS in the activation of AMPK during hypoxia, they employed ρ^0 cells in which mitochondria were repopulated (cybrid cells). One group was repopulated with normal mitochondria, whereas a second group was repopulated with mitochondria carrying a genetic deletion in the b cytochrome gene. This "rescue" was not sufficient to restore ATP production or oxygen consumption, but it did restore the ability to generate ROS at the Qo site of complex III. Interestingly, these cells recovered the ability to activate AMPK in hypoxia. While AMPK activation is normally attributed to increases in [AMP] during a bioenergetic crisis, the recovery of ROS production without the recovery of ATP generation in the cybrid cells indicates that ATP depletion was not responsible for the activation of AMPK. Collectively, these findings are consistent with the existence of a mitochondrial oxygen sensing mechanism at complex III that signals cellular hypoxia by releasing ROS signals to the cytosol. That system then participates in the activation of AMP kinase during hypoxia through an ROS-dependent and AMP-independent manner.

More recent studies reveal that ROS signals during hypoxia can activate AMPK in diverse cell types by triggering the entry of extracellular calcium. During hypoxia, ROS signals initiate release of intracellular calcium from the endoplasmic reticulum (ER). The resulting decrease in ER calcium leads to the oligomerization of stromal interaction molecule 1 (STIM1), the ER calcium sensor, which organizes the calcium release-activated calcium (CRAC) channels at sites where ER and plasma membranes associate. Knockdown of CaMKKβ abolishes the AMPK response, indicating that hypoxia can trigger AMPK activation in the apparent absence of increased [AMP] through ROS-dependent CRAC channel activation, leading to increases in cytosolic calcium that activate the AMPK upstream kinase CaMKKβ [116].

In summary, multiple lines of evidence provide support for the idea that molecular O_2 interacts with mitochondrial complex III in a manner that causes an increased release of ROS to the intermembrane space. This O_2 sensing function represents a "second function" of the complex, which also functions as a central component of the electron transport/proton translocation system that drives oxidative phosphorylation. To be sure, acceptance of this model has been hindered by the paradoxical finding that ROS production, at least in some subcellular compartments, occurs despite the decrease in oxygen availability. Additional important details remain to be addressed with respect to this system. One issue pertains to the structural changes that might occur at complex III in response to changes in O_2 abundance. In that regard, several hypotheses have been suggested to explain the increase in ROS generation [117]. The "ubisemiquinone hypothesis" proposes that structural changes at complex III lead to an increase in the lifetime of the semiquinone radical at the Qo site, thereby increasing the probability of ROS generation. The "vectoral hypothesis" proposes that hypoxia-induced structural changes lead to a vectoral shift in the release of superoxide from the inner membrane toward the intermembrane space and away from the matrix compartment. That mechanism could explain how increases in intermembrane space ROS signals occur even if overall ROS generation decreases. The "oxygen access" hypothesis proposes that hypoxia-induced structural changes lead to an increased access of O_2 to the Qo pocket, allowing superoxide generation to increase despite a lower abundance of molecular oxygen in the membrane. A definitive test of this hypothesis will require studies involving genetic deletion of complex III function in vivo, where the oxygen sensing functions of intact tissues can be evaluated.

References

1. Leung DW, Cachianes G, Kuang WJ, et al. Vascular endothelial growth factor is a secreted angiogenic mitogen. Science. 1989;246(4935):1306–9.
2. May D, Gilon D, Djonov V, et al. Transgenic system for conditional induction and rescue of chronic myocardial hibernation provides insights into genomic programs of hibernation. Proc Natl Acad Sci USA. 2008;105(1):282–7.
3. Ferrara N, Carver-Moore K, Chen H, et al. Heterozygous embryonic lethality induced by targeted inactivation of the VEGF gene. Nature. 1996;380(6573):439–42.
4. Hardie DG, Hawley SA, Scott JW. AMP-activated protein kinase–development of the energy sensor concept. J Physiol. 2006;574(Pt 1):7–15.
5. Semenza GL. Involvement of hypoxia-inducible factor 1 in pulmonary pathophysiology. Chest. 2005;128(6):592S–4.
6. Voter WA, Gayeski TE. Determination of myoglobin saturation of frozen specimens using a reflecting cryospectrophotometer. Am J Physiol. 1995;269 (4 Pt 2):H1328–41.
7. Gayeski TE, Honig CR. Intracellular PO2 in individual cardiac myocytes in dogs, cats, rabbits, ferrets, and rats. Am J Physiol. 1991;260:H522–31.
8. Gayeski TE, Honig CR. Intracellular PO2 in long axis of individual fibers in working dog gracilis muscle. Am J Physiol. 1988;254:H1179–86.
9. Gayeski TE, Honig CR. O2 gradients from sarcolemma to cell interior in red muscle at maximal VO2. Am J Physiol. 1986;251:H789–99.
10. Honig CR, Gayeski TE. Comparison of intracellular PO2 and conditions for blood-tissue O2 transport in heart and working red skeletal muscle. Adv Exp Med Biol. 1987;215:309–21.
11. Gayeski TE, Connett RJ, Honig CR. Minimum intracellular PO2 for maximum cytochrome turnover in red muscle in situ. Am J Physiol. 1987;252:H906–15.
12. Clark Jr A, Clark PA, Connett RJ, et al. How large is the drop in PO2 between cytosol and mitochondrion? Am J Physiol. 1987;252:C583–7.

13. Takahashi E, Endoh H, Xu ZL, et al. Direct estimation of intracellular PO2 gradients in a single cardiomyocyte of the rat. Adv Exp Med Biol. 1998;454:409–13.
14. Takahashi E, Sato K, Endoh H, et al. Direct observation of radial intracellular PO2 gradients in a single cardiomyocyte of the rat. Am J Physiol. 1998;275(1 Pt 2):H225–33.
15. Jones DP, Mason HS. Gradients of O_2 concentration in hepatocytes. J Biol Chem. 1978;253:4874–80.
16. Goldwasser E, Jacobson LO, Fried W, et al. Studies on erythropoiesis. V. The effect of cobalt on the production of erythropoietin. Blood. 1958;13:55–60.
17. Goldberg MA, Glass GA, Cunningham JM, et al. The regulated expression of erythropoietin by two human hepatoma cell lines. Proc Natl Acad Sci USA. 1987;84(22):7972–6.
18. Goldberg MA, Dunning SP, Bunn HF. Regulation of the erythropoietin gene: evidence that the oxygen sensor is a heme protein. Science. 1988;242:1412–5.
19. Fandrey J, Bunn HF. In vivo and in vitro regulation of erythropoietin mRNA: measurement by competitive polymerase chain reaction. Blood. 1993;81(3):617–23.
20. Horiguchi H, Bunn HF. Erythropoietin induction in Hep3B cells is not affected by inhibition of heme biosynthesis. Biochim Biophys Acta Mol Cell Res. 2000;1495(3):231–6.
21. Babior BM. The leukocyte NADPH oxidase. Isr Med Assoc J. 2002;4(11):1023–4.
22. Babior BM, Lambeth JD, Nauseef W. The neutrophil NADPH oxidase. Arch Biochem Biophys. 2002;397(2):342–4.
23. Wolin MS, Ahmad M, Gao Q, et al. Cytosolic NAD(P)H regulation of redox signaling and vascular oxygen sensing. Antioxid Redox Signal. 2007;9(6):671–8.
24. Muller G, Morawietz H. NAD(P)H oxidase and endothelial dysfunction. Horm Metab Res. 2009;41(2):152–8.
25. Kummer W, Acker H. Immunohistochemical demonstration of four subunits of neutrophil NAD(P)H oxidase in type I cells of carotid body. J Appl Physiol. 1995;78(5):1904–9.
26. Griendling KK, Sorescu D, Ushio-Fukai M. NAD(P)H oxidase—role in cardiovascular biology and disease. Circ Res. 2000;86(5):494–501.
27. Babior BM. The NADPH oxidase of endothelial cells. IUBMB Life. 2000;50(4–5):267–9.
28. Archer SL, Reeve HL, Michelakis E, et al. O_2 sensing is preserved in mice lacking the gp91 phox subunit of NADPH oxidase. Proc Natl Acad Sci USA. 1999;96:7944–9.
29. Weissmann N, Zeller S, Schafer RU, et al. Impact of mitochondria and NADPH oxidases on acute and sustained hypoxic pulmonary vasoconstriction. Am J Respir Cell Mol Biol. 2006;34(4):505–13.
30. Ganfornina MD, Lopez-Barneo J. Single K^+ channels in membrane patches of arterial chemoreceptor cells are modulated by O_2 tension. Proc Natl Acad Sci USA. 1991;88(7):2927–30.
31. Ganfornina MD, Lopez-Barneo J. Potassium channel types in arterial chemoreceptor cells and their selective modulation by oxygen. J Gen Physiol. 1992;100(3):401–26.
32. Peers C. Hypoxic suppression of K+ currents in type I carotid body cells: selective effect on the Ca2(+)-activated K+ current. Neurosci Lett. 1990;119(2):253–6.
33. Wyatt CN, Peers C. Ca(2+)-activated K+ channels in isolated type I cells of the neonatal rat carotid body. J Physiol. 1995;483(Pt 3):559–65.
34. Perez-Garcia MT, Colinas O, Miguel-Velado E, et al. Characterization of the Kv channels of mouse carotid body chemoreceptor cells and their role in oxygen sensing. J Physiol (Lond). 2004;557(2):457–71.
35. Weir EK, Lopez-Barneo J, Buckler KJ, et al. Mechanisms of disease—acute oxygen-sensing mechanisms. N Engl J Med. 2005;353(19):2042–55.
36. Williams SEJ, Wootton P, Mason HS, et al. Hemoxygenase-2 is an oxygen sensor for a calcium-sensitive potassium channel. Science. 2004;306(5704):2093–7.
37. Ortega-Saenz P, Pascual A, Gomez-Diaz R, et al. Acute oxygen sensing in heme oxygenase-2 null mice. J Gen Physiol. 2006;128(4):405–11.
38. Semenza GL, Roth PH, Fang H-M, et al. Transcriptional regulation of genes encoding glycolytic enzymes by hypoxia-inducible factor 1. J Biol Chem. 1994;269:23757–63.
39. Semenza GL, Wang GL. A nuclear factor induced by hypoxia via de novo protein synthesis binds to the human erythropoietin gene enhancer at a site required for transcriptional activation. Mol Cell Biol. 1992;12(12):5447–54.
40. Semenza GL, Nejfelt MK, Chi SM, et al. Hypoxia-inducible nuclear factors bind to an enhancer element located 3' to the human erythropoietin gene. Proc Natl Acad Sci USA. 1991;88(13):5680–4.
41. Maxwell PH, Pugh CW, Ratcliffe PJ. Inducible operation of the erythropoietin 3' enhancer in multiple cell lines: evidence for a widespread oxygen-sensing mechanism. Proc Natl Acad Sci USA. 1993;90:2423–7.
42. Marx J. How cells endure low oxygen. Science. 2004;303:1454–6.
43. Iyer NV, Kotch LE, Agani F, et al. Cellular and developmental control of O2 homeostasis by hypoxia- inducible factor 1 alpha. Genes Dev. 1998;12(2):149–62.
44. Maltepe E, Schmidt JV, Baunoch D, et al. Abnormal angiogenesis and responses to glucose and oxygen deprivation in mice lacking the protein ARNT. Nature. 1997;386(6623):403–7.
45. Maltepe E, Simon MC. Oxygen, genes, and development: an analysis of the role of hypoxic gene regulation during murine vascular development. J Mol Med. 1998;76(6):391–401.
46. Wang GL, Jiang B-H, Rue EA, et al. Hypoxia-inducible factor 1 is a basic-helix-loop-helix-PAS

heterodimer regulated by cellular O_2 tension. Proc Natl Acad Sci USA. 1995;92:5510–4.
47. Jiang BH, Semenza GL, Bauer C, et al. Hypoxia-inducible factor 1 levels vary exponentially over a physiologically relevant range of O_2 tension. Am J Physiol. 1996;271(4 Pt 1):C1172–80.
48. Huang LE, Gu J, Schau M, et al. Regulation of hypoxia-inducible factor 1α is mediated by an O_2-dependent degradation domain via the ubiquitin-proteasome pathway. Proc Natl Acad Sci USA. 1998;95(14):7987–92.
49. Maxwell PH, Wiesener MS, Chang GW, et al. The tumour suppressor protein VHL targets hypoxia-inducible factors for oxygen-dependent proteolysis. Nature. 1999;399(6733):271–5.
50. Semenza GL. HIF-1, O2, and the 3 PHDs: how animal cells signal hypoxia to the nucleus. Cell. 2001;107(1):1–3.
51. Takeda K, Ho VC, Takeda H, et al. Placental but not heart defects are associated with elevated hypoxia-inducible factor alpha levels in mice lacking prolyl hydroxylase domain protein. Mol Cell Biol. 2006;26(22):8336–46.
52. Jaakkola P, Mole DR, Tian YM, et al. Targeting of HIF-alpha to the von Hippel-Lindau ubiquitylation complex by O2-regulated prolyl hydroxylation. Science. 2001;292(5516):468–72.
53. Ivan M, Kondo K, Yang H, et al. HIFalpha targeted for VHL-mediated destruction by proline hydroxylation: implications for O2 sensing. Science. 2001; 292(5516):464–8.
54. Carmeliet P, Dor Y, Herbert JM, et al. Role of HIF-1alpha in hypoxia-mediated apoptosis, cell proliferation and tumour angiogenesis. Nature. 1998;394(6692):485–90.
55. Mole DR, Maxwell PH, Pugh CW, et al. Regulation of HIF by the von Hippel-Lindau tumour suppressor: implications for cellular oxygen sensing. IUBMB Life. 2001;52(1–2):43–7.
56. Lando D, Peet DJ, Whelan DA, et al. Asparagine hydroxylation of the HIF transactivation domain: a hypoxic switch. Science. 2002;295(5556):858–61.
57. Lando D, Peet DJ, Gorman JJ, et al. FIH-1 is an asparaginyl hydroxylase enzyme that regulates the transcriptional activity of hypoxia-inducible factor. Genes Dev. 2002;16(12):1466–71.
58. Masson N, Singleton RS, Sekirnik R, et al. The FIH hydroxylase is a cellular peroxide sensor that modulates HIF transcriptional activity. EMBO Rep. 2012;13(3):251–7.
59. Hagen T, Taylor CT, Lam F, et al. Redistribution of intracellular oxygen in hypoxia by nitric oxide: effect on HIF1alpha. Science. 2003;302(5652):1975–8.
60. Cassina A, Radi R. Differential inhibitory action of nitric oxide and peroxynitrite on mitochondrial electron transport. Arch Biochem Biophys. 1996;328(2):309–16.
61. Castello PR, David PS, McClure T, et al. Mitochondrial cytochrome oxidase produces nitric oxide under hypoxic conditions: implications for oxygen sensing and hypoxic signaling in eukaryotes. Cell Metab. 2006;3(4):277–87.
62. Poyton RO, Castello PR, Ball KA, et al. Mitochondria and hypoxic signaling: a new view. Ann N Y Acad Sci. 2009;1177:48–56.
63. Kadenbach B, Barth J, Akgun R, et al. Regulation of mitochondrial energy generation in health and disease. Biochim Biophys Acta. 1995;1271:103–9.
64. Kadenbach B, Stroh A, Ungibauer M, et al. Isozymes of cytochrome-c oxidase: characterization and isolation from different tissues. Methods Enzymol. 1986;126:32–45.
65. Huttemann M, Kadenbach B, Grossman LI. Mammalian subunit IV isoforms of cytochrome c oxidase. Gene. 2001;267(1):111–23.
66. Brunori M, Wilson MT. Electron transfer and proton pumping in cytochrome oxidase. Biochimie. 1995;77:668–76.
67. Mills E, Jobsis FF. Mitochondrial respiratory chain of carotid body and chemoreceptor response to changes in oxygen tension. J Neurophysiol. 1972;35(4):405–28.
68. Nair PK, Buerk DG, Whalen WJ, et al. Two cytochrome oxygen consumption model and mechanism for carotid body chemoreception. Adv Exp Med Biol. 1986;200:293–300.
69. Nair PK, Buerk DG, Whalen WJ. Cat carotid body oxygen metabolism and chemoreception described by a two-cytochrome model. Am J Physiol. 1986;250(2 Pt 2):H202–7.
70. Buerk DG, Nair PK, Whalen WJ. Two-cytochrome metabolic model for carotid body PtiO2 and chemosensitivity changes after hemorrhage. J Appl Physiol. 1989;67(1):60–8.
71. Eyzaguirre C, Zapata P. Perspectives in carotid body research. J Appl Physiol. 1984;57(4):931–57.
72. Cooper CE. The steady-state kinetics of cytochrome c oxidation by cytochrome oxidase. Biochim Biophys Acta. 1990;1017:187–203.
73. Joels N, Neil E. The action of high tensions of carbon monoxide on the carotid chemoreceptors. Arch Int Pharm Ther. 1962;130:528–34.
74. Lahiri S, Iturriaga R, Mokashi A, et al. CO reveals dual mechanisms of O2 chemoreception in the cat carotid body. Respir Physiol. 1993;94(2):227–40.
75. Lahiri S. Chromophores in O2 chemoreception: the carotid body model. News Physiol Sci. 1994;9: 161–5.
76. Jensen PK. Antimycin-insensitive oxidation of succinate and reduced nicotinamide-adenine dinucleotide in electron-tranport particles. Biochim Biophys Acta. 1966;122:157–66.
77. Boveris A, Oshino N, Chance B. The cellular production of hydrogen peroxide. Biochem J. 1972;128(3):617–30.
78. Ambrosio G, Zweier JL, Duilio C, et al. Evidence that mitochondrial respiration is a source of potentially toxic oxygen free radicals in intact rabbit

hearts subjected to ischemia and reflow. J Biol Chem. 1993;268(25):18532–41.
79. Becker LB, Vanden Hoek TL, Shao ZH, et al. Generation of superoxide in cardiomyocytes during ischemia before reperfusion. Am J Physiol. 1999;277(6 Pt 2):H2240–6.
80. Di Lisa F, Menabo R, Canton M, et al. The role of mitochondria in the salvage and the injury of the ischemic myocardium. Biochim Biophys Acta. 1998;1366(1–2):69–78.
81. Ferrari R. The role of mitochondria in ischemic heart disease. J Cardiovasc Pharmacol. 1996;28 Suppl 1:S1–10.
82. Turrens JF, Beconi M, Barilla J, et al. Mitochondrial generation of oxygen radicals during reoxygenation of ischemic tissues. Free Radic Res Commun. 1991;12–13(Pt 2):681–9.
83. Vanden Hoek TL, Li C, Shao Z, et al. Significant levels of oxidants are generated by isolated cardiomyocytes during ischemia prior to reperfusion. J Mol Cell Cardiol. 1997;29:2571–83.
84. Freeman BA, Crapo JD. Hyperoxia increases oxygen radical production in rat lungs and lung mitochondria. J Biol Chem. 1981;256(21):10986–92.
85. Shimada H, Hirai K, Simamura E, et al. Mitochondrial NADH-quinone oxidoreductase of the outer membrane is responsible for paraquat cytotoxicity in rat livers. Arch Biochem Biophys. 1998;351(1):75–81.
86. Boveris A, Cadenas E. Mitochondrial production of hydrogen peroxide regulation by nitric oxide and the role of ubisemiquinone. IUBMB Life. 2000;50(4–5): 245–50.
87. Chandel NS, Maltepe E, Goldwasser E, et al. Mitochondrial reactive oxygen species trigger hypoxia-induced transcription. Proc Natl Acad Sci USA. 1998;95:11715–20.
88. Chandel NS, Schumacker PT. Cellular oxygen sensing by mitochondria: old questions, new insight. J Appl Physiol. 2000;88(5):1880–9.
89. Jones DP, Shan X, Park Y. Coordinated multisite regulation of cellular energy metabolism. Annu Rev Nutr. 1992;12:327–43.
90. Turrens JF, Boveris A. Generation of superoxide anion by the NADH dehydrogenase of bovine heart mitochondria. Biochem J. 1980;191:421–7.
91. Votyakova TV, Reynolds IJ. DeltaPsi(m)-Dependent and -independent production of reactive oxygen species by rat brain mitochondria. J Neurochem. 2001;79(2):266–77.
92. Genova ML, Ventura B, Giuliano G, et al. The site of production of superoxide radical in mitochondrial Complex I is not a bound ubisemiquinone but presumably iron-sulfur cluster N2. FEBS Lett. 2001;505(3):364–8.
93. Li Y, Trush MA. Diphenyleneiodonium, an NAD(P) H oxidase inhibitor, also potently inhibits mitochondrial reactive oxygen species production. Biochem Biophys Res Commun. 1998;253(2):295–9.
94. Misra HP, Fridovich I. The univalent reduction of oxygen by reduced flavins and quinones. J Biol Chem. 1972;247:188–92.
95. Turrens JF, Alexandre A, Lehninger AL. Ubisemiquinone is the electron donor for superoxide formation by complex III of heart mitochondria. Arch Biochem Biophys. 1985;237:408–14.
96. Garcia-Ruiz C, Colell A, Morales A, et al. Role of oxidative stress generated from the mitochondrial electron transport chain and mitochondrial glutathione status in loss of mitochondrial function and activation of transcription factor nuclear factor-kappa B: studies with isolated mitochondria and rat hepatocytes. Mol Pharmacol. 1995;48(5):825–34.
97. Kwong LK, Sohal RS. Substrate and site specificity of hydrogen peroxide generation in mouse mitochondria. Arch Biochem Biophys. 1998;350:118–26.
98. Garcia-Ruiz C, Colell A, Mari M, et al. Direct effect of ceramide on the mitochondrial electron transport chain leads to generation of reactive oxygen species. Role of mitochondrial glutathione. J Biol Chem. 1997;272(17):11369–77.
99. Quillet-Mary A, Jaffrezou JP, Mansat V, et al. Implication of mitochondrial hydrogen peroxide generation in ceramide-induced apoptosis. J Biol Chem. 1997;272(34):21388–95.
100. Gille L, Nohl H. The ubiquinol/bc1 redox couple regulates mitochondrial oxygen radical formation. Arch Biochem Biophys. 2001;388(1):34–8.
101. Zhang L, Yu LD, Yu CA. Generation of superoxide anion by succinate-cytochrome c reductase from bovine heart mitochondria. J Biol Chem. 1998;273(51):33972–6.
102. Guzy RD, Hoyos B, Robin E, et al. Mitochondrial complex III is required for hypoxia-induced ROS production and cellular oxygen sensing. Cell Metab. 2005;1(6):401–8.
103. Mansfield KD, Guzy RD, Pan Y, et al. Mitochondrial dysfunction resulting from loss of cytochrome c impairs cellular oxygen sensing and hypoxic HIF-alpha activation. Cell Metab. 2005;1(6):393–9.
104. Waypa GB, Marks JD, Guzy R, et al. Hypoxia triggers subcellular compartmental redox signaling in vascular smooth muscle cells. Circ Res. 2010;106(3): 526–35.
105. Remington SJ. Fluorescent proteins: maturation, photochemistry and photophysics. Curr Opin Struct Biol. 2006;16(6):714–21.
106. Hanson GT, Aggeler R, Oglesbee D, et al. Investigating mitochondrial redox potential with redox-sensitive green fluorescent protein indicators. J Biol Chem. 2004;279(13):13044–53.
107. Waypa GB, Guzy R, Mungai PT, et al. Increases in mitochondrial reactive oxygen species trigger hypoxia-induced calcium responses in pulmonary artery smooth muscle cells. Circ Res. 2006;99(9): 970–8.
108. Waypa GB, Marks JD, Mack MM, et al. Mitochondrial reactive oxygen species trigger calcium

increases during hypoxia in pulmonary arterial myocytes. Circ Res. 2002;91(8):719–26.
109. Waypa GB, Chandel NS, Schumacker PT. Model for hypoxic pulmonary vasoconstriction involving mitochondrial oxygen sensing. Circ Res. 2001; 88(12):1259–66.
110. Jung HJ, Shim JS, Lee J, et al. Terpestacin inhibits tumor angiogenesis by targeting UQCRB of mitochondrial complex III and suppressing hypoxia-induced reactive oxygen species production and cellular oxygen sensing. J Biol Chem. 2010;285(15):11584–95.
111. King MP, Attardi G. Human cells lacking mtDNA: repopulation with exogenous mitochondria by complementation. Science. 1989;246:500–3.
112. Vaux EC, Metzen E, Yeates KM, et al. Regulation of hypoxia-inducible factor is preserved in the absence of a functioning mitochondrial respiratory chain. Blood. 2001;98(2):296–302.
113. Chandel NS, McClintock DS, Feliciano CE, et al. Reactive oxygen species generated at mitochondrial Complex III stabilize HIF-1-alpha during hypoxia: a mechanism of O2 sensing. J Biol Chem. 2000;275:25130–8.
114. Gusarova GA, Dada LA, Kelly AM, et al. Alpha1-AMP-activated protein kinase regulates hypoxia-induced Na, K-ATPase endocytosis via direct phosphorylation of protein kinase C zeta. Mol Cell Biol. 2009;29(13):3455–64.
115. Emerling BM, Weinberg F, Snyder C, et al. Hypoxic activation of AMPK is dependent on mitochondrial ROS but independent of an increase in AMP/ATP ratio. Free Radic Biol Med. 2009;46(10):1386–91.
116. Mungai PT, Waypa GB, Jairaman A, et al. Hypoxia triggers AMPK activation through ROS-mediated activation of CRAC channels. Mol Cell Biol. 2011;31(17):3531–45.
117. Guzy RD, Schumacker PT. Oxygen sensing by mitochondria at complex III: the paradox of increased reactive oxygen species during hypoxia. Exp Physiol. 2006;91(5):807–19.

Cellular and Molecular Defenses Against Hypoxia

2

Stilla Frede and Joachim Fandrey

Abstract

The ability to cope with hypoxia is essential for the development and survival of all vertebrate species. Cellular hypoxia occurs when oxygen demand exceeds oxygen supply. The development of cellular hypoxia depends both on the type of tissue and partial pressure oxygen (PO_2) in the tissue, because cells can vary extremely in their physiologic oxygen demand. Under normal conditions up to 90 % of the available oxygen is consumed by mitochondria to yield ATP through oxidative phosphorylation. Thus, oxygen deprivation will always lead to fundamental changes in cell metabolism and function. Different physiological systems have evolved for adaptation to conditions of low oxygen. Acute adaptation includes increased ventilation, changes in metabolism, protection against hypoxia-induced cell death and vasodilation. Adaptations to chronic hypoxia are characterized by the aim to restore oxygen delivery to the tissue. This is achieved by improving the oxygen transport capacity by increasing haemoglobin and red blood cell mass. Furthermore, long-term adaptation includes the remodelling of existing vessels as well as the formation of new vessels to increase blood and concurrently oxygen supply. Most of the hypoxic adaptations implicate gene expression driven by transcription factors especially activated under hypoxic conditions, such as *H*ypoxia-*i*nducible *f*actor 1 (HIF-1). HIF-1 is the best characterized regulators of cellular responses to hypoxia and products of its target genes are involved in all phases of hypoxic adaptation. In addition, for certain tissues a variety of HIF-1 independent molecular changes contributing to the cellular hypoxic response have been described including activation of NF-κB, CREB, and Notch. Particular attention is devoted to a potential cross talk between the oxygen-dependent HIF-1 regulators prolyl-hydroxylase (PHDs) and these HIF-1 independent hypoxia regulated factors which are known to be critical for survival under general stress conditions.

S. Frede, Ph.D. • J. Fandrey, M.D. (✉)
Institut für Physiologie, Universität Duisburg-Essen, Hufelandstr. 55, 45122 Essen, Germany
e-mail: stilla.frede@ukb.uni-bonn.de; joachim.fandrey@uni-due.de

E.R. Swenson and P. Bärtsch (eds.), *High Altitude: Human Adaptation to Hypoxia*,
DOI 10.1007/978-1-4614-8772-2_2, © Springer Science+Business Media New York 2014

Abbreviations

AMPK	AMP-activated protein kinase
ARNT	Arylhydrocarbon nuclear receptor translocator
bHLH domain	Basic-helix-loop-helix domain
BNIP3	BCL2/adenovirus E1B 19 kDa protein-interacting protein 3
CREB	Cyclic AMP response element binding protein
eEF	Eukaryotic elongation factor
eIF	Eukaryotic initiation factor
Epo	Erythropoietin
FIH-1	Factor inhibiting HIF
GLUT	Glucose transporter
HIF	Hypoxia-inducible factor
HNF-4	Hepatic nuclear factor
HRE	Hypoxia response element
ICD	Intracellular domain
IKKβ	I-kappa B kinase β
IL-3	Interleukin 3
IRE	Iron response element
IκB	Inhibitor of NF-κB
LDHA	Lactate dehydrogenase A
MAPK	Mitogen-activated protein kinase
mTOR	Mammalian target of Rapamycin
NF-κB	Nuclear factor kappa B
NLS	Nuclear localization sequence
NO	Nitric oxide
PAS	Per-ARNT-SIM domain
PDK1	Pyruvate dehydrogenase kinase-1
PER	"Period"—drosophila protein
PHD	Prolyl-hydroxylase
PI3	Phosphoinositid-3-kinase
PKA	Protein kinase A
PML	Tumor suppressor promyelocytic leukemia protein
RNS	Reactive nitrogen species
ROS	Reactive oxygen species
SIM	"Similar"—drosophila protein
UPR	Unfolded protein response
VEGF	Vascular endothelial growth factor

Effects of Hypoxia on Transcription

Hypoxia exerts dramatic effects on the cellular transcriptome [1]. In general, transcription is down-regulated under hypoxic conditions to conserve energy. Despite this general response, genes encoding proteins indispensable to overcome the hypoxic situation are specifically up-regulated. Activation of HIF-1 is a common and direct hypoxic event while activation of other oxygen-regulated transcription factors is more tissue specific, and depends on the duration and degree of hypoxia. Especially in tissues with normally high PO_2 like the lung, the PO_2 might not fall below the threshold necessary for HIF activation even under severely hypoxic conditions. Hypoxia responsive transcription factors other than HIF-1 are believed to regulate hypoxic gene expression in those tissues [2, 3]. In this context it is important to note that many of the genes involved in hypoxic adaptation are controlled by different hypoxia responsive transcription factors. In the following section activation and regulation of those transcription factors which are known to be directly involved in acute and chronic hypoxic adaptations are described in more detail. A comprehensive overview of hypoxia responsive transcription factors in general was provided by Cummins and Taylor [4]. In addition, the very recent review by Semenza especially focussed on the role of HIFs in physiology and pathophysiology [5].

Hypoxia-Inducible Transcription Factors: HIFs

*H*ypoxia-*i*nducible transcription *f*actors HIFs [6–8] are the best characterized transcription factors activated under hypoxic conditions; they are crucial to directly linking reduced oxygen supply with changes in gene expression. HIF consists as heterodimer of an α- and a β-subunit highly conserved across the animal kingdom. Both subunits belong to the family of basic-helix-loop-helix (bHLH)/PAS transcription factors. The first proteins of this transcription factor family were *P*ER, *A*RNT, and *S*IM, and the activity of HIF transcription factors is mainly regulated by the abundance of their α-subunit [7]. Three different α-subunits have been described by now: HIF-1α, HIF-2α, and HIF-3α. While HIF-1α seems to be ubiquitously expressed, HIF-2α displays a more tissue and cell-specific expression pattern [9, 10]. Despite similar activity in vitro, HIF-1α and HIF-2α have some important non-redundant, non-overlapping functions in the regulation of gene expression in vivo [11]. While HIF-1α is crucial for most short-term adaptations to reduced oxygen supply like the metabolic switch from oxidative phosphorylation as the main energy source to glycolysis, HIF-2α seems to be involved in mid—to long-term adaptations to hypoxia like increased erythropoiesis and angiogenesis [12]. The regulation, function, and tissue distribution of HIF-3α is incompletely understood, but distinct splice variants of HIF-3α were identified to potentially act as an endogenous HIF-1α antagonist [13, 14]. HIF-1α and HIF-2α subunits are regulated in a very similar way [15]: both genes are constitutively expressed and translated into the respective proteins, but the proteins are targeted for immediate degradation under normoxic PO_2 [16]. The degradation is initiated by hydroxylation of specific proline residues within the oxygen-dependent degradation domain of HIFα carried out by HIF-specific prolyl hydroxylases (PHDs). PHDs belong to a family of 2-oxo-glutarate-dependent dioxygenases. Three different PHD isoforms have been described to date with PHD2 being primarily important for HIF-1α hydroxylation under normoxia. The enzymatic activity of PHDs depends on molecular oxygen which qualifies these enzymes for cellular oxygen sensors [16, 17]. Furthermore, 2-oxo-glutarate, vitamin C, and ferrous iron (Fe^{2+}) were identified as indispensable cofactors. Hydroxylated HIF-α is recognized by the von Hippel–Lindau tumor suppressor protein (VHL). VHL then recruits components of the E3-ubiquitin ligase complex to polyubiquitinate HIFα which serves as a recognition signal for proteasomal degradation of HIFαs [18]. In addition, hydroxylation of an asparagine residue in the transactivation domain of HIFα by a hydroxylase called FIH (*F*actor *I*nhibiting *H*IF) inhibits the recruitment of transcriptional coactivators [19, 20].

In hypoxia, as a consequence of oxygen deficiency, proline and asparagine hydroxylation are inhibited, HIFα-subunits become stabilized, translocate into the nucleus where they dimerize with the common β-subunit to form the HIF-1 complex. HIF-1β is ubiquitously expressed and unaffected by changes in cellular oxygen concentration.

The HIF-1 heterodimers specifically bind to motifs termed hypoxia response elements (HREs) located within enhancer or promoter regions of target genes to either induce or suppress gene expression. Originally identified as the transcription factor responsible factor for the hypoxia-induced transcription of the red cell producing hormone erythropoietin [6], HIF-1 has been demonstrated to potentially control up to several hundreds (i.e., ~2–5 % of genome) of hypoxia responsive genes in humans [1]. HIF-1 target genes typically fall into three main categories according to the functions of their target proteins [21]. All gene products should serve to restore energy and O_2 homeostasis by either cellular or systemic means:

(a) By increasing anaerobic energy production via stimulated glycolytic substrate flux
(b) By protecting against hypoxia-induced cell death
(c) By improving tissue oxygenation via stimulation of vasodilation, angiogenesis, and erythropoiesis

Screening the human genome for potential HIF-binding sites (HIF-1 and HIF-2) by Schödel and coworkers revealed [22] at least up to 500 typical HIF-binding motifs indicating the importance of HIFs for the above mentioned cellular and systemic responses to hypoxia. Interestingly, some genes like hepcidin [23] are down-regulated under hypoxia, nevertheless a direct HIF-1/2-dependent repression of genes has not been documented until now.

An overview of different HIF targets is shown in Fig. 2.1.

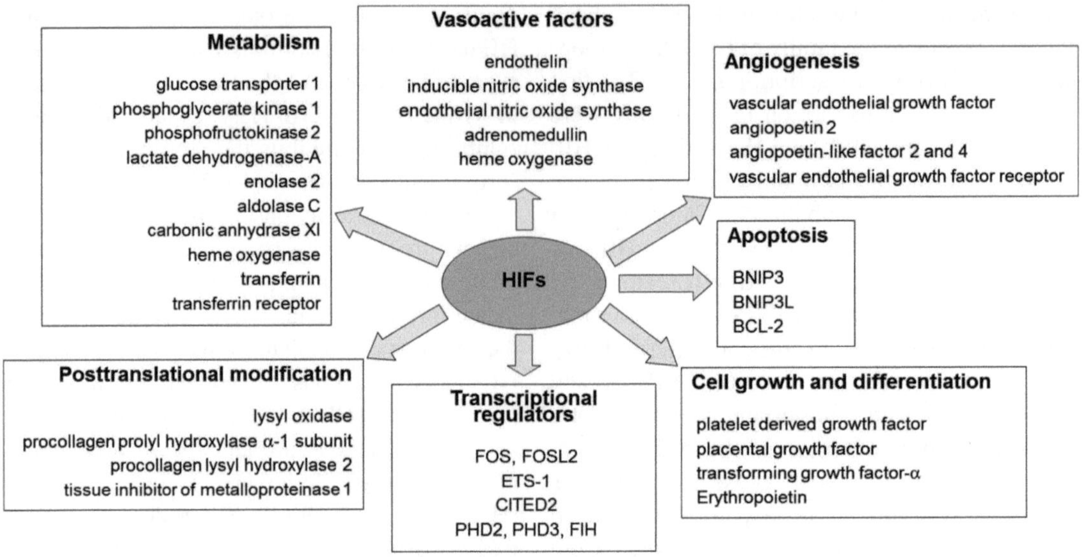

Fig. 2.1 Overview of HIF target genes involved in acute and chronic adaptations to hypoxia

Nuclear Factor Kappa B: NF-κB

During the last decade, a variety of transcription factors described to be involved in cellular stress response, proliferation, or protection from cell death have been shown to be regulated by changes in cellular oxygen partial pressure as well [24, 25]. Nuclear factor kappa B (NF-κB) was originally described as a transcription factor necessary for the expression of the kappa-light chain of immune globulins in B-lymphocytes [26]. Today, NF-κB is recognized as the master transcriptional mediator of the inflammatory response. NF-κB is indispensable for the expression of a wide variety of inflammatory cytokines, is important for cell survival and development [27] and moreover was identified as an upstream regulator of HIF-1α gene expression [28–30]. The NF-κB transcription factor superfamily comprises proteins with a highly conserved Rel homology domain [31]. Five members of the family are identified to date: p65, cRel, and RelB are the transcriptionally active members, whereas p50 and p52 derived from the precursor proteins p105 and p100 respectively. The most common transcriptionally active dimer is the complex consisting of p65 and p50. In the absence of a stimulus NF-κB is bound to the repressor molecule IκB in the cytosol. This binding covers the nuclear localization sequence (NLS) resulting in a sequestration of NF-κB in the cytosol. Upon stimulation e.g., by inflammatory cytokines, reactive oxygen species (ROS), or bacterial lipopolysaccharides, IκB is phosphorylated at serine residues by specific upstream kinases, which targets IκB for ubiquitination and proteasomal degradation. Degradation of IκB unmasks the NLS of NF-κB enabling translocation of the transcription factor into the nucleus [31]. Very recently it was shown that IκBα directly interacts with FIH resulting in sequestering of FIH in the cytosol and enhanced HIF-1α activity [32].

Although hypoxic activation of NF-κB was described in different cell types [25, 33], the underlying molecular mechanisms are not completely understood. Recently, an interesting link between the regulation of HIF and NF-κB has been reported: The IκB kinase IKKβ, an upstream regulator in the activation cascade of NF-κB was identified to be hydroxylated by PHD1 and PHD2 in normoxia. This results in suppression of IKKβ activity. In contrast, during hypoxia the reduced hydroxylation of IKKβ increases its activity which leads to enhanced NF-κB activation [34].

In addition, it was proposed that ROS increasingly leaking out of the mitochondria during hypoxia [35] could to activate NF-κB under this condition [36]. Especially under conditions of severe hypoxia (O_2<0.1 %) or ischemia NF-κB activation rather than HIF activation was demonstrated to be responsible for the adhesion and infiltration of monocytic cells [37]. A number of the HIF target genes listed in Fig. 2.1 contain NF-κB binding sites as well. Hypoxic activation of NF-κB represents therefore an additional pathway to enhance HIF-induced gene expression. But most importantly upregulation of genes lacking a hypoxia responsive element might be enabled under hypoxic conditions through NF-κB activation. Therefore, the NF-κB transcription factor family plays an important role in integrating hypoxic and inflammatory stimuli to overcome hypoxic stress.

Cyclic AMP Response Element Binding Protein

The cyclic AMP response element binding protein cyclic AMP response element binding (CREB) belongs to a family of leucine zipper transcription factors that are regulated by changes in intracellular cAMP or Ca^{2+} levels [38]. CREB target genes share some functions with typical HIF-1 target genes since they are involved in cell metabolism, apoptosis, and inflammation. In contrast to the immediate effects of HIF activation hypoxic CREB activation seems to have be protective effects only when induced by hypoxic/ischemic preconditioning. Hypoxic activation of CREB is achieved by phosphorylation of Ser133 by protein kinase A (PKA) [39]. In general CREB acts as an activator of transcription. In contrast to HIF-1 activation the hypoxic activation of CREB seems to be restricted to the following tissues. Especially in neuronal tissue CREB was shown to activate the anti-apoptotic Bcl-2 resulting in neuroprotection under hypoxic conditions [40]. In the hypoxic lung CREB activation was detected under conditions of moderate hypoxia and contributed in part by upregulation of endothelin-1 and vascular endothelial growth factor (VEGF) to the increase in pulmonary vascular resistance and remodelling [2]. Hypoxia-induced phosphorylation of CREB targets CREB for proteasomal degradation resulting in de-repression and activation of the respective target genes, e.g., VEGF, endothelin 1, tissue plasminogen activator, or VEGF receptor 1 [41, 42].

Effects of Hypoxia on Translation

In response to hypoxic conditions, cells reduce their overall rate of messenger RNA (mRNA) translation [43]. However, translation of some individual mRNAs is stimulated under hypoxia. The ability of cells to regulate translation during hypoxia is important for their survival. First, reduction of translation can save up to 70 % of cellular ATP consumption depending on the cell type [44]. Second, reducing translation will result in the preferential loss of proteins with a short half-life. This is of importance e.g., for the balance between short-lived anti-apoptotic and long-lived pro-apoptotic proteins [45]. Hypoxia causes the loss of polyribosome complexes [46]. Under normoxic conditions, translation is initiated by the assembling of the eukaryotic initiation factor 4F (eIF4F) to the cap structure of mRNAs [47], which consists of three subunits: the cap-binding protein eIF4E, the scaffolding protein eIF4G, and the helicase eIF4α. Binding of eIF4F facilitates the recruitment of the 40s ribosomal subunit, the initiation factor eIF3, and the ternary complex eIF2-GTP-tRNA. The assembly of the latter depends on the exchange of GDP or GTP catalyzed by eIF2B. Under hypoxic conditions the α-subunit of eIF2B becomes phosphorylated resulting in the inhibition of its catalytic activity [48]. Other important intracellular signalling pathways integrating metabolic changes, hypoxia and cellular stress signals are the mTOR (mammalian target of rapamycin kinase) and the unfolded protein response (UPR). mTOR activation in general is inhibited under hypoxic condition, but a number of recent publication suggest a potential interaction between HIF-1α and mTOR [49]. The exact cellular mechanisms underlying this paradox of

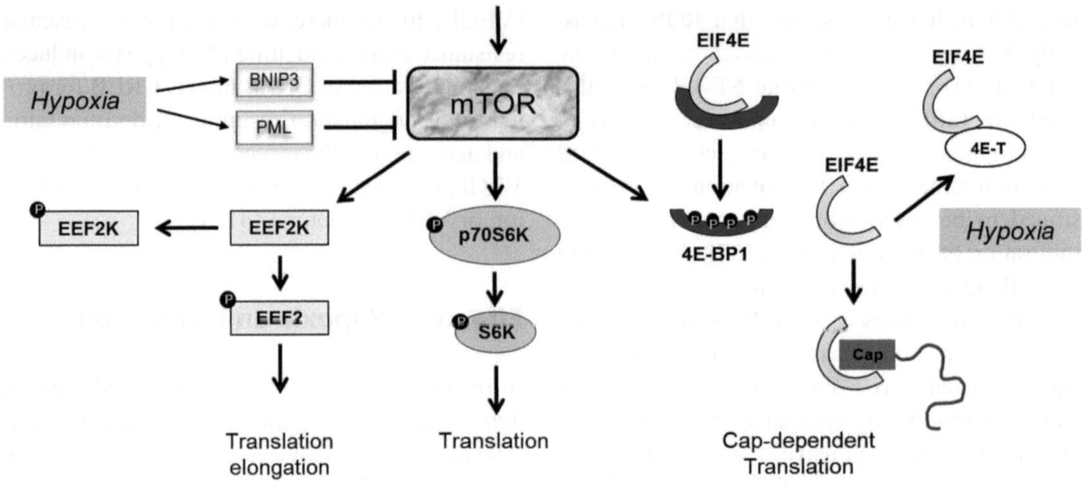

Fig. 2.2 Inhibition of the mammalian target of rapamycin signalling (mTOR) by hypoxia. Under normoxic conditions mTOR activation induces translation by three different mechanisms: activation of elongation via the eukaryotic elongation factor (2EEF2), by activation of the ribosomal protein S6 kinase (p70S6K) or by phosphorylation of the eukaryotic translation initiation factor 4E (eIF4E)-binding protein 1(4E-BP1) which results in cap-dependent translation of mRNA. Hypoxia inhibits mTOR indirectly by activation of BNIP3 or the tumor suppressor protein PML. A second mode of hypoxic inactivation of translation is the sequestering of eIF4E and its nuclear import factor 4E-T to the nucleus or cytoplasmic processing proteins (adapted from [50])

an increased mTOR-dependent translation of HIF-1α under hypoxic conditions still await elucidation. As recently reviewed, the impact of mTOR on translation of proteins under hypoxic conditions critically depends on the degree and duration of hypoxia [50]. Some hypoxia-induced proteins like HIF-1α or VEGF evade the requirement for eIF4F formation at the 5′cap structure via an internal ribosomal entry site [51] enabling an efficient translation of these proteins under hypoxic conditions. A simplified cartoon of hypoxia on the mTOR-dependent translation is given in Fig. 2.2.

Acute Cellular Responses to Hypoxia

Metabolic Changes

Most critical for the acute response to hypoxia is the adaption of cellular generation and consumption of energy-rich phosphates to avoid cell death, minimize cellular injury, and maintain important functions. The most efficient metabolic pathway for the generation of energy-equivalents in form of ATP is through oxidative phosphorylation of glucose (38 mol ATP/mol glucose). This process depends upon the availability of glucose and molecular oxygen. Under hypoxic conditions, however, cells are forced to shift their generation of ATP from oxidative phosphorylation to glycolysis (2 mol ATP/mol glucose). Two main cellular pathways are involved in the cells' response to acute hypoxic stress: The AMPK (AMP-activated protein kinase) pathway and the HIF-pathway.

The AMPK system appears to play a key role in maintaining the energy balance at the whole body level [52]. When ATP levels decline as consequence of cellular hypoxia or increase in cellular ATP consumption, AMP levels rise and AMPK is activated which promotes catabolic and inhibits anabolic processes. Alterations in the cellular AMP: ATP ratio is sensed directly by the AMPK, which therefore acts as a cellular energy sensor [52]. As consequence in many different cells glucose uptake is enhanced via increased expression and translocation of glucose transporter (GLUT) 4, or via increase in GLUT1 activity [53, 54]. Glycolysis and fatty acid oxidation are up-regulated whereas protein and glycogen synthesis are down-regulated [55, 56]. Moreover, AMPK exerts direct effects on the activity of different transcription factors like p300 and HNF-4 [57, 58] whereby the actual energy status of the cell is directly coupled to gene expression. Under conditions of acute hypoxia accumulation and activation of HIF-1 is primarily caused by the lack of molecular oxygen, resulting in inhibition of the activity of PHDs, in particular PHD2. In addition several intracellular signalling molecules merge in modulating or inhibiting PHD activity. The shift from aerobic to anaerobic metabolism first requires the upregulation of glycolysis. While in normoxia the primary function of glycolysis is to feed the Krebs cycle with pyruvate, glycolysis down to lactate to generate ATP becomes the main source of energy under hypoxic conditions. A number of important glycolytic enzymes are well characterized HIF-1 targets and are induced within minutes after the onset of hypoxia [59]. Moreover, the entry of pyruvate to the Krebs cycle is inhibited by HIF-dependent upregulation of PDK-1 (pyruvate dehydrogenase kinase-1) [60]. The HIF-1-dependent upregulation of lactate dehydrogenase A (LDHA) leads to an increased conversion of pyruvate to lactate. Both mechanisms contribute to reduced activity of the Krebs cycle under hypoxic conditions which also causes a decrease of succinate and fumarate levels. However, PHDs are sensitive to changes in 2-oxo-glutarate, succinate, or fumarate [61]. Deficiency in these co-substrates of PHDs mimics a hypoxic cell response [62–64].

Impact of Reactive Oxygen Species

The question whether the ROS are increased or decreased under conditions of hypoxia and what the role of ROS in HIF-1 signalling might be, provoked a lot of controversy during the last decades [35, 65, 66]. This question is discussed in detail in Chap. 1. On the one hand lack of molecular oxygen as final electron acceptor in the mitochondria might promote the formation of ROS, leaking out of mitochondria and induce HIF-1 accumulation. ROS and also reactive nitrogen species (RNS) were found to inhibit PHD activity resulting in an accumulation of HIF-1α [67–69]. Oxidation of Fe^{2+} to Fe^{3+} as part of a Fenton reaction has been proposed to be the underlying cellular mechanisms for this inhibition because PHD activity requires ferrous iron (Fe^{2+}) [70]. On the other hand due to the extreme high oxygen affinity of mitochondrial cytochrome C oxidase mitochondria act as cellular oxygen traps. Reactive oxygen and nitrogen species have be described to inhibit cytochrome C oxidase which was suggested to result in a shift of oxygen away from mitochondria to other cell compartments to increase PHD activity [71, 72].

Effects of Hypoxia on Cell Motility, Invasiveness, and Differentiation

The asparagines hydroxylase FIH-1 was originally described to modify HIF-1α transactivation [19, 73]. Hydroxylation of the asparagine residue 803 in the transactivation domain of HIF-1α prevents binding of the transcriptional coactivator CBP/p300 leading to inactivation of HIF-1 [74]. Recently it was shown, that FIH-1 can also hydroxylate asparagine residues in many other proteins containing ankyrin repeats [75]. One of this ankyrin repeat proteins interacting with FIH-1 is the Notch receptor intracellular domain (Notch-ICD) [76, 77]. Notch signalling plays an important role in cell differentiation, motility, and invasiveness [76, 78]. Hydroxylation of the Notch ICD by FIH-1 reduces Notch activity but concurrently derepresses HIF-1α transactivation. Dependent on the degree of hypoxia Notch ICD can function as a sequestering molecule for FIH-1 preventing HIF-1α hydroxylation. A similar role was recently assigned to IκBα [32].

Chronic Responses to Hypoxia

Adaptations to chronic hypoxia are characterized by the aim to restore oxygen delivery to the tissue. This is achieved by increasing the oxygen transport capacity by elevating the levels of haemoglobin and red blood cell mass. Furthermore, long-term adaptation includes the remodelling of existing vessels as well as the formation of new vessels to increase blood and concurrently oxygen supply.

Role of Erythropoietin

Hypoxic induction of erythropoietin (Epo) became the paradigm for oxygen-dependent gene expression. Originally, Epo was identified as the key hormone regulating the differentiation and proliferation of erythroid progenitors in the bone marrow [79]. Epo active in erythropoiesis is mainly produced in interstitial peritubular fibroblasts in the adult kidney and in fetal hepatocytes. Severe tissue hypoxia is able to induce an up to 1,000-fold upregulation of Epo expression in the kidney [80], predominantly by recruiting Epo producing cells [81]. Interestingly, Epo production is switched from the liver before birth to the kidneys after birth [82]. So far, the factors that control this developmental switch of Epo production have not been identified.

During the past years Epo production was found in various other tissues like brain, heart, and the reproductive tract. Epo synthesized in these organs does not have erythropoietic functions, but probably acts in a paracrine manner as a cell and tissue protective factor [83]. In particular, Epo may protect neuronal cells from hypoxic and ischemic injury [84] and induces neuronal growth and differentiation. In contrast to signalling through an Epo receptor homodimer to stimulate erythropoiesis, Epo might exert its protective effects by activation of a heterodimeric receptor consisting of the classical Epo receptor and the common β chain of the IL-3 receptor [85]. Likewise, Epo was shown to prevent cardiomyocytes from apoptosis and ischemia/reperfusion injury [86]. Several mechanisms were discussed to be involved in Epo-dependent cardiac protection. Epo is known to activate PI3/Akt kinases as well as MAPK-dependent signalling pathways. Burger and coworkers demonstrated that Epo-dependent activation of these pathways resulted in the upregulation of endothelial NO-synthase with NO as the key anti-apoptotic factor [87].

Recently, Epo was reported to increase hypoxia-induced stimulation of ventilation in mice and men [88, 89]. In these studies the role of Epo in connecting the acute ventilatory and the chronic erythropoietic response became evident.

Tissue Protection by Hypoxic Preconditioning

Hypoxic preconditioning is achieved by a spontaneous or experimentally induced episode of short hypoxia that protects the tissue from a subsequent, more severe hypoxic or ischemic insult. In 1986, Murry et al. [90] first described the principle of ischemic preconditioning. Several cycles of ischemia followed by short episodes of reperfusion significantly reduced the infarct size after a subsequent myocardial ischemia. Very recent studies with patients suffering from cardiac angina corroborate these experimental data, that injury-induced ischemic events protect these patients from severe myocardial infarction [91]. Ischemic preconditioning is not confined to the heart. In contrast, it appears that it is a powerful protective mechanism against ischemic injury that has been described for a variety of organs, including the brain, spinal cord, retina, liver, lung, and skeletal muscle [92]. Ischemic preconditioning has both immediate and delayed protective effects, the importance of which varies between species and organ systems. The exact molecular mechanisms of both protective components yet have to be exactly defined, but the involvement of hypoxia-induced transcription factors as well as changes in energy consumption have been discussed [92].

Regulation of Iron Metabolism as Protective Mechanism

Iron is necessary for a multitude of biological processes like catalyzing essential enzymatic reactions and is essential as a cofactor for oxygen-binding proteins in all living organisms. Iron is required for heme formation and—if not available in sufficient amounts—is the most common limiting factor for erythropoiesis. Iron metabolism, oxygen homeostasis, and erythropoiesis are consequently strongly linked. Both, iron excess and iron shortage have important consequences, therefore the use and storage of iron must be tightly regulated. The HIF-1 system has been identified to be influenced by the systemic iron status in a dual manner. First, iron deficiency directly reduces the activity of the iron-dependent PHDs resulting in increased HIF-α accumulation [17, 93]. Second, anemia caused by iron deficiency leads to tissue hypoxia which again increases HIF-α and HIF-1-dependent gene expression. Transferrin, the transferrin receptor, ceruloplasmin and heme oxygenase-1, which are responsible for iron transport and uptake, oxidation and recycling respectively have been identified as HIF-1 target genes [94–96].

Direct regulation of mRNAs encoding proteins involved in iron metabolism via iron response elements (IRE) located in the 3′ or 5′ untranslated regions of target mRNAs was described [97]. IREs can be occupied by binding of iron regulatory proteins (IRP) which may modify the efficiency of translation or the stability of these mRNAs thus matching iron availability and iron usage. Interestingly, IRPs are regulated by proteasomal degradation requiring 2-oxo-glutarate, iron and oxygen-dependent PHDs [98]. Hereby, cellular oxygen concentration can directly influence the regulation of iron metabolism.

Induction of Angiogenesis and Vascular Remodelling

Chronic adaptation to hypoxia may also improve blood and oxygen supply by either increasing the growth of new capillaries (angiogenesis) or by remodelling existing vessels [99]. Angiogenesis is a complex process, involving multiple proteins expressed by different cell types. VEGF is the most prominent HIF target gene involved in vascular biology [100]. In tumors, VEGF mRNA expression is significantly enhanced in perinecrotic regions of very low PO_2, suggesting a mechanism by which a hypoxic microenvironment stimulates angiogenesis [101]. Both tumor cells, but as recently shown tumor infiltrating macrophages, release VEGF to promote a functional vascular architecture [102]. Apart from VEGF numerous other factors are involved in the different steps initiating angiogenesis, most of which were identified as HIF-1 targets [21]. While de novo vessel formation seems to be critically dependent on the activity of HIF target genes, the role of HIFs in the preservation and remodelling of existing vessels is less clear. However, the therapeutic increase of HIF-1α under ischemic and hypoxic conditions was found to increase the number of blood vessels and regional oxygen supply [103].

Conclusions

All cellular and molecular defenses against hypoxia aim to restore energy and O_2 homeostasis by either cellular or systemic means to protect against hypoxia-induced cell death and maintain cellular function. In acute, severe hypoxia, when oxygen concentrations at tissue levels drop below 0.1 % (~1 mmHg) cells down-regulate oxygen consuming processes like transcription and translation, excepting those proteins which are indispensable for rescue from hypoxia. These proteins enable cells to switch to anaerobic energy production. This fall in tissue oxygenation is unlikely to occur at high altitudes but maybe expected in ischemic tissue. When hypoxia persists systemic responses are required to improve oxygen delivery to the tissue through vasodilation, erythropoiesis, and angiogenesis. The organ and tissue-specific adaptations all engage activation of HIFs. It therefore appears most appropriate to designate HIFs as the "master regulator of oxygen homeostasis" [104].

References

1. Manalo DJ, Rowan A, Lavoie T, et al. Transcriptional regulation of vascular endothelial cell responses to hypoxia by HIF-1. Blood. 2005;105(2):659–69.
2. Leonard MO, Howell K, Madden SF, et al. Hypoxia selectively activates the CREB family of transcription

factors in the in vivo lung. Am J Respir Crit Care Med. 2008;178(9):977–83.
3. Eltzschig HK, Kohler D, Eckle T, Kong T, Robson SC, Colgan SP. Central role of Sp1-regulated CD39 in hypoxia/ischemia protection. Blood. 2009;113(1): 224–32.
4. Cummins E, Taylor C. Hypoxia-responsive transcription factors. Pflugers Arch. 2005;450(6): 363–71.
5. Semenza GL. Hypoxia-inducible factors in physiology and medicine. Cell. 2012;148(3):399–408.
6. Wang GL, Jiang BH, Rue EA, Semenza GL. Hypoxia-inducible factor 1 is a basic-helix-loop-helix-PAS heterodimer regulated by cellular O_2 tension. Proc Natl Acad Sci USA. 1995;92(12): 5510–4.
7. Schofield CJ, Ratcliffe PJ. Oxygen sensing by HIF hydroxylases. Nat Rev Mol Cell Biol. 2004;5(5): 343–54.
8. Bunn HF, Poyton RO. Oxygen sensing and molecular adaptation to hypoxia. Physiol Rev. 1996;76(3): 839–85.
9. Talks KL, Turley H, Gatter KC, et al. The expression and distribution of the hypoxia-inducible factors HIF-1α and HIF-2α in normal human tissues, cancers, and tumor-associated macrophages. Am J Pathol. 2000;157(2):411–21.
10. Wiesener MS, Jurgensen JS, Rosenberger C, et al. Widespread, hypoxia-inducible expression of HIF-2alpha in distinct cell populations of different organs. FASEB J. 2003;17(2):271–3.
11. Pugh CW, Ratcliffe PJ. Regulation of angiogenesis by hypoxia: role of the HIF system. Nat Med. 2003; 9(6):677–84.
12. Frede S, Freitag P, Geuting L, Konietzny R, Fandrey J. Oxygen-regulated expression of the erythropoietin gene in the human renal cell line REPC. Blood. 2011;117(18):4905–14.
13. Maynard MA, Evans AJ, Shi W, Kim WY, Liu FF, Ohh M. Dominant-negative HIF-3 alpha 4 suppresses VHL-null renal cell carcinoma progression. Cell Cycle. 2007;6(22):2810–6.
14. Makino Y, Cao R, Svensson K, et al. Inhibitory PAS domain protein is a negative regulator of hypoxia-inducible gene expression. Nature. 2001;414(6863): 550–4.
15. O'Rourke JF, Tian YM, Ratcliffe PJ, Pugh CW. Oxygen-regulated and transactivating domains in endothelial PAS protein 1: comparison with hypoxia-inducible factor-1alpha. J Biol Chem. 1999;274(4): 2060–71.
16. Jaakkola P, Mole DR, Tian YM, et al. Targeting of HIF-alpha to the von Hippel-Lindau ubiquitylation complex by O_2-regulated prolyl hydroxylation. Science. 2001;292(5516):468–72.
17. Epstein AC, Gleadle JM, McNeill LA, et al. C. elegans EGL-9 and mammalian homologs define a family of dioxygenases that regulate HIF by prolyl hydroxylation. Cell. 2001;107(1):43–54.
18. Maxwell P. The von Hippel-Lindau gene product is necessary for oxgyen-dependent proteolysis of hypoxia-inducible factor [alpha] subunits. Nature. 1999;399:271–5.
19. Mahon PC, Hirota K, Semenza GL. FIH-1: a novel protein that interacts with HIF-1alpha and VHL to mediate repression of HIF-1 transcriptional activity. Genes Dev. 2001;15(20):2675–86.
20. Lando D, Peet DJ, Gorman JJ, Whelan DA, Whitelaw ML, Bruick RK. FIH-1 is an asparaginyl hydroxylase enzyme that regulates the transcriptional activity of hypoxia-inducible factor. Genes Dev. 2002;16(12):1466–71.
21. Mole DR, Blancher C, Copley RR, et al. Genome-wide association of hypoxia-inducible factor (HIF)-1{alpha} and HIF-2{alpha} DNA binding with expression profiling of hypoxia-inducible transcripts. J Biol Chem. 2009;284(25):16767–75.
22. Schödel J, Oikonomopoulos S, Ragoussis J, Pugh CW, Ratcliffe PJ, Mole DR. High-resolution genome-wide mapping of HIF-binding sites by ChIP-seq. Blood. 2011;117(23):e207–17.
23. Hintze KJ, McClung JP (2011) Hepcidin: a critical regulator of iron metabolism during hypoxia. Adv Hematol 2011:510304.
24. Koong AC, Chen EY, Mivechi NF, Denko NC, Stambrook P, Giaccia AJ. Hypoxic activation of nuclear factor-{kappa}B is mediated by a Ras and Raf signaling pathway and does not involve MAP kinase (ERK1 or ERK2). Cancer Res. 1994;54(20): 5273–9.
25. Chandel NS, Trzyna WC, McClintock DS, Schumacker PT. Role of oxidants in NF-κB activation and TNF-α Gene transcription induced by hypoxia and endotoxin. J Immunol. 2000;165(2): 1013–21.
26. Sen R, Baltimore D. Inducibility of κ immunoglobulin enhancer-binding protein NF-κB by a posttranslational mechanism. Cell. 1986;47(6):921–8.
27. Vallabhapurapu S, Karin M. Regulation and function of NF-κB transcription factors in the immune system. Annu Rev Immunol. 2009;27(1):693–733.
28. Frede S, Stockmann C, Freitag P, Fandrey J. Activation of HIF-1 by bacterial lipopolysaccharides in human monocytes requires NF-κB. Biochem J. 2006;396:517–27.
29. Frede S, Stockmann C, Winning S, Freitag P, Fandrey J. Hypoxia-inducible factor (HIF) 1α accumulation and HIF target gene expression are impaired after induction of endotoxin tolerance. J Immunol. 2009;182(10):6470–6.
30. Rius J, Guma M, Schachtrup C, et al. NF-κB links innate immunity to the hypoxic response through transcriptional regulation of HIF-1α. Nature. 2008; 453(7196):807–11.
31. Baeuerle PA, Baltimore D. NF-[kappa]B: ten years after. Cell. 1996;87(1):13–20.
32. Shin DH, Li SH, Yang SW, Lee BL, Lee MK, Park JW. Inhibitor of nuclear factor-kappaB alpha

derepresses hypoxia-inducible factor-1 during moderate hypoxia by sequestering factor inhibiting hypoxia-inducible factor from hypoxia-inducible factor 1alpha. FEBS J. 2009;276(13):3470–80.
33. Guo G, Bhat NR. Hypoxia/reoxygenation differentially modulates NF-kappaB activation and iNOS expression in astrocytes and microglia. Antioxid Redox Signal. 2006;8(5–6):911–8.
34. Cummins EP, Berra E, Comerford KM, et al. Prolyl hydroxylase-1 negatively regulates IκB kinase-beta, giving insight into hypoxia-induced NFκB activity. PNAS. 2006;103(48):18154–9.
35. Chandel NS, Maltepe E, Goldwasser E, Mathieu CE, Simon MC, Schumacker PT. Mitochondrial reactive oxygen species trigger hypoxia-induced transcription. Proc Natl Acad Sci USA. 1998;95(20): 11715–20.
36. Oliver KM, Taylor CT, Cummins EP. Hypoxia. Regulation of NFkappaB signalling during inflammation: the role of hydroxylases. Arthritis Res Ther. 2009;11(1):215.
37. Winning S, Splettstoesser F, Fandrey J, Frede S. Acute hypoxia induces HIF-Independent monocyte adhesion to endothelial cells through increased intercellular adhesion molecule-1 expression: the role of hypoxic inhibition of prolyl hydroxylase activity for the induction of NF-{kappa}B. J Immunol. 2010; 185(3):1786–93.
38. Mayr B, Montminy M. Transcriptional regulation by the phosphorylation-dependent factor CREB. Nat Rev Mol Cell Biol. 2001;2(8):599–609.
39. Beitner-Johnson D, Millhorn DE. Hypoxia induces phosphorylation of the cyclic AMP response element-binding protein by a novel signaling mechanism. J Biol Chem. 1998;273(31):19834–9.
40. Delivoria-Papadopoulos M, Ashraf QM, Mishra OP. Differential expression of apoptotic proteins following hypoxia-induced CREB phosphorylation in the cerebral cortex of newborn piglets. Neurochem Res. 2007;32(7):1256–63.
41. Leonard MO, O'Reilly S, Comerford KM, Taylor CT. Identification of cyclic AMP response element-binding protein-dependent transcriptional responses in hypoxia by microarray analysis. Methods Enzymol. 2004;381:511–24.
42. Taylor CT, Furuta GT, Synnestvedt K, Colgan SP. Phosphorylation-dependent targeting of cAMP response element binding protein to the ubiquitin/proteasome pathway in hypoxia. Proc Natl Acad Sci USA. 2000;97(22):12091–6.
43. Pettersen EO, Juul NO, Ronning OW. Regulation of protein metabolism of human cells during and after acute hypoxia. Cancer Res. 1986;46(9):4346–51.
44. Hochachka PW, Buck LT, Doll CJ, Land SC. Unifying theory of hypoxia tolerance: molecular/metabolic defense and rescue mechanisms for surviving oxygen lack. Proc Natl Acad Sci USA. 1996;93(18):9493–8.
45. Holley CL, Olson MR, Colon-Ramos DA, Kornbluth S. Reaper eliminates IAP proteins through stimulated IAP degradation and generalized translational inhibition. Nat Cell Biol. 2002;4(6):439–44.
46. Koritzinsky M, Magagnin MG, van den Beucken T, et al. Gene expression during acute and prolonged hypoxia is regulated by distinct mechanisms of translational control. EMBO J. 2006;25(5): 1114–25.
47. Gingras AC, Raught B, Sonenberg N. Control of translation by the target of rapamycin proteins. Prog Mol Subcell Biol. 2001;27:143–74.
48. Sonenberg N, Hinnebusch AG. Regulation of translation initiation in eukaryotes: mechanisms and biological targets. Cell. 2009;136(4):731–45.
49. Land SC, Tee AR. Hypoxia-inducible factor 1{alpha} is regulated by the mammalian target of rapamycin (mTOR) via an mTOR signaling motif. J Biol Chem. 2007;282(28):20534–43.
50. Wouters BG, Koritzinsky M. Hypoxia signalling through mTOR and the unfolded protein response in cancer. Nat Rev Cancer. 2008;8(11):851–64.
51. Lang KJ, Kappel A, Goodall GJ. Hypoxia-inducible factor-1alpha mRNA contains an internal ribosome entry site that allows efficient translation during normoxia and hypoxia. Mol Biol Cell. 2002;13(5): 1792–801.
52. Towler MC, Hardie DG. AMP-activated protein kinase in metabolic control and insulin signaling. Circ Res. 2007;100(3):328–41.
53. Kurth-Kraczek EJ, Hirshman MF, Goodyear LJ, Winder WW. 5' AMP-activated protein kinase activation causes GLUT4 translocation in skeletal muscle. Diabetes. 1999;48(8):1667–71.
54. Barnes K, Ingram JC, Porras OH, et al. Activation of GLUT1 by metabolic and osmotic stress: potential involvement of AMP-activated protein kinase (AMPK). J Cell Sci. 2002;115(11):2433–42.
55. Horman S, Browne G, Krause U, et al. Activation of AMP-activated protein kinase leads to the phosphorylation of elongation factor 2 and an inhibition of protein synthesis. Curr Biol. 2002;12(16):1419–23.
56. Carling D, Hardie DG. The substrate and sequence specificity of the AMP-activated protein kinase. Phosphorylation of glycogen synthase and phosphorylase kinase. Biochim Biophys Acta. 1989; 1012(1):81–6.
57. Hong YH, Varanasi US, Yang W, Leff T. AMP-activated protein kinase regulates HNF4{alpha} transcriptional activity by inhibiting dimer formation and decreasing protein stability. J Biol Chem. 2003;278(30):27495–501.
58. Yang W, Hong YH, Shen XQ, Frankowski C, Camp HS, Leff T. Regulation of transcription by AMP-activated protein kinase. Phosphorylation of p300 blocks its interaction with nuclear receptors. J Biol Chem. 2001;276(42):38341–4.
59. Seagroves TN, Ryan HE, Lu H, et al. Transcription factor HIF-1 Is a necessary mediator of the pasteur effect in mammalian cells. Mol Cell Biol. 2001; 21(10):3436–44.

60. Papandreou I, Cairns RA, Fontana L, Lim AL, Denko NC. HIF-1 mediates adaptation to hypoxia by actively downregulating mitochondrial oxygen consumption. Cell Metab. 2006;3:187–97.
61. Isaacs JS, Jung YJ, Mole DR, et al. HIF overexpression correlates with biallelic loss of fumarate hydratase in renal cancer: novel role of fumarate in regulation of HIF stability. Cancer Cell. 2005;8(2):143–53.
62. Lu H, Forbes RA, Verma A. Hypoxia-inducible factor 1 activation by aerobic glycolysis implicates the Warburg effect in carcinogenesis. J Biol Chem. 2002;277(26):23111–5.
63. Dalgard CL, Lu H, Mohyeldin A, Verma A. Endogenous 2-oxoacids differentially regulate expression of oxygen sensors. Biochem J. 2004; 380(Pt 2):419–24.
64. Selak MA, Armour SM, MacKenzie ED, et al. Succinate links TCA cycle dysfunction to oncogenesis by inhibiting HIF-alpha prolyl hydroxylase. Cancer Cell. 2005;7(1):77–85.
65. Fandrey J, Frede S, Jelkmann W. Role of hydrogen peroxide in hypoxia-induced erythropoietin production. Biochem J. 1994;303(Pt 2):507–10.
66. Guzy RD, Hoyos B, Robin E, et al. Mitochondrial complex III is required for hypoxia-induced ROS production and cellular oxygen sensing. Cell Metab. 2005;1(6):401–8.
67. Huang LE, Arany Z, Livingston DM, Bunn HF. Activation of hypoxia-inducible transcription factor depends primarily upon redox-sensitive stabilization of its α-subunit. J Biol Chem. 1996;271(50):32253–9.
68. Berchner-Pfannschmidt U, Yamac H, Trinidad B, Fandrey J. Nitric oxide modulates oxygen sensing by hypoxia-inducible factor 1-dependent Induction of prolyl hydroxylase 2. J Biol Chem. 2007;282(3): 1788–96.
69. Tug S, Reyes BD, Fandrey J, Berchner-Pfannschmidt U. Non-hypoxic activation of the negative regulatory feedback loop of prolyl-hydroxylase oxygen sensors. Biochem Biophys Res Commun. 2009;384(4): 519–23.
70. Gerald D, Berra E, Frapart YM, et al. JunD reduces tumor angiogenesis by protecting cells from oxidative stress. Cell. 2004;118(6):781–94.
71. Malis CD, Bonventre JV. Mechanism of calcium potentiation of oxygen free radical injury to renal mitochondria. A model for post-ischemic and toxic mitochondrial damage. J Biol Chem. 1986;261(30): 14201–8.
72. Hagen T, Taylor CT, Lam F, Moncada S. Redistribution of intracellular oxygen in hypoxia by nitric oxide: effect on HIF1α. Science. 2003;302(5652):1975–8.
73. Lando D, Peet DJ, Whelan DA, Gorman JJ, Whitelaw ML. Asparagine hydroxylation of the HIF transactivation domain a hypoxic switch. Science. 2002; 295(5556):858–61.
74. Fandrey J, Gorr TA, Gassmann M. Regulating cellular oxygen sensing by hydroxylation. Cardiovasc Res. 2006;71(4):642–51.
75. Coleman ML, McDonough MA, Hewitson KS, et al. Asparaginyl hydroxylation of the Notch ankyrin repeat domain by factor inhibiting hypoxia-inducible factor. J Biol Chem. 2007;282(33):24027–38.
76. Zheng X, Linke S, Dias JM, et al. Interaction with factor inhibiting HIF-1 defines an additional mode of cross-coupling between the Notch and hypoxia signaling pathways. Proc Natl Acad Sci USA. 2008; 105(9):3368–73.
77. Wilkins SE, Hyvarinen J, Chicher J, et al. Differences in hydroxylation and binding of Notch and HIF-1alpha demonstrate substrate selectivity for factor inhibiting HIF-1 (FIH-1). Int J Biochem Cell Biol. 2009;41(7):1563–71.
78. Keith B, Simon MC. Hypoxia-inducible factors, stem cells, and cancer. Cell. 2007;129(3):465–72.
79. Stockmann C, Fandrey J. Hypoxia-induced erythropoietin production: a paradigm for oxygen-regulated gene expression. Clin Exp Pharmacol Physiol. 2006;33(10):968–79.
80. Fandrey J, Bunn HF. In vivo and in vitro regulation of erythropoietin mRNA: measurement by competitive polymerase chain reaction. Blood. 1993;81(3): 617–23.
81. Koury ST, Bondurant MC, Koury MJ. Localization of erythropoietin synthesizing cells in murine kidneys by in situ hybridization. Blood. 1988;71(2): 524–7.
82. Fandrey J. Oxygen-dependent and tissue-specific regulation of erythropoietin gene expression. Am J Physiol Regul Integr Comp Physiol. 2004;286(6):R977–88.
83. Stevens RD, Bhardwaj A. Eryrthropoietin and the promise of ischemic multiorgan protection. Crit Care Med. 2008;36(8):2446–7.
84. Gassmann M, Heinicke K, Soliz J, Ogunshola OO. Non-erythroid functions of erythropoietin. Adv Exp Med Biol. 2003;543:323–30.
85. Brines M, Grasso G, Fiordaliso F, et al. Erythropoietin mediates tissue protection through an erythropoietin and common {beta}-subunit heteroreceptor. Proc Natl Acad Sci USA. 2004;101(41):14907–12.
86. Prunier F, Pottier P, Clairand R, et al. Chronic erythropoietin treatment decreases post-infarct myocardial damage in rats without venous thrombogenic effect. Cardiology. 2009;112(2):129–34.
87. Burger D, Lei M, Geoghegan-Morphet N, Lu X, Xenocostas A, Feng Q. Erythropoietin protects cardiomyocytes from apoptosis via up-regulation of endothelial nitric oxide synthase. Cardiovasc Res. 2006;72(1):51–9.
88. Soliz J, Joseph V, Soulage C, et al. Erythropoietin regulates hypoxic ventilation in mice by interacting with brainstem and carotid bodies. J Physiol. 2005; 568(Pt 2):559–71.
89. Soliz J, Thomsen JJ, Soulage C, Lundby C, Gassmann M. Sex-dependent regulation of hypoxic ventilation in mice and humans is mediated by erythropoietin. Am J Physiol Regul Integr Comp Physiol. 2009; 296(6):R1837–46.

90. Murry CE, Jennings RB, Reimer KA. Preconditioning with ischemia: a delay of lethal cell injury in ischemic myocardium. Circulation. 1986;74(5): 1124–36.
91. Czibik G, Wu Z, Berne GP, et al. Human adaptation to ischemia by preconditioning or unstable angina: involvement of nuclear factor kappa B, but not hypoxia-inducible factor 1 alpha in the heart. Eur J Cardiothorac Surg. 2008;34(5):976–84.
92. Fraisl P, Aragones J, Carmeliet P. Inhibition of oxygen sensors as a therapeutic strategy for ischaemic and inflammatory disease. Nat Rev Drug Discov. 2009;8(2):139–52.
93. Wang GL, Semenza GL. Desferrioxamine induces erythropoietin gene expression and hypoxia-inducible factor 1 DNA-binding activity: implications for models of hypoxia signal transduction. Blood. 1993;82(12):3610–5.
94. Rolfs A, Kvietikova I, Gassmann M, Wenger RH. Oxygen-regulated transferrin expression is mediated by hypoxia-inducible factor-1. J Biol Chem. 1997; 272(32):20055–62.
95. Mukhopadhyay CK, Mazumder B, Fox PL. Role of hypoxia-inducible factor-1 in transcriptional activation of ceruloplasmin by iron deficiency. J Biol Chem. 2000;275(28):21048–54.
96. Lee PJ, Jiang BH, Chin BY, et al. Hypoxia-inducible factor-1 mediates transcriptional activation of the heme oxygenase-1 gene in response to hypoxia. J Biol Chem. 1997;272(9):5375–81.
97. Peyssonnaux C. Regulation of iron homeostasis by the hypoxia-inducible transcription factors (HIFs). J Clin Invest. 2007;117(7):1926–32.
98. Wang F, Sekine H, Kikuchi Y, et al. HIF-1[alpha]-prolyl hydroxylase: molecular target of nitric oxide in the hypoxic signal transduction pathway. Biochem Biophys Res Commun. 2002;295(3):657–62.
99. Huang Y, Giordano FJ. Chapter 13 Oxygen as a direct and indirect biological determinant in the vasculature. Methods Enzymol. 2008;444:285–304.
100. Gleadle JM, Ratcliffe PJ. Induction of hypoxia-inducible factor-1, erythropoietin, vascular endothelial growth factor, and glucose transporter-1 by hypoxia: evidence against a regulatory role for Src kinase. Blood. 1997;89(2):503–9.
101. Plate KH, Breier G, Weich HA, Risau W. Vascular endothelial growth factor is a potential tumour angiogenesis factor in human gliomas in vivo. Nature. 1992;359(6398):845–8.
102. Stockmann C, Doedens A, Weidemann A, et al. Deletion of vascular endothelial growth factor in myeloid cells accelerates tumorigenesis. Nature. 2008;456(7223):814–8.
103. Vincent KA, Shyu KG, Luo Y, et al. Angiogenesis is induced in a rabbit model of hindlimb ischemia by naked DNA encoding an HIF-1{alpha}/VP16 hybrid transcription factor. Circulation. 2000;102(18):2255–61.
104. Semenza GL. Hypoxia-inducible factor 1: master regulator of O_2 homeostasis. Curr Opin Genet Dev. 1998;8(5):588–94.

Control of Breathing

3

Luc J. Teppema and Remco R. Berendsen

Abstract

Upon rapid ascent to high altitude or acute exposure to hypoxia ventilation rises rapidly (the acute hypoxic ventilatory response, initiated by the carotid bodies containing oxygen sensors), followed immediately by a secondary roll-off that partly is caused by the hyperventilation-induced hypocapnia. When the hypoxia is sustained, ventilation starts to rise gradually within the next few hours, that, depending on the species and altitude (degree of hypoxia), may last from hours to months. This phenomenon is known as ventilatory acclimatization to high altitude (VAH). Many studies have shown that this ventilatory adaptation is not due to a gradual acidosis in blood or extracellular brainstem fluid bathing the central chemoreceptors. One of the most remarkable changes is a considerable increase in the ventilatory response to hypoxia that is manifest already after ~4 h and that may continue to augment during the next days–weeks. This adaptive change is thought to underlie VAH and is caused by many plastic morphological and biochemical changes in both the carotid bodies and the central nervous system. An important orchestrator of these plastic changes is the transcription factor, hypoxia inducible factor 1 (HIF-1) that induces the transcription of many genes encoding substances indispensable for the adaptation of the organism to sustained hypoxia. This chapter describes the most important of these and summarizes the state of the art on the mechanisms underlying the plastic changes in carotid bodies and central nervous system.

Ventilatory Acclimatization to High Altitude and the Increase in the Hypoxic Ventilatory Response

At high altitude the uptake of a given amount of oxygen requires a higher level of pulmonary ventilation than at sea level due to the lower partial

L.J. Teppema Ph.D. (✉) • R.R. Berendsen, M.D.
Department of Anesthesiology,
Leiden University Medical Center,
P.O. Box 9600, 2300 RC Leiden, The Netherlands
e-mail: l.j.s.m.teppema@lumc.nl

E.R. Swenson and P. Bärtsch (eds.), *High Altitude: Human Adaptation to Hypoxia*,
DOI 10.1007/978-1-4614-8772-2_3, © Springer Science+Business Media New York 2014

pressure of oxygen. An increase in ventilation is one of the first adaptations to a hypoxic environment. This hypoxic ventilatory response (HVR) initiated by the peripheral chemoreceptors of the carotid bodies occurs in several phases, depending on the pattern, duration, and intensity of the hypoxic stimulus. The ventilatory response to acute poikilocapnic hypoxia, 30–60 min in duration, is characterized by a rapid increase in ventilation (decrease in arterial PCO_2) that, depending on the intensity of the hypoxic stimulus, is followed by a secondary decrease (roll-off), caused by the fall in $PaCO_2$ and by a centrally mediated depression. Within the first few hours of sustained hypoxia ventilation starts to rise again and this can continue for hours, days, weeks, or even months depending on the species and on the altitude [1, 2]. This phenomenon during short-term acclimatization is known as ventilatory acclimatization to high altitude (VAH). Ventilatory deacclimatization is the gradual return of ventilation to its normoxic level with descent to low altitude, i.e., a continuing slow decline in hyperventilation upon return to normoxia (reviewed in [3, 4]). Long-term adaptation takes years and can be accompanied by a "blunting" of the HVR (see also [3, 4]). Evolutional adaptation to high altitude is covered by Beall in this volume (Chap. 19; see also [5]). The different time domains of the ventilatory response to hypoxia are schematically shown in Fig. 3.1.

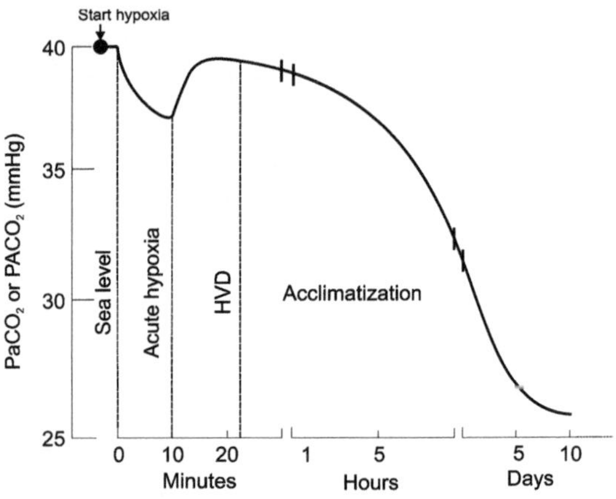

Fig. 3.1 Time course of the ventilatory response to hypoxia. The acute rise in ventilation (drop in PCO_2) is followed by a secondary roll-off with a rise in PCO_2 (hypoxic ventilatory depression HVD). During acclimatization there is a gradual drop in PCO_2 due to the rise in ventilation. Reproduced from reference [4].

Mechanisms Contributing to VAH

Smith et al. [4] have reviewed the processes underlying VAH and placed these in the context of the mechanical, diffusion, and other limitations imposed on the respiratory system by a hypoxic environment (see also [6, 7]). Additional studies in both animals and humans over the last 10–12 years have provided new insights into the way the carotid bodies and the central nervous system increase the HVR and contribute to the ventilatory adaptation in chronic hypoxia, and the purpose of this brief review is to summarize this new information. How crucial the HVR is for work performance at high altitude is covered in Chap. 16 of this volume. The separate important subject of the control of breathing during sleep at high altitude is taken up in Chap. 17.

Role of Carotid Bodies and Peripheral Chemoreceptors

Crucial Role of the Carotid Bodies in VAH

Major cell types in the carotid bodies are type I or glomus cells considered the oxygen sensing elements, and type II cells that have a glia-like appearance [8, 9]. Type I cells synapse with peripheral nerve endings of sensory neurons in the petrosal ganglion. These neurons send axonal projections to the caudal subnuclei of the NTS that act as the first relay station of carotid body afferent input to the medullary respiratory control center (see [40]).

Many studies in both animals and humans clearly indicate that VAH is associated with an increase in HVR. We consider the goat model developed by Bisgard and Forster (references in [4]) as representative for humans because studies in humans performed over the last 13 years by Robbins and coworkers have revealed remarkable similarities in the way goats and humans adapt to chronic hypoxia, at least in the initial stage. Goats show (a virtually complete) VAH (i.e., a stage wherein ventilation and $PaCO_2$ have reached a new steady-state level) and an increase in hypoxic sensitivity within 4 h of carotid body hypoxia. Similarities between humans and goats are: (a) an equally rapid increase in acute HVR after both poikilocapnic and isocapnic hypoxia [10]; (b) hypocapnia alone does also not evoke VAH in humans [11]; (c) an increased peripheral chemoreceptor sensitivity during prolonged hypoxia [12, 13]; (d) sympathetic and dopaminergic mechanisms are not involved in the HVR increase [14–16]. Further observations by Robbins and coworkers were: (a) hypocapnic hyperventilation per se cannot account for VAH or enhanced HVR [17]; (b) lower oxygen content, induced by phlebotomy or breathing low concentrations of CO, did not evoke VAH [11]; (c) keeping subjects at an end-tidal PO_2 as little as 10 Torr below resting for 5 days increases the HVR and peripheral CO_2 sensitivity, and maintaining end-tidal PO_2 10 Torr *above* resting has the reverse effects [18]; (d) resting ventilation in normoxia and hyperoxia consistently increase in the course of chronic poikilocapnic and isocapnic hypoxia; this is accompanied by an enhanced peripheral (hyperoxic) CO_2 sensitivity without changes in intercept of the (hyperoxic) CO_2 response curve, indicating an augmented carotid body sensitivity as the underlying mechanism for the hyperventilation [12, 13, 19]. Robbins and colleagues used dynamic end-tidal forcing (a technique enabling manipulation and control of end-tidal PO_2 and PCO_2 independently from changes in ventilation) and subjected their data to sophisticated model analysis to separate peripheral and central CO_2 sensitivity. Other workers have used modified rebreathing but with less consistent results, e.g., sometimes failing to find an increase of the HVR [20, 21]. Taken together all human studies on chronic hypoxia, both in the field and under simulated hypobaric circumstances (operation Everest II—[22]), what emerges is a clear picture of a gradual rise in ventilation and fall in end-tidal PCO_2, a rapid increase of the acute HVR, a relative short period of hyperventilation upon (short) return to normoxia and hyperoxia, and finally an increased peripheral CO_2 sensitivity [18]. The carotid bodies play a pivotal role in all these adaptations.

Other species such as (high altitude bred) guinea pigs [23] and mice [24–26] also clearly show a gradual rise in ventilation and an augmented HVR. In rats the findings were somewhat variable, some studies showing a relatively rapid return to normoxic rate of ventilation within 1 week (e.g., [27–29]), while others clearly demonstrated an increased ventilation (fall in PCO_2) and/or enhanced HVR after 2–7 weeks of acclimatization [30, 31]. Similar to goats, both rats and cats display a gradual rise in discharge frequency of the carotid sinus nerve in chronic hypoxia and an increase in the carotid body response to hypoxia [32–34].

Oxygen Sensing in the Carotid Bodies

The currently accepted framework of oxygen sensing is that hypoxia leads to membrane depolarization of type I cells caused by a reduced current through specific potassium channels, a subsequent influx of calcium ions and a release of neurotransmitters, of which ATP and acetylcholine are the most important in exciting afferent nerve endings that lie in close apposition to Type I cells (reviewed in [35, 36]). Recent studies have shown a coupling of cellular metabolism and potassium channel inhibition via AMP-activated protein kinase (AMPK—[37, 38]). A remarkable property of the carotid body is that it contains numerous neurotransmitters and modulators, the functions of which are far from clear. Prabhakar [39] has proposed a "push-pull" concept involving both excitatory (many of which are discussed below) and less well studied inhibitory transmitters. A variable balance between excitatory and inhibitory transmitters could act to produce a sustained activation in chronic hypoxia, instead of a situation in the absence of inhibitors, whereby overactivation would occur followed by an exhaustion of the response.

In animals, with chronic hypoxia (acclimatization to high altitude), the carotid body shows morphological and biochemical adaptations that result in a net increase in carotid sinus nerve activity and increased sensitivity to hypoxia in the acute stage of acclimatization. However, humans and several animal species with a history of generations at high altitude develop a blunting of the hypoxic response, often in the presence of enlarged carotid bodies.

When animals are exposed (and acclimatized) to chronic hypoxia the plastic changes occurring in the carotid bodies are fast, sometimes transient, and refer to both the oxygen-sensing candidates themselves but also to virtually all the local neurotransmission systems and modulators [40]. It would be interesting to know to what extent these animal data can be translated to human sea level residents who move to high altitude, but obviously invasive carotid body studies in humans are impossible. It is of interest, moreover, to consider potential effects of pharmacological agents used by many people staying at high altitude for shorter or longer periods in the light of what we know from animal studies because these drugs may influence their acclimatization, VAH and HVR via the carotid bodies or other respiratory structures.

Neurochemical and Membrane Ion Channel Adaptations

Dopamine is considered a predominantly inhibitory transmitter in the carotid bodies although not without controversies [8]. In the rat and cat, carotid body chronic hypoxia is associated with a rapid up-regulation of tyrosine hydroxylase (TH, the rate-limiting enzyme for dopamine synthesis) and an increase in dopamine turnover (references in [41–43]). In cats, rats, and rabbits dopamine is down-regulated in the initial stage of chronic hypoxia but up-regulated in later stages [24, 28, 29, 44–46]. Male mice with a targeted deletion of dopamine D_2 receptors failed to show VAH and an increase in HVR upon exposure to hypoxia for 2–8 days [25]. In goats and humans, however, no involvement of dopaminergic mechanisms in early acclimatization could be demonstrated [15, 47, 48]. Altogether, at least in the initial stages of acclimatization, changes in carotid body dopaminergic mechanisms alone are probably not responsible for (or even related to) the increase of the HVR and VAH, but in some species, in later stages of the adaptation, a modulating role of dopamine cannot be ruled out.

Norepinephrine. Data from cats suggest that down-regulation of carotid body noradrenergic/adrenoceptor mechanisms may contribute to the chronic hypoxia-induced increase of the HVR [49, 50]. In other species, elevated carotid body noradrenaline levels were reported (references in [8, 51]). In humans, exposure to high altitude leads to sympathetic activation and increased plasma catecholamine levels [52] but sympathetic/noradrenergic mechanisms do not appear to play a significant role in VAH [14, 16, 53]. Also, at least in the initial stage of acclimatization, changes in chemoreflex sensitivity and muscle sympathetic nerve activity may not run in parallel: chemoreflex sensitivity is enhanced after this period, but muscle sympathetic nerve activity is decreased (after an initial increase—[54]). After 4 weeks at 5,260 m, however, the picture may be quite different with a clear sympathetic neural overactivity [55].

Acetylcholine. A role of the cholinergic system in acclimatization is unclear. Acclimatized rats showed an augmented stimulation by acetylcholine of their carotid sinus nerve discharge on the one hand but failure of the nicotine receptor antagonist mecamylamine to cause any reduction in the hypoxia-induced increase in discharge on the other hand [56]. Studies in isolated type I cells from acclimatized rats indicate that chronic hypoxia may alter the balance between the cholinergic and dopaminergic systems to make the cells more responsive to O_2 [57]. Whereas mecamylamine failed to alter basal DA release from type I cells isolated from juvenile rats cultured in normoxia, the same concentration almost completely inhibited the release from cells cultured in hypoxia for ~12 days [57].

ATP. He et al. [58] suggested an altered balance between purinergic and acetylcholinergic transmission as a mechanism for VAH because in contrast to nicotinic antagonists, $P2X_2$ purinoceptor antagonists retained their ability to block the hypoxia-induced increase in nerve activity in acclimatized rats (in chronic hypoxic rats the inhibitory effect on neural discharge was larger).

Preparations from acclimatized but not control animals showed a prominent after-discharge in the carotid sinus nerve after removal of a brief hypoxic stimulus that was substantially reduced by purinergic blockers indicating that this after-discharge may be caused by prolonged ATP release after the removal of hypoxia [58]. It remains to be seen if this phenomenon may be relevant to the well-known continued hyperventilation in men and animals upon return from high altitude.

Nitric oxide, an inhibitory carotid body neurotransmitter, is up-regulated in chronic hypoxia ([59]; further references in [42]). In rats exposed to 380 Torr for up to 16 days, NOS inhibition caused a progressively greater stimulation of the carotid sinus nerve while the inhibitory effect of an NO donor also increased [60]. It is unknown how mice deficient in neuronal [61] or endothelial NOS [62] adapt to chronic hypoxia.

Endothelin-1. In the rat, exogenously applied endothelin-1 (ET-1) augments carotid sinus nerve discharge in situ and increases ventilation [63–65]. In chronic hypoxia, endothelin, along with ET_1—and ET_A receptors, is up-regulated in type I cells [33, 42, 65]. In the course of chronic hypoxia, selective inhibition of ET_A receptors progressively reduced the hypoxia-induced increase in carotid sinus nerve discharge in vitro from 11 % on day 0 (normoxia) to 50 % on day 16 [33]. Treating rats with the $ET_{A/B}$ receptor antagonist bosentan during exposure to 380 Torr for 14 days not only substantially diminished the hypoxic response of the carotid sinus nerve but also greatly reduced the carotid body enlargement and mitotic activity in type I cells [66]. Exposing healthy humans to hypoxia during 4 h increased their plasma [ET-1], while continuous infusion of ET-1 during such a hypoxic challenge augmented the chronic hypoxia-induced increase of their HVR [67].

Angiotensin II (ANG II), acting via AT_1 receptors, excites the carotid body (references in [68, 69]). In the rat, chronic hypoxia leads to a progressive increase in local angiotensinogen mRNA and protein, AT_1 receptor expression and angiotensin-converting enzyme (ACE) mRNA, as well as an enhanced chemoreceptor ANG II sensitivity [68, 69]. Chronic hypoxia also up-regulates AT_4 receptors that bind angiotensin IV, a pentapeptide metabolite of ANG II that raises intracellular $[Ca^{2+}]$ in glomus cells (and more so after chronic hypoxia—[70]). In rabbits with congestive heart failure, activation of AT_1 receptors by angiotensin II plays a key role in the sensitization of the carotid bodies [71]. Angiotensin II activates NADPH oxidase, which leads to increased carotid body superoxide anion production and enhanced O_2 sensitivity. Superoxide dismutase levels in the carotid bodies of these animals are down-regulated, while exogenous administration of this superoxide anion scavenger reverses the increase in carotid body sensitivity [71]. These interesting data, coupled to the known stimulating effect of ANG II on ventilation [72, 73], may be of relevance for human adaptation to high altitude for several reasons. Humans tend to increase their ACE levels upon ascent to high altitude [74]. Does this also contribute to the increase in HVR? In long-term adaptation, ACE levels normalize [75]. Low circulating ACE levels are of benefit in the physiological adaptation to high altitude [76, 77]. Does this play a role in "blunting" of the HVR in high altitude natives? Part of the variation in plasma ACE levels is explained by polymorphism of the *ACE* gene, with a prevalence of the insertion (I) allele (i.e., the presence of a 287-base fragment in intron 16) over the deletion allele (D) associated with lower ACE levels, better high altitude performance, and higher oxygen saturation during exercise (references in [5, 76, 78, 79]). In Peruvian Quechua, the ACE genotype does not appear to be associated with the magnitude of the isocapnic HVR [79]. However, a study in Caucasians showed a greater HVR in subjects homozygous for the I allele [80]. Individuals with the ACE I/I genotype had more success in reaching the summit of Mt Blanc [76].

Proinflammatory cytokines. Type I cells contain receptors for the interleukins IL-1β and IL-6 and for TNF-α [81, 82]. Type-I cells from chronic hypoxic rats displayed transcriptional up-regulation and increased protein expression of these proinflammatory cytokines and their receptors while they also showed increased $[Ca^{2+}_i]$

responses to these cytokines [83]. Liu et al. [34] showed up-regulation of carotid body IL-1β and TNF-α already after 1 day of hypoxia (exposure to 380 Torr) when the number of invaded macrophages was still low. After prolonged exposure of 28 days cytokine levels were normalized, except those for IL-6, which could be ascribed to increased production by type-I and -II cells. Finally, the chronic hypoxia-induced increase in hypoxic carotid sinus sensitivity was inhibited by dexamethasone and ibuprofen [34]. These intriguing data need further verification in humans: would NSAIDs and corticosteroids at high altitude influence the acclimatization process? That at least the initial phase of exposure to high altitude is associated with increased plasma levels of inflammatory cytokines may be relevant in this context; whether or not this occurs via a reactive oxygen species-dependent pathway remains controversial (references in [84]).

Ion channels and membrane currents. Several studies in type I cells from adult chronic hypoxic animals indicate that they manifest altered expression profiles and/or activity of O_2-sensitive potassium channels and other channel species such as sodium channels and gap junctions. For example, in rat type I cells chronic hypoxia leads to a decrease in current density in the membrane of type I cells, but not that caused by maxi-K channels (i.e., large conductance calcium-dependent potassium channels which are considered as oxygen sensitive channels in this species); this results in a larger contribution of these channels to the resting membrane potential and an increased hypoxia-induced catecholamine release thus contributing to the increase in hypoxic sensitivity in chronic hypoxic animals [85–90].

Role of HIFs in VAH. Generally, chronic hypoxia is accompanied by up-regulation of genes that promote tissue oxygenation, facilitate oxygen uptake by cells, and adapt cellular metabolism, e.g., by limiting and optimizing oxidative phosphorylation. The master plan of gene expression in chronic hypoxia is orchestrated by the hypoxia-inducible transcription factor HIF-1 [91, 92]. HIF-1 is a heterodimer consisting of the nuclear subunit HIF-1β and the cytosolic subunit HIF-1α. Both subunits are constitutively expressed but HIF-1α undergoes posttranscriptional modification under the influence of the PO_2. In normoxic conditions HIF-1α is subject to degradation but in hypoxia it is rescued from proteasomal breakdown, where after it is translocated to the nucleus where it dimerizes with HIF-1β to form the transcription factor HIF-1. Increased mitochondrial production of ROS is thought to play an important part in preventing the breakdown of HIF-1α in hypoxia ([93, 94]; see also Chap. 1). The dimer can now bind to the hypoxia response elements of target genes where it recruits co-activator proteins to promote transcription. In this way many target genes can be induced, e.g., those encoding erythropoietin, vascular endothelial growth factor, glucose transporters, ET-1, and many others [91].

By setting the level of gene products that influence the signal transduction and morphology of the carotid bodies and the further processing of afferent input by the central nervous system, HIF-1 can be considered an important modulator of the HVR. In the carotid bodies, HIF-1α (but also the paralogues HIF-2α and HIF-3α) is (are) up-regulated during chronic hypoxia, which has been related to several of the hypertrophic, vascular, and neurochemical adaptations discussed above [95, 96]. Generally, neurons, astrocytes, and endothelial cells in the central nervous system respond to chronic hypoxia with up-regulation of HIF-1α aimed at restoring tissue oxygen levels [97]. For example, induction of some target genes results in an increased synthesis of neuronal NOS [98, 99] and of TH in catecholaminergic cardiorespiratory neurons in the medulla [100]. For further references on the role of HIF-1α in chronic hypoxia we refer to [43].

That HIF-1α may play an important part in acclimatization is supported by data from both animals and humans. Mice with a heterozygous loss of function of the *HIF-1α* gene (*HIF-1α*$^{+/-}$) have a substantially reduced carotid sinus nerve response to acute hypoxia; do not show the normal pattern of VAH; and have pulmonary hypertension, right ventricular hypertrophy, and a delayed polycythemia [101, 102]. Humans with Chuvash polycythemia have increased HIF-1 levels due to impaired degradation of HIF-1α and show an abnormally high resting ventilation, an exaggerated HVR, and a low arterial PCO_2 [103]. This phenotype is reminiscent of VAH; also, these subjects have an elevated pulmonary arterial pressure and show exaggerated pulmonary vasoconstriction in acute hypoxia [103]. Chronic hypoxia not only increases the HVR but also augments pulmonary vascular O_2 sensitivity, and the latter phenomenon seems to depend on the iron status of the body [104]. An influence of the iron status on VAH and the (increase in) HVR could not be demonstrated but cannot be ruled out [104]. Also in relation to HIF-1, altered expression of the *Epo* gene, one of its many target genes, and of the Epo receptor may contribute to VAH (see also above).

In fact the involvement of HIF-1 in VAH will be rather complex because often paradigms with intermittent hypoxia will be superimposed on the chronic hypoxia itself. For example, sleeping at high altitude is characterized by periodic breathing frequently accompanied by apneas. As is the case in sleep apnea patients, the resultant intermittent hypoxia will up-regulate HIF-1α and is associated with generation of reactive oxygen species (reviewed in [105]).

More transcription factors than HIF-1 alone contribute to VAH [106]. For example, a role of the transcription factor activator protein-1 (AP-1, a heterodimer comprising proteins belonging to the c-Fos and c-Jun families) is indicated by the finding that mice lacking the *fos B* gene (a member of the *fos* family of immediate early genes), although having a normal HVR when kept in normoxia, do not exhibit VAH and an increase in HVR when exposed to chronic hypoxia [107]. The interesting question remains as to whether the up-regulation of various transcription factors, apart from sensitizing O_2 chemoreceptors, may also recruit oxygen sensing mechanisms that are silent in normoxia.

Role of reactive oxygen species (ROS) in VAH? A thus far unanswered question is whether during chronic hypoxia (CH) an altered balance between pro- and antioxidants plays a role in the increased carotid body sensitivity as in chronic intermittent hypoxia (CIH). In animals, CIH results in an augmented carotid body O_2 (not CO_2) sensitivity that can be prevented by an antioxidant treatment [101, 102, 108]. In contrast to wild types, heterozygous HIF-1$\alpha^{+/-}$ mice fail to respond to hypoxia and do neither accumulate ROS or augment their carotid body O_2 sensitivity after CIH. Mice with a haploinsufficiency for HIF-2α (*Epas 1*$^{+/-}$) resulting in a reduced expression of antioxidant enzymes, notably superoxide dismutase, show an augmented response to hypoxia, unstable breathing, and hypertension [108]. Chronic hypoxia in these animals has not been studied yet. These findings may suggest a role of ROS in the adaptation to high altitude hypoxia, but two points should be emphasized here. First, CH and CIH lead to different carotid body adaptations. For example, in contrast to CH, CIH is not accompanied by morphological changes in the carotid bodies and does not result in an appreciable increase in baseline ventilation [40]. Second, ROS are not directly involved in the process of oxygen sensing per se, and neither oxidizing nor reducing agents are able to modulate the hypoxia-induced catecholamine release by type I cells [40, 109]. Further studies are needed to study a possible role of ROS in VAH.

AMP-activated protein kinase (AMPK). AMPK is an important metabolic sensor that is activated by an increase in the AMP/ATP ratio that occurs in metabolic stress conditions, for example acute hypoxia [110, 111]. The enzyme suppresses ATP-consuming processes, while catabolic processes are up-regulated (references in [40]). AMPK is able to phosphorylate various types of O_2-sensitive potassium channels and as such is suggested to provide the link between oxygen sensing and cell metabolism [37, 38, 109]. In muscle cells, AMPK activation results in increased uptake of glucose and insulin sensitivity [111, 112]. Chronic hypoxia has similar effects but it is unclear whether this is also mediated by up-regulation of AMPK [113]. Further studies are needed to identify metabolic pathways that are linked to VAH and to determine the role of AMPK.

Hydrogen sulphate. Genetically engineered mice lacking cystathionine γ-ligase (CSE, the enzyme that is required for H_2S synthesis in the carotid body) have severely impaired sensitivity to hypoxia [114]. The identity, however, of H_2S as a mediator of oxygen sensitivity has been questioned because it also inhibits mitochondrial function (cytochrome *c*) raising intracellular Ca^{2+} and as yet has not been shown to inhibit O_2-sensitive potassium channels [109]. In addition, the levels of H_2S required to increase carotid body activity seem to be toxic concentrations known to increase ventilation and inhibit cytochrome *c* [115]. It is unknown how CSE knockout mice adapt to chronic hypoxia.

Morphological Changes in the Carotid Bodies

The carotid bodies undergo considerable morphological changes in chronic hypoxia. Typical vascular adaptations are remodeling and endothelial proliferation [51, 95, 116–118]. Earlier studies showed enlarged type I cells with increased numbers of mitochondria and enlarged dense core vesicles [119–122]. More recently, Bisgard and coworkers, using bromodeoxyuridine (BrdU), a uridine analog that is incorporated into cells undergoing cell division, reported hyperplasia of glomus cells from rats during the first 1–3 days of hypoxia but no further increase in labeling after 1 week [51, 123]. Carotid bodies from mice exposed to hypoxia for up to 20 days showed an increased BrdU incorporation after 1 day that was followed by a *delayed* rise in the amount of $BrdU^+/TH^+$ (i.e., dividing type I) cells from hypoxic day 5 on, even *before* a macroscopically recognizable increase in carotid body size; upon return to normoxia the size normalized, but 50 % of TH^+ cells were newly formed ([124], see also [117]). These intriguing observations led Pardal et al. [124] to suggest the existence of progenitor cells starting to divide before they differentiate into TH-containing type I cells. Type II cells were the predominant cell type to divide in the first 24–48 h; with progressing hypoxia almost all of them were replaced by intermediate progenitor cells and after about 1 week dividing TH^+ cells appeared [124].

Animals (such as guinea pig, dogs, cats, and dogs but not llama and alpaca) and humans living at high altitude possess enlarged carotid bodies [125–127]. Adaptation to *long-term* hypoxia is associated with glomus cell hyperplasia of the so-called light variant of type I cells, which is preceded by multiplication of the "dark" variant that may operate as the progenitor of the light variety [128, 129]. At least in the Andes, high altitude is associated with a relatively high incidence of chemodectomas in humans and animals [125, 126, 130]. It has been suggested that this type I cell hyperplasia, together with the diminished HVR reflects a genetic failure to adapt to high altitude (references in [131]). Hyperplasia during short-term adaptation, however, may well be associated with an increase in the hypoxic response.

In summary, the carotid bodies respond to chronic hypoxia with profound adaptive and possibly species-dependent changes in morphology, membrane properties, and expression of numerous stimulatory and inhibitory modulators/neurotransmitters. Under the orchestration of transcription factors such as HIF-1 all these simultaneous plastic changes may serve to keep the balance between excitatory and inhibitory influences within control limits. The question as to which of these adaptations are relevant for humans at high altitude remains unanswered to date. Also it might be anticipated that commonly used pharmacological agents such as ACE inhibitors, inhibitors of endothelin receptors, β-blockers, phosphodiesterase-5 inhibitors, and anti-inflammatory agents could influence the acclimatization process per se by their possible actions in the carotid body.

Role of Central Nervous System Changes to HVR and VAH

Although the peripheral chemoreceptors and carotid body play the dominant role in the VAH, changes in the CNS may have modulating effects on the final response. To investigate a possible adaptation (facilitation) of the processing of afferent carotid body input by the central nervous system, Dwinell and Powell [132] compared the effect of electrical carotid sinus nerve stimulation on phrenic nerve activity in control and acclimatized rats (380 Torr 1 week) and indeed found a large increase in the central nervous system gain ("CNS gain") of a standard electrical input from this nerve.

This phenomenon may also exist in other species (e.g., humans, ponies—for references see [133, 134]), but in the cat it is not manifest, at least after a hypoxic exposure lasting 48 h [32]. In goats, NaCN (assumed to cause maximal carotid body stimulation) caused equal effects on ventilation before and after 4 h of isocapnic or poikilocapnic hypoxia [135], but this does not preclude changes in CNS gain during prolonged hypoxia in this species. The potential mechanisms for this CNS facilitation and gain are discussed below.

Altered balance of peripheral and central dopaminergic systems. In the NTS, dopamine, acting via dopamine-D_2 receptors, has a stimulatory effect on ventilation [136], so local up-regulation of TH during chronic hypoxia as observed in rats [137] may facilitate respiratory output. In mice, the stimulatory effect on ventilation of the lipophilic dopamine antagonist droperidol in normoxia was transformed into an inhibitory one after chronic hypoxia [138]. Mice with a deletion of the *dopamine-D_2 receptor* gene did not manifest any VAH after 8 days of acclimatization [25, 136]. Thus, it was suggested that a shift in the relative contribution of the peripheral and central dopaminergic systems may contribute to VAH [136].

Up-regulation of NMDA receptors. Processing of afferent carotid body input by the caudal NTS occurs via glutamatergic (NMDA and non-NMDA) receptors [139, 140]. In chronic hypoxic rats, the dorsocaudal brain stem showed time-dependent changes in the expression of individual NMDA receptor subunits that were related to the increase in the acute HVR [141]. Up-regulation of NMDA receptors was observed in the medulla of chronic hypoxic mice [142]. In chronic hypoxic rats, the NMDA antagonist MK-801 had a larger inhibitory effect on hypoxic ventilation than in unacclimatized controls, also suggesting up-regulation of NMDA receptors [143].

PDGF-β receptors. When activated, platelet-derived growth factor-β receptors reduce NMDA receptor-mediated ionic currents (references in [139]). Down-regulation of PDGF-β receptors during chronic hypoxia was suggested to contribute to VAH because (1) PDGF-β injection in the NTS decreased the acute hypoxic response in the rat [144]; (2) mutant mice heterozygous for the PDGF-β receptor showed a reduced hypoxic ventilatory decline [139]; (3) rats acclimatized to 10 % O_2 during 2 weeks showed down-regulation of PDGF-β receptors in the dorsocaudal brainstem that was significantly correlated with VAH, while the attenuation of the hypoxic response by injection of PDGF-β in the NTS decreased over time and in fact was absent from day 7 on [145]. Mice lacking the PDGF receptor gene show a substantially reduced hypoxic ventilatory decline but VAH in these mice has not been studied yet [146].

NO-related mechanisms. NO is an important central excitatory mediator of the HVR. Chronic hypoxic mice had higher neuronal NOS levels and NO production in their medulla that seemed related to VAH [142]. Up-regulation of neuronal NOS was also observed in peripheral and central neurons of rats exposed to 12–24 h hypobaric hypoxia [99]. A hypoxia-induced rise in intracellular NO in the NTS may be one of the secondary consequences of NMDA receptor activation [139]. Thus, a feed-forward mechanism may exist in the NTS between NMDA receptor activation and NO release to enhance the CNS gain [147]. Rats exposed to chronic hypoxia and receiving a continuous intracerebroventricular infusion of the NOS inhibitor L-NAME displayed a lower normoxic ventilation (frequency) than controls after exposure indicating a reduced VAH. However, their HVR was indistinguishable from acclimatized control animals that had received no NOS inhibitor [148]. Thus, up-regulation of NO in the medulla during chronic hypoxia may contribute to VAH but the scarce data do not support a contribution to the increased HVR.

Erythropoietin (*Epo*). Mice overexpressing human Epo in the brain without elevation of plasma Epo and hematocrit (Tg21 mice) were reported to show larger increases in ventilation and HVR in chronic hypoxia, even after carotid sinus nerve section [149]. Female (but not male) mice with increased peripheral and central Epo expression (Tg6 mice, with increased hematocrit) were reported to have an augmented HVR [150]. However, another study in Tg6 mice (sex not mentioned) reported no difference in HVR with wild types; after acclimatization, VAH and an augmented HVR were clearly present in wild types but totally absent in the Tg6 genotype and this was suggested to be related to up-regulated carotid body dopaminergic mechanisms [26]. Finally, infusion into the brain of the soluble Epo receptor—that normally is down-regulated in chronic hypoxia thus increasing Epo bioavailability—reversed VAH in normal mice [151]. These studies in mice should be interpreted with caution because they were obtained with body plethysmography, hypoxic challenges were performed in poikilocapnic conditions, arterial blood gases were not available, and ventilation levels were low relative to CO_2 production.

In humans Epo concentration reaches a peak after 2 days of chronic hypoxia [152]. An intermittent hypoxia paradigm in male subjects for 4 days caused a rise in plasma [Epo] that was correlated with the increase in HVR [153]. Acute infusion of Epo in humans was reported to increase the HVR in women but not males, but this study is seriously flawed by the extremely low reported minute ventilation in these subjects and by the fact that the authors treated the *in*spired O_2 fraction as the ventilatory stimulus [150]. In conclusion, at this stage a role of Epo in VAH is not clear and more studies in acclimatized humans are needed to establish its potential role in the increase of the HVR.

Central oxygen-sensing neurons. Intrinsic oxygen sensors in the ventrolateral medulla of rats displayed enhanced sensitivity in vitro in the initial phase (4–5 days) of chronic hypoxia that in later stages (9–10 days) of acclimatization was offset [154]. Whether these neurons could also become more sensitive to synaptic input further enhancing CNS gain by a synergistic action with low PO_2 is unknown. In rats, chronic hypoxia for 10 days induced heme oxygenase 1 (but not heme oxygenase 2) in the rostroventrolateral medulla which has been suggested to facilitate respiratory output via modulation of potassium ion currents by CO generated by this enzyme [155]. Oxygen sensors may thus contribute to VAH but further studies elucidating the mechanism(s) of central oxygen sensing are needed to confirm this.

Expression profile of ion channels in the NTS. It has been reported that during chronic hypoxia outward currents through specific ATP-dependent potassium channels in NTS neurons receiving monosynaptic input from the carotid bodies are reduced. This would lead to an increased excitability of these neurons and an enhanced CNS gain [156].

Hypoadditive O_2–CO_2 interaction. Studies in the rat indicate a negative O_2–CO_2 interaction, i.e., the lower the central $PaCO_2$ the larger the stimulatory effect of hypoxia on ventilation [157, 158]. In this species this negative O_2–CO_2 interaction, suggested to be due to a negative interaction between peripheral and central chemoreceptors [158], could increase the CNS gain and thus play a role in VAH. A negative O_2–CO_2 interaction, however, has not proven to exist in humans.

Role of Carotid Bodies in CO_2 Sensitivity

Several studies have reported an augmented ventilatory response to CO_2 after chronic hypoxia or acclimatization to high altitude. Robbins and colleagues [159] exposed subjects to 8 h of isocapnic hypoxia prior to an incremental exercise protocol and found an increase in slope of the CO_2 response curve that was mainly attributed to augmented carotid body sensitivity, although a central contribution could not be excluded. In earlier studies, Mathew et al. [160] and Schoene and colleagues [22] had shown an increased ventilatory CO_2 sensitivity after acclimatization at 3,500 and 7,315 m, respectively. More recently, in humans following a gradual ascent and acclimatization to 5,500 m, Ainslie and co-workers [161], employing modified rebreathing, reported a 30 % increase in slope of the hyperoxic CO_2 response curve that they attributed to the central chemoreceptors; note however that during hyperoxia the carotid body is not silenced and that during the rebreathing procedure that starts with inhaling from a bag containing 7 % CO_2 and 93 % O_2 the influence of oxygen on CO_2 sensitivity progressively increases—see [40]. At the same time these authors also found an increased cerebrovascular CO_2 sensitivity, both in the hypocapnic and hypercapnic range. What causes this apparent increase in central ventilator CO_2 sensitivity? By itself, a higher cerebrovascular CO_2 sensitivity would tend to reduce the slope of the CO_2 response curve so this tendency must be overruled by other mechanisms. Peripheral CO_2 sensitivity will be augmented by all the adaptive changes in the carotid bodies that are discussed above and that could explain the findings of Robbins and coworkers. And certainly, an increase in central gain (section "Role of Carotid Bodies in CO_2 Sensitivity") would also augment the peripheral response to systemic changes in PCO_2.

Recent data obtained in awake dogs with one denervated carotid body and the other one perfused separately from the systemic circulation showed that when this carotid body was kept hypoxic central CO_2 sensitivity could be higher by as much as 200 % than control, while the opposite, a reduction by almost 80 %, was seen when it was perfused with hyperoxic/hypercapnic blood [162]. If humans would possess similar anatomical connections between the peripheral and central chemoreceptors as the rat [163–165] and employ a similar cross-talk between the two as suggested in this awake dog model, VAH, in a condition of combined hypoxia/hypocapnia, could, at least partly, be attributed to an increase in central chemosensitivity. Note, however, that Robbins and colleagues showed an augmented CO_2 sensitivity after both eucapnic and hypocapnic hypoxia, i.e., a combined hypocapnia and hypoxia is not a prerequisite for the increased CO_2 sensitivity to become manifest. In addition, in humans there are no convincing data showing an interaction between the peripheral and central chemoreceptors other than simply additive (references in [40]). And finally, in the dog model a combined selective carotid body hypoxia/hypocapnia, a condition similar to high altitude in which the low CO_2 would tend to reduce carotid body output by the local O_2–CO_2 interaction, was not tested.

Role of Acid–Base Status and Central Chemoreceptors?

As reviewed by Smith et al. in the first edition of this volume [4], earlier studies have failed to demonstrate a causal relationship between ventilatory adaptation and arterial, cerebrospinal fluid or brain interstitial pH. Changes in brain intracellular pH could also not account for VAH [166, 167].

A thus far unsettled issue concerns the role of lactate in VAH. Combined central hypoxia and hypocapnia during acclimatization results in an increase in brain lactate production combined with alkalosis, but the gradual alleviation of this alkalosis in time does not run in parallel with the gradual rise in ventilation, arguing against this original hypothesis of Severinghaus and also not supporting a role of a decrease in strong ion difference affecting CSF pH as suggested by Javaheri [4, 168, 169].

In evaluating the role of lactate on the HVR, a distinction should be made between the acute response to hypoxia (AHR) initiated by the carotid bodies (i.e., stimulation followed by a secondary roll-off) and the chronic response that is relevant to VAH during which both the carotid bodies and central mechanisms are influencing ventilation. With regard to the carotid bodies different effects are reported in various animal models. For example, in awake goats dichloroacetate (DCA), that reduces lactate production by activating the enzyme pyruvate dehydrogenase, increases the AHR, which may be hard to reconcile with known stimulating effects of lactate on type I cells [170] and with the smaller stimulatory effect on ventilation of intravenous lactate in carotid body denervated compared to intact rats [171]. Note, however, that DCA-induced effects may also be a secondary consequence of increased synthesis of Acetyl CoA and downstream intermediates of the citric acid cycle, rather than being caused by a decrease in lactate concentration per se. In other words, DCA may not be the most appropriate pharmacological tool to study the effects of lactate. In humans low dose DCA does not influence the AHR or HVD, but this has been suggested to be due to too low an administered dose [172]. With regard to central effects of lactate and its potential influence during VAH we refer to a frequently cited study in anesthetized, artificially ventilated, and carotid body denervated cats in which DCA prevented the occurrence of central hypoxic depression and acidosis during CO hypoxia, which was interpreted as an inhibitory effect of lactate on ventilation during hypoxia. Whether these data are relevant for VAH in conscious humans and animals is doubtful because these animals were kept under anesthesia and isocapnic, their carotid bodies were denervated, and the duration of the hypoxic challenge was only 15 min [173]. In awake goats, a time-dependent alleviation of lactate production in the brain was proposed as a mechanism underlying VAH, but also in this study a low dose DCA was used, lactate concentrations were not measured, while in addition the carotid bodies were intact so that peripheral effects of changes in lactate could not be excluded [174]. In the rat, intravenous infusion of lactate to achieve arterial levels up to 20 mM increased ventilation independent from changes in arterial PCO_2 [175]. Finally, DCA, in the maximally effective dose to inhibit lactate synthesis, greatly reduced the ventilatory response to poikilocapnic hypoxia measured over 150 min [172], suggesting that lactate may contribute to VAH rather than acting as an inhibitory stimulus. In conclusion, more studies are needed to determine how and whether CNS and peripheral lactate changes at high altitude in rest and during exercise may have a potential role in breathing at high altitude.

Despite the lack of a causal relationship between changes in brain ECF/CSF pH and ventilation during VAH, the absence of VAH after carotid body denervation or isolated central hypocapnic hypoxia in several species, and the fact that carotid body hypoxia alone is able to elicit VAH (references in section "Role of Central Nervous System Changes to HVR and VAH"), a significant role of the central chemoreceptors in the acclimatization process cannot be ruled out. In the rat, the retrotrapezoid nucleus RTN in the rostroventrolateral medulla, an important site containing specific respiratory CO_2 chemoreceptors [163], possesses afferent and efferent connections with the subnuclei of the nucleus of the solitary tract NTS where carotid body afferents terminate [164]. In carotid body intact but not in denervated rats, CO_2 sensitive neurons in the RTN are strongly activated by systemic hypoxia via a glutamatergic pathway from the commissural NTS [165]. Thus, during chronic hypoxia these neurons having extensive projections to nearby respiratory neurons may receive a time-dependent progressively increasing input from the carotid bodies and may thus appear to be an indispensable link in VAH. Recently, a canine model was developed in which one carotid body of awake dogs (the other carotid body being denervated) can be selectively perfused independently of the systemic circulation. With this model it is possible to study possible cross-talk between peripheral and central chemoreceptors and this may yield new insights into the mechanism of VAH [162]. This is briefly discussed in section "Role of Central Nervous System Changes to HVR and VAH."

Conclusion

The HVR and VAH are important determinants of success at high altitude. A subject's HVR at sea level may have a predictive value for best performance at high altitude (e.g., [176, 177]) but this is not necessarily always the case (e.g., [178, 179]). Obviously, what matters is the magnitude of ventilation at high altitude that for a number of reasons will be the result of more than just the immediate effect of inspired low oxygen, such as low barometric pressure, diffusion, and mechanical limitations, the O_2–CO_2 interaction in the carotid bodies (the lower the PCO_2 the lower carotid body sensitivity to changes in PaO_2), a possible peripheral–central interaction (the lower the PaO_2 the higher central CO_2 sensitivity) and, last but not least, the impressive amount of plastic changes in the carotid bodies and central nervous system. These adaptations are likely governed by transcription factors with HIF-1 as the master regulator of hypoxic responses, but may take some time to fully develop.

Further studies are needed to determine the precise mechanisms by which carotid body sensitivity already is enhanced after a *few* hours of hypoxia and whether these are also orchestrated by HIF-1-related gene products.

Apart from an augmented HVR, the response to hypercapnia is also increased after high altitude adaptation. The entire respiratory system operates at higher gain, at the background of a low arterial PO_2 and PCO_2. Especially a high gain and a low PCO_2 lead to breathing instability and are responsible for the periodic breathing during sleep commonly observed at high altitude ([180] and Chap. 17).

High altitude is associated with oxidative stress from various sources [181–184]. There is little information on the consequences of antioxidants supplementation on the HVR at high altitude. This issue may be of relevance and deserves to be studied in more detail. First, the balance between antioxidants and prooxidants (ROS in particular) plays an important part in the stabilization of HIF-1α [93, 94] and thus in the induction of HIF-1 target genes encoding modulators of the HVR. Second, ROS may play a role in cytokine production at high altitude and are involved in the signal transduction cascade upon activation of the AT-1 receptor. And finally, in various circumstances antioxidants have proven to increase the HVR or to reverse its reduction by acetazolamide and other pharmacological agents [185–189]. In fact changes in redox state may influence carotid body sensitivity via several pathways (references in [190, 191]). Antioxidant supplementation will reduce the oxidative stress and may improve muscle performance at high altitude (e.g., [192, 193]) but the effects on the HVR may be complex and warrant further study.

References

1. West JB. Rate of ventilatory acclimatization to extreme altitude. Respir Physiol. 1988;743:323–33.
2. West JB. Improving oxygenation at high altitude, acclimatization and O_2 enrichment. High Alt Med Biol. 2003;43:389–98.
3. Bisgard GE, Forster HV. Ventilatory responses to acute and chronic hypoxia. In: Frely MJ, Blatteis CM, editors. Handbook of physiology, environmental physiology. New York: Oxford University Press; 1996. p. 1207–39.
4. Smith CA, Dempsey JA, Hornbein TF. Control of breathing at high altitude. In: Hornbein TF, Schoene RB, editors. High altitude. An exploration of human adaptation. New York: Marcel Dekker; 2001. p. 139–73.
5. Hochachka PW, Rupert JL, Monge C. Adaptation and conservation of physiological systems in the evolution of human hypoxia tolerance. Comp Biochem Physiol A Mol Integr Physiol. 1999;1241:1–17.
6. West JB. Respiratory and circulatory control at high altitudes. J Exp Biol. 1982;100:147–57.
7. Schoene RB. Limits of respiration at high altitude. Clin Chest Med. 2005;263:405–14, vi.
8. Gonzalez C, Almaraz L, Obeso A, Rigual R. Carotid body chemoreceptors, from natural stimuli to sensory discharges. Physiol Rev. 1994;744:829–98.
9. Porzionato A, Macchi V, Parenti A, De Caro R. Trophic factors in the carotid body. Int Rev Cell Mol Biol. 2008;269:1–58.
10. Howard LS, Robbins PA. Alterations in respiratory control during 8 h of isocapnic and poikilocapnic hypoxia in humans. J Appl Physiol. 1995;783: 1098–107.
11. Ren X, Dorrington KL, Robbins PA. Respiratory control in humans after 8 h of lowered arterial PO2, hemodilution, or carboxyhemoglobinemia. J Appl Physiol. 2001;904:1189–95.
12. Fatemian M, Robbins PA. Human ventilatory response to CO_2 after 8 h of isocapnic or poikilocapnic hypoxia. J Appl Physiol. 1998;855:1922–8.
13. Fatemian M, Robbins PA. Selected contribution, chemoreflex responses to CO_2 before and after an 8-h exposure to hypoxia in humans. J Appl Physiol. 2001;904:1607–14.
14. Clar C, Dorrington KL, Robbins PA. Ventilatory effects of 8 h of isocapnic hypoxia with and without beta-blockade in humans. J Appl Physiol. 1999; 866:1897–904.
15. Pedersen ME, Dorrington KL, Robbins PA. Effects of dopamine and domperidone on ventilatory sensitivity to hypoxia after 8 h of isocapnic hypoxia. J Appl Physiol. 1999;861:222–9.
16. Liu C, Smith TG, Balanos GM, Brooks J, Crosby A, Herigstad M, et al. Lack of involvement of the autonomic nervous system in early ventilatory and pulmonary vascular acclimatization to hypoxia in humans. J Physiol. 2007;579:215–25.
17. Ren X, Robbins PA. Ventilatory responses to hypercapnia and hypoxia after 6 h passive hyperventilation in humans. J Physiol. 1999;514:885–94.
18. Donoghue S, Fatemian M, Balanos GM, Crosby A, Liu C, O'Connor D, et al. Ventilatory acclimatization in response to very small changes in PO_2 in humans. J Appl Physiol. 2005;985:1587–91.
19. Tansley JG, Clar C, Pedersen ME, Robbins PA. Human ventilatory response to acute hyperoxia during and after 8 h of both isocapnic and poikilocapnic hypoxia. J Appl Physiol. 1997;822:513–9.

20. Somogyi RB, Preiss D, Vesely A, Fisher JA, Duffin J. Changes in respiratory control after 5 days at altitude. Respir Physiol Neurobiol. 2005;145:41–52.
21. Ainslie PN, Burgess KR. Cardiorespiratory and cerebrovascular responses to hyperoxic and hypoxic rebreathing, effects of acclimatization to high altitude. Respir Physiol Neurobiol. 2008;161:201–9.
22. Schoene RB, Roach RC, Hackett PH, Sutton JR, Cymerman A, Houston CS. Operation Everest II, ventilatory adaptation during gradual decompression to extreme altitude. Med Sci Sports Exerc. 1990; 226:804–10.
23. Yilmaz C, Hogg DC, Ravikumar P, Hsia CC. Ventilatory acclimatization in awake guinea pigs raised at high altitude. Respir Physiol Neurobiol. 2005;145:235–42.
24. Huey KA, Brown IP, Jordan MC, Powell FL. Changes in dopamine D2-receptor modulation of the hypoxic ventilatory response with chronic hypoxia. Respir Physiol. 2000;123:177–87.
25. Huey KA, Low MJ, Kelly MA, Juarez R, Szewczak JM, Powell FL. Ventilatory responses to acute and chronic hypoxia in mice, effects of dopamine D2 receptors. J Appl Physiol. 2000;89:1142–50.
26. Villafuerte FC, Cardenas-Alayza R, Macarlupu JL, Monge C, Leon-Velarde F. Ventilatory response to acute hypoxia in transgenic mice over-expressing erythropoietin, effect of acclimation to 3-week hypobaric hypoxia. Respir Physiol Neurobiol. 2007;158:243–50.
27. Wach RA, Bee D, Barer GR. Dopamine and ventilatory effects of hypoxia and almitrine in chronically hypoxic rats. J Appl Physiol. 1989;671:186–92.
28. Bee D, Pallot DJ. Acute hypoxic ventilation, carotid body cell division, and dopamine content during early hypoxia in rats. J Appl Physiol. 1995;795: 1504–11.
29. Gamboa J, Macarlupu JL, Rivera-Chira M, Monge C, Leon-Velarde F. Effect of domperidone on ventilation and polycythemia after 5 weeks of chronic hypoxia in rats. Respir Physiol Neurobiol. 2003;1351:1–8.
30. Olson Jr EB, Dempsey JA. Rat as a model for humanlike ventilatory adaptation to chronic hypoxia. J Appl Physiol. 1978;445:763–9.
31. Aaron EA, Powell FL. Effect of chronic hypoxia on hypoxic ventilatory response in awake rats. J Appl Physiol. 1993;744:1635–40.
32. Vizek M, Pickett CK, Weil JV. Increased carotid body hypoxic sensitivity during acclimatization to hypobaric hypoxia. J Appl Physiol. 1987;636: 2403–10.
33. Chen J, He L, Dinger B, Stensaas L, Fidone S. Role of endothelin and endothelin A-type receptor in adaptation of the carotid body to chronic hypoxia. Am J Physiol Lung Cell Mol Physiol. 2002;2826: L1314–23.
34. Liu X, He L, Stensaas L, Dinger B, Fidone S. Adaptation to chronic hypoxia involves immune cell invasion and increased expression of inflammatory cytokines in rat carotid body. Am J Physiol Lung Cell Mol Physiol. 2009;2962:L158–66.
35. Weir EK, Lopez-Barneo J, Buckler KJ, Archer SL. Acute oxygen-sensing mechanisms. N Engl J Med. 2005;353:2042–55.
36. Lopez-Barneo J, Ortega-Saenz P, Pardal R, Pascual A, Piruat JI. Carotid body oxygen sensing. Eur Respir J. 2008;325:1386–98.
37. Evans AM. AMP-activated protein kinase underpins hypoxic pulmonary vasoconstriction and carotid body excitation by hypoxia in mammals. Exp Physiol. 2006;915:821–7.
38. Wyatt CN, Mustard KJ, Pearson SA, Dallas ML, Atkinson L, Kumar P, et al. AMP-activated protein kinase mediates carotid body excitation by hypoxia. J Biol Chem. 2007;282:8092–8.
39. Prabhakar NR. Novel partners and mechanisms in oxygen sensing. Exp Physiol. 2006;915:801.
40. Teppema LJ, Dahan A. The ventilatory response to hypoxia in mammals, mechanisms, measurement, and analysis. Physiol Rev. 2010;902:675–754.
41. Wang ZZ, Dinger B, Fidone SJ, Stensaas LJ. Changes in tyrosine hydroxylase and substance P immunoreactivity in the cat carotid body following chronic hypoxia and denervation. Neuroscience. 1998;834:1273–81.
42. Bisgard GE. Carotid body mechanisms in acclimatization to hypoxia. Respir Physiol. 2000;1212–3: 237–46.
43. Powell FL, Fu Z. HIF-1 and ventilatory acclimatization to chronic hypoxia. Respir Physiol Neurobiol. 2008;164:282–7.
44. Tatsumi K, Pickett CK, Weil JV. Possible role of dopamine in ventilatory acclimatization to high altitude. Respir Physiol. 1995;991:63–73.
45. Huey KA, Powell FL. Time-dependent changes in dopamine D2-receptor mRNA in the arterial chemoreflex pathway with chronic hypoxia. Brain Res Mol Brain Res. 2000;752:264–70.
46. Gonzalez-Guerrero PR, Rigual R, Gonzalez C. Effects of chronic hypoxia on opioid peptide and catecholamine levels and on the release of dopamine in the rabbit carotid body. J Neurochem. 1993;605:1769–76.
47. Janssen PL, Dwinell MR, Pizarro J, Bisgard GE. Intracarotid dopamine infusion does not prevent acclimatization to hypoxia. Respir Physiol. 1998;1111:33–43.
48. Janssen PL, O'Halloran KD, Pizarro J, Dwinell MR, Bisgard GE. Carotid body dopaminergic mechanisms are functional after acclimatization to hypoxia in goats. Respir Physiol. 1998;1111:25–32.
49. Cao H, Kuo YR, Prabhakar NR. Absence of chemoreceptor inhibition by alpha-2 adrenergic receptor agonist in cats exposed to low PO_2. FASEB J. 1991;5:A1118.
50. Prabhakar NR, Kou YR. Inhibitory sympathetic action on the carotid body responses to sustained hypoxia. Respir Physiol. 1994;951:67–79.
51. Wang ZY, Bisgard GE. Chronic hypoxia-induced morphological and neurochemical changes in the carotid body. Microsc Res Tech. 2002;593:168–77.

52. Bartsch P, Gibbs JS. Effect of altitude on the heart and the lungs. Circulation. 2007;116:2191–202.
53. Moore LG, Cymerman A, Huang SY, McCullough RE, McCullough RG, Rock PB, et al. Propranolol blocks metabolic rate increase but not ventilatory acclimatization to 4300 m. Respir Physiol. 1987; 702:195–204.
54. Tamisier R, Hunt BE, Gilmartin GS, Curley M, Anand A, Weiss JW. Hemodynamics and muscle sympathetic nerve activity after 8 h of sustained hypoxia in healthy humans. Am J Physiol Heart Circ Physiol. 2007;2935:H3027–35.
55. Hansen J, Sander M. Sympathetic neural overactivity in healthy humans after prolonged exposure to hypobaric hypoxia. J Physiol. 2003;546(Pt 3):921–9.
56. He L, Dinger B, Fidone S. Effect of chronic hypoxia on cholinergic chemotransmission in rat carotid body. J Appl Physiol. 2005;982:614–9.
57. Jackson A, Nurse CA. Role of acetylcholine receptors and dopamine transporter in regulation of extracellular dopamine in rat carotid body cultures grown in chronic hypoxia or nicotine. J Neurochem. 1998; 702:653–62.
58. He L, Chen J, Dinger B, Stensaas L, Fidone S. Effect of chronic hypoxia on purinergic synaptic transmission in rat carotid body. J Appl Physiol. 2006;1001: 157–62.
59. Ye JS, Tipoe GL, Fung PC, Fung ML. Augmentation of hypoxia-induced nitric oxide generation in the rat carotid body adapted to chronic hypoxia, an involvement of constitutive and inducible nitric oxide synthases. Pflugers Arch. 2002;4441–2:178–85.
60. He L, Chen J, Liu X, Dinger B, Fidone S. Enhanced nitric oxide-mediated chemoreceptor inhibition and altered cyclic GMP signaling in rat carotid body following chronic hypoxia. Am J Physiol Lung Cell Mol Physiol. 2007;2936:L1463–8.
61. Kline DD, Overholt JL, Prabhakar NR. Mutant mice deficient in NOS-1 exhibit attenuated long-term facilitation and short-term potentiation in breathing. J Physiol. 2002;539(Pt 1):309–15.
62. Kline DD, Yang T, Premkumar DR, Thomas AJ, Prabhakar NR. Blunted respiratory responses to hypoxia in mutant mice deficient in nitric oxide synthase-3. J Appl Physiol. 2000;884:1496–508.
63. McQueen DS, Dashwood MR, Cobb VJ, Bond SM, Marr CG, Spyer KM. Endothelins and rat carotid body, autoradiographic and functional pharmacological studies. J Auton Nerv Syst. 1995;532–3:115–25.
64. Rey S, Iturriaga R. Endothelins and nitric oxide, vasoactive modulators of carotid body chemoreception. Curr Neurovasc Res. 2004;15:465–73.
65. Chen Y, Tipoe GL, Liong E, Leung S, Lam SY, Iwase R, et al. Chronic hypoxia enhances endothelin-1-induced intracellular calcium elevation in rat carotid body chemoreceptors and up-regulates ETA receptor expression. Pflugers Arch. 2002;4434:565–73.
66. Chen J, He L, Liu X, Dinger B, Stensaas L, Fidone S. Effect of the endothelin receptor antagonist bosentan on chronic hypoxia-induced morphological and physiological changes in rat carotid body. Am J Physiol Lung Cell Mol Physiol. 2007;2925: L1257–62.
67. Talbot NP, Balanos GM, Robbins PA, Dorrington KL. Intravenous endothelin-1 and ventilatory sensitivity to hypoxia in humans. Adv Exp Med Biol. 2008;605:57–62.
68. Leung PS, Fung ML, Tam MS. Renin-angiotensin system in the carotid body. Int J Biochem Cell Biol. 2003;356:847–54.
69. Lam SY, Fung ML, Leung PS. Regulation of the angiotensin-converting enzyme activity by a time-course hypoxia in the carotid body. J Appl Physiol. 2004;962:809–13.
70. Fung ML, Lam SY, Wong TP, Tjong YW, Leung PS. Carotid body AT4 receptor expression and its upregulation in chronic hypoxia. Open Cardiovasc Med J. 2007;1:1–7.
71. Ding Y, Li YL, Zimmerman MC, Davisson RL, Schultz HD. Role of CuZn superoxide dismutase on carotid body function in heart failure rabbits. Cardiovasc Res. 2009;814:678–85.
72. Potter EK, McCloskey DI. Respiratory stimulation by angiotensin II. Respir Physiol. 1979;363:367–73.
73. Ohtake PJ, Jennings DB. Angiotensin II stimulates respiration in awake dogs and antagonizes baroreceptor inhibition. Respir Physiol. 1993;912–3:335–51.
74. Woods DR, Montgomery HE. Angiotensin-converting enzyme and genetics at high altitude. High Alt Med Biol. 2001;22:201–10.
75. Milledge JS, Catley DM, Blume FD, West JB. Renin, angiotensin-converting enzyme, and aldosterone in humans on Mount Everest. J Appl Physiol. 1983;554:1109–12.
76. Tsianos G, Eleftheriou KI, Hawe E, Woolrich L, Watt M, Watt I, et al. Performance at altitude and angiotensin I-converting enzyme genotype. Eur J Appl Physiol. 2005;935–6:630–3.
77. Hochachka PW, Beatty CL, Burelle Y, Trump ME, McKenzie DC, Matheson GO. The lactate paradox in human high-altitude physiological performance. News Physiol Sci. 2002;17:122–6.
78. Strohl KP. Lessons in hypoxic adaptation from high-altitude populations. Sleep Breath. 2008;122: 115–21.
79. Bigham AW, Kiyamu M, Leon-Velarde F, Parra EJ, Rivera-Ch M, Shriver MD, et al. Angiotensin-converting enzyme genotype and arterial oxygen saturation at high altitude in Peruvian Quechua. High Alt Med Biol. 2008;92:167–78.
80. Patel S, Woods DR, Macleod NJ, Brown A, Patel KR, Montgomery HE, et al. Angiotensin-converting enzyme genotype and the ventilatory response to exertional hypoxia. Eur Respir J. 2003;225:755–60.
81. Wang X, Wang BR, Duan XL, Zhang P, Ding YQ, Jia Y, et al. Strong expression of interleukin-1 receptor type I in the rat carotid body. J Histochem Cytochem. 2002;5012:1677–84.

82. Wang X, Zhang XJ, Xu Z, Li X, Li GL, Ju G, et al. Morphological evidence for existence of IL-6 receptor alpha in the glomus cells of rat carotid body. Anat Rec A Discov Mol Cell Evol Biol. 2006;2883: 292–6.
83. Lam SY, Tipoe GL, Liong EC, Fung ML. Chronic hypoxia upregulates the expression and function of proinflammatory cytokines in the rat carotid body. Histochem Cell Biol. 2008;1303:549–59.
84. Hagobian TA, Jacobs KA, Subudhi AW, Fattor JA, Rock PB, Muza SR, et al. Cytokine responses at high altitude, effects of exercise and antioxidants at 4300 m. Med Sci Sports Exerc. 2006;382:276–85.
85. Carpenter E, Bee D, Peers C. Ionic currents in carotid body type I cells isolated from normoxic and chronically hypoxic adult rats. Brain Res. 1998; 8111–2:79–87.
86. Stea A, Jackson A, Nurse CA. Hypoxia and N6, O2′-dibutyryladenosine 3′,5′-cyclic monophosphate, but not nerve growth factor, induce Na^+ channels and hypertrophy in chromaffin-like arterial chemoreceptors. Proc Natl Acad Sci U S A. 1992;8920: 9469–73.
87. Stea A, Jackson A, Macintyre L, Nurse CA. Long-term modulation of inward currents in O_2 chemoreceptors by chronic hypoxia and cyclic AMP in vitro. J Neurosci. 1995;153(Pt 2):2192–202.
88. Chen J, He L, Dinger B, Stensaas L, Fidone S. Chronic hypoxia upregulates connexin43 expression in rat carotid body and petrosal ganglion. J Appl Physiol. 2002;924:1480–6.
89. Kaab S, Miguel-Velado E, Lopez-Lopez JR, Perez-Garcia MT. Down regulation of Kv3.4 channels by chronic hypoxia increases acute oxygen sensitivity in rabbit carotid body. J Physiol. 2005;566(Pt 2): 395–408.
90. Caceres AI, Obeso A, Gonzalez C, Rocher A. Molecular identification and functional role of voltage-gated sodium channels in rat carotid body chemoreceptor cells. Regulation of expression by chronic hypoxia in vivo. J Neurochem. 2007;1021: 231–45.
91. Semenza GL, Shimoda LA, Prabhakar NR. Regulation of gene expression by HIF-1. Novartis Found Symp. 2006;272:2–8.
92. Rocha S. Gene regulation under low oxygen, holding your breath for transcription. Trends Biochem Sci. 2007;32:389–97.
93. Chandel NS, Budinger GR. The cellular basis for diverse responses to oxygen. Free Radic Biol Med. 2007;422:165–74.
94. Semenza GL. Life with oxygen. Science. 2007; 318(5847):62–4.
95. Tipoe GL, Fung ML. Expression of HIF-1alpha, VEGF and VEGF receptors in the carotid body of chronically hypoxic rat. Respir Physiol Neurobiol. 2003;1382–3:143–54.
96. Lam SY, Tipoe GL, Liong EC, Fung ML. Differential expressions and roles of hypoxia-inducible factor-1alpha, -2alpha and -3alpha in the rat carotid body during chronic and intermittent hypoxia. Histol Histopathol. 2008;233:271–80.
97. Chavez JC, Agani F, Pichiule P, LaManna JC. Expression of hypoxia-inducible factor-1alpha in the brain of rats during chronic hypoxia. J Appl Physiol. 2000;895:1937–42.
98. Ward ME, Toporsian M, Scott JA, Teoh H, Govindaraju V, Quan A, et al. Hypoxia induces a functionally significant and translationally efficient neuronal NO synthase mRNA variant. J Clin Invest. 2005;11511:3128–39.
99. Prabhakar NR, Pieramici SF, Premkumar DR, Kumar GK, Kalaria RN. Activation of nitric oxide synthase gene expression by hypoxia in central and peripheral neurons. Brain Res Mol Brain Res. 1996;431–2:341–6.
100. Pascual O, Denavit-Saubie M, Dumas S, Kietzmann T, Ghilini G, Mallet J, et al. Selective cardiorespiratory and catecholaminergic areas express the hypoxia-inducible factor-1alpha HIF-1alpha under in vivo hypoxia in rat brainstem. Eur J Neurosci. 2001;1412:1981–91.
101. Yu AY, Shimoda LA, Iyer NV, Huso DL, Sun X, McWilliams R, et al. Impaired physiological responses to chronic hypoxia in mice partially deficient for hypoxia-inducible factor 1alpha. J Clin Invest. 1999;1035:691–6.
102. Kline DD, Peng YJ, Manalo DJ, Semenza GL, Prabhakar NR. Defective carotid body function and impaired ventilatory responses to chronic hypoxia in mice partially deficient for hypoxia-inducible factor 1 alpha. Proc Natl Acad Sci U S A. 2002;992: 821–6.
103. Smith TG, Brooks JT, Balanos GM, Lappin TR, Layton DM, Leedham DL, et al. Mutation of von Hippel-Lindau tumour suppressor and human cardiopulmonary physiology. PLoS Med. 2006; 37:e290.
104. Smith TG, Balanos GM, Croft QP, Talbot NP, Dorrington KL, Ratcliffe PJ, et al. The increase in pulmonary arterial pressure caused by hypoxia depends on iron status. J Physiol. 2008;586: 5999–6005.
105. Nanduri J, Yuan G, Kumar GK, Semenza GL, Prabhakar NR. Transcriptional responses to intermittent hypoxia. Respir Physiol Neurobiol. 2008; 164:277–81.
106. Kenneth NS, Rocha S. Regulation of gene expression by hypoxia. Biochem J. 2008;4141:19–29.
107. Malik MT, Peng YJ, Kline DD, Adhikary G, Prabhakar NR. Impaired ventilatory acclimatization to hypoxia in mice lacking the immediate early gene fos B. Respir Physiol Neurobiol. 2005;1451:23–31.
108. Prabhakar NR, Semenza GL. Gaseous messengers in oxygen sensing. J Mol Med (Berl). 2012;903: 265–72.
109. Peers C, Wyatt CN, Evans AM. Mechanisms for acute oxygen sensing in the carotid body. Respir Physiol Neurobiol. 2010;1743:292–8.
110. Mu J, Brozinick Jr JT, Valladares O, Bucan M, Birnbaum MJ. A role for AMP-activated protein kinase in contraction- and hypoxia-regulated glucose transport in skeletal muscle. Mol Cell. 2001; 75:1085–94.

111. Fujii N, Jessen N, Goodyear LJ. AMP-activated protein kinase and the regulation of glucose transport. Am J Physiol Endocrinol Metab. 2006;2915: E867–77.
112. Fisher JS, Nolte LA, Kawanaka K, Han DH, Jones TE, Holloszy JO. Glucose transport rate and glycogen synthase activity both limit skeletal muscle glycogen accumulation. Am J Physiol Endocrinol Metab. 2002;2826:E1214–21.
113. Gamboa JL, Garcia-Cazarin ML, Andrade FH. Chronic hypoxia increases insulin-stimulated glucose uptake in mouse soleus muscle. Am J Physiol Regul Integr Comp Physiol. 2011;3001:R85–91.
114. Peng YJ, Nanduri J, Raghuraman G, Souvannakitti D, Gadalla MM, Kumar GK, et al. H_2S mediates O_2 sensing in the carotid body. Proc Natl Acad Sci U S A. 2010;10723:10719–24.
115. Haouzi P, Bell H, Van de Louw A. Hypoxia-induced arterial chemoreceptor stimulation and hydrogen sulfide, too much or too little? Respir Physiol Neurobiol. 2011;1792–3:97–102.
116. Kusakabe T, Hirakawa H, Oikawa S, Matsuda H, Kawakami T, Takenaka T, et al. Morphological changes in the rat carotid body 1, 2, 4, and 8 weeks after the termination of chronically hypocapnic hypoxia. Histol Histopathol. 2004;194:1133–40.
117. Kusakabe T, Matsuda H, Hayashida Y. Hypoxic adaptation of the rat carotid body. Histol Histopathol. 2005;203:987–97.
118. Clarke JA, Daly MB, Marshall JM, Ead HW, Hennessy EM. Quantitative studies of the vasculature of the carotid body in the chronically hypoxic rat. Braz J Med Biol Res. 2000;333:331–40.
119. Laidler P, Kay JM. A quantitative morphological study of the carotid bodies of rats living at a simulated altitude of 4300 metres. J Pathol. 1975;1173: 183–91.
120. Heath D, Edwards C, Winson M, Smith P. Effects on the right ventricle, pulmonary vasculature, and carotid bodies of the rat of exposure to, and recovery from, simulated high altitude. Thorax. 1973;281:24–8.
121. McGregor KH, Gil J, Lahiri S. A morphometric study of the carotid body in chronically hypoxic rats. J Appl Physiol. 1984;575:1430–8.
122. Pequignot JM, Hellstrom S, Johansson C. Intact and sympathectomized carotid bodies of long-term hypoxic rats, a morphometric ultrastructural study. J Neurocytol. 1984;133:481–93.
123. Wang ZY, Olson Jr EB, Bjorling DE, Mitchell GS, Bisgard GE. Sustained hypoxia-induced proliferation of carotid body type I cells in rats. J Appl Physiol. 2008;1043:803–8.
124. Pardal R, Ortega-Saenz P, Duran R, Lopez-Barneo J. Glia-like stem cells sustain physiologic neurogenesis in the adult mammalian carotid body. Cell. 2007;1312:364–77.
125. Arias-Stella J, Bustos F. Chronic hypoxia and chemodectomas in bovines at high altitudes. Arch Pathol Lab Med. 1976;10012:636–9.
126. Arias-Stella J, Valcarcel J. Chief cell hyperplasia in the human carotid body at high altitudes, physiologic and pathologic significance. Hum Pathol. 1976;74:361–73.
127. Edwards C, Heath D, Harris P, Castillo Y, Kruger H, Arias-Stella J. The carotid body in animals at high altitude. J Pathol. 1971;1044:231–8.
128. Heath D, Smith P, Fitch R, Harris P. Comparative pathology of the enlarged carotid body. J Comp Pathol. 1985;952:259–71.
129. Smith P, Heath D, Fitch R, Hurst G, Moore D, Weitzenblum E. Effects on the rabbit carotid body of stimulation by almitrine, natural high altitude, and experimental normobaric hypoxia. J Pathol. 1986; 1492:143–53.
130. Saldana MJ, Salem LE, Travezan R. High altitude hypoxia and chemodectomas. Hum Pathol. 1973; 42:251–63.
131. Monge C, Leon-Velarde F. Physiological adaptation to high altitude, oxygen transport in mammals and birds. Physiol Rev. 1991;714:1135–72.
132. Dwinell MR, Powell FL. Chronic hypoxia enhances the phrenic nerve response to arterial chemoreceptor stimulation in anesthetized rats. J Appl Physiol. 1999;872:817–23.
133. Dempsey JA, Forster HV. Mediation of ventilatory adaptations. Physiol Rev. 1982;621:262–346.
134. Powell FL, Huey KA, Dwinell MR. Central nervous system mechanisms of ventilatory acclimatization to hypoxia. Respir Physiol. 2000;1212–3:223–36.
135. Engwall MJ, Bisgard GE. Ventilatory responses to chemoreceptor stimulation after hypoxic acclimatization in awake goats. J Appl Physiol. 1990;694: 1236–43.
136. Huey KA, Szewczak JM, Powell FL. Dopaminergic mechanisms of neural plasticity in respiratory control, transgenic approaches. Respir Physiol Neurobiol. 2003;1352–3:133–44.
137. Schmitt P, Soulier V, Pequignot JM, Pujol JF, Avit-Saubie M. Ventilatory acclimatization to chronic hypoxia, relationship to noradrenaline metabolism in the rat solitary complex. J Physiol. 1994;477(Pt 2):331–7.
138. Olson LG, Saunders NA. Effect of a dopamine antagonist on ventilation during sustained hypoxia in mice. J Appl Physiol. 1987;623:1222–6.
139. Gozal D, Gozal E, Simakajornboon N. Signaling pathways of the acute hypoxic ventilatory response in the nucleus tractus solitarius. Respir Physiol. 2000;1212–3:209–21.
140. Reeves SR, Carter ES, Guo SZ, Gozal D. Calcium/calmodulin-dependent kinase II mediates critical components of the hypoxic ventilatory response within the nucleus of the solitary tract in adult rats. Am J Physiol Regul Integr Comp Physiol. 2005; 2893:R871–6.
141. Reeves SR, Gozal E, Guo SZ, Sachleben Jr LR, Brittian KR, Lipton AJ, et al. Effect of long-term intermittent and sustained hypoxia on hypoxic

ventilatory and metabolic responses in the adult rat. J Appl Physiol. 2003;955:1767–74.
142. El Hasnaoui-Saadani R, Alayza RC, Launay T, Pichon A, Quidu P, Beaudry M, et al. Brain stem NO modulates ventilatory acclimatization to hypoxia in mice. J Appl Physiol. 2007;1035:1506–12.
143. Reid SG, Powell FL. Effects of chronic hypoxia on MK-801-induced changes in the acute hypoxic ventilatory response. J Appl Physiol. 2005;996: 2108–14.
144. Gozal D, Simakajornboon N, Czapla MA, Xue YD, Gozal E, Vlasic V, et al. Brainstem activation of platelet-derived growth factor-beta receptor modulates the late phase of the hypoxic ventilatory response. J Neurochem. 2000;741:310–9.
145. Alea OA, Czapla MA, Lasky JA, Simakajornboon N, Gozal E, Gozal D. PDGF-beta receptor expression and ventilatory acclimatization to hypoxia in the rat. Am J Physiol Regul Integr Comp Physiol. 2000;2795:R1625–33.
146. Tsunekawa S, Ohi Y, Ishii Y, Sasahara M, Haji A. Hypoxic ventilatory response in platelet-derived growth factor receptor-beta-knockout mice. J Pharmacol Sci. 2009;110:270–5.
147. Ogawa H, Mizusawa A, Kikuchi Y, Hida W, Miki H, Shirato K. Nitric oxide as a retrograde messenger in the nucleus tractus solitarii of rats during hypoxia. J Physiol. 1995;486(Pt 2):495–504.
148. Schwenke DO, Pearson JT, Kangawa K, Shirai M. Does central nitric oxide chronically modulate the acute hypoxic ventilatory response in conscious rats? Acta Physiol (Oxf). 1864;2006:309–18.
149. Soliz J, Joseph V, Soulage C, Becskei C, Vogel J, Pequignot JM, et al. Erythropoietin regulates hypoxic ventilation in mice by interacting with brainstem and carotid bodies. J Physiol. 2005;568 (Pt 2):559–71.
150. Soliz J, Thomsen JJ, Soulage C, Lundby C, Gassmann M. Sex-dependent regulation of hypoxic ventilation in mice and humans is mediated by erythropoietin. Am J Physiol Regul Integr Comp Physiol. 2009;2966:R1837–46.
151. Soliz J, Gassmann M, Joseph V. Soluble erythropoietin receptor is present in the mouse brain and is required for the ventilatory acclimatization to hypoxia. J Physiol. 2007;583(Pt 1):329–36.
152. Robach P, Fulla Y, Westerterp KR, Richalet JP. Comparative response of EPO and soluble transferrin receptor at high altitude. Med Sci Sports Exerc. 2004;36:1493–8.
153. Brugniaux JV, Pialoux V, Foster GE, Duggan CTC, Eliasziw M, Hanly PJ, et al. Effects of intermittent hypoxia on erythropoietin, soluble erythropoietin receptor and ventilation in humans. Eur Respir J. 2011;374:880–7.
154. Nolan PC, Waldrop TG. In vitro responses of VLM neurons to hypoxia after normobaric hypoxic acclimatization. Respir Physiol. 1996;1051–2:23–33.
155. Mazza E, Thakkar-Varia S, Tozzi CA, Neubauer JA. Expression of heme oxygenase in the oxygen-sensing regions of the rostral ventrolateral medulla. J Appl Physiol. 2001;911:379–85.
156. Zhang W, Carreno FR, Cunningham JT, Mifflin SW. Chronic sustained and intermittent hypoxia reduce the function of ATP-sensitive potassium channels in the nucleus of the solitary tract. Am J Physiol Regul Integr Comp Physiol. 2008;295: R1555–62.
157. Cragg PA, Drysdale DB. Interaction of hypoxia and hypercapnia on ventilation, tidal volume and respiratory frequency in the anaesthetized rat. J Physiol. 1983;341:477–93.
158. Day TA, Wilson RJ. Brainstem PCO_2 modulates phrenic responses to specific carotid body hypoxia in an in situ dual perfused rat preparation. J Physiol. 2007;578(Pt 3):843–57.
159. Herigstad M, Fatemian M, Robbins PA. Respiratory control during air-breathing exercise in humans following an 8 h exposure to hypoxia. Respir Physiol Neurobiol. 2008;1623:169–75.
160. Mathew L, Gopinath PM, Purkayastha SS, Sen GJ, Nayar HS. Chemoreceptor sensitivity in adaptation to high altitude. Aviat Space Environ Med. 1983; 542:121–6.
161. Fan JL, Burgess KR, Basnyat R, Thomas KN, Peebles KC, Lucas SJ, et al. Influence of high altitude on cerebrovascular and ventilatory responsiveness to CO_2. J Physiol. 2010;588(Pt 3):539–49.
162. Blain GM, Smith CA, Henderson KS, Dempsey JA. Contribution of the carotid body chemoreceptors to eupneic ventilation in the intact, unanesthetized dog. J Appl Physiol. 2009;1065:1564–73.
163. Mulkey DK, Stornetta RL, Weston MC, Simmons JR, Parker A, Bayliss DA, et al. Respiratory control by ventral surface chemoreceptor neurons in rats. Nat Neurosci. 2004;712:1360–9.
164. Rosin DL, Chang DA, Guyenet PG. Afferent and efferent connections of the rat retrotrapezoid nucleus. J Comp Neurol. 2006;4991:64–89.
165. Takakura AC, Moreira TS, Colombari E, West GH, Stornetta RL, Guyenet PG. Peripheral chemoreceptor inputs to retrotrapezoid nucleus RTN. CO_2-sensitive neurons in rats. J Physiol. 2006;572(Pt 2):503–23.
166. Musch TI, Dempsey JA, Smith CA, Mitchell GS, Bateman NT. Metabolic acids and [H+] regulation in brain tissue during acclimatization to chronic hypoxia. J Appl Physiol. 1983;555:1486–95.
167. Goldberg SV, Schoene RB, Haynor D, Trimble B, Swenson ER, Morrison JB, et al. Brain tissue pH and ventilatory acclimatization to high altitude. J Appl Physiol. 1992;721:58–63.
168. Javaheri S, Corbett W, Wagner K, Adams JM. Quantitative cerebrospinal fluid acid-base balance in acute respiratory alkalosis. Am J Respir Crit Care Med. 1994;1501:78–82.
169. Severinghaus JW, Mitchell RA, Richardson BW, Singer MM. Respiratory control at high altitude suggesting active transport regulation of CSF pH. J Appl Physiol. 1963;18:1155–66.

170. Monti-Bloch L, Abudara V, Eyzaguirre C. Electrical communication between glomus cells of the rat carotid body. Brain Res. 1993;6221–2:119–31.
171. Lee LY, Morton RF, Lundberg JM. Pulmonary chemoreflexes elicited by intravenous injection of lactic acid in anesthetized rats. J Appl Physiol. 1996;816:2349–57.
172. Gargaglioni LH, Bicego KC, Steiner AA, Branco LG. Lactate as a modulator of hypoxia-induced hyperventilation. Respir Physiol Neurobiol. 2003;1381:37–44.
173. Neubauer JA, Simone A, Edelman NH. Role of brain lactic acidosis in hypoxic depression of respiration. J Appl Physiol. 1988;653:1324–31.
174. Aaron EA, Forster HV, Lowry TF, Korducki MJ, Ohtake PJ. Effect of dichloroacetate on $PaCO_2$ responses to hypoxia in awake goats. J Appl Physiol. 1996;801:176–81.
175. Hardarson T, Skarphedinsson JO, Sveinsson T. Importance of the lactate anion in control of breathing. J Appl Physiol. 1998;842:411–6.
176. Schoene RB. Control of ventilation in climbers to extreme altitude. J Appl Physiol. 1982;534:886–90.
177. Ogawa T, Hayashi K, Ichinose M, Nishiyasu T. Relationship between resting ventilatory chemosensitivity and maximal oxygen uptake in moderate hypobaric hypoxia. J Appl Physiol. 2007;1034:1221–6.
178. Milledge JS. The ventilatory response to hypoxia, how much is good for a mountaineer? Postgrad Med J. 1987;63737:169–72.
179. Milledge JS, Beeley JM, Broome J, Luff N, Pelling M, Smith D. Acute mountain sickness susceptibility, fitness and hypoxic ventilatory response. Eur Respir J. 1991;48:1000–3.
180. Dempsey JA, Smith CA, Przybylowski T, Chenuel B, Xie A, Nakayama H, et al. The ventilatory responsiveness to CO_2 below eupnoea as a determinant of ventilatory stability in sleep. J Physiol. 2004;560(Pt 1):1–11.
181. Bailey DM, Ainslie PN, Jackson SK, Richardson RS, Ghatei M. Evidence against redox regulation of energy homoeostasis in humans at high altitude. Clin Sci (Lond). 2004;1076:589–600.
182. Bartsch P, Bailey DM, Berger MM, Knauth M, Baumgartner RW. Acute mountain sickness, controversies and advances. High Alt Med Biol. 2004;52:110–24.
183. Vij AG, Dutta R, Satija NK. Acclimatization to oxidative stress at high altitude. High Alt Med Biol. 2005;64:301–10.
184. Magalhaes J, Ascensao A, Marques F, Soares JM, Ferreira R, Neuparth MJ, et al. Effect of a high-altitude expedition to a Himalayan peak Pumori, 7,161 m on plasma and erythrocyte antioxidant profile. Eur J Appl Physiol. 2005;935–6:726–32.
185. Pokorski M, Marczak M. Ascorbic acid enhances hypoxic ventilatory reactivity in elderly subjects. J Int Med Res. 2003;315:448–57.
186. Hildebrandt W, Alexander S, Bartsch P, Droge W. Effect of N-acetyl-cysteine on the hypoxic ventilatory response and erythropoietin production, linkage between plasma thiol redox state and O_2 chemosensitivity. Blood. 2002;995:1552–5.
187. Teppema LJ, Bijl H, Romberg RR, Dahan A. Antioxidants reverse depression of the hypoxic ventilatory response by acetazolamide in man. J Physiol. 2006;572(Pt 3):849–56.
188. Teppema LJ, Nieuwenhuijs D, Sarton E, Romberg R, Olievier CN, Ward DS, et al. Antioxidants prevent depression of the acute hypoxic ventilatory response by subanaesthetic halothane in men. J Physiol. 2002;544(Pt 3):931–8.
189. Teppema LJ, Romberg RR, Dahan A. Antioxidants reverse reduction of the human hypoxic ventilatory response by subanesthetic isoflurane. Anesthesiology. 2005;1024:747–53.
190. Gonzalez C, Sanz-Alfayate G, Agapito MT, Gomez-Nino A, Rocher A, Obeso A. Significance of ROS in oxygen sensing in cell systems with sensitivity to physiological hypoxia. Respir Physiol Neurobiol. 2002;1321:17–41.
191. Gonzalez C, Agapito MT, Rocher A, Gonzalez-Martin MC, Vega-Agapito V, Gomez-Nino A, et al. Chemoreception in the context of the general biology of ROS. Respir Physiol Neurobiol. 2007;1571:30–44.
192. Vats P, Singh VK, Singh SN, Singh SB. Glutathione metabolism under high-altitude stress and effect of antioxidant supplementation. Aviat Space Environ Med. 2008;7912:1106–11.
193. Subudhi AW, Jacobs KA, Hagobian TA, Fattor JA, Muza SR, Fulco CS, et al. Changes in ventilatory threshold at high altitude, effect of antioxidants. Med Sci Sports Exerc. 2006;388:1425–31.

Lung Function and Gas Exchange

4

Andrew M. Luks and Susan R. Hopkins

Abstract

As the first step in the path for oxygen transport to the body, the response of the respiratory system is critical to maintaining an adequate level of function at high altitude. Environmental hypoxia, combined with cold and heavy exercise all contribute to considerable stress to the lung. Acute altitude exposure results in hypoxia, and a rapid and sustained increase in alveolar ventilation, with an associated fall in alveolar and arterial partial pressure of carbon dioxide. In addition cardiac output and pulmonary vascular pressures also increase. These changes along with any occurring as a result of altitude illness have the potential to alter pulmonary function and gas exchange. In addition, acclimatization to hypoxia of hours to days duration results in additional physiological changes overlying the acute changes. This review explores the changes in pulmonary function, as well as changes in pulmonary gas exchange at rest or during exercise that occur within the first hours to weeks of hypoxic exposure following ascent to high altitude.

Introduction

Travelers to high altitude are exposed to low barometric pressure and ambient oxygen tensions, setting in motion a series of physiologic responses that determine adaptation to these environmental conditions. The respiratory system is one of the primary systems involved in these responses. Important changes occur in both pulmonary mechanics and gas exchange that, in turn, affect physiologic parameters, such as arterial oxygenation and acid–base status, and clinical outcomes such as exercise tolerance and development of acute altitude illness.

A.M. Luks, M.D. (✉)
Division of Pulmonary and Critical Care Medicine, Harborview Medical Center, University of Washington, 325 Ninth Avenue, Box 359762, Seattle, WA 98104, USA
e-mail: aluks@u.washington.edu

S.R. Hopkins, M.D., Ph.D. (✉)
Division of Physiology 0623A, Departments of Medicine and Radiology, University of California, San Diego, 9500 Gilman Drive, Box 0623, La Jolla, CA 92093-0623, USA
e-mail: shopkins@ucsd.edu

E.R. Swenson and P. Bärtsch (eds.), *High Altitude: Human Adaptation to Hypoxia*,
DOI 10.1007/978-1-4614-8772-2_4, © Springer Science+Business Media New York 2014

In this chapter, we review several components of the respiratory system's response to high altitude that may affect such outcomes. In the first section, we review the changes in pulmonary function that occur within the first hours to weeks of hypoxic exposure including changes in spirometry and lung volumes, diffusing capacity, closing volume and closing capacity, airways resistance, respiratory muscle strength and the work of breathing, while in the second section of the chapter, we review important changes in pulmonary gas exchange that occur at rest or during exercise 48 or more hours following ascent to high altitude. Both of these topics were reviewed extensively in the previous edition of this volume [80, 121]. In the space that follows, we provide a concise review of the material covered in the previous edition and highlight the more recent studies where applicable.

Pulmonary Function

Before reviewing the changes in pulmonary function with hypoxic exposure, it is important to recognize that for many of the topics discussed below, there is considerable variability in the reported results, not only in the magnitude but also in the direction of the observed changes. The most likely reason for this variability is the significant methodological differences across studies which make comparisons very difficult: Some studies are hypobaric chamber studies (usually of short duration) while others are field studies (usually of longer exposure), and additionally there is variation in measurement techniques, the altitudes at which data was collected, the ascent rate and time spent at that altitude, and the activities of the participants during their sojourn (rest or exercise). In addition, sample sizes in many of the studies are small, which might affect statistical power and the ability to detect differences between groups or at different altitudes. Each of these factors has the ability to affect changes in the measured variables. For example, if studies are conducted at different altitudes, the degree of hypoxia will vary, as will the subsequent physiologic responses. The magnitude of hypoxic pulmonary vasoconstriction, for example, demonstrates a dose–response relationship (discussed in Chaps. 5 and 22). This might affect the degree of pulmonary capillary blood flow and, therefore, alter diffusing capacity measurements. Similarly, the time elapsed between ascent and data collection may affect the subjects' acclimatization to hypoxia, which, in turn, may either magnify or diminish the observed physiologic responses.

Spirometry and Lung Volumes

Of all the different measures of pulmonary function, spirometry has the greatest amount of data, as numerous studies have examined changes in spirometric parameters in hypoxia. The majority of these studies reveal that vital capacity (VC) decreases with hypoxic exposure, whether the exposure is acute [42, 101, 113] or subacute [77, 130] or whether it was in a hypobaric chamber [39, 92] or in the field [76, 89, 101]. Depending upon the parameter studied, various mechanisms have been proposed for the observed changes including decreased respiratory muscle strength [29], hypocapnia [110], increased interstitial fluid accumulation [77], and increased pulmonary blood volume [130]. It should be noted, however, that two more recent studies did not reveal changes in this parameter. In their study of 262 recreational climbers ascending over 24 h to 4,559 m, Cremona et al. [26] showed no changes in forced vital capacity (FVC) or the Forced Expiratory Volume in one second (FEV_1) and forced expiratory flow at 25–75 % of vital capacity ($FEF_{25-75\,\%}$) when measured within 1–2 h of arrival at the target elevation, while Dehnert et al. [31] studied 34 climbers using a similar ascent profile and showed that the FVC did not change upon initial ascent to 4,559 nor over the following 2 days at that altitude. The reason for the differences between these two studies and the others cited above is not clear but may relate to methodological differences such as the elapsed time following ascent before measurements were conducted or difference in altitude at which measurements were performed in this entire cohort of studies.

Although there are some inconsistencies in the data regarding FVC, all previous studies consistently show that peak expiratory flow (PEF) rates increase at high altitude, with some differences in the magnitude of the observed change between studies. Deboeck et al. [29], for example, reported an 11–20 % increase from sea level values at 4,267 m while Pollard et al. [89] noted a 25 % increase in PEF at 5,300 m. Part of this variability may be a function of the altitude at which PEF has been measured, as Mason [77] noted a 9 % increase in PEF at 2,800 m and a 16 % increase at 5,300 m in their study of 46 subjects trekking to Everest Base Camp. Fasano et al. [38] have shown similar changes in PEF with increasing altitude. These results imply that the changes in air density that occur with ascent to higher elevation are responsible for the observed changes in PEF. However, some studies [29] have noted that the magnitude of the observed change in PEF is less than that expected based on the changes in air density alone, thereby suggesting that factors, such as cold-induced bronchoconstriction or respiratory muscle fatigue (discussed further below), might mitigate the effect of air density.

Like the PEF, maximum voluntary ventilation (MVV) is also consistently improved across studies in both the hypobaric chamber [24, 39] and in the field [27, 40, 90]. Only a single study [7] has noted a decrease in this parameter on initial arrival at high altitude and continued declines with subsequent gains in elevation. As with changes in PEF, the improvement in MVV seen in the majority of studies is likely due to changes in air density as the increases in MVV are also seen in subjects breathing oxygen-enriched gas mixtures while exposed to hypobaric conditions [24, 39]. The consistent improvement in MVV is somewhat surprising, however, as this maneuver is dependent on muscle strength, as well as subject effort. As indicated below, some studies suggest that maximum inspiratory and expiratory pressure may decline at high altitude and the high incidence of acute mountain sickness (AMS) with acute exposure [50, 53] makes it likely that many study participants may not feel well enough to expend the full effort necessary for this test.

Decreased air density, however, does not seem to have a positive impact on another important measure of pulmonary function, the FEV_1; the majority of studies demonstrate that FEV_1 does not change in either a hypobaric chamber [130] or the field [26, 29, 77, 89]. There is some inconsistency in this data, as well, however. Gautier et al. [42], for example, reported a small increase in FEV_1 following ascent to 3,457 m while Dehnert et al. [31] demonstrated an increase in this parameter following ascent to 4,559 m which persisted over the following 2 days at that elevation. On the other hand, Mansell [76] reported a small, nonsignificant decrease with ascent to 5,366 m and Basu et al. [7] noted a noted no change on initial ascent to 3,110 m but a decrease on subsequent days at that elevation and with further gains in elevation. Given the consistent changes in PEF and MVV, it is somewhat surprising that the data regarding FEV_1 has also not shown consistent changes. If decreased air density has a large effect on pulmonary mechanics at altitude, one would expect FEV_1 to consistently move in the same direction as these other parameters.

Lung volumes have been measured less frequently at high altitude than spirometric parameters. This likely reflects the fact that this measurement requires large, bulky equipment, such as a body plethysmograph, which is more difficult to transport to the field environment than equipment necessary to measure standard spirometric parameters. The data regarding changes in total lung capacity (TLC) are mixed with older hypobaric chamber and field studies showing either an increase in this parameter [22, 76, 98, 113] and or no change [42, 44]. More recently, Dehnert et al. [31], used body plethysmography in 34 healthy volunteers who ascended to 4,559 m in less than 24 h, and found no changes in TLC relative to sea level values either upon arrival or on the following 2 days at this elevation. Similar findings were reported by Pellegrino et al. [85] in their study of 14 largely healthy (one subject had a history of seasonal allergic asthma while another had a history of asthma) male volunteers who ascended from sea level to 4,559 m over a period of 3 days.

As noted above, most studies suggest that vital capacity decreases upon ascent to high altitude. Given the equivocal changes in TLC, the decrease in vital capacity should, therefore, derive from an increase in the residual volume (RV). The majority of studies [22, 62, 98, 113] do, in fact, reveal an increase in RV, although Gautier et al. [42] did not find changes in this parameter over 6 days at an altitude of 3,457 m. The increase in residual volume has been attributed to various factors, the most common of which is an increase in extravascular fluid accumulation leading to peribronchial cuffing, premature airway closure, and an increase in the closing volume [62]. This issue is considered in greater detail in the next section.

Closing Volume and Closing Capacity

Closing *volume* refers to the volume of gas in excess of residual volume when small airway closure begins during a maximal exhalation, while closing *capacity* is defined as the sum of the closing volume and the residual volume. These variables have been assessed infrequently at high altitude with mixed results. In several early hypobaric chamber [22] and field studies [62] (discussed in greater detail in the previous edition of this book) individuals were noted to have an increase in closing capacity following hypoxic exposure, although in the field study by Jaeger et al. [62] the increase was only found at 36 and 72 h after arrival at high altitude and was not seen upon initial ascent. At the same time, however, Gray et al. [46] found no change in closing volume when measured by single-breath nitrogen washout in healthy subjects after either 4 h in a hypobaric chamber (equivalent to 4,880 m) or following 7 days at an altitude of 5,335 m.

More recent studies have shown similarly mixed results. Using the single-breath nitrogen washout method, Dehnert et al. [31], found no change in closing volume relative to sea level over a 3-day period following ascent to 4,559 m while two other studies, using similar ascent profiles to the same elevation, found that closing volume did, in fact, increase. Senn et al. [101] studied 26 unacclimatized lowlanders who ascended from 490 to 4,559 m and showed an increase in closing volume from 0.35 ± 0.14 to 0.44 ± 0.11 L above residual volume within 3 h of arrival at the peak altitude. There were no further changes in mean closing volume between the first and second day at this altitude, although there was significant inter-individual variability in these changes. In 9 of the 21 subjects who continued to have increased closing volume on Day 2, diffusing capacity for carbon monoxide (DLCO) was reduced. The decrease in DLCO could possibly be attributed to interstitial and alveolar edema formation, but arguing against this pulmonary artery pressures, which have been shown to correlate with blood and protein levels in the bronchoalveolar lavage fluid of climbers who have rapidly ascended to the same elevation [111], remained the same as the remaining 12 subjects whose DLCO did not change.

Using a similar ascent profile, Cremona et al. [26] measured closing volume in 262 climbers and noted that among those climbers with radiographic or physical exam evidence of high altitude pulmonary edema (HAPE), closing volume was increased following ascent compared to the low-altitude measurements (Fig. 4.1). They argued that this likely occurs because interstitial fluid accumulation causes peribronchial cuffing that, in turn, narrows small airways and leads to premature airway closure on exhalation. Of greater note was the fact that there was also a statistically significant increase in closing volume between low and high altitude among those individuals without evidence of HAPE (Fig. 4.1). In total, 74 % of these climbers had increased closing volume, which, the authors argue, suggests that may asymptomatic people might be developing subclinical pulmonary edema. Given that clinically evident HAPE affects only 0.2–15 % of climbers to high altitude depending on the rate of ascent and ultimate altitude attained [6, 50, 103], this represents a startlingly high value. Although this is, by far, the largest study of pulmonary function conducted at altitude, the conclusions regarding subclinical edema must be viewed with caution, as there were several important methodological issues. First, it is unclear if the observed

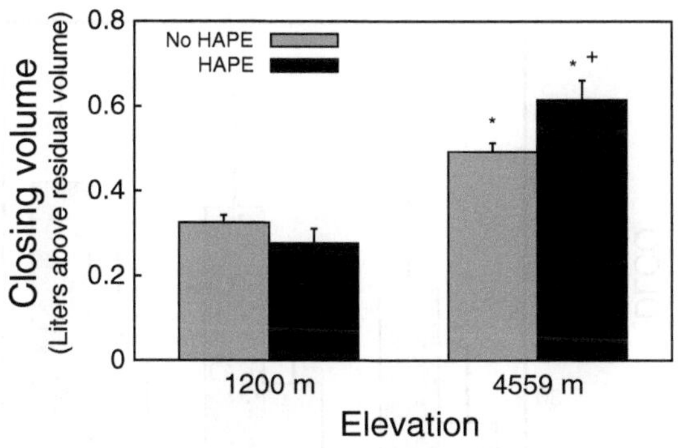

Fig. 4.1 Closing volume (mean ± SE) at 1,200 and 4,559 m in individuals with (*black bars*, *N* = 37) and without (*grey bars*, *N* = 197) clinical and radiographic evidence of high altitude pulmonary edema (HAPE) upon arrival at 4,559 m. * denotes significant differences between the two elevations while + denotes significant differences between groups (drawn with data from [26])

changes were due to the hypoxia itself, hypocapnia, or the effect of exercise. Most of the subjects were studied within 1–2 h of climbing to 4,559 m, an important feature of the study design given that exercise causes ventilation-perfusion inequality in up to 70 % of individuals (discussed further in the gas exchange section of this chapter). Second, subjects did not undergo bronchodilator response testing to determine if the observed closing volume changes could be attributed to cold- or exercise-induced bronchoconstriction rather than interstitial fluid accumulation, an issue that also arises with the day 1 measurements in the study by Senn et al. [101]. Third, closing volume measurements are a surrogate measure for extravascular lung water and were not compared directly to a different, perhaps, gold standard measurement of this variable. This is particularly important considering other studies such as that of Snyder et al. [104] who measured extravascular lung water in healthy volunteers using computed tomography, which allows direct visualization of extravascular lung water, following 17 h of hypoxia (F_IO_2 0.125) and actually found a decrease in this parameter. Finally, unlike the three other studies, which used the single-breath nitrogen washout method in order to measure closing volume, these authors determined closing volume by assessing changes in the intra-breath respiratory exchange ratio and did not correlate their measurements with those of the more standard nitrogen washout method. Given these issues and the discrepant results in the other studies mentioned above, it is difficult to draw firm conclusions about changes in closing volume with acute exposure to high altitude or make any inferences about closing volumes and extravascular lung water.

Diffusion Capacity

Further insight into the incidence of subclinical pulmonary edema might be gained by examining another important marker of pulmonary function at high altitude, the diffusing capacity for carbon monoxide (DLCO). To the extent that individuals develop subclinical or overt pulmonary edema, one would expect this parameter to decline with hypoxic exposure. As with many of the other parameters reviewed thus far, however, the data regarding the effects of high altitude on diffusing capacity are inconsistent and do not provide adequate insight into this question. In early studies, for example, Guleria et al. [49] showed an increase in DLCO on days 2, 5, and 10 following ascent to 3,658 m, while Weiskopf and Severinghaus showed a decrease in this parameter after 3 days at 4,340 m [128] and Kreuzer and Van Lookeran Campagne [65] found no difference after 7–10 days at 4,559 m. While measurements were conducted at rest in these studies, West [132] also found no change in this parameter during exercise at sea level, 4,700 and 5,800 m. More recent studies have also failed to show consistent changes in DLCO. In their study of 26 healthy individuals after 1 day at 4,559 m, for example, Senn et al. [101] found that DLCO decreased slightly when adjusted for the lower ambient PO_2. Dehnert et al. [31] also adjusted for the lower PO_2 in their study of healthy volunteers at the same elevation, yet found that DLCO increased on days 1, 2, and 3 following ascent to 4,559 m. Most recently, Agostoni et al. [1] reported an increase in DLCO when adjusted for hemoglobin and the low PO_2 following a 10-day trek to and 2 weeks of rest at 5,400 m, although whether these results can be compared to those of

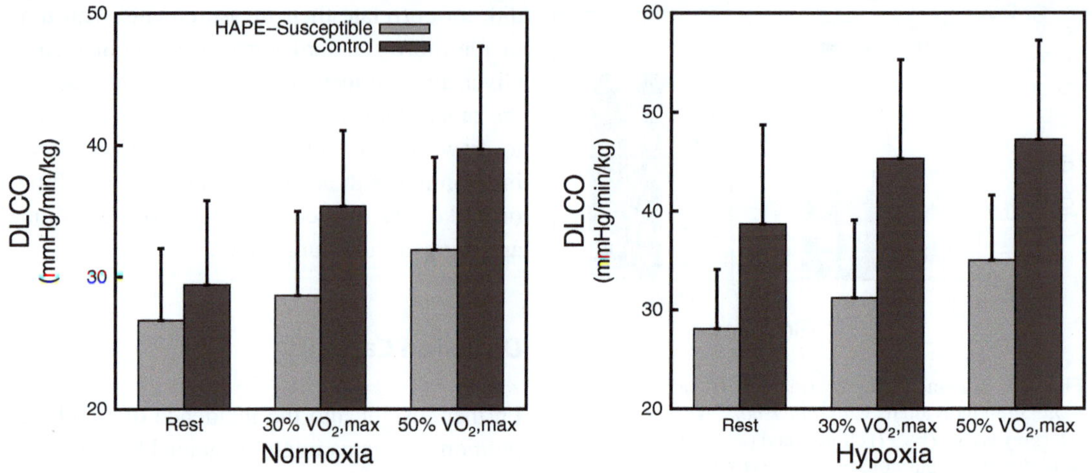

Fig. 4.2 DLCO measurements in HAPE-susceptible subjects (*black bars*) and controls (*grey bars*) at rest and while cycling at 30 and 50 % of VO_2 max in normoxia and hypoxia. Values are presented as mean + SD (created using data from [105])

Senn et al. and Dehnert et al. is not clear given the marked difference in the time frame (several days vs. several weeks) of the measurements relevant to the ascent to the target altitude.

As with many of the other measures of pulmonary function discussed in this chapter, the reasons for these differences are not clear. For this particular parameter, with the exception of the data from Agostoni et al. [1] the studies were all conducted following quick ascents to relatively similar altitudes (3,600–4,559 m) so the observed differences cannot be simply attributed to differences in ascent rate or the altitude reached. The more recent studies also adjusted DLCO measurements for the lower ambient oxygen tensions, an important adjustment given that the lower oxygen tensions will lead to higher carbon monoxide affinity for hemoglobin which could affect measured values.

An interesting fact to consider regarding DLCO measurements is that there may be important inter-individual differences in this parameter at altitude. Steinacker et al. [105] provided evidence of such variability in their study of eight HAPE-susceptible and five HAPE-resistant individuals during exercise at sea level in normoxia and hypoxia ($F_IO_2 = 0.14$). In ambient air and hypoxia, DLCO increased during exercise at 30 and 50 % of maximal oxygen consumption ($\dot{V}O_2$ max) compared to resting values but the magnitude of change was significantly lower in the HAPE-susceptible individuals (Fig. 4.2). This may reflect differences in the ability to recruit and distend the pulmonary vasculature between these two groups of patients that affects the total alveolar-capillary surface area for carbon monoxide uptake.

Ge et al. [43] has noted similar differences between individuals with and without AMS following ascent from 2,260 to 4,700 m, with smaller increases in DLCO seen in the AMS group and decreased values in those individuals with the highest AMS scores. While this result suggests that people with AMS may have subclinical interstitial edema, Dehnert et al. [31] were unable to replicate these results in subjects with and without AMS following ascent to 4,559 m. The reasons for these discrepant results are not clear but may relate to differences in ascent profile as the subjects studied by Ge et al. resided at 2,260 m while those examined by Dehnert et al. began their ascent from sea level.

Airways Resistance

Because air density decreases with increasing altitude, one would expect airways resistance to be lower at high altitude than at sea level. This effect may be mitigated or enhanced, however, by other features of the high altitude environment. For example, hypoxia has been shown in some studies to increase bronchial hyperresponsiveness [28, 35], although, as discussed further below, Pellegrino et al. [85] recently showed no effect of high altitude on airway responsiveness. Similarly, the hypocapnia that develops as a result of the hypoxic ventilatory response can increase airway resistance [82, 116]. At the same time as these factors might lead to increased resistance, however, acute hypoxic exposure leads to increased sympathoadrenal activity [78], which may, in turn, promote bronchodilation.

Field studies have not demonstrated a consistent effect of acute high altitude exposure on airways resistance. On the one hand, several earlier studies reported a decrease in airways resistance. Gautier et al. [42], for example, measured airways resistance and total pulmonary resistance during a 6-day sojourn to 3,457 m and found nonsignificant decreases in both variables on day 1 but significant decreases on day 2 that persisted through the remainder of the sojourn. Mansell et al. [76] noted a 29 % decreased in total pulmonary resistance after 9–30 days at 5,366 m while Cruz [27] noted a 6 % decrease in airways resistance in unacclimatized lowlanders with acute exposure to 4,350 m. More recently, however, two studies reported no change in this parameter between low and high altitude. Dehnert et al. [31] measured resistance using body plethysmography and found no difference following ascent to 4,559 m in less than 24 h while Pellegrino et al. [85] reported similar findings using impulse oscillometry to measure resistance following ascent to the same elevation over a period of 3 days. One potential explanation for the discrepancy between these and the earlier studies may be the difference in time frames over which the measurements took place relative to the ascent to high altitude. The studies by Gautier et al. and Mansell et al. were conducted after at least 6 days at the target altitude during which time acclimatization is well underway, while resistance was measured in all of the other studies within 4 days of arrival. Differences in duration at altitude might, for example, affect sympathetic nervous system activity [61] which could, in turn, affect overall bronchial tone.

It is interesting to note that in the studies by Gautier et al. [42] and Cruz [27], however, the observed changes in airways resistance were different than that expected based on changes in air density alone; Gautier et al. [42], for example, found agreement between the observed and predicted resistance on day one, but less resistance than predicted on days 2–6 at high altitude, a finding which suggests the presence of additional bronchodilation at high altitude. Cruz [27], on the other hand, noted that resistance fell by only 7 % at high altitude as compared with the 17 % predicted decrease based on air density alone. This finding suggests that other factors, such as hypoxia, hypocapnia, or cold temperature may be mediating some degree of bronchoconstriction that counteracts the effects of the changes in air density. The reasons for the discrepant results between these two studies are not clear. They may result from differences in methodology for measuring airways resistance and the fact that Gautier et al. measured resistance changes over a 6-day period whereas Cruz's measurements were only performed on the first day at high altitude.

Airway Hyperresponsiveness

Related to the issue of airway resistance is the question of what happens to airway hyperresponsiveness following ascent to high altitude. This question has been infrequently examined but the available data suggests that airway hyperresponsiveness decreases in this setting. Cogo et al. [23] examined responses to hypoosmolar aerosol and methacholine in 11 adults with a history of mild asthma at 5,050 m and found a decrease in airway responsiveness that they attributed to higher levels of cortisol and catecholamines. The same group reported similar findings in a study done at 4,559 m [2]. Pellegrino et al. [85] also examined this issue and found a smaller increase in airway resistance following methacholine inhalation at high altitude when compared to sea level with preservation of the bronchodilatory effect of deep breathing following methacholine administration. They attributed these findings to the reduced gas density at high altitude, hyperventilation and the effect of increased elastic recoil associated with a proposed increase in extravascular lung water and decrease in lung compliance.

Respiratory Muscle Strength

With acute high altitude exposure, resting minute ventilation increases due to the hypoxic ventilatory response [93]. Similarly, for any given work rate during exercise, minute ventilation will be higher at high altitude than at sea level [91]. In light of these increased ventilatory requirements, it is reasonable to question whether the hypoxic conditions at high altitude adversely affect the individual's ability to generate and sustain these increased levels of minute ventilation.

The limited available data reveal conflicting evidence about the effect of high altitude on resting maximum and minimum inspiratory pressures. Deboeck et al. [29] for example, exposed individuals to the equivalent of 4,267 m in a hypobaric chamber and found that maximum inspiratory pressure (MIP) and maximum expiratory pressure (MEP) were decreased relative to baseline values following 1 h of hypoxia and remained depressed over the course of the 12-h exposure (Fig. 4.3). They also found that MIP and MEP strongly correlated with vital capacity and, as a result, argued that these changes in respiratory muscle strength were responsible for the decrease in vital capacity at altitude. Similarly, Fasono et al. [38] studied eight normal individuals during and following an ascent to 4,559 m and documented decreases in MIP and MEP at each step in the ascent that correlated with the decline in oxygen saturations. With acclimatization and increasing time at peak altitude, MIP and MEP improved but did not return to sea level baseline values, suggesting that muscle strength improves as acclimatization occurs. These results, however, do not agree with those seen in other studies. Forte et al. [40], for example, demonstrated no change in MIP or MEP during acute exposure

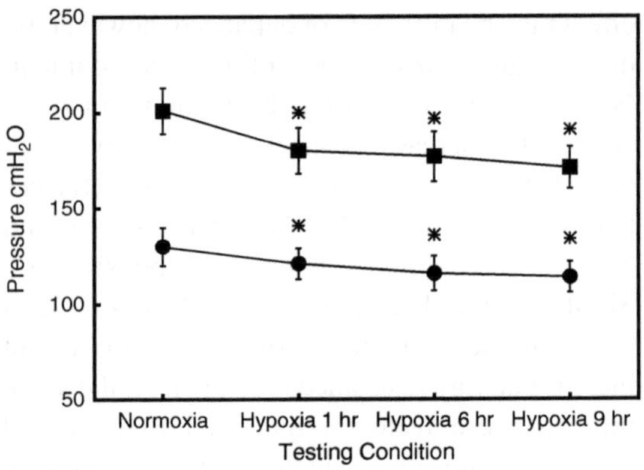

Fig. 4.3 Mean ± SE maximum inspiratory pressure (*squares*) and maximum expiratory pressure (*circles*) during a 12-h exposure to hypobaric hypoxia equivalent to 4,267 m. Fifteen subjects were tested in normoxia and at 1 h of hypoxia while 14 and 13 subjects were tested at 6 and 12 h, respectively. * denotes $P<0.05$ relative to normoxia (created using data from [29])

to 4,300 m in the field or to a barometric pressure of 460 Torr in a hypobaric chamber, while Sharma and Brown [102] found that the effect of high altitude on MIP and MEP varied based on the altitude at which they were measured and the time spent at that altitude. At 3,450 m, MIP and MEP were slightly increased upon initial arrival but fell over the subsequent 2 days at that elevation, while after 3 days at 5,350 m, both MIP and MEP rose relative to the values measured at 3,450 m and remained largely unchanged over the subsequent 3 weeks at the higher elevation. Why MIP and MEP improved with increasing elevation in this study, while declining with increasing elevation in the study by Fasano et al. [38] is not clear but may be due to the longer duration of stay at high elevation which, in turn, allowed more time for acclimatization and improved tissue oxygen delivery which may, in turn, lead to recovery of muscle strength.

The above studies were all performed with subjects at rest. Additional evidence suggests that the respiratory muscle function may also be further impaired during exercise in hypoxia. Gudjonsdottir et al. [48], for example, demonstrated that the force generating capacity of the diaphragm at maximal exercise was reduced and diaphragm recovery from fatigue was prolonged at an altitude of 3,325 m compared to sea level.

Similarly, Babcock et al. [5] exercised normal individuals at 85 % of their VO_2 max to exhaustion in normoxia and hypoxia ($F_IO_2=0.15$) and demonstrated that while similar degrees of diaphragm fatigue occur under both conditions, the time to exhaustion and peak diaphragm fatigue was markedly reduced at high altitude. Similar to Gudjonsdottir et al. [48] they also found that diaphragm recovery, as measured by bilateral phrenic nerve stimulation was prolonged in hypoxia compared to normoxia; whereas full recovery occurred by 60 min in all individuals in normoxia, the response to phrenic nerve stimulation was still reduced at 90 min post exercise in hypoxia. The consistency in results between these exercise studies stands in contrast to the conflicting data from studies conducted with subjects at rest.

Lung Compliance

Several studies have assessed dynamic and static lung compliance at high altitude. Most studies show no significant changes in dynamic lung compliance [27, 42, 67, 76], although in a considerably more recent study, Pellegrino et al. [85] reported a decrease in dynamic compliance following ascent to 4,559 m. The assessments of static compliance have not yielded consistent results. On the one hand, many studies suggest that compliance remains relatively constant in the initial days to weeks at high altitude. Gautier et al. [42], for example, studied nine healthy individuals during a 6-day sojourn to 3,457 m and demonstrated that while the static pressure-volume relationships shifted to the left between days 1 and 5 at high altitude there was no change in the slope of these curves and static lung compliance measured at 60–70 % of TLC remained unchanged. Similarly, Mansell et al. [76] studied seven healthy men between days 9 and 30 at 5,366 m and also noted leftward and upward shifts in the static pressure-volume curves but no changes in static compliance during the high altitude sojourn.

While the results of the studies by Gautier et al. and Mansell et al. agree with those of other assessments of static compliance at high altitude

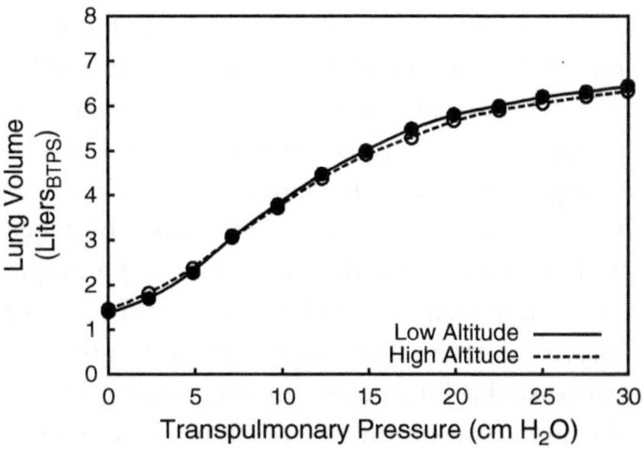

Fig. 4.4 Expiratory limb pressure-volume curves for low (200–875 m, *solid line*) and high altitude (3,000–4,300 m, *dotted line*). Each point is the mean of observations obtained at 0, 36, and 72 h of the low and high altitude phases of the study (redrawn with permission from [62])

[27, 67] several studies have yielded discrepant results. Jaeger et al. [62], for example, studied 25 soldiers during field exercises between 3,000 and 4,300 m and reported clockwise rotation of the pressure-volume curves at high altitude relative to the low-altitude curve, a result that is suggestive of decreased compliance at high altitude. The authors attributed this change to increased interstitial fluid accumulation and increased thoracic blood volume. The magnitude of the observed change in this study, however, is questionable, as close inspection of their pressure-volume curves reveal that the compliance curves measured at low and high altitude were not markedly different from each other (Fig. 4.4). Kronenberg et al. [66] have also reported a decrease in static lung compliance in their study of four healthy individuals at 3,800 m. Static lung compliance fell from 176 ± 8 to 141 ± 8 mL/cm H_2O when measured after 72 h at high elevation. Using a similar methodology as these authors, Raymond et al. [94] measured static lung compliance on four individuals over 10 days at 4,343 m and found the opposite result: non-statistically significant increases in static compliance relative to sea level values on days 2–8 and a statistically significant increase on day 10. Most recently, Pellegrino et al. noted a decrease in both static inspiratory and expiratory compliance at 4,559 m compared to sea level which they attribute to extravascular lung water accumulation.

The reasons for the discrepancies in these various studies are not clear but also probably relate to differences in study design. In the studies by Gautier et al. [42] and Mansell et al. [76], for example, the subjects remained largely at rest during their stay at altitude. In the study by Jaeger et al. [62], however, the soldiers were participating in field exercises and, to the extent that prolonged exercise at high altitude promotes extravascular fluid accumulation, this might have affected measurements of lung compliance. Similarly, in the study by Kronenberg et al. [66] subjects were flown to high altitude while breathing supplemental oxygen; while in the study by Raymond et al. [94] individuals breathed room air during a slower ascent by automobile.

Work of Breathing

The overall work of breathing in a given individual is a function of both the elastic and resistive work of breathing. Lung compliance affects the elastic work of breathing and, as noted above, studies on this parameter in resting subjects have not found significant changes at high altitude. As a result, one may not expect large changes in the elastic work of breathing. To the extent that interstitial fluid accumulation occurs, however, particularly with exercise, compliance may decrease and the elastic work of breathing may get worse. Resistive work of breathing is largely a function of airway resistance, which, as noted earlier, decreases with high altitude exposure. This would lead one to expect decreased resistive work of breathing, but as emphasized earlier, if there is significant cold- or exercise-induced bronchoconstriction, airway resistance and, therefore, resistive work may actually increase.

Only a few studies have examined how these theoretical issues are borne out at high altitude. In his study of six low-altitude natives exposed to an altitude of 4,350 m, Cruz [27] measured the work of breathing at rest and showed a statistically significant increase in the work of breathing from 0.033 ± 0.002 kgm/L at sea level to 0.039 ± 0.002 kgm/L at high altitude. There were small increases in both elastic (0.025 ± 0.002 to 0.028 ± 0.002 kgm/L) and resistive work (0.008 ± 0.001 to 0.011 ± 0.001 kgm/L) but only the increase in resistive work was statistically significant. The increase in resistive work is somewhat surprising because Cruz measured a 7 % decrease in airways resistance in the same study, although the measured decrease was lower than that predicted from the changes in air density alone.

While the study by Cruz examined individuals at rest, Cibella et al. [21] examined changes in the mechanical power of breathing (Wrs) in four individuals during submaximal exercise at sea level and following 1 month at 5,050 m. In one individual, Wrs at a given level of minute ventilation was lower at high altitude than at sea level, while in the other three subjects, the relationship between Wrs and minute ventilation was unchanged between the two environments (Fig. 4.5). In addition, they showed that the oxygen cost of breathing at altitude is a function of mechanical efficiency. If mechanical efficiency is as low as 5 %, the oxygen cost of breathing amounts to 26 % of VO_2 max whereas if efficiency rises to 20 %, the oxygen cost of breathing is only 6.5 % of VO_2 max.

Other studies have of the work of breathing during exercise have yielded divergent findings. Thoden et al. [114], for example, reported an increase in Wrs for any given minute ventilation in seven individuals during exercise at 3,100 m while Petite et al. [86] reported a decrease in Wrs for any given minute ventilation in two subjects at an equivalent altitude of 7,500 m in a hypobaric chamber and Mognoni et al. [81] reported a 20 % decrease in dynamic respiratory work in three subjects in a field study at 3,500 m.

One of the important differences between these studies that may account for the discrepant results is the duration of time spent at altitude prior to the measurements. In the study by Cibella et al. [21], for example, measurements were made after 1 month at 5,050 m. As a result, it is entirely possible that the measured work of breathing is different than what one would measure before acclimatization had occurred or before any effects of cold- or

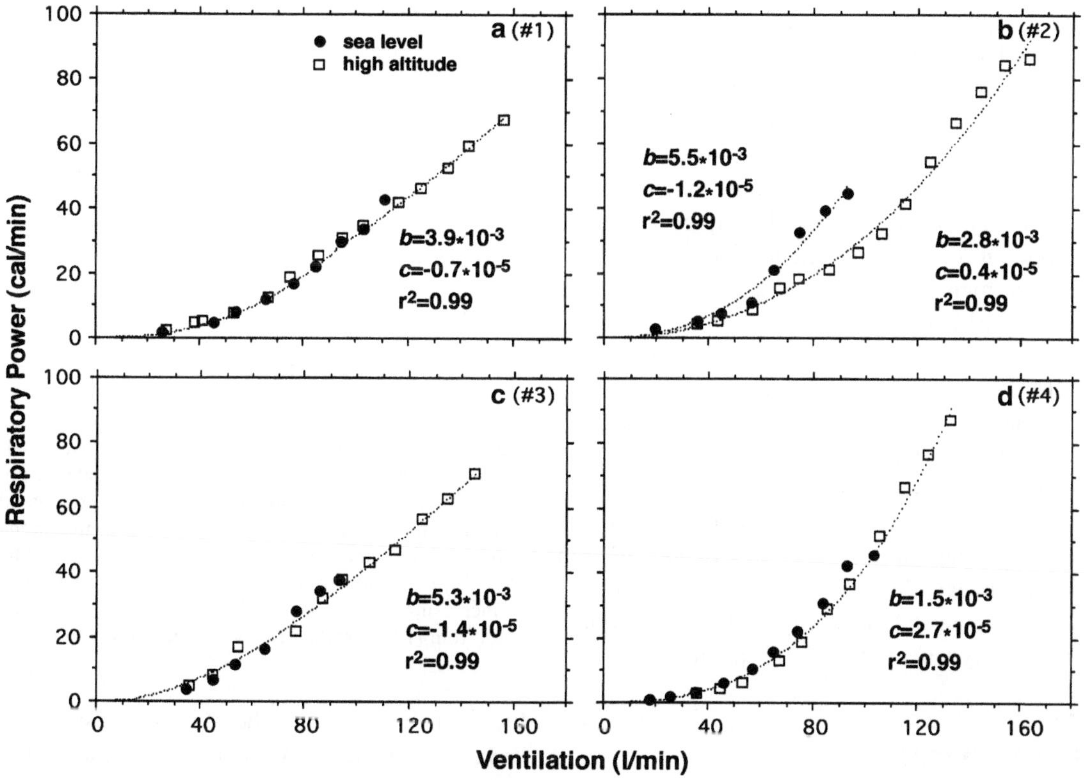

Fig. 4.5 Respiratory power as a function of minute ventilation. Panels **a–d** represent data from four individual subjects. Curves were calculated according to the equation: $\dot{W}_{rs} = bV_E^2 + c\dot{V}_E^3$ where b and c are constants (reprinted with permission from [21])

exercise-induced bronchoconstriction on resistive work or interstitial fluid accumulation on elastic work may have dissipated.

Extravascular Lung Water at High Altitude

A topic of considerable debate within the field of high altitude medicine and physiology is whether individuals who ascend to high altitude develop increased extravascular lung water in the absence of any pulmonary symptoms. Concern has also been raised as to whether those who develop this problem, often referred to as subclinical pulmonary edema, may be at increased risk for progression to overt HAPE [26]. Many of the changes in pulmonary function noted above such as the decrease in vital capacity and observed increases closing volume are invoked as evidence of this phenomenon. Drawing conclusions about this issue from the pulmonary function data described above, however, is problematic in several respects. First, with the exception of vital capacity and PEF rates, there is significant variability across studies in observed changes in many of the parameters cited as evidence of extravascular lung water accumulation, such as TLC and closing volume. Second, many of the observed changes in pulmonary function parameters that are cited as support for extravascular lung water accumulation can alternately be explained by other phenomena such as cold air hyperpnea, hypocapnia, heavy exercise, and mild respiratory muscle fatigue [110]. Finally, the changes in pulmonary function are, at best, surrogate, or indirect, measures for extravascular lung water accumulation. In none of these studies are the observed changes in pulmonary function correlated with findings on more direct assessment of extravascular lung water such as magnetic resonance imaging or computed tomography. In fact, studies using these techniques have not consistently shown an increase in extravascular lung water with hypoxia and/or exercise [110]. Given these issues, no definitive conclusions can really be made about the presence of extravascular lung water based on the pulmonary function data. The question is an important one to investigate but more direct measurements will need to be made using other techniques in order to resolve it.

The many parameters discussed above are reflective of changes in lung mechanics and function that occur with exposure to high altitude. It is these changes that will affect the ability of the respiratory system to deliver oxygen to the alveoli. In the second half of the chapter, we consider the next step of the oxygen cascade and, in particular, examine how hypoxic exposure affects gas exchange and the ability of the lung to transfer that oxygen into the blood stream.

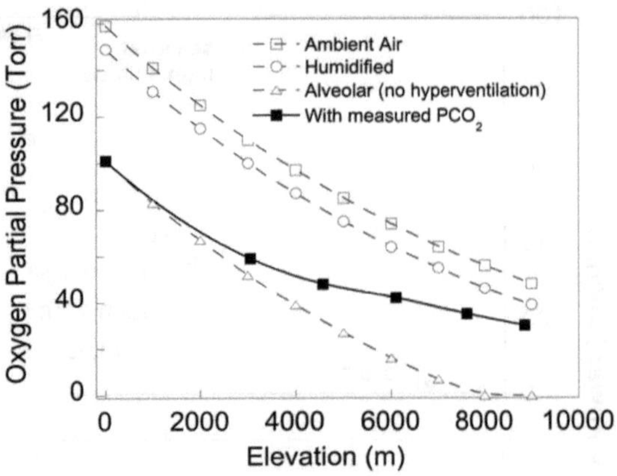

Fig. 4.6 Effect of increasing elevation and corresponding reduction in barometric pressure (calculated from [133]) on the partial pressure of oxygen as it progressed from dry ambient air, becomes warmed and wetted in the upper airways, and then mixed with alveolar gas. The *closed squares* show the effect of hyperventilation induced by increasing alveolar hypoxia is to reduce carbon dioxide (data from [121]), and buffer the fall in alveolar PO_2

Gas Exchange

Overview

There is a considerable body of work documenting the effects of acute hypoxia of a few minutes to hours duration on pulmonary gas exchange. However, very few travelers arrive at high altitude without any acclimatization at all, and many are present at high altitude for days, weeks, or years. Thus, most are at least partially acclimatized. The focus of this section of this chapter will be to integrate new information related to factors affecting pulmonary gas exchange in acclimatized subjects with that published in the previous edition [121].

Oxygen Availability

The primary physiological challenge at high altitude is the reduction in oxygen availability resulting from reduced barometric pressure [12]. Although the fractional concentration of oxygen remains constant at 0.21, with increasing elevation (and corresponding decline in barometric pressure) the partial pressure of oxygen (PO_2) falls. After inspiration, a further reduction in ambient PO_2 occurs because of warming and humidification of the inspired air, and then a further reduction occurs as it mixes into the resident gas of the gas exchange portions of the lung whose PO_2 is lower as a result of ongoing uptake into blood and associated CO_2 evolution into the alveolar space. Figure 4.6 shows the progressive decline as elevation increases, of PO_2 in ambient air, in the upper airways after being warmed and wetted, and in the alveoli. From this figure two points are particularly important: (1) While the absolute value of the saturation vapor pressure of water is a constant 47 Torr at 37 °C (body temperature), it is a progressively larger percentage of alveolar gas as elevation increases; (2) The effect of hyperventilation induced by stimulation of peripheral chemoreceptors reduces the alveolar CO_2 (P_ACO_2) and raises alveolar O_2 (P_AO_2), particularly above ~1,500 m and this buffers the decline in alveolar oxygen (P_AO_2). Thus, this second point highlights the importance of ventilatory sensitivity to hypoxia (covered in Chap. 3 of this book) as an important determinant of arterial oxygenation at altitude.

This is because of the effects on the arterial partial pressure of oxygen (PaO_2) and also because hyperventilation-induced respiratory alkalosis shifts the oxygen-hemoglobin dissociation curve, facilitating oxygen loading at the lung [75, 125, 137].

Barometric Pressure Variation

While barometric pressure plays a key role in determining pulmonary gas exchange and subsequent oxygen delivery at all elevations, at extremely high altitudes small fluctuations are of particular importance. An equation [133] predicting barometric pressure within 1 % of actual measured values for many locations of interest at high altitude within latitudes of 15° (in all seasons) and 30° (in the summer) is:

$$P_B(Torr) = \exp\left(6.63268 - 0.1112h - 0.00149h^2\right)$$

where h is the altitude in kilometers.

This equation predicts a barometric pressure of 252.7 Torr on the summit of Mt. Everest which agrees closely with the measured value of 253 Torr [134, 136]. This summit is of particular interest because the physiologic conditions at that elevation approach the limits of human survival. This equation also predicts a barometric pressure of 269 Torr at 8,400 m, a value also very close to that measured in the field for that elevation [47].

Although the primary effect of hypobaria at altitude is a reduction in inspired PO_2, there are a few reports of specific effects of decreased barometric pressure *per se* on respiration and other physiological variables. For example, humans exposed to simulated altitude in a hypobaric chamber have been shown to have decreased ventilation compared to the same inspired PO_2 at sea level [70, 71]; however this is not universally seen. In addition, AMS has been reported to be more severe in individuals subjected to hypobaria than when they are exposed to the equivalent level of normobaric hypoxia [97]. Also, hypobaric hypoxia has been shown to have some subtly different effects on plasma volumes and fluid balance compared to normobaric hypoxia [71]. In a systematic study comparing pulmonary gas exchange in normobaric and hypobaric hypoxia, hypobaric hypoxia resulted in a significantly lower arterial partial pressure of oxygen (PaO_2), saturation of arterial blood with oxygen (SaO_2) and arterial partial pressure of carbon dioxide ($PaCO_2$) although these differences were small (~4–8 %) and confounded by marked intersubject variability [99]. Under the majority of research situations, the effects are similar enough that data from both hypobaric and normobaric hypoxia can be generalized, since changes resulting from hypoxia *per se* are so much greater that those from hypobaria.

Resting Pulmonary Gas Exchange

Arterial Blood Gases

The previous edition of this book [121] extensively documented the results of several studies evaluating pulmonary gas exchange at altitude. Clearly as altitude increases, hypoxemia worsens as the alveolar hypoxia becomes more severe. The net result is that approaching the summit of Everest (or equivalent altitude in an altitude chamber) resting PAO_2 stabilizes at ~32–38 Torr and alveolar PCO_2 reaches 7–11 Torr [47, 125, 136]. Although the extreme altitudes of Mount Everest are mentioned as an example, it can be appreciated from Fig. 4.6 that PAO_2 changes remarkably little with increasing elevation in acclimatized subjects decreasing only ~25 Torr between 3,000 and 8,000 m. This decrease in PAO_2 represents a decrease of about 5 Torr per 1,000 m increase in elevation above 3,000 m, and is a substantial mitigation of hypoxia by hyperventilation.

The effects of the increased elevation and associated increase in alveolar ventilation on resting arterial blood gases can be seen in Fig. 4.7, which shows data from direct sampling of arterial blood from field studies with in subjects acclimatized for a period of days to weeks [6, 9, 25, 47, 74, 88, 96, 122, 135]. Resting PaO_2 decreases, at an average rate of ~4 Torr/1,000 m of altitude increase, although there is a suggestion that the rate of decline lessens at elevations above 4,000 m. The resting $PaCO_2$ is decreased

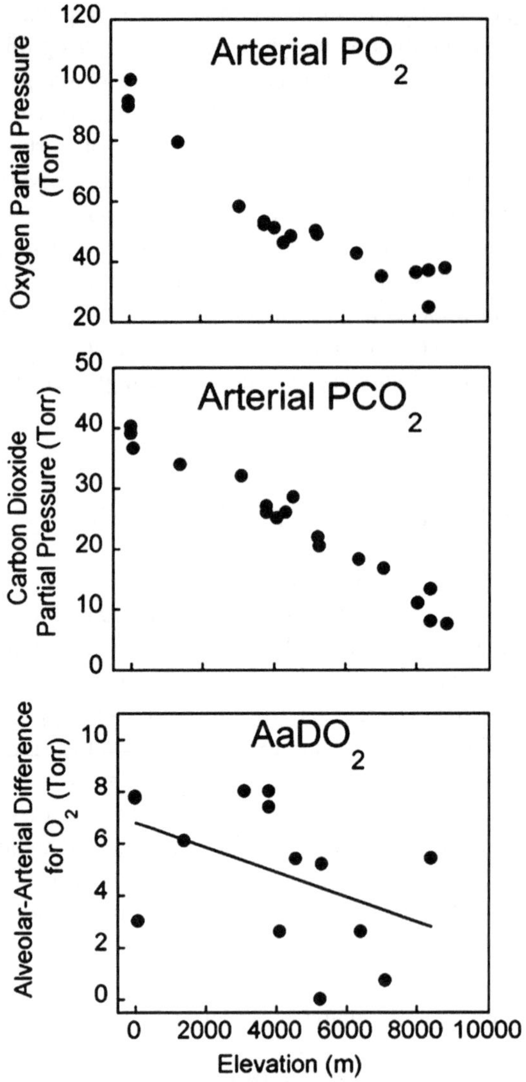

Fig. 4.7 The effect of increasing altitude on arterial PO_2, PCO_2 and the alveolar-arterial difference for O_2. The rate of fall in arterial PO_2 (*top*) with increasing elevation is reduced by increased alveolar ventilation resulting in a decreased arterial CO_2 (*middle*) and increased alveolar PO_2 as seen in Fig. 4.6, as well as an improvement in gas exchange efficiency, resulting in a reduction in the alveolar-arterial difference (*bottom*). See text for details. Data compiled from [6, 9, 25, 47, 74, 88, 96, 122, 135] and represent mean data from several subjects at each elevation

with increasing elevation, reflecting increased stimulation of carotid chemoreceptors. At the highest elevations the rate of fall in $PaCO_2$ may increase substantially particularly in individuals with a brisk hypoxic ventilatory response (see [135] for details) although this may not be a generalized characteristic.

Gas Exchange Efficiency: The Alveolar-Arterial Difference

The effect of increasing altitude on the efficiency of resting pulmonary gas exchange, as measured by the alveolar-arterial difference for oxygen ($AaDO_2$), is interesting. Resting $AaDO_2$ decreases with increasing altitude and fall in alveolar PO_2, and the fall in PaO_2 with increasing elevation is further buffered [45]. The reasons for this decrease can be found in examining the relative contributions from pulmonary diffusion limitation of oxygen transport, ventilation-perfusion heterogeneity, and shunt [131] to the $AaDO_2$. This complex issue reflecting the interaction of many potentially interdependent factors has been discussed extensively in the previous version of this chapter [121], and is summarized here.

Ventilation-Perfusion Inequality

Although ventilation-perfusion inequality may be increased at high altitude (discussed below), this does not necessarily mean that the effect on pulmonary gas exchange and the $AaDO_2$ is increased compared to sea level. In fact, modeling studies [121] show that the effect of a given (unchanging) amount of ventilation-perfusion inequality on the $AaDO_2$ is reduced at altitude (Fig. 4.8). The reason for this is threefold: In hypoxia, more of the ventilation-perfusion ratios of the individual lung subunits fall on the steep part of the oxygen-hemoglobin dissociation curve, and consequently the range of PO_2 values from low to high ventilation-perfusion ratio parts of the lung is less, reducing the difference between the alveolar and arterial PO_2. Secondly since regions of relatively high ventilation-perfusion ratio are on the steep part of the dissociation curve, small changes in PO_2 result in large changes in content in these lung regions, compared to normoxia. Finally, the overall ventilation of the lung is increased relative to cardiac output [121] and thus the overall ventilation-perfusion ratio of the lung is increased by hyperventilation, such that units of relatively low

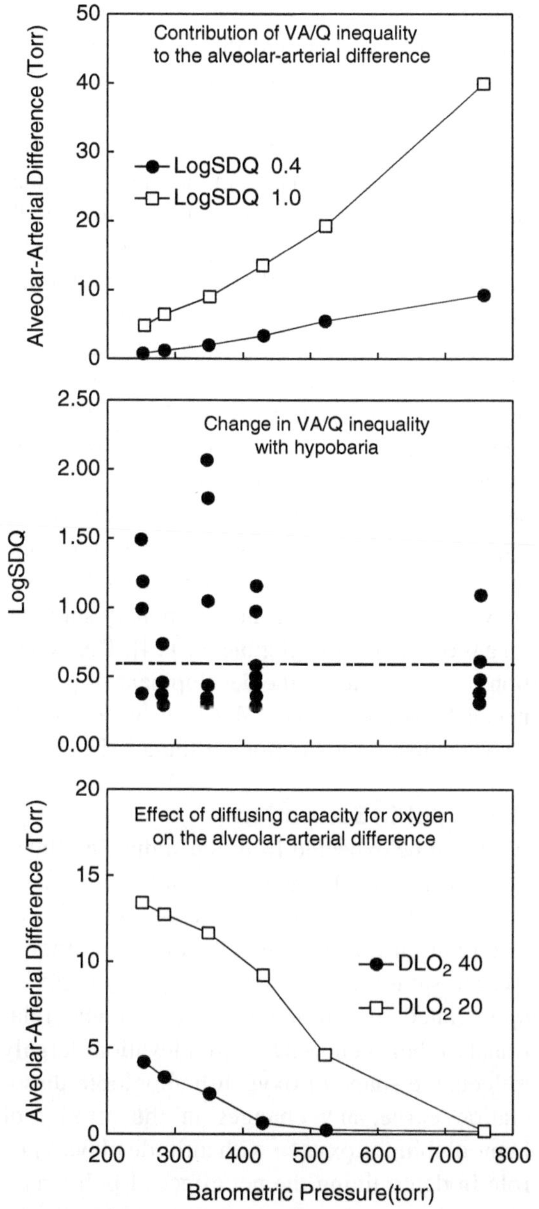

Fig. 4.8 The effect of decreasing barometric pressure on the ventilation-perfusion inequality and diffusion limitation components of the $AaDO_2$. The contribution of a given degree of ventilation-perfusion inequality (*top*, LogSDQ, the log standard deviation of the perfusion inequality) to the $AaDO_2$, decreases with decreasing barometric pressure. The effect of decreasing barometric pressure on the LogSDQ is variable between individuals (*middle*) with the *dotted line* indicating the upper bounds of normal sea level values. The net result is a reduction in the $AaDO_2$ attributable to ventilation-perfusion inequality. However, the effect of decreasing barometric pressure in the face of a fixed diffusing capacity for oxygen (DLO_2, *bottom*) is a marked increase in this component of the $AaDO_2$. Figures simplified and redrawn from [121]

ventilation-perfusion ratio occur at a higher ventilation (and thus PO_2) than if alveolar ventilation remained unchanged.

Other factors affect the extent of ventilation-perfusion inequality itself. There is a reduction in gas density with increasing altitude (discussed in the first part of this chapter) and this, combined with increased alveolar ventilation, has been suggested to make the distribution of inspired gas in the lung more uniform. However, this has not been shown experimentally [20, 138]. Depending on the interaction between convective and diffusive processes in the heterogeneous lung, resident alveolar gas is thought to mix differently with inspired gas depending on position within the tracheal bronchial tree, offsetting effects that result from reductions in density alone [20, 83]. Similarly sympathetically mediated increases in cardiac output along with hypoxia-induced increases in pulmonary arterial pressure are expected to cause dilation and recruitment of the capillary bed [17, 18] increasing the uniformity of the distribution of perfusion. Thus, since the distribution of both ventilation and perfusion are expected to be more uniform at high altitude, one might similarly expect ventilation-perfusion matching to be more uniform.

The picture is still more complicated, as interstitial pulmonary edema compressing small airways and blood vessels can disrupt ventilation-perfusion matching [100]. Since the increase in pulmonary arterial pressure with hypoxia will increase the driving pressure for fluid efflux these changes might counterbalance other factors increasing uniformity. In keeping with this idea, chamber studies have shown that the overall extent of resting ventilation-perfusion inequality is increased with increasing altitude (Fig. 4.8), consistent with the development of interstitial edema, and is notable for the extent of inter-subject variability. Interestingly in individuals who have previously developed HAPE, the spatial heterogeneity of pulmonary blood flow is increased in response to acute hypoxia. These changes occurred within 5–10 min of hypoxia exposure [32, 55], suggesting that in these subjects hypoxic pulmonary vasoconstriction is uneven, providing an additional stimulus for the development of patchy edema and increased ventilation-perfusion inequality. However, in healthy normal subjects, the spatial uniformity of pulmonary perfusion is either unchanged [4, 55] or minimally increased [32].

All of this information together implies that the effect of ventilation-perfusion inequality on the $AaDO_2$ at altitude will reflect a balance between the extent of the increase in ventilation-perfusion inequality, if any, and the net effect of the lungs position on the oxygen-hemoglobin dissociation curve. Thus in individuals who experience an increase in ventilation-perfusion inequality at altitude, this may be a particularly important contributor to the alveolar-arterial difference if pulmonary edema occurs.

Diffusion Limitation

In contrast to the situation for ventilation-perfusion inequality, the contribution of a fixed degree of alveolar-capillary diffusion limitation to the resting $AaDO_2$ increases with decreasing barometric pressure. This is illustrated in Fig. 4.8, where notably diffusion limitation is not present at rest at sea level—consistent with the range of values assumed for DLO_2. It can be appreciated that if diffusing capacity is unchanged in hypoxia, as modeled in this figure, an increasing amount of the resting $AaDO_2$ would be explained by diffusion limitation. However since the $AaDO_2$ falls with increasing altitude, this is more consistent with ventilation-perfusion inequality being the major contributor to the resting gas exchange inefficiency at altitude [9, 88, 124]. Diffusion limitation likely does not contribute more to the resting $AaDO_2$ likely because of an increased diffusing capacity in hypoxia [9, 51, 52, 124]. This is confirmed by multiple inert gas elimination studies showing the greater contribution of ventilation-perfusion inequality than diffusion limitation to the resting $AaDO_2$ [9, 124]. Thus in healthy normal subjects, diffusion limitation generally does not occur at rest. However there is some evidence that at extremely high altitudes it may contribute a small amount in some individuals [124], although this is difficult to prove conclusively.

Shunt

Intrapulmonary shunting, the passage of mixed venous blood through the pulmonary circulation without contact with ventilated regions of the lung [131] has attracted renewed attention recently as a cause of gas exchange impairment particular during hypoxic exercise as well as in some individuals at rest in hypoxia [37, 73, 106]. This topic has been debated recently and is presently quite controversial [58, 72]. Agitated saline contrast echocardiography studies demonstrate that microbubbles appear in the left heart after intravenous injection into a peripheral vein [37, 73, 106]. In a similar fashion to that for ventilation-perfusion inequality the contribution of a fixed level of shunt to pulmonary gas exchange decreases as hypoxia worsens [118]. However, both at rest and during exercise, inert gas elimination studies have shown that the effect of intrapulmonary shunting on gas exchange is negligible [9, 124]. The exception to this would be the development of pulmonary edema and alveolar flooding, which would also lead to an intrapulmonary shunt [95].

Hemoglobin Saturation

As might be expected from the somewhat linear decrement in PaO_2, resting arterial oxygen saturation falls with increasing hypobaric hypoxia. The relationship between elevation and arterial oxygen saturation is given in Fig. 4.9. It can be appreciated that although the curvilinear relationship between SaO_2 and elevation largely reflects the shape of oxygen-hemoglobin dissociation curve, any changes in the affinity of hemoglobin for oxygen with altitude plays a key role in determining the net effect of pulmonary gas exchange on oxygen delivery. The importance of hemoglobin affinity is supported from comparative studies of high altitude animals. Perhaps one of the most robust features observed across widely divergent species that are adapted to hypoxia is a high affinity of hemoglobin for oxygen (i.e., a low P_{50}) [112, 126, 127]. These observations are consistent with gas exchange models that predict that high hemoglobin-O_2 affinity (low P_{50}) will increase O_2 uptake at extremely high altitudes where pulmonary diffusion limitation becomes increasingly

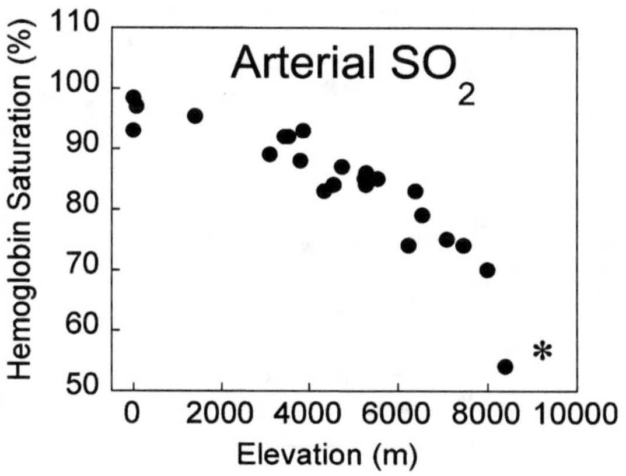

Fig. 4.9 Effect of increasing elevation on resting arterial oxygen saturation. Data are mean data for several subjects from field studies in acclimatized lowlanders and compiled from [6, 9, 25, 47, 74, 84, 88, 96, 122, 135]. The * indicates the data from the highest arterial blood samples recorded in the field [47]. This particular value may be an outlier because of previous use of supplementary oxygen, or because of rapid ascent disrupting gas exchange, see text for details

important [10]. It has previously been shown that [68, 69] that 2,3-diphosphoglycerate concentrations are increased with altitude exposure, resulting in a rightward shift in the dissociation curve and, thereby decreasing the affinity of hemoglobin for oxygen. This is counterbalanced by hyperventilation and respiratory alkalosis acting to shift the dissociation curve to the left. A recent study reported the changes in standard P_{50} and in vivo P_{50} for five climbers during a simulated ascent of Mt. Everest [125]. Although standard P_{50} rose from 28.1 to 33.1 Torr, the in vivo P_{50} was remarkably constant, remaining close to sea level values and even decreasing slightly at the barometric pressure equivalent to the summit. An analysis of the effects of P_{50} on oxygen delivery and VO_2 max [119, 120] suggests that, at least in humans, any given shift in the oxygen-hemoglobin dissociation curve is counterbalanced by opposing factors in the lungs and tissues. The net result is such that these shifts have little overall effect on gas exchange and oxygen delivery, although regional effects (i.e., lung vs. muscle) may be important [119, 120, 125].

Differences Between Field Studies and Chamber Studies

It is possible that chamber studies simulating altitude by hypobaric hypoxia may differ from field studies conducted at high altitude, even though the barometric pressures may be equivalent and the hypoxic stimulus may be similar. The proposed mechanisms for this include differing ascent profiles, use of supplemental oxygen, increased sympathetic activation, cold temperature, etc. [47]. Since the ability to make measurements at terrestrial high altitude is severely limited at very high elevations, this is worth discussion. Pulmonary gas exchange is highly variable between individuals, even in healthy subjects, and many such potential differences in the observations between chamber and field studies may be due to inter-subject variability. Figure 4.10 shows resting arterial blood gases and hemoglobin saturation obtained by arterial sampling at various barometric pressures during a simulated ascent [124] and an actual ascent of Mt. Everest [47]. Over a wide range of altitudes the results are remarkably similar, and both show substantial inter-subject variability as evidenced by the large standard deviations for PaO_2 and hemoglobin saturation, particularly at very high altitudes. At the very extreme altitudes the differences between the two studies likely represent recent use of supplemental oxygen acting to depress ventilation and thus PAO_2 and PaO_2 at the time the field measurements were made [47]. Thus it seems reasonable to combine the results from chamber studies with field studies in understanding the various contributors to the $AaDO_2$, particularly during exercise, where the field measurements are extremely limited.

Pulmonary Gas Exchange During Exercise

Arterial Blood Gases

Having reviewed that factors affecting pulmonary gas exchange at rest in hypobaric hypoxia, the corresponding changes during exercise can now be considered. Figure 4.11 (top panel) shows PaO_2 at different altitudes in acclimatized normal subjects exercising at intensities up to VO_2 max. In normal subjects at sea level (in this case the

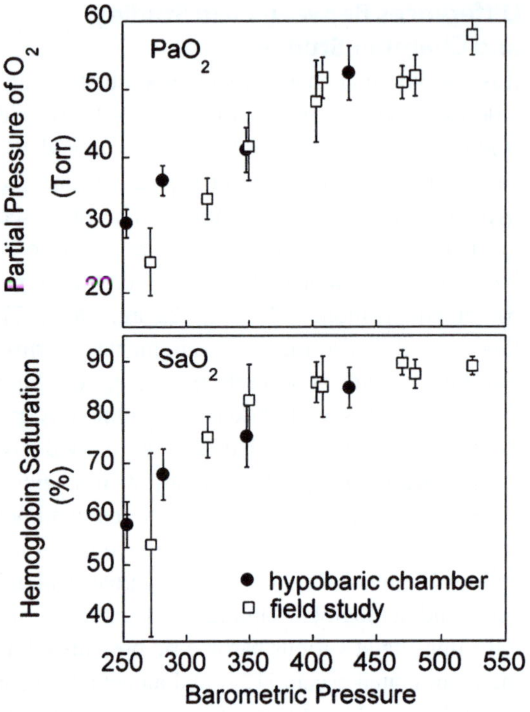

Fig. 4.10 Comparison of resting values for arterial PO_2 (*top*) and saturation (*bottom*) comparing chamber studies to field studies. Error bars are ± SD. At all except the lowest barometric pressures the data overlie one another. The discrepancies at the lowest barometric pressures likely are a result of more complete acclimatization in the controlled environment of a hypobaric chamber. Data compiled from [9, 33, 47, 74, 108, 122]

measured Pbar was 755 Torr), the PaO_2 is maintained near resting levels, albeit at the expense of increased alveolar ventilation and an increase in the $AaDO_2$ (Fig. 4.11, middle panel). In highly trained individuals, PaO_2 may fall particularly at high levels of work [34] and this is more common in individuals with a high $\dot{V}O_2$ max [54]. However during exercise at altitude in acclimatized subjects, virtually all individuals experience a fall in PaO_2 from resting values. While the absolute decrements in PaO_2 from rest to exercise appear relatively constant (Fig. 4.11), averaging ~5 Torr, they form a greater percentage of an ever diminishing baseline. As a result oxygen delivery has the potential to become increasingly compromised (or conversely improved) by even small changes in alveolar ventilation (and thus alveolar PO_2) and/or gas exchange efficiency ($AaDO_2$).

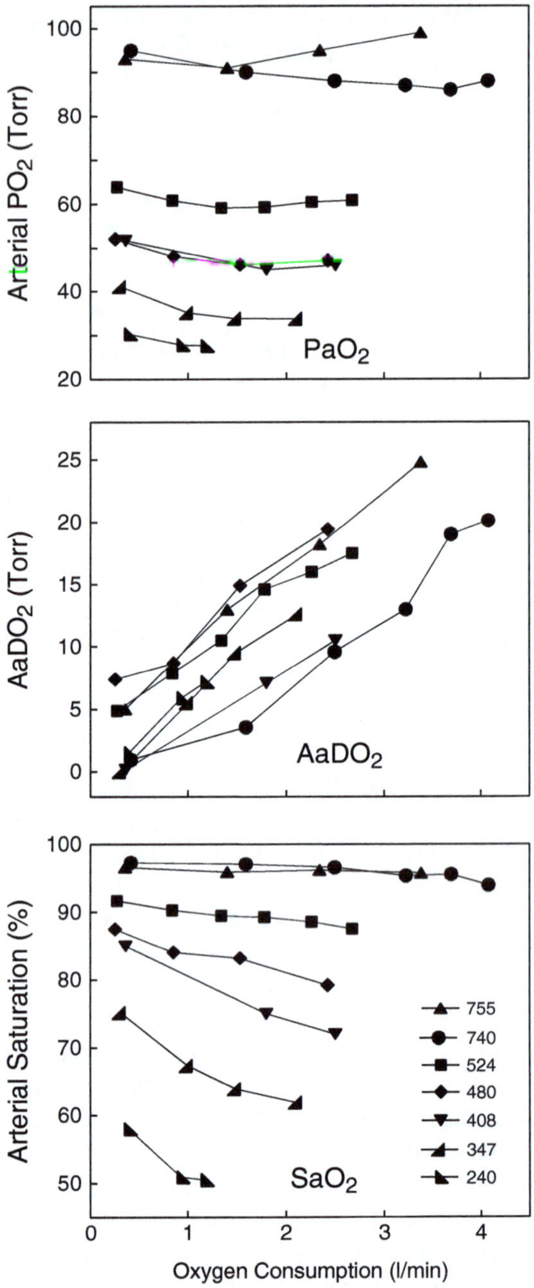

Fig. 4.11 Effect of barometric pressure on arterial PO_2 (*top*), $AaDO_2$ (*middle*), and arterial saturation (*bottom*) during exercise. Data are mean data for several subjects. At sea level, PaO_2 and arterial saturation are maintained in most subjects during exercise except in a minority at very high levels of exercise. As barometric pressure decreases PaO_2 falls during exercise, despite an overall unchanged $AaDO_2$, and SaO_2 drops precipitously. Data compiled from [9, 33, 107, 108, 122, 124]

Gas Exchange Efficiency

The preceding discussion on the relative contributions of ventilation-perfusion matching and diffusion limitation to the $AaDO_2$ might lead one to think that since both ventilation-perfusion matching [41] and pulmonary diffusion limitation [115] are expected to increase with increasing exercise intensity, that the $AaDO_2$ might similarly be expected to further increase with when the altitude is also increased. While this is true for subjects exercising in acute hypoxia [51, 115], this is not the case for acclimatized subjects. Figure 4.11, middle panel, shows the remarkable variability in the $AaDO_2$ between groups of subjects exercising at differing altitudes. While at each barometric pressure the $AaDO_2$ increases with increasing exercise intensity, there is no apparent additional effect of altitude. The same is true even in chamber studies when the same group of subjects is repeatedly studied at different barometric pressures under standardized conditions [124]. In addition to inter-subject variability, these findings can likely be explained by the relative contributions of pulmonary diffusion limitation and ventilation-perfusion inequality to the $AaDO_2$ during exercise, and the differing effects that altitude has on each of them, discussed further below.

Ventilation-Perfusion Inequality

As previously mentioned, the extent of resting ventilation-perfusion inequality increases with increasing hypobaric hypoxia, although the magnitude of the increase varies between subjects and depends on ascent profile, with faster ascent associated with greater ventilation-perfusion inequality. Ventilation-perfusion inequality increases with increasing exercise intensity in normoxia [57] and is also further increased in hypoxic exercise [41, 123]. Figure 4.12 shows the effect of a simulated ascent of Mt. Everest [124] on the LogSD$\dot{Q}$, an index of ventilation-perfusion heterogeneity derived from the multiple inert gas elimination technique. It is evident from Fig. 4.12, that at a barometric pressure equivalent to the summit of Mt. Everest, significant resting and exercising ventilation-perfusion inequality is present. The data from both 348 Torr (~6,100 m) and 253 Torr (~8,850 m, Everest

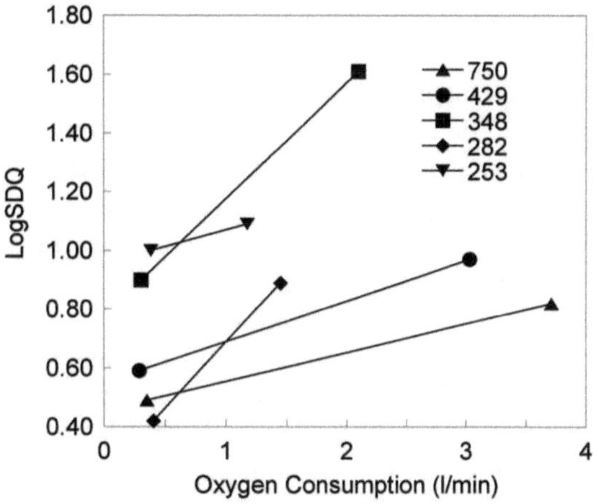

Fig. 4.12 The effect of barometric pressure on ventilation-perfusion inequality as measured by the LogSD$\dot{Q}$, at rest and during maximal exercise. Ventilation-perfusion inequality is increased by exercise and is increased further by high altitude. See text for details. Redrawn and simplified from [124]

Summit) are elevated above that seen at other barometric pressures, possibly reflecting the rapid ascent profile and the development of pulmonary edema (see below) before these data were collected.

There is considerable evidence that increased ventilation-perfusion inequality during exercise results from the development of interstitial pulmonary edema, including the following: (1) An increase in lung density is observed in subjects after exercise at moderate altitude [3]; (2) Ventilation-perfusion inequality is increased in acute hypoxia [41], and decreased by breathing 100 % O_2, which would alter capillary pressure and fluid filtration; (3) Prolonged exercise results in a progressive increase in ventilation-perfusion inequality [56], which persists in the recovery period after heavy exercise, when ventilation and cardiac output are back to resting values [100] consistent with fluid accumulation; (4) Subjects who have a history of HAPE also show larger increases in exercise-induced ventilation-perfusion inequality compared to subjects without a history of HAPE [88]; (5) A large scale study, described earlier in this chapter evaluated climbers at 1,200 and 4,559 m and found evidence for an increase in closing volume in 74 % of climbers who lacked clinical evidence of edema [26]. Increased closing volume is taken as evidence of interstitial edema, as peribronchial fluid forms around and compresses small airways, leading to air trapping. Together, these findings suggest interstitial edema as the most likely cause of ventilation-perfusion inequality and this is consistent with the increases in ventilation-perfusion inequality with rapid simulated ascent described in the preceding paragraph. However this is still an area of active investigation and by no means a resolved issue (108).

The extent of ventilation-perfusion inequality during exercise may also be altered by drugs commonly taken for altitude illness. In exercising normoxic horses, a species noted for very high exercising pulmonary arterial pressures, chronic acetazolamide (Diamox®) administration reduces pulmonary transvascular fluid flux [117]. In human subjects exercising in normoxia and acute hypoxia, acetazolamide has been shown to increase PaO_2, and decrease the $AaDO_2$ as well the extent of ventilation-perfusion inequality compared to placebo [64]. The mechanism for this is speculated to be release of hypoxic pulmonary vasoconstriction and reductions in pulmonary arterial pressure, although other effects of acetazolamide (such as diuresis) on the pulmonary circulation cannot be eliminated [109]. In keeping with this idea, sildenafil, a phosphodiesterase inhibitor that causes pulmonary vasodilation, has been shown to reduce pulmonary arterial pressure and the $AaDO_2$ with an increase in PaO_2, in acclimatized subjects exercising at 4,350 m [96] although this is not a consistent finding in all studies. However few studies report arterial blood gases and importantly, ventilation-perfusion inequality has not been directly measured in this context.

Thus, ventilation-perfusion inequality as a contributor to inefficient gas exchange is likely present in most individuals exercising at altitude. However as mentioned earlier, the net effect of ventilation-perfusion inequality on the $AaDO_2$ for a given individual will depend on the overall location of the lung (i.e., the P_AO_2) on the oxygen-hemoglobin dissociation curve. Figure 4.13 top panel shows the effect of exercise on the $AaDO_2$ attributable to ventilation-perfusion inequality in five subjects during a simulated ascent of Mount Everest [124]. It can be appreciated that although the extent of ventilation-perfusion inequality increases markedly, the effect on the $AaDO_2$ decreases to an even greater extent as more of the lung moves on to the steepest part of the dissociation curve. The combined result is that ventilation-perfusion inequality contributes almost nothing to the exercising $AaDO_2$ by the time the summit barometric pressure is reached. Thus, in contrast to the resting data, ventilation-perfusion inequality is a relatively minor contributor to gas exchange inefficiency during exercise at altitude.

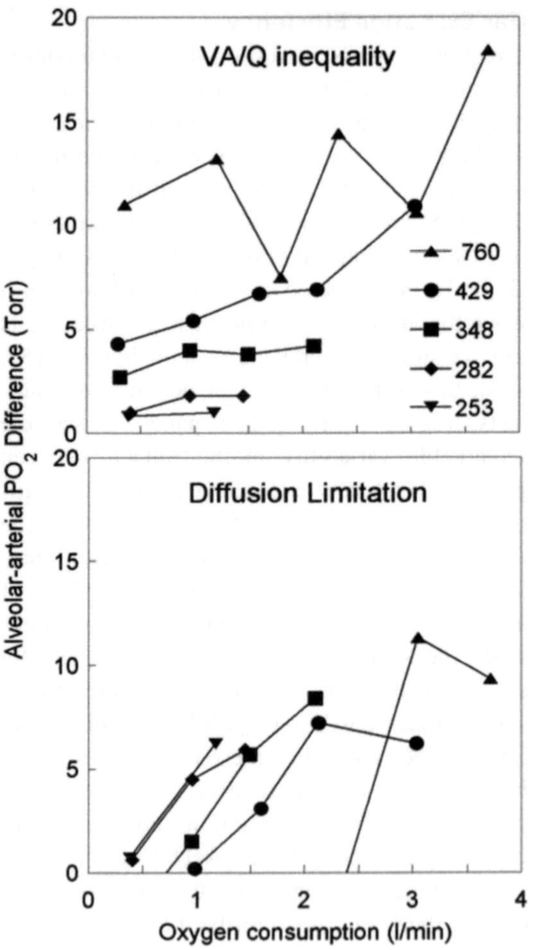

Fig. 4.13 The effect of exercise at different barometric pressures on the contribution of ventilation-perfusion inequality (*top*) and diffusion limitation (*bottom*) to the $AaDO_2$. At lower barometric pressures, as is the case for rest, the contribution of ventilation-perfusion inequality to the $AaDO_2$ becomes almost zero; however the contribution of diffusion limitation becomes increasingly important, even at very low rates of oxygen consumption elicited in extreme hypobaric hypoxia. Data from [124]

Diffusion Limitation

It was previously shown in Fig. 4.8 that the contribution from diffusion limitation to the $AaDO_2$ for a fixed DLO_2 increases with increasing elevation. During hypoxic exercise, the DLO_2 would be expected to increase above resting values as pulmonary capillaries become maximally recruited, but this is not sufficient to prevent diffusion limitation. The bottom panel of Fig. 4.13 shows the amount of the $AaDO_2$ that can be attributed to diffusion limitation of oxygen equilibrium during exercise at various simulated altitudes. It can be appreciated that, in contrast to ventilation-perfusion inequality, diffusion limitation becomes an increasingly larger contributor to the $AaDO_2$ during exercise, despite an overall increase in DLO_2. The reason that diffusion of oxygen from the alveolus into the blood is increasingly problematic at altitude is found in characteristics of the oxygen-hemoglobin dissociation curve (see the previous version of this chapter [121] for a complete explanation). The diffusion of oxygen can be modeled [87] as:

$$PAO_2 - \mathrm{Pc}'\,\mathrm{O}_2 = \left(PAO_2 - P\overline{v}O_2\right)\exp\left(-DLO_2 / \mathrm{b}O_2 \times \dot{Q}\right)$$

where $PAO_2 \cdot Pc'O_2$, and PvO_2 are the alveolar, capillary, and mixed venous partial pressures of O_2 respectively, DLO_2 is the diffusion capacity of the lung for oxygen, βO_2 is the instantaneous slope of the O_2-hemoglobin dissociation curve, and Q is cardiac output. As elevation increases, the ratio of $DLO_2/\beta\dot{Q}$ falls since the slope of the O_2-hemoglobin dissociation curve (i.e., β) increases in hypoxia. Thus, diffusion limitation becomes a progressively more important contributor to the $AaDO_2$ the altitude increases. The contribution of diffusion limitation to the $AaDO_2$ during exercise has been shown to decrease with acclimatization [9] compared to acute hypoxia, presumably on the basis of reduced $\dot{Q}$ and a consequent increase in $DLO_2/\beta\dot{Q}$ Interestingly, in addition to the effects on ventilation-perfusion matching described above, acetazolamide has also been shown to reduce diffusion limitation during acute hypoxic exercise, likely by reducing β and improving diffusion equilibrium [64], although it is unknown if the effect is sustained with chronic ingestion at altitude.

Special Considerations

Although the focus of this review is on pulmonary gas exchange in lowland individuals who travel and acclimatize to high altitude, there are two additional issues relevant to pulmonary gas exchange at altitude worth further discussion: Intermittent exposure to hypoxia (either occupationally or as part of training for athletic competition) and the effects of lifelong residence at high altitude, as these represent opposite ends of the exposure spectrum.

High Altitude Peoples

The exercise performance in hypoxia of high altitude peoples is well documented (see [8, 139] and Chap. 19), and investigation of potential mechanisms has attracted considerable interest. It is well known that pulmonary function is enhanced in lifelong residents of high altitude, who show larger lung volumes and increased diffusing capacity for carbon monoxide [19, 30, 36]. A substantial difficulty in conducting such research relates to teasing out which attributes are a result of lifelong altitude residence and which are related to the long-term genetic effects of natural selection in a population. One approach is to compare genetic lowlanders and highlanders, both born and raised at high altitude and compare physiologic responses compared to acclimatized lowland sojourners [15]. Interestingly, using this comparison, some high altitude Andean Aymara have been shown to have a greater oxygen saturation at rest and during exercise than lifelong high altitude resident people of European ancestry who did not differ from fully acclimatized sojorners born and raised at low altitude [13]. In addition, in a subset of Aymara born and raised at low altitude, who showed marked desaturation of hemoglobin at high altitude, the extent of the decrement was related to the degree of genetic admixture [14]. In a similar population, the angiotensin converting enzyme insertion deletion polymorphism-I/I allele is associated with higher arterial oxygen saturations both at rest and during exercise in a manner independent of a relationship with the hypoxic ventilatory response [11]. Thus, this evidence points to a genetic component to the maintenance of arterial oxygenation at high altitude.

However, the extent of genetic control of SaO_2 at altitude likely varies between high altitude populations. There are significant differences between Tibetan and Andean populations of high altitude peoples, with the result that Tibetans maintain higher resting ventilations, lower hemoglobin concentrations, and, paradoxically, reduced arterial oxygen saturations compared to Andean populations. Of particular note, Tibetans resident at high altitude do not show differences in saturation from Han Chinese (a lowland population) who have been born and raised at high altitude [129]. This reduction in SaO_2, affecting oxygen delivery in Tibetans, is likely compensated for by increased capillary density (reviewed in [8]). These phenotypic differences, among others, combined with population heterogeneity, and therefore more genetic variance in Tibetan high altitude peoples have lead some authors to suggest that in Tibetan populations there is greater potential for ongoing natural selection of traits beneficial for high altitude residence [8] than in Andean populations.

Exposure to High Altitude During Growth and Development

In addition to the genetic factors previously discussed, lifelong residence and hypoxic exposure at high altitude has a significant effect on pulmonary gas exchange. A series of recent animal experiments have shown that high altitude residence during somatic growth and development results in increased lung volumes and increased alveolar tissue volume and surface area [59, 79, 140]. These changes persist for years after return to low elevation [60] and are not found in mature animals transported to the same elevation for the same duration [63]. The effect of high altitude residence during early life was to improve pulmonary gas exchange in hypoxia, by reducing the portion of the $AaDO_2$ due to diffusion limitation, with a corresponding increase in arterial oxygenation. There was no change in the extent of ventilation-perfusion inequality in these animals compared to low-altitude controls [60]. Thus these studies identify high altitude residence during somatic growth and development as a powerful stimulus for lung growth affecting diffusive oxygen transport in the lung. However the effect of high altitude residence on lung growth and development may be additionally augmented in Tibetan or Andean peoples [16].

The Effect of Lifelong High Altitude Exposure on Pulmonary Gas Exchange

The net effect of high altitude residence in Andean people is to improve the efficiency of pulmonary gas exchange compared to acclimatized lowlanders [74, 122]. During exercise at 4,100 m, the $AaDO_2$ is less in high altitude Bolivian Aymara compared to acclimatized lowland controls. Even when these individuals are acutely exposed to higher altitudes (5,300 m), increased gas exchange efficiency compared to lowlanders is still observed [122]. Although the effect of acclimatization in lowlanders is to reduce the differences between lowlanders and high altitude natives, these differences do not entirely disappear [74]. The mechanism of the reduction in the $AaDO_2$ in lowlanders with acclimatization is likely as result of more complete diffusion equilibrium during exercise as cardiac output falls with prolonged exposure to altitude improving $DLO_2/\beta Q$ [9]. This improved gas exchange efficiency is likely a result of enhanced DLO_2 as it is not explained by differences in β or cardiac output between acclimatized lowlanders and Andeans [122] and is consistent with the animal data described previously. Enhanced arterial saturation was not a function of increased exercise ventilation in the high altitude natives as the ventilatory equivalent for oxygen was lower in the highlanders at the outset and similar between groups after the lowlanders acclimatized [74].

Future Directions

While earlier studies focused on characterizing changes in pulmonary function in normal individuals at high altitude, much of the more recent literature has examined changes in pulmonary function as a means to determine whether or not individuals develop subclinical pulmonary edema upon ascent [31, 85, 101]. This question is an important one with implications for advising people engaged in climbing, trekking, or other pursuits at high altitude but research into this question will need to move beyond the use of pulmonary function testing as surrogate measures of the presence of subclinical edema and focus on direct measurements that quantify extravascular lung water. Beyond this particular question, the literature needs to devote more attention to changes in pulmonary function in people with underlying cardiopulmonary diseases such as chronic obstructive pulmonary disease and asthma following exposure to acute hypoxia. With improvements in the quality of medical care, many of these individuals are engaging in a wider variety of activities than previously thought possible and further information about how pulmonary function in people with such disease changes at high altitude will facilitate better pre-travel counseling and planning for these groups.

Despite considerable indirect evidence, the role in interstitial edema in the development of ventilation-perfusion inequality has not been definitively established. The longstanding questions of what causes ventilation-perfusion inequality with exercise, and why does the extent of ventilation-perfusion inequality increase with hypoxic exercise have not yet been definitively answered. In light of recent work, the effect that intrapulmonary shunting has on pulmonary gas exchange and how this is affected by hypoxia merits further investigation. In particular it will be important to resolve the information from gas exchange techniques that consistently document only a minimal contribution to the $AaDO_2$ from intrapulmonary shunting with data from imaging techniques, showing that the passage of agitated saline contrast is induced by hypoxia. Finally the degree to which genetic and environmental influences interact and affect pulmonary function and gas exchange in individuals born and raised at high altitude both in high altitude native peoples as those of lowland ancestry is an unanswered question.

References

1. Agostoni P, Swenson ER, Bussotti M, Revera M, Meriggi P, Faini A, et al. High altitude exposure of three weeks duration increases lung diffusing capacity in humans. J Appl Physiol. 2011;110(6): 1564–71.
2. Allegra L, Cogo A, Legnani D, Diano PL, Fasano V, Negretto GG. High altitude exposure reduces bronchial responsiveness to hypo-osmolar aerosol in lowland asthmatics. Eur Respir J. 1995;8(11): 1842–6.
3. Anholm JD, Milne EN, Stark P, Bourne JC, Friedman P. Radiographic evidence of interstitial pulmonary edema after exercise at altitude. J Appl Physiol. 1999;86(2):503–9.
4. Arai TJ, Henderson AC, Dubowitz DJ, Levin DL, Friedman PJ, Buxton RB, et al. Hypoxic pulmonary vasoconstriction does not contribute to pulmonary blood flow heterogeneity in normoxia in normal supine humans. J Appl Physiol. 2009;106(4): 1057–64.
5. Babcock MA, Johnson BD, Pegelow DF, Suman OE, Griffin D, Dempsey JA. Hypoxic effects on exercise-induced diaphragmatic fatigue in normal healthy humans. J Appl Physiol. 1995;78(1):82–92.
6. Bartsch P, Maggiorini M, Mairbaurl H, Vock P, Swenson ER. Pulmonary extravascular fluid accumulation in climbers. Lancet. 2002;360(9332):571. author reply -2.
7. Basu CK, Selvamurthy W, Bhaumick G, Gautam RK, Sawhney RC. Respiratory changes during initial days of acclimatization to increasing altitudes. Aviat Space Environ Med. 1996;67(1):40–5.
8. Beall CM. Two routes to functional adaptation: Tibetan and Andean high-altitude natives. Proc Natl Acad Sci U S A. 2007;104 Suppl 1:8655–60.
9. Bebout DE, Storey D, Roca J, Hogan MC, Poole DC, Gonzales-Camerena R, et al. Effects of altitude acclimatization on pulmonary gas exchange during exercise. J Appl Physiol. 1989;67(6):2286–95.
10. Bencowitz HZ, Wagner PD, West JB. Effect of change in P50 on exercise tolerance at high altitude: a theoretical study. J Appl Physiol. 1982;53:1487.
11. Bigham AW, Kiyamu M, Leon-Velarde F, Parra EJ, Rivera-Ch M, Shriver MD, et al. Angiotensin-converting enzyme genotype and arterial oxygen saturation at high altitude in Peruvian Quechua. High Alt Med Biol. 2008;9(2):167–78.
12. Bouverot P, Farner DS, Heinrich B, Johansen K, Langer H, Neuweiler G, et al. Adaptation to altitude-hypoxia in vertebrates. Berlin: Springer; 1985.
13. Brutsaert TD, Araoz M, Soria R, Spielvogel H, Haas JD. Higher arterial oxygen saturation during submaximal exercise in Bolivian Aymara compared to European sojourners and Europeans born and raised at high altitude. Am J Phys Anthropol. 2000; 113(2):169–81.

14. Brutsaert TD, Parra EJ, Shriver MD, Gamboa A, Palacios JA, Rivera M, et al. Spanish genetic admixture is associated with larger V(O2) max decrement from sea level to 4338 m in Peruvian Quechua. J Appl Physiol. 2003;95(2):519–28.
15. Brutsaert TD, Roach R, Wagner PD, Hackett PH. Genetic and environmental adaptation in high altitude natives: conceptual, methodological and statistical concerns. Hypoxia: from genes to the bedside. New York: Kluwer Academic/Plenum Publishers; 2001. p. 133.
16. Brutsaert TD, Soria R, Caceres E, Spielvogel H, Haas JD. Effect of developmental and ancestral high altitude exposure on chest morphology and pulmonary function in Andean and European/North American natives. Am J Hum Biol. 1999;11(3):383–95.
17. Capen R, Latham L, Wagner WJ. Diffusing capacity of the lung during hypoxia: role of capillary recruitment. J Appl Physiol. 1981;50(1):165–71.
18. Capen R, Wagner WJ. Intrapulmonary blood flow redistribution during hypoxia increases gas exchange surface area. J Appl Physiol. 1982;52(6):1575–80.
19. Cerny FC, Dempsey JA, Reddan WG. Pulmonary gas exchange in nonnative residents of high altitude. J Clin Invest. 1973;52(12):2993–9.
20. Christopherson SK, Hlastala MP. Pulmonary gas exchange during altered density gas breathing. J Appl Physiol. 1982;52(1):221–5.
21. Cibella F, Cuttitta G, Romano S, Grassi B, Bonsignore G, Milic-Emili J. Respiratory energetics during exercise at high altitude. J Appl Physiol. 1999;86(6):1785–92.
22. Coates G, Gray G, Mansell A, Nahmias C, Powles A, Sutton J, et al. Changes in lung volume, lung density, and distribution of ventilation during hypobaric decompression. J Appl Physiol. 1979;46(4):752–5.
23. Cogo A, Basnyat B, Legnani D, Allegra L. Bronchial asthma and airway hyperresponsiveness at high altitude. Respiration. 1997;64(6):444–9.
24. Cotes JE. Ventilatory capacity at altitude and its relation to mask design. Proc R Soc Lond B Biol Sci. 1954;143(910):32–9.
25. Crapo RO, Jensen RL, Hegewald M, Tashkin DP. Arterial blood gas reference values for sea level and an altitude of 1,400 meters. Am J Respir Crit Care Med. 1999;160(5 Pt 1):1525–31.
26. Cremona G, Asnaghi R, Baderna P, Brunetto A, Brutsaert T, Cavallaro C, et al. Pulmonary extravascular fluid accumulation in recreational climbers: a prospective study. Lancet. 2002;359(9303):303–9.
27. Cruz JC. Mechanics of breathing in high altitude and sea level subjects. Respir Physiol. 1973;17(2):146–61.
28. Dagg KD, Thomson LJ, Clayton RA, Ramsay SG, Thomson NC. Effect of acute alterations in inspired oxygen tension on methacholine induced bronchoconstriction in patients with asthma. Thorax. 1997;52(5):453–7.
29. Deboeck G, Moraine JJ, Naeije R. Respiratory muscle strength may explain hypoxia-induced decrease in vital capacity. Med Sci Sports Exerc. 2005;37(5):754–8.
30. DeGraff Jr AC, Grover RF, Johnson Jr RL, Hammond Jr JW, Miller JM. Diffusing capacity of the lung in Caucasians native to 3,100 m. J Appl Physiol. 1970;29(1):71–6.
31. Dehnert C, Luks AM, Schendler G, Menold E, Berger MM, Mairbaurl H, et al. No evidence for interstitial lung oedema by extensive pulmonary function testing at 4,559 m. Eur Respir J. 2010; 35(4):812–20.
32. Dehnert C, Risse F, Ley S, Kuder TA, Buhmann R, Puderbach M, et al. Magnetic resonance imaging of uneven pulmonary perfusion in hypoxia in humans. Am J Respir Crit Care Med. 2006;174(10):1132–8.
33. Dempsey JA, Reddan WG, Birnbaum ML, Forster HV, Thoden JS, Grover RF, et al. Effects of acute through life-long hypoxic exposure on exercise pulmonary gas exchange. Respir Physiol. 1971;13(1): 62–89.
34. Dempsey JA, Wagner PD. Exercise-induced arterial hypoxemia. J Appl Physiol. 1999;87(6):1997–2006.
35. Denjean A, Roux C, Herve P, Bonniot JP, Comoy E, Duroux P, et al. Mild isocapnic hypoxia enhances the bronchial response to methacholine in asthmatic subjects. Am Rev Respir Dis. 1988;138(4):789–93.
36. Droma T, McCullough RG, McCullough RE, Zhuang JG, Cymerman A, Sun SF, et al. Increased vital and total lung capacities in Tibetan compared to Han residents of Lhasa (3,658 m). Am J Phys Anthropol. 1991;86(3):341–51.
37. Eldridge MW, Dempsey JA, Haverkamp HC, Lovering AT, Hokanson JS. Exercise-induced intrapulmonary arteriovenous shunting in healthy humans. J Appl Physiol. 2004;97(3):797–805.
38. Fasano V, Paolucci E, Pomidori L, Cogo A. High-altitude exposure reduces inspiratory muscle strength. Int J Sports Med. 2007;28(5):426–30.
39. Finkelstein S, Tomashefski JF, Shillito FH. Pulmonary mechanics at altitude in normal and obstructive lung disease patients. Aerosp Med. 1965;36:880–4.
40. Forte Jr VA, Leith DE, Muza SR, Fulco CS, Cymerman A. Ventilatory capacities at sea level and high altitude. Aviat Space Environ Med. 1997;68(6):488–93.
41. Gale GE, Torre BJ, Moon RE, Saltzman HA, Wagner PD. Ventilation-perfusion inequality in normal humans during exercise at sea level and simulated altitude. J Appl Physiol. 1985;58(3):978–88.
42. Gautier H, Peslin R, Grassino A, Milic-Emili J, Hannhart B, Powell E, et al. Mechanical properties of the lungs during acclimatization to altitude. J Appl Physiol. 1982;52(6):1407–15.
43. Ge RL, Matsuzawa Y, Takeoka M, Kubo K, Sekiguchi M, Kobayashi T. Low pulmonary diffusing capacity in subjects with acute mountain sickness. Chest. 1997;111(1):58–64.
44. Goldstein RS, Zamel N, Rebuck AS. Absence of effects of hypoxia on small airway function in humans. J Appl Physiol. 1979;47(2):251–6.
45. Gray BA, Blalock JM. Interpretation of the alveolar-arterial oxygen difference in patients with hypercapnia. Am Rev Respir Dis. 1991;143(1):4–8.

46. Gray GW, Rennie ID, Houston CS, Bryan AC. Phase IV volume of the single-breath nitrogen washout curve on exposure to altitude. J Appl Physiol. 1973;35(2):227–30.
47. Grocott MP, Martin DS, Levett DZ, McMorrow R, Windsor J, Montgomery HE. Arterial blood gases and oxygen content in climbers on Mount Everest. N Engl J Med. 2009;360(2):140–9.
48. Gudjonsdottir M, Appendini L, Baderna P, Purro A, Patessio A, Vilianis G, et al. Diaphragm fatigue during exercise at high altitude: the role of hypoxia and workload. Eur Respir J. 2001;17(4):674–80.
49. Guleria JS, Pande JN, Sethi PK, Roy SB. Pulmonary diffusing capacity at high altitude. J Appl Physiol. 1971;31(4):536–43.
50. Hackett PH, Rennie D. The incidence, importance, and prophylaxis of acute mountain sickness. Lancet. 1976;2(7996):1149–55.
51. Hammond MD, Gale GE, Kapitan KS, Ries A, Wagner PD. Pulmonary gas exchange in humans during normobaric hypoxic exercise. J Appl Physiol. 1986;61(5):1749–57.
52. Hammond MD, Hempleman SC. Oxygen diffusing capacity estimates derived from measured VA/Q distributions in man. Respir Physiol. 1987;69:129–47.
53. Honigman B, Theis MK, Koziol-McLain J, Roach R, Yip R, Houston C, et al. Acute mountain sickness in a general tourist population at moderate altitudes. Ann Intern Med. 1993;118(8):587–92.
54. Hopkins SR. Exercise induced arterial hypoxemia: the role of ventilation-perfusion inequality and pulmonary diffusion limitation. Adv Exp Med Biol. 2006;588:17–30.
55. Hopkins SR, Garg J, Bolar DS, Balouch J, Levin DL. Pulmonary blood flow heterogeneity during hypoxia and high-altitude pulmonary edema. Am J Respir Crit Care Med. 2005;171(1):83–7.
56. Hopkins SR, Gavin TP, Siafakas NM, Haseler LJ, Olfert IM, Wagner H, et al. Effect of prolonged, heavy exercise on pulmonary gas exchange in athletes. J Appl Physiol. 1998;85(4):1523–32.
57. Hopkins SR, McKenzie DC, Schoene RB, Glenny R, Robertson HT. Pulmonary gas exchange during exercise in athletes I: ventilation-perfusion mismatch and diffusion limitation. J Appl Physiol. 1994;77(2):912–7.
58. Hopkins SR, Olfert IM, Wagner PD. Point: exercise-induced intrapulmonary shunting is imaginary. J Appl Physiol. 2009;107(3):993–4.
59. Hsia CC, Carbayo JJ, Yan X, Bellotto DJ. Enhanced alveolar growth and remodeling in Guinea pigs raised at high altitude. Respir Physiol Neurobiol. 2005;147(1):105–15.
60. Hsia CC, Johnson Jr RL, McDonough P, Dane DM, Hurst MD, Fehmel JL, et al. Residence at 3,800-m altitude for 5 mo in growing dogs enhances lung diffusing capacity for oxygen that persists at least 2.5 years. J Appl Physiol. 2007;102(4):1448–55.
61. Hughson RL, Yamamoto Y, McCullough RE, Sutton JR, Reeves JT. Sympathetic and parasympathetic indicators of heart rate control at altitude studied by spectral analysis. J Appl Physiol. 1994;77(6):2537–42.
62. Jaeger JJ, Sylvester JT, Cymerman A, Berberich JJ, Denniston JC, Maher JT. Evidence for increased intrathoracic fluid volume in man at high altitude. J Appl Physiol. 1979;47(4):670–6.
63. Johnson Jr RL, Cassidy SS, Grover RF, Schutte JE, Epstein RH. Functional capacities of lungs and thorax in beagles after prolonged residence at 3,100 m. J Appl Physiol. 1985;59(6):1773–82.
64. Jonk AM, van den Berg IP, Olfert IM, Wray DW, Arai T, Hopkins SR, et al. Effect of acetazolamide on pulmonary and muscle gas exchange during normoxic and hypoxic exercise. J Physiol. 2007;579(Pt 3):909–21.
65. Kreuzer F, Van Lookeren CP. Resting pulmonary diffusing capacity for CO and O2 at high altitude. J Appl Physiol. 1965;20(3):519–24.
66. Kronenberg RS, Safar P, Lee J, Wright F, Noble W, Wahrenbrock E, et al. Pulmonary artery pressure and alveolar gas exchange in man during acclimatization to 12,470 ft. J Clin Invest. 1971;50(4):827–37.
67. Lefrancois R, Gautier H, Pasquis P. Mecanique ventilatoire chez l'homme a haute altitude. CR Soc Biol. 1969;163:2037–42.
68. Lenfant C, Torrance J, English E, Finch CA, Reynafarje C, Ramos J, et al. Effect of altitude on oxygen binding by hemoglobin and on organic phosphate levels. J Clin Invest. 1968;47(12):2652–6.
69. Lenfant C, Torrance JD, Reynafarje C. Shift of the O2-Hb dissociation curve at altitude: mechanism and effect. J Appl Physiol. 1971;30(5):625–31.
70. Loeppky JA, Icenogle M, Scotto P, Robergs R, Hinghofer-Szalkay H, Roach RC. Ventilation during simulated altitude, normobaric hypoxia and normoxic hypobaria. Respir Physiol. 1997;107(3):231–9.
71. Loeppky JA, Roach RC, Maes D, Hinghofer-Szalkay H, Roessler A, Gates L, et al. Role of hypobaria in fluid balance response to hypoxia. High Alt Med Biol. 2005;6(1):60–71.
72. Lovering AT, Eldridge MW, Stickland MK. Counterpoint: exercise-induced intrapulmonary shunting is real. J Appl Physiol. 2009;107(3):994–7 [Research Support, Non-U.S. Gov't].
73. Lovering AT, Romer LM, Haverkamp HC, Pegelow DF, Hokanson JS, Eldridge MW. Intrapulmonary shunting and pulmonary gas exchange during normoxic and hypoxic exercise in healthy humans. J Appl Physiol. 2008;104(5):1418–25.
74. Lundby C, Calbet JA, van Hall G, Saltin B, Sander M. Pulmonary gas exchange at maximal exercise in Danish lowlanders during 8 wk of acclimatization to 4,100 m and in high-altitude Aymara natives. Am J Physiol Regul Integr Comp Physiol. 2004;287(5):R1202–8.
75. Mairbaurl H, Schobersberger W, Oelz O, Bartsch P, Eckardt KU, Bauer C. Unchanged in vivo P50 at high altitude despite decreased erythrocyte age and

elevated 2,3-diphosphoglycerate. J Appl Physiol. 1990;68(3):1186–94.
76. Mansell A, Powles A, Sutton J. Changes in pulmonary PV characteristics of human subjects at an altitude of 5,366 m. J Appl Physiol. 1980;49(1):79–83.
77. Mason NP, Barry PW, Pollard AJ, Collier DJ, Taub NA, Miller MR, et al. Serial changes in spirometry during an ascent to 5,300 m in the Nepalese Himalayas. High Alt Med Biol. 2000;1(3):185–95.
78. Mazzeo RS, Wolfel EE, Butterfield GE, Reeves JT. Sympathetic response during 21 days at high altitude (4,300 m) as determined by urinary and arterial catecholamines. Metabolism. 1994;43(10):1226–32.
79. McDonough P, Dane DM, Hsia CC, Yilmaz C, Johnson Jr RL. Long-term enhancement of pulmonary gas exchange after high-altitude residence during maturation. J Appl Physiol. 2006;100(2):474–81.
80. Milic-Emili J, Kayser B, Gautier H. Mechanics of breathing. In: Hornbein TF, Schoene RB, editors. High altitude: an exploration of human adaptation. New York: Marcel Dekker; 2001.
81. Mognoni P, Saibene F, Veicsteinas A. Ventilatory work during exercise at high altitude. Int J Sports Med. 1982;3(1):33–6.
82. Newhouse MT, Becklake MR, Macklem PT, McGregor M. Effect of alterations in end-tidal Co2 tension on flow resistance. J Appl Physiol. 1964;19:745–9.
83. Paiva M, Engel LA. Pulmonary interdependence of gas transport. J Appl Physiol. 1979;47(2):296–305.
84. Peacock AJ, Jones PL. Gas exchange at extreme altitude: results from the British 40th Anniversary Everest Expedition. Eur Respir J. 1997;10(7):1439–44.
85. Pellegrino R, Pompilio P, Quaranta M, Aliverti A, Kayser B, Miserocchi G, et al. Airway responses to methacholine and exercise at high altitude in healthy lowlanders. J Appl Physiol. 2010;108(2):256–65.
86. Petit JM, Milic-Emili G, Troquet J. Travail dynamique pulmonaire et altitude. Rev Med Aeronaut. 1963;2:276–9.
87. Piiper J. Apparent increase of the O2 diffusing capacity with increased O2 uptake in inhomogeneous lungs: theory. Respir Physiol. 1969;6(2): 209–18.
88. Podolsky A, Eldridge MW, Richardson RS, Knight DR, Johnson EC, Hopkins SR, et al. Exercise-induced VA/Q inequality in subjects with prior high-altitude pulmonary edema. J Appl Physiol. 1996; 81(2):922–32.
89. Pollard AJ, Mason NP, Barry PW, Pollard RC, Collier DJ, Fraser RS, et al. Effect of altitude on spirometric parameters and the performance of peak flow meters. Thorax. 1996;51(2):175–8.
90. Pugh LG. Muscular exercise on Mount Everest. J Physiol. 1958;141(2):233–61.
91. Pugh LG, Gill MB, Lahiri S, Milledge JS, Ward MP, West JB. Muscular exercise at great altitudes. J Appl Physiol. 1964;19:431–40.
92. Rahn H, Hammond D. Vital capacity at reduced barometric pressure. J Appl Physiol. 1952;4(9): 715–24.
93. Rahn H, Otis AB. Man's respiratory response during and after acclimatization to high altitude. Am J Physiol. 1946;157:445–559.
94. Raymond L, Severinghaus JW. Static pulmonary compliance of man during altitude hypoxia. J Appl Physiol. 1971;31(5):785–7.
95. Reyes A, Roca J, Rodriguez-Roisin R, Torres A, Ussetti P, Wagner PD. Effect of almitrine on ventilation-perfusion distribution in adult respiratory distress syndrome. Am Rev Respir Dis. 1988;137(5): 1062–7.
96. Richalet JP, Gratadour P, Robach P, Pham I, Dechaux M, Joncquiert-Latarjet A, et al. Sildenafil inhibits altitude-induced hypoxemia and pulmonary hypertension. Am J Respir Crit Care Med. 2005;171(3): 275–81.
97. Roach RC, Loeppky JA, Icenogle MV. Acute mountain sickness: increased severity during simulated altitude compared with normobaric hypoxia. J Appl Physiol. 1996;81(5):1908–10.
98. Saunders NA, Betts MF, Pengelly LD, Rebuck AS. Changes in lung mechanics induced by acute isocapnic hypoxia. J Appl Physiol. 1977;42(3):413–9.
99. Savourey G, Launay JC, Besnard Y, Guinet A, Travers S. Normo- and hypobaric hypoxia: are there any physiological differences? Eur J Appl Physiol. 2003;89(2):122–6.
100. Schaffartzik W, Poole DC, Derion T, Tsukimoto K, Hogan MC, Arcos JP, et al. VA/Q distribution during heavy exercise and recovery in humans: implications for pulmonary edema. J Appl Physiol. 1992;72(5): 1657–67.
101. Senn O, Clarenbach CF, Fischler M, Thalmann R, Brunner-La Rocca H, Egger P, et al. Do changes in lung function predict high-altitude pulmonary edema at an early stage? Med Sci Sports Exerc. 2006;38(9):1565–70.
102. Sharma S, Brown B. Spirometry and respiratory muscle function during ascent to higher altitudes. Lung. 2007;185(2):113–21.
103. Singh I, Kapila CC, Khanna PK, Nanda RB, Rao BD. High-altitude pulmonary oedema. Lancet. 1965; 191:229–34.
104. Snyder EM, Beck KC, Hulsebus ML, Breen JF, Hoffman EA, Johnson BD. Short-term hypoxic exposure at rest and during exercise reduces lung water in healthy humans. J Appl Physiol. 2006; 101(6):1623–32.
105. Steinacker JM, Tobias P, Menhold E, Reissnecker S, Hohenhaus E, Liu Y, et al. Lung diffusing capacity and exercise in subjects with previous high altitude pulmonary oedema. Eur Respir J. 1998;11(3): 643–50.
106. Stickland MK, Lovering AT. Exercise-induced intrapulmonary arteriovenous shunting and pulmonary gas exchange. Exerc Sport Sci Rev. 2006;34(3): 99–106.
107. Stickland MK, Welsh RC, Haykowsky MJ, Petersen SR, Anderson WD, Taylor DA, et al. Intra-pulmonary shunt and pulmonary gas exchange during exercise in humans. J Physiol. 2004;561(Pt 1):321–9.
108. Sutton JR, Reeves JT, Wagner PD, Groves BM, Cymerman A, Malconian MK, et al. Operation Everest II: oxygen transport during exercise at extreme simulated altitude. J Appl Physiol. 1988; 64(4):1309–21.

109. Swenson ER. Carbonic anhydrase inhibitors and hypoxic pulmonary vasoconstriction. Respir Physiol Neurobiol. 2006;151(2–3):209–16.
110. Swenson ER. Most climbers do not develop subclinical interstitial pulmonary edema. High Alt Med Biol. 2011;12(2):125–8.
111. Swenson ER, Maggiorini M, Mongovin S, Gibbs JS, Greve I, Mairbaurl H, et al. Pathogenesis of high-altitude pulmonary edema: inflammation is not an etiologic factor. JAMA. 2002;287(17):2228–35.
112. Tenney SM. Functional differences in mammalian hemoglobin affinity for oxygen. In: Sutton JR, Houston CS, Coates MD, editors. Hypoxia and the brain. Burlington: Queen City Printers; 1995. p. 57–68.
113. Tenney SM, Rahn H, Stroud RC, Mithoefer JC. Adoption to high altitude: changes in lung volumes during the first seven days at Mt. Evans, Colorado. J Appl Physiol. 1953;5(10):607–13.
114. Thoden JS, Dempsey JA, Reddan WG, Birnbaum ML, Forster HV, Grover RF, et al. Ventilatory work during steady-state response to exercise. Fed Proc. 1969;28(3):1316–21.
115. Torre-Bueno JR, Wagner PD, Saltzman HA, Gale GE, Moon RE. Diffusion limitation in normal humans during exercise at sea level and simulated altitude. J Appl Physiol. 1985;58(3):989–95.
116. van den Elshout FJ, van Herwaarden CL, Folgering HT. Effects of hypercapnia and hypocapnia on respiratory resistance in normal and asthmatic subjects. Thorax. 1991;46(1):28–32.
117. Vengust M, Staempfli H, Viel L, Heigenhauser G. Effects of chronic acetazolamide administration on fluid flux from the pulmonary vasculature at rest and during exercise in horses. Equine Vet J Suppl. 2006;36:508–15.
118. Vogiatzis I, Zakynthinos S, Boushel R, Athanasopoulos D, Guenette JA, Wagner H, et al. The contribution of intrapulmonary shunts to the alveolar-to-arterial oxygen difference during exercise is very small. J Physiol. 2008;586(9):2381–91.
119. Wagner PD. A theoretical analysis of factors determining VO2 MAX at sea level and altitude. Respir Physiol. 1996;106(3):329–43.
120. Wagner PD. Insensitivity of VO2max to hemoglobin-P50 as sea level and altitude. Respir Physiol. 1997;107(3):205–12.
121. Wagner PD. Gas exchange. In: Hornbein TF, Schoene R, editors. High altitude an exploration of human adaptation. New York: Marcel Dekker; 2001. p. 199–234.
122. Wagner PD, Araoz M, Boushel R, Calbet JA, Jessen B, Radegran G, et al. Pulmonary gas exchange and acid-base state at 5,260 m in high-altitude Bolivians and acclimatized lowlanders. J Appl Physiol. 2002;92(4):1393–400.
123. Wagner PD, Gale GE, Moon RE, Torre BJ, Stolp BW, Saltzman HA. Pulmonary gas exchange in humans exercising at sea level and simulated altitude. J Appl Physiol. 1986;61(1):260–70.
124. Wagner PD, Sutton JR, Reeves JT, Cymerman A, Groves BM, Malconian MK. Operation Everest II: pulmonary gas exchange during a simulated ascent of Mt. Everest. J Appl Physiol. 1987;63(6): 2348–59.
125. Wagner PD, Wagner HE, Groves BM, Cymerman A, Houston CS. Hemoglobin P(50) during a simulated ascent of Mt. Everest, Operation Everest II. High Alt Med Biol. 2007;8(1):32–42.
126. Weber RE. Hemoglobin adaptations to hypoxia and altitude-the phylogenetic perspective. In: Sutton JR, Houston CS, Coates MD, editors. Hypoxia and the brain. Burlington: Queen City Printers; 1995. p. 31–44.
127. Weber RE. High-altitude adaptations in vertebrate hemoglobins. Respir Physiol Neurobiol. 2007; 158(2–3): 132–42.
128. Weiskopf RB, Severinghaus JW. Diffusing capacity of the lung for CO in man during acute acclimation to 14,246 ft. J Appl Physiol. 1972;32(3):285–9.
129. Weitz CA, Garruto RM. A comparative analysis of arterial oxygen saturation among Tibetans and Han born and raised at high altitude. High Alt Med Biol. 2007;8(1):13–26.
130. Welsh CH, Wagner PD, Reeves JT, Lynch D, Cink TM, Armstrong J, et al. Operation Everest. II: spirometric and radiographic changes in acclimatized humans at simulated high altitudes. Am Rev Respir Dis. 1993;147(5):1239–44.
131. West J. Respiratory physiology-the essentials. 5th ed. Baltimore: Williams & Wilkins; 1995. p. 39–43.
132. West JB. Diffusing capacity of the lung for carbon monoxide at high altitude. J Appl Physiol. 1962;17: 421–6.
133. West JB. Prediction of barometric pressures at high altitude with the use of model atmospheres. J Appl Physiol. 1996;81(4):1850.
134. West JB. Barometric pressures on Mt. Everest: new data and physiological significance. J Appl Physiol. 1999;86(3):1062–6.
135. West JB, Hackett PH, Maret KH, Milledge JS, Peters Jr RH, Pizzo CJ, et al. Pulmonary gas exchange on the summit of Mount Everest. J Appl Physiol. 1983;55(3):678–87.
136. West JB, Lahiri S, Maret KH, Peters Jr RM, Pizzo CJ. Barometric pressures at extreme altitudes on Mt. Everest: physiological significance. J Appl Physiol. 1983;54(5):1188–94.
137. Winslow RM, Monge CC, Statham NJ, Gibson CG, Charache S, Whittembury J, et al. Variability of oxygen affinity of blood: human subjects native to high altitude. J Appl Physiol. 1981;51(6):1411–6.
138. Wood LD, Bryan AC, Bau SK, Weng TR, Levison H. Effect of increased gas density on pulmonary gas exchange in man. J Appl Physiol. 1976;41(2): 206–10.
139. Wu T, Li S, Ward MP. Tibetans at extreme altitude. Wilderness Environ Med. 2005;16(1):47–54.
140. Yilmaz C, Dane DM, Hsia CC. Alveolar diffusion-perfusion interactions during high-altitude residence in Guinea pigs. J Appl Physiol. 2007;102(6):2179–85.

Pulmonary Circulation

5

Marco Maggiorini, Peter Bärtsch,
and Erik R. Swenson

Abstract

Increased pulmonary vascular resistance (PVR) and pulmonary artery pressure (Ppa) upon ascent to high altitude universally occur in humans and other mammals, although the magnitude can vary almost five-fold among individuals, across species, and with time at altitude from years to many generations. The mechanisms that lead to hypoxic pulmonary hypertension are numerous and have differing contributions to the pressure measured at any one time. The first and best characterized response is acute hypoxic pulmonary vasoconstriction (HPV). Its physiologic relevance or advantage, if any, in healthy humans moving to altitude remains uncertain, but when it is very excessive it can lead within days to high altitude pulmonary edema (HAPE). Sustained very high Ppa over weeks and months can lead to congestive right heart failure of high altitude also known as subacute mountain sickness (SMS) or "cor pulmonale or acute right heart failure of high altitude". With longer durations of high pressure in the pulmonary circulation there can be within days sufficient remodelling of the vasculature to protect the pulmonary capillaries, hence avoid HAPE, and within weeks compensatory changes in the right heart to overcome the

M. Maggiorini, M.D. (✉)
Medical Intensive Care Unit,
University Hospital, University of Zurich,
Zurich, Switzerland
e-mail: klinmax@usz.uzh.ch

P. Bärtsch, M.D. (✉)
Division of Sports Medicine, Department
of Internal Medicine, Medical University Clinic,
University of Heidelberg, Heidelberg, Germany
e-mail: peter.bartsch@med.uni-heidelberg.de

E.R. Swenson, M.D.
VA Puget Sound Health Care System, University
of Washington, Seattle, WA, USA
e-mail: eswenson@u.washington.edu

E.R. Swenson and P. Bärtsch (eds.), *High Altitude: Human Adaptation to Hypoxia*,
DOI 10.1007/978-1-4614-8772-2_5,

higher pulmonary vascular resistance and thus avoid SMS. This condition of high altitude pulmonary hypertension (HAPH) is now recognized as a maladaptation to long-term residency at high altitude in some but not all high altitude populations. In this chapter, we will discuss the different aspects of the pulmonary circulation at altitude from the normal physiological response to environmental hypoxia to its contribution to the pathophysiology of high altitude related diseases and review in depth the work of the last decade in these areas.

Introduction

Increased pulmonary vascular resistance (PVR) and pulmonary artery pressure (Ppa) upon ascent to high altitude universally occur in humans and other mammals, although the magnitude can vary almost fivefold among individuals, across species, and with time at altitude from years to many generations. Furthermore, the mechanisms that lead to hypoxic pulmonary hypertension are numerous and have differing contributions to the pressure measured at any one time [1, 2]. The first and best characterized response is the von Euler-Liljestrand reflex or acute hypoxic pulmonary vasoconstriction (HPV). Its physiologic relevance or advantage, if any, in healthy humans moving to altitude remains uncertain. It has been postulated that moderately elevated pulmonary artery pressure (Ppa) optimizes systemic oxygen delivery by increasing blood flow into areas of lung with relatively lower blood flow, thus recruiting a greater fraction of the total alveolar capillary surface area for gas exchange [3]. Because the magnitude of HPV is so variable and those most highly adapted to high altitude (such as natives of the Tibetan plateau and high altitude mammals) have the least pulmonary hypertension, the argument is not a very compelling. It would appear that the oxygen sensitivity of the lung vasculature or HPV evolved along with hypercapnic vasoconstriction as a mechanism to divert blood flow away from poorly ventilated lung regions with localized airway or airspace pathology in post-fetal life [4] as a survival advantage in pneumonia or thoracic trauma, or in very late gestation to aid in reducing PVR (i.e., oxygen-dependent pulmonary vasodilation rather than HPV) with birth and assumption of air breathing in order to eliminate the 75–90 % right to left shunting past the lungs of venous blood in utero [5].

An excessive rise in Ppa with high altitude exposure is clearly disadvantageous and leads either acutely to high altitude pulmonary edema (HAPE) [6, 7], (taken up in detail in Chap. 21), or within weeks or months to congestive right heart failure of high altitude also known as subacute mountain sickness (SMS) [8, 9]. This latter disease, which might more appropriately be termed "cor pulmonale or acute right heart failure of high altitude," was first described in cattle taken to summertime high pastures (>3,000 m) in the Rocky Mountains and named "Brisket Disease" [10]. In this chapter, we will discuss the different aspects of the pulmonary circulation at altitude from the normal physiological response to environmental hypoxia to its contribution to the pathophysiology of high altitude-related diseases and review in depth the work of the last decade in these areas.

Physiological Response of the Pulmonary Circulation at High Altitude

Acute Hypoxic Pulmonary Vasoconstriction

HPV is a complex process with elements of its expression arising from multiple points in the neuro-cardiopulmonary axis, with variation in intensity and mechanisms over time [11]. In addition to the intrinsic hypoxic response of pulmonary vessels that can be elicited in isolated pulmonary vascular smooth muscle cells and vessels, there are numerous extrinsic modulating influences sensitive to oxygen in vivo that include the vascular endothelium, red cells, chemoreceptors, autonomic nervous system, and lung innervation. The response of the pulmonary circulation to environmental hypoxia is characterized largely by contraction of smooth muscle cells within small pulmonary arterioles and veins of a diameter less than 900 μm, the veins accounting approximately for 20 % of the total increase in PVR [12, 13]. Although it is not generally thought that hypoxia acts at the microvascular or acinar level, pulmonary capillary endothelial cells respond to hypoxia with membrane depolarization [14]. As yet, no evidence has been found for capillary constriction [15] despite evidence that other vasoconstrictors are active at this level and in surrounding parenchymal perivascular cells that contain actin and myosin microfilaments [16].

HPV in intact animals and humans appears to be fully expressed within several hours and has several temporal components. The first occurs with 5 min with a half time of about 90 s Morrell et al. [17]; Deem et al. [18]; Talbot et al. [19]. A second phase of greater pressure elevation is evident and plateaus at 2 h [19]. In studies of isolated pulmonary arteries, lungs or vascular smooth muscle cells a third phase taking upward of 8 h has been shown [20]. The mechanisms behind these differing time phases has not been well studied, but the isolated vessel studies suggest the first phase is intrinsic smooth muscle contraction with the later phases representing the summation of numerous other modulating influences acting on the smooth muscle [11] as discussed below. All of these differing hypoxic responses are fully and immediately reversible with return to normoxia.

In contrast, with sustained hypoxia there is remodeling of the vasculature with hypertrophy of smooth muscle and endothelial hyperplasia that leads to more fixed resistance to blood flow and a relative loss of acute hypoxic sensitivity as will be discussed below.

HPV at the Level of the Vascular Smooth Muscle

There are several mechanisms involved in HPV that are activated in parallel or sequentially, leading to the critical increase of intracellular calcium and/or an enhanced calcium sensitivity of the actin-myosin that initiates contraction [2]. Intracellular calcium concentration is increased by hypoxia-mediated inhibition of several potassium channels, leading to membrane depolarization and extracellular calcium entry through L-type channels, and a release of calcium from the sarcoplasmic reticulum (SR), with consequent further influx through store-operated calcium channels (SOCC). In addition, sensitivity to calcium of the contractile elements is enhanced via a hypoxia-induced increase in Rho-kinase activity [21]. The change in oxygen tension that stimulates these components of HPV is signaled by an alteration in the redox status of the smooth muscle cells [2, 11, 22, 23]. Whether an increase or a decrease of reactive oxygen species (ROS) is responsible for HPV signal transduction is still under debate, but a strong case is emerging that hypoxia increases mitochondrial ROS generation (see Chap. 1) as an upstream signal for HPV. It is clear that high altitude exposure increases stable circulating markers of ROS production and persons with higher HPV appear to generate more ROS and less bioactive NO species across the lung [24].

Endothelium-Dependent Modulation of HPV

The pulmonary vascular endothelium generates a variety of vasoactive mediators that act in a paracrine fashion on the surrounding vascular smooth muscle cells. These include nitric oxide and prostacyclin as vasodilators and endothelin-1 as both a vasoconstrictor via binding to endothelin A receptors and a vasodilator by binding to endothelin B receptors causing NO release. Isolated human PA endothelial cells exposed to 3 % oxygen produce more hydrogen peroxide and thus may also be a source for ROS that initiate HPV

[25]. The endothelium also produces carbon monoxide (CO) via heme-oxygenase-2 [26], which is up-regulated by hypoxia [27]. CO dilates vessels by activating guanylate cyclase to generate cyclic GMP in a manner similar to NO. Hydrogen sulfide (H_2S), a strong reducing agent, generated in hypoxia is vasoconstricting in the pulmonary circulation by several as yet not fully quantified mechanisms [28]. It should be noted that many of these "gaso-transmitters" alter the concentrations of each other, making it difficult to assess the contribution of each to HPV modulation [29, 30].

As mentioned above, pulmonary microvascular endothelial cells respond to hypoxia with membrane depolarization. Wang et al. [14] have suggested that the capillary vascular endothelium owing to its intimate proximity to alveolar gas is the site where rapid and efficient oxygen sensing occurs to initiate HPV. They propose that their membrane depolarization is propagated in a retrograde fashion to the upstream resistance vessels via inter-endothelial cell connexin 40-dependent gap junctions. The evidence for this novel concept is supported by lesser HPV, and greater V/Q mismatching and hypoxemia when connexin 40-deficient mice are exposed to hypoxia or regional ventilation reduction.

Erythrocyte-Dependent Modulation of HPV

Red cells may contribute to HPV and pulmonary pressures in several ways. Although hypoxia-mediated decrease in deformability might reduce flow and increase measured vascular resistance [31, 32], direct measurements of human and other mammalian red cells over a range of PO_2 from 120 to 47 mmHg show no evidence of significant deformability changes with hypoxia [33]. With elevations in hematocrit with altitude, pulmonary vascular pressures increase partly due to increased blood viscosity and direct increases in lung vascular resistance as shown by hemodilution studies at high altitude in patients with chronic mountain sickness (CMS) [34] and in animal studies [35]. Red cell-mediated changes in PVR with hypoxia represent a balance between those effects that are vasodilating and others that are vasoconstricting. Direct endothelial cell NO scavenging by oxyhemoglobin HPV [36] and ROS generation by hypoxic red cells [37] will enhance HPV. In contrast the oxylabile behavior of red cells and hemoglobin that lead to vasodilating s-nitrosothiol release [38] and NO generation from nitrite with hemoglobin desaturation [39] will blunt HPV. Additionally deoxygenated red cells also release ATP which activates endothelial cell NO production via purinergic receptor binding [40] and so act in a vasodilating fashion. Thus parallel with the various and sometimes competing interactions of endothelial cell vasoactive mediators on HPV, the contribution of red cells is similarly complicated and the net result on PVR may vary depending on the degree and duration of hypoxia.

Neurohumoral-Dependent Modulation of HPV

The lung vasculature is innervated by sympathetic noradrenergic fibers from the large conduit arteries and veins down to 50 μm vessels in larger species such as man and dogs, but much less so in smaller species [41]. In addition to the release of norepinephrine with sympathetic activation causing vasoconstriction via alpha-1 adrenergic receptors on vascular smooth muscle, other opposing vasodilating neurotransmitters such as neuropeptide Y and vasoactive intestinal peptide can be released [41]. Additionally there is opposing vasodilating parasympathetic innervation that is NO dependent [42]. Hypoxia, sensed at the peripheral chemoreceptors which project afferents to the medullary cardiovascular control areas in the brain stem in addition to the respiratory control center, activates both parasympathetic and sympathetic outflow to the lung. Denervation of the carotid bodies and loss of afferent input from the peripheral chemoreceptors increases HPV [43, 44]. The efferent arc of this response is not well defined but is conveyed by the vagus nerve because vagotomy reduces HPV [45, 46]. Studies using atropine and propranolol suggest that vasodilating parasympathetic activity is more dominant than sympathetic activity in HPV inhibition [46, 47]. Other data suggest a stronger sympathetic contribution [48]. However, not all studies find evidence for neural modulation of HPV [49, 50]. The reason for this discrepancy is not clear, but those studies finding no effect on HPV have employed receptor blocking drugs rather than neural pathway interruption. It is entirely possible that peripheral chemoreceptor-mediated modulation of HPV may involve other neurotransmitter release via the lung innervation besides catecholaminergic or cholinergic agonists as described above. In humans, the finding that a stronger hypoxic ventilatory response (almost wholly a peripheral chemoreceptor-mediated response) is associated with weaker HPV supports the majority of the animal work [51].

Given that persons with exaggerated HPV are at greatest risk for HAPE [52], it would be interesting to determine which of the many processes contributing to HPV are most important in this subset of vulnerable people. In regard to neurohumoral mediation of HPV, HAPE-susceptibility may stem from greater generalized sympathetic nervous system activation to hypoxia [53, 54]. Furthermore owing to lesser hypoxic peripheral chemoreceptor responsiveness [55, 56], HAPE-susceptible subjects will breathe less at any given altitude or ambient PIO_2. As a result they will have lower alveolar, bronchial arterial, and mixed venous PO_2, all of which set the PO_2 sensed by the pulmonary vasculature, although alveolar PO_2 is most dominant [11].

The pulmonary vasculature expresses adrenergic and cholinergic receptors, as well as other receptors, such as for thyroxine angiotensin, adenosine, natriuretic peptides, and estrogen, to name only a few. Thus it can respond to circulating vasoactive mediators with dilation by epinephrine via beta-2 receptors [41], estrogens [57] and natriuretic peptides [58], and constriction with angiotensin [59], adenosine [60] and thyroxine [61]. The full neurohumoral response of the lung vascular response to hypoxia is often neglected in discussions of HPV, and while admittedly difficult to study, better understanding is much needed.

Other Modulating Influences on HPV

In addition to the many mechanisms involved in the control of pulmonary vascular tone in the hypoxic environment; individual genetic background [62, 63], a history of familial susceptibility [64–66], and environmental factors such as cold, intensity of exercise activity and other stressors [52] (Fig. 5.3) also contribute. pH status and carbon dioxide have a considerable influence on HPV with alkalosis and hypocapnia both diminishing HPV in animals and humans [67]. Lastly, with even low-grade infection or inflammation circulating and locally produced inflammatory products such as leukotrienes, thromboxane, and cytokines, such as tumor necrosis factor, interleukin-6 [68–71], or activation of their receptors in the vasculature [72] appear to modulate HPV (both negatively and positively) in animal studies, although there are no studies on humans. Sepsis leads to order of magnitude increases in NO production by up-regulation of the inducible NO synthase isoform [73] with virtual suppression of HPV.

Hypoxia Inducible Factors and HPV

The study of HPV continues to identify new sensing, signaling and effector mechanisms an pathways; the most recent is the role of hypoxia inducible factors (HIFs) (discussed in Chaps. 1 and 2). Iron is emerging as a critical element in HPV and pulmonary vascular changes with hypoxia. Iron supplementation and iron chelation reduce and increase HPV respectively [74, 75], possibly via altered HIF metabolism [76] involving prolyl hydroxylases, the O_2 sensitive enzymes that degrade HIF and require iron. In two rat strains with differing pulmonary hypoxic responses, HIF-1 activity, and HIF-mediated protein expression were higher in the strain with greater pulmonary hypertension [77]. In contrast, mice with heterozygous HIF 1-alpha deficiency have weaker acute and chronic hypoxic responses in isolated pulmonary vascular smooth myocytes and pulmonary vessels than wild-type mice [78, 79]. Interestingly, carotid body sensitivity to hypoxia in these same HIF 1-alpha deficient heterozygote mice is depressed [80], although this does not appear to diminish the hypoxic ventilatory response. Further supporting pharmacological evidence for HIF-1alpha mediation of HPV was demonstrated in mice by reduction in hypoxic pulmonary hypertension [81] and with increased ventilation in humans during hypoxic exercise [82] with digoxin, a known inhibitor of HIF-1alpha transcriptional activity [83]. At present it is not clear how HIF-dependent gene transcription affects HPV, but it likely involves alterations in pulmonary vascular smooth muscle calcium signaling [81].

Chronic Hypoxic Vasoconstriction and Remodeling

With chronic high altitude exposure of many weeks to years lung vasculature resistance increases and there is a loss of immediate vasorelaxation to oxygen [84]. Medial hypertrophy of pulmonary arteries is the characteristic histological pattern of chronic pulmonary hypertension of high altitude [85]. Vascular remodeling with vessel hypertrophy is due to an inhibition of the normal rate of apoptosis and greater stimulation of endothelial and smooth muscle proliferation. Hypoxia-mediated inhibition of potassium channel activity and increase in cytosolic potassium and calcium concentrations which figure prominently in acute HPV persist and participate in proliferative signaling.

Each of the several cell types in the vasculature (endothelial, smooth muscle, and adventitial fibroblast) undergo temporal- and location-specific changes in proliferation, extracellular matrix production, growth factor, and cytokine expression [86]. Furthermore, the vasculature may contain progenitor cells that arise from the vessel itself or migrate in out of a pool of circulating progenitor cells. Additionally, circulating monocytic leukocytes and mononuclear fibroblasts are recruited into the vessel wall by chemotactic signals generated by hypoxia such as endothelin-1, monocyte chemoattractant protein-1, vascular endothelial growth factor-A and tumor growth factor-B, and there exert paracrine proliferative effects on resident vessel wall cells [87, 88]. ROS [89, 90] and other pro-inflammatory mediators generated by hypoxic cells appear to be central in signaling a reinforcing cascade of gene transcription and growth factor release [91]. Nuclear factor of activated T cells isoform 3 (NFATc3), which is activated by ET 1 binding to endothelin receptor-A and Rho kinase, is another transcription factor that drives pulmonary vascular smooth muscle hypertrophy. Its down-regulation by gene deletion in mice reduces pulmonary artery wall thickness with intermittent and chronic hypoxia [92, 93].

HIF-signaling may contribute to this pro-inflammatory and proliferative state, because in addition to its role in hypoxic survival, HIF drives gene transcription of many inflammatory proteins [94]. This may in part involve HIF-mediated up-regulation of newly discovered micro RNAs or hypoxamirs, which regulate the expression of a large number of genes involved in growth control (both positive and negative) of pulmonary endothelial and smooth muscle cells. In cell studies with hypoxia, these include miRNA-210 [95] miRNA-328 [96], miRNA-21 [97], miRNA-17 [98]. Single-stranded complementary DNA antagonists of miRNA-17, miRNA-21 (antagomirs), or overexpression of miRNA-328 in mice reduce pulmonary artery hypertension in chronically hypoxia [96–98]. Beyond a single study of climbers ascending to just over 8,000 m in which a variety of miRNAs were altered in blood [99], including miRNA-21, little is known about these hypoxamirs in humans, but given their ubiquity across the animal kingdom, it is likely that they play a role in HPV and high altitude pulmonary hypertension (HAPH) in humans and may represent therapeutic targets. Lending overall support to the critical role of HIF and its up-regulation with hypoxia on HPV; mice that are heterozygously deficient for HIF-2 alpha [100]; and HIF-1 alpha [78, 79] develop less vascular remodeling and pulmonary hypertension with chronic hypoxia. Additionally, humans with a HIF-2 alpha gain of function mutation have greater HPV and slightly higher normoxic pulmonary pressures than controls [101].

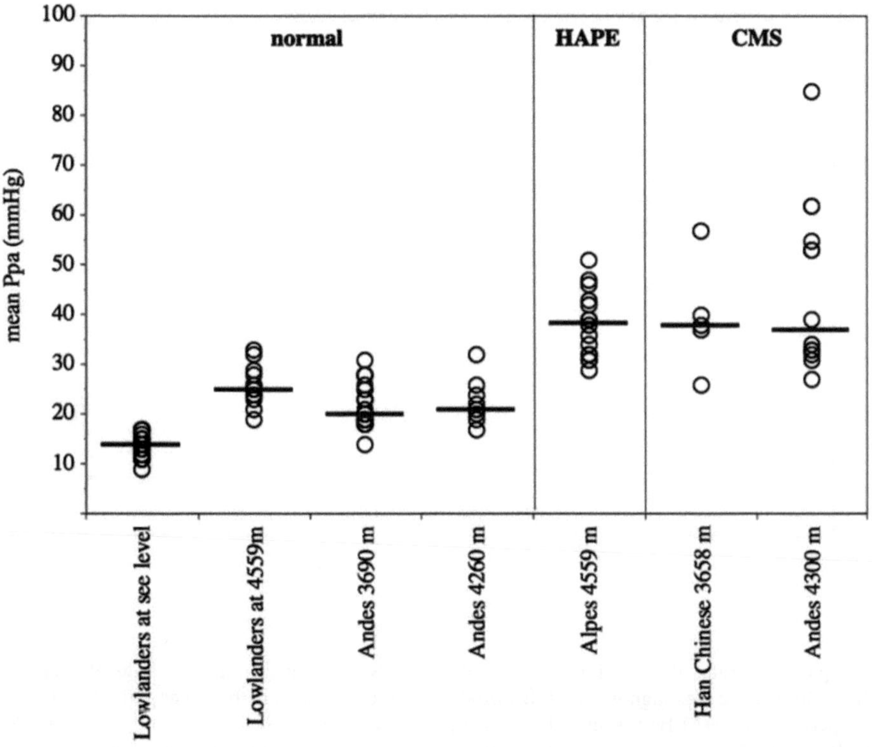

Fig. 5.1 The figure shows individual mean pulmonary artery pressure (Ppa) values of sea level residents and high altitude residents (3,500–4,300 m) in the Himalayas and in South America. In its *left panel* we report the mean Ppa values of healthy natives [105, 106] and *right panel* values of low altitude residents susceptible to HAPE [104, 167] and residents of high altitude with right heart failure of high altitude [123, 134]. The – indicates median Ppa value for each individual group of subjects. The figures indicate that subjects susceptible to high altitude pulmonary edema (HAPE) and chronic mountain sickness (CMS) have significantly higher Ppa values compared to a control population

Pulmonary Hemodynamics in Humans at High Altitude

In healthy lowlanders at altitudes between 3,800 and 4,600 m, invasively assessed resting mean pulmonary artery pressure (Ppa) ranges between 15 and 35 mmHg (average 25 mmHg) and the systolic Ppa between 27 and 48 mmHg (average 37 mmHg) [102–104] (Fig. 5.1). Mean Ppa at altitudes above 5,000 m during exposure in a hypobaric chamber (Operation Everest II) in healthy and partially acclimatized volunteers was 24 mmHg at 6,100 m (barometric pressure 347 mmHg) and 34 mmHg at 7,620 m (282 mmHg). At both altitudes physical exercise significantly increased mean Ppa to 41 and 54 mmHg respectively [84].

Mean resting Ppa has been reported to be lowest in Tibetans compared to high altitude Han Chinese residents, South- and North-American natives (Fig. 5.1). At similar altitudes between 3,658 and 3,950 m mean Ppa was on average 14 mmHg in Tibetans of Lhasa, 28 mmHg in Han Chinese residents of the Qinghai Province [105] and 20 mmHg among natives of South America [106]. In Leadville, Colorado mean Ppa in 28 healthy men living at 3,100 m averaged 25 mmHg [107]. The absence of highly muscularized pulmonary arterioles in natives of Ladakh, India [108] and some but not all natives of the Andes [109] fits well with the data on Ppa. All these findings are indicative of a likely evolutionary genetic adaptation (down-regulation) of hypoxic pulmonary vasoreactivity among populations

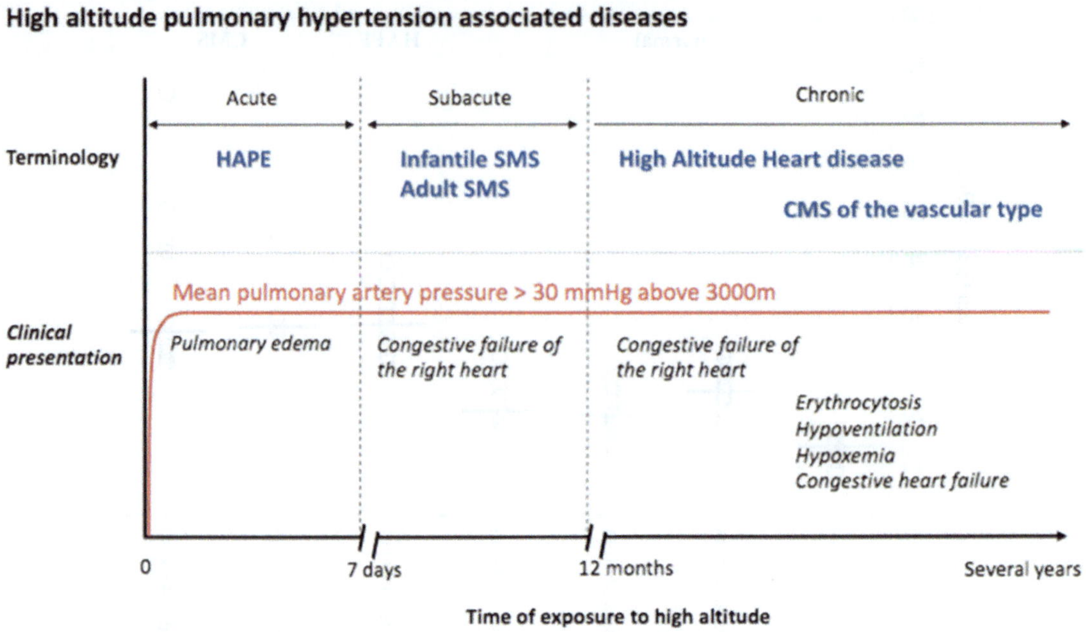

Fig. 5.2 The figure illustrates the relationship between exposure to high altitude and the diagnosis of high altitude pulmonary hypertension (HAPH)-associated diseases, which are HAPE, infantile and adult subacute mountain sickness (SMS), high altitude heart disease (Asia) and CMS. The major clinical findings of the different diseases are summarized in the lower part of the figure. It is of note that CMS may present with and without HAPH. Excessive erythrocytosis, hypoventilation, and severe hypoxemia are characteristic signs of CMS

living at different altitudes for different periods of time.

High Altitude Pulmonary Hypertension

HAPH in residents of high altitude has been defined in a consensus statement as a clinical condition above 2,500 m associated with a mean Ppa of more than 30 mmHg or a systolic Ppa above 50 mmHg measured at the altitude of residence [110]. It should be noted that no distinction in this definition was made for the actual altitude of residence except to exclude those living below 2,500 m. Because pulmonary artery pressures increase with higher altitude, these thresholds will progressively overestimate true pulmonary vascular pathology at greater altitudes. Not included also within this definition are residents of low altitude developing either HAPE within the first days at high altitude, high altitude congestive right heart failure SMS after several weeks at high altitude, or those with excessive erythrocytosis or CMS, which is discussed later in the chapter and more fully in Chap. 22. Since invasive pulmonary hemodynamic measurements clearly indicate that in HAPE-susceptible persons mean Ppa at altitudes above 3,000 m exceeds 30 mmHg [104, 111–116] Fig. 5.1 and that this is likely true for persons developing SMS [117], we think it reasonable to include HAPE and SMS as clinical conditions associated with HAPH. In the following sections we will refer to HAPH in the lowlander and resident of high altitude and illustrate their possible common pathophysiological mechanisms. Figure 5.2 shows the spectrum of HAPH from acute exposure to high altitude leading to HAPE within a few days, to longer exposure causing SMS and right heart failure in several weeks to more chronic exposure also causing right heart failure (high altitude heart disease) in newcomers to altitude, and to CMS in high altitude natives (those born and raised at high altitude).

In conclusion elevated pulmonary artery pressure in persons with HAPH may be initially the result of an imbalance between vasoconstrictors and vasodilators of the pulmonary circulation

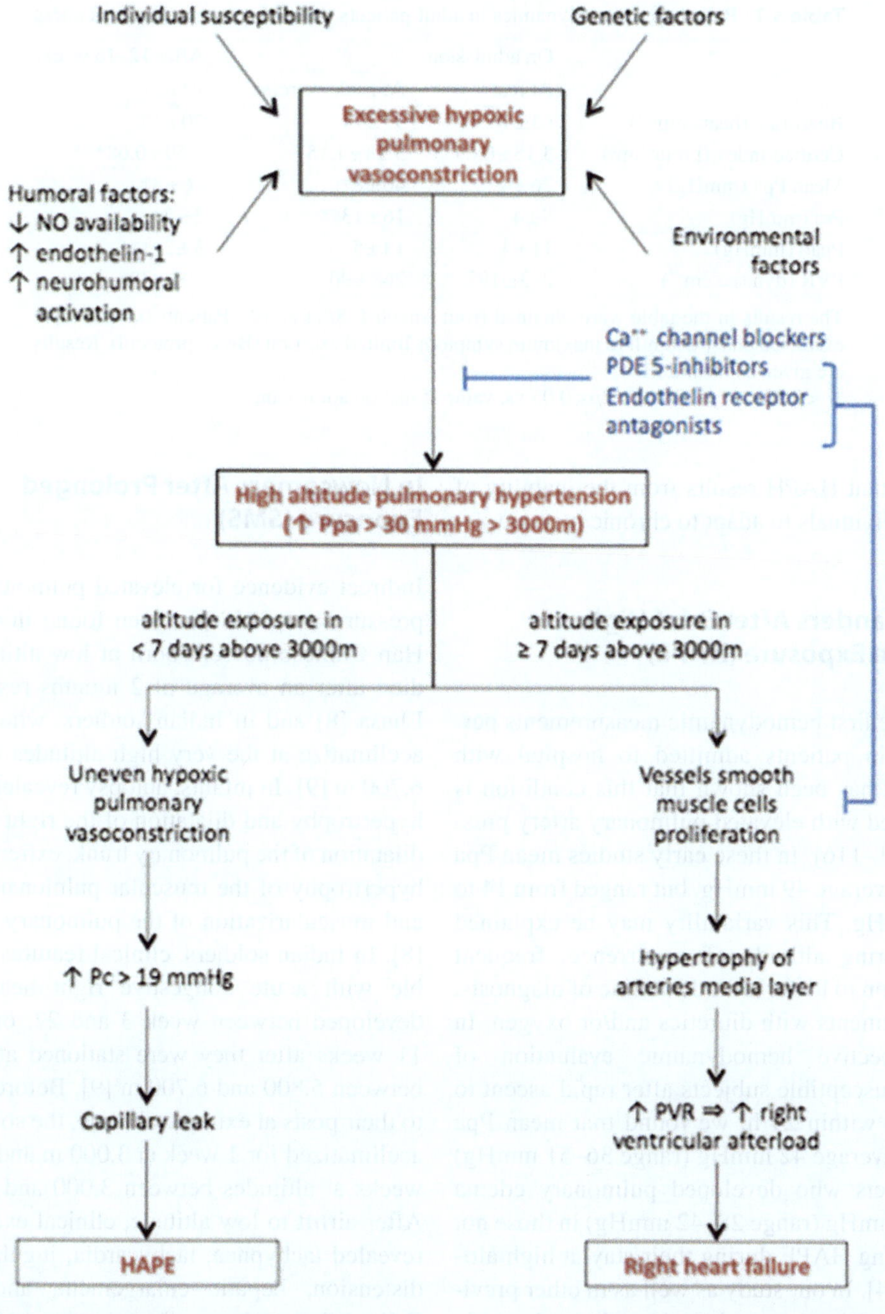

Fig. 5.3 The figure illustrated the common pathophysiological mechanism between HAPE and right heart failure of high altitude. An excessive hypoxic pulmonary vasoconstriction (HPV) leading to an out of proportion increase in pulmonary artery pressure (Ppa) at high altitude is the common denominator between these high altitude diseases. Different genetic factors, individual susceptibility, and environmental factors may be at the origin of the excessive HPV

increasing vascular resistance. With more sustained hypoxia some of these same mediators along with multiple other signaling and effector mechanisms initiate and sustain a remodeling of the pulmonary vascular bed to further increase PVR. As depicted in Fig. 5.3 these events in some lead to HAPE with acute exposure and right heart failure with more chronic exposure to high altitude. Reversibility of increased PVR several weeks after return to low altitude underlines the

Table 5.1 Pulmonary hemodynamics in adult patients with subacute mountain sickness

	On admission		After 12–16 weeks
	At rest	At peak exercise	At rest
Heart rate (beats.min^{-1})	63±10	99±14*	70±12
Cardiac Index (l.min^{-1}.m^{2})	3.15±0.69	5.28±1.15*	3.59±0.68***
Mean Ppa (mmHg)	26±4	40±8*	16±3*
Pra (mmHg)	8±4	16±13**	2±2*
Ppao (mmHg)	11±4	13±5	8±2***
PVR (dyn.sec.cm^{-5})	273±197	268±90	107±38*

The results in the table were obtained from Anand I. S. et al. [9]. Patients (n=21) were exercised on a treadmill to maximum symptom-limited exertion (Bruce protocol). Results are given as mean±SD

*p<0.001, **p<0.01, ***p<0.05 vs. value at rest on admission

concept that HAPH results from the inability of some individuals to adapt to chronic hypoxia.

In Lowlanders After Brief High Altitude Exposure (HAPE)

Since the first hemodynamic measurements performed in patients admitted to hospital with HAPE it has been shown that this condition is associated with elevated pulmonary artery pressure [111–116]. In these early studies mean Ppa was on average 49 mmHg, but ranged from 14 to 117 mmHg. This variability may be explained by differing altitude of occurrence, frequent evacuation to lower altitude at time of diagnosis, and treatments with diuretics and/or oxygen. In a prospective hemodynamic evaluation of HAPE-susceptible subjects after rapid ascent to 4,559 m within 24 h, we found that mean Ppa was on average 42 mmHg (range 36–51 mmHg) in subjects who developed pulmonary edema and 33 mmHg (range 28–42 mmHg) in those not developing HAPE during their stay at high altitude [104]. In our study as well as in other previous investigations, invasive hemodynamic measurements performed at high altitude consistently show that left atrial pressure, as assessed by transient segmental pulmonary artery occlusion (pulmonary artery wedge pressure) is normal both in healthy subjects and in those with HAPE. These results effectively rule out any element of left sided ventricular dysfunction as a cause for elevated pulmonary artery pressure.

In Newcomers After Prolonged Exposure (SMS)

Indirect evidence for elevated pulmonary artery pressure and SMS has been found in infants of Han Chinese descent born at low altitude, who died after an average of 2 months residence in Lhasa [8] and in Indian soldiers, who failed to acclimatize at the very high altitudes of 5,800–6,700 m [9]. In infants, autopsy revealed massive hypertrophy and dilatation of the right ventricle, dilatation of the pulmonary trunk, extreme medial hypertrophy of the muscular pulmonary arteries and muscularization of the pulmonary arterioles [8]. In Indian soldiers, clinical features compatible with acute congestive right heart failure developed between week 3 and 22, on average 11 weeks after they were stationed at altitudes between 5,800 and 6,700 m [9]. Before trekking to their posts at extreme altitude, the soldiers had acclimatized for 1 week at 3,000 m and then 1–3 weeks at altitudes between 3,000 and 4,500 m. After airlift to low altitude, clinical examination revealed tachypnea, tachycardia, jugular venous distension, hepatic enlargement, and ascites. Echocardiography confirmed enlargement of the right ventricle and showed normal left ventricular dimensions and ejection fraction. On admission, mean Ppa was on average 26 mmHg at rest and increased to 40 mmHg (see table 5.1) during mild exercise. After 12–16 weeks mean Ppa decreased on average to 16 mmHg. Pulmonary artery occlusion pressure was within normal range (Table 5.1). There was no severe hypoxemia and/or excessive polycythemia in these soldiers.

Taken together, these studies suggest that in lowlanders excessive pulmonary hypertension is not a transient phenomenon and may continue with remodeling of the small weakly muscularized and nonmuscular pulmonary vessels within hours of exposure to hypoxia [118–121]. Thus gradual ascent to high altitude may lead to a more homogeneous distribution of HPV and protection of pulmonary capillaries from elevated pressure and flow, so that the possibility of HAPE is only an early risk. However, when exaggerated pronounced thickening of the pulmonary artery media occurs, PVR increases enough to engender right ventricular failure.

Right heart failure or cor pulmonale may occur in lowlanders within a few weeks or months after ascent to high altitude [8, 117, 122]. The considerable dilatation and hypertrophy of the right ventricle seen with echocardiography [123] and autopsy [8, 123], without any hemodynamic evidence of left ventricular failure, clearly indicate that the clinical symptoms and signs observed within weeks and months after exposure to high altitude are independent of age and due to failure of the right ventricle secondary to HAPH.

Additional factors may also contribute to the development of right heart failure of high altitude. In both Indian soldiers with and without cor pulmonale exposure to altitudes above 5,800 m increased total body water and sodium [9, 124]. In the latter group exposure to extreme altitude decreased the effective renal blood flow and increased plasma norepinephrine and aldosterone concentrations, while renin plasma activity did not change [124]. These findings suggest that activation of the neurohumoral system may contribute to edema and ascites formation in subjects susceptible to acute right heart failure of high altitude. In Indian soldiers polycythemia was moderate (averaged hemoglobin concentration <20 g/dL) and hence unlikely to be an important factor in this setting.

In Highlanders Returning from low Altitude (re-Entry HAPE)

HAPE is not only confined to lowlanders after rapid ascent to high altitude, but has also been reported in residents of high altitude returning to their homes after days or weeks at low altitude. Of 113 high altitude residents developing HAPE after reascent 10 % spent less than 7 days, 55 % 14–20 days and 19 % more than 27 days at low altitude (ref). Re-entry HAPE has also been described in a Tibetan with CMS after a sojourn of 12 days at low altitude [125]. Susceptibility for re-entry HAPE appears to run in families and affect predominantly children [126]. Studies from Leadville, Colorado have shown that children descending to Denver, Colorado developed re-entry HAPE after staying only 3–6 days at low altitude [65]. Hemodynamic measurements in Denver (1,600 m) showed, compared to controls matched for age and residency, that in these HAPE-susceptible children mean Ppa was higher while breathing ambient air (22 vs. 16 mmHg) and even more in hypoxia (56 vs. 18 mmHg) [127].

In Highlanders: Idiopathic or Associated with CMS

An incomplete transition from the fetal to mature pattern of pulmonary circulation in newborns at high altitude or a loss of acclimatization while living at high altitude are possible mechanisms leading HAPH in natives of high altitude [85, 128]. Infants born at altitudes between 3,500 and 4,500 m may show persistence of near systemic-like Ppa values for several weeks following birth [129]. There is evidence that at least in some cases elevated Ppa persist during infancy [130, 131] leading, within the first few months of life, either to cor pulmonale with fatal right heart failure [8] or to a compensated chronic pulmonary hypertension of high altitude in adulthood with the typical histological pattern [132, 133]. Pulmonary hemodynamic features of such patients were reported in natives of the Andes [134], most of them examined at an altitude of 4,300 m, and Han Chinese who developed the disease after having lived between 11 and 36 years in Lhasa (3,658 m) [123]. Mean Ppa averaged 45 mmHg in South Americans and 40 mmHg in Han Chinese (Fig. 5.1). The right atrial pressure, pulmonary artery occlusion pressure (wedge pressure) and cardiac output were within normal in both patient groups. Recent echocardiographic examinations performed on high altitude residents of Cerro del Pasco with CMS confirm these results and indicate that these patients have usually normal left ventricular function [135]. Whether increased Ppa is a cause or a consequence of the disease has not been established.

Development of right heart failure or cor pulmonale may occur in residents of high altitude of different ethnic origin in Tibet [123, 128, 131], Kyrgyzstan [136] and the Andes [85]. The considerable dilatation and hypertrophy of the right ventricle seen with echocardiography [135, 137] and at autopsy [8, 123], without evidence of left ventricular failure, indicate that long term exposure to high altitude in all these different populations are independent of age and due to failure of the right ventricle secondary to HAPH.

Patients with CMS or Monge's disease (see Chap. 22) generally have a milder degree of pulmonary hypertension of CMS than the populations described above and it is not certain that this condition should necessarily be considered a variant of HAPH. The high pulmonary pressures are in part due to the high viscosity of polycythemic blood [138]. This has been demonstrated by hemodilution studies in CMS patients where considerable reductions in Ppa and PVR are observed [34, 139]. Applying the 21 % decrease in PVR found by [34] following a 16 % reduction in hematocrit (from 65 to 54 %) to the recent data of Groepenhoff et al. [140] in which a group of healthy high altitude controls were also studied (Hct 72 vs. 54 %), yields an almost equivalent PVR in the CMS patients to the controls after this correction. A similar finding is revealed using the data of Hoffman et al. [138]. Right heart failure is not omnipresent in patients with CMS [110, 128, 140–142] and appears to occur only as a much later manifestation of the disease after many years (Fig. 5.2). In fact, patients with CMS and much greater pulmonary hypertension and right heart failure may simply constitute those at the far right hand of a broad distribution of hypoxic pulmonary artery pressures, such as is seen in lowlanders [143].

Genetic Basis of High Altitude Pulmonary Hypertension

A genetic background for susceptibility to HAPH is largely assumed by studies in patients with HAPE (reviewed in Chap. 21) as suggested by anecdotal observations that HAPE runs in families [64–66] and more recently by candidate gene analyses, particularly involving genes encoding for NO, endothelin, and the renin-angiotensin system. Nonetheless, Ppa is lowest in Tibetans when compared to other populations resident at high altitude [85, 142, 144] and evidence is accumulating that with the same degree of inspired hypoxia residents of low altitude have lower exhaled NO concentrations compared to those living at high altitude [145, 146]. The higher endogenous NO production found in Ladakhi natives [147] and Sherpas living at high altitude [148] is attributed to an overrepresentation of the wild-type alleles G and 4b in the gene for endothelial nitric oxide synthase (eNOS) and linked to the observation that the combination of the wild-type genotypes GG and BB were associated high altitude adaptation. However, these results seem to be confined to a population of Asian-Indian origin. By contrast, in high altitude Amerindian natives eNOS gene polymorphism was not found to be associated with residency at high altitude or CMS [63]. The 287 bp Ala insertion (I)/deletion (ID) polymorphism of the ACE gene has been studied in several populations. An overrepresentation of I allele (genotypes II and I) was found in Kyrgyz highlands subjects with is associated with HAPH [136]. This observation could not be confirmed in the Andes where, compared to lowlanders of the same area, the genotype II and I were overrepresented in both high altitude residents without and with CMS [63, 149]. In Amerindians the I/I genotype was associated with good athletic performance and the ability to maintain high arterial oxygen saturation while altitude increases and the D/D genotype with higher hemoglobin levels [63, 149]. Other polymorphisms of genes that might modify HPV have been also investigated: angiotensin receptor-1 and 2, endothelin-1, vascular endothelial growth factor, and BMPR-2 gene. However none of them were found to be related to HAPH [150–153].

It must be appreciated that the power of these association studies to definitively address possible causation is severely limited by the small numbers of subjects in every report. The only studies with more adequate power are three independent investigations that find a polymorphism in the HIF-2alpha gene is much more expressed in Tibetans, who as group have lower pulmonary artery pressures, greater NO production and lower hematocrits [154]. Finally, we cannot exclude that variance between populations of different ethnic origin may simply reflect differences in adaptation due to the duration of time that these populations have resided in the mountains.

Prevention and Treatment of High Altitude Pulmonary Hypertension

There are no well-established treatments for pulmonary hypertension of high altitude in addition to descent. In residents of low altitude with HAPE, in which treatment is discussed in Chap. 21, Ppa normalizes within days after descent to low altitude [114]. In those with SMS, symptoms resolve within weeks at sea level [9]. In residents of high altitude who migrate to low altitude because of right heart failure and/or CMS, symptoms resolve and Ppa normalizes within weeks of residency at low altitude [155]. These results suggest that HAPH is a relatively benign disease, which resolves spontaneously when environmental hypoxemia is relieved.

However, there are patients with HAPH who may not want to descend to live at low altitude. In these circumstances HAPH can be reduced or treated by the use of pulmonary vasodilators. In a series of children with re-entry HAPE and persisting HAPH weeks after HAPE resolution while residing at an altitude of 3,050 m, treatment with the calcium channel blocker diltiazem significantly decreased Ppa and improved exercise tolerance [156]. In residents of high altitude with and without CMS, partial reversibility of elevated Ppa with nifedipine at the altitude of La Paz (3,600 m) was successful in 23 of 31 high altitude residents with moderate to severe pulmonary hypertension without signs of right heart failure [157]. More recently, the phosphodiesterase 5 inhibitor, tadalafil, has been successfully used in the treatment of HAPH [158]. However, because of limited experience and the occurrence of headache [159] further studies are needed before general recommendation can be given. Newer options are emerging, such as mimicking NO-mediated pulmonary vasodilation with administration of soluble guanylate cyclase activators [160]. The Rho-kinase inhibitor, fasudil, was found to lower pulmonary artery pressure and PVR in patients with HAPH at 3,200–3,600 m [161]. Agents in both of these classes are in late stage human testing. Lastly, acetazolamide, a proven respiratory stimulant at high altitude [162] and inhibitor of acute HPV [163, 164] improves symptoms of CMS and slightly decreases PVR after 24 weeks of treatment [165, 166].

Taken together, the results of these studies indicate that for the prevention and treatment of HAPH pulmonary vasodilators, particularly calcium channel blockers are the first choice, because they are cheap and available in the most countries.

References

1. Grover RF, Wagner WW, McMurtry IF, Reeves JT. Pulmonary circulation. In: Handbook of physiology: the cardiovascular system. Peripheral circulation and organ blood flow. Sect. 2, vol III, pt. 1, chapter 4. Besthesda, MD; 1985. p. 103–36.
2. Sommer N, Dietrich A, Schermuly RT, et al. Regulation of hypoxic pulmonary vasoconstriction: basic mechanisms. Eur Respir J. 2008;32:1639–51.
3. Wagner Jr WW, Latham LP, Capen RL. Capillary recruitment during airway hypoxia: role of pulmonary artery pressure. J Appl Physiol. 1979;47:383–7.
4. Naeije R, Brimioulle S. Physiology in medicine: importance of hypoxic pulmonary vasoconstriction in maintaining arterial oxygenation during acute respiratory failure. Crit Care. 2001;5:67–71.
5. Morin FC, Egan EA. Pulmonary hemodynamics in fetal lambs during development at normal and increased oxygen tension. J Appl Physiol. 1992;73:213–8.

6. Maggiorini M. High altitude-induced pulmonary oedema. Cardiovasc Res. 2006;72:41–50.
7. Dehnert C, Berger MM, Mairbaurl H, Bartsch P. High altitude pulmonary edema: a pressure-induced leak. Respir Physiol Neurobiol. 2007;158:266–73.
8. Sui GJ, Liu Y, Cheng XS, et al. Subacute infantile mountain sickness. J Pathol. 1988;155:161–70.
9. Anand IS, Malhotra RM, Chandrashekhar Y, et al. Adult subacute mountain sickness—a syndrome of congestive heart failure in man at very high altitude. Lancet. 1990;335:561–5.
10. Hecht HH, Kuida H, Lange RL, et al. Brisket disease. Am J Med. 1962;32:171–83.
11. Sylvester JT, Shimoda LA, Aaronson PI, Ward JP. Hypoxic pulmonary vasoconstriction. Physiol Rev. 2012;92:367–520.
12. Hakim TS, Michel RP, Minami H, Chang HK. Site of pulmonary hypoxic vasoconstriction studied with arterial and venous occlusion. J Appl Physiol. 1983;54:1298–302.
13. Audi SH, Dawson CA, Rickaby DA, Linehan JH. Localization of the sites of pulmonary vasomotion by use of arterial and venous occlusion. J Appl Physiol. 1991;70:2126–36.
14. Wang L, Yin J, Nickles HT, Ranke H, et al. Hypoxic pulmonary vasoconstriction requires connexin 40-mediated endothelial signal conduction. J Clin Invest. 2012;122:4218–30.
15. Conhaim RL, Burt Olson Jr E, Vidruk EH, et al. Acute hypoxia does not alter inter-alveolar perfusion distribution in unanesthetized rats. Respir Physiol Neurobiol. 2008;160:277–83.
16. Watson KE, Dovi WF, Conhaim RL. Evidence for active control of perfusion within lung microvessels. J Appl Physiol. 2012;112:48–53.
17. Morrell NW, Nijran KS, Biggs T, Seed WA. Magnitude and time course of acute hypoxic pulmonary vasoconstriction in man. Respir Physiol. 1995;100:271–81.
18. Deem S, Hedges RG, Kerr ME, Swenson ER. Acetazolamide reduces hypoxic pulmonary vasoconstriction in isolated perfused rabbit lungs. Respir Physiol. 2000;123:109–19.
19. Talbot NP, Balanos GM, Dorrington KL, Robbins PA. Two temporal components within the human pulmonary vascular response to approximately 2 h of isocapnic hypoxia. J Appl Physiol. 2005;98:1125–39.
20. Aaronson PI, Robertson TP, Ward JP. Endothelium-derived mediators and hypoxic pulmonary vasoconstriction. Respir Physiol Neurobiol. 2002;132:107–20.
21. Weigand L, Shimoda LA, Sylvester JT. Enhancement of myofilament calcium sensitivity by acute hypoxia in rat distal pulmonary arteries. Am J Physiol. 2011;301:L380–738.
22. Weir EK, Olschewski A. Role of ion channels in acute and chronic responses of the pulmonary vasculature to hypoxia. Cardiovasc Res. 2006;71:630–41.
23. Schumacker PT. Lung cell hypoxia: role of mitochondrial reactive oxygen species signaling in triggering responses. Proc Am Thorac Soc. 2011;8:477–784.
24. Bailey DM, Dehnert C, Luks AM, et al. High-altitude pulmonary hypertension is associated with a free radical-mediated reduction in pulmonary nitric oxide bioavailability. J Physiol. 2010;588:4837–47.
25. Irwin DC, McCord JM, Nozik-Grayck E, et al. A potential role for reactive oxygen species and the HIF-1alpha-VEGF pathway in hypoxia-induced pulmonary vascular leak. Free Radic Biol Med. 2009;47:55–61.
26. Zhang F, Kaide JI, Yang L, et al. CO modulates pulmonary vascular response to acute hypoxia: relation to endothelin. Am J Physiol. 2004;286:H137–44.
27. Motterlini R, Foresti R, Bassi R, et al. Endothelial heme oxygenase-1 induction by hypoxia. Modulation by inducible nitric-oxide synthase and S-nitrosothiols. J Biol Chem. 2000;275:13613–20.
28. Madden JA, Ahlf SB, Dantuma MW, et al. Precursors and inhibitors of hydrogen sulfide synthesis affect acute hypoxic pulmonary vasoconstriction in the intact lung. J Appl Physiol. 2012;112:411–8.
29. Skovgaard N, Gouliaev A, Aalling M, Simonsen U. The role of endogenous H2S in cardiovascular physiology. Curr Pharm Biotechnol. 2011;12:1385–93.
30. Evans AM, Hardie DG, Peers C, Mahmoud A. Hypoxic pulmonary vasoconstriction: mechanisms of oxygen-sensing. Curr Opin Anaesthesiol. 2011;24:13–20.
31. Palareti G, Coccheri S, Poggi M, et al. Changes in the rheologic properties of blood after a high altitude expedition. Angiology. 1984;35:451–8.
32. Hakim TS, Macek AS. Role of erythrocyte deformability in the acute hypoxic pressor response in the pulmonary vasculature. Respir Physiol. 1988; 72:95–107.
33. Kaniewski WS, Hakim TS, Freedman JC. Cellular deformability of normoxic and hypoxic mammalian red blood cells. Biorheology. 1994;31:91–101.
34. Manier G, Guenard H, Castaing Y, Varene N, Vargas E. Pulmonary gas exchange in Andean natives with excessive polycythemia—effect of hemodilution. J Appl Physiol. 1988;65:2107–17.
35. Kerbaul F, Van der Linden P, Pierre S, et al. Prevention of hemodilution-induced inhibition of hypoxic pulmonary vasoconstriction by N-acetylcysteine in dogs. Anesth Analg. 2004;99:547–51.
36. Deem S, Swenson ER, Alberts MK, et al. Red-blood-cell augmentation of hypoxic pulmonary vasoconstriction: hematocrit dependence and the importance of nitric oxide. Am J Respir Crit Care Med. 1998;157:1181–6.
37. Kiefmann R, Rifkind JM, Nagababu E, Bhattacharya J. Red blood cells induce hypoxic lung inflammation. Blood. 2008;111:5205–14.
38. Stamler JS, Singel DJ, Loscalzo J. Biochemistry of nitric oxide and its redox-activated forms. Science. 1992;258:1898–902.
39. Crawford JH, Isbell TS, Huang Z, et al. Hypoxia, red blood cells, and nitrite regulate NO-dependent hypoxic vasodilation. Blood. 2006;107:566–74.
40. Sprague RS, Olearczyk JJ, Spence DM, et al. Extracellular ATP signaling in the rabbit lung: eryth-

rocytes as determinants of vascular resistance. Am J Physiol. 2003;285:H693–700.
41. Kummer W. Pulmonary vascular innervation and its role in responses to hypoxia: size matters! Proc Am Thorac Soc. 2011;8:471–6.
42. McMahon TJ, Hood JS, Kadowitz PJ. Pulmonary vasodilator response to vagal stimulation is blocked by N omega-nitro-L-arginine methyl ester in the cat. Circ Res. 1992;70:364–9.
43. Naeije R, Lejeune P, Leeman M, et al. Pulmonary vascular responses to surgical chemodenervation and chemical sympathectomy in dogs. J Appl Physiol. 1989;66:42–50.
44. Levitzky MG, Newell JC, Dutton RE. Effect of chemoreceptor denervation on the pulmonary vascular response to atelectasis. Respir Physiol. 1978;35:43–51.
45. Chapleau MW, Wilson LB, Gregory TJ, Levitzky MG. Chemoreceptor stimulation interferes with regional hypoxic pulmonary vasoconstriction. Respir Physiol. 1988;71:185–200.
46. Wilson LB, Levitzky MG. Chemoreflex blunting of hypoxic pulmonary vasoconstriction is vagally mediated. J Appl Physiol. 1989;66:782–91.
47. Marshall JM. Peripheral chemoreceptors and cardiovascular regulation. Physiol Rev. 1994;74:543–94.
48. Lejeune P, Vachiery JL, Leeman M, et al. Absence of parasympathetic control of pulmonary vascular pressure-flow plots in hyperoxic and hypoxic dogs. Respir Physiol. 1989;78(2):123–33. PubMed PMID: 2609023. Epub 1989/11/01. eng.
49. Lodato RF, Michael JR, Murray PA. Absence of neural modulation of hypoxic pulmonary vasoconstriction in conscious dogs. J Appl Physiol. 1988;65(4):1481–7.
50. Liu C, Smith TG, Balanos GM, et al. Lack of involvement of the autonomic nervous system in early ventilatory and pulmonary vascular acclimatization to hypoxia in humans. J Physiol. 2007;579:215–25.
51. Alberts TJ, Swenson ER. Peripheral chemoreceptor responsivness and hypoxic pulmonary asoconstriction in humans. Am J Respir Crit Care Med. 2007;183:A5033.
52. Bartsch P, Mairbaurl H, Maggiorini M, Swenson ER. Physiological aspects of high-altitude pulmonary edema. J Appl Physiol. 2005;98:1101–10.
53. Duplain H, Vollenweider L, Delabays A, et al. Augmented sympathetic activation during short-term hypoxia and high-altitude exposure in subjects susceptible to high-altitude pulmonary edema. Circulation. 1999;99:1713–8.
54. Koyama S, Kobayashi T, Kubo K, et al. The increased sympathoadrenal activity in patients with high altitude pulmonary edema is centrally mediated. Jpn J Med. 1988;27:10–6.
55. Hohenhaus E, Paul A, McCullough RE, et al. Ventilatory and pulmonary vascular response to hypoxia and susceptibility to high altitude pulmonary oedema. Eur Respir J. 1995;8:1825–33.
56. Selland MA, Stelzner TJ, Stevens T, et al. Pulmonary function and hypoxic ventilatory response in subjects susceptible to high-altitude pulmonary edema. Chest. 1993;103:111–6.
57. Lahm T, Albrecht M, Fisher AJ, et al. 17beta-Estradiol attenuates hypoxic pulmonary hypertension via estrogen receptor-mediated effects. Am J Respir Crit Care Med. 2012;185:965–80.
58. Cargill RI, Lipworth BJ. Atrial natriuretic peptide and brain natriuretic peptide in cor pulmonale. Hemodynamic and endocrine effects. Chest. 1996;110:1220–5.
59. Cargill RI, Lipworth BJ. Acute effects of hypoxaemia and angiotensin II in the human pulmonary vascular bed. Pulm Pharmacol. 1994;7:305–10.
60. Thomas T, Marshall JM. The role of adenosine in hypoxic pulmonary vasoconstriction in the anaesthetized rat. Exp Physiol. 1993;78:541–3.
61. Herget J, Frydrychova M, Kawikova I, McMurtry IF. Thyroxine treatment increases the hypoxic pulmonary vasoconstriction in isolated lungs from thyroidectomized rats. Bull Eur Physiopathol Respir. 1987;23:217–21.
62. Stobdan T, Karar J, Pasha MA. High altitude adaptation: genetic perspectives. High Alt Med Biol. 2008;9:140–7.
63. Leon-Velarde F, Mejia O. Gene expression in chronic high altitude diseases. High Alt Med Biol. 2008;9:130–9.
64. Fred HL, Schmith AM, Bates T, Hecht HH. Acute pulmonary edema of altitude. Circulation. 1962;25:929–37.
65. Scoggin CH, Hyers TM, Reeves JT, Grover RF. High altitude pulmonary edema in the children and young adults of Leadville, Colorado. N Engl J Med. 1977;297:1269–71.
66. Lorenzo VF, Yang Y, Simonson TS, et al. Genetic adaptation to extreme hypoxia: study of high-altitude pulmonary edema in a three-generation Han Chinese family. Blood Cells Mol Dis. 2009;43:221–5.
67. Balanos GM, Talbot NP, Dorrington KL, Robbins PA. Human pulmonary vascular response to 4 h of hypercapnia and hypocapnia measured using Doppler echocardiography. J Appl Physiol. 2003;94:1543–51.
68. Tsai BM, Wang M, Pitcher JM, et al. Hypoxic pulmonary vasoconstriction and pulmonary artery tissue cytokine expression are mediated by protein kinase C. Am J Physiol. 2004;287:L1215–9.
69. Petersen B, Austen KF, Bloch KD, et al. Cysteinyl leukotrienes impair hypoxic pulmonary vasoconstriction in endotoxemic mice. Anesthesiology. 2011;115:804–11.
70. Johnson D, Hurst T, Wilson T, et al. NG-monomethyl-L-arginine does not restore loss of hypoxic pulmonary vasoconstriction induced by TNF-alpha. J Appl Physiol. 1993;75:618–25.
71. Savale L, Tu L, Rideau D, Izziki M, Maitre B, Adnot S, et al. Impact of interleukin-6 on hypoxia-induced

pulmonary hypertension and lung inflammation in mice. Respir Res. 2009;10:6–10.
72. Petersen B, Bloch KD, Ichinose F, et al. Activation of Toll-like receptor 2 impairs hypoxic pulmonary vasoconstriction in mice. Am J Physiol. 2008;294: L300–8.
73. De Cruz SJ, Kenyon NJ, Sandrock CE. Bench-to-bedside review: the role of nitric oxide in sepsis. Expert Rev Respir Med. 2009;3:511–21.
74. Smith TG, Balanos GM, Croft QP, et al. The increase in pulmonary arterial pressure caused by hypoxia depends on iron status. J Physiol. 2008;586: 5999–6005.
75. Smith TG, Talbot NP, Privat C, et al. Effects of iron supplementation and depletion on hypoxic pulmonary hypertension: two randomized controlled trials. JAMA. 2009;302:1444–50.
76. Knowles HJ, Raval RR, Harris AL, Ratcliffe PJ. Effect of ascorbate on the activity of hypoxia-inducible factor in cancer cells. Cancer Res. 2003;63:1764–8.
77. Engebretsen BJ, Irwin D, Valdez ME. e tal. Acute hypobaric hypoxia (5486 m) induces greater pulmonary HIF-1 activation in hilltop compared to madison rats. High Alt Med Biol. 2007;8:312–21.
78. Shimoda LA, Manalo DJ, Sham JS, et al. Partial HIF-1alpha deficiency impairs pulmonary arterial myocyte electrophysiological responses to hypoxia. Am J Physiol. 2001;281:L202–8.
79. Yu AY, Shimoda LA, Iyer NV, et al. Impaired physiological responses to chronic hypoxia in mice partially deficient for hypoxia-inducible factor 1alpha. J Clin Invest. 1999;103:691–6.
80. Kline DD, Peng YJ, Manalo DJ, et al. Defective carotid body function and impaired ventilatory responses to chronic hypoxia in mice partially deficient for hypoxia-inducible factor 1 alpha. Proc Natl Acad Sci. 2002;99:821–6.
81. Abud EM, Maylor J, Undem C, et al. Digoxin inhibits development of hypoxic pulmonary hypertension in mice. Proc Natl Acad Sci. 2012;109:1239–44.
82. Janssen C, Lheureux O, Beloka S, et al. Digoxin increases peripheral chemosensitivity and the ventilatory response to exercise in normal subjects. Clin Exp Pharmacol Physiol. 2010;37:303–8.
83. Zhang H, Qian DZ, Tan YS, et al. Digoxin and other cardiac glycosides inhibit HIF-1alpha synthesis and block tumor growth. Proc Natl Acad Sci. 2008;105:19579–86.
84. Groves BM, Reeves JT, Sutton JR, et al. Operation Everest II: elevated high altitude pulmonary resistance unresponsive to oxygen. J Appl Physiol. 1987;63:521–30.
85. Penaloza D, Arias-Stella J. The heart and pulmonary circulation at high altitudes: healthy highlanders and chronic mountain sickness. Circulation. 2007;115: 1132–46.
86. Stenmark KR, Fagan KA, Frid MG. Hypoxia-induced pulmonary vascular remodeling: cellular and molecular mechanisms. Circ Res. 2006;99: 675–91.
87. Stenmark KR, Davie NJ, Reeves JT, Frid MG. Hypoxia, leukocytes, and the pulmonary circulation. J Appl Physiol. 2005;98:715–21.
88. Li M, Riddle SR, Frid MG, et al. Emergence of fibroblasts with a proinflammatory epigenetically altered phenotype in severe hypoxic pulmonary hypertension. J Immunol. 2012;187:2711–22.
89. Hartney T, Birari R, Venkataraman S, et al. Xanthine oxidase-derived ROS upregulate Egr-1 via ERK1/2 in PA smooth muscle cells; model to test impact of extracellular ROS in chronic hypoxia. PLoS One. 2011;6:e27531.
90. Nozik-Grayck E, Stenmark KR. Role of reactive oxygen species in chronic hypoxia-induced pulmonary hypertension and vascular remodeling. Adv Exp Med Biol. 2007;618:101–12.
91. Cavasin MA, Demos-Davies K, Horn TR, et al. Selective class I histone deacetylase inhibition suppresses hypoxia-induced cardiopulmonary remodeling through an antiproliferative mechanism. Circ Res. 2012;110:739–48.
92. de Frutos S, Spangler R, Alo D, et al. NFATc3 mediates chronic hypoxia-induced pulmonary arterial remodeling with alpha-actin up-regulation. J Biol Chem. 2007;282:15081–9.
93. de Frutos S, Caldwell E, Nitta CH, et al. NFATc3 contributes to intermittent hypoxia-induced arterial remodeling in mice. Am J Physiol. 2010;299: H356–63.
94. Eltzschig HK, Carmeliet P. Hypoxia and inflammation. N Engl J Med. 2012;364:656–65.
95. Chan YC, Banerjee J, Choi SY, Sen CK. miR-210: the master hypoxamir. Microcirculation. 2012;19:215–23.
96. Guo L, Qiu Z, Wei L, et al. The microRNA-328 regulates hypoxic pulmonary hypertension by targeting at insulin growth factor 1 receptor and L-type calcium channel-alpha1C. Hypertension. 2012;59: 1006–13.
97. Sarkar J, Gou D, Turaka P, et al. MicroRNA-21 plays a role in hypoxia-mediated pulmonary artery smooth muscle cell proliferation and migration. Am J Physiol. 2010;299:L861–71.
98. Pullamsetti SS, Doebele C, Fischer A, et al. Inhibition of microRNA-17 improves lung and heart function in experimental pulmonary hypertension. Am J Respir Crit Care Med. 2012;185:409–19.
99. Chen F, Zhang W, Liang Y, et al. Transcriptome and network changes in climbers at extreme altitudes. PLoS One. 2012;7:e31645.
100. Brusselmans K, Compernolle V, Tjwa M, et al. Heterozygous deficiency of hypoxia-inducible factor-2alpha protects mice against pulmonary hypertension and right ventricular dysfunction during prolonged hypoxia. J Clin Invest. 2003;111: 1519–27.
101. Formenti F, Beer PA, Croft QP, et al. Cardiopulmonary function in two human disorders of the hypoxia-inducible factor (HIF) pathway: von Hippel-Lindau disease and HIF-2alpha gain-of-function mutation. FASEB J. 2012;25:2001–11.

102. Kronenberg RG, Safar P, Wright F, et al. Pulmonary artery pressure and alveolar gas exchange in men during acclimatization to 12,470 ft. J Clin Invest. 1971;50:827–37.
103. Vogel JHK, Goss GE, Mori M, Brammell HL. Pulmonary circulation in normal man with acute exposure to high altitude (14.260 feet). Circulation. 1966;43(suppl III):3–233.
104. Maggiorini M, Mélot C, Pierre S, et al. High altitude pulmonary edema is initially caused by an increase in capillary pressure. Circulation. 2001;103:2078–83.
105. Groves BM, Droma T, Sutton JR, McCullough RG, McCullough RE, Zhuang J, et al. Minimal hypoxic pulmonary hypertension in normal Tibetans at 3,658 m. J Appl Physiol. 1993;74(1):312–8.
106. Hultgren HN, Kelly J, Miller H. Pulmonary circulation in acclimatized mean at high altitude. J Appl Physiol. 1965;20(2):233–8.
107. Vogel J, Weaver W, Rose R, et al. Pulmonary hypertension on exertion in normal man living at 10,150 feet (Leadville Colorado). Med Thorac. 1962;19:461–77.
108. Gupta ML, Rao KS, Anand IS, et al. Lack of smooth muscle in the small pulmonary arteries of the native Ladakhi. Is the Himalayan highlander adapted? Am Rev Respir Dis. 1992;145:1201–14.
109. Wagenvoort CA, Wagenvoort N. Hypoxic pulmonary vascular lesions in man at high altitude and in patients with chronic respiratory disease. Pathol Microbiol. 1973;39:276–82.
110. Leon-Velarde F, Maggiorini M, Reeves JT, et al. Consensus statement on chronic and subacute high altitude diseases. High Alt Med Biol. 2005;6:147–57.
111. Antezana G, Leguia G, Guzman A. Hemodynamic study of high altitude pulmonary edema (12'000 ft). In: Brendel W, Zink R, editors. High altitude physiology and medicine. New York: Springer; 1982. p. 232–41.
112. Hultgren NH, Lopez CE, Lundberg E, Miller H. Physiologic studies of pulmonary edema at high altitude. Circulation. 1964;29:393–408.
113. Kobayashi T, Koyama S, Kubo K, et al. Clinical features of patients with high altitude pulmonary edema in Japan. Chest. 1987;92:814–21.
114. Koitzumi T, Kawashima A, Kubo K, et al. Radiographic and hemodynamic changes during recovery from high altitude pulmonary edema. Intern Med. 1994;33:525–8.
115. Roy BS, Guleria JS, Khanna PK, et al. Haemodynamic studies in high altitude pulmonary edema. Br Heart J. 1969;31:52–8.
116. Penaloza D, Sime F. Circulatory dynamics during high altitude pulmonary edema. Am J Cardiol. 1969;23:368–78.
117. Anand IS, Wu T. Syndromes of subacute mountain sickness. High Alt Med Biol. 2004;5:156–70.
118. Sobin SS, Tremer HM, Hardy JD, Chiodi HP. Changes in arteriole in acute and chronic hypoxic pulmonary hypertension and recovery in rat. J Appl Physiol. 1983;55:1445–55.
119. Rabinovitch M, Gamble W, Miettinen O, Reid L. Age and sex influence on pulmonary hypertension of chronic hypoxia and on recovery. Am J Physiol. 1981;240:H62–72.
120. Rabinovitch M, Konstam MA, Gamble WJ, et al. Changes in pulmonary blood flow affect vascular response to chronic hypoxia in rats. Circ Res. 1983;52:432–41.
121. Berg JT, Breen EC, Fu Z, et al. Alveolar hypoxia increases gene expression of extracellular matrix proteins and platelet-derived growth factor-B in lung parenchyma. Am J Respir Crit Care Med. 1998;158:1920–8.
122. Ge RL, Helun G. Current concept of chronic mountain sickness: pulmonary hypertension-related high-altitude heart disease. Wilderness Environ Med. 2001;12:190–4.
123. Pei SX, Chen XJ, Si Ren BZ, et al. Chronic mountain sickness in Tibet. Q J Med. 1989;71:555–74.
124. Anand IS, Chandrashekhar Y, Rao SK, et al. Body fluid compartments, renal blood flow, and hormones at 6,000 m in normal subjects. J Appl Physiol. 1993;74:1234–9.
125. Wu T. A Tibetan with chronic mountain sickness followed by high altitude pulmonary edema on reentry. High Alt Med Biol. 2004;5:190–4.
126. Hultgren HN, Marticorena EA. High altitude pulmonary edema. Epidemiologic observations in Peru. Chest. 1978;74:372–6.
127. Fasules JW, Wiggins JW, Wolfe RR. Increased lung vasoreactivity in children from Leadville, Colorado, after recovery from high altitude. Circulation. 1985;72:957–62.
128. Wu TY. Chronic mountain sickness on the Qinghai-Tibetan plateau. Chin Med J. 2005;118: 161–8.
129. Gamboa R, Marticorena E. Pulmonary arterial pressure in newborn infants in high altitude. Arch Inst Biol Andina. 1971;4:55–66.
130. Sime F, Banchero N, Penaloza D, et al. Pulmonary hypertension in children born and living at high altitudes. Am J Cardiol. 1963;11:143–9.
131. Ge RL, Ma RY, Bao HH, et al. Changes of cardiac structure and function in pediatric patients with high altitude pulmonary hypertension in Tibet. High Alt Med Biol. 2009;10:247–52.
132. Arias-Stella J, Saldaña M. The terminal portion of the pulmonary arterial tree in people native to high altitude. Circulation. 1963;28:915–25.
133. Gamboa R, Marticorena E. The ductus arteriosus in the newborn infant at high altitude. Vasa. 1972;1: 192–5.
134. Hultgren H. Chronic mountain sickness. In: Hultgren H, editor. High altitude medicine. San Francisco, CA: Hultgren Publication; 1997. p. 348–67.
135. Maignan M, Rivera-Ch M, Privat C, Leon-Velarde F, et al. Pulmonary pressure and cardiac function in chronic mountain sickness patients. Chest. 2009;135: 499–504.
136. Aldashev AA, Sarybaev AS, Sydykov AS, et al. Characterization of high-altitude pulmonary hypertension in the Kyrgyz: association with angiotensin-converting enzyme genotype. Am J Respir Crit Care Med. 2002;166:1396–402.

137. Kojonazarov BK, Imanov BZ, Amatov TA, et al. Noninvasive and invasive evaluation of pulmonary arterial pressure in highlanders. Eur Respir J. 2007; 29:352–6.
138. Hoffman JI. Pulmonary vascular resistance and viscosity: the forgotten factor. Pediatr Cardiol. 2011;32: 557–61.
139. Winslow RM, Monge CC, Brown EG, et al. Effects of hemodilution on O2 transport in high-altitude polycythemia. J Appl Physiol. 1985;59:1495–502.
140. Groepenhoff H, Overbeek MJ, Mule M, et al. Exercise pathophysiology in patients with chronic mountain sickness. Chest. 2012;142:877–84.
141. Vargas E, Spielvogel H. Chronic mountain sickness, optimal hemoglobin, and heart disease. High Alt Med Biol. 2006;7:138–49.
142. Maggiorini M, Leon-Velarde F. High-altitude pulmonary hypertension: a pathophysiological entity to different diseases. Eur Respir J. 2003;22:1019–25.
143. Grunig E, Weissmann S, Ehlken N, et al. Stress Doppler echocardiography in relatives of patients with idiopathic and familial pulmonary arterial hypertension: results of a multicenter European analysis of pulmonary artery pressure response to exercise and hypoxia. Circulation. 2009;119: 1747–57.
144. Wu T, Kayser B. High altitude adaptation in Tibetans. High Alt Med Biol. 2006;7:193–208.
145. Beall CM, Decker MJ, Brittenham GM, et al. An Ethiopian pattern of human adaptation to high-altitude hypoxia. Proc Natl Acad Sci U S A. 2002;99: 17215–8.
146. Beall CM, Laskowski D, Strohl KP, et al. Pulmonary nitric oxide in mountain dwellers. Nature. 2001;414:411–2.
147. Ahsan A, Norboo T, Baig MA, Qadar Pasha MA. Simultaneous selection of the wild-type genotypes of the G894T and 4B/4A polymorphisms of NOS3 associate with high-altitude adaptation. Ann Hum Genet. 2005;69:260–7.
148. Droma Y, Hanaoka M, Basnyat B, et al. Genetic contribution of the endothelial nitric oxide synthase gene to high altitude adaptation in sherpas. High Alt Med Biol. 2006;7:209–20.
149. Mejia OM, Prchal JT, Leon-Velarde F, et al. Genetic association analysis of chronic mountain sickness in an Andean high-altitude population. Haematologica. 2005;90:13–9.
150. Mortimer H, Patel S, Peacock AJ. The genetic basis of high-altitude pulmonary oedema. Pharmacol Ther. 2004;101:183–92.
151. Bushuev VI, Miasnikova GY, Sergueeva AI, et al. Endothelin-1, vascular endothelial growth factor and systolic pulmonary artery pressure in patients with Chuvash polycythemia. Haematologica. 2006;91: 744–9.
152. Hotta J, Hanaoka M, Droma Y, et al. Polymorphisms of renin-angiotensin system genes with high-altitude pulmonary edema in Japanese subjects. Chest. 2004;126:825–30.
153. Rupert JL, Koehle MS. Evidence for a genetic basis for altitude-related illness. High Alt Med Biol. 2006;7:150–67.
154. van Patot MC, Gassmann M. Hypoxia: adapting to high altitude by mutating EPAS-1, the gene encoding HIF-2alpha. High Alt Med Biol. 2011;12:157–67.
155. Penaloza D, Sime F. Chronic cor pulmonale due to loss of altitude acclimatization (chronic mountain sickness). Am J Med. 1971;50:728–43.
156. Das BB, Wolfe RR, Chan KC, et al. High-altitude pulmonary edema in children with underlying cardiopulmonary disorders and pulmonary hypertension living at altitude. Arch Pediatr Adolesc Med. 2004;158:1170–6.
157. Antezana AM, Antezana G, Aparicio O, et al. Pulmonary hypertension in high-altitude chronic hypoxia: response to nifedipine. Eur Respir J. 1998;12:1181–5.
158. Aldashev AA, Kojonazarov BK, Amatov TA, et al. Phosphodiesterase type 5 and high altitude pulmonary hypertension. Thorax. 2005;60:683–7.
159. Maggiorini M, Brunner-La Rocca HP, Peth S, et al. Both tadalafil and dexamethasone may reduce the incidence of high-altitude pulmonary edema: a randomized trial. Ann Intern Med. 2006;145:497–506.
160. Ghofrani HA, Grimminger F. Soluble guanylate cyclase stimulation: an emerging option in pulmonary hypertension therapy. Eur Respir Rev. 2009;18:35–41.
161. Kojonazarov B, Myrzaakhmatova A, Sooronbaev T, et al. Effects of fasudil in patients with high-altitude pulmonary hypertension. Eur Respir J. 2012;39: 496–8.
162. Swenson ER, Teppema LJ. Prevention of acute mountain sickness by acetazolamide: as yet an unfinished story. J Appl Physiol. 2007;102:1305–7.
163. Swenson ER. Carbonic anhydrase inhibitors and hypoxic pulmonary vasoconstriction. Respir Physiol Neurobiol. 2006;151:209–16.
164. Teppema LJ, Balanos GM, Steinback CD, et al. Effects of acetazolamide on ventilatory, cerebrovascular, and pulmonary vascular responses to hypoxia. Am J Respir Crit Care Med. 2007;175:277–81.
165. Rivera-Ch M, Huicho L, Bouchet P, et al. Effect of acetazolamide on ventilatory response in subjects with chronic mountain sickness. Respir Physiol Neurobiol. 2008;162:184–9.
166. Richalet JP, Rivera-Ch M, Maignan M, et al. Acetazolamide for Monge's disease: efficiency and tolerance of 6-month treatment. Am J Respir Crit Care Med. 2008;177:1370–6.

Cardiovascular System

6

Heart and Systemic Circulation

Aaron L. Baggish, Eugene E. Wolfel, and Benjamin D. Levine

Abstract

The cardiovascular response to hypoxia is a dynamic process that evolves over the days, weeks, and years of prolonged exposure. In both acute and sustained hypoxia, coordinated changes in cardiac pump function and systemic vascular function work to facilitate adequate end-organ perfusion. During acute hypoxia, sudden reductions in oxygen supply are compensated for by increased oxygen delivery (cardiac output) which preserves or at least minimizes reductions in conductive oxygen transport. With prolonged exposure and sustained hypoxia, gradual adaptations restore cardiovascular function towards normoxic levels. Exercise in the hypoxic environment represents a unique physiologic stress that has important implication for healthy high-altitude travelers and patients with preexisting cardiovascular conditions.

Introduction

The major role of the cardiovascular system is to supply adequate oxygen and substrate to meet the metabolic demands of the body. Both the central (cardiac pump) and peripheral (regional vascular beds) components of the cardiovascular system are challenged by the hypoxic environment. The magnitude of this challenge is intensified during hypoxic exercise when oxygen demand increases considerably. The cardiovascular response to hypoxic exposure appears to be driven largely by stimulation of the autonomic nervous system which facilitates changes in cardiac pump function (heart rate and stroke volume) and systemic vascular function (regional flow redistribution) to ensure that organ-specific metabolic demands are met.

A.L. Baggish, M.D.
Division of Cardiology, Massachusetts General Hospital, Boston, MA, USA

E.E. Wolfel, M.D.
University of Colorado Health Sciences, Denver, CO, USA

University of Texas Southwestern Medical Center, Dallas, TX, USA

B.D. Levine, M.D. (✉)
University of Texas Southwestern Medical Center, Dallas, TX, USA

Institute for Exercise and Environmental Medicine, Texas Health Presbyterian Hospital Dallas, Dallas, TX, USA
e-mail: BenjaminLevine@TexasHealth.org

E.R. Swenson and P. Bärtsch (eds.), *High Altitude: Human Adaptation to Hypoxia*, DOI 10.1007/978-1-4614-8772-2_6, © Springer Science+Business Media New York 2014

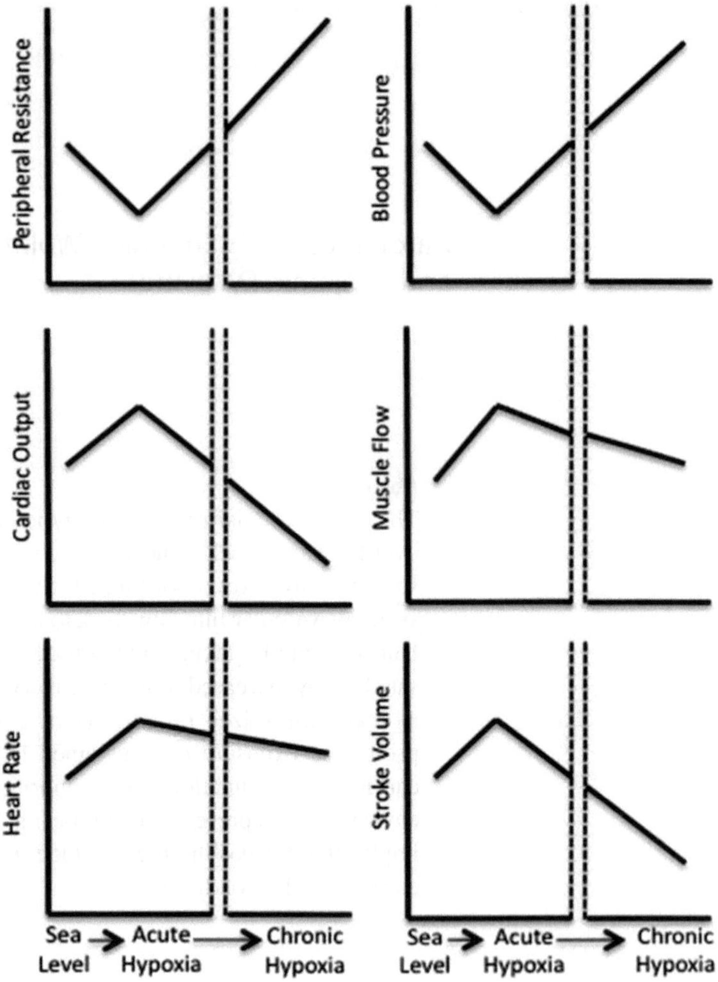

Fig. 6.1 Schematic of the cardiovascular hemodynamic changes that occur with acute and chronic hypoxia compared to sea level. Heart rate, cardiac output, muscle flow, and stroke volume initially rise with acute hypoxia but decrease to levels at or below sea level with sustained exposure. Blood pressure and vascular resistance initially fall with acute hypoxia but rise progressively with prolonged hypoxia

The cardiovascular response to hypoxia is a dynamic process that begins immediately upon exposure and evolves over the days, weeks, and years of prolonged exposure. Although the temporal nature of this response has inherent interindividual variability, it is useful to consider the response in two distinct phases: *Acute hypoxia* and *Sustained hypoxia*. During acute hypoxia, sudden reductions in oxygen supply are compensated for by increases in oxygen delivery (cardiac output) which appear aimed at preserving or at least minimizing reductions in conductive oxygen transport. With prolonged exposure, the hyperdynamic state of acute hypoxia abates and sustained hypoxia is marked by gradual adaptations that attempt to restore cardiovascular function towards normoxic levels. A qualitative summary of the acute and chronic effects of hypoxia on the cardiovascular system is presented in Fig. 6.1. Cardiovascular function during resting conditions and exercise in the settings of both acute hypoxia and sustained hypoxia will be reviewed. This review will outline historically important fundamental concepts, describe recent advances in knowledge, and propose important areas of future work.

Acute Hypoxia

Acute hypoxia decreases PaO_2 and arterial oxygen content leading to activation of the sympathetic nervous system. This activation is the key element of the acute hypoxic response of the cardiovascular system. The carotid bodies appear to respond to partial pressure of oxygen while local vascular beds seem to be regulated by arterial oxygen content. With respect to heart function, increased sympathetic activity (with or without vagal withdrawal) leads to a rise in heart rate and thus an increase in cardiac output. However, peripheral vasoconstriction, the expected vascular response to increased sympathetic tone, appears to be blunted by the release of local vasodilatory substances constituting a phenomenon equivalent to the "functional sympatholysis" that occurs during sustained endurance exercise. Thus, cardiac output increases while systemic vascular resistance and blood pressure decrease transiently. Acute responses occur within minutes of exposure to a hypoxic environment, and in the context of this section, are generally considered to reflect the first few hours–days of hypoxia.

With acute exposure to normobaric or hypobaric hypoxia, the partial pressure of oxygen in the air, alveoli, and blood is reduced. This results in oxyhemoglobin desaturation and a reduction in arterial oxygen content. To facilitate preserved tissue oxygen delivery, the cardiovascular system must increase systemic blood flow. Classic studies by Grollman on Pikes Peak (4,300 m) demonstrated that there is an approximately 40 % increase in resting cardiac output within the first few days of ascent to high altitude [1]. Similar observations have been made under more rigorously controlled conditions in the laboratory, with equivalent degrees of normobaric [2, 3] or hypobaric hypoxia [4–6]. The magnitude of the increase in cardiac output is proportional to the degree of hypoxia, with no apparent increase at altitudes below 700 m [4] and augmentation nearing 75 % at altitudes of 5,000 m [4–6]. This response appears to be independent of gender [4, 7] or age [8].

Cardiac Function During Acute Hypoxia

Heart Rate

The increase in cardiac output during acute hypoxia is due primarily to elevation in heart rate [4–6, 9]. Resting heart rate is a function of both the intrinsic heart rate and the balance of sympathetic and parasympathetic neural activity. Intrinsic heart rate does not appear to change significantly with acute hypoxia (i.e., no change in the face of combined adrenergic and vagal blockade [10]) suggesting that autonomic mechanisms must be responsible.

The evidence for sympathetic activation during acute hypoxia is compelling. Indirect measures of sympathetic function such as plasma and urinary catecholamines levels have shown a consistent elevation in both norepinephrine and epinephrine concentrations with acute altitude exposure [11–13]. However, such data, especially with respect to urinary measurements, must be interpreted cautiously as clearance of these neurotransmitters increases prominently during acute hypoxia [14]. Frequency analysis of heart rate variability during acute hypoxia has been used as surrogate measure of autonomic nervous system function, yet interpretation of these data has proven challenging due to inadequate control for confounding variables that may affect heart rate variability during acute hypoxia (for summary and review, see [15] including ventilatory volume and respiratory rate [16], cardiac volume and mechanical function [17], arterial stiffness and pulse amplification, arterial baroreflex function [18], and sinus node responsiveness [19]). Direct measurements of efferent postganglionic sympathetic activity to skeletal muscle [20] have confirmed a direct relationship between the magnitudes of hypoxia and sympathetic activation [21, 22]. Although such nerve activity has not been documented in cardiac tissue responsible for heart rate control, it is highly likely that increases in sympathetic activity underlie the global stimulation of the circulatory system in the face of acute hypoxia.

However, sympathetic activation does not appear to explain all of the increase in heart rate and cardiac output acute during acute hypoxia. For example, beta-blockade alone does not completely abolish the heart rate response [9]. This finding led to the notion that simultaneous vagal withdrawal might complement sympathetic activation. Evidence for vagal withdrawal during hypoxia comes from two lines of evidence: (a) R-R interval shortening occurs extremely rapidly after brief periods of hypoxia which are too transient for full sympathetic nervous system activation [23]; this R-R shortening occurred almost exclusively during inspiration which strongly suggests that vagal withdrawal is driven largely by increases in ventilation. (b) vagal blockade with atropine in combination with beta-blockade prevented any further increase in resting heart rate with acute hypoxia [10] confirming at least some role of vagal withdrawal in the acute HR response. In summary, acute hypoxia, via stimulation of peripheral chemoreceptors [24] and the subsequent coupling of sympathetic activation and vagal withdrawal, results in a significant increase in resting heart rate and cardiac output.

The specific central neural pathways involved in the recognition of acute hypoxia and the subsequent adjustment of autonomic activity have been an area of important recent work and have been reviewed recently [25]. Early work demonstrated that afferent fibers from both baroreflex and chemoreflex receptor populations synapse closely together in the nucleus tractus solitarius (NTS) [26]. It is now apparent that multiple sites within the brain including the thalamus, hypothalamus, pons, and medulla are oxygen sensitive and contribute to the augmentation of both sympathetic and respiratory activity. The exact neural pathways, cellular mechanisms, and relevant central nervous system neurotransmitters which mediate the autonomic response to hypoxia are being actively explored [25].

Nitric oxide seems to be a strong candidate for a key neurotransmitter modulating CV control: for example, microinjections of an endogenous nitric oxide synthase inhibitor into the NTS attenuated heart rate responses to pharmacologic autonomic nervous system stimulation [27]. In contrast, brain stem activity of the neurotransmitter NMDA and its coupling to the protein Kinase C [28] and calcium/calmodulin dependent kinase II [29] pathways have recently been identified as key effectors of the acute hypoxic ventilatory response. The exact nature of the anatomic and neurophysiologic "cross-talk" between cardiovascular and respiratory reflexes remains to be determined.

Stroke Volume

Changes in stroke volume play a minor role in the cardiovascular response to acute hypoxia. Augmentation of myocardial contraction velocity [30] and resultant reduction in end-systolic volume appear to be offset by a reduction in end-diastolic volume, particularly when the measurements are taken within 1–2 h after exposure to acute hypoxia [4–6]. Importantly, even severe acute hypoxia does not appear to be associated with depression of myocardial contractile function in normal hearts [31]. Although sustained hypoxia equivalent to 10,000 m altitude will result in heart failure in dogs if adrenergic compensatory mechanisms are inhibited [32], there is no evidence that this finding has any relevance to the human cardiac response to acute hypoxia.

Recent experimental work has documented several mechanisms by which the human heart compensates for arterial hypoxemia thus preserving contractile function. The most important of these is coronary vasodilation and increased myocardial blood flow [33]. In healthy individuals, this increase is quite robust with a 24 % increase in resting coronary flow at a simulated altitude of 4,500 m [34]. Additionally, there is some preliminary evidence that myocardial metabolism may shift from highly oxygen-dependent free fatty acid metabolism towards increased use of glucose during acute hypoxia [35]. Preliminary animal data further suggest that hypoxia-induced changes in metabolism may differ between the left and right ventricles, possibly because of the differential changes in afterload to the right and left heart with acute hypoxia [36]. It is also possible that respiratory alkalosis causes Ca^{2+} sensitization of cardiac fibers, thus improving mechanical efficiency and reducing the O_2 cost of contraction [37]. Thus, it seems likely that myocardial contractile function and thus stroke volume are unchanged during acute hypoxia due to compensatory hyperperfusion and shifts in intracellular energy substrate utilization.

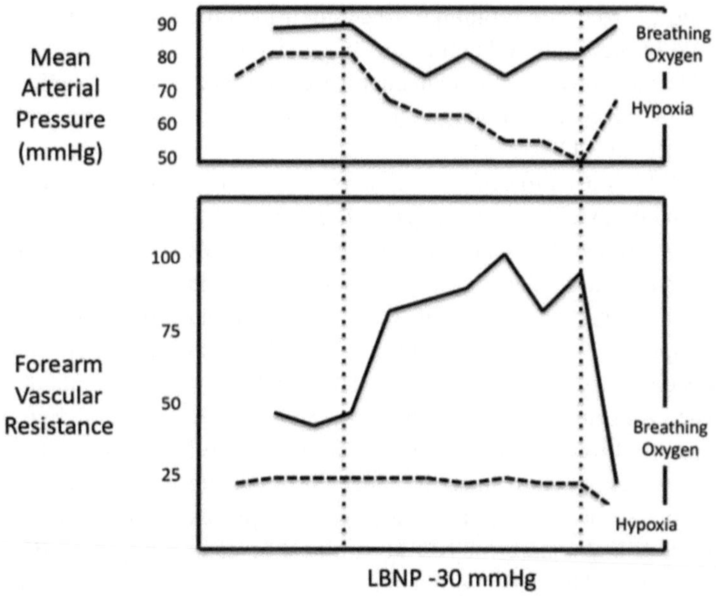

Fig. 6.2 Hypoxia prevents the rise in forearm vascular resistance with lower body negative pressure (LBNP) and the ability to maintain blood pressure is compromised (data from [50]) common to both exercise and hypoxia. During exercise, the mechanism of functional sympatholysis appears to be muscular contraction-induced opening of ATP-sensitive potassium channels by endothelium-derived hyperpolarizing factor mediated, at least in part, by nitric oxide

Systemic Circulatory Changes—Acute Hypoxia

Activation of the sympathetic nervous system generally results in peripheral vasoconstriction. However, with the sympathetic activation of acute hypoxia, this vasoconstriction is absent in all vascular beds except the pulmonary circulation [38–40]. This capacity of local control mechanisms to override sympathetic vasoconstriction has been well described during normoxic exercise and has been termed "functional sympatholysis" [41, 42]. This concept is illustrated in Fig. 6.2. The combination of global sympathetic nervous system activation with a resultant increase in cardiac output and simultaneous selective peripheral vasodilation resulting in distribution of flow to vascular beds with the greatest metabolic demand appears to be an important adaptive strategy to handle the challenge of vigorous exercise. The mechanism of functional sympatholysis during exercise appears to be muscular contraction-induced opening of adenosine triphosphate (ATP)-sensitive channels by endothelium-derived hyperpolarizing factor [43] in part mediated by nitric oxide [44].

Although mechanisms remain unclear, there is strong experimental evidence to support the presence of "functional sympatholysis" during acute hypoxia [45]. For example, leg blood flow during acute hypoxia precisely matches the decrease in arterial oxygen content, keeping O_2 delivery constant both at rest and during exercise [39, 46]. In the coronary arteries, acute hypoxia causes neurally mediated vasodilation [47]. There is no evidence that a similar mechanism exists in the peripheral circulation and the observation that beta-adrenergic blockade in the forearm does not alter the vasodilator response to hypoxia [9] argues against a neural mechanism for the sympatholysis of acute hypoxia.

The extraordinary precision of the adjustment of blood flow to *O_2 content* (as opposed to pO_2) [48, 49] suggests that the red cell or its contents play an intimate role in oxygen sensing. One prominent hypothesis proposed by Ellsworth et al. suggests that the metabolic activity of the red cell is a major mediator of vascular tone during acute hypoxia [50]. Specifically, they demonstrated that deoxygenated blood cells release ATP which binds to P2y receptors on the vascular endothelium and stimulates vasodilation [51]. An elegant series of studies from this group has since confirmed the role of the red blood cell as an "oxygen sensor" [52] and have clarified mechanisms by which acute hypoxia stimulates the red blood cell release of ATP [53–55].

Additional molecular mediators independent of ATP have also been described and may provide alternative and/or redundant mechanisms for hypoxia-mediated vasodilation. Clearly nitric oxide plays an important role in facilitating peripheral hypoxic vasodilation [56] which depends upon concomitant local production of adenosine and its metabolites [57, 58]. Originally proposed by Mark Gladwin and colleagues [59] and recently confirmed by others [60], nitrite reductase activity of deoxyhemoglobin appears to be a key source for NO at the local level in animals and humans, with a greater effect during conditions where hemoglobin is deoxygenated such as exercise [59] and hypoxia [60]. More recently, several groups have shown that hypoxia-mediated nitrite reductase activity resides in the vessel wall and that this may be more substantial than deoxyhemoglobin-mediated nitrite reduction [61, 62]. Regardless of mechanism, increasing the availability of nitrite by increasing dietary intake of nitrate (via beet root juice) has recently been demonstrated to improve exercise performance in hypoxia [63]. In contrast, Stamler and colleagues have published extensively on the concept that a conformational change in an S-nitrosothiol (SNO) residue on the beta-93 cysteine on the hemoglobin molecule (SNO-Hb) causes an allosteric conversion from an oxygenated "relaxed" (R-state) to a deoxygenated "tight" (T-state) of SNO-Hb thereby releasing O_2 and transferring functional vasodilatory NO to the cell membrane. This group has recently extended this work by demonstrating that (a) RBC-mediated relaxation is absent when SNO Hb is depleted; and (b) hypoxic vasodilation is present even when the endothelium has been removed, and in the presence of NO synthase inhibition [64]. However like most physiological mechanisms with substantial redundancy SNO–hemoglobin is not essential for hypoxia-mediated vasodilation by red blood cells [65]. Finally, an elegant animal study by Coney and Marshall provides compelling evidence that neuropeptide Y (via the neuropeptide Y_1 receptor) and catecholamines (via the alpha-2 adrenoreceptors) contribute to hypoxic sympatholysis [66]. Although these multiple hypotheses are often presented by their proponents as conflicting, they are not mutually exclusive and in the intact, exercising, hypoxic human, it may be that all the above mechanisms play some role. Further human study is required to determine quantitatively exactly how these biochemical pathways interact to vivo responses to acute hypoxia.

Ultimately, the balance between sympathetic vasoconstriction and hypoxic vasodilation determines arterial blood pressure during acute hypoxia. Blood pressure during acute hypoxia depends upon several factors including the absolute altitude achieved, the rapidity of the acclimatization responses, and the timing of the blood pressure measurement. When blood pressure measurements are made very early—within the first hour of exposure—most studies show a decrease in total peripheral resistance with a small but measurable reduction in blood pressure [4, 8, 67]. However, when measurements are made even a few hours after exposure, the balance shifts toward predominant sympathetic vasoconstriction with the potential for resultant moderate hypertension [68]. This shift may be due to increases in minute ventilation or plasma hemoglobin concentration which increase arterial oxygen content and thus attenuate some of the hypoxic peripheral vasodilation. A key question that remains is whether changes in the partial pressure of arterial oxygen (PaO_2) or the arterial oxygen content dictate blood pressure during acute hypoxia? While it has been shown that the receptors in the carotid body respond directly to PaO_2 [69], it is also clear that organs like the kidney respond to arterial oxygen content such that anemia with a normal PaO_2 results in vasodilation and synthesis of erythropoietin. This distinction may be critically important in understanding the vascular response to both acute and chronic hypoxia and is an important area of future research.

Summary and Key Unanswered Questions—Acute Hypoxia

The cardiovascular systems play a crucial role in the maintenance of peripheral tissue oxygenation during acute hypoxia. Increased sympathetic nervous system activity results in increases in heart rate and cardiac output which augments the delivery of oxygen to the central organs and peripheral tissues. Acute hypoxia stimulates numerous local peripheral vasodilating mechanisms which lead to reduced systemic vascular resistance and a modest but significant transient reduction in arterial blood pressure.

1. Which biochemical mediators/pathways contribute to hypoxic "functional sympatholysis" and what are their relative contributions to this process during acute hypoxic exposure?
2. Do the cardiovascular and ventilatory systems share common central nervous system control that facilitates matching of ventilation and oxygen transport during hypoxia?
3. Is the sympathetic hyperactivity necessary for successful acute hypoxia exposure or can it precipitate high-altitude illness (see Chaps. 20, 21, and 22) or symptoms of underlying cardiovascular disease?

Sustained Hypoxia

As hypoxia becomes sustained, arterial oxygen content (Cao_2) increases secondary to hemoconcentration, ventilatory acclimatization, and increases in red cell mass. These changes lead to an abatement of the hyperdynamic state of the circulatory system. Ventricular contractile function remains normal but reductions in plasma volume and ventricular filling lead to decreased stoke volume as predicted by the Frank–Starling mechanism. Persistent sympathetic hyperactivity leads to down-regulation of cardiac beta receptors and increased vagal activity which combine to promote a reduction in resting heart rate. These heart rate and stroke volume reductions lead to an overall fall in the initial acute increase in cardiac output. In the systemic circulation blood pressure gradually rises as the peripheral mechanisms responsible for local vasodilation diminished. In sum, these changes lead to increased transit time of blood flow across most organ systems and thus improved delivery of oxygen.

In contrast to the acute hypoxic response, in which the cardiovascular system plays a primary role in restoring oxygen transport toward normal, the behavior of the cardiovascular system during sustained exposure reflects adaptation in other organ systems. Ventilatory acclimatization (see Chap. 3), a key adaptive response characterized by increases in alveolar ventilation and thus PaO_2, shifts the oxyhemoglobin dissociation curve to the left. In addition, sustained hypoxic exposure leads to erythropoiesis and gradual restoration of plasma volume which result in increased blood volume. Together, these adaptations substantially increase the CaO_2 after the first few days (increased Hgb concentration and O_2 binding) to weeks (increased red cell mass) of sustained hypoxic exposure. The cardiovascular system appears to respond to increased CaO_2 with changes in heart rate, stroke volume, intrinsic cardiac function, cardiac output, and systemic vascular resistance. Data from relevant studies are summarized in Tables 6.1 and 6.2 and are reviewed below.

Cardiac Function—Sustained Hypoxia

Heart Rate

Resting heart rate remains elevated at altitudes greater than 3,000 m during sustained exposure. The magnitude of resting heart rate elevation is dependent on the severity of sustained hypoxia [31, 70–78]. The greater the altitude, the higher the resting heart rate, with little evidence for a decrease toward sea level values over 2–3 weeks of exposure (Fig 6.1, and Table 6.2).

The sympathetic nervous system continues to increase resting heart rate after prolonged exposure to hypoxia. Support for this notion comes from the observation that beta blockade prior to ascent prevents most (but not all) of the increase in resting heart rate with sustained altitude exposure [79, 80]. However, beta blockers administered after acclimatization do not reduce heart rate to the same magnitude [81]. This intriguing

Table 6.1 Effects of sustained hypoxia on resting cardiac hemodynamics

Altitude (m)	*n*	Age (year)	Duration (days)	Method	Response (%)[a] CO	SV	HR	References
3,658	50	21–34	10	Imped.	−26*	−36*	+18*	[77]
3,800	5	20–73	21–28	CO_2 rb	−4	−11	+9	[76]
4,300	8	18–24	21	Dye dilut.	−8	−20	+18	[84]
4,350	4	20–22	10	Dye dilut.	−21	−22	+15	[140]
3,100	8	23–32	10	Fick	−7	No Δ	No Δ	[81]
4,300	4	20–21	14	Dye dilut.	+28	−23	+31	[156]
4,300	12	21–28	21	Dye dilut.	−17	−30	+11	[82, 87]

Rb Rebreathing, *dye dilute.* indocyanine green dye method, *imped.* impedance, *CO* cardiac output, *SV* stroke volume, *HR* heart rate
$*p<0.05$ versus sea level
[a]% changes in responses from sea level

Table 6.2 Resting hemodynamics in healthy men at 4,300 m—Pikes Peak, Colorado

($n=12$)	Sea level	Acute hypoxia	Sustained hypoxia
Heart rate (bpm)	72±5	80±4*	80±4*
Cardiac output (L/min)	6.6±0.6	6.7±0.6	5.5±0.6*
Stroke volume (mL)	91±6	84±6	70±7*, **
VO_2 (mL/min)	290±17	316±14	376±14*
a-v O_2 (vol%)	4.4±0.5	4.7±0.8	6.8±0.8*, **
CaO_2 (vol%)	18.4±0.3	15.3±0.3*	18.3±0.4**
O_2 delivery (L O_2/min)	1,230±120	1,010±90*	1,015±96*
MAP (mmHg)	88±3	86±3	107±3*,**
SVR (dynes-sec-cm^{-3})	1,176±116	1,162±126	1,772±204*,**

MAP mean arterial pressure
$*p<0.05$ versus sea level; $**p<0.05$ versus acute hypoxia
The increase in a-v O_2 with sustained hypoxia was related to the decrease in cardiac output (80 %) and the increase in VO_2 (20 %)
[a]Sea level studies performed in Palo Alto, California
[b]Acute hypoxia data obtained within 4 h of arrival at 4,300 m, Pikes Peak
[c]Sustained hypoxia studies performed at 21 days on Pikes Peak
Sources: [82, 87]

finding suggesting that the tachycardia of acute exposure confers some degree of resistance to subsequent sympathetic blockade. The mechanism(s) underlying this relative resistance to sympathetic blockade after acute exposure remain uncertain, though alteration in vagal responsiveness likely plays a role.

Although the sympathetic nervous system is of undeniable importance to the sustained increase in heart rate with prolonged hypoxia, one perplexing issue deserves mention. It is well accepted that persistent sympathetic hyperactivity in heart failure causes down-regulation of cardiac beta receptors [82] and thus a reduced chronotropic response to catecholamines. Such a down-regulation has been well described in animals and humans after acclimatization to high altitude [83–86]. Cardiac receptor down-regulation during sustained hypoxia would favor a shift in heart rate back toward resting sea level values. However, observational data suggest that this does not fully occur raising a question about the physiologic importance of adrenergic receptor down-regulation during sustained hypoxia. Moreover, inhibition of persistent vagal activity by glycopyrrolate at maximal exercise after successful acclimatization fully returns maximal heart rate to normal [87], suggesting that beta receptor down-regulation may be of limited functional importance at maximal exercise.

The mechanism underlying this persistent heart rate elevation during sustained hypoxia remains incompletely understood. Decrease in plasma volume, re-setting of chemo/barorecpetor sensitivity, and persistent vagal withdrawal (at rest) may each contribute. Plasma volume reduction is a well-known consequence of sustained hypoxia [88, 89] and has important implications with respect to reduction in exercise capacity [90]. This relative arterial under-filling may unload the arterial baroreceptors and favor persistent sympathetic activity [91, 92]. Resetting of the chemo/baroreceptor threshold appears to also be a factor. Hansen et al. studied acclimatized lowlanders with saline infusion and supplemental oxygen after 4 weeks at 5,260 m (Chacaltaya, Bolivia) [93]. Although both interventions were expected to attenuate sympathetic activity, these maneuvers had only minimal effect on sympathetic activity as measured by peroneal nerve microneurography. Although vagal withdrawal is a potential contributor, recent work by Boushel et al. suggests otherwise as an *increase* in vagal activity at rest (greater increase in HR after glycopyrrolate after acclimatization) was demonstrated among lowlanders following 9 weeks at ≥5,260 m [87].

Studies examining the effects of further increasing altitude (i.e., superimposing a more severe hypoxic stimulus upon prior acclimatization) have shown that resting heart rate continues to increase with each increment in altitude. For example, in the American Medical Research Expedition to Mount Everest (AMREE), resting heart rate increased 20 beats per minute from sea level to 6,300 m [71]. Similar data have been obtained in a hypobaric chamber study, confirming that hypobaric hypoxia per se, rather than the stresses of mountaineering, is responsible for this tachycardia [31]. Thus, even in the setting of prior acclimatization, the ability to respond to a further increase in hypoxia with cardioacceleration remains present.

Stroke Volume

Despite the increases in resting heart rate that accompany sustained hypoxia, stroke volume and cardiac output are consistently depressed below sea level values after acclimatization to high altitude [9, 76, 94]. Reductions in stroke volume can be seen as early as a few hours after exposure to hypobaric hypoxia during submaximal exercise [5] and reductions at rest have been seen after only 2 days at high altitude [70]. There appears to be some continued decline over the first week, after which stroke volume stabilizes for a given altitude [74]. However, with progressively increasing altitude there appears to be further reductions in stroke volume [31], although differences in measurement techniques make precise determination of the effect of progressive prolonged hypoxia on stroke volume difficult.

Although these results are consistent across studies performed at moderate high altitude, important differences are observed if hypoxia is incremental and more extreme, as seen in a typical climbing expedition. For example, in the studies performed on Pikes Peak [75, 79], although arterial oxygen content increased over the course of acclimatization, cardiac output actually continued to decrease so that conductive O_2 transport fell to levels below sea level. Interestingly, there appears to be some rebound increase in cardiac output at extreme high altitude. In Operation Everest II (Table 6.3), resting cardiac output actually began to increase at altitudes above 7,000 m [31, 95, 96]. Though absolute resting cardiac output remained well below sea-level values, this increase at extreme simulated high altitude was intriguing. Increases in cardiac output, and thus higher conductive O_2 delivery, at very low PiO_2 come at the expense of reduced capillary transit time and an overall reduction in oxygen diffusion capacity. The in vivo balance between these factors ultimately determines the actual net O_2 flux. At present, a mechanism for the sensing and effecting matching of diffusion time to diffusion gradients is unknown and remains an area of important future work.

Since stroke volume is end-diastolic volume minus end-systolic volume, it will be reduced only if end-*systolic* volume is *increased* due to impaired contractile function, or if end-*diastolic* volume is *reduced* due to decreased left ventricular filling. Each of these possibilities will be considered in turn.

Table 6.3 Resting hemodynamic measurements with progressive high altitude—Operation Everest II

	Sea level	6,100 m	7,260 m	8,848 m
Heart rate (bpm)	64±4	86±6*	95±6*	99±6*
Cardiac output (L/min)	6.3±1.2	5.0±1.1	7.3±1.5	8.6±0.8
Stroke volume (mL)	107±8	72±6*	69±10*	81±11*
CaO_2 (vol%)	17.9±1.2	15.7±2.0*	13.6±1.7*	11.8±1.9*
a-v O_2 (vol%)	5.7±0.5	6.4±1.1	5.7±1.0	4.6±0.2
O_2 delivery (L O_2/min)	1,125±144	786±220*	992±155	1,018±152
VO_2 (mL/min)	360±20	306±25	406±16	386±17
MAP (mmHg)	96±3	96±3	90±2	96±9
SVR (dynes-sec-cm^{-3})	1,219±200	1,536±48*	986±107*	893±300*

Simulated altitude where PIO_2=63 Torr (6,100 m); 49 Torr (7,620 m); 43 Torr (8,840 m)
*$p<0.05$ versus sea level
Source: [43, 102, 103]

Cardiac Contraction (Systolic Function)

Early studies evaluating the hemodynamic responses to sustained hypoxia suggested that decreased cardiac contractile function contributed to the reductions in stroke volume [74, 97, 98]. However, the measurement techniques used in these studies were extremely load dependent and misleading in the setting of a prominent reduction in plasma and LV end-diastolic volumes (LVEDVs). Subsequently, Fowles and Hultgren normalized LV ejection indices to left ventricular end-diastolic dimensions and demonstrated that contractile function was actually enhanced rather than impaired for a given preload [73], similar to observations made after diuretic treatment [99].

The most convincing evidence that cardiac function is well preserved during sustained hypoxia comes from the Operation Everest II project. In this study, two-dimensional echocardiograms were performed at progressively greater simulated altitudes up to and including that equivalent to the summit of Mount Everest [100]. As noted above, stroke volume was decreased with chronic altitude exposure, even at 8,848 m. However, at each altitude, end-systolic volume was smaller than at sea level and all indices of left ventricular systolic function (left ventricular ejection fraction, mean normalized systolic ejection rate, peak systolic pressure/end-systolic volume) were either unchanged or improved across the altitude spectrum. Simultaneous invasive measurements of cardiac filling pressure allowed the construction of Starling function curves and demonstrated that high-altitude curves are superimposable on those obtained at sea level [31]. These data confirm that in normal individuals, LV contractile function is not diminished during sustained exposure to high altitude and is thus not a contributor to reduced stroke volume. Left ventricular pressure–volume relationships are highlighted in Fig. 6.3.

The mechanisms responsible for preserved myocardial contractile function during severe hypoxia are not well understood. Candidate mechanisms underlying the "hypoxia-tolerant" myocardium include heightened sympathetic activity [12, 101], a shift in myocardial metabolism, with increased ATP yield per molecule of oxygen [102], augmentation of coronary artery blood flow [103], transcriptional changes in mitochondrial and nuclear genes [104], and generation of NO by deoxymyoglobin [105]. Recently, sustained hypoxia has been shown to induce various cellular regulatory factors that preserve mitochondrial function such as hypoxia-inducible factor [106, 107], nuclear factor (NF)-kB [108], heat shock proteins [109, 110], and mitochondrial nitric oxide synthase [111] in various animal and tissue culture models. Although none of these mechanisms have been definitely shown to occur in humans at high altitude, these studies present compelling data that changes at the cellular level do occur with sustained hypoxia. A combination of these

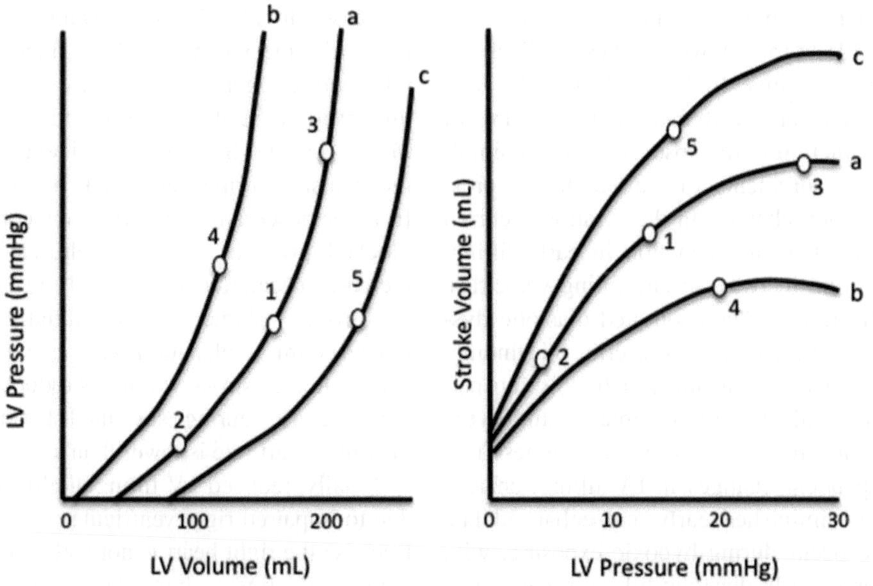

Fig. 6.3 Left ventricular pressure–volume relationships. Point 1 on *curve* represents the normal supine position on the pressure–volume (*left*) and Starling (*right*) *curves*. Acute dehydration would cause a shift to point 2, to a less stiff region (smaller slope) of the *P–V curve* and a large fall in stroke volume (SV) during orthostasis. In contrast, hyperhydration would shift to point 3, where changes in SV would be modest despite large changes in LVEDP. For a given left ventricular volume, changes in the underlying *P–V* relationship could result in an increase in cardiac compliance (*curve c*) or decrease in compliance (*curve b*) due to intrinsic or extrinsic factors

molecular and cellular adaptations could result in a more energy-efficient, functioning myocardium that is relatively protected from functionally relevant tissue ischemia.

Cardiac Relaxation (Diastolic Function)

The reduction in stroke volume during sustained hypoxia must therefore be due to a reduction in LVEDV. LVEDV is determined by filling during ventricular diastole and in practice, decreased LVEDV has been universally observed during sustained hypoxia [73, 74, 100, 112]. Diastolic filling is a dynamic and complex process comprised of distinct hemodynamic phases. Diastole is practically divided into three phases: early filling, diastasis, and late filling. At sea level in the healthy heart, the majority (~80–90 %) of total LVEDV is accounted for by early diastolic filling while late filling, generated by atrial contraction, accounts for the remaining total volume. Determinants of overall LV filling include *preload* (left atrial pressure), *intrinsic myocardial factors* (active myocyte relaxation, atrioventricular coupling, heart rate, and the dynamic establishment of atrioventricular and intraventricular pressure gradients), and *extrinsic cardiac factors* (pericardial and pulmonary mechanical constraint). In theory, sustained hypoxia could lead to changes in LV filling via alterations in any of these factors.

Previous studies examining the impact of altitude exposure on diastolic function have yielded mixed results. The OE II investigators provided important measurements of pulmonary capillary wedge pressure during progressive altitude exposure with consistent documentation of low left atrial pressure [31]. This finding suggests that alterations in LV relaxation, if present, do not result in significant rise in this surrogate measure of left atrial pressure.

Reductions in early LV diastolic filling by Doppler have been identified by older [113] as well as more recent studies [114]. For example, decrease in E/A-wave transmitral flow has been noted in a relatively large number of healthy subjects after rapid ascent to 4,559 m [114] or short-term exposure to normobaric hypoxia ($FiO_2=0.12$) [115]. However, these indices are exquisitely sensitive to LV preload [116] and must be interpreted with caution as measures of diastolic function in the absence of direct simultaneous measurements of cardiac filling pressures. Of note, changes in these ratios occurred due to simultaneous reduction in early filling/relaxation and increase in late filling velocities. Although the term "compensated diastolic dysfunction" has been coined to describe this finding, the evidence that diastolic function is intrinsically abnormal at altitude remains unproven. In aggregate, these recent studies suggest that phase-dependent changes in LV filling, characterized by diminished early and enhanced late filling, do occur during hypoxic exposure, with reductions in early diastolic filling in part offset by increased atrial stroke volume.

The primary mechanism for diminished early LV filling and subsequent stroke volume reduction during *sustained hypoxia* is a drop in plasma volume. Numerous studies have documented that plasma volume is reduced by 20–30 % within the first few hours to days at altitude [117, 118]. When this reduction in plasma volume is prevented with CO_2 breathing, the reduction in stroke volume is also prevented [119]. Over time, a gradual increase in red cell mass occurs, such that blood volume gradually increases. For example, in most of the studies from Pikes Peak there was approximately a 15–20 % decrease in plasma volume throughout the period of residence and a smaller (2–5 %) decrease in blood volume after 3 weeks at 4,300 m due to the progressive increase in red cell mass [75, 120]. The observation that stroke volume reduction at altitude is seen early in the acclimatization period and does not appear to be totally ameliorated by the gradual increase in blood volume over time suggests that plasma volume reduction may not fully explain this phenomenon. Further, experimental intravenous volume supplementation has been shown to increase but not fully restore stroke volumes at altitude [74].

The increase in heart rate observed during sustained hypoxia may further contribute to decreases in LVEDV. For example, in models using chronic pacing to obtain increased heart rate there are generally decreases in stroke volume even in the absence of LV function deterioration [121]. Supportive evidence for this mechanism during sustained hypoxia comes from the observation that when subjects are beta-blocked prior to ascent to altitude [79], the increase in heart rate is prevented and the decrease in stroke volume is substantially reduced. However, this explanation is at best only a partial contributor, as stroke volume is reduced at maximal exercise during sustained hypoxia when maximal heart rate is lower than at sea level.

Finally, reduced LV filling at altitude may be due to impaired right ventricular function [122]. Because the right heart is not well suited to geometric modeling, volume data similar to that of the LV are not available for the right ventricle at altitude. The role of the right ventricle may be particularly important because hypoxia causes pulmonary vasoconstriction and sustained pulmonary hypertension at altitude [95], resulting in numerous electrocardiographic reports of right ventricular strain or overload [71, 123, 124]. Because the right ventricle is comparatively preload dependent and does not have the contractile reserve available to the left ventricle, it may not be able to perform optimally against acute or subacute increases in afterload, especially when plasma volume and thus preload are reduced [125]. Anecdotal reports of acute right ventricular dilatation have been made in climbers with high-altitude pulmonary edema, and RV dilatation has been documented by echocardiography even in well subjects at extreme altitude in OE II [100], in a group of Japanese climbers acclimatized in the field to altitudes above 6,000 m [112], and in ultra-endurance athletes competing in a high-altitude endurance race [126]. The potential significance of the right ventricle for supporting LV filling has been highlighted in some patients with congestive heart failure [127]. In this study, patients with heart failure, RV dilation, and depressed LV stroke volume were subjected to lower body negative pressure with pooling of blood in the lower part of the body. Contrary to what occurs in normal individuals, these patients had an increase rather than a decrease in cardiac output, associated with an increase in LVEDV and SV. Thus, in some patients RV dilation may impair LV filling through ventricular interdependence and pericardial constraint with shift of the interventricular septum from right to left. Although such an effect is theoretically possible during sustained hypoxia, it may not play a major role in most individuals because LV filling pressure when measured has been uniformly low, providing evidence against pericardial constraint. The absence of a correlation between estimated changes in pulmonary artery pressure and Doppler indices of LV filling is consistent with, but does not refute an important contribution of RV volume and ventricular interdependence as a determinant of LV filling volume at high altitude [128]. Better measures of RV and LV size, perhaps with MRI, or modern 3D echo techniques and conformation during unloading at altitude will be necessary to confirm or disprove this hypothesis.

In sum, during sustained hypoxia, diastolic LV filling and thus LVEDV appear to be reduced by reductions in preload and thus diminished early diastolic filling. Augmentation of late filling due to enhanced atrial contraction appears to minimize but not fully offset this reduction in early filling. Whether additional factors such as myocardial ischemia or ventricular interdependence contribute to reduce early filling remains uncertain but the former appears to be of little relevance in the healthy heart.

Systemic Circulatory Changes—Sustained Hypoxia

General Systemic Vascular Responses

The systemic circulation is comprised of the heart, central vessel, and regional vascular beds. Although the local peripheral circulation appears to be precisely regulated by arterial oxygen content, the central circulation, known to be exquisitely sensitive to changes in autonomic nervous system function, appears to behave differently. This observation led to two important questions that puzzled investigators for a long period of time: Does sympathetic activation persist or diminish during sustained hypoxia over the course of prolonged exposure to high altitude and what role does it play in the control of vascular tone? It now appears that sympathetic nervous system activity, as measured by circulating catecholamine neurotransmitters or muscle sympathetic nervous system activity (MSNA), stays elevated and may actually continue to increase during sustained hypoxia [22, 93].

Regional Vascular Responses

Since oxygen delivery (blood flow × CaO_2) seems to be precisely regulated at a local level during acute hypoxia, it is reasonable to assume that the changes in peripheral blood flow with sustained hypoxia would also mirror the changes in Ca_2, and in fact that is the case [129]. The largest body of work examining the effect of sustained hypoxia on peripheral blood flow has been performed at 4,300 m on the summit of Pikes Peak. Leg blood flow was measured using the thermodilution technique under both acute and sustained hypoxia (~3 weeks) at rest and during exercise [75, 79, 130]. In general, these studies used a common paradigm: baseline measurements at sea level, followed by acute exposure to 4,300 m, and then repeat measurements after sustained hypoxia at 4,300 m. Similar findings have been reported for the nonexercising forearm of men at the same high altitude [131], i.e., a progressive increase in forearm vascular resistance over 7 days of acclimatization to 4,300 m (Fig. 6.4). Moreover, regardless of whether the upper or lower extremity was studied, and independent of metabolic activity, the degree of peripheral vasoconstriction during acclimatization is related to elevated urinary and arterial norepinephrine levels, indicating that enhanced sympathetic stimulation was likely playing an important role [12, 131]. Based on these experiments, the following conclusions can be drawn: (1) Peripheral skeletal muscle blood flow is consistently elevated with acute hypoxia compared to sea level, thereby maintaining oxygen delivery. In the presence of

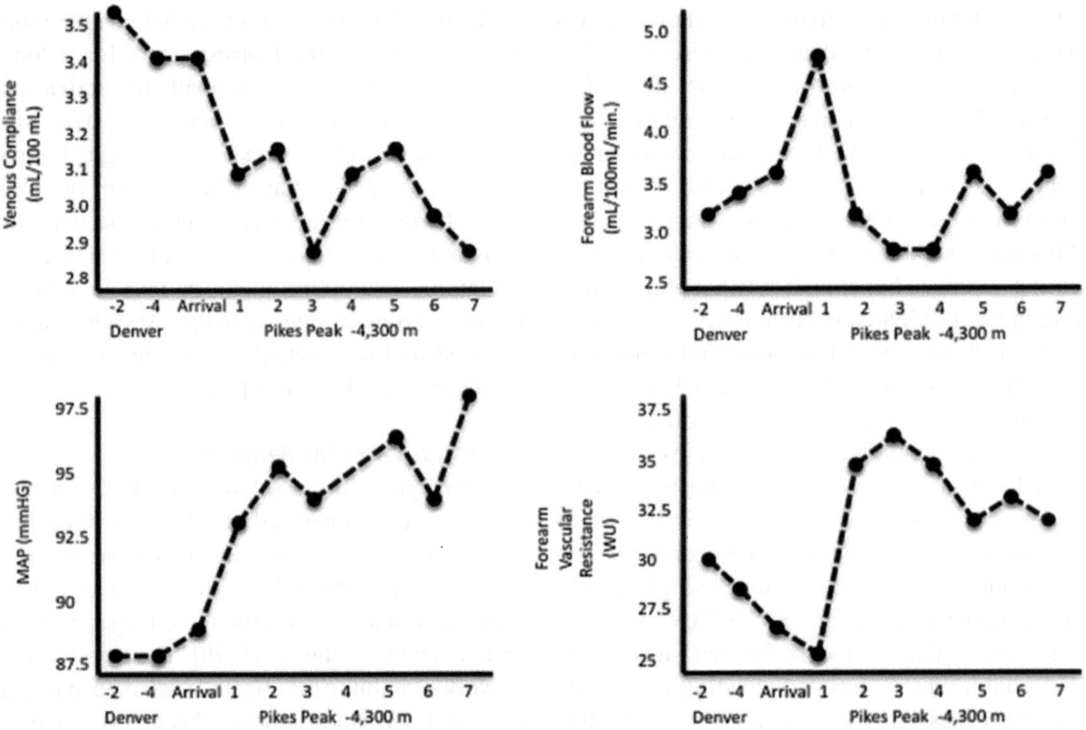

Fig. 6.4 Sustained hypoxia at 4,300 m results in a reduction in forearm compliance (*upper left*) and blood flow (*upper right*) along with increases in mean arterial pressure (MAP) (*lower left*) and forearm vascular resistance (*lower right*). Subjects were residents of 1,600 m (Denver, CO) (figures redrawn from original data in [133])

elevated sympathetic nervous system function, this vasodilation appears to represent a form of function sympatholysis. (2) Sustained hypoxia, during which there are increases in arterial oxygen content due both to hemoconcentration and increased red cell mass, is associated with a decrease in peripheral skeletal muscle blood flow and a concomitant increase in oxygen extraction. (3) The addition of supplemental oxygen to acclimatized subjects results in a further reduction in leg blood flow, confirming the tight coupling of leg blood flow to oxygen content during sustained hypoxia. Intriguingly, hypocapnia might be playing an important role in mediating changes in limb blood flow and vascular resistance, as forearm flow and resistance were unaltered at altitude under conditions of normocapnic hypoxia [132]. These data suggest a potentially significant interaction between ventilatory and cardiovascular chemoreflexes.

Finally, there is evidence that the changes in peripheral blood flow that are induced by sustained hypoxia remain present for some degree of time after return to sea level. After a sojourn at altitude ranging from 3,500 to 8,400 m, climbers were found to have a 26–34 % reduction in muscle blood flow, determined by ^{133}Xe clearance, during submaximal exercise [133]. Thus, even with restoration of normal PaO_2, the surfeit of oxygen availability continues to regulate oxygen delivery via alterations in limb blood flow.

Thus, the peripheral circulation develops adaptive responses with sustained hypoxia that maintain tissue oxygen uptake and substrate delivery. Both at rest and during exercise, in a small (forearm) or a large (legs) muscle bed, progressive vasoconstriction appears to occur during acclimatization as arterial oxygen content increases and a persistent sympathetic activation is unmasked by withdrawal of functional sympatholysis.

Coronary Circulation

The coronary circulation has a number of features that distinguish it from the peripheral circulation and may influence its adaptation to hypoxia. Coronary blood flow is tightly coupled to myocardial oxygen demand, which in turn is a function of three main factors: heart rate, contractility, and wall stress. In addition, oxygen extraction is extremely high in the myocardium even at rest, and thus most of the adaptive range of myocardial oxygen demand must be met by alterations in coronary blood flow. For normal young individuals without coronary artery disease, this vasodilator reserve is adequate to meet myocardial oxygen demands across a wide spectrum of conditions. Whether myocardial oxygen demand is increased, decreased, or unchanged during sustained hypoxia is not clear and depends on the complex interplay of these factors, some of which are increased (heart rate, contractility, blood pressure, afterload) while others are maintained or decreased (preload). Data from OE II [123] and the AMREE [71] suggest that extreme hypoxia is well tolerated by the healthy individuals with normal coronary vasculature despite extreme hypoxia, as evidenced by the absence of ischemic symptoms or ECG evidence of ischemia in subjects exercising maximally ($VO_2 = 1.17$ L/min) at 8,848 m despite extraordinary metabolic derangements including an SaO_2 as low as 49 % and pH of 7.52.

Similar to the peripheral circulation, coronary blood flow increases in proportion to the reduction in arterial oxygen content [134]. However, studies performed in both high-altitude residents and recently acclimatized lowlanders suggest that coronary blood flow decreases to levels lower than sea level with sustained hypoxia. For example, high-altitude Andean natives demonstrated a progressive decline in coronary blood flow at rest with increasing altitude, compared to sea level natives, reaching 30 % at the highest altitudes [135]. These findings were related to a lower heart rate, which contributed to a decreased myocardial O_2 demand as well as to an increased red cell mass, which allowed a relatively greater coronary arterial-venous O_2 difference. Similar reductions in coronary blood flow have also been observed in newly acclimatized young men after only 10 days at 3,100 m (Leadville, CO) [136]. Like high-altitude natives, these recently acclimatized lowlanders also had a 30 % decrease in coronary blood flow compared to sea level values, though without a concomitant reduction in heart rate and myocardial O_2 consumption. Although coronary arterial PO_2 was appropriately low, coronary O_2 extraction was increased as reflected by a decrease in coronary sinus O_2 content and saturation. Thus, it appeared that coronary blood flow is tightly regulated to maintain a constant myocardial O_2 tension at high altitude.

Systemic Blood Pressure

There has long been controversy regarding the impact of sustained hypoxia on systemic blood pressure. Most of the controversy, however, probably relates to the timing of blood pressure measurements. With acute exposure to altitude, blood pressure usually falls due to a decrease in systemic vascular resistance. However, with acclimatization stroke volume and cardiac output fall, even in the face of persistent tachycardia, and peripheral vascular resistance gradually rises as ventilatory and hematological acclimatization restore arterial oxygen content towards normal. This acclimatization response may occur relatively rapidly, particularly at moderate altitudes of 3,000–4,000 m. Thus, persistent sympathetic activation due to hypoxia is no longer offset by functional sympatholysis and locally mediated vasodilation, and blood pressure rises over time. Increases in systemic arterial pressure and vascular resistance have been noted in numerous studies of high-altitude acclimatization [75, 76, 79, 81, 130, 137–139]. Studies using β-blockers prior to ascent have not observed a complete reduction in the systemic blood pressure elevation associated with chronic hypoxia [68, 79], thus raising the importance of α-adrenergic vasoconstriction in this process. In support of this concept, healthy young women treated with the α-adrenergic blocker prazosin were found to have a decline in both diastolic and mean arterial pressure during exercise after 48 h of hypoxia in a hypobaric chamber simulating an altitude of 4,300 m when compared to the unblocked state [140]. However, administration of prazosin to women prior to a 12-day sojourn at 4,300 m blunted but did not fully abolish the rise in systemic blood pressure observed with ambulatory monitoring [141]. Thus, sympathetic blockade with either α- or β-blockers appears not to be sufficient to prevent the systemic pressor response at moderate high altitude. Recent studies of carvedilol in a small number of healthy volunteers showed a prominent reduction in blood pressure with this combined alpha/beta blocker, though there was still an increase in night time blood pressure [142]. Moreover, few data exist regarding the relative importance of other neurohormones such as angiotensin, aldosterone, or vasopressin, all of which may be involved in some degree with raising blood pressure during sustained high-altitude exposure. These mediators may be particularly important in sustaining the pressor response to sustained hypoxia, as acute administration of 35 % O_2 after acclimatization to 4,300 m did not return blood pressure to sea level values [130].

Although several studies at sustained hypoxia with altitudes between 3,000 and 5,000 m have reported increases in systemic arterial pressure, no increases in blood pressure or systemic vascular resistance have been observed at more extreme altitudes [95, 143]. At these extreme altitudes, the degree of hypoxia is profound and the acclimatization process cannot restore arterial oxygen content to normal. Thus, persistent peripheral vascular oxygen depletion likely serves a continued stimulus for peripheral vasodilation.

Few data exist regarding blood pressure after descent from sustained hypoxia. A study of 47 men who were taken from sea level to 3,658 m for a 10-day sojourn reported a progressive rise in resting diastolic blood pressure that was associated with an increase in total urinary catecholamines [138]. On return to sea level, diastolic blood pressure remained elevated for several days while urinary catecholamines returned to normoxic levels within the first 24 h. Insufficient data were available to provide a satisfying explanation and more research is needed in this area.

Finally, elevations in systemic arterial pressure do not seem to persist in lowlanders who remain at altitude for several months or longer. Although mechanisms remain speculative, this may be due at the eventual abatement of sympathetic activity or a compensatory up-regulation of circulating or local vasodilatory molecules with continued high-altitude residence. For example, urinary catecholamine levels have been reported to return to sea level values in subjects who remained at altitudes greater than 3,000 m for more than 90 days [144]. However, no direct data regarding sympathetic nerve activity are available in sea level natives over such a prolonged period of time. In addition, studies in high-altitude populations have shown that systemic hypertension is uncommon and blood pressures in highlanders tend to be lower than in sea level natives [145]. Thus, elevations in systemic arterial pressure appear to be part of the early (i.e., days to weeks) acclimatization process to high altitude associated with heightened sympathetic stimulation but may gradually abate over time as the sea level native becomes more like a high-altitude native. This represents an important area of future work.

Summary and Key Unanswered Questions—Sustained Hypoxia

Over a period of days to weeks, sustained hypoxia perpetuates and probably augments sympathetic hyperactivity. This results in persistent tachycardia relative to sea level. However, reductions in plasma volume result in a reduced stroke volume that ultimately, despite tachycardia, leads to a reduction in cardiac output to levels below sea level values. Coronary and peripheral oxygen transport is maintained via increases in CaO_2, and regional blood flow is precisely regulated to support delivery of oxygen and substrate to the tissues. With acclimatization, a gradual reduction in coronary and peripheral blood flow is associated with sympathetic activation that is now unopposed by regional vasodilation, leading to increases in total peripheral resistance and elevations in arterial blood pressure.

1. Are reductions in cardiac output with sustained hypoxia due fully to preload mediated changes in stroke volume or are other factors (i.e., central nervous system control, cellular mechanisms) contributory?

2. Is the systemic vasoconstriction with sustained hypoxia an adaptive or maladaptive response?
3. Do changes in myocardial metabolism govern coronary blood flow and are there long-term implications or altered myocardial substrate utilization?

Cardiovascular Function During Exercise at High Altitude

In comparison to sea level, submaximal exercise during acute hypoxia is associated with increased heart rate and preserved stroke volume leading to an increased cardiac output for any given absolute value of oxygen uptake. The same response has been shown with leg blood flow during acute hypoxia. However, with sustained hypoxia, submaximal exercise is associated with an increased heart rate but a decrease in stroke volume for any given absolute level of oxygen uptake. The usual sea level relationship between cardiac output and oxygen uptake is preserved in some but not all studies at moderate altitudes of 2,500–4,500 m, while other studies report a reduced cardiac output for a given oxygen uptake. With more extreme hypoxia, altitudes greater than 5,000 m, the cardiac output and oxygen uptake relationship is similar to sea level. Maximal exercise capacity is reduced to a similar degree with both acute and sustained hypoxia, although the mechanisms for the reduction differ depending on the duration of hypoxia. With acute hypoxia, heart rate, stroke volume, and cardiac output are generally maintained at sea level values with reductions in arterial oxygen content responsible for the decrease in exercise capacity. With sustained hypoxia, reduction in maximal exercise capacity is associated with a decreased maximal heart rate and cardiac output that is as yet not clearly explained. Whether the reduction in cardiac output at maximal exercise with acclimatization is responsible for the decrease in maximal oxygen uptake or is a consequence of it is one of the most intriguing questions in regard to human adaptation to high altitude.

Total metabolic demands of a human performing high intensity exercise may increase by 10–20-fold above resting conditions. Successful performance of such vigorous exercise requires a cardiovascular system that can supply the skeletal muscle with adequate substrate to meet these demands. The remarkable responsiveness of the cardiovascular system facilitates increases in cardiac output by nearly an order of magnitude during high levels of exercise. It has long been recognized that hypoxia presents a formidable challenge to exercise as decrements in exercise capacity have been documented universally in high-altitude environments. Impairment of exercise capacity at altitude is a multisystem process and the relative contribution of each organ system remains unclear. Key points related to cardiovascular function during exercise deserve emphasis and are discussed below (see also Chaps. 9 and 16).

Submaximal Exercise-Acute Hypoxia

When compared to normoxic exercise, submaximal exercise during acute hypoxia is associated with an increase in muscle blood flow, heart rate, and cardiac output for any given absolute work rate. These features enable the maintenance of sea-level oxygen uptake in the face of reduced arterial oxygen content [31, 71, 75, 79, 130] (Figs. 6.5 and 6.6). Resting heart rate increase is proportional to the severity of hypoxia, and changes in heart rate during submaximal exercise are similar to those of resting heart rate. Exposure to hypoxia of short duration (hours) is not associated with any reduction in stroke volume.

One simple explanation for the increase in heart rate during submaximal exercise is that the same *absolute* work rate in hypoxia represents a greater relative work rate (greater percentage of maximal oxygen uptake) compared with normoxia, and relative oxygen uptake appears to govern the cardiovascular responses to exercise [146]. Yet when workload is adjusted for % maximal oxygen uptake, heart rate is still somewhat greater in hypoxia than in normoxia [5]. Increases in sympathetic stimulation, especially

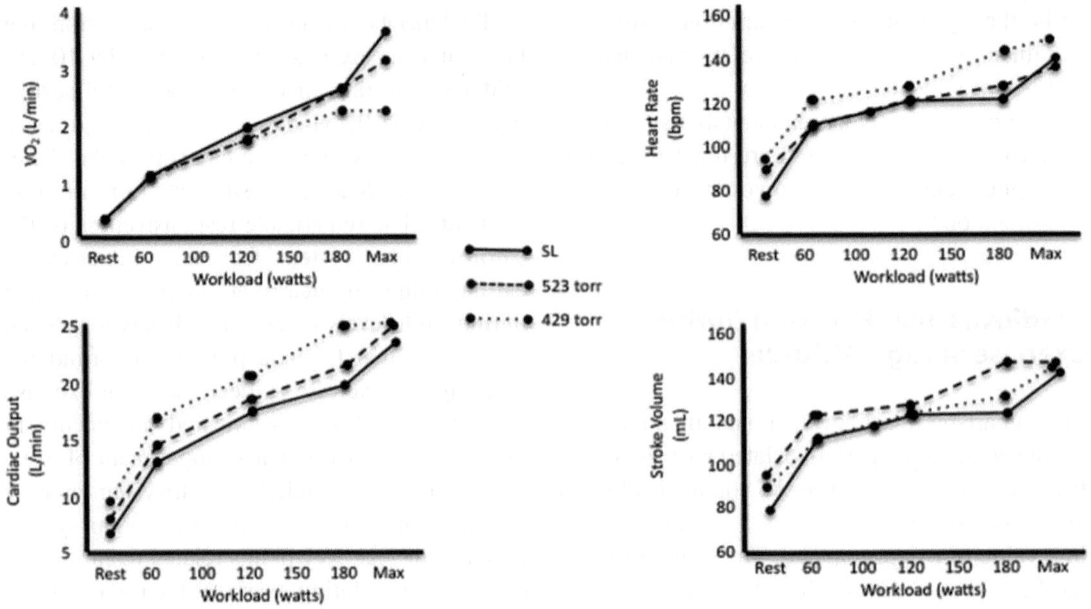

Fig. 6.5 Cardiac hemodynamic responses to acute hypoxia at simulated altitudes of 3,048 m (523 Torr) and 4,572 m (429 Torr). The VO_2 workload relationship is maintained as at sea level, while both heart rate and cardiac output increase in a progressive fashion with greater hypoxia at a given workload. Stroke volume remains at or above sea level values. Although maximal VO_2 is lower at sea level, both maximal heart rate and cardiac output are unchanged compared to sea level with less than 1 h of hypoxia exposure

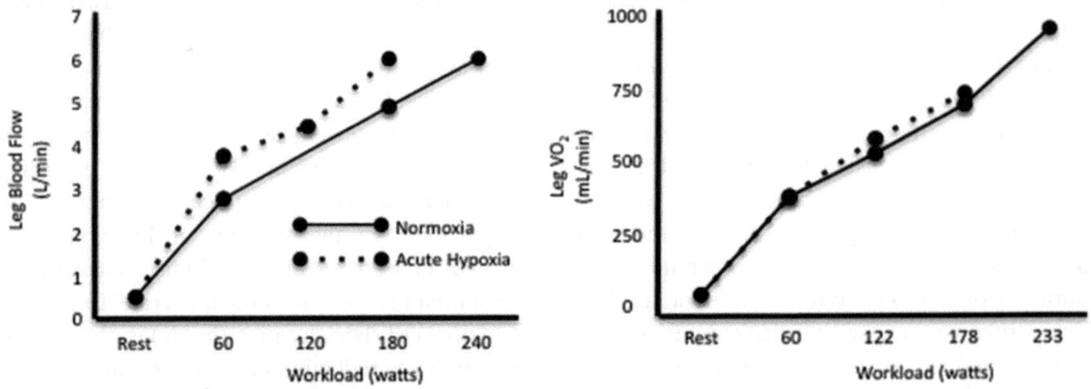

Fig. 6.6 Leg blood flow is greater for a given workload with acute hypobaric hypoxia compared to normoxia. During acute hypoxia, the relationship between leg VO_2 and workload is similar to that at sea level (figures drawn from data from [132])

the elevation in arterial epinephrine, may be partly responsible for this increase in heart rate [13]. However, studies in both humans and animals with β adrenergic blockade during acute hypoxia continue to show some rise in cardiac output and heart rate with no effect on the vasodilator effect of hypoxia [9]. Thus, peripheral factors involved in "functional sympatholysis" or alternative central mechanisms may play a major role in the augmentation of cardiac output during acute hypoxia by reducing peripheral resistance and afterload.

Table 6.4 Sustained hypoxia—submaximal exercise hemodynamic changes

Altitude (m)	*n*	Duration	Method	O_2 uptake at altitude, L/min (mL/kg/min)	Response (%)[a] CO	SV	HR	References
3,800	5	3–4 week	CO_2 rb	1.88 (23.3)	−7*	−14*	+9*	[76]
4,300	8	3 week	Dye dilut.	1.70 (23.0)	nc	−16*	+6*	[84]
3,100	8	10 days	Fick	1.60 (20.5)	−15	−11*	+4	[86]
4,300	4	2 week	Dye dilut.	1.57 (20.3)	nc	−15*	+8*	[84]
4,350	4	10 days	Dye dilut.	1.43 (19.6)	−23*	−32*	+24*	[140]
3,100	6	2–3 week	N_2O	1.62 (21.4)	nc	−8	+13	[85]
4,300	12	3 week	Dye dilut.	1.77 (24.7)	−19*	−26*	+8*	[82, 87]
6,100	6	3–4 week	Thermo/Fick	1.49 (19.8)	−13	−28*	+17*	[43]
7,620	6	1–2 week	Thermo/Fick	1.45 (19.1)	−7	−30*	+16*	[43]
8,840	6	20 min	Thermo/Fick	1.18 (15.5)	+8	−26*	+25	[43]

Rb Rebreathing, *dye dilut.* indocyanine green dye method, *thermo* thermodilution, *nc* no change, *CO* cardiac output, *SV* stroke volume, *HR* heart rate
*$p < 0.05$ versus sea level
[a]% changes from sea level, relates to same absolute oxygen uptake at sea level and altitude

Maximal Exercise—Acute Hypoxia

The majority of studies performed within the first hour of acute hypoxia have shown that maximal heart rate and cardiac output are unchanged from sea level. Maximal oxygen uptake is decreased because of the obligatory reduction in maximal convective oxygen transport caused by decreased availability of inspired oxygen [5, 6]. However at maximal exercise, tissue oxygen extraction is unable to compensate for the reduced convective oxygen transport and maximal oxygen uptake falls [5, 6]. Similar to acute hypoxia, intracardiac filling pressures remain at or below sea level values aside from the hypoxic vasoconstriction-related rise in pulmonary arterial pressures [6].

Submaximal Exercise—Sustained Hypoxia

The key difference between acute hypoxia and sustained hypoxia with respect to exercise is the reduction in stroke volume, heart rate, and cardiac output that develops with sustained hypoxic exposure. These reductions occur in response to adaptations in other organ systems such as increased ventilation, enhanced tissue oxygen extraction, and increases in hemoglobin concentration. Compared to acute hypoxia, there is less cardiac output and limb blood flow at a given absolute work rate, and conductive oxygen delivery is supported to a greater extent by the increase in oxygen content in arterial blood. However, the degree of hypobaric hypoxia (P_B), the duration of exposure, and the type of exercise, i.e., submaximal or maximal, influence the specific physiological changes in cardiovascular function during exercise over time at high altitude. Hemodynamics associated with submaximal exercise in sustained hypoxia is summarized in Table 6.4.

A strong and consistent direct relationship between cardiac output and oxygen consumption exists at sea level. As previously discussed, acute hypoxia is associated with an elevated cardiac output for any given oxygen uptake. With sustained hypoxia, the cardiac output/oxygen uptake relationship varies based on the degree of altitude or the severity of hypoxia. At altitudes >5,000 m, the relationship between cardiac output and oxygen uptake is similar to that at sea level [31, 72, 96, 143]. In contrast, at more moderate altitudes (3,100–4,500 m), where there is both less peripheral hypoxic vasodilation and less sympathetic activation, cardiac output may fall to levels that are actually less than those at sea level for the same oxygen uptake [74, 75, 79, 137, 147]. Studies examining climbers who have spent an extended period of time at high altitude have

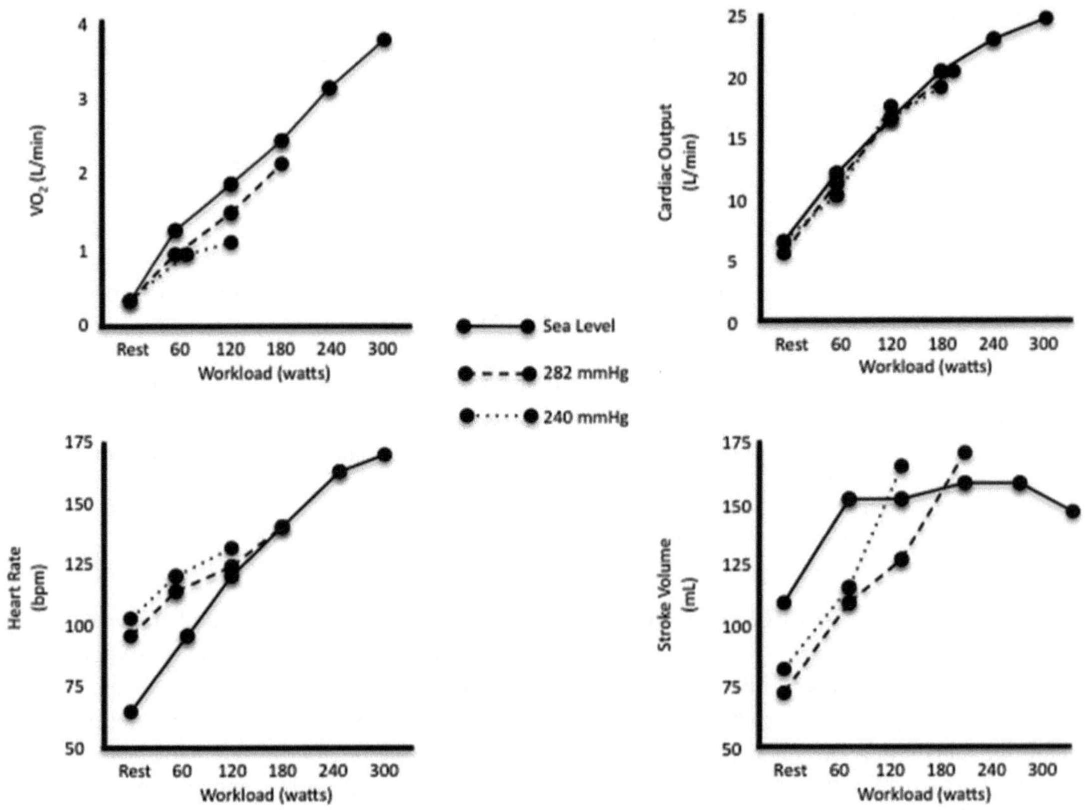

Fig. 6.7 Cardiac hemodynamic responses during exercise with progressive hypoxia exposure in the Operation Everest II study [62]. The VO_2–workload (*upper left*) and the cardiac output–workload relationships (*upper right*) are maintained during submaximal exercise at all altitudes. With progressive hypoxia, heart rate increases (*lower left*) and stroke volume decreases (*lower right*). The response of stroke volume to progressive exercise differs between sea level and extreme hypoxia

demonstrated consistent reductions in both stroke volume and cardiac output at submaximal workloads even after return to sea level, indicating that there may be a persistent alteration in the cardiac output–oxygen uptake relationship after sustained hypoxia [148, 149].

Although heart rate remains elevated at the same absolute or relative submaximal work rate during sustained hypoxia [72, 79, 150], stroke volume is typically lower than at sea level (Fig. 6.7). This serves as the key factor for the reduction in cardiac output during acclimatization. Similar to observations made at rest, pulmonary capillary wedge pressure is also decreased during submaximal exercise, indicating that reductions in cardiac filling or preload and not decreased contractility are responsible for the decrease in stroke volume. Thus, even with the contribution of contracting muscle to augment venous return to the heart (skeletal muscle pump) during exercise, central blood volume remains reduced during sustained hypoxia resulting in a decreased stroke volume.

One of the intriguing unanswered regulatory questions regarding the cardiovascular response to exercise during sustained hypoxia is what determines the interplay between central systemic blood flow, local peripheral blood flow, and oxygen delivery/extraction. Although at least some of the reduction in cardiac output may be offset (if not induced) by an increased oxygen-carrying capacity of the blood, total oxygen delivery is virtually always reduced compared to sea level, and there is a greater dependence on

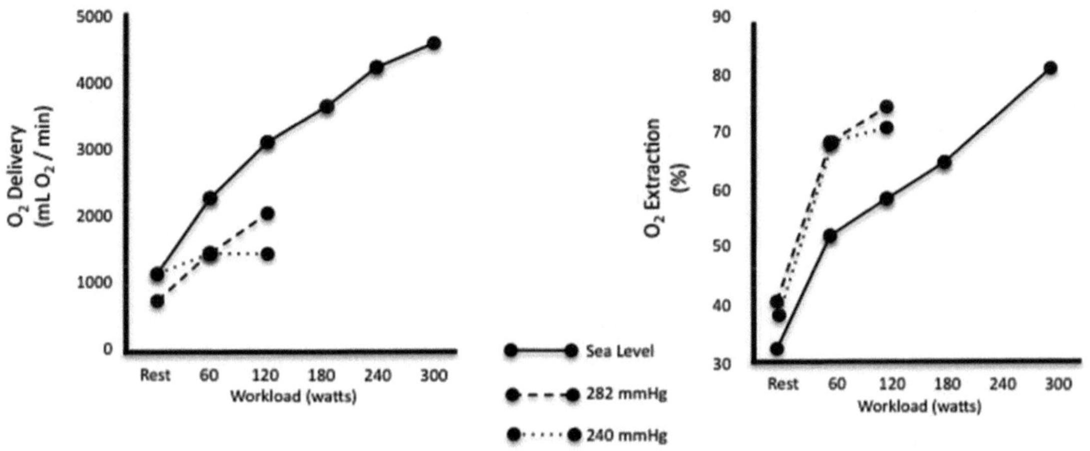

Fig. 6.8 With progressive hypoxia exposure, oxygen uptake is maintained by an increase in oxygen extraction (a-v O_2/CaO_2) as oxygen delivery is decreased compared to sea level (from [103])

peripheral extraction [96, 143] (Fig. 6.8). This response occurs in spite of substantial heart rate and blood flow reserve, which could, if more substantially engaged, restore oxygen delivery to normal levels [75, 79].

With prolonged sustained hypoxia, there appears to be a reduction in heart rate and thus cardiac output at any given level of submaximal exercise [75, 79]. The important observation that administration of supplemental O_2 during submaximal exercise in acclimatized subjects results in a further reduction in heart rate suggests that oxygen content of the blood is an important determinant of the heart rate response [31]. Both arms of the autonomic nervous system have been investigated in this context. Vagal activity does not appear to be responsible for reduced submaximal exercise heart rate during sustained hypoxia as heart rate increases with atropine to the same degree, both at sea level and during sustained hypoxia to 4,300 m [151]. However, the fact that acute beta-blockade at high altitude reduces submaximal exercise heart rate to near sea level values [81] emphasizes the importance of hypoxia-induced sympathetic activation in this adaptation. In sum, the reductions in submaximal exercise heart rate that occur during sustained hypoxia appear to rely on improved conductive oxygen transport (i.e., improved tissue delivery) and simultaneous decrease in the responsiveness to sympathetic stimulation as a result of down-regulation of cardiac β-adrenergic receptors [83, 84]. The mechanism(s) by which arterial oxygen content regulates both central and peripheral blood flow during exercise remains unknown. Endothelial nitric oxide release, as summarized in a comprehensive recent review [152], appears to be an important mediator of this phenomenon. As previously discussed, animal studies suggest that it is the oxygen content of red blood cells, endothelial cell activity, and circulating molecules that respond to hypoxia and dictate changes in blood flow and vascular resistance. More investigation is required to determine the sensing mechanism and factors responsible for this regulation of blood flow during submaximal exercise with sustained hypoxia.

Sustained Hypoxia—Maximal Exercise

Virtually all studies of maximal exercise during sustained hypoxia show a reduction in maximal cardiac output (Table 6.5). Reductions in both peak heart rate and stroke volume are responsible for the reduced cardiac output at maximal exercise to varying degrees depending on the absolute altitude. For example, at moderate altitudes of 3,000–4,000 m, reductions in stroke volume are predominant [76, 139, 150] and are similar in magnitude to those observed during submaximal exercise. However, at altitudes of ≥4,000 m, reductions in maximal heart rate appear to play an increasingly important role [31, 143, 153, 154].

Table 6.5 Sustained hypoxia—maximal exercise hemodynamic changes

Altitude (m)	*n*	Duration (days)	Method	O_2 uptake at altitude, L/min (mL/kg/min)	Response (%)[a] CO	SV	HR	References
4,300	8	3 week	Dye dilut.[b]	2.42 (32.7)	+4	+14*	−4	[84]
4,300	4	2 week	Dye dilut.	2.61 (33.9)	−22	−19*	−3	[156]
4,350	4	10 days	Dye dilute.	2.41 (33.0)	−29*	−24*	−7*	[140]
5,000	6	10 week	CO_2 rb[c]	Not available	−23*	24*	nc	[149]
6,100	6	3–4 week	Thermo/Fick	2.10 (27.8)	−16*	−4	−14*	[43]
7,620	6	1–2 week	Thermo/Fick	1.45 (19.2)	−31*	−30*	−23*	[43]
8,848[d]	6	20 min	Thermo/Fick	1.12 (14.8)	−30*	−14	−26*	[43]

Rb Rebreathing, *dye dilute.* indocyanine green dye method, *thermo* thermodilution, *nc* no change, *CO* cardiac output, *SV* stroke volume, *HR* heart rate

*$p<0.05$ versus sea level

[a]% changes from sea level

[b]Potential concerns about methodology of dye dilution (see [26])

[c]Studies done immediately after descent under more normoxic conditions

[d]Only 3 subjects studied at simulated 8,848 m altitude

This reduction in maximal cardiac output, compounded by decreased arterial oxygen content, results in a decrease in peak O_2 delivery. Thus, despite appropriate acclimatization, maximal oxygen uptake remains reduced compared to sea level and is similar to the reduction observed with acute hypoxia. Though reductions in conductive oxygen delivery are offset to some extent by increased tissue O_2 extraction, increased extraction is not sufficient to return aggregate tissue oxygen supply to sea level values. It remains unknown whether the reduction in cardiac output at maximal exercise is a primary cause of, or secondary to the reduction in maximal oxygen uptake at high altitude [155].

In contrast to the elevated heart rate at rest and during submaximal exercise, there is a consistent reduction in maximal heart rate at virtually all altitudes studied. The degree of reduction in heart rate is dependent on the duration and severity of sustained hypoxia with the greatest reductions at the most extreme altitudes. Numerous studies have reported reductions in maximal heart rate after 2–3 weeks residence at 4,300 m [75, 76, 78, 79, 139]. At altitudes of >5,800 m there is a 25 % reduction in maximal heart rate associated with a >60 % reduction in maximal oxygen uptake [31, 143, 153, 154]. The mechanism for this reduction in maximal heart rate is unclear but could be related to a direct depressant effect of hypoxia on either the sinus node or central cortical irradiation (central command, in practice, these different mechanisms have been very difficult to differentiate a secondary effect of reduced work capacity and therefore under control of exercising skeletal muscle, or specific alterations in autonomic nervous system activity associated with sustained hypoxia).

Studies with inhalation of supplemental oxygen to simulate normoxic conditions after sustained hypoxia have yielded mostly expected results. For example, at an altitude of 4,300 m, abrupt restoration of normoxic conditions in acclimatized subjects led to an increase in heart rate and cardiac output with a corollary increase in maximal work rate and oxygen uptake [156]. However, cardiac output was still less than at sea level. Similar findings were seen at an altitude of 5,800 m with improvement in heart rate and oxygen uptake at maximal exercise with oxygen inhalation [143]. In contrast, with more prolonged, progressive exposure to more extreme altitude, there was little restoration of heart rate and work capacity with supplemental oxygen [31]. However, changes in cardiovascular function at maximal work capacity during sustained hypoxia are difficult to interpret as differentiating the impact of hypoxia from that of simple deconditioning is challenging. More work is needed to further examine this issue.

In addition to the known blunting of the heart rate response to enhanced sympathetic activity with prolonged hypoxia, early studies suggested an additional enhancement of parasympathetic activity with chronic hypoxia, as maximal heart rate was restored to sea level values with atropine in recently acclimatized subjects to 4,600 m [157]. Moreover, in lifelong high-altitude residents, atropine produces a greater increase in exercise heart rate compared to recently acclimatized subjects [158]. Studies examining both β-adrenergic blockade and parasympathetic blockade in climbers residing at altitudes of >5,000 m suggest that the degree of parasympathetic modulation of the heart is intimately influenced by the extent of cardiac sympathetic activity [81]. In an elegant study by Savard et al. [81], climbers who had the greatest reduction in adrenergic responsiveness also had the greatest increase in maximal heart rate with atropine. Conversely, if sympathetic responsiveness was minimally reduced, then the response to atropine was less prominent. This concept was elegantly proven by Boushel et al. [87] who documented increases in heart rate during submaximal and maximal exercise after administration of glycopyrrolate in sustained hypoxic conditions. However, the glycopyrrolate-facilitated heart rate augmentation did not lead to increased cardiac output during exercise [87]. Interestingly, in all the studies where maximal heart rate was increased with atropine or glycopyrrolate, there was no improvement in maximal oxygen uptake, suggesting that the decreased maximal heart rate with acclimatization is unlikely to be a primary cause of the reduction in peak work capacity.

The vast majority of studies examining the link between peak exercise capacity and cardiac function have focused on the LV. To date, little information is available about the role of the RV in this context. In a study of patients born without a right ventricle who had undergone the Fontan operation (passive conduit directly from the systemic veins into the pulmonary artery) there were findings to suggest that the right ventricle may be an important determinant of peak exercise capacity [159]. Among individuals with no right ventricle, completely normal LV stroke volume and cardiac output response were observed at 3,100 m both at rest and with submaximal exercise. In contrast, there was a depression in stroke volume at maximal exercise, indicating that right ventricular dysfunction could contribute to the reductions in stroke volume during exercise with hypoxia. To what degree these findings are applicable to sustained hypoxia remains unclear and further work aimed at examining the role of the right ventricle in sustained hypoxia exercise is warranted.

Finally, a key unanswered question is the degree to which the reduction in maximal cardiac output at high altitude contributes to the reduction in maximal oxygen uptake. At present, most evidence points to the converse: that the reduction in oxygen availability/utilization is the primary determinant of systemic and regional blood flow [155]. Lines of evidence that support this hypothesis include: (1) the relationship between cardiac output and oxygen uptake remains similar to that at sea level, even at maximal exercise at extreme altitude; (2) the addition of supplemental oxygen results in an immediate increase in work rate, oxygen uptake, heart rate, and cardiac output; (3) increases or decreases in heart rate by pharmacological intervention fail to alter maximal oxygen uptake; and even autologous blood transfusion [160] or chronic erythropoietin injection [161] at altitude, which increases blood volume and red cell mass, fails to increase maximal oxygen uptake at high altitudes >4,000 m. Issues related to distribution of cardiac output and to the degree of regional autonomic activation during maximal exercise with sustained hypoxia remain to be elucidated before this hypothesis can be universally accepted.

Summary and Key Unanswered Questions—Exercise During Acute and Sustained Hypoxia

During acute hypoxia, simultaneous increases in central pump output and sympatholytic reductions in peripheral vascular resistance lead to higher cardiac output for any given oxygen uptake than that observed at sea level.

Presumably these responses ultimately maintain conductive oxygen transport while mechanisms responsible for increased skeletal muscle extraction are set into motion. With sustained hypoxia, cardiac output and limb blood flow are reduced in proportion to improvement in arterial oxygenation afforded by the acclimatization process. At maximal exercise, oxygen uptake is reduced but the relationship between cardiac output and oxygen uptake remains similar to sea level. The factors responsible for the reduction in cardiac output remain to be determined but there clearly is interplay between the autonomic nervous system and the degree of hypoxia which influences the cardiovascular responses to exercise with sustained hypoxia.

1. What are the mechanisms responsible for increased blood flow responses during hypoxic exercise and are they the same as those recently elucidated at rest?
2. What factors regulate peripheral oxygen extraction during exercise at both acute and sustained hypoxia?
3. Do the changes in cardiac output and peripheral blood flow play a primary or secondary role in the reduced exercise capacity during sustained hypoxia?

High-Altitude Residents and Populations

Exposure to high altitude over many years leads to alterations in hemodynamic, autonomic, metabolic, and coronary circulatory changes that may enhance tolerance to chronic hypoxia. Autonomic balance favors enhanced parasympathetic and reduced sympathetic activity in high-altitude natives, though definitive studies have not yet been done. An enhanced utilization of glucose by the heart in high-altitude natives appears to result in greater oxygen efficiency (more high-energy phosphate production per molecule of oxygen) and may be an important factor in the preservation of cardiac function with prolonged hypoxia. The significance of these findings for cardiovascular function has not been determined. There may be favorable changes in the coronary circulation of high-altitude natives that enhance myocardial blood flow and aid in myocardial protection. The combination of reduced coronary blood flow and enhanced vascularity may be a favorable adaptation to chronic hypoxia. The mechanisms for these coronary responses remain to be determined. Finally, the distinction between lifelong acclimatization to hypoxia and population-based, genetic adaptations of the high-altitude native has yet to be determined (see also Chap. 19).

There is increasing evidence that chronic high-altitude dwellers, especially those born at high altitude, differ from sea level dwellers exposed to chronic hypoxia. Key hemodynamic differences are summarized in Fig. 6.9. Early reports on the cardiovascular function of the Sherpas of Nepal suggested that this high-altitude population has unique physiological characteristics that explain their superior exercise capacity under conditions of extreme hypoxia [3]. These responses include a greater heart rate and cardiac output at maximal exercise along with lower minute ventilation, a higher PCO_2, and a normal blood pH compared to recently acclimatized lowlanders. One of the key differences between high-altitude natives and recently acclimatized lowlanders is the substantially greater red cell mass in the former. This difference appears to result in more prominent relative reductions in intracardiac volumes among high-altitude natives when compared to recently acclimatized lowlanders. For example, the higher the hematocrit with progressive altitude residence, the lower the cardiac output secondary to decreases in stroke volume [135].

When high-altitude residents are taken to sea level, their cardiac outputs and stroke volumes are lower than those of sea level natives despite increases in plasma volume [74, 98, 139]. With short-term residence at sea level, cardiac output and stroke volume continue to increase but may remain lower than in sea level natives. These differences in stroke volume are largely attributed to changes in plasma volume [6], though differences in fitness, diet, blood pressure, and other factors may be contributory. As such, definitive conclusions from short-term deacclimatization studies are inconclusive. The breathing of a

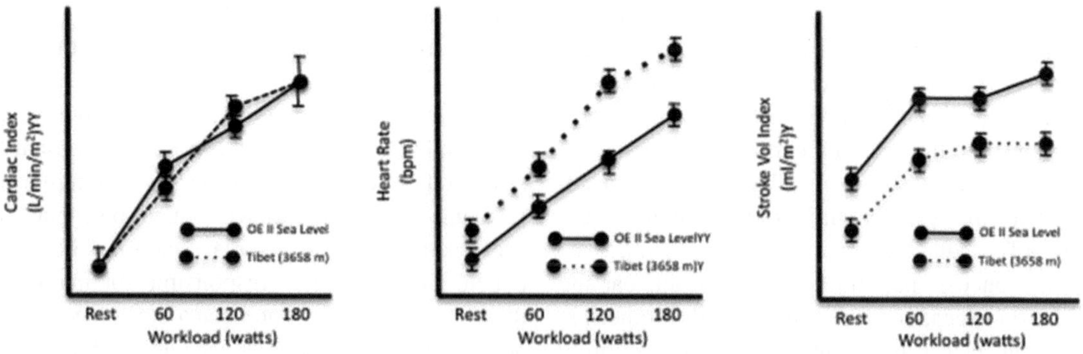

Fig. 6.9 Native Tibetans have higher heart rates and lower stroke volume indices at rest and during exercise compared to North American men. Cardiac index, however, is similar between the groups. VO_2 values were similar at identical workloads in both subject groups. Figures drawn from data on sea level subjects from Operation Everest II [43] and from lifelong residents of Tibet at 3,658 m (Groves et al. J Appl Physiol 1993; 74:312–318)

high-oxygen mixture to simulate sea level conditions while the subjects remained at altitude does not appear to result in any significant hemodynamic changes [157]. However, when cardiac hemodynamics were studied in high-altitude natives after 2 years of residence at sea level, parameters were similar to sea level natives, indicating that the majority of the cardiovascular changes with chronic hypoxia are reversible with prolonged exposure to normoxia. This observation argues against any permanent structural changes or fundamental genetic differences [162].

A key question regarding long-term exposure to chronic hypoxia is whether the dramatic sympathetic activation documented with short-term acclimatization persists or gradually abates. This response may be very different in individuals born in a hypoxic environment, compared with those who were born at sea level but have migrated to high altitudes. Studies with β-adrenergic blockade and atropine in Tibetans compared to newly acclimatized Han Chinese suggest greater parasympathetic activity in the high-altitude natives [158], manifested by greater increases in heart rate with atropine. Heart rate variability studies have also shown a greater respiratory sinus arrhythmia [163], though such differences simply may be a function of differences in respiratory rate and tidal volume. This is an important area for future research.

One of the most intriguing findings in high-altitude natives relates to alterations in myocardial metabolism. Study of Sherpa men using ^{31}P-magnetic resonance spectroscopy suggest a greater reliance on carbohydrate metabolism for energy sources by the resting myocardium. This study used ^{31}P-magnetic resonance spectroscopy to document lower ratios of phosphocreatine to ATP consistent with a threefold increase in free ADP concentration in the myocardium [164]. These metabolic conditions would accommodate the higher enzyme kinetic constants (K_M) of the enzymes phosphoglycerate kinase and pyruvate kinase, indicating enhanced capacity for carbohydrate metabolism. Similar findings were also reported in both Quechua and Sherpa subjects in whom an enhanced glucose uptake compared to sea level natives was observed with positron emission tomography [165]. It remains to be determined whether this alteration in myocardial substrate utilization is a phenotypic expression of an altered genetic myocardial adaptation in high-altitude natives or a common acclimatization response seen with more short-term exposure to high altitude.

Limited studies in high-altitude natives also suggest that there are changes in coronary anatomy and physiology that may be protective in the setting of chronic hypoxia. Resting coronary blood flow has been shown to decline progressively with increasing altitude residence [166]. However, myocardial oxygen consumption is also lower at rest in high-altitude natives, a finding that differs from studies of recently acclimatized newcomers to altitude [136].

Differences in methodology and the possible inclusion of patients with chronic mountain sickness make these data difficult to interpret. Pathological studies on high-altitude natives also demonstrate that lifelong exposure to hypoxia is associated with a more abundant coronary vascular bed with a greater density of peripheral branching of smaller coronary vessels [167]. The significance of these findings remains unclear and is hard to separate from population differences in diet, physical activity, and genetic susceptibility to coronary artery disease. However, it is possible that coronary adaptations to chronic hypoxia, similar to the increased collateralization seen with chronic myocardial ischemia [168], could result in more effective myocardial vascularization with greater surface area of oxygen diffusion. Such adaptations could be responsible for a possible protective effect of long-term residence at high altitude against coronary artery disease [169].

In addition, there may be significant differences among geographically distinct groups of chronic high-altitude dwellers. For example, Himalayan highlanders have been reported to have lower hemoglobin concentrations, higher plasma volumes, and a more robust hypoxic ventilatory response than residents of similar elevation from the South American Andes [170–172]. In addition, comparison of data from separate studies suggests that Andean natives may have higher pulmonary arterial pressures than Tibetan counterparts [173, 174]. It has been postulated that such geographically distinct patterns of high-altitude physiology may reflect differences in time intervals that these respective populations have spent at high altitude and or differences in underlying genetics [175]. For example, populations found in the Tibetan Plateau are thought to have migrated there more than 20,000 years ago while Andean high-altitude populations have spent only half that time at high altitude. To date, there are insufficient data to determine if high-altitude physiology truly differs with geography and if so, what underlies the differences observed across high-altitude populations. Direct comparative study of cardiovascular function at rest and during exercise across different native high-altitude populations is an important area of future work.

Summary and Key Unanswered Questions—High-Altitude Populations

In summary, there are various hemodynamic, autonomic, metabolic, and possibly coronary structural adaptations that occur in long-term residents at high altitude that may contribute to preservation of cardiovascular function. The time course and the factors responsible for many of these changes have not been determined. The heterogeneous responses to sustained high-altitude exposure may relate to different genetic susceptibility as well as to alterations in controlling mechanisms of cardiovascular function.

1. What is the time course for the various cardiovascular adaptations observed in high-altitude natives?
2. How do alterations in autonomic activity influence the cardiovascular adaptations to chronic hypoxia in high-altitude residents?
3. Are there actual genotypic changes that result from generations of chronic hypoxia exposure, or are there changes reversible with prolonged normoxic exposure?

Cardiovascular Disease at High Altitude—Clinical Correlations

Although the cardiovascular system appears to function normally at rest and during exercise at high altitude, the effect of hypoxia on cardiac function in patients with underlying cardiovascular disease has not been well defined. Current understanding of the impact of hypoxia on the manifestation and progression of cardiac disease is based on limited data.

In theory, exposure to hypoxia in patients with preexisting cardiovascular disease could precipitate new symptoms not present at sea level or could accentuate mild preexisting symptoms. Exposure to high altitude presents several stresses that may exacerbate the clinical manifestations of underlying cardiac disease. These include the degree of hypoxia itself, alkalosis, heightened sympathetic activity, elevated systemic blood pressure and heart rate, and increased blood viscosity. Exercise at altitude also may be an important factor, as any submaximal work load at

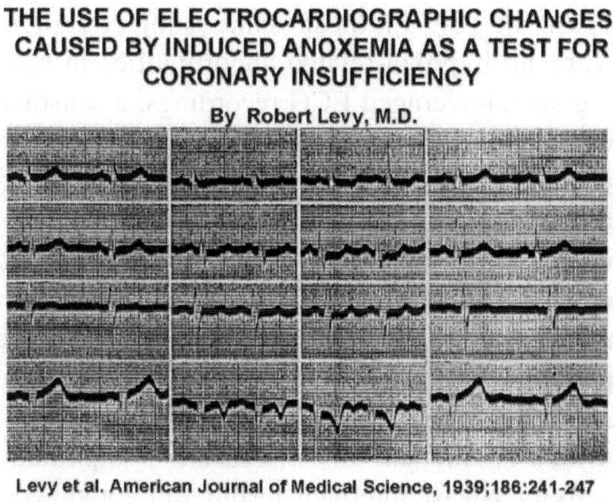

Fig. 6.10 Representative electrocardiographic tracings obtained during performance of the "Levy Test" during which patients are exposed to an altitude equivalent to 18,000 ft. in an attempt to diagnose coronary arterial flow insufficiency

altitude is at a higher percentage of maximal oxygen uptake than at sea level and thereby requires increased cardiac work. Clinical issues of particular concern include the provocation of myocardial ischemia and acute unstable coronary syndromes, the triggering of cardiac arrhythmias, and the precipitation or exacerbation of heart failure.

Myocardial Ischemia and Infarction

In patients with known coronary disease, exposure to the hypoxia of high altitude may result in myocardial ischemia, especially during exercise. Indeed, prior to the initiation of exercise testing, a typical test for the provocation of myocardial ischemia was the "Levy test" exposing patients to an altitude equivalent to 18,000 ft. (Fig. 6.10). There is surprisingly limited information available regarding the risk of ischemia in patients with coronary disease during acute or sustained hypoxia. Extreme hypoxia alone does not appear to cause any clinical or ECG evidence of ischemia in the absence of atherosclerotic disease [71, 123]. However, the vasomotor response of coronary arteries, especially via endothelial-dependent mechanisms, has been shown to be abnormal in patients with atherosclerotic coronary disease. For example, intracoronary infusion of acetylcholine, an endothelial-dependent vasodilator in normal vessels, causes vasoconstriction in diseased coronary vessels [176]. A similar observation has been noted during exercise at sea level in coronary patients where vasoconstriction, rather than vasodilation, can occur in diseased coronary segments [177]. Recently, Arbab-Zadeh et al. examined the coronary arterial response to hypoxia in patients with angiographically documented coronary atherosclerosis [103]. They found that the normal hypoxic coronary vasodilation failed to occur in arterial segments with preexisting narrowing. Of note, this finding was observed at moderate ($FiO_2=0.15$) and more severe ($FiO_2=0.10$) levels of hypoxemia. In addition, there are data documenting blunted exercise-induced myocardial flow reserve among patients with coronary disease [34]. These findings raise concern about the ability of individuals with preexisting atherosclerotic coronary disease to tolerate acute bouts of hypoxia such as those inherent to rapid ascents to high altitude. Clinical correlations will be required to address this concern.

Acute coronary syndromes including unstable angina and myocardial infarction are caused by disruption of atherosclerotic plaque and localized thrombus formation. Local factors that predispose to this syndrome include the presence of lipid-rich plaque, vasoconstriction, and thrombus formation [178]. Systemic factors are also important and include increased catecholamine levels, an abnormal coagulation and fibrinolysis profile, abnormal metabolic states (diabetes, homocysteinemia), and shear forces in the vessels as can occur with hypertension. Although there have not been any reported coagulation abnormalities with hypoxia [179], other factors such as hypertension, heightened sympathetic activity, and possibly coronary vasoconstriction could lead to plaque disruption in the susceptible patient with lipid-laden plaques while at altitude. A report examining some relevant cardiovascular parameters in an "at-risk" population provides some reassuring results. Schobersberger et al. reported the impact of a 3-week sojourn at 1,700 m on blood pressure, heart rate, glycemic parameters, and lipid metabolism in 22 males with metabolic syndrome [180]. In addition to complete freedom from adverse clinical events, these subjects were found to have favorable short-term cardiovascular responses including reduced heart rate and blood pressure during altitude exposure. Exercise may be a relevant factor as it has been found to be an important cause of plaque rupture at sea level [181].

Despite these theoretical reasons for increased cardiac risk at high altitude, the limited available data suggest that altitude exposure is generally well tolerated by patients with coronary disease and there is evidence that birth or long-term living at high altitude may protect against coronary heart disease [182]. Coronary patients residing at 1,600 m developed objective evidence of myocardial ischemia at a lower exercise workload upon initial arrival at 3,100 m [183]. However, after 5 days of acclimatization to the higher altitude, their ischemic pattern returned to their baseline level. In these partially altitude-acclimatized individuals, there was no change in the ischemic threshold (heart rate–systolic blood pressure product at the onset of ischemia) with acute or prolonged exposure to hypoxia. In contrast, sea level natives with coronary artery disease studied under conditions of acute hypoxia in a hypobaric chamber at a simulated altitude of 2,500 m did develop myocardial ischemia at a lower hemodynamic threshold compared to normoxic conditions [8]. However, these patients when studied after 5 days of residence at 2,500 m only developed ischemia at their sea level threshold. Thus, it appeared that the initial adverse response to acute hypoxia was reversible with even short-term acclimatization.

Cardiac Arrhythmias

There is limited information on the occurrence of cardiac arrhythmias at high altitude. In elderly patients with either an increased risk for or known coronary disease, there were no significant arrhythmias noted by either short-term ECG monitoring at rest or exercise electrocardiography with acute or more prolonged exposure to 2,500 m [8]. In these same subjects, there were no hypoxia-related abnormalities in resting signal-averaged ECG recordings, a sensitive marker for the presence of a myocardial electrophysiological substrate conducive for arrhythmia production. Recently, Gibelli et al. used microvolt T-wave alternans testing, a documented method of risk stratification for ventricular tachyarrhythmias [184], to examine arrhythmic potential in healthy climbers ($n=8$) during a climbing excursion to 8,150 m. Despite confirmed evidence of increased sympathetic activity, t-wave alternans testing did not suggest an increased risk of malignant rhythm events. One small study by Woods et al. used implantable loop recorders in healthy subjects ($n=9$) during a high-altitude trek in Nepal [185]. In this small group, palpitations were frequent and were found to correlate with underlying atrial flutter, sinus arrhythmia, nonconducted p-waves, and ST-segment depression. Aside from the Woods study, the majority of data argue against significant arrhythmia during acute or sustained hypoxia in individuals without preexisting heart disease. Further work will be required to resolve this issue definitively.

To date, no occurrence of significant cardiac arrhythmias has been reported in any study of heart disease patients at high altitude. However, most of these studies did not include patients with significant left ventricular systolic dysfunction or prior history of malignant arrhythmias. In one small study of patients with a reduced mean left ventricular ejection fraction of 39 %, no arrhythmias were reported during exercise at 2,500 m [186]. An interesting observation regarding the management of atrial fibrillation anticoagulation was recently provided by Van Patot and colleagues who documented an increased risk of subtherapeutic INR values among patients after ascent to altitude [187]. This observational study is noteworthy but requires clarification and was not sufficient to provide mechanistic interpretation. Further work at high altitudes with larger numbers of patients with a greater clinical likelihood of cardiac arrhythmias is required before any definitive statement can be made about the risks and clinical relevance of arrhythmia during altitude exposure.

Heart Failure

Although the myocardium functions normally under conditions of acute and sustained hypoxia, it is unclear how a heart with underlying LV systolic function would tolerate these conditions. In theory, the stimuli leading to increased cardiac output with acute hypoxia and the systemic vasoconstriction that occurs with more prolonged exposure could precipitate decompensated heart failure in patients with preexisting LV dysfunction. Although the usual cardiac response to prolonged hypoxia is a reduction in end-diastolic volume with a maintained or increased ejection fraction [100], echocardiographic studies in coronary subjects living in Denver, CO (1,600 m), who spent 5 days at 3,100 m demonstrated an increase in end-diastolic and end-systolic dimensions with a fall in LVEF from 51 % at 1,600 m to 37 % at 3,100 m [188]. These results are contradictory to those of another study at 2,500 m in which sea level residents with known coronary disease were studied after 5 days at altitude and were found to have no echocardiographic abnormalities [8]. Patients with known left ventricular dysfunction who were exercised at 2,500 m were found to have the same decrement in exercise capacity as normal age-matched subjects and no adverse cardiovascular events were noted after 2 days at moderate altitude in these patients [186].

Sea level residents with chronic heart failure have been studied while exercising at different degrees of acute normobaric hypoxia equivalent to altitudes varying from 1,000 to 3,000 m [189]. These patients, all of whom had reduced exercise capacity when compared to normal subjects, had a similar response to progressive hypoxia in regards to arterial oxygen saturation, exercise ventilation, heart rate, and blood pressure. In normal subjects there was a 3 % decrement in maximal exercise capacity for each 1,000 m in elevation. Heart failure patients with a sea level maximal uptake greater than 15 mL/kg/min had a similar decrement of 5 % per each 1,000 m. However, patients with maximal oxygen uptake less than 15 mL/kg/min at sea level had a more marked reduction in exercise capacity, with an 11 % reduction for each 1,000 m of elevation. This reduction in exercise capacity probably reflected the inability to increase cardiac output in response to acute reductions in arterial oxygenation. No data are available in more prolonged hypoxia exposure in heart failure patients. In addition, all these patients were on heart failure medications that may have modified the cardiovascular responses to hypoxia.

In summary, there is limited information on altitude tolerance in patients with cardiovascular disease. Hypoxia-induced coronary vasoconstriction is a particularly intriguing concept that needs to be explored. The limited data would suggest that moderate altitude exposure (2,500–3,100 m) is well tolerated by stable coronary patients with normal or moderately depressed ventricular function. Although difficult to obtain, more information is needed in this important area of high-altitude medicine.

Sudden Death in the High-Altitude Environment

Despite concerns elucidated in the previous sections, available data suggest that performance of vigorous physical activity in moderately high-altitude environments such as those typically obtained by recreational mountains is relatively safe. Specifically, Burtscher et al. reported estimated sudden cardiac death rates of 1:780,000 hiking hours and 1:1,630,000 skiing hours based on retrospective data from the Austrian Alps [190]. More recently, Ponchia et al. reported estimated cardiovascular event rates of 1 per 319 000 person-days of physical activity in the mountains, 1 sudden cardiac death per 980,000 person-days of physical activity, and 1 acute coronary syndrome per 2,895,000 person-days of physical activity in the mountains [191]. Careful examination of these data strongly suggests that increasing age, prior low activity exposure, and preexisting traditional cardiovascular risk factors are important determinants of cardiovascular event risk at high altitude. This notion is further substantiated by a recent report in which Burtscher et al. studied cases of sudden death in the mountains and found that victims were more likely to have had prior

MI, known coronary artery disease without prior MI, diabetes, hypercholesterolemia, and less prior experience with mountain sports activities than matched controls [192]. It should be noted that there are presently sparse data examining cardiovascular outcomes at high or extreme altitude.

Recommendations and Key Unanswered Questions—Clinical Correlations

Based on the above information, we suggest that patients with cardiac disease should be evaluated at sea level prior to ascent to altitude to insure a stable disease state and to document objective measures of exercise capacity. We suggest that patients with known coronary disease limit their physical activity during the first few days at altitude to allow the favorable effects of acclimatization to occur. Recommendations for patients with left ventricular dysfunction are more difficult to determine. However, it appears that patients with mild to moderate ventricular dysfunction should be able to tolerate moderate altitude without serious adverse consequences. In questionable cases, nocturnal oxygen supplementation should be considered and medications used at sea level should be continued at altitude. Patients with more significant cardiac impairment at sea level, especially with significant exercise limitation, should not travel to high altitude in the absence of supplemental oxygen to simulate normoxic conditions. An intriguing issue that deserves future attention is the use of normobaric hypoxic exercise testing as a routine component of prealtitude ascent risk stratification in patients with underlying cardiac disease.

1. Can hypoxia alone lead to instability of coronary vascular lesions and vasomotor tone with the precipitation of unstable coronary syndromes and possibly acute myocardial infarction?
2. Does exposure to acute or sustained hypoxia increase the likelihood of arrhythmias in patients with cardiovascular disease?
3. What are the implications of altitude exposure on patients with reduced ventricular function at sea level?

References

1. Grollman A. Physiological variations of the cardic output of man. VII. The effects of high altitude on the cardiac output and its related functions: an account of experiments conducted on the summit of Pikes Peak, Colorado. Am J Physiol. 1930;93:19–40.
2. Doyle JT, Wilson JS, Warren JV. The pulmonary vascular responses to short-term hypoxia in human subjects. Circulation. 1952;5(2):263–70.
3. Wescott RN, Fowler NO, Scott RC, Hauenstein VD, McGuire J. Anoxia and human pulmonary vascular resistance. J Clin Invest. 1951;19:957–70.
4. Vogel JA, Harris CW. Cardiopulmonary responses of resting man during early exposure to high altitude. J Appl Physiol. 1967;22(6):1124–8.
5. Stenberg J, Ekblom B, Messin R. Hemodynamic response to work at simulated altitude, 4,000 m. J Appl Physiol. 1966;21(5):1589–94.
6. Wagner PD, Gale GE, Moon RE, Torre-Bueno JR, Stolp BW, Saltzman HA. Pulmonary gas exchange in humans exercising at sea level and simulated altitude. J Appl Physiol. 1986;61(1):260–70.
7. Wagner JA, Miles DS, Horvath SM. Physiological adjustments of women to prolonged work during acute hypoxia. J Appl Physiol. 1980;49(3):367–73.
8. Levine BD, Zuckerman JH, deFilippi CR. Effect of high-altitude exposure in the elderly: the Tenth Mountain Division study. Circulation. 1997;96(4):1224–32.
9. Richardson DW, Kontos HA, Raper AJ, Patterson Jr JL. Modification by beta-adrenergic blockade of the circulatory respones to acute hypoxia in man. J Clin Invest. 1967;46(1):77–85.
10. Koller EA, Drechsel S, Hess T, Macherel P, Boutellier U. Effects of atropine and propranolol on the respiratory, circulatory, and ECG responses to high altitude in man. Eur J Appl Physiol Occup Physiol. 1988;57(2):163–72.
11. Cunningham WL, Becker EJ, Kreuzer F. Catecholamines in plasma and urine at high altitude. J Appl Physiol. 1965;20(4):607–10.
12. Mazzeo RS, Wolfel EE, Butterfield GE, Reeves JT. Sympathetic response during 21 days at high altitude (4,300 m) as determined by urinary and arterial catecholamines. Metabolism. 1994;43(10):1226–32.
13. Mazzeo RS, Bender PR, Brooks GA, Butterfield GE, Groves BM, Sutton JR, et al. Arterial catecholamine responses during exercise with acute and chronic high-altitude exposure. Am J Physiol. 1991;261(4 Pt 1):E419–24.
14. Leuenberger U, Gleeson K, Wroblewski K, Prophet S, Zelis R, Zwillich C, et al. Norepinephrine clearance is increased during acute hypoxemia in humans. Am J Physiol. 1991;261(5 Pt 2):H1659–64.
15. Perini R, Milesi S, Biancardi L, Veicsteinas A. Effects of high altitude acclimatization on heart rate

variability in resting humans. Eur J Appl Physiol Occup Physiol. 1996;73(6):521–8.
16. Brown TE, Beightol LA, Koh J, Eckberg DL. Important influence of respiration on human R-R interval power spectra is largely ignored. J Appl Physiol. 1993;75(5):2310–7.
17. Shibata S, Zhang R, Hastings J, Fu Q, Okazaki K, Iwasaki K, et al. Cascade model of ventricular-arterial coupling and arterial-cardiac baroreflex function for cardiovascular variability in humans. Am J Physiol Heart Circ Physiol. 2006;291(5): H2142–51.
18. Piepoli M, Sleight P, Leuzzi S, Valle F, Spadacini G, Passino C, et al. Origin of respiratory sinus arrhythmia in conscious humans. An important role for arterial carotid baroreceptors. Circulation. 1997; 95(7):1813–21.
19. Grassman E, Blomqvist CG. Absence of respiratory sinus arrhythmia: a manifestation of the sick sinus syndrome. Clin Cardiol. 1983;6(4):151–4.
20. Somers VK, Mark AL, Abboud FM. Interaction of baroreceptor and chemoreceptor reflex control of sympathetic nerve activity in normal humans. J Clin Invest. 1991;87(6):1953–7.
21. Saito M, Mano T, Iwase S, Koga K, Abe H, Yamazaki Y. Responses in muscle sympathetic activity to acute hypoxia in humans. J Appl Physiol. 1988;65(4): 1548–52.
22. Duplain H, Vollenweider L, Delabays A, Nicod P, Bartsch P, Scherrer U. Augmented sympathetic activation during short-term hypoxia and high-altitude exposure in subjects susceptible to high-altitude pulmonary edema. Circulation. 1999;99(13):1713–8.
23. Eckberg DL, Bastow III H, Scruby AE. Modulation of human sinus node function by systemic hypoxia. J Appl Physiol. 1982;52(3):570–7.
24. Kahler RL, Goldblatt A, Braunwald E. The effects of acute hypoxia on the systemic venous and arterial systems and on myocardial contractile force. J Clin Invest. 1962;41:1553–63.
25. Neubauer JA, Sunderram J. Oxygen-sensing neurons in the central nervous system. J Appl Physiol. 2004;96(1):367–74.
26. Mifflin SW, Felder RB. Synaptic mechanisms regulating cardiovascular afferent inputs to solitary tract nucleus. Am J Physiol. 1990;259(3 Pt 2):H653–61.
27. Dias AC, Vitela M, Colombari E, Mifflin SW. Nitric oxide modulation of glutamatergic, baroreflex, and cardiopulmonary transmission in the nucleus of the solitary tract. Am J Physiol Heart Circ Physiol. 2005;288(1):H256–62.
28. Reeves SR, Gozal D. Protein kinase C activity in the nucleus tractus solitarii is critically involved in the acute hypoxic ventilatory response, but is not required for intermittent hypoxia-induced phrenic long-term facilitation in adult rats. Exp Physiol. 2007;92(6):1057–66.
29. Reeves SR, Carter ES, Guo SZ, Gozal D. Calcium/calmodulin-dependent kinase II mediates critical components of the hypoxic ventilatory response within the nucleus of the solitary tract in adult rats. Am J Physiol Regul Integr Comp Physiol. 2005;289(3):R871–6.
30. Hultgren HN. The systemic circulation. In: High altitude medicine. Stanford, CA: Hultgren Publications; 1997:33–63.
31. Reeves JT, Groves BM, Sutton JR, Wagner PD, Cymerman A, Malconian MK, et al. Operation Everest II: preservation of cardiac function at extreme altitude. J Appl Physiol. 1987;63(2):531–9.
32. Pool PE, Covell JW, Chidsey CA, Braunwald E. Myocardial high energy phosphate stores in acutely induced hypoxic heart failure. Circ Res. 1966;19: 221–9.
33. Kaufmann PA, Schirlo C, Pavlicek V, Berthold T, Burger C, von Schulthess GK, et al. Increased myocardial blood flow during acute exposure to simulated altitudes. J Nucl Cardiol. 2001;8(2):158–64.
34. Wyss CA, Koepfli P, Fretz G, Seebauer M, Schirlo C, Kaufmann PA. Influence of altitude exposure on coronary flow reserve. Circulation. 2003;108(10): 1202–7.
35. Chen CH, Liu YF, Lee SD, Huang CY, Lee WC, Tsai YL, et al. Altitude hypoxia increases glucose uptake in human heart. High Alt Med Biol. 2009;10(1):83–6.
36. Adrogue JV, Sharma S, Ngumbela K, Essop MF, Taegtmeyer H. Acclimatization to chronic hypobaric hypoxia is associated with a differential transcriptional profile between the right and left ventricle. Mol Cell Biochem. 2005;278(1–2):71–8.
37. Onishi K, Sekioka K, Ishisu R, Tanaka H, Nakamura M, Ueda Y, et al. Decrease in oxygen cost of contractility during hypocapnic alkalosis in canine hearts. Am J Physiol. 1996;270(6 Pt 2):H1905–13.
38. Heistad DD, Abboud FM, Dickinson W. Richards lecture: circulatory adjustments to hypoxia. Circulation. 1980;61(3):463–70.
39. Rowell LB, Saltin B, Kiens B, Christensen NJ. Is peak quadriceps blood flow in humans even higher during exercise with hypoxemia? Am J Physiol. 1986;251(5 Pt 2):H1038–44.
40. Hartley LH, Vogel JA, Landowne M. Central, femoral, and brachial circulation during exercise in hypoxia. J Appl Physiol. 1973;34(1):87–90.
41. Remensnyder JP, Mitchell JH, Sarnoff SJ. Functional sympatholysis during muscular activity. Observations on influence of carotid sinus on oxygen uptake. Circ Res. 1962;11:370–80.
42. Hansen J, Thomas GD, Harris SA, Parsons WJ, Victor RG. Differential sympathetic neural control of oxygenation in resting and exercising human skeletal muscle. J Clin Invest. 1996;98(2):584–96.
43. Thomas GD, Hansen J, Victor RG. ATP-sensitive potassium channels mediate contraction-induced attenuation of sympathetic vasoconstriction in rat skeletal muscle. J Clin Invest. 1997;99(11):2602–9.
44. Thomas GD, Victor RG. Nitric oxide mediates contraction-induced attenuation of sympathetic vasoconstriction in rat skeletal muscle. J Physiol. 1998;506(Pt 3):817–26.

45. Kojonazarov BK, Imanov BZ, Amatov TA, Mirrakhimov MM, Naeije R, Wilkins MR, et al. Noninvasive and invasive evaluation of pulmonary arterial pressure in highlanders. Eur Respir J. 2007;29(2):352–6.
46. Knight DR, Schaffartzik W, Poole DC, Hogan MC, Bebout DE, Wagner PD. Effects of hyperoxia on maximal leg O2 supply and utilization in men. J Appl Physiol. 1993;75(6):2586–94.
47. Hackett JG, Abboud FM, Mark AL, Schmid PG, Heistad DD. Coronary vascular responses to stimulation of chemoreceptors and baroreceptors: evidence for reflex activation of vagal cholinergic innervation. Circ Res. 1972;31(1):8–17.
48. Singel DJ, Stamler JS. Chemical physiology of blood flow regulation by red blood cells: the role of nitric oxide and S-nitrosohemoglobin. Annu Rev Physiol. 2005;67:99–145.
49. Gonzalez-Alonso J, Mortensen SP, Dawson EA, Secher NH, Damsgaard R. Erythrocytes and the regulation of human skeletal muscle blood flow and oxygen delivery: role of erythrocyte count and oxygenation state of haemoglobin. J Physiol. 2006;572(Pt 1):295–305.
50. Ellsworth ML, Forrester T, Ellis CG, Dietrich HH. The erythrocyte as a regulator of vascular tone. Am J Physiol. 1995;269(6 Pt 2):H2155–61.
51. Rosenmeier JB, Yegutkin GG, Gonzalez-Alonso J. Activation of ATP/UTP-selective receptors increases blood flow and blunts sympathetic vasoconstriction in human skeletal muscle. J Physiol. 2008;586 (Pt 20):4993–5002.
52. Dietrich HH, Ellsworth ML, Sprague RS, Dacey Jr RG. Red blood cell regulation of microvascular tone through adenosine triphosphate. Am J Physiol Heart Circ Physiol. 2000;278(4):H1294–8.
53. Abraham EH, Sterling KM, Kim RJ, Salikhova AY, Huffman HB, Crockett MA, et al. Erythrocyte membrane ATP binding cassette (ABC) proteins: MRP1 and CFTR as well as CD39 (ecto-apyrase) involved in RBC ATP transport and elevated blood plasma ATP of cystic fibrosis. Blood Cells Mol Dis. 2001;27(1):165–80.
54. Sprague RS, Ellsworth ML, Stephenson AH, Kleinhenz ME, Lonigro AJ. Deformation-induced ATP release from red blood cells requires CFTR activity. Am J Physiol. 1998;275(5 Pt 2):H1726–32.
55. Sprague RS, Bowles EA, Hanson MS, DuFaux EA, Sridharan M, Adderley S, et al. Prostacyclin analogs stimulate receptor-mediated cAMP synthesis and ATP release from rabbit and human erythrocytes. Microcirculation. 2008;15(5):461–71.
56. Walsh MP, Marshall JM. The early effects of chronic hypoxia on the cardiovascular system in the rat: role of nitric oxide. J Physiol. 2006;575(Pt 1):263–75.
57. Ray CJ, Marshall JM. Measurement of nitric oxide release evoked by systemic hypoxia and adenosine from rat skeletal muscle in vivo. J Physiol. 2005;568(Pt 3):967–78.
58. Pyner S, Coney A, Marshall JM. The role of free radicals in the muscle vasodilatation of systemic hypoxia in the rat. Exp Physiol. 2003;88(6):733–40.
59. Cosby K, Partovi KS, Crawford JH, Patel RP, Reiter CD, Martyr S, et al. Nitrite reduction to nitric oxide by deoxyhemoglobin vasodilates the human circulation. Nat Med. 2003;9(12):1498–505.
60. Maher AR, Milsom AB, Gunaruwan P, Abozguia K, Ahmed I, Weaver RA, et al. Hypoxic modulation of exogenous nitrite-induced vasodilation in humans. Circulation. 2008;117(5):670–7.
61. Liu X, Srinivasan P, Collard E, Grajdeanu P, Lok K, Boyle SE, et al. Oxygen regulates the effective diffusion distance of nitric oxide in the aortic wall. Free Radic Biol Med. 2010;48(4):554–9.
62. Dalsgaard T, Simonsen U, Fago A. Nitrite-dependent vasodilation is facilitated by hypoxia and is independent of known NO-generating nitrite reductase activities. Am J Physiol. 2007;292(6):H3072–8.
63. Vanhatalo A, Fulford J, Bailey SJ, Blackwell JR, Winyard PG, Jones AM. Dietary nitrate reduces muscle metabolic perturbation and improves exercise tolerance in hypoxia. J Physiol. 2012;589 (Pt 22):5517–28.
64. Diesen DL, Hess DT, Stamler JS. Hypoxic vasodilation by red blood cells: evidence for an s-nitrosothiol-based signal. Circ Res. 2008;103(5):545–53.
65. Isbell TS, Sun CW, Wu LC, Teng X, Vitturi DA, Branch BG, et al. SNO-hemoglobin is not essential for red blood cell-dependent hypoxic vasodilation. Nat Med. 2008;14(7):773–7.
66. Coney AM, Marshall JM. Contribution of alpha2-adrenoceptors and Y1 neuropeptide Y receptors to the blunting of sympathetic vasoconstriction induced by systemic hypoxia in the rat. J Physiol. 2007;582(Pt 3):1349–59.
67. Hultgren HN. Circulatory responses to acute hypoxia in normal subjects at sea level. In: High altitude medicine. Stanford, CA: Hultgren Publications; 1997. p. 34.
68. Wolfel EE, Selland MA, Mazzeo RS, Reeves JT. Systemic hypertension at 4,300 m is related to sympathoadrenal activity. J Appl Physiol. 1994; 76(4):1643–50.
69. Nielsen AM, Bisgard GE, Vidruk EH. Carotid chemoreceptor activity during acute and sustained hypoxia in goats. J Appl Physiol. 1988;65(4):1796–802.
70. Klausen K. Cardiac output in man in rest and work during and after acclimatization to 3,800 m. J Appl Physiol. 1966;21(2):609–16.
71. Karliner JS, Sarnquist FF, Graber DJ, Peters Jr RM, West JB. The electrocardiogram at extreme altitude: experience on Mt. Everest. Am Heart J. 1985;109(3 Pt 1):505–13.
72. Houston CS, Riley RL. Respiratory and circulatory changes during acclimatization to high altitude. Am J Physiol. 1947;140:565–88.
73. Fowles RE, Hultgren HN. Left ventricular function at high altitude examined by systolic time intervals

and M-mode echocardiography. Am J Cardiol. 1983;52(7):862–6.
74. Alexander JK, Hartley LH, Modelski M, Grover RF. Reduction of stroke volume during exercise in man following ascent to 3,100 m altitude. J Appl Physiol. 1967;23(6):849–58.
75. Wolfel EE, Groves BM, Brooks GA, Butterfield GE, Mazzeo RS, Moore LG, et al. Oxygen transport during steady-state submaximal exercise in chronic hypoxia. J Appl Physiol. 1991;70(3):1129–36.
76. Vogel JA, Hansen JE, Harris CW. Cardiovascular responses in man during exhaustive work at sea level and high altitude. J Appl Physiol. 1967;23(4): 531–9.
77. Dempsey JA, Thomson JM, Forster HV, Cerny FC, Chosy LW. HbO2 dissociation in man during prolonged work in chronic hypoxia. J Appl Physiol. 1975;38(6):1022–9.
78. Reeves JT, Grover RF, Cohn JE. Regulation of ventilation during exercise at 10,200 ft in athletes born at low altitude. J Appl Physiol. 1967;22(3):546–54.
79. Wolfel EE, Selland MA, Cymerman A, Brooks GA, Butterfield GE, Mazzeo RS, et al. O2 extraction maintains O2 uptake during submaximal exercise with beta-adrenergic blockade at 4,300 m. J Appl Physiol. 1998;85(3):1092–102.
80. Moore LG, Cymerman A, Huang SY, McCullough RE, McCullough RG, Rock PB, et al. Propranolol blocks metabolic rate increase but not ventilatory acclimatization to 4300 m. Respir Physiol. 1987;70(2):195–204.
81. Savard GK, Areskog NH, Saltin B. Cardiovascular response to exercise in humans following acclimatization to extreme altitude. Acta Physiol Scand. 1995;154(4):499–509.
82. Bristow MR, Ginsburg R, Minobe W, Cubicciotti RS, Sageman WS, Lurie K, et al. Decreased catecholamine sensitivity and beta-adrenergic-receptor density in failing human hearts. N Engl J Med. 1982;307(4):205–11.
83. Voelkel NF, Hegstrand L, Reeves JT, McMurty IF, Molinoff PB. Effects of hypoxia on density of beta-adrenergic receptors. J Appl Physiol. 1981;50(2):363–6.
84. Kacimi R, Richalet JP, Corsin A, Abousahl I, Crozatier B. Hypoxia-induced downregulation of beta-adrenergic receptors in rat heart. J Appl Physiol. 1992;73(4):1377–82.
85. Richalet JP, Larmignat P, Rathat C, Keromes A, Baud P, Lhoste F. Decreased cardiac response to isoproterenol infusion in acute and chronic hypoxia. J Appl Physiol. 1988;65(5):1957–61.
86. Antezana AM, Kacimi R, Le Trong JL, Marchal M, Abousahl I, Dubray C, et al. Adrenergic status of humans during prolonged exposure to the altitude of 6,542 m. J Appl Physiol. 1994;76(3):1055–9.
87. Boushel R, Calbet JA, Radegran G, Sondergaard H, Wagner PD, Saltin B. Parasympathetic neural activity accounts for the lowering of exercise heart rate at high altitude. Circulation. 2001;104(15):1785–91.
88. Myhre LG, Dill DB, Hall FG, Brown DK. Blood volume changes during three-week residence at high altitude. Clin Chem. 1970;16(1):7–14.
89. Jain SC, Bardhan J, Swamy YV, Krishna B, Nayar HS. Body fluid compartments in humans during acute high-altitude exposure. Aviat Space Environ Med. 1980;51(3):234–6.
90. Robach P, Dechaux M, Jarrot S, Vaysse J, Schneider JC, Mason NP, et al. Operation Everest III: role of plasma volume expansion on VO(2)(max) during prolonged high-altitude exposure. J Appl Physiol. 2000;89(1):29–37.
91. Kubo SH, Clark M, Laragh JH, Borer JS, Cody RJ. Identification of normal neurohormonal activity in mild congestive heart failure and stimulating effect of upright posture and diuretics. Am J Cardiol. 1987;60(16):1322–8.
92. Thompson CA, Tatro DL, Ludwig DA, Convertino VA. Baroreflex responses to acute changes in blood volume in humans. Am J Physiol. 1990;259(4 Pt 2):R792–8.
93. Hansen J, Sander M. Sympathetic neural overactivity in healthy humans after prolonged exposure to hypobaric hypoxia. J Physiol. 2003;546(Pt 3):921–9.
94. Hoon RS, Balasubramanian V, Mathew OP, Tiwari SC, Sharma SC, Chadha KS. Effect of high-altitude exposure for 10 days on stroke volume and cardiac output. J Appl Physiol. 1977;42(5):722–7.
95. Groves BM, Reeves JT, Sutton JR, Wagner PD, Cymerman A, Malconian MK, et al. Operation Everest II: elevated high-altitude pulmonary resistance unresponsive to oxygen. J Appl Physiol. 1987;63(2):521–30.
96. Sutton JR, Reeves JT, Wagner PD, Groves BM, Cymerman A, Malconian MK, et al. Operation Everest II: oxygen transport during exercise at extreme simulated altitude. J Appl Physiol. 1988;64(4):1309–21.
97. Balasubramanian V, Behl A, Das GS, Wadhwa AK, Mathew OP, Hoon RS. Effect of digoxin and diuretics on high altitude left ventricular dysfunction. Circulation. 1978;57(6):1180–5.
98. Hartley LH, Alexander JK, Modelski M, Grover RF. Subnormal cardiac output at rest and during exercise in residents at 3,100 m altitude. J Appl Physiol. 1967;23(6):839–48.
99. Buch J, Egeblad H, Hansen PB, Kjaergard H, Waldorff S, Steiness E. Correlation between changes in systolic time intervals and left ventricular end-diastolic diameter after preload reduction. Non-invasive monitoring of pharmacological intervention. Br Heart J. 1980;44(6):668–71.
100. Suarez J, Alexander JK, Houston CS. Enhanced left ventricular systolic performance at high altitude during Operation Everest II. Am J Cardiol. 1987;60(1):137–42.
101. Webster KA, Bishopric NH. Molecular regulation of cardiac myocyte adaptations to chronic hypoxia. J Mol Cell Cardiol. 1992;24(7):741–51.

102. Mazer CD, Stanley WC, Hickey RF, Neese RA, Cason BA, Demas KA, et al. Myocardial metabolism during hypoxia: maintained lactate oxidation during increased glycolysis. Metabolism. 1990;39(9):913–8.
103. Arbab-Zadeh A, Levine BD, Trost JC, Lange RA, Keeley EC, Hillis LD, et al. The effect of acute hypoxemia on coronary arterial dimensions in patients with coronary artery disease. Cardiology. 2009;113(2):149–54.
104. Bunn HF, Poyton RO. Oxygen sensing and molecular adaptation to hypoxia. Physiol Rev. 1996;76(3):839–85.
105. Rassaf T, Flogel U, Drexhage C, Hendgen-Cotta U, Kelm M, Schrader J. Nitrite reductase function of deoxymyoglobin: oxygen sensor and regulator of cardiac energetics and function. Circ Res. 2007;100(12):1749–54.
106. Arany Z, Huang LE, Eckner R, Bhattacharya S, Jiang C, Goldberg MA, et al. An essential role for p300/CBP in the cellular response to hypoxia. Proc Natl Acad Sci USA. 1996;93(23):12969–73.
107. Tan T, Marin-Garcia J, Damle S, Weiss HR. Hypoxia-inducible factor-1 improves inotropic responses of cardiac myocytes in ageing heart without affecting mitochondrial activity. Exp Physiol. 2010;95(6):712–22.
108. Kacimi R, Long CS, Karliner JS. Chronic hypoxia modulates the interleukin-1beta-stimulated inducible nitric oxide synthase pathway in cardiac myocytes. Circulation. 1997;96(6):1937–43.
109. Lau S, Patnaik N, Sayen MR, Mestril R. Simultaneous overexpression of two stress proteins in rat cardiomyocytes and myogenic cells confers protection against ischemia-induced injury. Circulation. 1997;96(7):2287–94.
110. Zimmerman LH, Levine RA, Farber HW. Hypoxia induces a specific set of stress proteins in cultured endothelial cells. J Clin Invest. 1991;87(3):908–14.
111. Zaobornyj T, Valdez LB, Iglesias DE, Gasco M, Gonzales GF, Boveris A. Mitochondrial nitric oxide metabolism during rat heart adaptation to high altitude: effect of sildenafil, L-NAME, and L-arginine treatments. Am J Physiol. 2009;296(6):H1741–7.
112. Hirata K, Ban T, Jinnouchi Y, Kubo S. Echocardiographic assessment of left ventricular function and wall motion at high altitude in normal subjects. Am J Cardiol. 1991;68(17):1692–7.
113. Cargill RI, Kiely DG, Lipworth BJ. Adverse effects of hypoxaemia on diastolic filling in humans. Clin Sci (Lond). 1995;89(2):165–9.
114. Allemann Y, Rotter M, Hutter D, Lipp E, Sartori C, Scherrer U, et al. Impact of acute hypoxic pulmonary hypertension on LV diastolic function in healthy mountaineers at high altitude. Am J Physiol Heart Circ Physiol. 2004;286(3):H856–62.
115. Huez S, Retailleau K, Unger P, Pavelescu A, Vachiery JL, Derumeaux G, et al. Right and left ventricular adaptation to hypoxia: a tissue Doppler imaging study. Am J Physiol Heart Circ Physiol. 2005;289(4):H1391–8.
116. Prasad A, Popovic ZB, Arbab-Zadeh A, Fu Q, Palmer D, Dijk E, et al. The effects of aging and physical activity on Doppler measures of diastolic function. Am J Cardiol. 2007;99(12):1629–36.
117. Jung RC, Dill DB, Horton R, Horvath SM. Effects of age on plasma aldosterone levels and hemoconcentration at altitude. J Appl Physiol. 1971;31(4): 593–7.
118. Surks MI, Chinn KS, Matoush LR. Alterations in body composition in man after acute exposure to high altitude. J Appl Physiol. 1966;21(6):1741–6.
119. Grover RF, Reeves JT, Maher JT, McCullough RE, Cruz JC, Denniston JC, et al. Maintained stroke volume but impaired arterial oxygenation in man at high altitude with supplemental CO2. Circ Res. 1976;38(5):391–6.
120. Grover RF, Selland MA, McCullough RG, Dahms TE, Wolfel EE, Butterfield GE, et al. Beta-adrenergic blockade does not prevent polycythemia or decrease in plasma volume in men at 4300 m altitude. Eur J Appl Physiol Occup Physiol. 1998;77(3):264–70.
121. Ross Jr J, Linhart JW, Brauwald E. Effects of changing heart rate in man by electrical stimulation of the right atrium. Studies at rest, during exercise, and with isoproterenol. Circulation. 1965;32(4):549–58.
122. Naeije R. Physiological adaptation of the cardiovascular system to high altitude. Prog Cardiovas Dis. 2010;52(6):456–66.
123. Malconian M, Rock P, Hultgren H, Donner H, Cymerman A, Groves B, et al. The electrocardiogram at rest and exercise during a simulated ascent of Mt. Everest (Operation Everest II). Am J Cardiol. 1990;65(22):1475–80.
124. Das BK, Tewari SC, Parashar SK, Akhtar M, Grover DN, Ohri VC, et al. Electrocardiographic changes at high attitude. Indian Heart J. 1983;35(1):30–3.
125. Abel FL, Waldhausen JA. Effects of alterations in pulmonary vascular resistance on right ventricular function. J Thorac Cardiovasc Surg. 1967;54(6): 886–94.
126. Davila-Roman VG, Guest TM, Tuteur PG, Rowe WJ, Ladenson JH, Jaffe AS. Transient right but not left ventricular dysfunction after strenuous exercise at high altitude. J Am Coll Cardiol. 1997;30(2): 468–73.
127. Atherton JJ, Moore TD, Lele SS, Thomson HL, Galbraith AJ, Belenkie I, et al. Diastolic ventricular interaction in chronic heart failure. Lancet. 1997;349(9067):1720–4.
128. Bernheim AM, Kiencke S, Fischler M, Dorschner L, Debrunner J, Mairbaurl H, et al. Acute changes in pulmonary artery pressures due to exercise and exposure to high altitude do not cause left ventricular diastolic dysfunction. Chest. 2007;132(2):380–7.
129. Calbet JA, Radegran G, Boushel R, Sondergaard H, Saltin B, Wagner PD. Effect of blood haemoglobin concentration on V(O2, max) and cardiovascular function in lowlanders acclimatised to 5260 m. J Physiol. 2002;545(Pt 2):715–28.

130. Bender PR, Groves BM, McCullough RE, McCullough RG, Huang SY, Hamilton AJ, et al. Oxygen transport to exercising leg in chronic hypoxia. J Appl Physiol. 1988;65(6):2592–7.
131. Weil JV, Byrne-Quinn E, Battock DJ, Grover RF, Chidsey CA. Forearm circulation in man at high altitude. Clin Sci. 1971;40(3):235–46.
132. Cruz JC, Grover RF, Reeves JT, Maher JT, Cymerman A, Denniston JC. Sustained venoconstriction in man supplemented with CO2 at high altitude. J Appl Physiol. 1976;40(1):96–100.
133. Cerretelli P, Marconi C, Deriaz O, Giezendanner D. After effects of chronic hypoxia on cardiac output and muscle blood flow at rest and exercise. Eur J Appl Physiol Occup Physiol. 1984;53(2):92–6.
134. Hellems HK, Ord JW, Talmers FN, Christensen RC. Effects of hypoxia on coronary blood flow and myocardial metabolism in normal human subjects. Circulation. 1963;16:893–8.
135. Moret P, Covarrubias E, Coudert J, Duchosal F. Cardiocirculatory adaptation to chronic hypoxia: comparative study of coronary flow, myocardial oxygen consumption and efficiency between sea level and high altitude residents. Acta Cardiol. 1972;27(2):283–305.
136. Grover RF, Lufschanowski R, Alexander JK. Alterations in the coronary circulation of man following ascent to 3,100 m altitude. J Appl Physiol. 1976;41(6):832–8.
137. Klausen K. Man's acclimatization to altitude during the first week at 3.800 M. Schweiz Z Sportmed. 1966;14(1):246–53.
138. Sharma SC, Balasubramanian V, Mathew OP, Hoon RS. Serial studies of heart rate, blood pressure and urinary catecholamine excretion on acute induction to high altitude (3658m). Indian J Chest Dis Allied Sci. 1977;19(1):16–20.
139. Vogel JA, Hartley LH, Cruz JC. Cardiac output during exercise in altitude natives at sea level and high altitude. J Appl Physiol. 1974;36(2):173–6.
140. Zamudio S, Douglas M, Mazzeo RS, Wolfel EE, Young DA, Rock PB, et al. Women at altitude: forearm hemodynamics during acclimatization to 4,300 m with alpha(1)-adrenergic blockade. Am J Physiol Heart Circ Physiol. 2001;281(6):H2636–44.
141. Mazzeo RS, Carroll JD, Butterfield GE, Braun B, Rock PB, Wolfel EE, et al. Catecholamine responses to alpha-adrenergic blockade during exercise in women acutely exposed to altitude. J Appl Physiol. 2001;90(1):121–6.
142. Bilo G, Caldara G, Styczkiewicz K, Revera M, Lombardi C, Giglio A, et al. Effects of selective and nonselective beta-blockade on 24-h ambulatory blood pressure under hypobaric hypoxia at altitude. J Hypertens. 2011;29(2):380–7.
143. Pugh LG, Gill MB, Lahiri S, Milledge JS, Ward MP, West JB. Muscular exercise at great altitudes. J Appl Physiol. 1964;19:431–40.
144. Sharma SC, Hoon RS, Balasubramanian V, Chadha KS. Urinary catecholamine excretion in temporary residents of high altitude. J Appl Physiol. 1978;44(5):725–7.
145. Marticorena E, Ruiz L, Severino J, Galvez J, Penaloza D. Systemic blood pressure in white men born at sea level: changes after long residence at high altitudes. Am J Cardiol. 1969;23(3):364–8.
146. Lewis SF, Taylor WF, Graham RM, Pettinger WA, Schutte JE, Blomqvist CG. Cardiovascular responses to exercise as functions of absolute and relative work load. J Appl Physiol. 1983;54(5):1314–23.
147. Vogel JA, Hartley LH, Cruz JC, Hogan RP. Cardiac output during exercise in sea-level residents at sea level and high altitude. J Appl Physiol. 1974;36(2):169–72.
148. Ferretti G, Boutellier U, Pendergast DR, Moia C, Minetti AE, Howald H, et al. Oxygen transport system before and after exposure to chronic hypoxia. Int J Sports Med. 1990;11 Suppl 1:S15–20.
149. Steinacker JM, Liu Y, Boning D, Halder A, Maassen N, Thomas A, et al. Lung diffusion capacity, oxygen uptake, cardiac output and oxygen transport during exercise before and after an himalayan expedition. Eur J Appl Physiol Occup Physiol. 1996;74(1–2): 187–93.
150. Moore LG, Cymerman A, Huang SY, McCullough RE, McCullough RG, Rock PB, et al. Propranolol does not impair exercise oxygen uptake in normal men at high altitude. J Appl Physiol. 1986; 61(5):1935–41.
151. Grover RF, Weil JV, Reeves JT. Cardiovascular adaptation to exercise at high altitude. Exerc Sport Sci Rev. 1986;14:269–302.
152. Casey DP, Joyner MJ. Local control of skeletal muscle blood flow during exercise: influence of available oxygen. J Appl Physiol. 2011;111(6):1527–38.
153. Cerretelli P. Limiting factors to oxygen transport on Mount Everest. J Appl Physiol. 1976;40(5):658–67.
154. West JB, Boyer SJ, Graber DJ, Hackett PH, Maret KH, Milledge JS, et al. Maximal exercise at extreme altitudes on Mount Everest. J Appl Physiol. 1983;55(3):688–98.
155. Wagner PD. Why does the maximal exercise cardiac output fall during altitude residence at extreme altitudes and is it important? Proceedings of the 11th International Hypoxia Symposium, Jasper, Alberta; 1999.
156. Saltin B, Grover RF, Blomqvist CG, Hartley LH, JOhnson RL. Maximal oxygen uptake and cardiac output after 2 weeks at 4,300 m. J Appl Physiol. 1968;25:400–9.
157. Hartley LH, Vogel JA, Cruz JC. Reduction of maximal exercise heart rate at altitude and its reversal with atropine. J Appl Physiol. 1974;36(3):362–5.
158. Zhuang J, Droma T, Sutton JR, McCullough RE, McCullough RG, Groves BM, et al. Autonomic regulation of heart rate response to exercise in Tibetan and Han residents of Lhasa (3,658 m). J Appl Physiol. 1993;75(5):1968–73.
159. Garcia JA, McMinn SB, Zuckerman JH, Fixler DE, Levine BD. The role of the right ventricle during

hypobaric hypoxic exercise: insights from patients after the Fontan operation. Med Sci Sports Exerc. 1999;31(2):269–76.
160. Young AJ, Sawka MN, Muza SR, Boushel R, Lyons T, Rock PB, et al. Effects of erythrocyte infusion on VO2max at high altitude. J Appl Physiol. 1996;81(1):252–9.
161. Robach P, Calbet JA, Thomsen JJ, Boushel R, Mollard P, Rasmussen P, et al. The ergogenic effect of recombinant human erythropoietin on VO2max depends on the severity of arterial hypoxemia. PLoS One. 2008;3(8):e2996.
162. Sime F, Penaloza D, Ruiz L. Bradycardia, increased cardiac output, and reversal of pulmonary hypertension in altitude natives living at sea level. Br Heart J. 1971;33(5):647–57.
163. Passino C, Bernardi L, Spadacini G, Calciati A, Robergs R, Anand I, et al. Autonomic regulation of heart rate and peripheral circulation: comparison of high altitude and sea level residents. Clin Sci (Lond). 1996;91(Suppl):81–3.
164. Hochachka PW, Clark CM, Holden JE, Stanley C, Ugurbil K, Menon RS. 31P magnetic resonance spectroscopy of the Sherpa heart: a phosphocreatine/adenosine triphosphate signature of metabolic defense against hypobaric hypoxia. Proc Natl Acad Sci USA. 1996;93(3):1215–20.
165. Holden JE, Stone CK, Clark CM, Brown WD, Nickles RJ, Stanley C, et al. Enhanced cardiac metabolism of plasma glucose in high-altitude natives: adaptation against chronic hypoxia. J Appl Physiol. 1995;79(1):222–8.
166. Moret P. Cotronary blood flow and myocardial metabolism in man at hihg altitude. In: Porter R, Knight J, editors. High altitude physiology: cardiac and respiratory aspects. Edinburgh: Churchill Livingston; 1971. p. 131–48.
167. Arias-Stella J, Topilsky M. Anatomy of the coronary circulation at high al;titude. In: Porter R, Knight J, editors. High altitude physiology: cardiac and respiratory aspects. Edinburgh: Churchill Livingstone; 1971. p. 149–57.
168. Fuster V, Frye RL, Kennedy MA, Connolly DC, Mankin HT. The role of collateral circulation in the various coronary syndromes. Circulation. 1979;59(6):1137–44.
169. Mortimer Jr EA, Monson RR, MacMahon B. Reduction in mortality from coronary heart disease in men residing at high altitude. N Engl J Med. 1977;296(11):581–5.
170. Beall CM, Brittenham GM, Strohl KP, Blangero J, Williams-Blangero S, Goldstein MC, et al. Hemoglobin concentration of high-altitude Tibetans and Bolivian Aymara. Am J Phys Anthropol. 1998;106(3):385–400.
171. Beall CM, Strohl KP, Blangero J, Williams-Blangero S, Almasy LA, Decker MJ, et al. Ventilation and hypoxic ventilatory response of Tibetan and Aymara high altitude natives. Am J Phys Anthropol. 1997;104(4):427–47.
172. Winslow RM, Chapman KW, Gibson CC, Samaja M, Monge CC, Goldwasser E, et al. Different hematologic responses to hypoxia in Sherpas and Quechua Indians. J Appl Physiol. 1989;66(4):1561–9.
173. Hoit BD, Dalton ND, Erzurum SC, Laskowski D, Strohl KP, Beall CM. Nitric oxide and cardiopulmonary hemodynamics in Tibetan highlanders. J Appl Physiol. 2005;99(5):1796–801.
174. Antezana AM, Antezana G, Aparicio O, Noriega I, Velarde FL, Richalet JP. Pulmonary hypertension in high-altitude chronic hypoxia: response to nifedipine. Eur Respir J. 1998;12(5):1181–5.
175. Beall CM. Two routes to functional adaptation: Tibetan and Andean high-altitude natives. Proc Natl Acad Sci USA. 2007;104 Suppl 1:8655–60.
176. Ludmer PL, Selwyn AP, Shook TL, Wayne RR, Mudge GH, Alexander RW, et al. Paradoxical vasoconstriction induced by acetylcholine in atherosclerotic coronary arteries. N Engl J Med. 1986;315(17):1046–51.
177. Vita JA, Treasure CB, Yeung AC, Vekshtein VI, Fantasia GM, Fish RD, et al. Patients with evidence of coronary endothelial dysfunction as assessed by acetylcholine infusion demonstrate marked increase in sensitivity to constrictor effects of catecholamines. Circulation. 1992;85(4):1390–7.
178. Fuster V, Lewis A. Conner memorial lecture. Mechanisms leading to myocardial infarction: insights from studies of vascular biology. Circulation. 1994;90(4):2126–46.
179. Bartsch P, Waber U, Haeberli A, Maggiorini M, Kriemler S, Oelz O, et al. Enhanced fibrin formation in high-altitude pulmonary edema. J Appl Physiol. 1987;63(2):752–7.
180. Schobersberger W, Schmid P, Lechleitner M, von Duvillard SP, Hortnagl H, Gunga HC, et al. Austrian Moderate Altitude Study 2000 (AMAS 2000). The effects of moderate altitude (1,700 m) on cardiovascular and metabolic variables in patients with metabolic syndrome. Eur J Appl Physiol. 2003;88(6):506–14.
181. Burke AP, Farb A, Malcom GT, Liang Y, Smialek JE, Virmani R. Plaque rupture and sudden death related to exertion in men with coronary artery disease. JAMA. 1999;281(10):921–6.
182. Faeh D, Gutzwiller F, Bopp M. Lower mortality from coronary heart disease and stroke at higher altitudes in Switzerland. Circulation. 2009;120(6):495–501.
183. Morgan BJ, Alexander JK, Nicoli SA, Brammell HL. The patient with coronary disease at altitude: observations during acute exposure to 3,100 m. J Wilderness Med. 1990;1:147–53.
184. Gibelli G, Fantoni C, Anza C, Cattaneo P, Rossi A, Montenero AS, et al. Arrhythmic risk evaluation during exercise at high altitude in healthy subjects: role of microvolt T-wave alternans. Pacing Clin Electrophysiol. 2008;31(10):1277–83.
185. Woods DR, Allen S, Betts TR, Gardiner D, Montgomery H, Morgan JM, et al. High altitude arrhythmias. Cardiology. 2008;111(4):239–46.

186. Erdmann J, Sun KT, Masar P, Niederhauser H. Effects of exposure to altitude on men with coronary artery disease and impaired left ventricular function. Am J Cardiol. 1998;81(3):266–70.
187. Van Patot MC, Hill AE, Dingmann C, Gaul L, Fralick K, Christians U, et al. Risk of impaired coagulation in warfarin patients ascending to altitude (>2400 m). High Alt Med Biol. 2006;7(1):39–46.
188. Alexander JK, Abinader EG, Sharif DS, Morgan BJ, Brammell HL. Left ventricular function in coronary heart disease at high altitude. Circulation. 1988;78(Suppl II):2–6.
189. Agostoni P, Cattadori G, Guazzi M, Bussotti M, Conca C, Lomanto M, et al. Effects of simulated altitude-induced hypoxia on exercise capacity in patients with chronic heart failure. Am J Med. 2000;109(6):450–5.
190. Burtscher M, Philadelphy M, Likar R. Sudden cardiac death during mountain hiking and downhill skiing. N Engl J Med. 1993;329(23):1738–9.
191. Ponchia A, Biasin R, Tempesta T, Thiene M, Volta SD. Cardiovascular risk during physical activity in the mountains. J Cardiovasc Med (Hagerstown). 2006;7(2):129–35.
192. Burtscher M, Pachinger O, Schocke MF, Ulmer H. Risk factor profile for sudden cardiac death during mountain hiking. Int J Sports Med. 2007;28(7):621–4.

Cerebral Circulation and Brain

7

Philip N. Ainslie, Mark H. Wilson, and Christopher H.E. Imray

Abstract

It should be noted that Dr. J.W. Severinghaus was the expert author of this chapter (Cerebral circulation at high altitude) in the First Edition of this textbook. Moreover, in updating this area, we also amalgamate from the First Edition the detailed chapter titled "The High-Altitude Brain," which was authored by Drs. M.S. Raichle and T.F. Hornbein. Readers are directed to these original and elegant comprehensive reviews (Severinghaus: High altitude; exploration of human adaptation, New York, Basel; Raichle: High altitude; exploration of human adaptation, New York, Basel). Rather than reproduce this information here, and while summarizing some of this original material in the context of new findings within the last decade, we provide an update within the broad topic of the cerebral circulation and brain at high altitude. This chapter is comprised of seven sections. The introduction is followed by the major factors which regulate cerebral blood flow (CBF) are initially presented in order to emphasize the integrated mechanisms by which CBF is controlled. Next, detailed discussion is provided to examine the influence of exposure to high-altitude exposure on these aforementioned mechanisms which regulate CBF. We then briefly review recent advances in the understanding of neurological clinical syndromes that occur on exposure to high altitudes. The next two sections summarize the influence of high altitude on cognitive function and highlight other neurological-based symptoms and events that can occur upon exposure to high altitude. Finally, we suggest avenues for future research.

P.N. Ainslie, Ph.D. (✉)
Centre for Heart, Lung and Vascular Health, School of Health and Exercise Sciences, University of British Columbia, Okanagan Campus, 3333 University Way, Kelowna, BC, Canada
e-mail: philip.ainslie@ubc.ca

M.H. Wilson, B.Sc., MB, BChir F.R.C.S., F.I.M.C., F.R.G.S., M.R.C.A.
Department of Neurosurgery, Imperial Hospitals NHS Trust, London, UK
e-mail: acutebrain@googlemail.com

C.H.E. Imray, Ph.D., F.R.C.S., F.R.C.P., M.B.B.S. (✉)
Department of Vascular Surgery, Warwick Medical School, University Hospitals Coventry and Warwickshire NHS Trust, Coventry, West Midlands, UK
e-mail: chrisimray@aim.com; chrisimray@aol.com

E.R. Swenson and P. Bärtsch (eds.), *High Altitude: Human Adaptation to Hypoxia*,
DOI 10.1007/978-1-4614-8772-2_7, © Springer Science+Business Media New York 2014

Regulation of Cerebral Blood Flow

The oxygen supply to the brain depends on the arterial oxygen content and the CBF. The rate of oxygen consumption of the entire brain accounts for ~20 % of the resting total body oxygen consumption. The brain is the most oxygen-dependent organ in the body and many pathophysiological processes either cause or result in an interruption to its oxygen supply. For example, loss of consciousness occurs within minutes of acute exposure to elevations greater than 7,000 m. Remarkably, however, following acclimatization to high-altitude consciousness can be maintained—by some—during "oxygenless" ascents of Mt Everest (which stands at ~8,850 m). The mechanisms underlying these responses to high altitude upon CBF are complex and involve interactions of many physiological, metabolic and biochemical processes. The major factors underlying the extent of change in CBF during high-altitude-induced reductions in the partial pressure of arterial oxygen (PaO_2) are depicted in Fig. 7.1 and are further detailed in this chapter.

Background

The relative constancy of the overall craniospinal volume [1], the cerebrovascular anatomy, and the positional and delay differences of the cerebral circulation relative to the heart make the intracranial hemodynamics complex [1–5] (Fig. 7.1). Control of cerebral perfusion is linked closely to status of the intracranial volume; it comprises the arterial cerebrovascular bed, the large cerebral veins, and the processes associated with production and reabsorption of cerebral spinal fluid (CSF) [6]. According to Poiseuille's law [7], CBF is determined by the cerebral perfusion pressure and the cerebrovascular conductance, or its reciprocal, cerebrovascular resistance [8]. The cerebral perfusion pressure is the difference

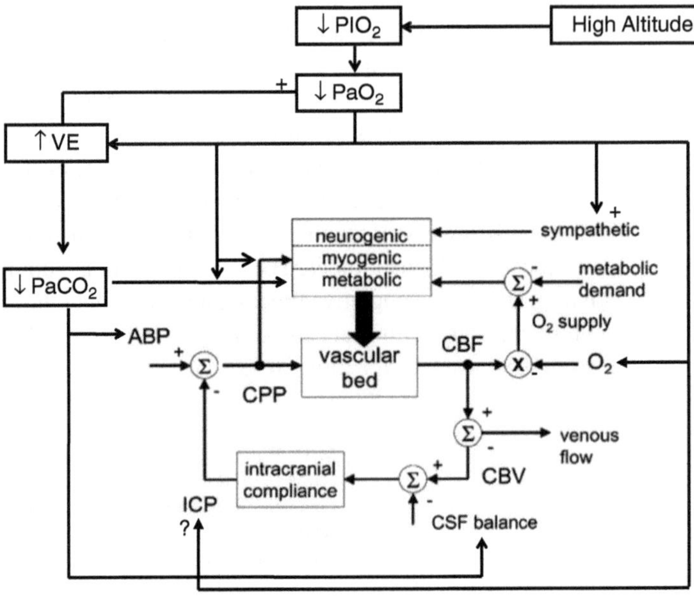

Fig. 7.1 Simplified block diagram of the main mechanisms responsible for the control of cerebral blood flow (CBF). The vascular bed corresponds to the passive properties of blood vessels, usually modeled as a combination of resistances and compliances. Modified from [3, 102]. *ABP* arterial blood pressure, *ICP* intracranial pressure, *CSF* cerebral spinal fluid, *CPP* cerebral perfusion pressure, *CBV* cerebral blood volume

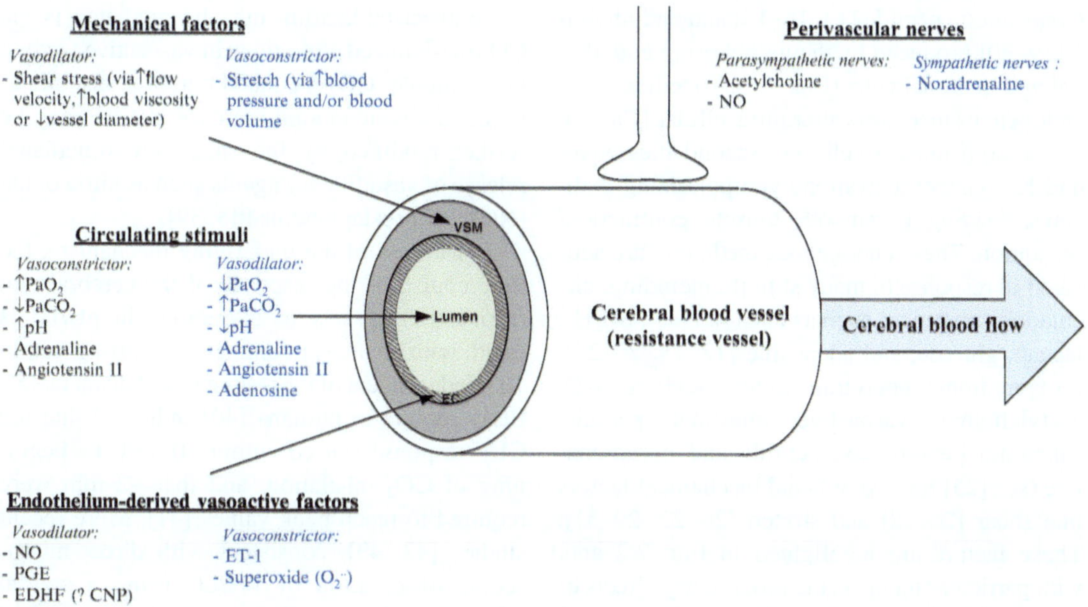

Fig. 7.2 The main factors influencing the diameter of cerebral resistance vessels. *VSM* vascular smooth muscle, *EC* endothelial cells, *NO* nitric oxide, *PaO_2* partial pressure of oxygen, *$PaCO_2$* partial pressure of carbon dioxide, *PGE* prostaglandins, *ET-1* endothelin-1, *EDHF* endothelial-derived hyperpolarizing factor, *CNP* C-natriuretic peptide. Modified from [252]. Note that angiotensin and adrenaline can cause vasoconstriction or vasodilation depending on which receptor they bind to (e.g., angiotensin-1 receptors (vasoconstriction) and angiotension-2 receptors (vasodilation); for adrenaline receptors, vasoconstriction is via the alpha receptors and vasodilation is via the beta receptors)

between blood pressure at the level of the circle of Willis and intracranial pressure; and intracranial pressure, in turn, encompasses central venous pressure and the pressures within the cerebral spinal fluid (Fig. 7.1). The resistance variable to flow has historically been suggested to occur mostly in the cerebral arteriolar and capillary bed [5, 9]. In turn, the large cerebral arteries and veins are believed to be noncompliant and act merely as a conduit for the pulsatile arterial flow from the aorta to the brain [10]. However, recently it has been reported in humans that cerebrovascular resistance is not solely modulated at the level of the arteriolar pial vessels but occurs also in the larger cerebral arteries (e.g., at the level of the internal carotid arteries) [11]. Nevertheless, CBF is dynamically adjusted to changes in the perfusion pressure, the metabolic activity of the brain, arterial oxygen content, acid–base status, humoral factors, and autonomic nerve activity [12]. Figure 7.1 is a simplified diagram of these main regulatory or active mechanisms that interact to maintain oxygen supply matched to cerebral metabolic demand and to protect the brain from changes in cerebral perfusion. This integrative model highlights the interdependence of CBF to many other variables through complex and likely nonlinear relationships. The highlighted areas represent the potential areas—both direct and indirect—in which reductions in PaO_2 may influence CBF (see Sect. 7.3.0).

At the vascular level, CBF is regulated by smooth muscle tone, which is under the influence of neural, chemical, metabolic, and physical factors (Figs. 7.1 and 7.2). Ultimately, it is the net effect of vasodilators and vasoconstrictors derived from the endothelium, neuronal innervations, perfusion pressure, and shear stress that determines CBF (Fig. 7.2). In the brain, vascular smooth muscle tone is closely controlled by endogenous substances such as nitric oxide (NO) [13] and prostanoids [14, 15], as well as

C-natriuretic peptide [13, 16–18] and endothelin-1 [19, 20] produced by the neighboring endothelial and neuronal cells (Fig. 7.2). Together, these endogenous mediators alter intracellular [Ca^{2+}] in the smooth muscle cells via second messenger and K^+ channel activation/hyperpolarizing pathways, leading to smooth muscle contraction/relaxation. These endogenous mediators are activated in response to many stimuli, including: circulating substances such as $PaCO_2$, PaO_2, pH/H^+, lactate, glucose, and adenosine [13, 14, 20–24]; postganglionic neurotransmitters such as NO, acetylcholine, vasoactive intestinal peptide, calcitonin gene-related peptide and noradrenaline (see [25] for review), and mechanical factors like shear [26–28] and stretch [20–22, 29–31]. These factors are highlighted in Fig. 7.2 and, with particular focus on the modulatory effects of hypoxia, are described in this chapter.

Arterial Blood Gases

Partial Pressure of Arterial Carbon Dioxide ($PaCO_2$): It is well established that the cerebral vasculature is profoundly affected by $PaCO_2$ [32]. The teleological relevance of the exquisite sensitivity of CBF to changes in $PaCO_2$ seems to be a vital homeostatic function that helps regulate and maintain central pH [33], which therefore affects the respiratory central chemoreceptor stimulus. The interaction between $PaCO_2$ and vasodilation/vasoconstriction are normally ascribed to occur at the level of the arterioles and the precapillary sphincters [5, 12, 34]. Increased CO_2 results in a relaxation of the vascular smooth muscle of all cerebral vessels, although the small vessels are the most responsive [35]. By contrast, the vasoconstrictor effect of hypocapnia is unaffected by vessel size [35]. The mechanism(s) behind these actions of CO_2 has not been entirely elucidated. Some evidence indicates that elevations in CO_2 and the concomitant change in pH activate potassium channels in the vascular smooth muscle [36, 37]. Thus, K^+ channels, through impact on the endothelium and vascular smooth muscle, play a role in coordinating vascular tone in upstream and downstream vessels. An alternative, or possible complimentary means by which cerebral vasodilatation may be mediated is via CO_2/pH-induced alterations in vasoactive factors. For example, CO_2-mediated cerebral vasodilatation and a concomitant increase in CBF may be further modified by the shear stress-mediated release of vasodilatory agents such as nitric oxide (NO) and prostaglandins [38, 39].

Regardless of the underlying mechanism, the time course of the response of the cerebral pial (resistance) vessels to alterations in pH/H^+ is rapid, with changes in diameter occurring within 10 s, independent of the resting vessel diameter [5]. Early reports in humans [40] indicated that the CBF response started within 30 s of the beginning of CO_2 inhalation, and that ~2 min were required to reach peak values [41]. More recent studies [42, 43], consistent with direct microscopic observation of vessels using a cranial window technique, incorporating transcranial Doppler (TCD) and more sophisticated methods to manipulate end-tidal gases, have demonstrated that the acute CBF response to step changes in CO_2 (including hypocapnia) in humans was much faster (delay of 6 s) than these previous reports.

Partial Pressure of Arterial Oxygen (PaO_2): The role of PaO_2 in the day-to-day regulation of CBF is minor, reflected in the findings that, depending on the prevailing $PaCO_2$ [44], a drop in PaO_2 below a certain threshold (<45 mmHg) is required before cerebral vasodilatation occurs [45]. However, while hypoxia per se is a cerebral vasodilator, reflected in a rise in CBF in proportion to the severity of isocapnic hypoxia [46, 47], under normal conditions the hypoxic-induced activation of peripheral chemoreceptor activity leads to hyperventilatory-induced lowering of $PaCO_2$ and subsequent cerebral vasoconstriction (Fig. 7.3). Therefore, the cerebrovascular bed receives conflicting signals during exposure to acute hypoxia. Recently it has been shown that, at least with severe acute isocapnic hypoxia, there is a disparate blood flow regulation to the brainstem and cortex such that blood flow to the brainstem is increased by a greater extent than that to the cortex [11]. The role of PaO_2 becomes critically important upon ascent to high altitude (see Sect. 7.3.0) and possibility during hypoxemia associated with chronic lung disease [48].

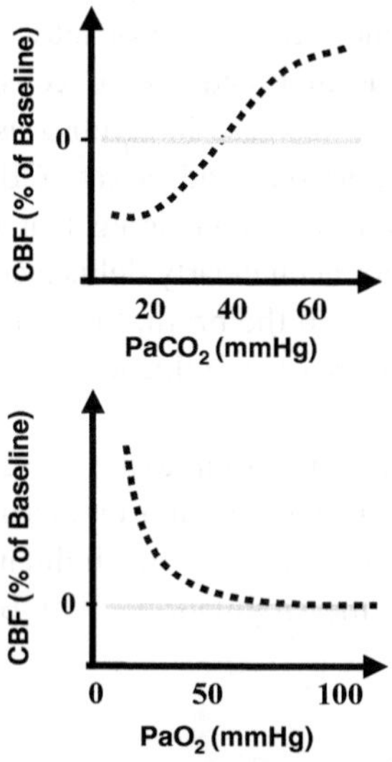

Fig. 7.3 Schematic of the influence of arterial PO_2 (PaO_2) and arterial PCO_2 ($PaCO_2$) on CBF. Upon ascent to high altitude, the fall in PaO_2 tends to cause vasodilatation especially at levels <45 mmHg; however, the drop in PaO_2 stimulates the peripherial chemoreceptor and a subsequent reflex hyperventilation. The reflex hyperventilation induces hypocapnia which subsequently causes cerebral vasoconstriction. The balance between the degree of hypoxia and hypocapnia, mediated by changes in ventilatory control, are critical determinants of CBF at high altitude. See Fig. 7.6 for further details

Cerebral Autoregulation

Cerebral autoregulation adjusts cerebral arteriolar caliber, or cerebrovascular resistance, to ensure that CBF levels are matched to metabolic needs, and it comprises two main components: static and dynamic. Although not perfect [49], static cerebral autoregulation keeps CBF relatively constant, over gradual and progressive changes in cerebral perfusion [50]. Dynamic cerebral autoregulation refers to the rapid regulation of cerebral blood flow (CBF) in response to changes in arterial blood pressure that occur in a few seconds [51]. Dynamic cerebral autoregulation may have different control mechanisms than static cerebral autoregulation [52], and data indicate that neural control of cerebral circulation may be more effective under dynamic than under steady-state conditions [51]. Although the actual range of blood pressure in which CBF is maintained is variable between subjects [53] and was originally based on steady-state CBF data points under several different conditions presented in previous publications [53], there have been a number of reports of the influence of hypoxia on cerebral autoregulation (see Sect. 7.3.2).

Sympathetic Nerve Activity

Although the cerebral circulation is richly innervated with sympathetic nerve fibers, the effect of sympathetic nerve activity on the regulation of CBF remains controversial [54]. The traditional view is that increases in sympathetic activity appear to have a limited effect on the cerebral vasculature and CBF of humans, particularly at rest [55, 56]. It seems likely that any potential influence of sympathetic nerve activity on regulating CBF is masked by the other more powerful regulatory influences on CBF—autoregulation, cerebrovascular CO_2 reactivity and, potentially, cardiac output [54, 57, 58]. Recently, however, elegant animal studies incorporating continuous recording of sympathetic nerve activity in the superior cervical ganglion [59] conclude that sympathetic nerve activity directed to cerebral vessels increases with acute hypertension, but not with hypotension, suggesting that it serves a protective function for the cerebral microcirculation, and not a regulatory role for maintenance of systemic arterial pressure. In addition to the influence of sympathetic nerve activity on cerebral autoregulation [51, 60], there are reports that sympathetic activity influences the reactivity of CBF to $PaCO_2$ [61, 62]. Thus, although the influence might be regarded as minor, hypoxic-induced elevations in sympathetic nerve activity have the potential to affect CBF by both direct and indirect (e.g., via systemically mediated changes in cardiac output being redistributed to the brain) mechanisms (Figs. 7.1 and 7.2: see Sect. 7.3.2).

Cerebral Metabolism

Cerebral oxygen consumption in normal, conscious, young humans is approximately 3.5 mL/100 g brain/min [63]. The rate of oxygen consumption of the entire brain of average weight (1,400 g) is therefore about 50 mL O_2/min. The magnitude of this rate can be appreciated more fully when it is compared with the average metabolic rate of the whole body—approximately 250 mL O_2/min in the basal state for a 70 kg male. Therefore, the brain, which represents only about 2 % of total body weight, accounts for ~20 % of the resting total body oxygen consumption.

Systemic Factors

Although the classical notion is that CBF is maintained over a range of blood pressure, it has more recently been established that CBF is also dependent on cardiac output ($\dot{Q}$) ([30, 64, 65], see [66] for review). Thus, because sympathetic nerve activity and heart rate are altered during O_2 and CO_2 reactivity testing, subsequent changes in $\dot{Q}$ may theoretically influence CBF. It should be acknowledged, however, that it is not clear whether these changes in $\dot{Q}$ would have direct effects on CBF independent of changes in cerebral perfusion pressure. While this idea is supported indirectly, it still remains counterintuitive because according to Poiseuille's law [7], blood flow is the quotient of perfusion pressure and vascular resistance and changes in $\dot{Q}$ are not variables of the equation [67]. Previous studies, however, have reported a positive relationship between $\dot{Q}$ and CBF at normothermia during rest and exercise [30, 68]. For example, increases or decreases in $\dot{Q}$ with volume expansion or application of lower body negative pressure altered CBF velocity respectively in a linear fashion when mean blood pressure and $PaCO_2$ were kept constant. Moreover, observations in patients with acute ischemic stroke and severe head injury showed that CBF in the affected brain regions correlated closely with changes in $\dot{Q}$ during intravascular volume expansion alone or in combination with induced hypertension [67]. Although the proposed mechanism(s) by which $\dot{Q}$ affects CBF are unknown, nonneural, flow-mediated mechanisms have been suggested [69]. For instance, sheer-stress related to increasing pulsatile pressure and/or blood flow may modulate the steady-state pressure-flow relationship of the cerebral circulation [70]. This may occur via the sheer-stress induced release of nitric oxide from the cerebrovascular endothelial cells, which in turn result in a decrease in cerebrovascular resistance in the arterioles and elevate CBF [67, 70]. Together, while acknowledging that the mechanisms have not been clearly delineated, assuming a redistribution to the brain, these findings suggest the existence of an influence of changes in $\dot{Q}$ on CBF.

In summary, the complexities of CBF regulation have been updated and emphasized. In the context of these factors, the influence of high altitude on the regulation of CBF is now considered.

Regulation of Cerebral Blood Flow at High Altitude

Alteration in Cerebral Perfusion at Altitude

Oxygen delivery to the brain is primarily determined by the oxygen carrying capacity of the blood and the CBF. Measurement of CBF can be achieved using a number of different techniques, each having certain advantages and also potential methodological flaws. While some techniques are better suited to use in the field such as TCD ultrasound and near infrared spectroscopy (NIRS), others give more detailed information but are restricted to use in a hospital or research institute such as magnetic resonance imaging. Our insights into cerebral perfusion are consequently determined by where, how, and when during the ascent to altitude the measurements have been made. It is important to note that cerebrovascular response to hypoxia can be broadly grouped into two time domains: (1) acute hypoxic response (seconds to hours); (2) short to long-term hypoxia (days to years), i.e., during acclimatization. Since the majority of the adaptive changes in CBF occur within 3–4 weeks upon exposure to a constant level of hypoxia, we will focus on the latter time domain of this response.

Time-dependent changes in CBF are clearly evident during acclimatization to high altitude (HA). Severinghaus et al. [71] used the Kety–Schmidt nitrous oxide washout method [72, 73] to analyze changes in CBF; they found a 24 % increase at 6–12 h and a 13 % increase at 3–5 days after arrival at 3,810 m. Jensen et al. [74] used radiolabeled xenon in 12 individuals who ascended from 150 m to 3,475 m, and found an increase in CBF of 24 %. Ascent from 3,200 m to between 4,785 and 5,430 m caused a further increase in CBF to 53 % above the readings at sea level. Using a variety of invasive and noninvasive methods, subsequent studies, in humans [75–82] and animals [83, 84] have confirmed and added to these initial observations by Severinghaus et al. [71] and Jensen et al. [74]. In essence, these studies collectively show that CBF reaches a peak approximately 2–3 days after arrival at HA and then returns to sea level values within 1–3 weeks (see Fig. 7.3). The subsequent onset of ventilatory acclimatization, coincides with the fall CBF due to an increase in PaO_2 and decrease in $PaCO_2$ due to hyperventilation [71, 75–77]. Therefore, over this time period at a constant altitude, the influence of the PaO_2-induced threshold for cerebral vasodilatation (<40–45 mmHg) is removed and the degree of hypocapnia is accentuated; both of these factors, along with elevations in hematocrit act to attenuate the initial rise in CBF. The time course of these ventilatory-induced changes in arterial blood gases, CBF velocity, blood pressure, and hematocrit are illustrated in Fig. 7.4, and other factors acting to determine the CBF response to high-altitude exposure are further considered. In a recent meta-analysis [85], it has been reported that CBF in Himalayan residents was slightly elevated compared with sea level subjects, and was 24 % higher compared with Andeans. After correction for hematocrit and arterial oxygen saturation, CBF remained ~20 % higher in Himalayans compared with Andeans. Although the need for experimental evidence is needed, an altered brain metabolism in Andeans, and/or increased nitric oxide

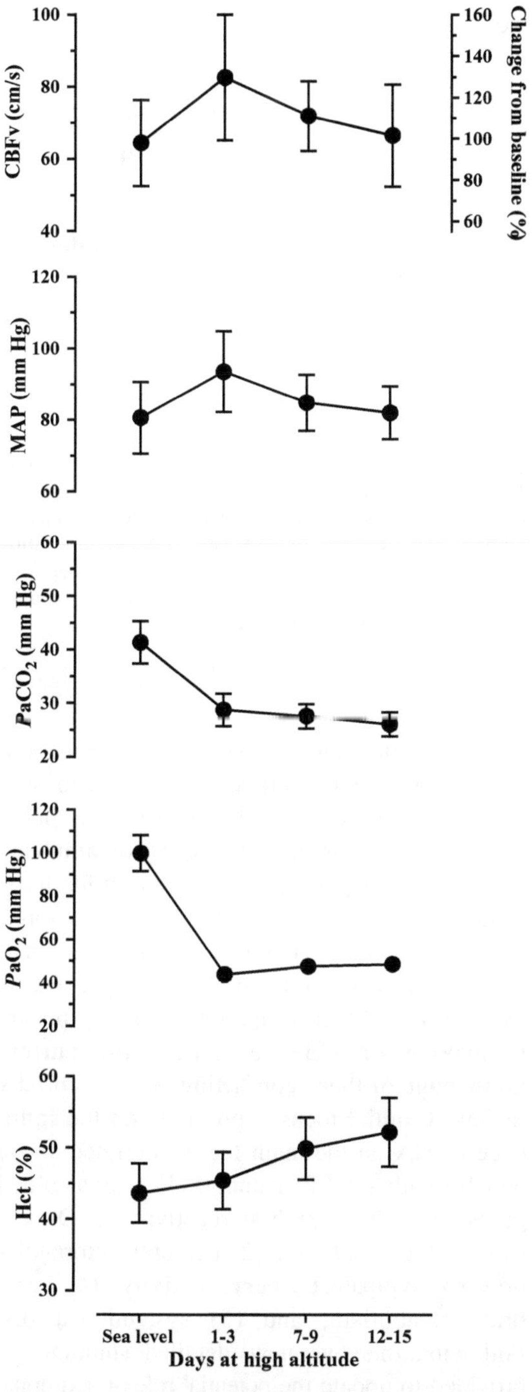

Fig. 7.4 Illustration of the time course in changes in CBF velocity in the middle cerebral artery (MCAv), mean arterial blood pressure (MAP), partial pressure of arterial PO_2 and PCO_2 (PaO_2 and $PaCO_2$) and hematocrit (HCT) at sea level and during various time points throughout a 15-day stay at 5,050 m (*Left-hand*). Data are mean ± SD; n = 14–18. Modified from [93]. Note: (1) ~25 % elevation in CBFv upon initial arrival at 5,050 m, coinciding with elevations in MAP, the greatest fall in PaO_2 and highest point of $PaCO_2$ before ventilatory adjustments occur; (2) progressive elevation in HCT during a 2-week stay at 5,050 m. See text for details of these multifactorial influences that likely determine the extent of CBF change

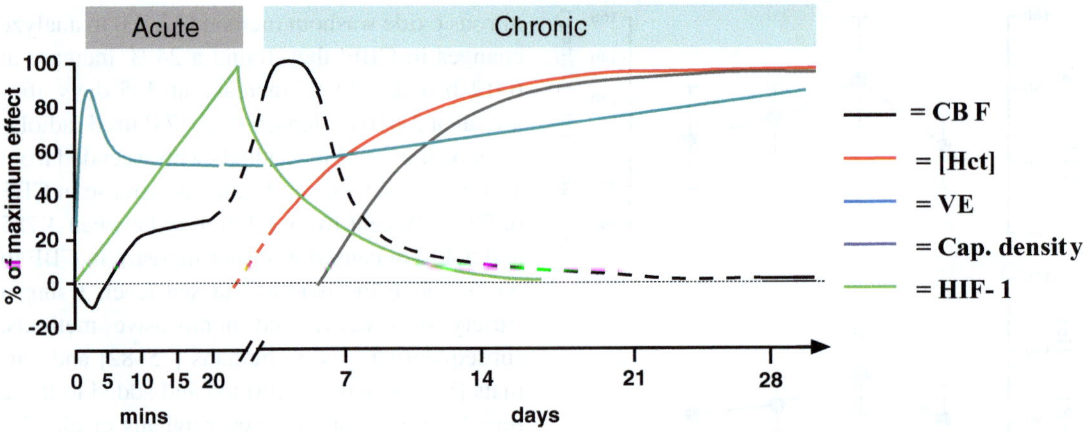

Fig. 7.5 Integrative changes in CBF, cerebral capillary density, hematocrit, and ventilation during poikilocapnic hypoxic exposure. During acute exposure to poikilocapnic hypoxia, the hypoxic-induced hyperventilation and subsequent hypocapnia causes cerebral vasoconstriction, thereby reducing CBF. Ventilatory decline associated with acute hypoxic exposure coincided with an increase in CBF. During chronic hypoxic exposure, the increase in CBF peaks after 1–2 days followed by a slow and progressive decline towards sea level baseline. In addition, this decline coincided with a steady increase in basal ventilation and elevation in cerebral capillary density from ~day 4. Furthermore, hematocrit concentration steadily increases with prolonged hypoxic exposure (Fig. 7.3). Modified from [139]

availability in Himalayans may have a role to explain this difference in CBF [85].

Regulation of CBF at High Altitude

The initial factors that determine the magnitude of change in CBF depend upon the relative strengths of four reflex mechanisms [86]: (1) hypoxic ventilatory response; (2) hypercapnic ventilatory response; (3) hypoxic cerebral vasodilatation, and (4) hypocapnic cerebral vasoconstriction. In addition to these reflex responses, which most likely adjust during the acclimatization process, CBF is also influenced by a myriad of other hypoxic-induced changes (e.g., capillary density (angiogenesis), adenosine, viscosity/haemtocrit, hypoxia-inducible factor, vascular endothelial growth factor (VEGF), free radicals, etc.); these factors are illustrated in Fig. 7.4, and the relative time course of some of these major events are depicted in Fig. 7.5. Following the first few days at HA, hematocrit begins to rise and, at least in animal studies, reaches ~80 % of maximum by 7 days (Xu et al. [84]). It should be noted, however, that in humans a similar time course of elevation in hematocrit does not occur until at least 2 weeks of exposure (Fig. 7.4). Differences in the experimental model (e.g., animal vs. human) and degree of simulated hypoxia may explain these apparent differences. Initial elevations in hypoxia-inducible factor-1, indicative of tissue hypoxia, fall to about half after 4 days and return to baseline by 3 weeks [87]. In the context of the influence of exposure to hypoxia on CBF regulation, the current knowledge of these conflicting factors is briefly reviewed. Initial focus is provided on the influence of HA on the main factors outlined above which regulate CBF; namely: (1) arterial blood gases (including cerebral reactivity to O_2 and CO_2 at high altitude); (2) cerebral autoregulation; (3) sympathetic nerve activity; (4) cerebral metabolism, and (5) systemic factors. Following this overview, detailed summary is provided to update the potential role of additional neuronal, hematological, and local factors which also act to regulate CBF at high altitude.

Respiratory factors: The relative importance of oxygen and carbon dioxide at altitude has been debated since the time of Bert [88] and Mosso

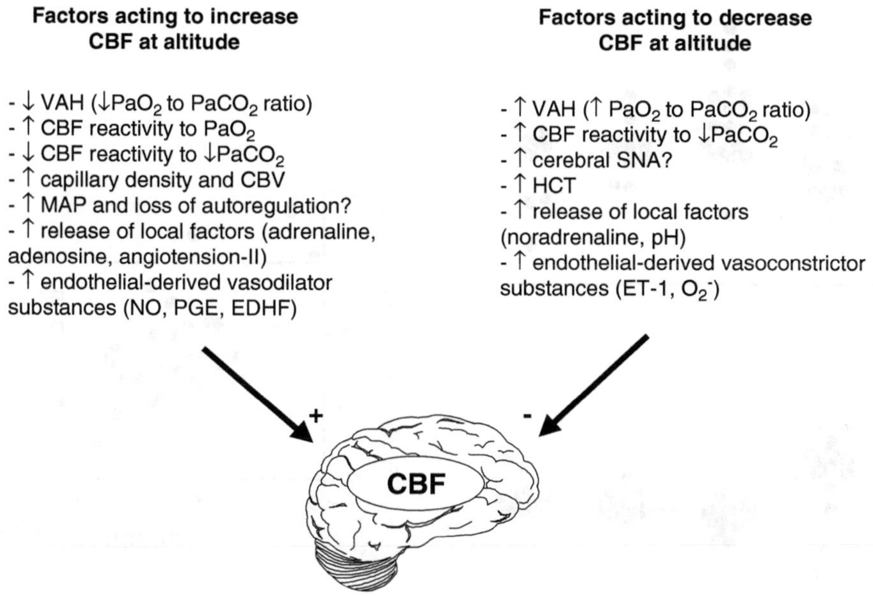

Fig. 7.6 Summary of the major factors acting to increase (*plus*) and decrease (*minus*) CBF during exposure to hypoxia. *VAH* ventilatory acclimatization to hypoxia. Data from sources cited in text

[89] over 100 years ago. Hypoxia causes vasodilatation and an increase in CBF [90] (Fig. 7.3). However, a break to this vasodilatation occurs because the hypoxic ventilatory response causes a fall in CO_2 and hence vasoconstriction [91]. The CBF response to isocapnic hypoxia is therefore greater than the response to poikilocapnic hypoxia [92]. During the first 1–3 days at HA, CBF is elevated via a greater relative degree of hypoxia compared with less severe level of hypocapnia. This balance of arterial blood gases in the regulation of CBF is highlighted in Fig. 7.6, and shows that a low PaO_2 to $PaCO_2$ ratio results in a greater degree of hypoxic vasodilatation for a given hypocapnic vasoconstriction [93]—this balance, prior to ventilatory acclimatization, accounts for ~30 % of the initial increase in CBF. Moreover, hypoxic-induced elevations in blood pressure (Fig. 7.4), coinciding with the greatest degree of hypoxia and loss of cerebral autoregulation (see Sect. 7.3.2) may also influence the initial increase in CBF. The process of ventilatory acclimatization results in a progressive rise of ventilation that increases PaO_2 and reduces $PaCO_2$ (reviewed in: [94]). Nevertheless, at a given altitude, those individuals who show a "brisk" ventilatory response will have a higher PaO_2 and lower $PaCO_2$ than those individuals who have a "blunted" response, and therefore are likely to have lower CBF (Fig. 7.7); a greater inhibitory response to the resultant hypocapnia may also accelerate this ventilatory drive to hypoxia. In addition, during ventilatory acclimatization, the associated change in CSF pH, and likely resulting alkalosis, has been suggested as another factor responsible for the falling CBF[1]; however, it should be noted that experimental evidence to directly link changes in CSF with CBF regulation have not been reported.

The balance of these changes in arterial blood gases, as mentioned, are time dependant (i.e., partly dependent on acid–base changes) and have a major influence on CBF. In addition, during ventilatory acclimatization, the associated change in CSF pH, and probable persisting but attenuated alkalosis, is likely to be another factor responsible for the falling CBF. In terms of the relative contribution of the two, because of this differential sensitivities to PaO_2 and $PaCO_2$ [11], it seems that $PaCO_2$ may be more important than PaO_2; however, an example where the chronic vasodilatory influences of hypoxia may

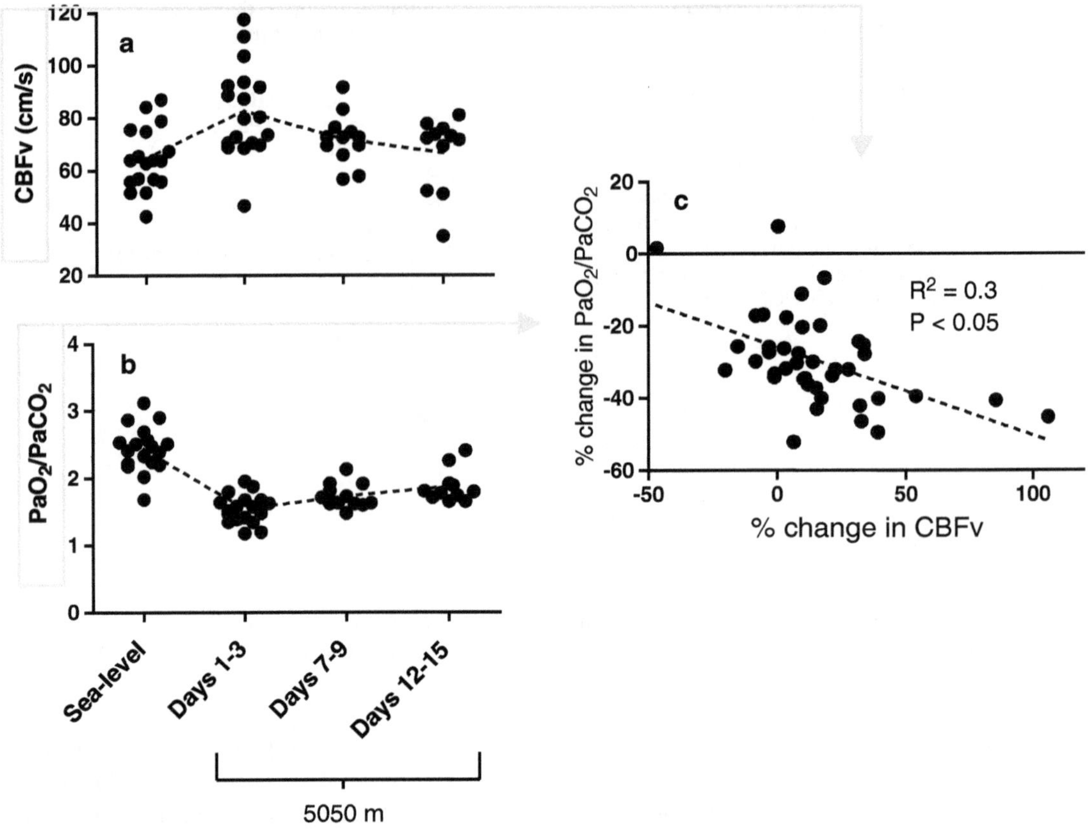

Fig. 7.7 Individual changes in CBF velocity (CBFv, (**a**)) and PaO_2 to PCO_2 ratio (**b**). A low PaO_2 to $PaCO_2$ ratio indicates (**c**) more hypoxic vasodilatation for a given hypocapnic vasoconstriction. Each point represents an individual response. Modified from [93]

outcompete the hypocapnic vasoconstriction is at extreme altitudes (>6,400 m) [95]. The benefit of this vasodilation reflects the need to maintain adequate cerebral oxygen delivery.

Cerebrovascular reactivity: The magnitude of change in CBF also depends upon the relative strengths of hypoxic vasodilatation and hypocapnic vasoconstriction. The significance of alterations in cerebrovascular reactivity to hypoxia and hypocapnia is related to whether or not the intrinsic ability of the cerebrovascular bed to dilate or constrict at HA is altered from that at sea level. Experimental studies that have examined the influence of exposure to HA on cerebral reactivity are limited and have produced variable results. For example, Jensen et al. [90] reported a 34 % increase in CBF response (TCD) to acute isocapnic hypoxia following 5 days at HA (3,810 m). They also observed increases in estimated CBF reactivity to CO_2, suggesting this increase in reactivity could be fully accounted for by the proportional and gradual changes in PCO_2 and cerebrospinal fluid [HCO_3^-] with acclimatization, which resulted in larger cerebrospinal fluid pH changes per mm Hg PCO_2 change. It should be noted that experimental data to support this possibility are not available and that it in order to clarify the mechanisms involved, knowledge of arterial-venous PCO_2 gradients and CO_2 flux across the brain would be needed. In contrast, other studies have reported that, when compared to sea level, CBF reactivity to hypercapnia is reduced [96–98] or unchanged [99] at HA and CBF reactivity to hypocapnia is either enhanced [97, 98, 100] or unchanged [96, 99]. In reconciling these divergent findings, it has been reported that subjects with acute mountain sickness

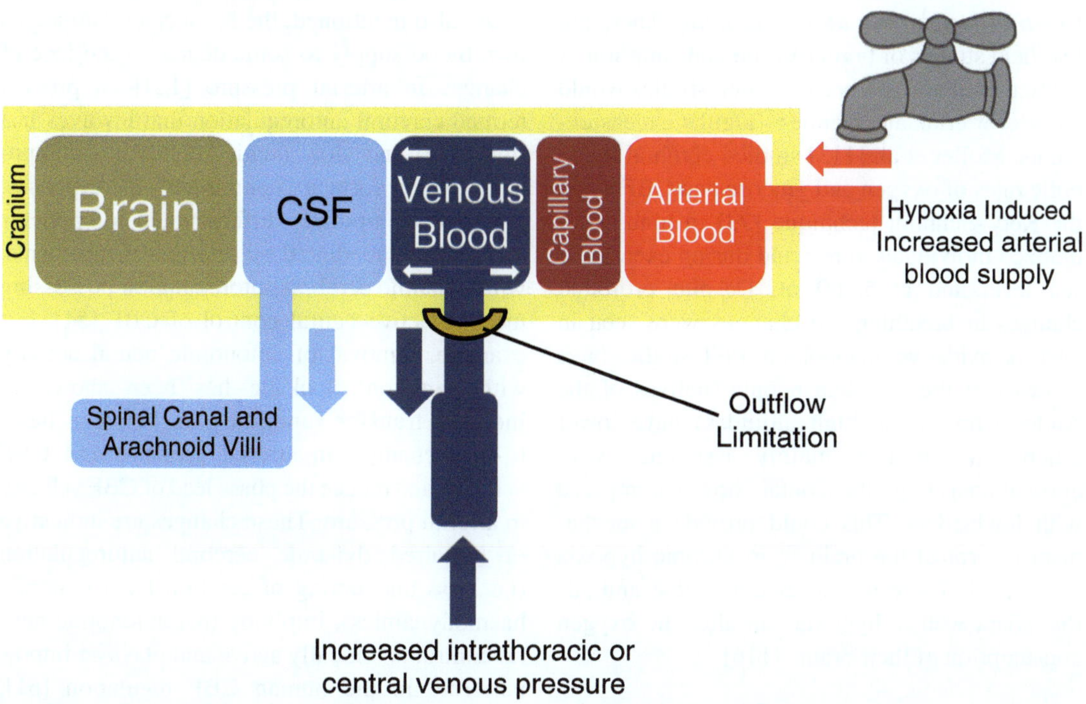

Fig. 7.8 Diagram of venous hypertension mechanism. *CSF* cerebrospinal fluid. Modified from [155]

(AMS) have greater cerebral hemodynamic responses to hypocapnia and a greater reduction in hypercapnic reactivity (compared with no-AMS subjects) [101]. Thus it seems like maladaption to altitude [100, 101] might lead to a differential change in CBF reactivity to hypocapnia compared to well acclimatized subjects [96, 97, 99] and HA natives [97, 101]. However, it is important to point out that differences in experimental protocol (length of hypoxic exposure), method of assessing CBF reactivity (steady-state vs. rebreathing [102]), degree and type of hypoxic exposure (simulated vs. HA), and limited sample size may explain the differences in the observed CBF reactivity following exposure to HA.

Cerebral autoregulation: Although the mechanism(s) for cerebral autoregulation have not been fully determined, they are most likely to involve a complex interplay of myogenic, metabolic, and neurogenic mechanisms [50]. As indicated (Figs. 7.1, 7.2, 7.3, 7.4, 7.5, 7.6, 7.7, and 7.8), many of these mechanisms are potentially influenced by hypoxia. There have been a number of studies that have examined the influence of high altitude on cerebral autoregulation. These studies indicate an impairment of cerebral autoregulation [96, 103, 104] in both newcomers to high altitude and in permanent high-altitude residents living above 4,000 m [105], especially in the presence of AMS [106–108]. Impairment in cerebral autoregulation, leading to over-perfusion and vasogenic edema subsequent to mechanical disruption of the blood–brain barrier, has been implicated [109] in the selective accumulation of intracellular cytotoxic and extracellular vasogenic edema observed in AMS [110]. However, the linkage of alterations in cerebral autoregulation with susceptibility to AMS is not supported in all studies [111, 112]. Moreover, the fact that permanent high-altitude residents living above 4,000 m [105] have an impairment in autoregulation in the absence of AMS may further support the notion that changes in autoregulation is not important in the pathophysiology of AMS. Differences in methods to assess static and dynamic cerebral autoregulation and our poor understanding of these methods and related mechanisms regulating autoregulation may explain these divergent findings.

Cerebral metabolism at high altitude: There are few field studies of brain oxygen consumption or extraction at altitude because such studies would require arterial and venous (jugular) measurements. Moller et al. [113] studied cerebral metabolic rates of oxygen and glucose and CBF using the Kety–Schmidt technique [73] in nine acclimatized individuals at rest and during exercise at sea level and at 5,260 m. Despite profound changes in breathing, no changes were seen in CBF or oxidative metabolism. PET studies have shown that the Quechua people (natives of the Andes who live at high altitudes) have lower glucose metabolism, mainly associated with cortical areas (e.g., the frontal cortex), compared with lowlanders. This could provide a mechanism to protect the brain from chronic hypoxia [114, 115]. There is also evidence that animals that adapt well to hypoxia can alter the oxygen consumption of their brains [116].

Autonomic control of CBF at high altitude: A number of studies have reported general elevations in sympathetic activity [117, 118] following ascent to high altitude. Although the cerebral vasculature has rich innervation by sympathetic nerves [119], the influence of adrenergic activity and high altitude on the cerebral circulation—particularly cerebral hemodynamics—is surprisingly unclear [120]. As mentioned, high altitude has been reported to elevate cerebrovascular reactivity to both hypocapnia and hypercapnia [81, 112]. In contrast, acute sympathetic activation (e.g., hand-grip, LBNP) at sea level generally leads to maintained cerebrovascular CO_2 reactivity [121, 122]. Moreover, at sea level, sympathetic deactivation (via ganglionic blockade) reduces cerebral reactivity to hypercapnia [123]; however, since blood pressure may influence CBF [124], this lowering of reactivity is likely confounded by an attenuated hypercapnia-induced pressure response [61, 123]. Thus, the effects of chronic high-altitude-induced sympathetic activation appear to manifest differently than acute sympathoexcitation with respect to cerebrovascular reactivity.

As also mentioned, the brain can modulate its own blood supply to some degree in the face of changes in arterial pressure [124]—a process termed cerebral autoregulation that involves fast (dynamic) and slow (static) regulatory components which form a "continuum" of responses. These two temporally differentiated components appear to be mediated by different mechanisms, with dynamic cerebral autoregulation possessing more effective neural control of CBF [51]. For example, removal of autonomic neural activity with ganglion blockade has been shown to increase transfer function gain between beat-to-beat changes in arterial pressure and CBF velocity, and reduce the phase lead of CBF velocity to arterial pressure. These changes are indicative of impaired dynamic cerebral autoregulation (i.e., less uncoupling of cerebral from systemic haemodynamics), implying that autonomic neural control is tonically active and plays an important role in the human CBF regulation [51]. Studies conducted at high altitude indicate that dynamic cerebral autoregulation is impaired upon acute exposure to hypoxia or high altitude [104, 107, 112, 125] and does not improve with partial acclimatization [126]. This deterioration in cerebral autoregulation is generally manifested by increased coherence and gain and/or reduced phase of the transfer function between blood pressure and MCAv within the optimal autoregulatory frequency range. An explanation of these findings may be that, despite favorable changes in hypoxia and hypocapnia that act to improve cerebral autoregulation [120], the autonomic nervous system may not function appropriately to effectively regulate CBF [126] due to the intense sympathetic hyperactivity at HA. Recently, the extent to which neural control governs any changes in cerebral autoregulation in humans at high altitude has been examined. Dynamic autoregulation and cerebrovascular reactivity was assessed at sea level and again at 5,050 m before and after combined α- and β-adrenergic blockade [127]. Despite elevations in CBF velocity reactivity to hypercapnia at 5,050 m, blockade reduced it comparably at sea level and HA—effects that were attributed to the hypotension and/or abolition of the hypercapnic-induced increase in MAP rather than the direct effects of sympathetic nerve activity. In contrast, this study revealed a functional role of sympathetic nerve activity in regulating some commonly used metrics of dynamic cerebral autoregulation. However, it should be noted that there is no "gold-standard" or normative data for the assessment of dynamic cerebral autoregulation; and the physiological and clinical implications for changes in dynamic cerebral autoregulation and related transfer function analysis metrics are still unclear [128]. For example, dynamic cerebral autoregulation incorporates both active cerebral autoregulation as well as Windkessel (i.e., compliance) components. Nevertheless, hypoxic-induced elevations in sympathetic nerve activity have the potential to affect CBF and its regulation (i.e., reactivity and autoregulation) by both direct and indirect (e.g., blood pressure) mechanisms (Fig. 7.1 and 7.2). Based on studies of sheep at simulated high altitude, it has been reported that increased sympathetic tone might increase cerebrovascular resistance, acting to attenuate CBF over time [129]. Although it would seem plausible that elevations in sympathetic nerve activity at high altitude would act to constrain elevations in CBF, experimental data is lacking to support or refute this possibility.

Additional Factors Affecting Vessel Tone and CBF

Neuronal activation: A major factor acting to increase CBF is the hypoxic-induced stimulation of neuronal pathways that originate or pass through the brain stem [130, 131]; these closely relate to the blood oxygen content [132, 133].

Hematocrit: CBF varies inversely with haemtocrit in many species under both acute (i.e., acute anaemia) and chronic (i.e., erythropoiesis) experimental conditions. While the mechanism(s) behind this relationship remains unclear, two possible explanations have been proposed: (1) alterations in plasma viscosity elicit shear stress-mediated compensatory changes in the cerebral vascular resistance in order to maintain CBF constant [28, 134–136]; (2) reductions in cerebral arterial content elicit vasodilatory responses in order to maintain bulk O_2 transport (e.g., CBF × cerebral arterial O_2 content) to the brain [27, 132, 133, 137].

Angiogenesis: Another mechanism acting to elevate CBF during exposure to high altitude is angiogenesis. Indeed, the induction of angiogenesis in the mammalian brain is one of the more dramatic adaptations to hypoxia, causing a near doubling of the capillary density that occurs between 1 and 3 weeks of exposure [138–140]. The cellular mechanisms through which hypoxia stimulate angiogenesis are now beginning to be understood (for review see: [141]). The fundamental importance of hypoxic-induced elevations in capillary density are related to an increase in cerebral blood volume [142, 143] and, potentially, subsequent cerebral vasodilatation. Although experimental evidence in humans is lacking, hypoxic-induced elevations in cerebral blood volume have been directly related to capillary density [144]. It is clear that hypoxia causes a cascade of local mechanisms that underline the reported hypoxic-induced vasodilatation; however, the normalization of CBF over time at HA indicates that the vasodilatory factors are balanced out by the various factors acting to attenuate CBF.

Adenosine: Brain adenosine levels increase rapidly within 30 s to 5 min hypoxia exposure which mirrors the increase in CBF [145] This mechanism is thought to be involved in the very fast arteriole distension of hypoxic vasodilatation and it is thought the adenosine expression must be close to arteriolar smooth muscle, e.g., from glial end feet.

Potassium (*K^+*): The hypoxia-induced rise in K^+ may also be involved in hypoxic vasodilatation. At least in single vessel studies, endothelial K^+ channel activation results in increased concentration of intracellular calcium which in turn results in release of nitric oxide [146]. However, it should be noted that respiratory alkalosis (as would be caused by high altitude) causes hypokalemia by stimulating intracellular uptake of K^+ and renal excretion. Thus, it is unclear if changes in K^+ may influence changes in CBF upon ascent to high altitude.

Nitric Oxide (*NO*): Nitric Oxide Synthase III (endothelial Nitric Oxide Synthase—eNOS) produces NO or endothelial-derived relaxing factor [147]. NO has a short half-life and rapidly diffuses to vascular smooth muscle where it interacts with Ca^{2+} modulation mediated by cGMP. This results in vasodilatation. Appenzeller et al. [148] used TCD changes in response to an exogenous NO donor in 9 altitude-native Ethiopians and 9 altitude-native Peruvians to assess how adapted to altitude each different race is. The circulation response to NO was minimal in Ethiopians at low altitude, while Peruvians had a large response. In contrast, at high altitude, Ethiopians had a large response and Peruvians had minimal. They concluded Peruvians were well adapted lowlanders while Ethiopians were highlanders adapted to that life. Nevertheless, since different NO donors and altitudes may affect the diameter of the MCA and therefore the CBF response, caution is needed in the interpretation of these findings.

Factors Affecting Vessel Permeability

Hydrostatic pressure: A number of studies [71, 74, 101] have speculated that the increased CBF at altitude would result in an increase in hydrostatic pressure. Such increases occur in other clinical conditions such as hypertensive encephalopathy and toxaemia of pregnancy. These conditions also cause a reversible increase in white matter T2 signal [149].

Venous Hypertension: Idiopathic (or Benign) Intracranial Hypertension has recently been demonstrated to be closely related to sinovenous outflow obstruction [150, 151]. This obstruction would directly increase hydrostatic pressures. Such anatomical variations that cause this clinically may only become apparent under the stressor of hypoxia and hence account in part for the predisposition some have for AMS or high-altitude cerebral edema (HACE) [145]. High-altitude pulmonary edema (HAPE) itself may increase venous pressures and hence trigger HACE. The presence of retinal [152] and cerebral hemorrhages [153] and autopsy evidence of antemortem venous thrombosis also suggests this is a possible mechanism [154]. Retinal venous distension [152], cerebral hemorrhages [153] and the increased venous blood demonstrated by near infrared spectroscopy and more recently by MRI imply that, in hypoxia, a relative venous insufficiency may exist (see: [155] for review). Similarly, there is increasing evidence that manifestations of the fluid shift during microgravity is of similar nature to idiopathic intracranial hypertension, which is thought to be primarily a venous insufficiency condition. The unique anthropomorphic adaptations of large brained biped humans with cerebral venous systems that have to cope with large changes in hydrostatic pressure may result in conditions of inflow/outflow mismatch. In addition, slight increases in central venous pressures (e.g., from hypoxia-induced pulmonary vasoconstriction) may further compromise venous outflow at altitude [155]. A simplified venous hypertension model is outlined in Fig. 7.8. In a steady state, flow in equals flow out. In clinical syndromes of outflow obstruction, flow in still equals flow out, but the pressure to drive the outflow is greater to overcome additional resistance. In hypoxia a number of mechanisms may contribute to increased intracranial blood (and in particular venous) volume, but all center on the limiting step of venous drainage [155]. Recently, it had been reported that arterial hypoxemia is associated with cerebral and retinal venous distension (as assessed from retinal venous distinction and MRI) whose magnitude correlates with symptoms of high-altitude headache [156]. This study provides the first evidence that restriction in cerebral venous outflow is associated with both retinal distension and symptoms of high-altitude headache [156]. Therefore, an inflow (i.e., CBF) and outflow (i.e., cerebral venous physiology) matching in the context of high altitude seems to be an important adjustment in the pathogenesis of AMS-related headache.

Role of pulmonary vascular vasoconstriction: Alterations in pulmonary vascular tone may also play a role in regulating cerebrovascular function at high altitude. It is well known that, because of hypoxic-induced pulmonary vasoconstriction, pulmonary systemic pressure increases during hypoxic exposure [157]. Elevations in pulmonary vascular resistance can also impair cardiac function, e.g., increase right ventricular end-diastolic and atrial pressure, and reduce left atrial filling [157]. In addition, elevations in right atrial pressure can lead to venous drainage obstruction in certain organs like the brain, resulting in increased cerebral venous pressure, cerebral blood volume, and subsequent cerebral edema at HA [145]. However, apart from severe cases of HAPE and Chronic Mountain Sickness (CMS), it is unlikely that hypoxic pulmonary vasoconstriction at high altitude leads to any degree of right heart insufficiency with its attendant elevation of right arterial pressure or failure to load the left side of the heart. Please refer to Chaps. 5, 6, and 21 for greater detail on the pulmonary circulation, heart, hypoxic pulmonary vasoconstriction, and HAPE.

Direct effects of hypoxia: Hypoxia itself can damage basal membrane structures [158]. Houston proposed that hypoxia suppresses the sodium-potassium pump in cell membranes leading to cell swelling [159]. Kallenberg et al. [110], finding of cytotoxic edema in AMS, support this possibility. A reduction in cellular PO_2 decreases the expression and activity of the Na^+/K^+ ATPase in several cell types, such as alveolar epithelial cells, endothelial cells, and neuronal cells (reviewed in: [145]). This might represent a mechanism that cells, possibly neurons, use to reduce energy expenditure in hypoxia.

Hypoxia-Inducible Factor (*HIF*): HIF-1 is a heterodimeric factor composed of HIF-1α and HIF-1β protein subunits. In normoxia HIF prolylhydroxylase breaks down HIF, but in hypoxia HIF prolyl-hydroxylase is inhibited and HIF-1α binds to promoter/enhancer elements causing increased transcription of classic hypoxia-inducible target genes. Such genes are involved in angiogenesis (VEGF) and erythropoesis (erythropoietin [160]). Atrial naturetic peptide and nitric oxide synthase are also induced [160].

Vascular Endothelial Growth Factor (*VEGF*): An explanation for hypoxic-induced increases in vascular permeability via the VEGF pathway was first suggested by Severinghaus [161]. Hypoxia induces VEGF expression and blockage of that expression with VEGF specific antibodies prevents vascular leakage in the brain. Dexamthasone appears to block VEGF expression and hence reverse hypoxia-induced brain edema [162]. Early studies had shown no correlation between VEGF levels and AMS [163]. Recently, however, the soluble VEGF receptor (sFlt-1) which can bind VEGF in the circulation and was not accounted for in earlier work has been studied [164]. In this study of 20 subjects who were driven to 4,300 m, subjects who developed AMS had lower sFlt-1 and hence significantly higher levels of free plasma VEGF on ascent than well subjects.

Free radicals: The study of free radicals is difficult due to their very short half-life. A number of studies however imply that neuroxidation and the subsequent inflammatory response may damage cerebrovascular endothelium [165–168]. In a recent study Bailey et al. found that there was a progressive increase in blood and CSF concentrations of free radicals (lipid-derived alkoxyl and alkyl species) and IL-6 during a 16–8 h simulated exposure to 4,600 m (12 % O_2) [165]. Although this induced a mild (0.6 % or 7 mL) increase in brain volume, no underlying morphological changes (e.g., edema) were seen on MRI.

Role of Cerebral Blood Flow in Altitude Illness

The role of CBF regulation in the pathophysiology of AMS and HACE is outlined in Chap. 20; thus, only brief aspects of this area are outlined below.

High-Altitude Cerebral Syndromes

Headache (*HAH*): The International Headache Society defined a high-altitude headache (HAH) as a headache associated with ascent to altitude that responds to simple analgesia [169]. The headache should have started within 24 h of ascent, resolve within 8 h of decent. The headache is bilateral, frontal or fronto-temporal and is of a dull or pressing quality, mild or moderate in intensity and aggravated by exertion, movement, straining, coughing, or bending. These are common but can be difficult to distinguish from a headache secondary to dehydration. Management principally consists of stopping ascent, analgesia, rehydration, and descending if no improvement. HAH would normally be expected to resolve within 10–15 min of supplementary oxygen, and this intervention can help to distinguish it from a common migraine.

Acute Mountain Sickness (*AMS*): Some consider that AMS is a progression of HAH. The Lake Louise Consensus Group defined AMS in 1993 as the presence of headache in an unacclimatized person who has recently arrived at an altitude above 2,500 m with one or more of the additional symptoms: gastrointestinal symptoms (anorexia, nausea, or vomiting), insomnia, dizziness, and lassitude or fatigue [170].

High-Altitude Cerebral Edema (*HACE*): HACE can be considered as the final encephalopathic and potentially life-threatening stage of cerebral altitude illness. There is often evidence of preceding AMS or HAPE. It is usually marked by disturbances of consciousness that may eventually progress to deep coma, psychiatric changes of varying degree, confusion and ataxia of gait [171]. There have been reports of HACE developing *de novo* in the absence of headache but this is rare [172].

Cerebral Perfusion and Altitude Illness

Some subjects are more tolerant of hypoxia than others resulting in variations in CBF changes. The less tolerant have more hypoxic drive, more hypocapnia, and more vasoconstriction. Harvey et al. found the addition of 3–5 % CO_2 to inspired air induced a feeling of well-being during high-altitude exposure [173] and also improves brain oxygenation [174, 175]. However chronic exposure to elevated CO_2 can worsen symptoms [176]. The claim that adding 3 % CO_2 to air has a beneficial effect was soon challenged by Bartsch et al. [177] who randomly allocated 20 mountaineers with AMS at 4,559 m to three treatment groups: (1) with 33 % O_2, (2) with 3 % CO_2 in air, and (3) an air control. Thirty-three percent O_2 significantly relieved symptoms of AMS and reduced CBF, but CO_2 addition did not significantly ameliorate AMS, despite the rise of PCO_2, ventilation, and alveolar PO_2.

Jensen et al. [74] used radiolabeled xenon in 12 individuals who ascended from 150 m to 3,475 m, and found an increase in CBF of 24 %. Ascent from 3,200 m to between 4,785 and 5,430 m caused a further increase in CBF to 53 % above the readings at sea level; however, there was no difference in CBF between individuals with and without AMS.

TCD is a noninvasive ultrasound technique that has been used to assess relative changes in CBF velocity at high altitudes. A 2 Mhz ultrasound probe is focused through the temporal bone window onto the middle cerebral artery measuring the velocity of the blood within the artery, which to date is assumed to represent CBF. Recent evidence highlights the need to interpret TCD ultrasound with caution during extremes of hypoxia [11, 95]; during such levels, dilation of the middle cerebral artery may occur, thus invaliding this approach to estimate CBF.

It has been suggested that the headache observed in subjects with AMS could be secondary to an increase in CBF [74, 101]. However Baumgartner et al. [178] found no significant change in velocity and no correlation with the development of AMS when they studied 10 subjects at 0, 3, and 6 h on decompression to 4,559 m. The absence of consistent early changes in middle cerebral artery velocity has also been corroborated by the results of studies of carotid and vertebral artery velocities [79, 179]. This is in part explained by the substantial interindividual differences that were observed. However, in contrast, 12–24 h after arrival at high altitudes (3,475–4,559 m) several Doppler studies found increases of 20–27 % in velocity in the middle cerebral artery [78–80]. These ultrasound findings are in accordance with the results of Severinghaus et al. [71] and Jensen et al. [74]. It may be that the observed delayed increase in middle cerebral artery velocity could explain the delay in the development of AMS and HACE.

Dyer et al. [180] quantified regional CBF using arterial spin labeling MRI during 30-min hypoxia in 12.5 % oxygen. They studied 6 AMS susceptible and 6 AMS resistant subjects. Steady-state whole-brain CBF increased in hypoxia ($p<0.005$), but did not differ between groups, and cerebral oxygen delivery remained constant. The percent change in CBF did not differ between brain regions or between groups. They found CBF increased in acute hypoxia, but was not different between white matter and grey matter, irrespective of AMS susceptibility.

It has been proposed that it is not the actual change in CBF itself but the loss of autoregulation that is implicated in the development of AMS; however, this linkage is not a universal finding (see Sect. 7.3.2).

Although there is some indirect evidence to suggest that exercise at altitude may increase the likelihood of an individual developing AMS and HACE [181], not all studies agree [182]. Imray et al. [183] assessed cerebral perfusion during exercise in nine individuals who ascended to 5,260 m, and found an increase in middle cerebral artery velocity with increasing altitude while subjects were at rest. At intermediate levels of exercise, there was an increase in middle cerebral artery velocity but at maximum exercise levels, a reduction of CBF velocity in the middle cerebral artery was observed. The pronounced rise in blood pressure and elevated middle cerebral artery flow could cause stress to the blood–brain barrier, possibly leading to vasogenic edema. Subudhi et al. [184] have reported similar results. However, a recent study indicated that mass cerebral oxygen delivery via CO_2-mediated vasodilation does not improve incremental exercise performance in normoxia or hypoxia, at least when accompanied by respiratory acidosis [185].

Cerebral Oxygenation

Cerebral NIRS has been shown to correlate with jugular venous bulb saturations in healthy volunteers undergoing isocapnic hypoxia [186]. The technique has also been validated with PET scanning [187] ^{133}Xe washout methods [188] and with internal carotid artery stump pressures [189]. Imray et al. [190] demonstrated a progressive fall in resting regional brain saturation (rSO_2) with altitude that did not correlate with AMS severity. Hadolt examined 17 volunteers trekking in Nepal in 2003 [191] and found that increases in altitude resulted in a significant fall in rSO_2. Imray et al. [175] subsequently demonstrated that administering small supplemental CO_2 actually increased rSO_2 (presumably by reducing hypocapnic-induced vasoconstriction). Imray et al. [183] studied 9 subjects using a recumbent bicycle up to 5,260 m. At sea level, sub maximal exercise resulted in an increase in rSO_2 from 68.4 to 70.9 %. At VO_2 max, this reduced slightly to 69.8 %. However, at each altitude above this, rSO_2 was reduced during increasing exercise. Similarly, Subudhi et al. [192] studied 13 cyclists exercising to VO_2 max under normoxic and acute hypoxic (12 % FiO_2) conditions. NIRS monitored left vastus lateralis and frontal cerebral cortex. In normoxia, they demonstrated that frontal cortex rSO_2 increased from 25 to 75 % of VO_2 max then fell towards 100 % of VO_2 max. During hypoxia however, they found cerebral rSO_2 dropped across all work levels. From this they concluded that cerebral oxygenation could limit incremental exercise performance at altitude. Using multichannel near infrared cerebral spectroscopy, it has been suggested that prefrontal, premotor, and motor cortex deoxygenation during high-intensity exercise contributed to an integrative decision to stop exercise [193]. Vuyk et al. [194] demonstrated that acetazolomide helped to maintain cerebral oxygenation (as measured using NIRS during exercise in a group of 16) up to 5,700 m. Wolff [72], in a reanalysis of results from the paper by Severinghaus et al. [71], demonstrated that despite an approximate 20 % reduction in arterial oxygen content at 3,810 m altitude, increases in CBF were able to offset any significant fall in cerebral oxygen delivery. Although these findings indicate that brain oxygen saturation decreases with altitude and to an even greater degree with exercise at altitude, they highlight that at extreme altitude (>6,400 m), cerebral vasodilation occurs resulting in maintained cerebral oxygenation [95].

Influence of High Altitude on Cognitive Function

Background

Ascent to altitude results in a subtle decline in some of our more sophisticated cognitive skills. The mountain environment is potentially dangerous and the complex decisions needed for safe travel require a normal functioning brain. Some of the tragedies that have occurred at high altitude, such as the events on high on Everest in 1996, may have resulted from a critical loss of judgment [195, 196]. The mortality on descent from the summit of Everest is three times greater for those climbers not using oxygen than in those who did, which lends further support to the possibility [196, 197]. More recently a study into all the deaths on Mount Everest between 1921 and 2006 found that of the 212 deaths, 94 occurred above 8,000 m [198]. Most of these deaths occurred on descent. Cognitive impairment, ataxia, profound fatigue, late summit times, and the tendency to fall behind companions were all features seen in non-survivors. Respiratory distress, nausea, vomiting, and headache, common indicators of altitude illness at 2,500–5,000 m, were rarely noted in the non-survivors above 8,000 m.

Neuropsychological Effects of Acute Exposure

Nothing caught the attention of the public regarding the potential dangers of acute exposure to high altitude more than the exploits of the early balloonists. On 5th of September 1862, two British balloonists, James Glaisher and Henry Coxwell, took off from Wolverhampton in the West Midlands, UK and ascended to over 8,800 m. Glaisher reported paralysis of his arms and legs and sudden loss of vision before losing consciousness. Coxwell lost the use of his hands and could only open the valve to initiate balloon descent by pulling the cord with his teeth. After landing, they were able to walk seven miles to Cold Weston near Ludlow, suggesting they had no residual neurological deficits [199–201]. Thirteen years later, the first deaths attributable to acute high-altitude exposure were reported when three Frenchmen lost consciousness ascending through 7,000 m. Hypoxia had dulled their minds sufficiently they had forgotten to use the oxygen Paul Bert had provided. Only Tissandier survived to tell the tale of the fateful flight of the Zenith [202].

These early accounts relate to the acute neurological effects of hypobaric hypoxia. The first detailed clinical descriptions of the consequences of slower ascent were given by Thomas Ravenhill in 1913 while working as a medical officer in the mines of northern Chile. He provided a classification of high-altitude illness and described the features of both high-altitude cerebral and pulmonary edema [203, 204]. Seminal among the early studies is the account published by McFarland in 1937. Using a variety of assessments of sensory, motor, and cognitive function McFarland found that individuals taken rapidly (over hours) to altitudes between 15,000 and 16,500 ft. exhibited impairment of both simple and complex psychological performance [205].

With the development of the jet engine and improvements in aircraft design, including cabin pressurization after World War II, rapid ascent to extremely high altitude became possible. Luft et al. studied the effects of rapid decompression to 55,000 ft. from 30,000 ft. in subjects breathing 100 % oxygen [206]. There appeared to be a minimum latency of between 15 and 17 s prior to the loss of consciousness regardless of the altitude achieved. It is clearly difficult to study, but cerebral function appeared to unimpaired for several seconds after the oxygen supply in the surrounding capillaries reached its lowest level. Interestingly, humans born and living at altitude have significantly longer time to unconsciousness than do their sea level counterparts [207].

Chronic adaptation (60 days) to high altitude appears to enable some individuals to function surprisingly well in extreme hypoxia as evidenced by femoral arterial blood gas sampling at 8,400 m on the descent from the summit of Everest [208]. The mean PaO_2 in the four subjects breathing ambient air (20 min after the relief of oxygen) was 24.6 mmHg (3.28 kPa), with a range of 19.1–29.5 mmHg (2.55–3.93 kPa).

Neuropsychological Effects of More Sustained Hypobaric Hypoxia

Greene, the physician for the 1933 Everest Expedition, quoted Hingston's observation on Everest in 1924: "Though the mind is clear, yet there was a disinclination for effort. It was far more pleasant to sit about than to do a job of work that required thought… Though mental work is a burden at high altitudes, yet with an effort it can be done….." [209].

A number of HA studies have shown impairment of arithmetic [210], memory and metamemory [211, 212], language, perception, learning, cognitive flexibility, and psychomotor skills [213, 214]. Increases in reaction time [215, 216] and P300 [217] and a slowing of pupil constriction [218] have also been seen, indicative of a fundamental slowing of neuronal processing. As with other symptoms, the neuropsychological changes observed are related to the rate of ascent and the altitude [214]. As a result there are large differences between studies where different ascent protocols have been applied. In addition neuropsychological performance is susceptible to the influence of fatigue and also anxiety, both of which occur in climbing to high altitudes [219]. A 6 % oxygen enrichment at a simulated altitude of 5,000 m reversed many of these observations [220].

Sleep disturbance either through periodic breathing and waking between periods of apnoea [221] or secondary to disturbance by tent companions is common and may contribute to daytime neuropsychological impairment.

It is also important to distinguish between neuropsychological changes due to hypobaric hypoxia and changes due to AMS. The literature suggests AMS has little effect on short-term memory, but a significant impairment in conceptual tasks. Those at altitude who do not develop AMS tend to have a greater short-term memory impairment and scarce alteration in conceptual tasks [215, 222].

Perception: No threshold change has been found for auditory stimuli; however, there is a rise in threshold for detecting visual stimuli when dark adapted [223]. There are mixed results regarding changes in color perception, but any change, if present, is of minimal significance [213].

Memory: Short-term memory has been shown to decline at 4,500 m and is especially noticeable above 6,000 m [213, 224]. Long-term memory appears to be preserved [225]. Both animal and human [215, 226] studies have compared controlled and automatic processing tasks and the results imply that hypobaric hypoxia causes a reduced capacity to learn rather than to retrieve. Spatial memory has been found to become impaired between 3,800 and 5,000 m [227]. Cognitive flexibility has been assessed using Stroop Color and Word test or Wisconsin Card Sorting test. It has been shown to be significantly impaired in world class mountain climbers even months after their last ascent [228].

Motor skills: Motor speed and precision are reduced compared to sea level [224, 229]. The finger taping test (FTT) and Purdue pegboard tests have been used to assess psychomotor changes but confounding variables in the field include fatigue and the cold. This may explain the wide variation in altitudes between different studies where dysfunction is detected (Berry et al. at 3,500 m [224] and Bolmont et al. at 8,000 m [216]). The American Medical Research Everest Expedition (AMREE-1981) demonstrated 15 of the 16 climbers had impaired FTT immediately after the expedition and 13 still had impairment a year on implying some damage may be long term [229].

Ataxia is a common feature of HACE, and this has been quantitatively examined using a wobble board [230]. It was found that a positive test gave a predictive value for AMS of 66.7 % at 4,650 m and 100 % at 5,005 m. Brain oxygenation was also found to correlate with stability on the wobble board (while peripheral saturations did not).

Psychological changes: New onset anxiety disorders are relatively common in trekkers to altitude [231]. These are often focused on health concerns. A heightened state of anxiety has been shown to offset some of these reductions in reaction time and psychomotor ability [202]. Mild auditory and visual hallucinations (e.g., a "third man") have been described in climbers at very high altitude [204, 232].

Cerebral dysfunction above 8,000 m: Climbing above 8,000 m either with or without supplementary oxygen is potentially dangerous. There is a higher attrition rate among western climbers compared to sherpas descending from the summit of Everest (2.5 % vs. 0.2 %, $p<0.001$) [198]. On the standard Tibetan and Nepalese routes, 3.4 % and 1.7 % of climbers, respectively, did not return from the summit. A common feature among the non-survivors was gross cognitive impairment and ataxia. Since arterial oxygenation decreases during strenuous exercise at high altitude [183, 233], maximal hypoxemia would occur during the vigorous exertion just below the summit. Acute hypoxic cerebral dysfunction at such extreme altitude might involve pathophysiological mechanisms apart from HACE.

Evidence of long-term brain injury: Anooshiravani et al. did not detect any functional or structural alterations (using MRI) in a group of 8 climbers who had ascended a 6,000 m peak [234]. However, Garrido et al. [235, 236] have performed studies demonstrating increased signal intensity in periventricular, posterior parietal, and occipital cortex in five of nine climbers who had ascended above 7,000 m. Vichow-Robin spaces (CSF spaces around vessels) tend to be enlarged in regular climbers suggesting chronic hypoxia-induced brain atrophy [237]. The implications of these changes for long-term brain injury would seem possible but has yet to been examined in combination with imaging and a prospective design.

Other Neurological Syndromes That can Occur at High Altitude

Some, usually focal, events that can also occur at sea level (and which fall outside the usual definition of altitude sickness) seem to have an increased incidence at high altitudes [238].

Transient ischemic attacks: Transient ischemic attacks occur in a younger population at altitude than they do at sea level, which implies that they are not caused by atherosclerosis and might be related to vasospasm, hypocapnic vasoconstriction, or the putative prothrombotic effects of hypoxia. Right-to-left shunts increase with exercise and altitude, which means that embolisation through a patent foramen ovale might account for some of the focal neurological events or cases of migraine [239]. Such effects can also contribute to rare cases of cerebral infarction, to which dehydration and hypoxia-induced polycythemia might also contribute.

Transient global amnesia: Transient global amnesia has been reported [240, 241] and might occur through similar mechanisms in the limbic cortex to those that cause the focal events described above.

Migraine: Migraine can be difficult to distinguish from AMS, but migraine should be suspected when there is a history or if a headache is associated with focal neurology. HACE, however, can also present with focal neurological symptoms and must be the first diagnosis to be considered after recent ascent to altitude [242].

Cerebral venous thrombosis: Cerebral venous thrombosis is a common autopsy finding in patients who die at high altitudes [154, 243] and might be related to volume depletion and polycythemia [244]. Altitude can also trigger central venous thrombosis in individuals who are susceptible owing to familial thrombophilia [245–247].

Epilepsy: Whether seizure frequency increases at high altitudes is not known, although hypoxia and hyperventilatory hypocapnia are thought to be triggers [248].

High-altitude syncope: High-altitude syncope is thought to be a vasovagal phenomenon that is related to hypoxemia, although an increase in arrhythmias might be a contributory factor [249]. High-altitude syncope has been reported to occur at moderate altitude and rapidly resolves [250], even without descent. Recently, it has been reported that initial orthostatic tolerance and symptomatic responses to standing upright is well maintained at 5,050 m [251]. These findings indicated that, while there was a faster drop in BP with standing (likely reflecting a temporary mismatch between peripheral vascular resistance and cardiac output) the comparable drop in BP and comparable time for correction of the hypotension indicates that the arterial baroreflex reflex is well preserved at high altitude.

Cranial nerve palsies and other neurological conditions: Cranial nerve palsies are well recognized both in the presence and absence of AMS or HACE. The sixth cranial nerve is most commonly affected, perhaps owing to compression of the trunk from adjacent brain swelling. Facial and hypoglossal cranial nerve palsies have also been reported [238]. There is evidence that acute hypoxic cerebral dysfunction at extreme altitude might involve pathophysiological mechanisms apart from HACE [198]. Many neurological conditions that occur at sea level can also occur at high altitudes (e.g., subarachnoid hemorrhage), and mild hypoxia can unmask indolent conditions (e.g., cerebral tumors) [247].

Ophthalmological disturbances: Retinal hemorrhages, with or without the symptoms of AMS, are common. Unless the macula is involved, they are relatively benign and resolve spontaneously. The cause of the retinal hemorrhages is unknown, but could be due to increased blood flow or breakdown in the blood–retina barrier [152]. The monocular blindness of amaurosis fugax is believed to result from the blood supply to the retina being compromised, and the aetiology might be similar to the visual symptoms of transient ischemic attacks. Similarly, cortical blindness could be caused by vascular compromise to the visual cortex.

Conclusions and Suggestions for Future Research

We have attempted to examine the key mechanisms and their interactions which are involved with the regulation of CBF during exposure to the hypoxia of high altitude. The ICP, cerebrovascular, and MRI studies have given us a greater understanding of gross changes with hypobaric hypoxia, but there are still many questions as to the mechanisms of these changes on a vascular and cellular level. It should be noted that MRI scans of people suffering with non-chamber-induced AMS are not possible as currently no such scanners exist at altitude and scans performed on descent will not represent the acute changes that would have occurred higher; thus, until MRI scans are possible at high altitude, progression of research within this area will be hindered. However with the increasingly sophisticated MR and CT cerebral imaging modalities that are becoming available, major advances in our understanding are likely to follow. With the increasing technical advances in cerebral imaging techniques there is a temptation to suggest that the days of field-based research must be numbered. However, developments in both the portability and robustness of vascular ultrasound [11, 95], TCD [128] and NIRS [183] equipment have improved markedly and will inevitably lead to further insights in our understanding of the pathophysiological mechanisms of cerebral altitude illness. The development and refinement of vascular ultrasound, "functional" (i.e., multi-channel) TCD and NIRS measurements might help bridge the gap and provide insightful information.

It is clear that many aspects of CBF regulation and brain function at high altitude clearly warrant further investigation, with particular focus on integrative systems physiology and adequately powered for statistical analysis. For example, over the period of ventilatory acclimatization (weeks to months), how does hypoxia alter cerebral autoregulation, cerebrovascular reactivity; and is CBF in anyway regulated by sympathetic activity? What mechanisms underlie these changes? Are the observed alterations in CBF regulation the primary cause of, or contribute to AMS, high-altitude cerebral edema and central sleep apnea? Related to this, are there regional changes in CBF upon ascent to high altitude to maintain brainstem blood flow over that of the cortex? What are the interactions between hypoxic-induced elevations in pulmonary vascular resistance and CBF regulation? How may exercise influence CBF regulation at high altitude? Addressing some of these intriguing questions will not only provide new information by which CBF is regulated at high altitude, but will also provide insight into the understanding of cerebral hypoxia in the clinical setting of chronic lung disease and altitude illness.

References

1. Ursino M, Lodi CA. A simple mathematical model of the interaction between intracranial pressure and cerebral hemodynamics. J Appl Physiol. 1997;82(4):1256–69.
2. Panerai RB, Eames PJ, Potter JF. Multiple coherence of cerebral blood flow velocity in humans. Am J Physiol Heart Circ Physiol. 2006;291(1):H251–9.
3. Panerai RB. The critical closing pressure of the cerebral circulation. Med Eng Phys. 2003;25(8): 621–32.
4. Marmarou A, Shulman K, LaMorgese J. Compartmental analysis of compliance and outflow resistance of the cerebrospinal fluid system. J Neurosurg. 1975;43(5):523–34.
5. Edvinsson L, Krause DN. Cerebral blood flow and metabolism. 2nd ed. Philadelphia: Lippincott, Williams & Wilkins; 2002.
6. Grant DA, Franzini C, Wild J, Walker AM. Continuous measurement of blood flow in the superior sagittal sinus of the lamb. Am J Physiol. 1995 ;269(2 Pt 2):R274–9.
7. McDonald DA. Blood flow in arteries. http://www.amazon.com/Blood-Flow-Arteries-D-McDonald/dp/0713142138. 2nd ed. London: Edward Arnold; 1974.
8. Aaslid R. Cerebral hemodynamics. In: Newell DW, Aaslid R, editors. Transcranial Doppler. New York: Raven; 1992. p. 49–55.
9. Fog M. The relationship between the blood pressure and the tonic regulation of the pial arteries. J Neurol Psychiatry. 1938;1:187–97.
10. Olufsen MS, Nadim A, Lipsitz LA. Dynamics of cerebral blood flow regulation explained using a lumped parameter model. Am J Physiol Regul Integr Comp Physiol. 2002;282(2):R611–22.
11. Willie CKMD, Shaw AD, Smith KJ, Tzeng YC, Eves NE, Ikeda K, Graham J, Lewis NC, Day TA, Ainslie PN. Regional brain blood flow in man during acute changes in arterial blood gases. J Physiol. 2012;590:3261–75.
12. Busija DW, Heistad DD. Factors involved in the physiological regulation of the cerebral circulation. Rev Physiol Biochem Pharmacol. 1984;101:161–211.
13. Peebles KC, Richards AM, Celi L, McGrattan K, Murrell CJ, Ainslie PN. Human cerebral arteriovenous vasoactive exchange during alterations in arterial blood gases. J Appl Physiol. 2008;105(4):1060–8.
14. Xie A, Skatrud JB, Morgan B, Chenuel B, Khayat R, Reichmuth K, et al. Influence of cerebrovascular function on the hypercapnic ventilatory response in healthy humans. J Physiol. 2006;577(Pt 1):319–29.
15. Hortobagyi L, Kis B, Hrabak A, Horvath B, Huszty G, Schweer H, et al. Adaptation of the hypothalamic blood flow to chronic nitric oxide deficiency is independent of vasodilator prostanoids. Brain Res. 2007;1131(1):129–37.
16. Prickett TC, Yandle TG, Nicholls MG, Espiner EA, Richards AM. Identification of amino-terminal pro-C-type natriuretic peptide in human plasma. Biochem Biophys Res Commun. 2001;286(3): 513–7.
17. Konopacka A, Zielinska M, Albrecht J. Ammonia inhibits the C-type natriuretic peptide-dependent cyclic GMP synthesis and calcium accumulation in a rat brain endothelial cell line. Neurochem Int. 2008;52(6):1160–6.
18. Kobayashi H, Ueno S, Tsutsui M, Okazaki M, Uezono Y, Yanagihara N, et al. C-type natriuretic peptide increases cyclic GMP in rat cerebral microvessels in primary culture. Brain Res. 1994;648(2):324–6.
19. Hand MF, Haynes WG, Webb DJ. Reduced endogenous endothelin-1-mediated vascular tone in chronic renal failure. Kidney Int. 1999;55(2):613–20.
20. Henze D, Menzel M, Soukup J, Scharf A, Holz C, Nemeth N, et al. Endothelin-1 and cerebral blood flow in a porcine model. J Clin Neurosci. 2007;14(7):650–7.
21. Miekisiak G, Kulik T, Kusano Y, Kung D, Chen JF, Winn HR. Cerebral blood flow response in

adenosine 2a receptor knockout mice during transient hypoxic hypoxia. J Cereb Blood Flow Metab. 2008;28:1656–64.

22. Rubio R, Berne RM, Bockman EL, Curnish RR. Relationship between adenosine concentration and oxygen supply in rat brain. Am J Physiol. 1975;228(6):1896–902.
23. Iliff JJ, D'Ambrosio R, Ngai AC, Winn HR. Adenosine receptors mediate glutamate-evoked arteriolar dilation in the rat cerebral cortex. Am J Physiol Heart Circ Physiol. 2003;284(5):H1631–7.
24. Wang Q, Bryowsky J, Minshall RD, Pelligrino DA. Possible obligatory functions of cyclic nucleotides in hypercapnia-induced cerebral vasodilation in adult rats. Am J Physiol. 1999;276(2 Pt 2):H480–7.
25. Gulbenkian S, Uddman R, Edvinsson L. Neuronal messengers in the human cerebral circulation. Peptides. 2001;22(6):995–1007.
26. Rebel A, Ulatowski JA, Kwansa H, Bucci E, Koehler RC. Cerebrovascular response to decreased hematocrit: effect of cell-free hemoglobin, plasma viscosity, and CO2. Am J Physiol Heart Circ Physiol. 2003;285(4):H1600–8.
27. Hudak ML, Koehler RC, Rosenberg AA, Traystman RJ, Jones Jr MD. Effect of hematocrit on cerebral blood flow. Am J Physiol. 1986;251(1 Pt 2):H63–70.
28. Lenz C, Rebel A, Bucci E, van Ackem K, Kuschinsky W, Waschke KF. Lack of hypercapnic increase in cerebral blood flow at high blood viscosity in conscious blood-exchanged rats. Anesthesiology. 2001;95(2):408–15.
29. Osol G, Halpern W. Myogenic properties of cerebral blood vessels from normotensive and hypertensive rats. Am J Physiol. 1985;249(5 Pt 2):H914–21.
30. Ogoh S, Brothers RM, Barnes Q, Eubank WL, Hawkins MN, Purkayastha S, et al. The effect of changes in cardiac output on middle cerebral artery mean blood velocity at rest and during exercise. J Physiol. 2005;569(Pt 2):697–704.
31. White RP, Vallance P, Markus HS. Effect of inhibition of nitric oxide synthase on dynamic cerebral autoregulation in humans. Clin Sci (Lond). 2000;99(6):555–60.
32. Brian Jr JE, Faraci FM, Heistad DD. Recent insights into the regulation of cerebral circulation. Clin Exp Pharmacol Physiol. 1996;23(6–7):449–57.
33. Chesler M. Regulation and modulation of pH in the brain. Physiol Rev. 2003;83(4):1183–221.
34. Atkinson JL, Anderson RE, Sundt Jr TM. The effect of carbon dioxide on the diameter of brain capillaries. Brain Res. 1990;517(1–2):333–40.
35. Wei EP, Kontos HA, Patterson Jr JL. Dependence of pial arteriolar response to hypercapnia on vessel size. Am J Physiol. 1980;238(5):697–703.
36. Nelson MT, Quayle JM. Physiological roles and properties of potassium channels in arterial smooth muscle. Am J Physiol. 1995;268(4 Pt 1):C799–822.
37. Kinoshita H, Katusic ZS. Role of potassium channels in relaxations of isolated canine basilar arteries to acidosis. Stroke. 1997;28(2):433–7; discussion 7–8.
38. Parfenova H, Shibata M, Zuckerman S, Leffler CW. CO2 and cerebral circulation in newborn pigs: cyclic nucleotides and prostanoids in vascular regulation. Am J Physiol. 1994;266(4 Pt 2):H1494–501.
39. Leffler CW, Mirro R, Pharris LJ, Shibata M. Permissive role of prostacyclin in cerebral vasodilation to hypercapnia in newborn pigs. Am J Physiol. 1994;267(1 Pt 2):H285–91.
40. Shapiro E, Wasserman AJ, Patterson Jr JL. Human cerebrovascular response time to elevation of arterial carbon dioxide tension. Arch Neurol. 1965;13:130–8.
41. Ellingsen I, Hauge A, Nicolaysen G, Thoresen M, Walloe L. Changes in human cerebral blood flow due to step changes in PAO2 and PACO2. Acta Physiol Scand. 1987;129(2):157–63.
42. Poulin MJ, Liang PJ, Robbins PA. Dynamics of the cerebral blood flow response to step changes in end-tidal PCO2 and PO2 in humans. J Appl Physiol. 1996;81(3):1084–95.
43. Poulin MJ, Liang PJ, Robbins PA. Fast and slow components of cerebral blood flow response to step decreases in end-tidal PCO2 in humans. J Appl Physiol. 1998;85(2):388–97.
44. Ainslie PN, Poulin MJ. Ventilatory, cerebrovascular, and cardiovascular interactions in acute hypoxia: regulation by carbon dioxide. J Appl Physiol. 2004;97(1):149–59.
45. Gupta AK, Menon DK, Czosnyka M, Smielewski P, Jones JG. Thresholds for hypoxic cerebral vasodilation in volunteers. Anesth Analg. 1997;85(4):817–20.
46. Cohen PJ, Alexander SC, Smith TC, Reivich M, Wollman H. Effects of hypoxia and normocarbia on cerebral blood flow and metabolism in conscious man. J Appl Physiol. 1967;23(2):183–9.
47. Yang SP, Bergo GW, Krasney E, Krasney JA. Cerebral pressure-flow and metabolic responses to sustained hypoxia: effect of CO2. J Appl Physiol. 1994;76(1):303–13.
48. Bernardi L, Casucci G, Haider T, Brandstatter E, Pocecco E, Ehrenbourg I, et al. Autonomic and cerebro-vascular abnormalities in mild copd are worsened by chronic smoking. Eur Respir J. 2008;32:1458–65.
49. Lucas STY, Galvin S, Thomas KN, Ainslie PN. Influence of blood pressure on cerebral perfusion and oxygenation in healthy humans. Hypertension. 2010;55:698–705.
50. Paulson OB, Strandgaard S, Edvinsson L. Cerebral autoregulation. Cerebrovasc Brain Metab Rev. 1990;2(2):161–92.
51. Zhang R, Zuckerman JH, Iwasaki K, Wilson TE, Crandall CG, Levine BD. Autonomic neural control of dynamic cerebral autoregulation in humans. Circulation. 2002;106(14):1814–20.
52. Dawson SL, Blake MJ, Panerai RB, Potter JF. Dynamic but not static cerebral autoregulation is impaired in acute ischaemic stroke. Cerebrovasc Dis. 2000;10(2):126–32.
53. Lassen NA. Cerebral blood flow and oxygen consumption in man. Physiol Rev. 1959;39(2):183–238.

54. Van Lieshout JJ, Secher NH, Strandgaard S, Sigurdsson ST. Point: counterpoint: sympathetic activity does/does not influence cerebral blood flow. J Appl Physiol. 2008;26:1364–6.
55. Alm A, Bill A. The effect of stimulation of the cervical sympathetic chain on retinal oxygen tension and on uveal, retinal and cerebral blood flow in cats. Acta Physiol Scand. 1973;88(1):84–94.
56. Harper AM, Deshmukh VD, Rowan JO, Jennett WB. The influence of sympathetic nervous activity on cerebral blood flow. Arch Neurol. 1972;27(1): 1–6.
57. Levine BD, Zhang R. Comments on point: counterpoint: sympathetic activity does/does not influence cerebral blood flow. Autonomic control of the cerebral circulation is most important for dynamic cerebral autoregulation. J Appl Physiol. 2008;105(4): 1369–73.
58. Strandgaard S, Sigurdsson ST. Point: counterpoint: sympathetic activity does/does not influence cerebral blood flow. Counterpoint: sympathetic nerve activity does not influence cerebral blood flow. J Appl Physiol. 2008;105(4):1366–7; discussion 7–8.
59. Cassaglia PA, Griffiths RI, Walker AM. Sympathetic nerve activity in the superior cervical ganglia increases in response to imposed increases in arterial pressure. Am J Physiol Regul Integr Comp Physiol. 2008;294:1255–61.
60. Ogoh S, Brothers S, Raven PB. Autonomic neural control of the cerebral vasculature: acute hypotension. Stroke. 2008;39:1979–87.
61. Jordan J, Shannon JR, Diedrich A, Black B, Costa F, Robertson D, et al. Interaction of carbon dioxide and sympathetic nervous system activity in the regulation of cerebral perfusion in humans. Hypertension. 2000;36(3):383–8.
62. D'Alecy LG, Rose CJ, Sellers SA. Sympathetic modulation of hypercapnic cerebral vasodilation in dogs. Circ Res. 1979;45(6):771–85.
63. Rowell LB. Control of regional blood flow during dynamic exercise. In: Rowell LB, editor. Human cardiovascular control. New York: Oxford University Press; 1993. p. 204–54.
64. Ide K, Pott F, Van Lieshout JJ, Secher NH. Middle cerebral artery blood velocity depends on cardiac output during exercise with a large muscle mass. Acta Physiol Scand. 1998;162(1):13–20.
65. van Lieshout JJ, Pott F, Madsen PL, van Goudoever J, Secher NH. Muscle tensing during standing: effects on cerebral tissue oxygenation and cerebral artery blood velocity. Stroke. 2001;32(7):1546–51.
66. Secher NH, Seifert T, Van Lieshout JJ. Cerebral blood flow and metabolism during exercise: implications for fatigue. J Appl Physiol. 2008;104(1):306–14.
67. Treib J, Haass A, Koch D, Grauer MT, Schimrigk K. Influence of blood pressure and cardiac output on cerebral blood flow and autoregulation in acute stroke measured by TCD. Eur J Neurol. 1996;3: 539–43.
68. Ide K, Boushel R, Sorensen HM, Fernandes A, Cai Y, Pott F, et al. Middle cerebral artery blood velocity during exercise with beta-1 adrenergic and unilateral stellate ganglion blockade in humans. Acta Physiol Scand. 2000;170(1):33–8.
69. Zhang R, Levine BD. Autonomic ganglionic blockade does not prevent reduction in cerebral blood flow velocity during orthostasis in humans. Stroke. 2007;38(4):1238–44.
70. Rubanyi GM, Romero JC, Vanhoutte PM. Flow-induced release of endothelium-derived relaxing factor. Am J Physiol. 1986;250(6 Pt 2):H1145–9.
71. Severinghaus JW, Chiodi H, Eger II EI, Brandstater B, Hornbein TF. Cerebral blood flow in man at high altitude. Role of cerebrospinal fluid pH in normalization of flow in chronic hypocapnia. Circ Res. 1966;19(2):274–82.
72. Wolff CB. Cerebral blood flow and oxygen delivery at high altitude. High Alt Med Biol. 2000;1(1): 33–8.
73. Kety SS, Schmidt CF. The determination of cerebral blod flow in man by the use of nitrous oxide in low concentrations. Am J Physiol. 1945;143:53–66.
74. Jensen JB, Wright AD, Lassen NA, Harvey TC, Winterborn MH, Raichle ME, et al. Cerebral blood flow in acute mountain sickness. J Appl Physiol. 1990;69(2):430–3.
75. Milledge JS, Sorensen SC, Milledge JS, Sorensen SC. Cerebral arteriovenous oxygen difference in man native to high altitude. J Appl Physiol. 1972;32(5):687–9.
76. Moller K, Paulson OB, Hornbein TF, Colier WN, Paulson AS, Roach RC, et al. Unchanged cerebral blood flow and oxidative metabolism after acclimatization to high altitude. J Cereb Blood Flow Metab. [Research Support, Non-U.S. Gov't]. 2002;22(1): 118–26.
77. Roy SB, Guleria JS, Khanna PK, Talwar JR, Manchanda SC, Pande JN, et al. Immediate circulatory response to high altitude hypoxia in man. Nature. 1968;217(5134):1177–8.
78. Baumgartner RW, Bartsch P, Maggiorini M, Waber U, Oelz O. Enhanced cerebral blood flow in acute mountain sickness. Aviat Space Environ Med. 1994;65(8):726–9.
79. Huang SY, Moore LG, McCullough RE, McCullough RG, Micco AJ, Fulco C, et al. Internal carotid and vertebral arterial flow velocity in men at high altitude. J Appl Physiol. 1987;63(1):395–400.
80. Otis SM, Rossman ME, Schneider PA, Rush MP, Ringelstein EB. Relationship of cerebral blood flow regulation to acute mountain sickness. J Ultrasound Med. 1989;8(3):143–8.
81. Fan JL, Burgess KR, Basnyat R, Thomas KN, Peebles KC, Lucas SJ, et al. Influence of high altitude on cerebrovascular and ventilatory responsiveness to CO2. J Physiol. 2010;588(Pt 3):539–49.
82. Lucas SJ, Burgess KR, Thomas KN, Donnelly J, Peebles KC, Lucas RA, et al. Alterations in cerebral blood flow and cerebrovascular reactivity during 14 days at 5050 m. J Physiol. 2011;589(Pt 3):741–53.
83. LaManna JC, Vendel LM, Farrell RM, LaManna JC, Vendel LM, Farrell RM. Brain adaptation to chronic

hypobaric hypoxia in rats. J Appl Physiol. [Research Support, U.S. Gov't, P.H.S.]. 1992;72(6):2238–43.
84. Xu K, Puchowicz MA, LaManna JC, Xu K, Puchowicz MA, LaManna JC. Renormalization of regional brain blood flow during prolonged mild hypoxic exposure in rats. Brain Res. [Comparative Study]. 2004;1027(1–2):188–91.
85. Jansen GF, Basnyat B. Brain blood flow in Andean and Himalayan high-altitude populations: evidence of different traits for the same environmental constraint. J Cereb Blood Flow Metab. 2011;31(2): 706–14.
86. Severinghaus JW. Cerebral circulation at high altitude. In: Hornbein TF, Schoene RB, editors. High altitude; exploration of human adaptation. New York, Basel: M. Dekker, Inc.; 2001. p. 343–76.
87. Chavez JC, Agani F, Pichiule P, LaManna JC. Expression of hypoxia-inducible factor-1alpha in the brain of rats during chronic hypoxia. J Appl Physiol. 2000;89(5):1937–42.
88. La BP. Pression barometrique. Paris: Columbus Ohio College Book Company; 1878.
89. Mosso A. Life of man on the high alps. London: Fisher Unwin; 1898.
90. Jensen JB, Sperling B, Severinghaus JW, Lassen NA. Augmented hypoxic cerebral vasodilation in men during 5 days at 3,810 m altitude. J Appl Physiol. 1996;80(4):1214–8.
91. Brugniaux JV, Hodges AN, Hanly PJ, Poulin MJ. Cerebrovascular responses to altitude. Respir Physiol Neurobiol. 2007;158(2–3):212–23.
92. Ainslie PN, Poulin MJ. Respiratory, cerebrovascular and pressor responses to acute hypoxia: dependency on PET(CO2). Adv Exp Med Biol. 2004;551: 243–9.
93. Thomas KN, Burgess KR, Fan JL, Peebles KC, Lucas RAI, Cotter JD, et al., editors. Time course of changes in cerebral blood flow velocity at 5050, contribution of arterial blood gases, blood pressure and haematocrit. New Zealand Physiological Conference. Dunedin, New Zealand; 2008.
94. Dempsey JA, Forster HV. Mediation of ventilatory adaptations. Physiol Rev. 1982;62(1):262–346.
95. Wilson MH, Edsell ME, Davagnanam I, Hirani SP, Martin DS, Levett DZ, et al. Cerebral artery dilatation maintains cerebral oxygenation at extreme altitude and in acute hypoxia—an ultrasound and MRI study. J Cereb Blood Flow Metab. 2011;31(10): 2019–29.
96. Ainslie PN, Burgess K, Subedi P, Burgess KR. Alterations in cerebral dynamics at high altitude following partial acclimatization in humans: wakefulness and sleep. J Appl Physiol. 2007;102(2): 658–64.
97. Fan JL, Burgess KR, Thomas KN, Peebles KC, Lucas RAI, Cotter JD, et al., editors. Alterations in cerebral blood flow and ventilatory reactivity to CO2 at high altitude; implications in development of periodic breathing. New Zealand Physiological Socitey Conference. Dunedin, New Zealand; 2009.
98. Burgess KR, Lucas SJ, Shepherd K, Dawson A, Swart M, Thomas KN, et al. Worsening of central sleep apnea at high altitude–a role for cerebrovascular function. J Appl Physiol (1985). 2013 Apr;114(8): 1021–8.
99. Ainslie PN, Burgess KR. Cardiorespiratory and cerebrovascular responses to hyperoxic and hypoxic rebreathing: effects of acclimatization to high altitude. Respir Physiol Neurobiol. 2008;161(2):201–9.
100. Blaber AP, Hartley T, Pretorius PJ. Effect of acute exposure to 3,660 m altitude on orthostatic responses and tolerance. J Appl Physiol. 2003;95:591–601.
101. Jansen GF, Krins A, Basnyat B. Cerebral vasomotor reactivity at high altitude in humans. J Appl Physiol. 1999;86(2):681–6.
102. Ainslie PN, Duffin J. Integration of cerebrovascular CO2 reactivity and chemoreflex control of breathing: mechanisms of regulation, measurement and interpretation. Am J Physiol Regul Integr Comp Physiol. 2009;296:1473–95.
103. Jansen GF, Krins A, Basnyat B, Bosch A, Odoom JA. Cerebral autoregulation in subjects adapted and not adapted to high altitude. Stroke. 2000;31(10):2314–8.
104. Levine BD, Zhang R, Roach RC. Dynamic cerebral autoregulation at high altitude. Adv Exp Med Biol. 1999;474:319–22.
105. Jansen GF, Krins A, Basnyat B, Odoom JA, Ince C. Role of the altitude level on cerebral autoregulation in residents at high altitude. J Appl Physiol. 2007;103(2):518–23.
106. Van Osta A, Moraine JJ, Melot C, Mairbaurl H, Maggiorini M, Naeije R. Effects of high altitude exposure on cerebral hemodynamics in normal subjects. Stroke. 2005;36(3):557–60.
107. Bailey DM, Evans KA, James PE, McEneny J, Young IS, Fall L, et al. Altered free radical metabolism in acute mountain sickness: implications for dynamic cerebral autoregulation and blood–brain barrier function. J Physiol. 2009;587(Pt 1):73–85.
108. Cochand NJ, Wild M, Brugniaux JV, Davies PJ, Evans KA, Wise RG, et al. Sea-level assessment of dynamic cerebral autoregulation predicts susceptibility to acute mountain sickness at high altitude. Stroke. 2011;42(12):3628–30.
109. Roach RC, Hackett PH. Frontiers of hypoxia research: acute mountain sickness. J Exp Biol. 2001;204(Pt 18):3161–70.
110. Kallenberg K, Bailey DM, Christ S, Mohr A, Roukens R, Menold E, et al. Magnetic resonance imaging evidence of cytotoxic cerebral edema in acute mountain sickness. J Cereb Blood Flow Metab. 2007;27(5):1064–71.
111. Subudhi AW, Dimmen AC, Julian CG, Wilson MJ, Panerai RB, Roach RC. Effects of acetazolamide and dexamethasone on cerebral hemodynamics in hypoxia. J Appl Physiol. 2011;110(5):1219–25.
112. Subudhi AW, Panerai RB, Roach RC. Effects of hypobaric hypoxia on cerebral autoregulation. Stroke. 2010;41(4):641–6.

113. Moller K, Paulson OB, Hornbein TF, Colier WN, Paulson AS, Roach RC, et al. Unchanged cerebral blood flow and oxidative metabolism after acclimatization to high altitude. J Cereb Blood Flow Metab. 2002;22(1):118–26.
114. Hochachka PW, Clark CM, Brown WD, Stanley C, Stone CK, Nickles RJ, et al. The brain at high altitude: hypometabolism as a defense against chronic hypoxia? J Cereb Blood Flow Metab. 1994;14(4): 671–9.
115. Hochachka PW, Monge C. Evolution of human hypoxia tolerance physiology. Adv Exp Med Biol. 2000;475:25–43.
116. Curran-Everett DC, Iwamoto J, Meredith MP, Krasney JA. Intracranial pressures and O2 extraction in conscious sheep during 72 h of hypoxia. Am J Physiol. 1991;261(1 Pt 2):H103–9.
117. Hansen J, Sander M. Sympathetic neural overactivity in healthy humans after prolonged exposure to hypobaric hypoxia. J Physiol. 2003;546(Pt 3):921–9.
118. Saito M, Mano T, Iwase S, Koga K, Abe H, Yamazaki Y. Responses in muscle sympathetic activity to acute hypoxia in humans. J Appl Physiol. 1988;65(4): 1548–52.
119. Heistad DD, Marcus ML. Evidence that neural mechanisms do not have important effects on cerebral blood-flow. Circ Res. 1978;42(3):295–302.
120. Ainslie PN, Ogoh S. Regulation of cerebral blood flow in mammals during chronic hypoxia: a matter of balance. Exp Physiol. 2010;95(2):251–62.
121. Ainslie PN, Ashmead JC, Ide K, Morgan BJ, Poulin MJ. Differential responses to CO2 and sympathetic stimulation in the cerebral and femoral circulations in humans. J Physiol. 2005;566(Pt 2):613–24.
122. LeMarbre G, Stauber S, Khayat RN, Puleo DS, Skatrud JB, Morgan BJ. Baroreflex-induced sympathetic activation does not alter cerebrovascular CO2 responsiveness in humans. J Physiol. 2003;551(Pt 2):609–16.
123. Przybylowski T, Bangash MF, Reichmuth K, Morgan BJ, Skatrud JB, Dempsey JA. Mechanisms of the cerebrovascular response to apnoea in humans. J Physiol. 2003;548(1):323–32.
124. Lucas SJ, Tzeng YC, Galvin SD, Thomas KN, Ogoh S, Ainslie PN. Influence of changes in blood pressure on cerebral perfusion and oxygenation. Hypertension. 2010;55(3):698–705.
125. Ainslie PN, Ogoh S, Burgess K, Celi L, McGrattan K, Peebles K, et al. Differential effects of acute hypoxia and high altitude on cerebral blood flow velocity and dynamic cerebral autoregulation: alterations with hyperoxia. J Appl Physiol. 2008;104(2): 490–8.
126. Iwasaki K, Zhang R, Zuckerman JH, Ogawa Y, Hansen LH, Levine BD. Impaired dynamic cerebral autoregulation at extreme high altitude even after acclimatization. J Cereb Blood Flow Metab. 2011;31(1):283–92.
127. Ainslie PNLS, Fan M, Thomas KN, Cotter JD, Tzeng YC, Burgess KR. Influence of sympathoexcitation on cerebrovascular function and ventilatory control at high altitude. J Appl Physiol. 2012;113(7):1058–67.
128. Willie CK, Colino FL, Bailey DM, Tzeng YC, Binsted G, Jones LW, et al. Utility of transcranial Doppler ultrasound for the integrative assessment of cerebrovascular function. J Neurosci Methods. 2011;196(2):221–37.
129. Curran-Everett DC, Meredith MP, Krasney JA. Acclimatization to hypoxia alters cerebral convective and diffusive O2 delivery. Respir Physiol. 1992;88(3):355–71.
130. Golanov EV, Christensen JR, Reis DJ. Neurons of a limited subthalamic area mediate elevations in cortical cerebral blood flow evoked by hypoxia and excitation of neurons of the rostral ventrolateral medulla. J Neurosci. 2001;21(11):4032–41.
131. Golanov EV, Reis DJ. Contribution of oxygen-sensitive neurons of the rostral ventrolateral medulla to hypoxic cerebral vasodilatation in the rat. J Physiol. 1996;495(Pt 1):201–16.
132. Jones Jr MD, Traystman RJ, Simmons MA, Molteni RA. Effects of changes in arterial O2 content on cerebral blood flow in the lamb. Am J Physiol. 1981;240(2):H209–15.
133. Brown MM, Wade JP, Marshall J. Fundamental importance of arterial oxygen content in the regulation of cerebral blood flow in man. Brain. 1985;108(Pt 1):81–93.
134. Brown MM, Marshall J. Regulation of cerebral blood flow in response to changes in blood viscosity. Lancet. 1985;1(8429):604–9.
135. Waschke KF, Krieter H, Hagen G, Albrecht DM, Van Ackern K, Kuschinsky W. Lack of dependence of cerebral blood flow on blood viscosity after blood exchange with a Newtonian O2 carrier. J Cereb Blood Flow Metab. 1994;14(5):871–6.
136. Tomiyama Y, Brian Jr JE, Todd MM. Plasma viscosity and cerebral blood flow. Am J Physiol Heart Circ Physiol. 2000;279(4):H1949–54.
137. Ulatowski JA, Bucci E, Razynska A, Traystman RJ, Koehler RC. Cerebral blood flow during hypoxic hypoxia with plasma-based hemoglobin at reduced hematocrit. Am J Physiol. 1998;274(6 Pt 2): H1933–42.
138. Miller Jr AT, Hale DM. Increased vascularity of brain, heart, and skeletal muscle of polycythemic rats. Am J Physiol. 1970;219(3):702–4.
139. Xu K, Lamanna JC. Chronic hypoxia and the cerebral circulation. J Appl Physiol. 2006;100(2):725–30.
140. Boero JA, Ascher J, Arregui A, Rovainen C, Woolsey TA. Increased brain capillaries in chronic hypoxia. J Appl Physiol. 1999;86(4):1211–9.
141. Dore-Duffy P, LaManna JC. Physiologic angiodynamics in the brain. Antioxid Redox Signal. 2007;9(9):1363–71.

142. Julien-Dolbec C, Tropres I, Montigon O, Reutenauer H, Ziegler A, Decorps M, et al. Regional response of cerebral blood volume to graded hypoxic hypoxia in rat brain. Br J Anaesth. 2002;89(2):287–93.
143. Shockley RP, LaManna JC. Determination of rat cerebral cortical blood volume changes by capillary mean transit time analysis during hypoxia, hypercapnia and hyperventilation. Brain Res. 1988; 454(1–2):170–8.
144. Dunn JF, Roche MA, Springett R, Abajian M, Merlis J, Daghlian CP, et al. Monitoring angiogenesis in brain using steady-state quantification of DeltaR2 with MION infusion. Magn Reson Med. 2004;51(1):55–61.
145. Wilson MH, Newman S, Imray CH. The cerebral effects of ascent to high altitudes. Lancet Neurol. 2009;8(2):175–91.
146. Faraci FM, Heistad DD. Regulation of the cerebral circulation: role of endothelium and potassium channels. Physiol Rev. 1998;78(1):53–97.
147. Sanders DB, Kelley T, Larson D. The role of nitric oxide synthase/nitric oxide in vascular smooth muscle control. Perfusion. 2000;15(2):97–104.
148. Appenzeller O, Claydon VE, Gulli G, Qualls C, Slessarev M, Zenebe G, et al. Cerebral vasodilatation to exogenous NO is a measure of fitness for life at altitude. Stroke. 2006;37(7):1754–8.
149. Na SJ, Hong JM, Park JH, Chung TS, Lee KY. A case of reversible postpartum cytotoxic edema in preeclampsia. J Neurol Sci. 2004;221(1–2):83–7.
150. Farb RI, Vanek I, Scott JN, Mikulis DJ, Willinsky RA, Tomlinson G, et al. Idiopathic intracranial hypertension: the prevalence and morphology of sinovenous stenosis. Neurology. 2003;60(9):1418–24.
151. Owler BK, Parker G, Halmagyi GM, Johnston IH, Besser M, Pickard JD, et al. Cranial venous outflow obstruction and pseudotumor Cerebri syndrome. Adv Tech Stand Neurosurg. 2005;30:107–74.
152. Mader TH, Tabin G. Going to high altitude with preexisting ocular conditions. High Alt Med Biol. 2003;4(4):419–30.
153. Kallenberg K, Dehnert C, Dorfler A, Schellinger PD, Bailey DM, Knauth M, et al. Microhemorrhages in nonfatal high-altitude cerebral edema. J Cereb Blood Flow Metab. 2008;28(9):1635–42.
154. Dickinson J, Heath D, Gosney J, Williams D. Altitude-related deaths in seven trekkers in the Himalayas. Thorax. 1983;38(9):646–56.
155. Wilson MH, Imray CH, Hargens AR. The headache of high altitude and microgravity—similarities with clinical syndromes of cerebral venous hypertension. High Alt Med Biol. 2011;12(4):379–86.
156. Wilson MH, Davagnanam I, Holland G, Dattani R, Hirani S, Kolfschoten N, Strycharczuk L, Green C, Thornton J, Wright A, Bradwell A, Edsell M, Kitchen N, Holloway C, Clarke K, Grocott M, Montgomery H, Imray C. The cerebral venous system and anatomical predisposition to high altitude headache. Ann Neurol. 2012;73(3):381–9.
157. Bartsch P, Gibbs JS. Effect of altitude on the heart and the lungs. Circulation. 2007;116(19):2191–202.
158. Miserocchi G, Passi A, Negrini D, Del Fabbro M, De Luca G. Pulmonary interstitial pressure and tissue matrix structure in acute hypoxia. Am J Physiol Lung Cell Mol Physiol. 2001;280(5):L881–7.
159. Houston CS. Incidence of acute mountain sickness at intermediate altitudes. JAMA. 1989;261(24):3551–2.
160. Yamakawa M, Liu LX, Date T, Belanger AJ, Vincent KA, Akita GY, et al. Hypoxia-inducible factor-1 mediates activation of cultured vascular endothelial cells by inducing multiple angiogenic factors. Circ Res. 2003;93(7):664–73.
161. Severinghaus JW. Hypothesis: angiogenesis cytokines in high altitude cerebral oedema. Acta Anaesthesiol Scand Suppl. 1995;107:177–8.
162. Schoch HJ, Fischer S, Marti HH. Hypoxia-induced vascular endothelial growth factor expression causes vascular leakage in the brain. Brain. 2002;125(Pt 11):2549–57.
163. Dorward DA, Thompson AA, Baillie JK, MacDougall M, Hirani N. Change in plasma vascular endothelial growth factor during onset and recovery from acute mountain sickness. Respir Med. 2007;101(3):587–94.
164. Tissot van Patot MC, Leadbetter G, Keyes LE, Bendrick-Peart J, Beckey VE, Christians U, et al. Greater free plasma VEGF and lower soluble VEGF receptor-1 in acute mountain sickness. J Appl Physiol. 2005;98(5):1626–9.
165. Bailey DM, Roukens R, Knauth M, Kallenberg K, Christ S, Mohr A, et al. Free radical-mediated damage to barrier function is not associated with altered brain morphology in high-altitude headache. J Cereb Blood Flow Metab. 2006;26(1):99–111.
166. Bailey DM. Ascorbate, blood–brain barrier function and acute mountain sickness: a radical hypothesis. Wilderness Environ Med. 2004;15(3):231–3.
167. Bailey DM. Radical dioxygen: from gas to (unpaired!) electrons. Adv Exp Med Biol. 2003;543:201–21.
168. Bailey DM, Kleger GR, Holzgraefe M, Ballmer PE, Bartsch P. Pathophysiological significance of peroxidative stress, neuronal damage, and membrane permeability in acute mountain sickness. J Appl Physiol. 2004;96(4):1459–63.
169. Headache Classification Subcommittee of the International Headache Society. The International Classification of Headache Disorders: 2nd edition. Cephalalgia. 2004;24 Suppl 1:9–160.
170. Roach R, Bartsch P, Oelz O, Hackett P. Lake Louise AMS Scoring Consensus Committee. The Lake Louise acute mountain sickness scoring system. In: Coates G, editor. Hypoxia and molecular medicine. Burlington (VT): Charles S. Houston; 1993. p. 272–4.
171. Hackett PH, Roach RC. High altitude cerebral edema. High Alt Med Biol. 2004;5(2):136–46.
172. Thomassen O, Skaiaa SC. High-altitude cerebral edema with absence of headache. Wilderness Environ Med. 2007;18(1):45–7.

173. Harvey TC, Raichle ME, Winterborn MH, Jensen J, Lassen NA, Richardson NV, et al. Effect of carbon dioxide in acute mountain sickness: a rediscovery. Lancet. 1988;2(8612):639–41.
174. Imray CH, Clarke T, Forster PJ, Harvey TC, Hoar H, Walsh S, et al. Carbon dioxide contributes to the beneficial effect of pressurization in a portable hyperbaric chamber at high altitude. Clin Sci (Lond). 2001;100(2):151–7.
175. Imray CH, Walsh S, Clarke T, Tiivas C, Hoar H, Harvey TC, et al. Effects of breathing air containing 3% carbon dioxide, 35% oxygen or a mixture of 3% carbon dioxide/35% oxygen on cerebral and peripheral oxygenation at 150 m and 3459 m. Clin Sci (Lond). 2003;104(3):203–10.
176. Maher JT, Cymerman A, Reeves JT, Cruz JC, Denniston JC, Grover RF. Acute mountain sickness: increased severity in eucapnic hypoxia. Aviat Space Environ Med. 1975;46(6):826–9.
177. Bartsch P, Baumgartner RW, Waber U, Maggiorini M, Oelz O. Comparison of carbon-dioxide-enriched, oxygen-enriched, and normal air in treatment of acute mountain sickness. Lancet. 1990;336(8718): 772–5.
178. Baumgartner RW, Spyridopoulos I, Bartsch P, Maggiorini M, Oelz O. Acute mountain sickness is not related to cerebral blood flow: a decompression chamber study. J Appl Physiol. 1999;86(5):1578–82.
179. Reeves JT, Moore LG, McCullough RE, McCullough RG, Harrison G, Tranmer BI, et al. Headache at high altitude is not related to internal carotid arterial blood velocity. J Appl Physiol. 1985;59(3):909–15.
180. Dyer EA, Hopkins SR, Perthen JE, Buxton RB, Dubowitz DJ. Regional cerebral blood flow during acute hypoxia in individuals susceptible to acute mountain sickness. Respir Physiol Neurobiol. 2008;160(3):267–76.
181. Roach RCMD, Sandoval D, Robergs RA, Icenogle M, Hinghofer-Szalkay H, Lium D, Loeppky JA. Exercise exacerbates acute mountain sickness at simulated high altitude. J Appl Physiol. 2000;88(2):581–5.
182. Schommer KHM, Hotz L, Menold E, Bärtsch P, Berger MM. Exercise intensity typical of mountain climbing does not exacerbate acute mountain sickness in normobaric hypoxia. J Appl Physiol. 2012;113(7):1068–74.
183. Imray CH, Myers SD, Pattinson KT, Bradwell AR, Chan CW, Harris S, et al. Effect of exercise on cerebral perfusion in humans at high altitude. J Appl Physiol. 2005;99(2):699–706.
184. Subudhi AW, Dimmen AC, Roach RC. Effects of acute hypoxia on cerebral and muscle oxygenation during incremental exercise. J Appl Physiol. 2007;103(1):177–83.
185. Subudhi AW, Olin JT, Dimmen AC, Polaner DM, Kayser B, Roach RC. Does cerebral oxygen delivery limit incremental exercise performance? J Appl Physiol. 2011;111(6):1727–34.
186. Rostrup E, Law I, Pott F, Ide K, Knudsen GM. Cerebral hemodynamics measured with simultaneous PET and near-infrared spectroscopy in humans. Brain Res. 2002;954(2):183–93.
187. Henson LC, Calalang C, Temp JA, Ward DS. Accuracy of a cerebral oximeter in healthy volunteers under conditions of isocapnic hypoxia. Anesthesiology. 1998;88(1):58–65.
188. Bucher HU, Edwards AD, Lipp AE, Duc G. Comparison between near infrared spectroscopy and 133Xenon clearance for estimation of cerebral blood flow in critically ill preterm infants. Pediatr Res. 1993;33(1):56–60.
189. Williams IM, Vohra R, Farrell A, Picton AJ, Mortimer AJ, McCollum CN. Cerebral oxygen saturation, transcranial Doppler ultrasonography and stump pressure in carotid surgery. Br J Surg. 1994;81(7):960–4.
190. Imray CH, Barnett NJ, Walsh S, Clarke T, Morgan J, Hale D, et al. Near-infrared spectroscopy in the assessment of cerebral oxygenation at high altitude. Wilderness Environ Med. 1998;9(4):198–203.
191. Hadolt I, Litscher G. Noninvasive assessment of cerebral oxygenation during high altitude trekking in the Nepal Himalayas (2850–5600 m). Neurol Res. 2003;25(2):183–8.
192. Subudhi AW, Dimmen AC, Roach RC. Effects of acute hypoxia on cerebral and muscle oxygenation during incremental exercise. J Appl Phyisol. 2007;103:177–83.
193. Subudhi AW, Miramon BR, Granger ME, Roach RC. Frontal and motor cortex oxygenation during maximal exercise in normoxia and hypoxia. J Appl Physiol. 2009;106(4):1153–8.
194. Vuyk J, Van Den Bos J, Terhell K, De Bos R, Vletter A, Valk P, et al. Acetazolamide improves cerebral oxygenation during exercise at high altitude. High Alt Med Biol. 2006;7(4):290–301.
195. Krakauer J. Into thin air. New York: Villard; 1997.
196. Gw R. Decision making at extreme altitude; has anyone seen my excetutive function lately? Wilderness Environ Med. 2012;23(3):288–91.
197. Huey RB, Eguskitza X. Supplemental oxygen and mountaineer death rates on Everest and K2. JAMA. 2000;284(2):181.
198. Firth PG, Zheng H, Windsor JS, Sutherland AI, Imray CH, Moore GW, et al. Mortality on Mount Everest, 1921–2006: descriptive study. BMJ. 2008;337:a2654.
199. Rodway GW. Limb paralysis and visual changes during Glaisher and Coxwell's 1862 balloon ascent to over 8800 m. High Alt Med Biol. 2007;8(3):256–9.
200. West JB. Paralysis and blindness during a balloon ascent to high altitude. High Alt Med Biol. 2004;5(4):453–6.
201. Glaisher J. Notes of effects experienced during recent ballon ascents. Lancet. 1862;2:559–60.
202. Bert P. Barometric pressure. In: Hitchcock FA, editor. Researches in experimental physiology.

Columbus: College Book; 1878. p. 171–92, 704–7, 963–83.
203. West JB. T.H. Ravenhill and his contributions to mountain sickness. J Appl Physiol. 1996;80(3): 715–24.
204. Ravenhill TH. Some experiences of mountain sickness in the Andes. J Tropical Med Hyg. 1913;16: 313–20.
205. McFarland RA. Psycho-physiological studies at high altitude in the Andes. I. The effects of rapid ascents by aeroplane and train. J Comp Psychol. 1937;23:191–225.
206. Luft UC, Clamann HG, Opitz E. The latency of hypoxia on exposure to altitude above 50,000 feet. J Aviat Med. 1951;22(2):117–22; passim.
207. Velasquez T. Tolerance to acute anoxia in high altitude natives. J Appl Physiol. 1959;14(3):357–62.
208. Grocott MP, Martin DS, Levett DZ, McMorrow R, Windsor J, Montgomery HE. Arterial blood gases and oxygen content in climbers on Mount Everest. N Engl J Med. 2009;360(2):140–9.
209. Greene R. Mental performance in chronic anoxia. Br Med J. 1957;1(5026):1028–31.
210. Wu X, Li X, Han L, Wang T, Wei Y. Effects of acute moderate hypoxia on human performance of arithmetic. Space Med Med Eng (Beijing). 1998;11(6): 391–5.
211. Pelamatti G, Pascotto M, Semenza C. Verbal free recall in high altitude: proper names vs common names. Cortex. 2003;39(1):97–103.
212. Du JY, Li XY, Zhuang Y, Wu XY, Wang T. [Effects of acute mild and moderate hypoxia on human short memory]. Space Med Med Eng (Beijing). 1999;12(4):270–3.
213. Virues-Ortega J, Buela-Casal G, Garrido E, Alcazar B. Neuropsychological functioning associated with high-altitude exposure. Neuropsychol Rev. 2004;14(4):197–224.
214. Bouquet CA, Gardette B, Gortan C, Abraini JH. Psychomotor skills learning under chronic hypoxia. Neuroreport. 1999;10(14):3093–9.
215. Kramer AF, Coyne JT, Strayer DL. Cognitive function at high altitude. Hum Factors. 1993;35(2): 329–44.
216. Bolmont B, Bouquet C, Thullier F. Relationships of personality traits with performance in reaction time, psychomotor ability, and mental efficiency during a 31-day simulated climb of Mount Everest in a hypobaric chamber'. Percept Mot Skills. 2001;92(3 Pt 2): 1022–30.
217. Wesensten NJ, Crowley J, Balkin T, Kamimori G, Iwanyk E, Pearson N, et al. Effects of simulated high altitude exposure on long-latency event-related brain potentials and performance. Aviat Space Environ Med. 1993;64(1):30–6.
218. Wilson MH, Edsell M, Imray C, Wright A. Changes in pupil dynamics at high altitude—an observational study using a handheld pupillometer. High Alt Med Biol. 2008;9(4):319–25.
219. Bolmont B, Thullier F, Abraini JH. Relationships between mood states and performances in reaction time, psychomotor ability, and mental efficiency during a 31-day gradual decompression in a hypobaric chamber from sea level to 8848 m equivalent altitude. Physiol Behav. 2000;71(5):469–76.
220. Gerard AB, McElroy MK, Taylor MJ, Grant I, Powell FL, Holverda S, et al. Six percent oxygen enrichment of room air at simulated 5,000 m altitude improves neuropsychological function. High Alt Med Biol. 2000;1(1):51–61.
221. Reite M, Jackson D, Cahoon RL, Weil JV. Sleep physiology at high altitude. Electroencephalogr Clin Neurophysiol. 1975;38(5):463–71.
222. Forster PJ. Effect of different ascent profiles on performance at 4,200 m elevation. Aviat Space Environ Med. 1985;56(8):758–64.
223. Kobrick JL, Appleton B. Effects of extended hypoxia on visual performance and retinal vascular state. J Appl Physiol. 1971;31(3):357–62.
224. Berry DT, McConnell JW, Phillips BA, Carswell CM, Lamb DG, Prine BC. Isocapnic hypoxemia and neuropsychological functioning. J Clin Exp Neuropsychol. 1989;11(2):241–51.
225. Stobdan T, Karar J, Pasha MA. High altitude adaptation: genetic perspectives. High Alt Med Biol. 2008;9(2):140–7.
226. Nelson TO, Dunlosky J, White DM, Steinberg J, Townes BD, Anderson D. Cognition and metacognition at extreme altitudes on Mount Everest. J Exp Psychol Gen. 1990;119(4):367–74.
227. Nelson M. Psychological testing at high altitudes. Aviat Space Environ Med. 1982;53(2):122–6.
228. Regard M, Oelz O, Brugger P, Landis T. Persistent cognitive impairment in climbers after repeated exposure to extreme altitude. Neurology. 1989;39(2 Pt 1):210–3.
229. Hornbein TF, Townes BD, Schoene RB, Sutton JR, Houston CS. The cost to the central nervous system of climbing to extremely high altitude. N Engl J Med. 1989;321(25):1714–9.
230. Johnson BG, Simmons J, Wright AD, Hillenbrand P, Beazley ME, Sutton I, et al. Ataxia at altitude measured on a wobble board. Wilderness Environ Med. 2005;16(1):42–6.
231. Fagenholz PJ, Murray AF, Gutman JA, Findley JK, Harris NS. New-onset anxiety disorders at high altitude. Wilderness Environ Med. 2007;18(4):312–6.
232. Windsor JS. Voices in the air. BMJ. 2008;337:a2667.
233. Sutton JR, Reeves JT, Groves BM, Wagner PD, Alexander JK, Hultgren HN, et al. Oxygen transport and cardiovascular function at extreme altitude: lessons from Operation Everest II. Int J Sports Med. 1992;13 Suppl 1:S13–8.
234. Anooshiravani M, Dumont L, Mardirosoff C, Soto-Debeuf G, Delavelle J. Brain magnetic resonance imaging (MRI) and neurological changes after a single high altitude climb. Med Sci Sports Exerc. 1999;31(7):969–72.

235. Garrido E, Castello A, Ventura JL, Capdevila A, Rodriguez FA. Cortical atrophy and other brain magnetic resonance imaging (MRI) changes after extremely high-altitude climbs without oxygen. Int J Sports Med. 1993;14(4):232–4.
236. Garrido E, Segura R, Capdevila A, Aldoma J, Rodriguez FA, Javierra C, et al. New evidence from magnetic resonance imaging of brain changes after climbs at extreme altitude. Eur J Appl Physiol Occup Physiol. 1995;70(6):477–81.
237. Di Paola M, Bozzali M, Fadda L, Musicco M, Sabatini U, Caltagirone C. Reduced oxygen due to high-altitude exposure relates to atrophy in motor-function brain areas. Eur J Neurol. 2008;15(10):1050–7.
238. Basnyat B, Wu T, Gertsch JH. Neurological conditions at altitude that fall outside the usual definition of altitude sickness. High Alt Med Biol. 2004;5(2):171–9.
239. Imray CH, Pattinson KT, Myers S, Chan CW, Hoar H, Brearey S, et al. Intrapulmonary and intracardiac shunting with exercise at altitude. Wilderness Environ Med. 2008;19(3):199–204.
240. Litch JA, Bishop RA. Transient global amnesia at high altitude. N Engl J Med. 1999;340(18):1444.
241. Bucuk M, Tomic Z, Tuskan-Mohar L, Bonifacic D, Bralic M, Jurjevic A. Recurrent transient global amnesia at high altitude. High Alt Med Biol. 2008;9(3):239–40.
242. Clarke C. Acute mountain sickness: medical problems associated with acute and subacute exposure to hypobaric hypoxia. Postgrad Med J. 2006;82(973): 748–53.
243. Song SY, Asaji T, Tanizaki Y, Fujimaki T, Matsutani M, Okeda R. Cerebral thrombosis at altitude: its pathogenesis and the problems of prevention and treatment. Aviat Space Environ Med. 1986;57(1):71–6.
244. Zhou WD. Transient ischemic attack in young people in high altitude. Qinghai Med J. 1984;14:71–6.
245. Nair V, Mohapatro AK, Sreedhar M, Indrajeet IK, Tewari AK, Anand AC, et al. A case of hereditary protein S deficiency presenting with cerebral sinus venous thrombosis and deep vein thrombosis at high altitude. Acta Haematol. 2008;119(3): 158–61.
246. Boulos P, Kouroukis C, Blake G. Superior sagittal sinus thrombosis occurring at high altitude associated with protein C deficiency. Acta Haematol. 1999;102(2):104–6.
247. Baumgartner RW, Siegel AM, Hackett PH. Going high with preexisting neurological conditions. High Alt Med Biol. 2007;8(2):108–16.
248. Hackett P. High altitude and common medical conditions. An exploration of human adaptation. New York: Marcel Dekker; 2001. p. 839–85.
249. Woods DR, Allen S, Betts TR, Gardiner D, Montgomery H, Morgan JM, et al. High altitude arrhythmias. Cardiology. 2008;111(4):239–46.
250. Nicholas R, O'Meara PD, Calonge N. Is syncope related to moderate altitude exposure? JAMA. 1992;268(7):904–6.
251. Thomas KN, Lucas SJE, Burgess KR, Cotter JD, Fan JL, Peebles KC, et al. Initial orthostatic hypotension at high altitude. High Alt Med Biol. 2010;11:163–7.
252. Morgan BJ. Vascular consequences of intermittent hypoxia. In: Roach RC, Wagner PD, Hackett PH, editors. Hypoxia and the circulation. New York: Springer; 2007. p. 68–84.
253. Raichle ME, Hornbein TF. The high altitude brain. In: Hornbein TF, Schoene RB, editors. High altitude; exploration of human adaptation. New York, Basel: M. Dekker, Inc.; 2001.

Autonomic Nervous System

8

Mark J. Drinkhill, Roger Hainsworth, and Victoria E. Claydon

Abstract

All stresses to which the body is subjected induce widespread changes in neural activity and in hormonal secretions. Generally, these are protective in nature and the best known is the "fright and flight" reaction to danger. The responses to these stresses include increases in activity in sympathetic nerves, decreases in parasympathetic activity and increased secretions of various hormones including adrenaline and cortisol. Exposure to high altitude is another stress caused by the hypoxic environment as well as the low temperatures that frequently also occurs. This chapter is concerned with the effects of high altitude on the autonomic nervous system, both in lowlanders who visit high altitude locations and in people who are permanently resident there. The resulting changes in blood gases and often also in blood pressure induce reflex responses mediated largely through autonomic nerves. The role of the autonomic system in the adaptation as well as changes in the maladaptation syndromes of acute and chronic mountain sickness will be described.

Outline of the Autonomic Nervous System

The term "autonomic nervous system" was introduced by JN Langley a 100 years ago to refer to the nerves concerned mainly with the control of the internal environment of the body and bodily functions. Autonomic nerves predominantly control unconscious and involuntary functions, although some pain sensations, for example visceral distension and cardiac pain, are mediated through autonomic pathways. The autonomic nervous system is integrated with the rest of the central nervous system, and somatic afferents

M.J. Drinkhill, Ph.D. (✉)
R. Hainsworth, Ph.D.
Division of Cardiovascular and Neuronal Remodelling, Faculty of Medicine, University of Leeds, Leeds LS2 9JT, UK
e-mail: m.j.drinkhill@leeds.ac.uk; r.hainsworth@leeds.ac.uk

V.E. Claydon, Ph.D.
Department of Biomedical Physiology and Kinesiology, Faculty of Science, Simon Fraser University, 8888 University Drive, Burnaby, BC V5A 1S6, Canada
e-mail: victoria_claydon@sfu.ca

E.R. Swenson and P. Bärtsch (eds.), *High Altitude: Human Adaptation to Hypoxia*, DOI 10.1007/978-1-4614-8772-2_8,

can influence autonomic function and vice versa. Many regions in the brain, particularly the hypothalamus, affect autonomic activity.

The autonomic system is classically divided into the sympathetic and the parasympathetic divisions, the functions of which tend to be opposing. For example, heart rate is increased by sympathetic activity and decreased by parasympathetic activity. Both afferent and efferent nerves run in both divisions and reflexes with afferent activity in one division usually induce efferent responses in both, often reciprocally increasing activity in one and decreasing it in the other.

The two autonomic divisions do not function in a homogeneous manner. There are many examples of uneven responses. For example, stimulation of baroreceptors inhibits sympathetic activity to the heart and to blood vessels in muscle and abdominal viscera but has little effect on cutaneous vessels which are much more affected by thermal stimuli. Stimulation of cardiac atrial receptors increases sympathetic activity to the heart, decreases it to the kidney and has little effect on peripheral discharge [1]. This variation is important to consider when interpreting generalised effects.

Reflex Control of Autonomic Activity

Afferent activity in autonomic nerves from almost all regions of the body has been shown, using anaesthetised animal preparations, to induce widespread reflex responses in both divisions of the autonomic nervous system. However, the two reflexes that have received the most attention in humans, particularly in altitude and hypoxia research, are those arising from arterial baroreceptors and chemoreceptors. Accordingly, this chapter will primarily focus on the alterations in autonomic function that arise from stimulation of these reflexes.

Baroreceptor Reflexes

Baroreceptors sense the level of arterial blood pressure from the degree of stretch of the walls of certain arteries. Baroreceptors, situated at the origin of the internal carotid artery (carotid sinuses), are of particular importance as they are influenced by the blood pressure perfusing the brain. They are also accessible for study in humans. However, baroreceptors also exist elsewhere, including the aortic arch and the coronary arteries. They form the afferent limb of a negative feedback reflex control system. Increases in blood pressure increase the stretch of these receptors and they induce reflex decreases in sympathetic activity to blood vessels, causing decreases in vascular resistance and increases in capacitance, as well as decreases in sympathetic and increases in parasympathetic activity to the heart, leading to decreases in both rate and force of contraction.

Baroreceptors show phasic properties and are able to respond to changes in pulse pressure even when mean pressure does not change. In this way, they can respond to changes in cardiac output. An example of this occurs when moving from supine to standing, or especially during passive upright tilting, which leads to "pooling" of blood in dependent regions (legs and pelvic veins) and a consequent reduction in cardiac output. This and the associated tachycardia cause stroke volume and therefore pulse pressure to decrease, so that baroreceptors are stimulated less and consequently mean blood pressure does not fall and usually actually increases.

The characteristics of baroreceptors may be defined in terms of their sensitivity or gain (maximum slope of the stimulus–response curve), threshold (lowest pressure required to elicit a response), "set" point (usually the mid-point of their operating range) and saturation pressure (pressure above which no further responses can be obtained) (Fig. 8.1). Different receptors show different characteristics so that increasing pressure recruits more receptors. Also, different baroreceptors groups operate over different ranges: coronary baroreceptors have a lower pressure range than those in the aorta and carotid arteries [2].

Baroreceptors "reset" when subjected to sustained changes in pressure. This starts to occur in less than 30 min. They are therefore more effective in reducing short-term fluctuations in pressure, and their role in long-term control is unclear.

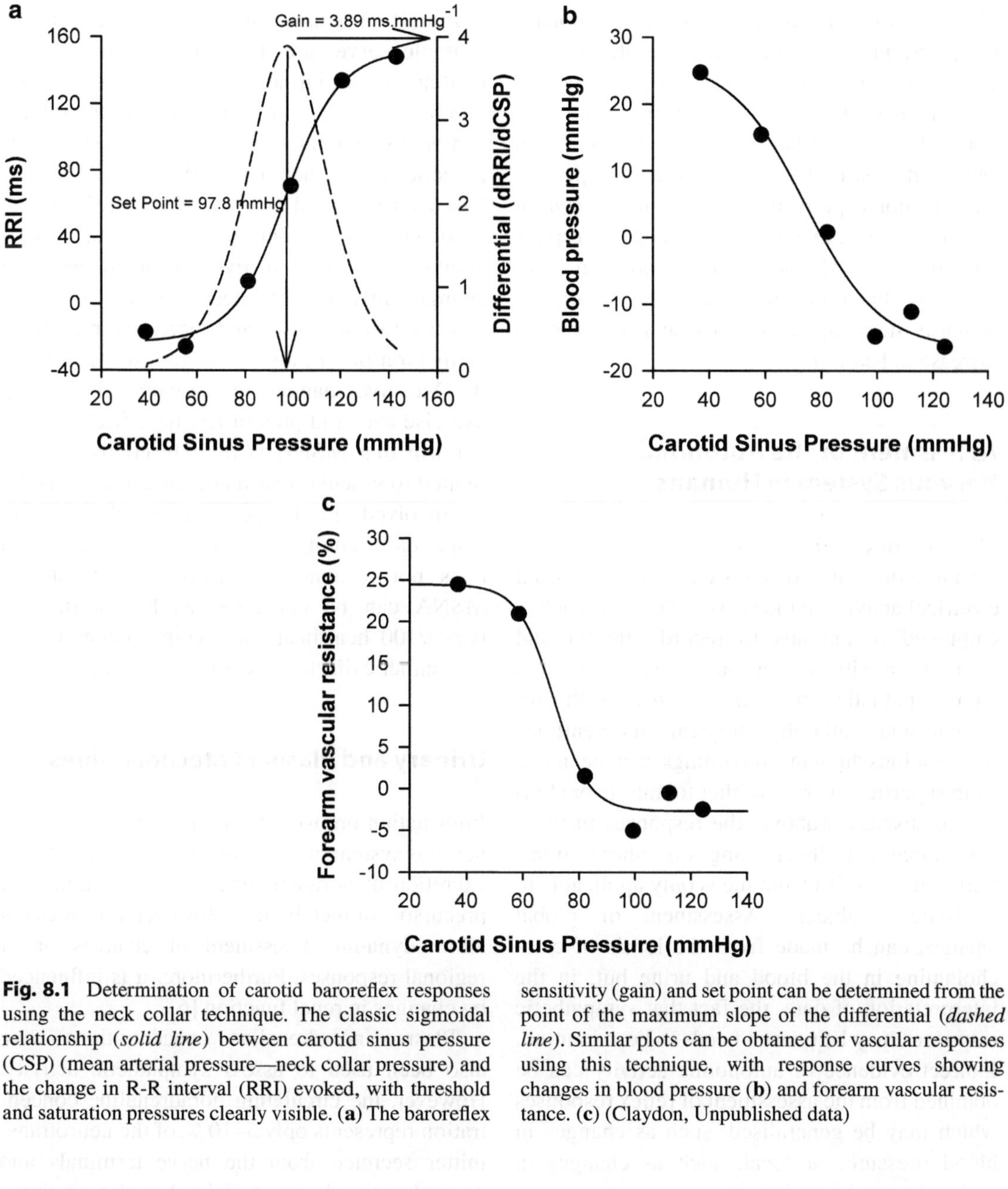

Fig. 8.1 Determination of carotid baroreflex responses using the neck collar technique. The classic sigmoidal relationship (*solid line*) between carotid sinus pressure (CSP) (mean arterial pressure—neck collar pressure) and the change in R-R interval (RRI) evoked, with threshold and saturation pressures clearly visible. (**a**) The baroreflex sensitivity (gain) and set point can be determined from the point of the maximum slope of the differential (*dashed line*). Similar plots can be obtained for vascular responses using this technique, with response curves showing changes in blood pressure (**b**) and forearm vascular resistance. (**c**) (Claydon, Unpublished data)

Peripheral Arterial Chemoreceptors

Chemoreceptors are found in the carotid and aortic bodies and are innervated, along with the adjacent baroreceptors, by the glossopharyngeal and vagus nerves (IX and X cranial nerves respectively). Their principal stimulus is hypoxia and this becomes increasingly effective below oxygen tensions of about 60 mmHg. However, they are also sensitive to arterial CO_2 and hydrogen ion levels. Stimulation of chemoreceptors increases the rate and depth of ventilation. The primary cardiovascular effects are bradycardia (vagal) and constriction of resistance and capacitance vessels (sympathetic). These responses can be demonstrated in anaesthetised controlled preparations, but in intact spontaneously breathing animals, including humans, the rate and depth of respiration increases, stimulating pulmonary receptors, and this largely abolishes the primary cardiovascular responses and usually induces a reflex increase in heart rate. A further complication is that the ventilatory response lowers PCO_2 and this decreases the stimulus both to the peripheral chemoreceptors themselves and to central chemoreceptors, thereby diminishing the respiratory effect. Overall, the effect of hypoxia in spontaneously breathing individuals is usually to increase the sympathetic drive and to cause an increase in heart rate.

Assessment of the Autonomic Nervous System in Humans

The only direct method of assessing activity in the autonomic nervous system is by recording electrical activity in the nerves. This approach is employed in animals to record afferent and efferent activity in various sympathetic and parasympathetic nerves in response to physiological and pathophysiological interventions. In conscious humans recordings can be made from superficial nerves (either to muscle or skin) but, as discussed above, the responses in these nerves may not reflect changes in other regions. Furthermore, this technique is only applicable to motionless subjects. Assessment of global changes can be made from the levels of catecholamine in the blood and urine but, in the interpretation of data, the fact that sympathetic responses may be uneven needs to born in mind. Indirect evidence of autonomic activity can be obtained from the assessment of reflex responses which may be generalised, such as changes in blood pressure, or local, such as changes in regional vascular resistance.

Microneurography

Microneurography in humans was developed by Hagbarth and Vallbo [3] and it provides information on regional sympathetic outflow. The technique involves the insertion of microelectrodes directly into a nerve and the recording of multiunit or single unit sympathetic activity. Suitable nerves are the peroneal, tibial, ulnar, median and radial nerves [4, 5] and these contain both afferent and efferent fibres. Recordings can be discriminated into muscle and skin sympathetic nerve activity (MSNA and SSNA) based on accepted criteria. MSNA reflects the vasoconstrictor activity to skeletal muscle vasculature and is sensitive to blood pressure changes; increases and decreases in blood pressure lead to decreases and increases in activity. Activation of cardiovascular reflexes, including the Valsalva manoeuvre, isometric hand grip exercise and cold pressor test have been used to confirm that MSNA is reflexly controlled and related to vascular resistance. In contrast, SSNA is involved in temperature regulation and responds to emotional and cortical effects, and does not respond to cardiovascular reflexes. MSNA can be expressed as bursts/min and bursts/100 heartbeats, allowing comparison of sympathetic discharge between individuals.

Urinary and Plasma Catecholamines

Information on overall activity of the adrenergic nervous system may be obtained from the urinary excretion of noradrenaline, adrenaline and their precursors or metabolites. However, this does not allow dynamic assessment of changes or of regional responses. Furthermore, it is influenced by changes in renal function [6].

Plasma noradrenaline concentrations have also been used to assess sympathetic activity. However, the circulating noradrenaline concentration represents only 5–10 % of the neurotransmitter secreted from the nerve terminals and depends not only on secretion but also on tissue clearance and reuptake processes, all of which may change [7–9]. Furthermore, peripheral venous blood sampling may be misleading since there is a wide variation in the release of noradrenaline between regions of the body [9]. Plasma noradrenaline values have poor inter- and intra-laboratory reproducibility [10].

A better method of assessing neurotransmitter release in humans is from the assessment of noradrenaline spillover using a radioisotope. This avoids the confounding influence of neurotransmitter clearance and permits assessment of release from specific target organs [11]. Tracer amounts of [^{3}H] noradrenaline are infused at a steady rate and, provided neuronal and extraneuronal uptake of catecholamines and blood flow remain stable, the rate at which neurally released noradrenaline enters the plasma should be directly proportional to the rate of neurotransmitter release from sympathetic nerve endings into the sympathetic cleft. Blood flow should be constant since the spillover of noradrenaline changes with flow, even if the stimulus to neuronal release is unchanged [12].

Assessment of Reflex Responses

Baroreceptor Reflex

Most methods to assess baroreflex sensitivity have concentrated on the cardiac responses because of their ease of measurement. However, in the control of blood pressure and blood flow it is the vascular component that is of greater importance.

Responses to Vasoactive Drugs

The cardiac component of the integrated baroreceptor reflex can be assessed from the responses to injections of vasoactive drugs such as phenylephrine and nitroprusside [13]. Baroreflex sensitivity is taken as the slope of the regression line fitting beat-to-beat changes in R-R interval against systolic blood pressure. This technique is widely used although it is invasive and allows only the heart rate responses to be investigated.

Neck Collar Method

An airtight collar round the neck overlying the carotid sinuses allows changes to be made in carotid transmural pressure. Negative pressures increase the stimulus to the baroreceptors; positive pressures decrease it. The advantage is that, unlike with vasoactive drugs, arterioles are not affected by the stimulus allowing responses of vascular resistance to be studied. This method is discussed elsewhere [14].

Spontaneous Heart Rate and Blood Pressure Variability

Heart rate variability reflects fluctuations rather than absolute levels of autonomic activity. Power spectral analysis separates the changes in R-R interval into its frequency components and quantifies them in terms of their relative powers. Three spectral components can normally be distinguished: high frequency (0.15–0.4 Hz), low frequency (0.04–0.15 Hz) and very low frequency (0–0.04 Hz) [15]. High frequency fluctuations are coupled with the respiratory cycle and believed to be caused predominantly by changes in vagal activity. Low frequency fluctuations involve cardiovascular reflexes and derive from changes in both cardiac sympathetic and vagal activities [16]. Very low frequency fluctuations are thought to represent long-term regulatory mechanisms related to thermoregulation or to humoral factors [17]. The powers of the different components and the total power of the entire spectrum are expressed in absolute units (ms^2). LF and HF components can also be expressed as normalised units. The LF/HF ratio is thought to represent the relative contributions of the two branches of the ANS [15]. Decreased heart rate variability is said to reflect enhanced sympathetic activity and decreased vagal activity [15].

More recently, this technique has been applied to beat-to-beat arterial blood pressure data to evaluate autonomic activation of vascular smooth muscle. The same three spectral components can be distinguished. High frequency oscillations are related to respiratory changes in intrathoracic pressure, and are not reflexly mediated. Low frequency oscillations are attributed to oscillations in sympathetic outflow to the vascular smooth muscle [18]. The very low frequency oscillations reflect influences on peripheral vascular tone and, based on recent animal studies, may represent myogenic responses mediated by L-type calcium channels [19]. This technique is facilitated by non-invasive beat-to-beat blood pressure monitoring devices.

Baroreflex sensitivity may be determined from spontaneous changes in heart rate and blood pressure using various analytical techniques. One involves identifying three or more

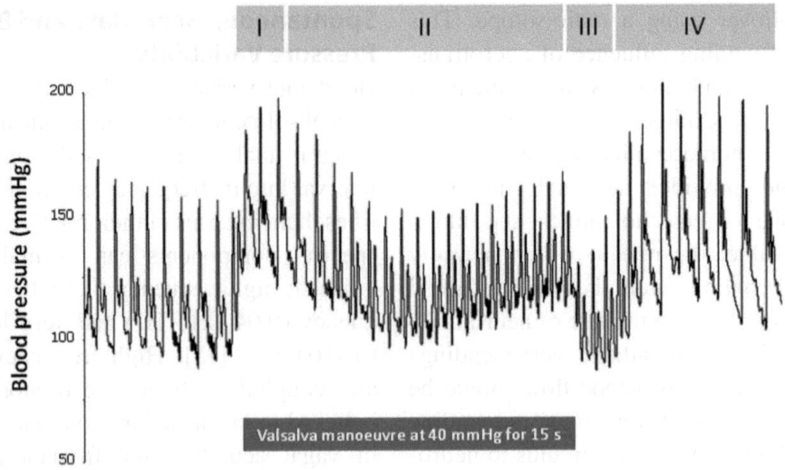

Fig. 8.2 Beat-to-beat blood pressure tracing obtained during a Valsalva manoeuvre. The four phases of the Valsalva can be clearly identified. In phase I, there is an increase in blood pressure with the onset of the strain. This progressively declines in early phase II. In late phase II, there is baroreflex compensation with restoration of blood pressure and tachycardia. When the Valsalva is released, blood pressure rapidly falls (phase III), followed by a blood pressure overshoot and baroreflex bradycardia (phase IV). (Claydon, Unpublished observations)

consecutive beats in which progressive increases/decreases in systolic blood pressure are followed by progressive lengthening/shortening in R-R interval [20]. The calculated slopes indicate baroreflex sensitivity. More recently cross-spectral analysis has been used to examine the relations between oscillations in R-R interval and oscillations in blood pressure in the different frequency components [21].

Other Tests of Autonomic Function

Several simple procedures may be used to assess autonomic control. These include the assessment of the *heart rate responses to standing* [22–24], which involves baroreflexes, cardiac and pulmonary reflexes as well as exercise responses. *The Valsalva manoeuvre* involves a forced expiratory effort and the determination of a series of changes in heart rate and blood pressure that are related to sympathetic and vagal control of the circulation (Fig. 8.2) [25, 26]. Other tests include: *slow deep breathing* which exaggerates respiratory sinus arrhythmia and is a test of vagal function [27–29]; *the cold pressor test*, in which sympathetic responses are determined to placing a hand or foot in water at temperatures of between 1 and 5 °C for periods of up to 6 min [30]; and *sustained handgrip*, a form of isometric exercise, which evokes increases in heart rate, and blood pressure [31].

The Autonomic Nervous System in Normobaric Hypoxia

The effects of hypoxia on autonomic function are complex due to interacting mechanisms including direct reflex effects of peripheral chemoreceptor stimulation, secondary effects of chemoreceptor-induced hyperventilation, influences of circulating hormones and the effects of hypoxia upon the heart and blood vessels, either directly or due to changes in receptor function [32–36].

Effects on the Cardiovascular System

Animal Studies

Results of cardiovascular responses to hypoxia have been inconsistent due to differences in species studied, experimental settings (such as the duration and severity of hypoxia) and whether animals were anaesthetised or conscious [35]. In spontaneously breathing anaesthetised rats, systemic hypoxia decreased blood pressure and increased heart rate [35]. In anaesthetised artificially ventilated cats, however, systemic hypoxia increased blood pressure [37]. In conscious dogs hypoxia increased both heart rate and blood pressure [38], while in rabbits, there was a decrease in heart rate and no change in blood pressure [39]. More recently, Sugimura et al. [40] gradually induced hypoxia and maintained it for 10 min, in conscious normal and spontaneously hypertensive rats (SHR). They reported initial increases in heart rate in both groups although they were not sustained in the hypertensive animals. Systolic blood pressure increased in both groups initially during the hypoxia, before falling, in the SHR, to significantly below baseline. The inconsistent effects of hypoxia on blood pressure suggest either that the effects on vascular tone are variable, or that any peripheral hypoxic vasoconstriction is compensated by other mechanisms. However, a direct effect of intermittent hypoxia on smooth muscle potassium channels associated with vasoconstriction has been noted in animal models [41]. Whether these effects of short-term intermittent hypoxia extrapolate to longer continuous exposures is unclear.

Human Studies

Richardson et al. [42] reported that hypoxia (breathing 7.5 % oxygen for 8 min) increased both heart rate and cardiac output but caused no change in blood pressure [42]. Cooper et al. [43] also obtained a significant increase in heart rate with 12 % oxygen for 10 min with no significant change in mean arterial pressure or forearm vascular resistance. Halliwill and Minson [44] also found that 12 % oxygen for 18 min caused a significant increase in heart rate, but they also reported a significant increase in blood pressure. Ziegler et al. [45], however, reported that breathing 15 % oxygen significantly increased heart rate and decreased blood pressure. Overall, the many studies of acute hypoxia in humans consistently showed increases in heart rate while the effect on blood pressure was more variable. Despite the resting tachycardia, there is reported to be a marked reduction in the maximal exercise heart rate, and this is attributed to increased parasympathetic activity [46]. Pharmacological blockade has been used to assess the mechanism of the hypoxic tachycardia. Propranolol caused only a slight reduction in the tachycardia, suggesting only a small sympathetic contribution [42]. Clar et al. [47] also showed that β-adrenergic receptor blockade had little effect on the heart rate response to 8 h of hypoxia. However, the sensitivity of the response was significantly reduced by muscarinic blockade with glycopyrrolate [48]. These findings suggest that an increase in sympathetic activity has only a minor role in the hypoxic tachycardia and that the principal mechanism is the withdrawal of parasympathetic tone.

Effects on Sympathetic Nerve Activity

Animal Studies

Acute hypoxia, in rats, cats and rabbits, increases sympathetic activity to the kidneys, adrenals, pulmonary circulation, heart and skeletal muscle [34, 37, 49–52]. This is in contrast to cutaneous sympathetic activity which either decreased during moderate hypoxia (5–8 % O_2) [52–54] or was unchanged [54, 55]. However, it did increase during severe hypoxia (0 or 3 % O_2) [54, 56].

Human Studies

Somers et al. [57] reported that breathing 14 % oxygen for 5 min increased heart rate but had no significant effect on blood pressure or MSNA, recorded from the peroneal nerve. However, increasing the hypoxic stress to 10 % oxygen for 5 min did result in a significant increase in MSNA, although blood pressure was still unchanged. The authors concluded that there was a threshold of hypoxia for sympathetic activation. Leuenberger et al. [58] also found that, when subjects breathed 10.5 % oxygen for 30 min, peroneal nerve activity increased despite no change in blood pressure.

Responses of Catecholamines

Results of plasma noradrenaline turnover in conscious dogs and rats indicate an increase in sympathetic activity with hypoxia [59, 60].

Results from studies in humans, however, have been inconsistent. Of 10 studies reviewed by Rostrup [56] only one, which involved 30 min of hypoxia at 10.5 %, demonstrated a significant increase in plasma noradrenaline, and this was accompanied by an increase in plasma adrenaline. Richardson et al. [42] found that breathing 7.5 % oxygen for 8 min actually decreased plasma noradrenaline, and increased adrenaline. These findings must be interpreted with caution as, not only were there differences in the degree and duration of the hypoxia, but there is also the problem that plasma catecholamine levels may change because hypoxia increases their clearance [58].

Heart Rate and Blood Pressure Variability

Animal Studies

Several studies have shown increases in the power of the low frequency component of blood pressure, which is thought to parallel sympathetic activity [61, 62]. Yasuma and Hayano [63] reported that progressive isocapnic hypoxia, in conscious dogs with pre-implanted telemetry devices, increased blood pressure and heart rate, decreased the high frequency component of the R-R interval, and increased the low frequency power of blood pressure. Sugimura et al. [40] also reported an increase in the low frequency component of blood pressure in both control rats and SHR during 5 min of progressive hypoxia. This subsequently decreased during 10 min of maintained hypoxia (10 % O_2), although it remained greater than the normoxic value. The high frequency component of the R-R interval was reduced in SHR but responses in control animals were more variable. These findings indicate reductions in cardiac vagal activity and increases in sympathetic activity to blood vessels. They also suggest that the inhibition of cardiac parasympathetic activity in acute hypoxia may be enhanced in hypertension.

Human Studies

Lucy et al. [64] reported that hypoxia decreased the high frequency component of heart rate variability but had no effect on the low frequency component implying that the tachycardia was due to vagal withdrawal alone. Nesterov [65] also found that hypoxia decreased high frequency power but they also found that the "sympathovagal balance" (LF/HF) significantly increased, suggesting that, in addition to vagal withdrawal, there was increased sympathetic activity. However, the interpretation of heart rate variability in these studies is difficult as it may be influenced by many variables such as ventilatory volume and rate, and arterial baroreflex function, all of which are altered by hypoxia [66, 67].

Baroreceptor Sensitivity

Halliwill and Minson [44, 68] reported that baroreflex sensitivity, assessed by the cardiac and MSNA responses to vasoactive drugs, was not affected by breathing 12 % oxygen. However, they did find that hypoxia resets the baroreflex control to higher pressures. Cooper et al. [43] investigated the effect of hypoxia on carotid baroreflex control of heart rate and forearm vascular resistance using the neck collar method. They also found no effect on the sensitivity of the heart rate response, but in contrast to earlier findings [44, 68] they also found the set point to be unaffected. There was no effect on either gain or set point of blood pressure control, although the sensitivity of the vascular resistance response was reduced with no change in set point (Fig. 8.3).

Summary of Effects of Normobaric Hypoxia

Effects of hypoxia in spontaneously breathing animals or humans depend on responses to chemoreceptor stimulation as well as the effects of the increased breathing. Most studies indicate that a sufficiently intense stimulus results in increased sympathetic and decreased vagal activity, resulting in tachycardia and vasoconstriction. This is supported by direct nerve recordings, changes in catecholamine activity and analysis of heart rate and blood pressure variability. The sensitivity of the vascular component of the baroreceptor reflex may be

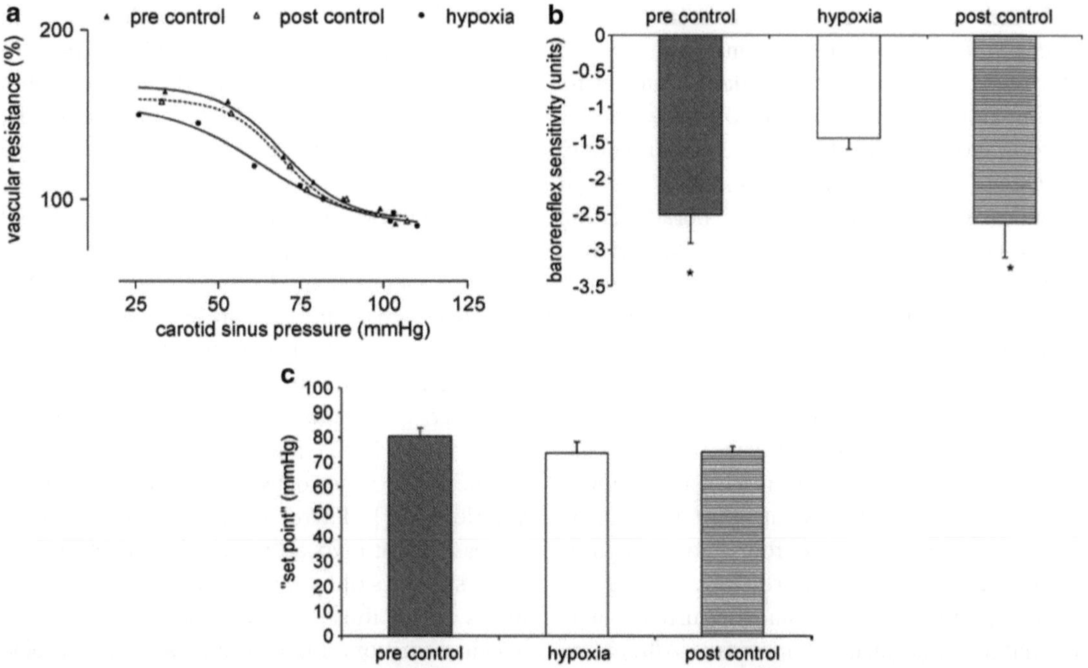

Fig. 8.3 Carotid baroreflex responses to hypoxia. Example carotid baroreflex curve showing vascular resistance responses to changes in CSP using the neck collar technique before (pre control), during and after (post control) hypoxia (12 % O_2 in N_2). (**a**) The slope of the relationship is reduced by hypoxia with no change in set point, indicating decreased vascular resistance baroreflex sensitivity without resetting. Group data are shown in (**b**) and (**c**). Significant difference from hypoxia denoted by *$p<0.05$ (adapted from Cooper et al. [43] J Physiol 568, 677–687)

decreased and this could contribute to the autonomic changes observed.

The Autonomic System in Hypobaric Hypoxia

Studies of the effects of hypobaric hypoxia on the autonomic nervous system have taken place either in hypobaric chambers or at high altitude locations. Interpretation of field studies is more complex since, in addition to hypoxia, other factors, including the cold, dehydration and (preceding) exercise may influence autonomic activity.

Effects on Sympathetic Nerve Activity

Saito et al. [69] reported a significant increase in MSNA during hypobaric hypoxia at a simulated altitude of 4,000 m, increasing further at 5,000 m. However, responses at 6,000 m were variable, with four of the ten subjects actually showing a decrease. Duplain et al. [70] tested healthy subjects at an altitude of 4,559 m and also reported an increase in peroneal nerve activity; this was greater in those with histories of high altitude pulmonary oedema. Hansen and Sander [71] observed a marked elevation in MSNA after 4 weeks of altitude exposure, which persisted for days after the return to sea level. Interestingly, the increase was little affected by administered oxygen (to eliminate chemoreflex activation) or saline infusion (to reduce baroreflex responses), and they suggested that chemoreflex activation by hypoxia, or baroreflex change by dehydration, accounted for only a small component of the sympathetic response, with the major mechanism unexplained.

Responses of Catecholamines

Subjects exposed to simulated altitudes of 3,000–4,750 m in hypobaric chambers showed a doubling of urinary adrenaline excretion but no change in urinary noradrenaline [72, 73].

Rather different results have been reported from altitude field studies. Cunningham et al. [74] observed a doubling of plasma and urine noradrenaline levels during 17 days at altitudes up to 4,560 m with little change in adrenaline. Mazzeo et al. [75] reported that in men after 4 h at altitude (4,300 m) plasma noradrenaline decreased by 36 %, but after 21 days it had increased to 52 % above sea level values. Arterial adrenaline values doubled following acute altitude exposure but declined to only 26 % above sea level values by day 21. Mazzeo et al. [76] also studied women at 4,300 m for 12 days and reported that both urinary noradrenaline and adrenaline excretion increased significantly after only 1 day at altitude. Plasma catecholamines were also found to be significantly elevated by the fourth day. During the 12-day period both plasma and urinary noradrenaline continued to increase. Adrenaline values however fell back to sea level values. Rostrup [56], however, reported that men exposed to 4,200 m had decreases in plasma noradrenaline and adrenaline on the second day but they recovered by day 7. Finally, Calbet [77] found that both plasma noradrenaline and adrenaline were elevated after 9 weeks of exposure to an altitude of 5,260 m, without a change in whole-body noradrenaline clearance. This increase persisted even after the arterial oxygen content had normalised with acclimatisation.

These differences between the chamber and the field studies may be attributed to experimental differences, including difficulties at altitude locations with the transport and analysis of samples [56]. Most studies show at least an initial increase in adrenaline levels, possibly partly due to the stress. Changes in noradrenaline levels were variable, but it should be remembered that hypoxia may increase renal clearance.

Hypoxia and Receptor Density

Animal Studies

Hypobaric hypoxia in rats for up to 15 days did not affect β-adrenergic receptor density, although by 21 days there was a 24 % reduction [78]. Favret et al. [79], however, showed that the density of rat right ventricular β-adrenoceptors was significantly elevated after just 30 min of hypoxia, but it subsequently declined to values below those found in normoxia.

The reported effects of simulated altitude exposure on the density of α_1-adrenergic receptors are inconsistent. Leon-Velarde et al. [80] reported a 66 % increase in left ventricular α_1-adrenergic receptor density in rats exposed to a simulated altitude of 5,500 m for 21 days. However, in another study rats exposed to 5,500 m for 15 days showed no change in α_1-adrenergic receptor density in either left or right ventricle [81]. Favret et al. [79] reported an increase in rat myocardial α_1-adrenoceptors over the first 3 days of hypoxia which then declined to levels below those in normoxia. Stimulation of α_1-adrenoceptors induces increases in cardiac contractility and has been suggested to be involved in the development of myocardial hypertrophy, although the evidence for this is controversial [82, 83].

Density of rat myocardial muscarinic receptors increases during simulated altitude exposure [79, 84]. The increase in the right ventricle peaked at 33 % above control after 1 day of hypoxia. The left ventricle showed a similar increase in receptor density which then continued to rise for 21 days, when it was 80 % higher [79]. This increase in muscarinic receptor density in response to hypoxia may reflect the reduction in parasympathetic (vagal) nervous activity to the heart suggested by heart rate variability studies.

Adenosine receptor density is decreased in the rat following a 30-day exposure to a simulated altitude of 5,500 m [84], possibly due to opening of K_{ATP} channels and impaired Ca^{2+} influx [85].

Human Studies

Indirect evidence in humans suggests that, as in animals, hypobaric hypoxia causes changes in receptor density. Richalet et al. [86] reported that the rate of infusion of isoprenaline required to increase heart rate by 25 beats min^{-1} increased with increasing exposure to altitude and this was attributed to down-regulation of the β-adrenergic receptors. Fischetti et al. [87] reported a reduction in platelet α_2-adrenergic receptor density after 4 weeks exposure to 5,050 m, that may indicate similar changes in the central nervous system [88]. This is important because these receptors play an important role in cardiovascular regulation [89]; stimulation of receptors in the ventrolateral medulla reduces sympathetic and increases parasympathetic outflow. A change in the density of these receptors may contribute to the effect of altitude on the autonomic system. However, at present, the evidence in humans is indirect and inconclusive.

Heart Rate Variability

Kanai et al. [90] reported that 2 h after arriving at an altitude of 2,700 m heart rate variability decreased in both high and low frequency bands, although the ratio of low to high frequency power increased. The increase in the ratio is believed to imply that the sympathetic system is dominant compared to the parasympathetic. Similar results were obtained by Cornolo et al. [91] who reported that exposure to an altitude of 4,350 m for 1–2 days reduced power in the high-frequency band but increased low-frequency power, thereby increasing the low to high frequency ratio. From this they concluded that acute exposure to hypoxia is associated with decreased parasympathetic and increased sympathetic tone. During acclimatisation, there was a progressive shift toward still higher sympathetic activity.

Baroreceptor Sensitivity

Sagawa et al. [92] exposed seven unacclimatised subjects to a simulated altitude of 4,300 m and stimulated carotid baroreceptors using a neck chamber. They found baroreceptor response curves showed no change in the set point but there was a 50 % reduction in the gain of the heart rate response. Other authors have examined changes in spontaneous baroreflex gain derived from R-R intervals and blood pressure during simulated high altitude and reported reductions of 35–43 % [93–96].

Summary of Effects of Hypobaric Hypoxia

As for normobaric hypoxia, most studies indicate an increase in sympathetic activity, although there may be a threshold for this, and in more severe hypoxia the change may be smaller. One report indicated that the increased sympathetic activity during hypobaric hypoxia was not prevented by oxygen administration [71], suggesting some other unknown mechanism might also be operating. Blood and urine analyses indicate increases in adrenaline levels with variable changes in noradrenaline, although renal clearance of these may change in hypoxia. Changes also occur in receptor density, particularly of β-receptors which are down-regulated. As seen during normobaric hypoxia, the sensitivity of the vascular limb of the baroreceptor reflex appears to be decreased.

Differences Between Hypobaric and Normobaric Hypoxia

Comparisons of effects of normobaric, hypobaric chamber and altitude location studies are complicated by experimental difficulties and differences in protocol, and carefully designed comparative studies are needed. Indeed, there remains considerable debate as to whether normobaric and hypobaric hypoxia produce equivalent physiological responses [97]. Studies examining these effects are lacking, and often fail to control for other environmental factors that could influence autonomic responses, such as alterations in ambient temperature. Protocol discrepancies (duration and level of altitude exposure) are common. There is a suggestion that the initial responses of oxygen saturation and symptoms of hypoxia are influenced by the protocol employed, but these differences were lost after 5 min [98]. Also, Savourey et al. [99]

observed initially higher respiratory frequencies and lower tidal volumes in hypobaric hypoxia than in the equivalent normobaric hypoxia, with lower end tidal oxygen and carbon dioxide levels, and lower oxygen saturations after 5 min of exposure. These differences were lost after 30 min of exposure. An earlier study by the same group showed greater hypoxemia, hypocapnia, alkalosis and lower arterial oxygen saturations (although no effect on pulse oximetry) with hypobaric compared to normobaric hypoxia [100]. Comparisons in rats exposed to hypobaric or nomobaric hypoxia for 5, 10 and 14 days found no differences between groups [101]. Blood gases were also similar in ducks exposed to equivalent hypobaric and normobaric hypoxia, although tidal volume and respiratory frequencies were greater in hypobaric hypoxia [102]. There are reports that nitric oxide metabolism may also be differentially affected by hypobaric and normobaric hypoxia [103], although this has been contested [104]. However, in general any physiological differences between normobaric and hypobaric exposures are thought to be too small to be clinically relevant [97, 105].

The Autonomic Nervous System in High Altitude Residents

The most extensively studied altitude locations are in the Andes, Himalayas, and Ethiopian highlands. The various populations differ in their adaptation strategies, and some develop the maladaptation syndrome, chronic mountain sickness (CMS) which is discussed further in Chap. 22.

Chronic Mountain Sickness

The incidence of CMS varies in different populations, occurring in 52 % of males living at 4,100 m in the Andes. It also occurs in Himalayan populations. However, there are no reports of CMS at similar altitudes in Ethiopia [106, 107]. Incidence of CMS is higher in males and increases with age and increasing altitude. The features are extreme polycythaemia and profound hypoxia (greater than expected for the altitude), finger clubbing, congested conjunctivae, pulmonary hypertension, right ventricular hypertrophy, and cyanosis [94]. The symptoms include headache, dizziness, confusion, anorexia, fatigue, exercise intolerance, acral paraesthesia, and ultimately symptoms of heart failure. CMS may be fatal. It completely resolves following descent to sea level, and reappears upon return to altitude [94].

Many of the features of CMS suggest changes in autonomic control. Different altitude dwelling populations adopt different adaptive responses in autonomic control (see below), and these may underlie the differing susceptibilities to CMS. It has been argued that CMS is merely the manifestation of minor impairments in renal, cardiovascular or respiratory function that are not severe enough to present symptoms during normoxia; hence the recovery following descent to sea level [108].

Effects on the Cardiovascular System

Resting blood pressures tend to be elevated in Andean altitude dwellers, particularly those with CMS [109]. Heart rate is significantly higher in altitude dwellers at their resident altitude than following descent to sea level [109]. This effect is more pronounced in those with CMS who also have increased arterial wall stiffness [110], increased circulating catecholamine levels, particularly noradrenaline [111] and altered baroreflex control (see below). This tendency to higher blood pressures and sympathetic activation may predispose to the increased cardiovascular morbidity in these populations. There is no elevation of resting heart rates or blood pressures in Ethiopian altitude dwellers in whom CMS does not occur [106].

Interestingly, there is recent evidence that, despite the prevailing hypoxia and altered resting autonomic function in healthy Andean high altitude residents, the autonomic responses to exercise training are not impaired [112].

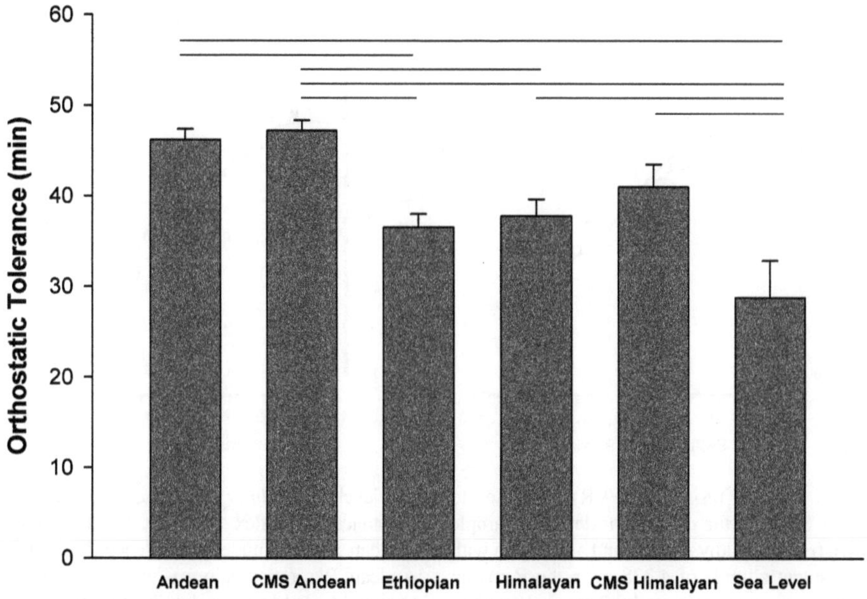

Fig. 8.4 Orthostatic tolerance (measured as time to presyncope) in permanent residents at high altitude compared to sea level residents. Orthostatic tolerance was determined using a combined test of head-upright tilting and lower body negative pressure. Data are shown for healthy Andean altitude dwellers, Andeans with Chronic Mountain Sickness (CMS), Ethiopian and Himalayan altitude residents (including those with CMS), and sea level controls. There are no data for Ethiopians with CMS because we could not find evidence of CMS in the Ethiopian highlands. Significant differences between groups ($p < 0.05$) are denoted by the *black bars* (adapted from Claydon et al. [109] Exp Physiol 89, 565–571 and unpublished observations)

Orthostatic Tolerance

Orthostatic tolerance is the ability to maintain an adequate blood pressure and consciousness during a gravitational stress that decreases return of blood to the heart. It may be quantified in terms of the stress required to induce syncope, or fainting. It is high in many high altitude populations tested [109, 113] compared to sea- level dwellers (Fig. 8.4) [114]. It is particularly high in Andeans, including those with CMS, despite a tendency for vasoconstriction to be less, and this may be attributed to their large red cell volumes [109] and possibly also to better autoregulation of cerebral blood flow [115].

Effects on Sympathetic Nerve Activity

Although in low altitude dwellers exposure to hypoxia results in an increased MSNA that persists for days after descent to sea level [71], there are no MSNA data from permanent high altitude residents.

Changes in Catecholamine Levels

Gamboa et al. [111] reported higher plasma levels of noradrenaline, dopamine and the noradrenaline metabolite dihydroxyphenylglycol (DHPG) in Andeans at their resident altitude compared to data obtained following descent to sea level, suggesting sympathetic activation at altitude. Other studies suggest that, unlike following acute exposure to hypoxia where there is down-regulation of β-adrenoceptors in response to sympathetic overactivity [86], in Andeans chronically exposed to hypoxia there is no such change [116], and no evidence for adrenergic desensitivity in either healthy altitude dwellers or those with CMS.

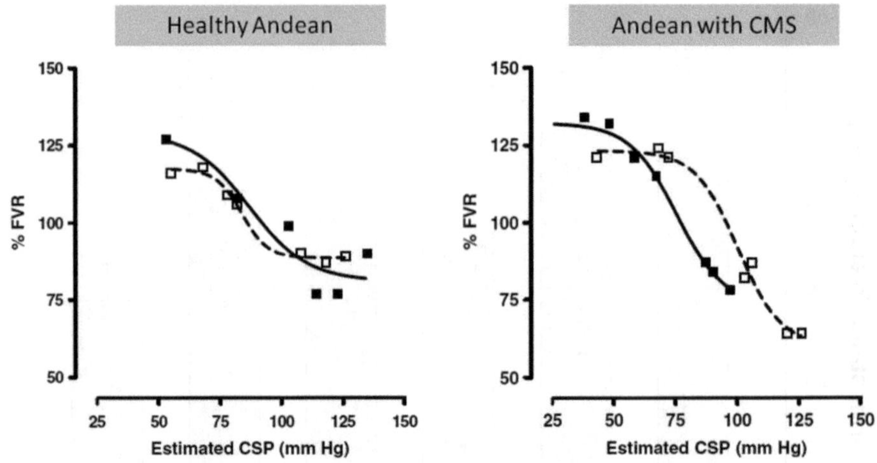

Fig. 8.5 Forearm vascular resistance (FVR) responses to manipulation of CSP using the neck collar device. Example traces are shown from a healthy Andean and an Andean with CMS at their resident altitude of 4,338 m in the Peruvian Andes (*dashed line*, *open squares*) and following descent to sea level (*solid line*, *closed squares*). Increased vascular resistance baroreflex set point can be seen in the CMS Andean at his altitude of residence, but not in the healthy Andean. Baroreflex gain was unaffected in both individuals (adapted from Moore et al. [120] Exp Physiol 91, 907–913)

Heart Rate and Blood Pressure Variability

Supine low frequency oscillations in R-R interval are reported to be low in altitude dwellers [117]. Highlanders with CMS also have low supine low frequency oscillations in systolic arterial pressure [117]. This suggests less sympathetic activity, implying that their tendency to increased blood pressure may be more related to their high blood viscosity. Orthostatic stress provoked the expected changes in heart rate and blood pressure variability in Ethiopians, but not in Andeans [117]. Interestingly, in those with CMS the low frequency oscillations in blood pressure while supine were greatly reduced, but increased markedly when upright [117], indicating that although sympathetic activity was low when supine, it increased to high levels when upright. Vascular resistance responses, however, were blunted [115], suggesting a mismatch between sympathetic activation and the effector response.

Baroreflex Control

Cross-spectral analysis of cardiac baroreflex control in altitude residents from the Andes and Ethiopia revealed reduced coherence between blood pressure and R-R interval and increased baroreflex delay. Cardiac baroreflex sensitivity was reduced only in subjects with CMS [117]. These data suggest impaired cardiac baroreflex control in altitude dwellers, particularly those with CMS.

In Andean high altitude dwellers baroreflex sensitivity was also shown to be reduced using spontaneous sequence analyses [118]. The reduction in baroreflex sensitivity correlated with haematocrit, being reduced more in those with CMS. Baroreflex sensitivity increased following 1 h of supplemental oxygen therapy, or slow frequency breathing. Interestingly, the same group reported *enhanced* baroreflex sensitivity and reduced sympathetic activation in Himalayan high altitude dwellers [119], further underscoring the different adaptive processes.

The above studies have focused on cardiac responses, whereas vascular resistance is arguably more important. Moore et al. [120] reported that the sensitivity of vascular resistance responses to carotid baroreceptor stimulation, using the neck collar technique, was similar in healthy Andeans and Andeans with CMS, both when studied at their resident altitude and 24 h after descent to sea level.

However, the set point was higher in the CMS subjects at altitude and it decreased following descent to sea level (Fig. 8.5). This provides a mechanism in CMS subjects linking sympathetic activation and hypertension to the hypoxia of altitude. The observation that alterations in baroreflex sensitivity, for both cardiac and vascular components, are rapidly reversed by normoxia is compatible with the known reversal of symptoms and signs of CMS upon descent to sea level.

Summary of Autonomic Changes in High Altitude Residents

Most altitude dwellers adapt well to their environment, but some develop CMS characterised by very high haematocrits and excessive hypoxia. Altitude residents tend to have high heart rates and blood pressures and indirect evidence, from catecholamine levels and heart rate and blood pressure variabilities, indicate increased sympathetic activation, particularly in CMS subjects. Unlike in unadapted subjects, there is no down-regulation of β receptors. Tolerance to orthostatic stress is high and correlates with haematocrit rather than with sympathetic responses. In CMS patients cardiac baroreceptor sensitivity may be decreased and the vascular response shifted to higher levels, possibly explaining their susceptibility to hypertension.

Conclusion

In this chapter, we have shown that exposure to high altitude induces changes in activity in autonomic nerves. Changes are seen in both visitors and in permanent high altitude residents. However, despite the extensive knowledge of these changes there are still many unknowns. For example, due to several confounding variables, it is not yet clear as to the exact differences between the effects of hypoxia in normobaria, chamber hypobaria and high altitude exposure. Also, although various reflex responses have been studied in the different groups (lowlanders, healthy highlanders and highlanders with chronic mountain sickness), there is relatively little information on the changes in the autonomic system itself.

References

1. Karim F, Kidd C, Malpus CM, Penna PE. The effects of stimulation of the left atrial receptors on sympathetic efferent nerve activity. J Physiol. 1972;227(1):243–60.
2. McMahon NC, Drinkhill MJ, Hainsworth R. Vascular responses to stimulation of carotid, aortic and coronary artery baroreceptors with pulsatile and non-pulsatile pressures in anaesthetized dogs. Exp Physiol. 1996;81(6):969–81.
3. Hagbarth KE, Vallbo AB. Discharge characteristics of human muscle afferents during muscle stretch and contraction. Exp Neurol. 1968;22(4):674–94.
4. Huggett RJ, Scott EM, Gilbey SG, Stoker JB, Mackintosh AF, Mary DA. Impact of type 2 diabetes mellitus on sympathetic neural mechanisms in hypertension. Circulation. 2003;108(25):3097–101.
5. Wallin BG. Human sympathetic nerve activity and blood pressure regulation. Clin Exp Hypertens A. 1989;11 Suppl 1:91–101.
6. Esler M. Clinical application of noradrenaline spillover methodology: delineation of regional human sympathetic nervous responses. Pharmacol Toxicol. 1993;73(5):243–53.
7. Bravo EL, Tarazi RC. Plasma catecholamines in clinical investigation: a useful index or a meaningless number? J Lab Clin Med. 1982;100(2):155–60.
8. Folkow B. Physiological aspects of primary hypertension. Physiol Rev. 1982;62(2):347–504.
9. Esler M, Jennings G, Lambert G, Meredith I, Horne M, Eisenhofer G. Overflow of catecholamine neurotransmitters to the circulation: source, fate, and functions. Physiol Rev. 1990;70(4):963–85.
10. Grassi G, Bolla G, Seravalle G, Turri C, Lanfranchi A, Mancia G. Comparison between reproducibility and sensitivity of muscle sympathetic nerve traffic and plasma noradrenaline in man. Clin Sci (Lond). 1997;92(3):285–9.
11. Esler M, Jennings G, Korner P, et al. Assessment of human sympathetic nervous system activity from measurements of norepinephrine turnover. Hypertension. 1988;11(1):3–20.
12. Grossman E, Chang PC, Hoffman A, Tamrat M, Kopin IJ, Goldstein DS. Tracer norepinephrine kinetics: dependence on regional blood flow and the site of infusion. Am J Physiol. 1991;260(5 Pt 2):R946–52.
13. Smyth HS, Sleight P, Pickering GW. Reflex regulation of arterial pressure during sleep in man. A quantitative method of assessing baroreflex sensitivity. Circ Res. 1969;24(1):109–21.

14. Cooper VL, Hainsworth R. Carotid baroreflex testing using the neck collar device. Clin Auton Res. 2009;19(2):102–12.
15. Task Force of the European Society of Cardiology and the North American Society of Pacing and Electrophysiology. Heart rate variability: standards of measurement, physiological interpretation and clinical use. Circulation. 1996;93(5):1043–65.
16. Akselrod S, Gordon D, Madwed JB, Snidman NC, Shannon DC, Cohen RJ. Hemodynamic regulation: investigation by spectral analysis. Am J Physiol. 1985;249(4 Pt 2):H867–75.
17. Kitney RI, Rompelman O. Thermal entrainment patterns in heart rate variability [proceedings]. J Physiol. 1977;270(1):41P–2.
18. Baselli G, Cerutti S, Civardi S, et al. Spectral and cross-spectral analysis of heart rate and arterial blood pressure variability signals. Comput Biomed Res. 1986;19(6):520–34.
19. Langager AM, Hammerberg BE, Rotella DL, Stauss HM. Very low-frequency blood pressure variability depends on voltage-gated L-type Ca2+ channels in conscious rats. Am J Physiol Heart Circ Physiol. 2007;292(3):H1321–7.
20. Bertinieri G, di RM, Cavallazzi A, Ferrari AU, Pedotti A, Mancia G. A new approach to analysis of the arterial baroreflex. J Hypertens Suppl. 1985;3(3): S79–81.
21. Fauvel JP, Cerutti C, Mpio I, Ducher M. Aging process on spectrally determined spontaneous baroreflex sensitivity: a 5-year prospective study. Hypertension. 2007;50(3):543–6.
22. Ewing DJ, Hume L, Campbell IW, Murray A, Neilson JM, Clarke BF. Autonomic mechanisms in the initial heart rate response to standing. J Appl Physiol. 1980;49(5):809–14.
23. Ewing DJ, Campbell IW, Clarke BF. Heart-rate response to standing as a test for automatic neuropathy. Br Med J. 1978;1(6128):1700.
24. Borst C, Wieling W, van Brederode JF, Hond A, de Rijk LG, Dunning AJ. Mechanisms of initial heart rate response to postural change. Am J Physiol. 1982;243(5):H676–81.
25. Levin AB. A simple test of cardiac function based upon the heart rate changes induced by the Valsalva maneuver. Am J Cardiol. 1966;18(1):90–9.
26. Ewing DJ, Clarke BF. Diagnosis and management of diabetic autonomic neuropathy. Br Med J (Clin Res Ed). 1982;285(6346):916–8.
27. Wheeler T, Watkins PJ. Cardiac denervation in diabetes. Br Med J. 1973;4(5892):584–6.
28. Angelone A, Coulter Jr NA. Respiratory sinus arrhythmia: a frequency dependent phenomenon. J Appl Physiol. 1964;19:479–82.
29. Bennett T, Fentem PH, Fitton D, Hampton JR, Hosking DJ, Riggott PA. Assessment of vagal control of the heart in diabetes. Measures of R-R interval variation under different conditions. Br Heart J. 1977;39(1):25–8.
30. Hines EA, Brown G. A standard stimulus for measuring vasomotor reactions: its application in the study of hypertension. Proc Staff Meet Mayo Clinic. 1932;332–5.
31. Seals DR, Victor RG, Mark AL. Plasma norepinephrine and muscle sympathetic discharge during rhythmic exercise in humans. J Appl Physiol. 1988;65(2): 940–4.
32. Murasato Y, Hirakawa H, Harada Y, Nakamura T, Hayashida Y. Effects of systemic hypoxia on R-R interval and blood pressure variabilities in conscious rats. Am J Physiol. 1998;275(3 Pt 2):H797–804.
33. Hirakawa H, Nakamura T, Hayashida Y. Effect of carbon dioxide on autonomic cardiovascular responses to systemic hypoxia in conscious rats. Am J Physiol. 1997;273(2 Pt 2):R747–54.
34. Fukuda Y, Sato A, Suzuki A, Trzebski A. Autonomic nerve and cardiovascular responses to changing blood oxygen and carbon dioxide levels in the rat. J Auton Nerv Syst. 1989;28(1):61–74.
35. Marshall JM, Metcalfe JD. Analysis of the cardiovascular changes induced in the rat by graded levels of systemic hypoxia. J Physiol. 1988;407:385–403.
36. Kawaguchi T, Tsubone H, Hori M, Ozaki H, Kuwahara M. Cardiovascular and autonomic nervous functions during acclimatization to hypoxia in conscious rats. Auton Neurosci. 2005;117(2): 97–104.
37. Shirai M, Matsukawa K, Nishiura N, Kawaguchi AT, Ninomiya I. Changes in efferent pulmonary sympathetic nerve activity during systemic hypoxia in anesthetized cats. Am J Physiol. 1995;269(6 Pt 2): R1404–9.
38. Hammill SC, Wagner Jr WW, Latham LP, Frost WW, Weil JV. Autonomic cardiovascular control during hypoxia in the dog. Circ Res. 1979;44(4): 569–75.
39. Chalmers JP, Isbister JP, Korner PI, Mok HY. The role of the sympathetic nervous system in the circulatory response of the rabbit to arterial hypoxia. J Physiol. 1965;181(1):175–91.
40. Sugimura M, Hirose Y, Hanamoto H, et al. Influence of acute progressive hypoxia on cardiovascular variability in conscious spontaneously hypertensive rats. Auton Neurosci. 2008;141(1–2):94–103.
41. Jackson-Weaver O, Paredes DA, Gonzalez Bosc LV, Walker BR, Kanagy NL. Intermittent hypoxia in rats increases myogenic tone through loss of hydrogen sulfide activation of large-conductance Ca(2+)-activated potassium channels. Circ Res. 2011; 108(12):1439–47.
42. Richardson DW, Kontos HA, Raper AJ, Patterson Jr JL. Modification by beta-adrenergic blockade of the circulatory respones to acute hypoxia in man. J Clin Invest. 1967;46(1):77–85.
43. Cooper VL, Pearson SB, Bowker CM, Elliott MW, Hainsworth R. Interaction of chemoreceptor and baroreceptor reflexes by hypoxia and hypercapnia—a mechanism for promoting hypertension in obstructive sleep apnoea. J Physiol. 2005;568(Pt 2):677–87.

44. Halliwill JR, Minson CT. Effect of hypoxia on arterial baroreflex control of heart rate and muscle sympathetic nerve activity in humans. J Appl Physiol. 2002;93(3):857–64.
45. Ziegler MG, Nelesen RA, Mills PJ, et al. The effect of hypoxia on baroreflexes and pressor sensitivity in sleep apnea and hypertension. Sleep. 1995;18(10): 859–65.
46. Boushel R, Calbet JA, Radegran G, Sondergaard H, Wagner PD, Saltin B. Parasympathetic neural activity accounts for the lowering of exercise heart rate at high altitude. Circulation. 2001;104(15):1785–91.
47. Clar C, Dorrington KL, Fatemian M, Robbins PA. Cardiovascular effects of 8 h of isocapnic hypoxia with and without beta-blockade in humans. Exp Physiol. 2000;85(5):557–65.
48. Clar C, Dorrington KL, Fatemian M, Robbins PA. Effects of 8 h of isocapnic hypoxia with and without muscarinic blockade on ventilation and heart rate in humans. Exp Physiol. 2001;86(4):529–38.
49. Malpas SC, Ninomiya I. Effect of asphyxia on the frequency and amplitude modulation of synchronized renal nerve activity in the cat. J Auton Nerv Syst. 1992;40(3):199–205.
50. Downing SE, Siegel JH. Baroreceptor and chemoreceptor influences on sympathetic discharge to the heart. Am J Physiol. 1963;204:471–9.
51. Biesold D, Kurosawa M, Sato A, Trzebski A. Hypoxia and hypercapnia increase the sympathoadrenal medullary functions in anesthetized, artificially ventilated rats. Jpn J Physiol. 1989;39(4):511–22.
52. Janig GM. Effects of sytemic hypoxia and hypercapnia on cutaneous and muscle vasoconstrictor neurones to the cat's hindlimb. Pflugers Arch. 1977; 368(1–2):71–81.
53. Blumberg H, Janig W, Rieckmann C, Szulczyk P. Baroreceptor and chemoreceptor reflexes in postganglionic neurones supplying skeletal muscle and hairy skin. J Auton Nerv Syst. 1980;2(3):223–40.
54. Iriki M, Walther OE, Pleschka K, Simon E. Regional cutaneous and visceral sympathetic activity during asphyxia in the anesthetized rabbit. Pflugers Arch. 1971;322(2):167–82.
55. Johnson C, Hudson S, Marshall J. Responses evoked in single sympathetic nerve fibres of the rat tail artery by systemic hypoxia are dependent on core temperature. J Physiol. 2007;584(Pt 1):221–33.
56. Rostrup M. Catecholamines, hypoxia and high altitude. Acta Physiol Scand. 1998;162(3):389–99.
57. Somers VK, Mark AL, Zavala DC, Abboud FM. Contrasting effects of hypoxia and hypercapnia on ventilation and sympathetic activity in humans. J Appl Physiol. 1989;67(5):2101–6.
58. Leuenberger U, Gleeson K, Wroblewski K, et al. Norepinephrine clearance is increased during acute hypoxemia in humans. Am J Physiol. 1991;261(5 Pt 2):H1659–64.
59. Rose Jr CE, Althaus JA, Kaiser DL, Miller ED, Carey RM. Acute hypoxemia and hypercapnia: increase in plasma catecholamines in conscious dogs. Am J Physiol. 1983;245(6):H924–9.
60. Kregel KC. Alterations in autonomic adjustments to acute hypoxia in conscious rats with aging. J Appl Physiol. 1996;80(2):540–6.
61. Pagani M, Lombardi F, Guzzetti S, et al. Power spectral analysis of heart rate and arterial pressure variabilities as a marker of sympatho-vagal interaction in man and conscious dog. Circ Res. 1986;59(2): 178–93.
62. Brown DR, Brown LV, Patwardhan A, Randall DC. Sympathetic activity and blood pressure are tightly coupled at 0.4 Hz in conscious rats. Am J Physiol. 1994;267(5 Pt 2):R1378–84.
63. Yasuma F, Hayano JI. Impact of acute hypoxia on heart rate and blood pressure variability in conscious dogs. Am J Physiol Heart Circ Physiol. 2000;279(5): H2344–9.
64. Lucy SD, Hughson RL, Kowalchuk JM, Paterson DH, Cunningham DA. Body position and cardiac dynamic and chronotropic responses to steady-state isocapnic hypoxaemia in humans. Exp Physiol. 2000;85(2):227–37.
65. Nesterov SV. [Autonomic regulation of the heart rate in humans under conditions of acute experimental hypoxia]. Fiziol Cheloveka. 2005;31(1):82–7.
66. Piepoli M, Sleight P, Leuzzi S, et al. Origin of respiratory sinus arrhythmia in conscious humans. An important role for arterial carotid baroreceptors. Circulation. 1997;95(7):1813–21.
67. Brown TE, Beightol LA, Koh J, Eckberg DL. Important influence of respiration on human R-R interval power spectra is largely ignored. J Appl Physiol. 1993;75(5):2310–7.
68. Halliwill JR, Minson CT. Cardiovagal regulation during combined hypoxic and orthostatic stress: fainters vs. nonfainters. J Appl Physiol. 2005;98(3): 1050–6.
69. Saito M, Mano T, Iwase S, Koga K, Abe H, Yamazaki Y. Responses in muscle sympathetic activity to acute hypoxia in humans. J Appl Physiol. 1988;65(4): 1548–52.
70. Duplain H, Vollenweider L, Delabays A, Nicod P, Bartsch P, Scherrer U. Augmented sympathetic activation during short-term hypoxia and high-altitude exposure in subjects susceptible to high-altitude pulmonary edema. Circulation. 1999;99(13): 1713–8.
71. Hansen J, Sander M. Sympathetic neural overactivity in healthy humans after prolonged exposure to hypobaric hypoxia. J Physiol. 2003;546(Pt 3):921–9.
72. Becker EJ, Kreuzer F. Sympathoadrenal response to hypoxia. Pflugers Arch. 1968;304(1):1–10.
73. Johnson TS, Rock PB, Young JB, Fulco CS, Trad LA. Hemodynamic and sympathoadrenal responses to altitude in humans: effect of dexamethasone. Aviat Space Environ Med. 1988;59(3):208–12.
74. Cunningham WL, Becker EJ, Kreuzer F. Catecholamines in plasma and urine at high altitude. J Appl Physiol. 1965;20(4):607–10.

75. Mazzeo RS, Bender PR, Brooks GA, et al. Arterial catecholamine responses during exercise with acute and chronic high-altitude exposure. Am J Physiol. 1991;261(4 Pt 1):E419–24.
76. Mazzeo RS, Child A, Butterfield GE, Mawson JT, Zamudio S, Moore LG. Catecholamine response during 12 days of high-altitude exposure (4, 300 m) in women. J Appl Physiol. 1998;84(4):1151–7.
77. Calbet JA. Chronic hypoxia increases blood pressure and noradrenaline spillover in healthy humans. J Physiol. 2003;551(Pt 1):379–86.
78. Kacimi R, Richalet JP, Corsin A, Abousahl I, Crozatier B. Hypoxia-induced downregulation of beta-adrenergic receptors in rat heart. J Appl Physiol. 1992;73(4):1377–82.
79. Favret F, Richalet JP, Henderson KK, Germack R, Gonzalez NC. Myocardial adrenergic and cholinergic receptor function in hypoxia: correlation with O(2) transport in exercise. Am J Physiol Regul Integr Comp Physiol. 2001;280(3):R730–8.
80. Leon-Velarde F, Bourin MC, Germack R, Mohammadi K, Crozatier B, Richalet JP. Differential alterations in cardiac adrenergic signaling in chronic hypoxia or norepinephrine infusion. Am J Physiol Regul Integr Comp Physiol. 2001;280(1):R274–81.
81. Morel OE, Buvry A, Le CP, et al. Effects of nifedipine-induced pulmonary vasodilatation on cardiac receptors and protein kinase C isoforms in the chronically hypoxic rat. Pflugers Arch. 2003;446(3): 356–64.
82. Tamai J, Hori M, Kagiya T, et al. Role of alpha 1-adrenoceptor activity in progression of cardiac hypertrophy in guinea pig hearts with pressure overload. Cardiovasc Res. 1989;23(4):315–22.
83. Martinez ML, Fernandez-Tome P, Lopez-Miranda V, Colado MI, Delgado C. Modulation of adrenergic receptors during left ventricular hypertrophy development and after regression by captopril. J Cardiovasc Pharmacol. 1999;34(4):505–11.
84. Kacimi R, Richalet JP, Crozatier B. Hypoxia-induced differential modulation of adenosinergic and muscarinic receptors in rat heart. J Appl Physiol. 1993;75(3):1123–8.
85. Coney AM, Marshall JM. Contribution of alpha2-adrenoceptors and Y1 neuropeptide Y receptors to the blunting of sympathetic vasoconstriction induced by systemic hypoxia in the rat. J Physiol. 2007;582 (Pt 3):1349–59.
86. Richalet JP, Larmignat P, Rathat C, Keromes A, Baud P, Lhoste F. Decreased cardiac response to isoproterenol infusion in acute and chronic hypoxia. J Appl Physiol. 1988;65(5):1957–61.
87. Fischetti F, Fabris B, Zaccaria M, et al. Effects of prolonged high-altitude exposure on peripheral adrenergic receptors in young healthy volunteers. Eur J Appl Physiol. 2000;82(5–6):439–45.
88. Piletz JE, Andorn AC, Unnerstall JR, Halaris A. Binding of [3H]-p-aminoclonidine to alpha 2-adrenoceptor states plus a non-adrenergic site on human platelet plasma membranes. Biochem Pharmacol. 1991;42(3):569–84.
89. Gavras I, Manolis AJ, Gavras H. The alpha2 -adrenergic receptors in hypertension and heart failure: experimental and clinical studies. J Hypertens. 2001; 19(12):2115–24.
90. Kanai M, Nishihara F, Shiga T, Shimada H, Saito S. Alterations in autonomic nervous control of heart rate among tourists at 2700 and 3700 m above sea level. Wilderness Environ Med. 2001;12(1):8–12.
91. Cornolo J, Mollard P, Brugniaux JV, Robach P, Richalet JP. Autonomic control of the cardiovascular system during acclimatization to high altitude: effects of sildenafil. J Appl Physiol. 2004;97(3): 935–40.
92. Sagawa S, Torii R, Nagaya K, Wada F, Endo Y, Shiraki K. Carotid baroreflex control of heart rate during acute exposure to simulated altitudes of 3,800 m and 4,300 m. Am J Physiol. 1997;273(4 Pt 2):R1219–23.
93. Sevre K, Bendz B, Hanko E, et al. Reduced autonomic activity during stepwise exposure to high altitude. Acta Physiol Scand. 2001;173(4):409–17.
94. Monge CC, Whittembury J. Chronic mountain sickness. Johns Hopkins Med J. 1976;139:S87–9.
95. Blaber AP, Hartley T, Pretorius PJ. Effect of acute exposure to 3660 m altitude on orthostatic responses and tolerance. J Appl Physiol. 2003;95(2):591–601.
96. Roche F, Reynaud C, Garet M, Pichot V, Costes F, Barthelemy JC. Cardiac baroreflex control in humans during and immediately after brief exposure to simulated high altitude. Clin Physiol Funct Imaging. 2002;22(5):301–6.
97. Millet GP, Faiss R, Pialoux V, Mounier R, Brugniaux JV. Point: counterpoint hypobaric hypoxia induces/does not induce different responses than normobaric hypoxia. J Appl Physiol. 2012;112(10):1788–94.
98. Self DA, Mandella JG, Prinzo OV, Forster EM, Shaffstall RM. Physiological equivalence of normobaric and hypobaric exposures of humans to 25,000 feet (7620 m). Aviat Space Environ Med. 2011; 82(2):97–103.
99. Savourey G, Launay JC, Besnard Y, et al. Normo or hypobaric hypoxic tests: propositions for the determination of the individual susceptibility to altitude illnesses. Eur J Appl Physiol. 2007;100(2):193–205.
100. Savourey G, Launay JC, Besnard Y, Guinet A, Travers S. Normo- and hypobaric hypoxia: are there any physiological differences? Eur J Appl Physiol. 2003;89(2):122–6.
101. Sheedy W, Thompson JS, Morice AH. A comparison of pathophysiological changes during hypobaric and normobaric hypoxia in rats. Respiration. 1996;63(4): 217–22.
102. Shams H, Powell FL, Hempleman SC. Effects of normobaric and hypobaric hypoxia on ventilation and arterial blood gases in ducks. Respir Physiol. 1990;80(2–3):163–70.
103. Hemmingsson T, Linnarsson D. Lower exhaled nitric oxide in hypobaric than in normobaric acute hypoxia. Respir Physiol Neurobiol. 2009;169(1):74–7.

104. Beall CM, Strohl KP, Laskowski D, Hutte R, Erzurum SC. "Lower exhaled nitric oxide in acute hypobaric than in normobaric hypoxia" by T Hemmingsson and D. Linnarsson [Respir. Physiol. Neurobiol. 169 (2009) 74–77]. Respir Physiol Neurobiol. 2010;170(3):211–2.
105. Kupper T, Milledge JS, Hillebrandt D, et al. Work in hypoxic conditions—consensus statement of the Medical Commission of the Union Internationale des Associations d'Alpinisme (UIAA MedCom). Ann Occup Hyg. 2011;55(4):369–86.
106. Claydon VE, Gulli G, Slessarev M, et al. Cerebrovascular responses to hypoxia and hypocapnia in Ethiopian high altitude dwellers. Stroke. 2008; 39:336–42.
107. Appenzeller O, Claydon VE, Gulli G, et al. Cerebral vasodilatation to exogenous NO is a measure of fitness for life at altitude. Stroke. 2006;37(7):1754–8.
108. Zubieta-Castillo Sr G, Zubieta-Calleja Jr GR, Zubieta-Calleja L. Chronic mountain sickness: the reaction of physical disorders to chronic hypoxia. J Physiol Pharmacol. 2006;57 Suppl 4:431–42.
109. Claydon VE, Norcliffe LJ, Moore JP, et al. Orthostatic tolerance and blood volumes in Andean high altitude dwellers. Exp Physiol. 2004;89(5): 565–71.
110. Otsuka K, Norboo T, Otsuka Y, et al. Chronoecological health watch of arterial stiffness and neuro-cardio-pulmonary function in elderly community at high altitude (3524 m), compared with Japanese town. Biomed Pharmacother. 2005;59 Suppl 1:S58–67.
111. Gamboa A, Gamboa JL, Holmes C, et al. Plasma catecholamines and blood volume in native Andeans during hypoxia and normoxia. Clin Auton Res. 2006;16(1):40–5.
112. Cornolo J, Brugniaux JV, Macarlupu JL, Privat C, Leon-Velarde F, Richalet JP. Autonomic adaptations in andean trained participants to a 4220-m altitude marathon. Med Sci Sports Exerc. 2005;37(12): 2148–53.
113. Malhotra MS, Selvamurthy W, Purkayastha SS, Mukherjee AK, Mathew L, Dua GL. Responses of the autonomic nervous system during acclimatization to high altitude in man. Aviat Space Environ Med. 1976;47(10):1076–9.
114. el Bedawi KM, Hainsworth R. Combined head-up tilt and lower body suction: a test of orthostatic tolerance. Clin Auton Res. 1994;4(1–2):41–7.
115. Claydon VE, Norcliffe LJ, Moore JP, et al. Cardiovascular responses to orthostatic stress in healthy altitude dwellers, and altitude residents with chronic mountain sickness. Exp Physiol. 2005;90(1): 103–10.
116. Antezana AM, Richalet JP, Antezana G, Spielvogel H, Kacimi R. Adrenergic system in high altitude residents. Int J Sports Med. 1992;13 Suppl 1: S96–100.
117. Gulli G, Claydon VE, Slessarev M, et al. Autonomic regulation during orthostatic stress in highlanders: comparison with sea-level residents. Exp Physiol. 2007;92(2):427–35.
118. Keyl C, Schneider A, Gamboa A, et al. Autonomic cardiovascular function in high-altitude Andean natives with chronic mountain sickness. J Appl Physiol. 2003;94(1):213–9.
119. Bernardi L. Heart rate and cardiovascular variability at high altitude. Conf Proc IEEE Eng Med Biol Soc. 2007;2007:6679–81.
120. Moore JP, Claydon VE, Norcliffe LJ, et al. Carotid baroreflex regulation of vascular resistance in high altitude Andean natives with and without chronic mountain sickness. Exp Physiol. 2006;91: 907–13.

Skeletal Muscle Tissue Changes with Hypoxia

9

Hans Hoppeler, Matthias Mueller, and Michael Vogt

Abstract

This review summarizes results of research into the effect on skeletal muscle tissue of prolonged exposure to high (3,000–5,500 m) and extreme altitude (>5,500 m). There is consensual evidence that continued sojourn at these altitudes has a number of negative consequences to muscle tissue. There is a loss of muscle mass related to a decrease of individual muscle fiber cross-sectional area. There is also a relative and absolute decrease in muscle oxidative capacity which manifests itself as a decrease in mitochondrial volume as well as a decrease in oxidative enzyme activities. The capillary to fiber ratio is maintained in hypoxia with the consequence that, without capillary neoformation, the oxygen supply of remaining mitochondria is improved. There is further a massive increase in lipofuscin, a lipid peroxidation product. Hypoxia activates defensive cellular mechanisms, among them the well-characterized response to the hypoxic master gene HIF (hypoxia-inducible factor). Reactive oxygen species (ROS) abound under hypoxic conditions and are further responsible for the orchestration of the hypoxia response. The permanent hypoxic stress of living at high altitude has led to a number of disparate but effective phylogenetic adaptations in native high-altitude populations, Tibetans and Quechua. When hypoxia is used as an adjunct limited to exercise training sessions, skeletal muscle tissue responds with a specific molecular signature. The functional consequences of which may offer benefits for competition at altitude.

Background

The influence of high-altitude/hypoxic conditions on muscle tissue has been of long-term scientific interest, one of the obvious reasons being that whole body and muscle performance decreases with increasing altitude [1]. Early reports on

H. Hoppeler, M.D. (✉) • M. Mueller, Ph.D.
M. Vogt, Ph.D.
Department of Anatomy, University of Bern,
Baltzerstrasse 2, CH 3000 Bern 9, Switzerland
e-mail: hoppeler@ana.unibe.ch

E.R. Swenson and P. Bärtsch (eds.), *High Altitude: Human Adaptation to Hypoxia*,
DOI 10.1007/978-1-4614-8772-2_9,

muscle tissue adaptations to hypoxia included observations of Valdivia [2] who reported skeletal muscle capillarity in guinea pigs native to the Andes to be 30 % larger than that of animals raised at sea level. In a landmark paper, Reynafarje [3] compared data from muscle biopsies from sartorius muscle in nine young subjects native to Cerro de Pasco (Peru 4,400 m) to those obtained from nine age-matched controls from Lima (0 m). In high-altitude natives, he found cytochrome c and myoglobin concentrations to be significantly higher by 26 % and 16 %, respectively, while lactate dehydrogenase activity was similar. These papers as well as a number of further studies in animals and humans supported the contention of hypoxia as an important stressor of muscle tissue favoring an enhanced oxidative capacity of muscle tissue (i.e., more capillaries and mitochondria) and associated changes such as an increased myoglobin concentration. Hochachka [4] carried out research on myocardium in a number of high-altitude species (llama, taruca, and alpaca) and found oxidative enzyme activities (citrate synthase and 3-hydroxyacyl-CoA dehydrogenase) to have an elevated scaling in line with previous findings. They argued that "…elevated oxidative enzyme activities allow increased maximum flux capacity of aerobic metabolism. This in turn calls for physiological adjustments in O_2 transfer systems; flux limits of the former must be matched by flux limits of the latter. Only then can an acceptably high scope for aerobic activity be achieved despite reduced O_2 availability." They called this then commonly held viewpoint "an interpretive hypothesis." The explanatory power of the view that higher oxidative capacities were the necessary consequence of short- and long-term adaptations to hypoxia was called into question by a review of Banchero [5]. In a critical analysis of published data on capillary and mitochondrial adaptations to hypoxia, he criticized that many studies supposedly investigating hypoxia were not controlled for concomitant exercise and cold exposure. The latter two stresses had previously been shown to be responsible for distinct elevations of muscle oxidative capacity (see [6, 7]) and are often associated with living at altitude.

The notion of hypoxia as an important environmental cue for an organism was eventually substantiated by the discovery of the hypoxia-inducible factor 1 (HIF-1; see [8]). HIF-1 has subsequently been shown to be a master gene involved in oxygen sensing in skeletal muscle tissue and many other mammalian cells [9, 10]. HIF-1 functions as a transcription factor for many genes coding for proteins involved in oxygen transfer by enhancing erythropoiesis and improving vascularization. HIF-1 also interferes with metabolic pathways, favoring anaerobic, glycolytic metabolism as well as glucose uptake and utilization. The discovery of the transcription factor HIF-1 and the demonstration of its involvement in multiple regulatory ways related to enhance cellular and organismal function under hypoxic conditions have given hypoxia research a solid mechanistic foundation. In the context of this review, we will refer to HIF-1 where appropriate, but we will not discuss HIF-1 function and regulation which is explored in Chaps. 1 and 2.

The quest into the decrease of human exercise capacity and VO_2 max and thus the "limiting factors" with altitude has stimulated many physiologists and motivated a large number of studies in real and simulated altitude. An intriguing finding in need of an explanation was the fact that as already noted by Pugh [1] supplying oxygen to (Caucasian) subjects chronically adapted to hypoxia in Everest base camp did not bring their exercise performance or their VO_2 max back to the values observed at sea level (see [11, 12]). Cerretelli [11] noted that in ten base-camp-adapted subjects, VO_2 max at 5,350 m was reduced by 30 % from sea-level control values. Supplying oxygen to these subjects brought their VO_2 max back to 92 % of sea level despite the fact that cardiac oxygen delivery massively surpassed sea-level values mainly due to a 40 % increase in hematocrit under these conditions. This finding was particularly noteworthy in the light of the generally prevailing notion of a monofactorial cardiac limitation of VO_2 max. Cerretelli [11] thus suggested that "…the limit to VO_2 max at altitude is at least in part peripheral." Based on this conviction, Cerretelli sought collaboration with the Department of Anatomy in

Bern. In Bern, we had used morphometric techniques to analyze human skeletal muscle tissue seeking relationships between muscle structure and muscle performance capacity [13]. The ensuing cooperation between the two labs lasted for over 2 decades and resulted in a number of publications focused on structural modifications of muscle tissue with altitude both in lowland and high-altitude populations [14]. In the context of the current review, we will refer to high altitude as elevations of 3,000–5,500 m, as extreme altitude elevations >5,500 m [15]. Comprehensive summary of structural findings of relevance to long-term hypoxia is given in Fig. 9.1.

Muscle Changes at Altitude in Lowlanders

Body and Muscle Mass: It has been a consistent finding that high-altitude exposure over normal expedition periods of 6 weeks results in weight losses of the order of 5–10 % with a reduction of quadriceps cross-sectional area of the order of 10 % [16]. It was subsequently shown that good acclimation procedures and optimal housing and nutritional conditions can mostly prevent loss of body and muscle mass [17, 18] and that malabsorption does not seem to be a major issue [19]. According to Westerterp [20] energy balance can be sustained to altitudes up to 5,000 m. Above that, loss of appetite and mountain sickness prevent maintenance of energy balance with malabsorption playing a minor role. More recent research shows the problem of body and muscle mass loss at altitude to be more deeply rooted. Reactive oxygen species (ROS), cytokines, hormones, neurotransmitters, receptors, and ligands all have been implied to be potentially involved in metabolic control at altitude [21, 22]. The involvement of the hypoxia master gene, HIF-1, has been shown for the regulation of the leptin gene at least in cell cultures [23]. Leptin seems to be upregulated in people responding with appetite loss upon high-altitude exposure [24] but seems to be unchanged in elite climbers at high altitude [25]. The latter study rather points to a number of endocrine adaptations of importance in high-altitude weight loss. The issue of appetite and weight loss at altitude thus remains currently unresolved but is likely multifactorial (see Chaps. 13 and 15).

Muscle Fiber Size and Muscle Fiber Types: In line with the reduction of muscle cross-sectional area, we found a 20 % loss of muscle fiber cross-sectional area in 14 mountaineers after 8 weeks of exposure to Everest base camp conditions [26]. Similar decreases in arm and leg muscle cross-sectional area and in type I and type II fiber cross-sectional area are also reported for five subjects of Operation Everest II for which an ascent to Mt. Everest was simulated in a hypobaric chamber over a 40-day period [27]. More recently muscle biopsies of the arm muscle biceps brachii and the thigh muscle vastus lateralis were obtained from seven active and eight less active subjects before and after sojourn for 75 days at or above 5,250 m [28]. Muscle fiber size is reported to have decreased by 15 % independent of muscle and activity level. It is thus reasonably well established that the total quantity of skeletal muscle tissue suffers when humans expose themselves to altitudes in excess of 5,000 m for longer periods of time. The reason for muscle loss remains debated. A recent report indicates that in early exposure to hypoxia (4,500 m), mTOR (mammalian target of rapamycin) a key regulator of translation of mRNA in muscle and other tissues is decreased [29]. This finding is in line with current concepts of regulators of muscle mass as the increase of translation of myofibrillar mRNAs via the mTOR pathway is considered responsible for muscle hypertrophy with strength training [30]. Holm et al. [31] found whole body amino acid turnover to be increased with myocontractile protein synthesis rate increased and sarcoplasmic protein synthesis rate unchanged in humans exposed for 1 week to 4,600 m. These recent results continue to leave the question as to the mechanism of altitude muscle mass loss open. More comprehensive future studies involving high-throughput metabolomics and transcriptomic and proteomic studies may help to resolve existing controversies (see also below) [32].

Muscle fiber type distribution does not seem to be affected by high-altitude/hypoxia exposure

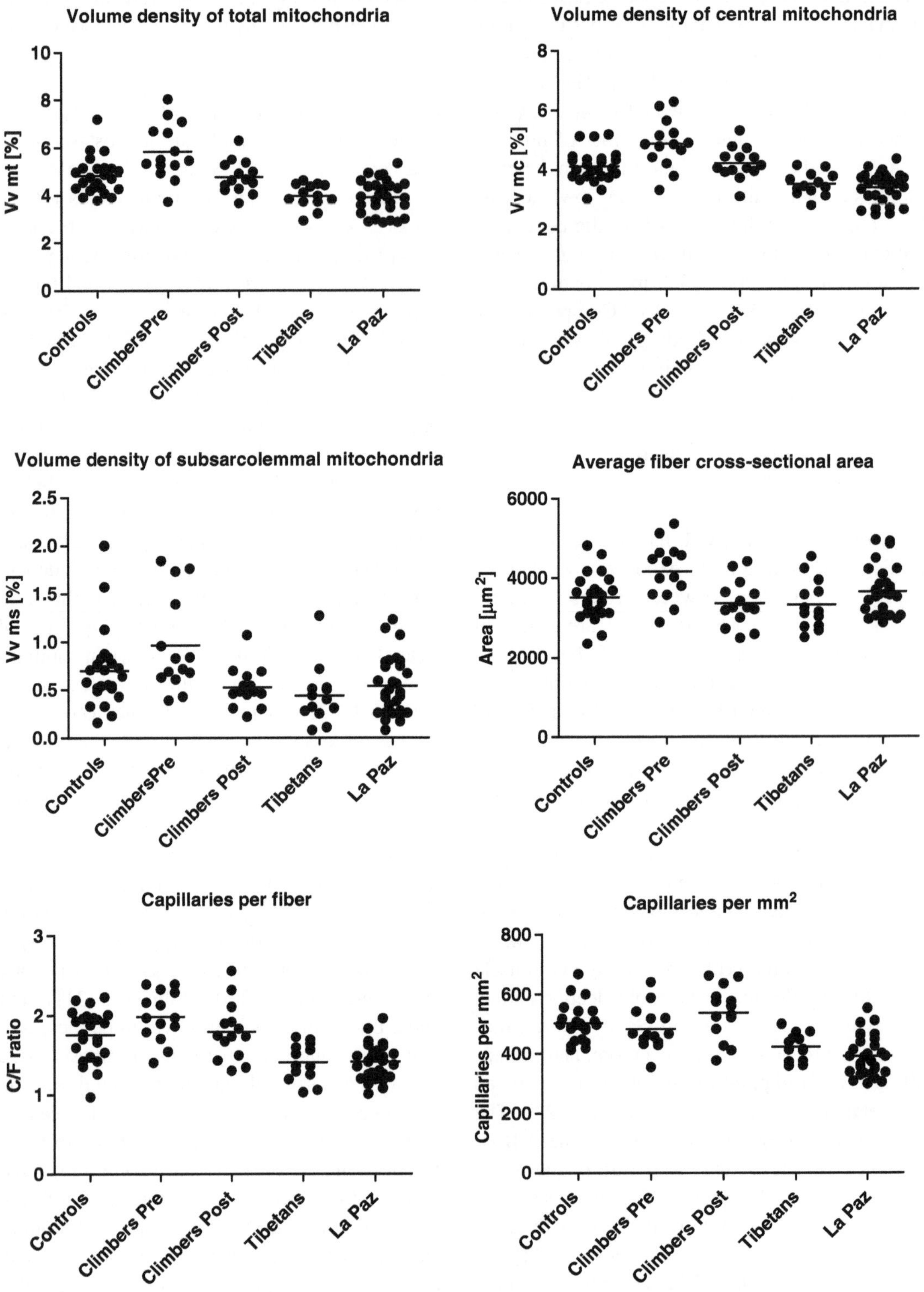

Fig. 9.1 Morphometric variables obtained from biopsies of various populations of subjects analyzed with identical ultrastructural morphometric techniques (mean and individual data points)

in humans [33] or in mice [34]. By contrast, the results of a recent study on seven male, untrained mountaineers, exposed to 5,000 m for 23 days during an expedition to Manaslu, indicate both a shift to type I fibers combined with an increase in mitochondrial content and fiber size [35]. These rather unexpected findings may be related to the fact that the untrained subjects involved in this study may have undergone typical endurance-type adaptations over the expedition period.

With long-term hypoxia such as seen in chronic obstructive pulmonary disease (COPD), it has been shown to shift fiber type distribution in vastus lateralis towards the most glycolytic fiber types IIb/x with a concomitant reduction in fiber size of all fiber types [36–38]. The latter finding could be due to the disease-related hypoxia, hypoactivity, or the chronic inflammatory state of COPD.

Muscle Capillarity: Muscle capillarity has consistently found to be increased by 9–20 % with high-altitude exposure in human biopsy studies, with longer exposure times showing larger increases [16, 27, 28, 33, 39]. In all cases the observed increase in capillary density (or capillary length per fiber volume) was due to a constant capillary to fiber ratio (C/F ratio) combined with the previously discussed decrease in fiber size. From this it can be concluded that high-altitude exposure does not lead to true capillary neoformation; it is rather found that the same capillary network supplies a smaller fiber volume at least in humans. This is a surprising finding as one would expect HIF-1, upregulated in hypoxia, to promote a host of factors involved in angiogenesis. All the more, other factors such as an increase in reactive oxygen species at altitude should support HIF action [40]. There is evidence of a lack in capillary neoformation and an attenuated gene expression of vascular endothelial growth factor (VEGF) and its receptors in gastrocnemius muscle of rats exposed to normobaric hypoxia simulating an altitude of 4,000 m for 8 weeks [41]. Likewise, VEGF mRNA has not found to be increased in humans after acclimatization to 4,100 m for 8 weeks [42]. Animal experiments would indicate that the capacity to increase capillary density and aerobic enzymes at altitude is in principle maintained at least under certain experimental conditions in some species (see [43]). Data from humans exposed to altitudes in excess of 5,000 m do not seem to support this notion [28]. There is currently no good explanation as to why the angiogenic program in human skeletal muscle tissue of subjects with long-term exposure to high or extreme altitude does not seem to be invoked, as it may be in other tissues such as the brain [44].

Muscle Mitochondria: Muscle oxidative capacity estimated by biochemical methods (i.e., citrate synthase, CS, or succinate dehydrogenase, SDH, activity) was reduced by some 20 % in seven subjects of a Swiss Himalayan expedition after return to sea level [45] as well as in five subjects after Operation Everest II [27, 33]. Somewhat smaller decreases in CS and 3-hydroxyacyl-CoA-dehydrogenase (HAD) activities are reported for nine subjects that were exposed to an altitude of 3,450 m for 6–44 weeks [39]. We found the mitochondrial volume density in vastus lateralis to be reduced by close to 20 % with the smaller fraction of subsarcolemmal mitochondria more reduced (−43 %) than the larger fraction of interfibrillar mitochondria (−13 %) in 14 subjects after return from Everest base camp [26]. Mizuno et al. [28] found no significant changes in aerobic enzyme activities (CS and HAD) in the leg muscle biopsies of their 15 subjects after 75 days at 5,250 m. These findings clearly refute Hochachka's interpretive hypothesis mentioned above [4]. In almost all instances, aerobic capacity of skeletal muscle tissue after sojourn at extreme altitude is compromised. We must further consider that the decrease of muscle mass and the decrease in aerobic enzyme activities or mitochondrial volume densities are multiplicative. For the 14 subjects on return from Everest, this meant a decrease in leg muscle oxidative capacity (absolute volume of mitochondria) of the order of 30 %. As a consequence and not surprisingly, we found that VO_2 max of these subjects was also significantly reduced from 4.1 to 3.9 l/min. A further finding of note of this study was the fact that the subjects with the highest mitochondrial volume densities suffered most from high-altitude exposure. The exercise training-related difference in mitochondrial volume density of sedentary and

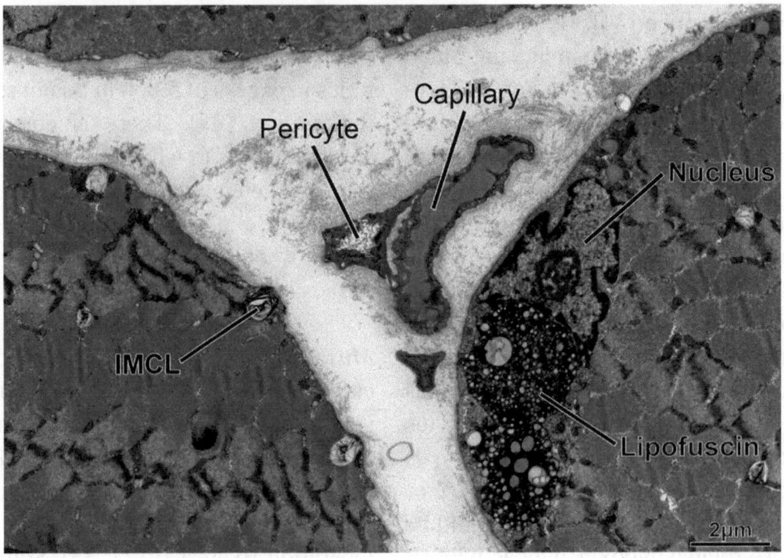

Fig. 9.2 Electron micrograph showing lipofuscin in human skeletal muscle

trained subjects was lost with all subjects having similar mitochondrial volume densities after return from base camp [26]. It cannot be excluded that the mitochondrial loss in trained subjects might at least in part be due to detraining related to the inability to maintain sea-level training volumes at high altitude.

The finding of a decrease of skeletal muscle mitochondrial volume and oxidative enzyme activities with prolonged high-altitude exposure in humans is surprising. There is evidence from studies in which endurance exercise training was carried out in hypoxia (discussed below)—that hypoxia per se activates HIF as well as a set of mitochondrial genes when hypoxia is present only during the limited time of an exercise session [46, 47]. In addition, there is increasing evidence that hypoxia influences mitochondrial genes not only via the HIF pathway but also in interaction with ROS [48–50]. We have to admit that we currently lack a mechanistic understanding of the failure of the hypoxic HIF and ROS induction to augment mitochondriogenesis. This failure is particularly noteworthy as other known HIF and ROS actions, such as a shift away from fatty acid to carbohydrate metabolism in muscle, work as expected [51–53].

Lipofuscin in Muscle: An interesting additional structural aspect of post-expeditional muscle tissue deterioration is an accretion of lipofuscin (Fig. 9.2) [54]. Lipofuscin is considered to be a degradation product of lipid peroxidation accumulating over a lifetime in postmitotic tissues such as nerve cells and cardiac and skeletal myocytes. A major stress of high-altitude exposure is the excess formation of ROS in mitochondria [55]. Our finding of a more than threefold increase in muscle lipofuscin content over an expedition period is the likely consequence of ROS formation [54]. This massive change in lipofuscin content can be seen as a consequence of the activation of macroautophagy and chaperone-mediated autophagy, necessary to deal with the bulk of ROS-damaged proteins [56, 57]. The accumulated lipofuscin is thought to diminish the efficiency of lysosomal degradation, with the process being akin to accelerated cellular senescence. Lipofuscin in skeletal muscle tissue is quite conspicuously elevated in elderly subjects, whereby lipofuscin is almost exclusively found in subsarcolemmal regions adjacent to the nucleus and close to capillaries (Hoppeler's unpublished observations). COPD patients have also shown to have increased levels of lipofuscin in their quadriceps muscles compared to healthy controls [58]. In these patients lipofuscin inclusions have been shown to be more abundant in the oxidative type I fibers with the suggestion that oxidative damage mediated by muscle tissue neutrophils could play a role in oxidative damage [59]. The long-term fate and biological relevance of lipofuscin accumulations after exposure to extreme altitude of lowlanders remain to be determined.

Myoglobin: Hurtado [60] noted a higher content of myoglobin in muscles of dogs native to high altitude. This was substantiated by Reynafarje [3] on biopsies obtained from miners living at 4,400 m. The functional role of myoglobin in muscle has been subject to controversy; however, it seems clearly established that myoglobin serves a supportive role in facilitating oxygen flux to mitochondria [61] and as a scavenger of nitric oxide [62]. The interrelationship of skeletal muscle myoglobin concentration, exercise, and hypoxia has remained controversial with studies on humans and various mammalian species showing inconsistent results (see [43]). It was shown that the myoglobin gene has no hypoxia-responsive element (HRE; [63]). Recent translational studies indicate that exercise together with hypoxia can regulate myoglobin gene expression via calcium transients and translocation of nuclear factor of activated T cells (NFAT; [64]). Given the distinct species specificity of changes of myoglobin concentrations in skeletal muscles with exercise [65, 66], it remains to be seen how the results of the elegant study in mice [64] are applicable to humans.

Acid–Base Regulation and Buffer Capacity: In eight lowland subjects exposed for 8 weeks to 4,100 m, there was a 40–70 % increases in some muscle proteins involved in acid–base balance ($Na+-HCO_3$- co-transporter; the membrane-bound carbonic anhydrase, CA IV), while others decreased (CA XIV) or remained unchanged (cytosolic CA II and III; [67]). Observed changes were smaller in muscle tissue than in erythrocytes. It is suggested that these changes could influence lactate, HCO_3-, and H^+ fluxes from muscle to blood; however, the functional relevance of these changes is currently unclear. There is also a report on a small increase in muscle buffer capacity (5 % and 9 % in vastus lateralis and biceps brachii, respectively) in 15 subjects exposed to 5,250 m for 75 days [28].

Lowlanders exposed to altitudes between 5,000 and 6,000 m, such as those typically experienced during mountaineering expeditions to the Himalayas, experience a loss in body mass mostly related to a loss of appetite that prevents maintenance of energy balance. A loss in muscle mass and muscle fiber size is also observed possibly in connection with a suppression of the mTOR pathway regulating mRNA translation of myofibrillar proteins. Despite of the activation of the HIF and ROS (redox) pathways in hypoxia, mitochondrial volume and muscle oxidative capacity are reduced while capillary supply is maintained. An accumulation of lipofuscin can be taken as proof of increased lipid peroxidation/degradation related to the increase of ROS at high altitude. Overall, we thus see a significant degradation of muscle tissue with an ensuing loss of muscle function in lowlanders exposed to altitudes in excess of 5,000 m.

Muscle Structure in Permanent High-Altitude Residents

During a research stay in La Paz in 1992, we had the opportunity to study the morphology of muscle tissue in a group of young untrained students of mixed ethnic background which had lived permanently in La Paz (3,600–4,000 m) with no stay (>1 month) at lower altitudes in the 3 years preceding the study [68]. Figure 9.1 compares biopsy data of these to those of lowland controls of similar age and background [69–71]. The comparison indicates that fiber size and fiber type distribution are similar in lowlanders and in permanent high-altitude dwellers. Muscle mitochondrial content as well as capillary to fiber ratio is reduced by some 20 % in high-altitude residents. This indicates that both indices of oxygen demand and of oxygen supply are similarly reduced. This finding is in contrast to the early report of Reynafarje [3] that indicated an increased oxidative capacity of skeletal muscle in permanent high-altitude residents. It has been suggested that the discrepancy of these findings stems from the fact that Reynafarje [3] compared an active miner population to sedentary laboratory personnel and that the effect noted by Reynafarje was thus not due to hypoxia but rather to a difference in the level of activity [39]. This observation is in line with the finding that a strenuous standard endurance exercise protocol of 6 weeks leads to similar relative improvements of an approx. 30 % increase in mitochondrial volume density in both lowland and highland inhabitants [68, 69]. In this context it has also been noted that the intramyocellular lipid stores in highland subjects are only half those of lowlanders. Moreover, there is only a marginal increase of these intramyocellular substrate stores with endurance exercise training [68], whereas there is a more than twofold increase in this parameter with an analogous training intervention in lowlanders [69]. It is possible that this structural reflection of myocellular substrate preference is a consequence of HIF activation and its multiple actions on metabolic pathways. The low oxidative capacity with a commensurately low capillary supply of skeletal muscle tissue of high-altitude residents puts permanent high-altitude residents at a disadvantage when exposed to normoxia. Favier et al. [72] demonstrated that VO_2 max increased by a mere 8.2 % in high-altitude residents when subjected to acute normoxia, less than half of what would be expected if acclimatized lowlanders would be subjected to a similar manipulation [14].

Muscle Tissue of Tibetans and Quechua

High-altitude populations offer fascinating options to study genetic selection in humans as a recognized major stressor, hypoxia, is present in permanence. In a temporal sequence, hypoxia first elicits an acute response (O_2 sensors in the carotid body, EPO production, etc.) aimed at defending homeostasis. Some of the acute responses triggered by hypoxia function as molecular signals inducing specific gene expressional changes in different tissues which favor cell, tissue, and organismal function under hypoxia. HIF-1 is recognized to be one of the key players of cellular hypoxia response. In a second step, organisms thus use "phenotypic plasticity" to acclimatize to hypoxic conditions. Given enough time, genetic variability of key traits, and selection pressure, phylogenetic adaptations can occur [73]. It is assumed that the Tibetan plateau has been inhabited for ≈22,000 years (1,100 generations) while the Andean altiplano has a much shorter colonization ≈11,000 years (550 generations; [74]). In principle both time spans offer sufficient opportunity for natural selection to act [75] (and Chap. 19). Interestingly, Tibetans and Quechua have evolved differently to successfully cope with living at altitudes around 4,000 m [75]. Most prominently, Tibetans have an increased hypoxic ventilatory response (HVR) absent in Quechua. By contrast, Quechua show elevated erythropoietin levels and hemoglobin concentrations, absent in Tibetans (see [75]). While Andean highlanders are susceptible to chronic mountain sickness, this condition is rare in Tibetans [76]. Van Patot and Gassmann [77] have argued recently that modifications in the EPAS-1 gene, encoding HIF-2α, could be responsible for some of the differences observed between highland populations.

With regard to skeletal muscle tissue, there is limited data from both high-altitude populations. Assuming typical lowlanders to have on average some 50 % type I (slow-oxidative) fibers, there is a tendency in Tibetans (59 % type I, $n=8$) and Quechua (68 % type I, $n=3$) to have an elevated number of type I fibers [78, 79]. However, all values observed in highland populations fall within the range usually observed in untrained and trained lowlanders. Fiber cross-sectional area is generally found to be at the lower end of the spectrum seen in lowlanders [80]. In a comparison of second-generation Tibetans living in Kathmandu (1,300 m) with subjects of lowland origin, the Tibetans were found to have significantly smaller muscle fibers [78]. Tibetans are found to have mitochondrial volume densities reduced by about 20 % when compared to lowland subjects of similar age and training status [80]. Likewise, second-generation Tibetans were also found to have significantly reduced mitochondrial densities and commensurately lower citrate synthase activities than Nepalese with lowland origin [78]. Generally in line with these findings, Quechua type I fibers are reported to have lower oxidative and higher glycolytic enzyme activities (malate dehydrogenase and lactate dehydrogenase, respectively [79]). Capillary density in Tibetans is found to be in the same range as that of comparable lowland populations [78, 80]. As a consequence, Tibetans have a relatively better capillary supply of muscle mitochondria; i.e., the same capillary network supplies a smaller oxidative capacity. Together with a reduced oxidative capacity, we also find significantly reduced quantities of intramyocellular lipids [78, 80]. This finding is compatible with the contention of a shift away from lipid metabolism and is corroborated by a significantly smaller activity of 3-hydroxyacyl-CoA dehydrogenase (HAD, a key enzyme of lipid metabolism) in Tibetans as compared to lowland Nepalese [78]. Relating VO_2 peak to mitochondrial volumes, it becomes apparent that Tibetans reach a higher ratio of peak oxygen consumption to mitochondria volume than lowland populations [78, 80]. The structural findings reported above may be responsible for the fact that Tibetans lose up to 20 % less of their aerobic performance capacity when asked to perform in a hypoxic environment compared to lowlanders [81].

A reanalysis of available Tibetan muscle tissue with proteomic, biochemical, and morphometric techniques has yielded a number of additional features of skeletal muscle tissue, some of which hint at phylogenetic adaptations of Tibetans to their high-altitude habitat [82]. Both first- and second-generation Tibetans show a significant upregulation of myoglobin, more than twofold in Tibetans previously exposed to high altitude. An increased content of myoglobin can be seen as advantageous for oxygen diffusion in muscle tissue and thus contributing towards maintaining muscle aerobic metabolism in hypoxia (see above). In Tibetans returning from Everest base camp, glutathione S-transferase (an antioxidant enzyme involved with lipid peroxidation) was upregulated over three fold, possibly related to the increased capacity of Tibetan muscle tissue to deal with ROS damage. We found lipofuscin content in these muscle biopsies barely elevated above values seen in Caucasian climbers before high-altitude exposure.

Exercise Training in Hypoxia

This review has so far been concerned with humans continuously exposed to high or extreme altitude for durations in excess of 1 month. The combined evidence indicates that a number of defensive mechanisms are activated, indicating that elevations above 5,000 m are not a desirable long-term environment for our species. It has long been noted that sojourn at moderate altitudes (2,000–3,000 m) has measurable physiological effects such as a HIF-1-mediated increase in erythropoietin and elevation in red blood cell counts. This is seen as advantage in endurance-oriented sports (see [83, 84]). In the athletic environment "live high—train low," i.e., spending some 12–14 h in hypoxia equivalent to at least 2,500 m but training under normoxic conditions has been shown to be effective in certain sports (see [85] and Chap. 16). We have used the alternative strategy "train high—live low" with the rationale to enhance the training stress (mainly on muscle) by using normobaric hypoxia during intense training sessions while recovering in ambient normoxia [86]. This was done in view of the negative effects on muscle tissue of permanent hypoxia exposure. We have recently critically reviewed the literature comprising 27 controlled studies with regard to the "train high—live low" paradigm in untrained and trained subjects [87]. In sedentary subjects it is evident that the main stressor is "exercise training," whether exercise is carried out in a normoxic or hypoxic environment. In nine studies of trained subjects, four report an added benefit of performing some of the training load under hypoxic conditions. With regard to muscle tissue, it is of note that hypoxic training induces a significant upregulation of a number of genes, among them HIF-1 and genes related to carbohydrate metabolism, mitochondrial biogenesis, pH regulation, and oxidative stress, compared to training in normoxia [47] (see also [88, 89]). While "train high—live low" cannot be shown to be an advantage for competition at sea level, it may be of advantage for competition at altitude when an athlete's usual training environment is near sea level (0–500 m).

In conclusion we find that long-term sojourn at altitudes >5,000 m has mainly negative consequences to muscle tissue as evidenced by a decrease in fiber size, a loss of mitochondria, and an increased accumulation of lipid peroxidation products. Tissues (including muscle) activate a defensive program in hypoxia of which the HIF-1 response is currently well described but not likely the only mechanism of protection. Further confirmation of the stress of permanent living at high altitude (3,000–5,500 m) comes from the demonstration of disparate but effective phylogenetic adaptations of high-altitude populations, Tibetans and Quechua. Finally, using hypoxia during exercise sessions has been shown to lead to a specific molecular signature of the muscle tissue response of potential benefit for competition at altitude. Further understanding of interactions of hypoxia with the human phenotype will come from the combination of careful classical physiology with genomic data and data obtained with high-throughput technology to assess changes in the transcriptome, proteome, and metabolome.

References

1. Pugh LG, Gill MB, Lahiri S, et al. Muscular exercise at great altitudes. J Appl Physiol. 1964;19:431–40.
2. Valdivia E. Total capillary bed in striated muscles of guinea pigs native to the Peruvian mountains. Am J Physiol. 1958;194:585–9.
3. Reynafarje B. Myoglobin content and enzymatic activity of muscle and altitude adaptation. J Appl Physiol. 1962;17:301–5.
4. Hochachka PW, Stanley C, Merkt J, et al. Metabolic meaning of elevated levels of oxidative enzymes in high altitude adapted animals: an interpretive hypothesis. Respir Physiol. 1983;52:303–13.
5. Banchero N. Cardiovascular responses to chronic hypoxia. Annu Rev Physiol. 1987;49:465–76.
6. Holloszy JO, Booth FW. Biochemical adaptations to endurance exercise in muscle. Annu Rev Physiol. 1976;38:273–91.
7. Himms-Hagen J, Cerf J, Desautels M, et al. Thermogenic mechanisms and their control. Experientia Suppl. 1978;32:119–34.
8. Semenza GL, Wang GL. A nuclear factor induced by hypoxia via de novo protein synthesis binds to the human erythropoietin gene enhancer at a site required for transcriptional activation. Mol Cell Biol. 1992;12:5447–54.
9. Semenza GL, Shimoda LA, Prabhakar NR. Regulation of gene expression by HIF-1. Novartis Found Symp. 2006;272:2–8. discussion 8–14, 33–6.
10. Semenza GL. Oxygen homeostasis. Wiley Interdiscip Rev Syst Biol Med. 2010;2:336–61.
11. Cerretelli P. Limiting factors to oxygen transport on Mount Everest. J Appl Physiol. 1976;40:658–67.
12. Ferretti G. Limiting factors to oxygen transport on Mount Everest 30 years after: a critique of Paolo Cerretelli's contribution to the study of altitude physiology. Eur J Appl Physiol. 2003;90:344–50.
13. Hoppeler H, Luthi P, Claassen H, et al. The ultrastructure of the normal human skeletal muscle. A morphometric analysis on untrained men, women and well-trained orienteers. Pflugers Arch. 1973;344:217–32.
14. Cerretelli P, Hoppeler H. Morphologic and metabolic response to chronic hypoxia: the muscle system. In: Fregly MJ, Blatteis CM, editors. Handbook of physiology, section 4, environmental physiology, vol. 2. New York: Oxford University Press; 1996. p. 1155–81.
15. Bartsch P, Saltin B. General introduction to altitude adaptation and mountain sickness. Scand J Med Sci Sports. 2008;18 Suppl 1:1–10.
16. Hoppeler H, Howald H, Cerretelli P. Human muscle structure after exposure to extreme altitude. Experientia. 1990;46:1185–7.
17. Butterfield GE, Gates J, Fleming S, et al. Increased energy intake minimizes weight loss in men at high altitude. J Appl Physiol. 1992;72:1741–8.
18. Kayser B, Narici M, Milesi S, et al. Body composition and maximum alactic anaerobic performance during a one month stay at high altitude. Int J Sports Med. 1993;14:244–7.
19. Kayser B, Acheson K, Decombaz J, et al. Protein absorption and energy digestibility at high altitude. J Appl Physiol. 1992;73:2425–31.
20. Westerterp KR, Meijer EP, Rubbens M, et al. Operation Everest III: energy and water balance. Pflugers Arch. 2000;439:483–8.
21. Tschop M, Morrison KM. Weight loss at high altitude. Adv Exp Med Biol. 2001;502:237–47.
22. Raguso CA, Guinot SL, Janssens JP, et al. Chronic hypoxia: common traits between chronic obstructive pulmonary disease and altitude. Curr Opin Clin Nutr Metab Care. 2004;7:411–7.

23. Grosfeld A, Zilberfarb V, Turban S, et al. Hypoxia increases leptin expression in human PAZ6 adipose cells. Diabetologia. 2002;45:527–30.
24. Tschop M, Strasburger CJ, Hartmann G, et al. Raised leptin concentrations at high altitude associated with loss of appetite. Lancet. 1998;352:1119–20.
25. Benso A, Broglio F, Aimaretti G, et al. Endocrine and metabolic responses to extreme altitude and physical exercise in climbers. Eur J Endocrinol. 2007;157: 733–40.
26. Hoppeler H, Kleinert E, Schlegel C, et al. Morphological adaptations of human skeletal muscle to chronic hypoxia. Int J Sports Med. 1990;11 Suppl 1:S3–9.
27. MacDougall JD, Green HJ, Sutton JR, et al. Operation Everest II: structural adaptations in skeletal muscle in response to extreme simulated altitude. Acta Physiol Scand. 1991;142:421–7.
28. Mizuno M, Savard GK, Areskog NH, et al. Skeletal muscle adaptations to prolonged exposure to extreme altitude: a role of physical activity? High Alt Med Biol. 2008;9:311–7.
29. Vigano A, Ripamonti M, De Palma S, et al. Proteins modulation in human skeletal muscle in the early phase of adaptation to hypobaric hypoxia. Proteomics. 2008;8:4668–79.
30. Baar K, Nader G, Bodine S. Resistance exercise, muscle loading/unloading and the control of muscle mass. Essays Biochem. 2006;42:61–74.
31. Holm L, Haslund ML, Robach P, et al. Skeletal muscle myofibrillar and sarcoplasmic protein synthesis rates are affected differently by altitude-induced hypoxia in native lowlanders. PLoS One. 2010;5:e15606.
32. Murray AJ. Metabolic adaptation of skeletal muscle to high altitude hypoxia: how new technologies could resolve the controversies. Genome Med. 2009;1:117.
33. Green HJ, Sutton JR, Cymerman A, et al. Operation Everest II: adaptations in human skeletal muscle. J Appl Physiol. 1989;66:2454–61.
34. Takahashi H, Kikuchi K, Nakayama H. Effect of chronic hypoxia on skeletal muscle fiber type in adult male rats. Ann Physiol Anthropol. 1992;11:625–30.
35. Doria C, Toniolo L, Verratti V, et al. Improved VO2 uptake kinetics and shift in muscle fiber type in high-altitude trekkers. J Appl Physiol. 2011;111:1597–605.
36. Whittom F, Jobin J, Simard PM, et al. Histochemical and morphological characteristics of the vastus lateralis muscle in patients with chronic obstructive pulmonary disease. Med Sci Sports Exerc. 1998;30:1467–74.
37. Pereira MC, Isayama RN, Seabra JC, et al. Distribution and morphometry of skeletal muscle fibers in patients with chronic obstructive pulmonary disease and chronic hypoxemia. Muscle Nerve. 2004;30:796–8.
38. Gosker HR, Zeegers MP, Wouters EF, et al. Muscle fibre type shifting in the vastus lateralis of patients with COPD is associated with disease severity: a systematic review and meta-analysis. Thorax. 2007;62: 944–9.
39. Saltin B, Nygaard E, Rasmussen B. Skeletal muscle adaptation in man following prolonged exposure to high altitude (abstract). Acta Physiol Scand. 1998; 109:31A.
40. Breen E, Tang K, Olfert M, et al. Skeletal muscle capillarity during hypoxia: VEGF and its activation. High Alt Med Biol. 2008;9:158–66.
41. Olfert IM, Breen EC, Mathieu-Costello O, et al. Skeletal muscle capillarity and angiogenic mRNA levels after exercise training in normoxia and chronic hypoxia. J Appl Physiol. 2001;91:1176–84.
42. Lundby C, Pilegaard H, Andersen JL, et al. Acclimatization to 4100 m does not change capillary density or mRNA expression of potential angiogenesis regulatory factors in human skeletal muscle. J Exp Biol. 2004;207:3865–71.
43. Mathieu-Costello O. Muscle adaptation to altitude: tissue capillarity and capacity for aerobic metabolism. High Alt Med Biol. 2001;2:413–25.
44. Schoch HJ, Fischer S, Marti HH. Hypoxia-induced vascular endothelial growth factor expression causes vascular leakage in the brain. Brain. 2002;125:2549–57.
45. Howald H, Pette D, Simoneau JA, et al. Effect of chronic hypoxia on muscle enzyme activities. Int J Sports Med. 1990;11 Suppl 1:S10–4.
46. Vogt M, Puntschart A, Geiser J, et al. Molecular adaptations in human skeletal muscle to endurance training under simulated hypoxic conditions. J Appl Physiol. 2001;91:173–82.
47. Zoll J, Ponsot E, Dufour S, et al. Exercise training in normobaric hypoxia in endurance runners. III. Muscular adjustments of selected gene transcripts. J Appl Physiol. 2006;100:1258–66.
48. Chavez A, Miranda LF, Pichiule P, et al. Mitochondria and hypoxia-induced gene expression mediated by hypoxia-inducible factors. Ann N Y Acad Sci. 2008; 1147:312–20.
49. Taylor CT. Mitochondria and cellular oxygen sensing in the HIF pathway. Biochem J. 2008;409:19–26.
50. Semenza GL. Regulation of oxygen homeostasis by hypoxia-inducible factor 1. Physiology (Bethesda). 2009;24:97–106.
51. Zhang JZ, Behrooz A, Ismail-Beigi F. Regulation of glucose transport by hypoxia. Am J Kidney Dis. 1999;34:189–202.
52. Webster KA. Evolution of the coordinate regulation of glycolytic enzyme genes by hypoxia. J Exp Biol. 2003;206:2911–22.
53. Katz A. Modulation of glucose transport in skeletal muscle by reactive oxygen species. J Appl Physiol. 2007;102:1671–6.
54. Martinelli M, Winterhalder R, Cerretelli P, et al. Muscle lipofuscin content and satellite cell volume is increased after high altitude exposure in humans. Experientia. 1990;46:672–6.
55. Askew EW. Work at high altitude and oxidative stress: antioxidant nutrients. Toxicology. 2002;180:107–19.
56. Terman A, Brunk UT. Lipofuscin. Int J Biochem Cell Biol. 2004;36:1400–4.
57. Rajawat YS, Hilioti Z, Bossis I. Aging: central role for autophagy and the lysosomal degradative system. Ageing Res Rev. 2009;8:199–213.

58. Allaire J, Maltais F, LeBlanc P, et al. Lipofuscin accumulation in the vastus lateralis muscle in patients with chronic obstructive pulmonary disease. Muscle Nerve. 2002;25:383–9.
59. Koechlin C, Maltais F, Saey D, et al. Hypoxaemia enhances peripheral muscle oxidative stress in chronic obstructive pulmonary disease. Thorax. 2005;60: 834–41.
60. Hurtado A, Rotta A, Merino C, et al. Studies of my hemoglobin at high altitudes. Am J Med Sci. 1937; 194:708–13.
61. Conley KE, Ordway GA, Richardson RS. Deciphering the mysteries of myoglobin in striated muscle. Acta Physiol Scand. 2000;168:623–34.
62. Garry DJ, Kanatous SB, Mammen PP. Emerging roles for myoglobin in the heart. Trends Cardiovasc Med. 2003;13:111–6.
63. Wystub S, Ebner B, Fuchs C, et al. Interspecies comparison of neuroglobin, cytoglobin and myoglobin: sequence evolution and candidate regulatory elements. Cytogenet Genome Res. 2004;105:65–78.
64. Kanatous SB, Mammen PP, Rosenberg PB, et al. Hypoxia reprograms calcium signaling and regulates myoglobin expression. Am J Physiol Cell Physiol. 2009;296:C393–402.
65. Masuda K, Okazaki K, Kuno S, et al. Endurance training under 2500-m hypoxia does not increase myoglobin content in human skeletal muscle. Eur J Appl Physiol. 2001;85:486–90.
66. Pattengale PK, Holloszy JO. Augmentation of skeletal muscle myoglobin by a program of treadmill running. Am J Physiol. 1967;213:783–5.
67. Juel C, Lundby C, Sander M, et al. Human skeletal muscle and erythrocyte proteins involved in acid–base homeostasis: adaptations to chronic hypoxia. J Physiol. 2003;548:639–48.
68. Desplanches D, Hoppeler H, Tuscher L, et al. Muscle tissue adaptations of high-altitude natives to training in chronic hypoxia or acute normoxia. J Appl Physiol. 1996;81:1946–51.
69. Hoppeler H, Howald H, Conley K, et al. Endurance training in humans: aerobic capacity and structure of skeletal muscle. J Appl Physiol. 1985;59:320–7.
70. Rosler K, Hoppeler H, Conley KE, et al. Transfer effects in endurance exercise. Adaptations in trained and untrained muscles. Eur J Appl Physiol Occup Physiol. 1985;54:355–62.
71. Elder GC, Bradbury K, Roberts R. Variability of fiber type distributions within human muscles. J Appl Physiol. 1982;53:1473–80.
72. Favier R, Spielvogel H, Desplanches D, et al. Maximal exercise performance in chronic hypoxia and acute normoxia in high-altitude natives. J Appl Physiol. 1995;78:1868–74.
73. Hochachka PW, Gunga HC, Kirsch K. Our ancestral physiological phenotype: an adaptation for hypoxia tolerance and for endurance performance? Proc Natl Acad Sci U S A. 1998;95:1915–20.
74. Aldenderfer MS. Moving up the world: archeologists seek to understand how and when people came to occupy the Andean and Tibetan plateaus. Am Sci. 2003;91:542–9.
75. Beall CM. Detecting natural selection in high-altitude human populations. Respir Physiol Neurobiol. 2007;158:161–71.
76. Wu TY. Chronic mountain sickness on the Qinghai-Tibetan plateau. Chin Med J (Engl). 2005;118:161–8.
77. van Patot MC, Gassmann M. Hypoxia: adapting to high altitude by mutating EPAS-1, the gene encoding HIF-2alpha. High Alt Med Biol. 2011;12:157–67.
78. Kayser B, Hoppeler H, Desplanches D, et al. Muscle ultrastructure and biochemistry of lowland Tibetans. J Appl Physiol. 1996;81:419–25.
79. Rosser BW, Hochachka PW. Metabolic capacity of muscle fibers from high-altitude natives. Eur J Appl Physiol Occup Physiol. 1993;67:513–7.
80. Kayser B, Hoppeler H, Claassen H, et al. Muscle structure and performance capacity of Himalayan Sherpas. J Appl Physiol. 1991;70:1938–42.
81. Marconi C, Marzorati M, Grassi B, et al. Second generation Tibetan lowlanders acclimatize to high altitude more quickly than Caucasians. J Physiol. 2004; 556:661–71.
82. Gelfi C, De Palma S, Ripamonti M, et al. New aspects of altitude adaptation in Tibetans: a proteomic approach. FASEB J. 2004;18:612–4.
83. Levine BD. Intermittent hypoxic training: fact and fancy. High Alt Med Biol. 2002;3:177–93.
84. Bonetti DL, Hopkins WG. Sea-level exercise performance following adaptation to hypoxia: a meta-analysis. Sports Med. 2009;39:107–27.
85. Stray-Gundersen J, Levine BD. Live high, train low at natural altitude. Scand J Med Sci Sports. 2008;18 Suppl 1:21–8.
86. Desplanches D, Hoppeler H, Linossier MT, et al. Effects of training in normoxia and normobaric hypoxia on human muscle ultrastructure. Pflugers Arch. 1993;425:263–7.
87. Hoppeler H, Klossner S, Vogt M. Training in hypoxia and its effects on skeletal muscle tissue. Scand J Med Sci Sports. 2008;18 Suppl 1:38–49.
88. Hoppeler H, Vogt M. Hypoxia training for sea-level performance. Training high-living low. Adv Exp Med Biol. 2001;502:61–73.
89. Mounier R, Pialoux V, Roels B, et al. Effect of intermittent hypoxic training on HIF gene expression in human skeletal muscle and leukocytes. Eur J Appl Physiol. 2009;105:515–24.

Blood and Haemostasis

10

James S. Milledge and Peter Bärtsch

Abstract

This chapter deals with oxygen-carrying capacity of blood and haemostasis. It focuses on the effects of intermittent exposure to hypoxia on red blood cell mass, particularly in the various settings used by athletes. Furthermore, mechanisms of neocytolysis occurring on descent from high altitude are discussed, as well as mechanisms accounting for a rapid decline of erythropoietin serum levels during persistent hypoxia, whilst increased erythropoiesis is maintained. Different strategies of adaptation to chronic hypoxia with regard to haemoglobin and oxygenation of the Andean, Ethiopian and Tibetan populations are discussed in the light of recent findings of genetic mutations in these populations. The review on haemostasis includes studies on platelets, blood coagulation and fibrinolysis in acute and chronic exposure to hypoxia and the discussion of potential mechanisms of activation of blood coagulation at altitudes above 4,000 m.

Oxygen-Carrying Capacity of Blood

Introduction

The blood has many components and functions, but in this chapter we will focus only on its oxygen-carrying capacity (in Part I) and on haemostasis (Part II).

In an earlier review on this topic, Robert Grover and Peter Bärtsch outlined the importance of blood in the transport of oxygen [1]. They pointed out the considerable individual variation in haematocrit even at sea level and its change with age and difference between men and women.

J.S. Milledge, MBChB., M.D., F.R.C.P.
Center for Altitude, Space and Extreme Environment Medicine UCL, University College London, London, UK
e-mail: jim@medex.org.uk

P. Bärtsch, M.D (✉)
Division of Sports Medicine, Department of Internal Medicine, Medical University Clinic, University of Heidelberg, Heidelberg, Germany
e-mail: peter.bartsch@med.uni-heidelberg.de

E.R. Swenson and P. Bärtsch (eds.), *High Altitude: Human Adaptation to Hypoxia*,
DOI 10.1007/978-1-4614-8772-2_10, © Springer Science+Business Media New York 2014

The increase in red blood cell count is probably the best-known response to hypoxia first shown by Viault in 1890. This is initially due mainly to a reduction in plasma volume (haemoconcentration) as a diuretic response to acute hypoxia and possibly a reduction in fluid balance or increased insensible water loss (see Chap. 11). With sustained hypoxia, there is a slow increase in red cell mass (RCM). Again the increase in hematocrit, Hb and RCM is very variable. It might be thought that this variability is partly due to the differing hypoxic ventilatory response (HVR) resulting in variability of oxygen saturation in subjects at the same altitude and hence O_2 supply to the kidney, but Chapman et al. [2] found no correlation between HVR and erythropoietin (EPO) levels. The increase in RCM is driven by an increase in EPO. However, the increase in serum EPO concentration falls away in 3–10 days, to levels only marginally above control whilst the RCM continues to rise for 3–6 months. The time course of these changes and the effect of PaO_2 on the rise in RCM were illustrated, again showing considerable individual variability. Note that the mechanism for hypoxia turning on the EPO gene via hypoxia-inducible factor (HIF) is discussed in Chap. 2.

Oxygen-haemoglobin affinity was discussed and the effects of changes in 2,3-diphosphoglycerate. The conclusion was that small changes in 2,3-DPG are unimportant and that the in vivo oxygen dissociation curve is probably little different from that at sea level, at least at rest and up to altitudes of 4,000–5,000 m. At extreme altitude the under-corrected respiratory alkalosis will shift the curve to the left, probably a beneficial effect in enhancing lung oxygen uptake to a greater extent than retarding oxygen release in the tissues [3].

The blood of high-altitude residents was then considered. The higher the altitude of residence, the higher is the haematocrit and haemoglobin concentration [Hb]. However, populations differ in their response, Andean (and USA) high-altitude residents having higher levels of haematocrit than Himalayan (Tibetan) residents. These populations are at risk of excessive polycythaemia and chronic mountain sickness (CMS) (Chap. 22 of this edition). CMS resolves if the patient descends to low altitude and also responds to respiratory stimulants.

Finally, the value of increased haematocrit was discussed. It was concluded that a modest increase to 50–53 % is beneficial in increasing oxygen delivery but probably much higher levels are not due, perhaps, to increased viscosity and less diffusive equilibration in the lungs and tissues [4].

Intermittent Hypoxia and Erythrocytosis

In the last 10–15 years, there has been much interest in the use of intermittent hypoxia in the sports science community as an aid to training. This stems from the use of altitude training to improve performance at sea level. The disadvantage of training at altitude is that the intensity of training has to be reduced. In order to overcome this problem, Levine and Stray-Gundersen introduced the concept of "live high, train low". It was hoped that by training at low altitude, the intensity of training could be maintained whilst by living at high altitude, for most of the day and night, the advantage of increasing the RCM would be gained. The hopes for benefits in both haematocrit and exercise performance have largely been disappointing as reviewed in Chap. 16.

Actually going to altitude and commuting to low altitude each day to train are impractical for most athletes, and so the possibility of simulating altitude in a chamber or tent or by simply breathing a low-oxygen mixture seemed to offer a cheap alternative. There have been a large number of studies exploring this possibility. Some have looked at the effect of training under hypoxic conditions for an hour or so a day for some weeks. Others have simulated "live high, train low" by providing hypoxia during the night and training in normoxia in the day. Others have used an intervention of switching frequently from hypoxia to normoxia for periods of an hour or two daily for a week or weeks. Again, as for living in hypobaric hypoxia, simulation by normobaric hypoxia has been disappointing in terms of enhanced performance and significant erythropoiesis (reviewed in

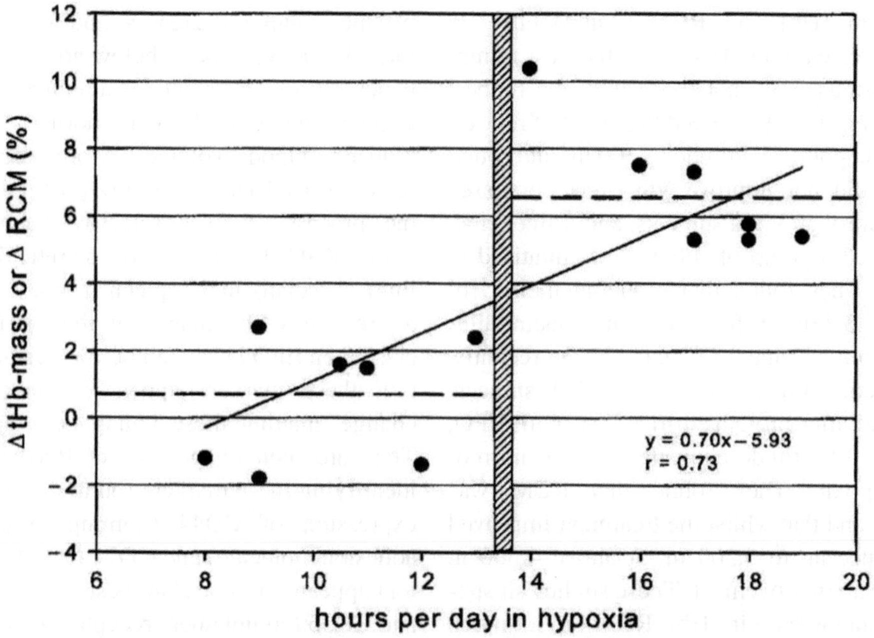

Fig. 10.1 Percentage change in Hb mass or RCM during "live high, train low" studies, in relation to hours of exposure each day. There appears to be a step change in response of RCM; those studies using >13–14 h per day showing a response whilst <13–14 h showed no significant response. The data comes from a review of some 14 studies; see original paper for details (From Schmidt and Prommer [6] with permission)

Chap. 16). Most recently, a double-blind placebo-controlled trial of normobaric "live high, train low" found no effect on either exercise performance or RCM [5].

However, enough studies have measured RCM to make a judgement on the effect of intermittent hypoxia on RCM. Schmidt and Prommer [6] have reviewed these studies and conclude that to effect a significant increase in RCM:

1. The altitude must be above 2,100 m (or equivalent).
2. The duration of hypoxia must be greater than 13–14 h per day.
3. There was little or no effect of altitude above the threshold, although the maximum altitude of these studies was 3,500 m (Fig. 10.1).

They also found that the total time at altitude had little effect but the days at altitude in the studies reviewed varied only from 12 to 31 days. Gore et al. [7] studied subjects who were exposed to an altitude equivalent to 4,000–5,500 m but for only 3 h a day, 5 days a week for 4 weeks. They found that even using this very high altitude, the rise in RCM was not significant (1.0 %).

Red Cell Mass and Performance at Altitude and Sea Level

In an earlier review, Grover pointed out that there is considerable variation in the various red cell indices, [Hb] and RCM [1]. In normal daily living, this does not seem to matter. Moderate decrease in [Hb] is compensated for by an increase in cardiac output. The lower viscosity of anaemic blood reduces the cardiac work, within certain limits. However, when the system is put under pressure as in athletic trials, especially middle-distance races, it has been shown that increasing the RCM (doping) certainly improves performance at sea level, i.e. VO_2 max and time to exhaustion [8]. At altitude, although performance increases to some extent with acclimatisation, other processes are at work besides an increase in [Hb] and RCM. Anecdotally, in climbing expeditions those with the best climbing performance are not those with the highest [Hb]. Young et al. [9] studied the effect of red cell infusion on going to altitude (4,300 m) and found that this did not improve VO_2 max compared with a control group of subjects, and Calbet et al. [10] studied a group of subjects acclimatised to 5,260 m. They found that reducing their [Hb] from 185 to 136 g/L by isovolaemic haemodilution did not change VO_2 max. More recently, Robach and colleagues [11] studied subjects before and after prolonged treatment with EPO, at a range of altitude equivalents using normobaric hypoxia. They found that RCM was increased and that whilst the treatment improved performance up to 3,500 m, at above 4,500 m there was no improvement. These studies all suggest that an increase in [Hb], RCM or Hct much above the normal sea-level values does not improve performance above about 4,000 m as predicted in modelling studies [4]. So presumably, performance at altitude above this level is not limited by oxygen delivery.

Haemoglobin Concentration on Descent from Altitude: Neocytolysis

On descent from altitude, arterial oxygen saturation will return to the normal 96–98 %, and this, together with the now raised [Hb], would be expected to inhibit EPO secretion. Milledge and Cotes [12] reported that serum EPO levels were indeed depressed to 66 % of control values 8 and 20 h after descent to 1,200 m following 2 months at or above 4,500 m. Risso et al. [13] found EPO levels to be only 25 % of control values 6 days after descent from 53 days at or above 4,500 m. The low EPO levels would be expected to reduce the rate of red cell production, but a study by Rice et al. [14] of haematological values of high-altitude residents descending to sea level found also evidence of a process called neocytolysis.

Neocytolysis is the process by which young red blood cells are selectively recalled by the reticuloendothelial system, allowing rapid reduction in RCM. The process was first observed in astronauts [15]. On entering microgravity, their plasma volume decreases and the hematocrit rises; EPO levels fall to below normal, triggering neocytolysis. This results in a reduction of RCM, bringing the [Hb] back to normal but with a reduced blood volume. More recent work reviewed by Rice and Alfrey [16] indicates that the process involves selective haemolysis of young RBCs by the cells of the reticuloendothelium, especially in the spleen. These young RBCs are recognised by changes in their surface markers. When EPO levels fall below a critical threshold, these markers appear to undergo rapid change, making these young cells appear old. They are then phagocytosed. Recent work on identifying these markers found that a decreased expression of CD44 (homing-associated cell adhesion molecule) and CD71 (transferrin receptor) appeared to correlate best with neocytolysis and CD35 (complement receptor) less so [17].

This mechanism was confirmed by Risso et al. [13] in the study cited above, showing that after a 6-day descent to sea level, EPO levels were 25 % of control values. Using density separation, they separated red cells into young, middle-aged and old subsets. Young and middle-aged subsets had almost disappeared after descent, representing only 0.19 and 1.90 % of all cells, whilst old cells increased from 29.5 to 97.9 %. The remaining red cells had acquired a senescent-like phenotype. The same team, in a follow-up study, showed that RBCs of mountaineers after return to sea level, following an altitude stay, showed a lower expression and fragmentation of β-actin. This suggested a change in the membrane skeleton of the RBCs. Interestingly they had found similar changes in the RBCs of patients with β-thalassaemia [18].

Low levels of EPO also trigger changes in the endothelial cells of the spleen, making the capillaries more permeable to young RBCs [19]. These changes are unique to the splenic endothelium.

For athletes who hope to increase their RCM and hence performance for sea-level events by altitude training, the reality of neocytolysis will mean that RCM is reduced even faster than would be calculated on the basis of normal red cell survival.

Other situations in which neocytolysis is involved include the neonate born with a high

[Hb] level, inappropriate for the now increased oxygen saturation thus depressing EPO levels and triggering neocytolysis. Also in renal failure patients where the Low EPO levels in renal failure also cause neocytolysis, which contributes further to the anaemia in these patients [20]. EPO therapy is improved in this situation, if small frequent doses are given to avoid a dip in serum EPO levels to below the critical value.

The Erythropoietin Paradox

On arrival at altitude, serum EPO levels rise within hours. The mechanisms involved in oxygen sensing and induction of a whole host of hormones, including EPO, is via hypoxia-inducible factor (HIF). These have been the subject of exciting new developments in the field of hypoxia. They are discussed in Chap. 1 and also by [21] so will not be reviewed here; except to mention that it now seems that HIF-2 alpha rather than HIF-1 alpha is the important isoform for upregulating the EPO gene. It is the predominant isoform in the interstitial cells [21] of the kidney [22]. EPO levels reached a peak on about the second or third day at altitude and then, even if the subject remains at that altitude, fall away in 5–10 days to levels only marginally above control (Fig. 10 [1]). However, the RCM continues to rise for 3–6 months (Fig. 6, [1]). This well-known, but unexpected, situation I have called the EPO paradox.

In an earlier review [1], Grover suggests that initially EPO production (increased due to hypoxia) exceeds consumption and the serum level rises but that after 3 days consumption has increased to exceed production and hence levels in the blood fall. However, there is no evidence for this mechanism. There is no direct way of measuring production or consumption of EPO. However, Tan et al. [23] measured EPO mRNA, in kidney and liver, in rats made hypoxic for up to 30 h. They also measured serum EPO. Figure 10.2 shows their results. It will be seen that EPO mRNA levels closely followed serum EPO concentration.

If mRNA can be taken as an index of EPO production, then the fall in serum EPO, with continued hypoxia, is accompanied by a fall in production.

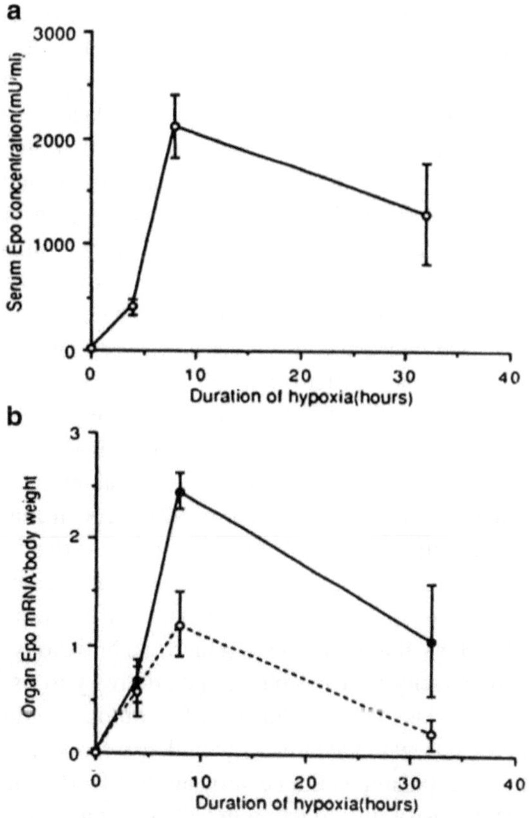

Fig. 10.2 Time course of response of EPO to hypoxia (7.5 % O) in rats. (**a**) Serum EPO from start of hypoxia. (**b**) EPO mRNA levels from renal (*solid line*) and hepatic (*dotted line*). Similar time courses suggest that serum EPO concentration is dependent upon production, as indexed by mRNA levels (From Tan et al. [23] with permission)

If we discard the idea of continued high production and consumption of EPO, we are left with two questions:

1. What is the mechanism that lowers production (and hence serum levels) of EPO in the face of continued tissue hypoxia?
2. What is the mechanism of continued red cell production, erythrocytosis, in the face of near-normal (sea level) EPO levels?

Reduction in EPO Levels at Altitude

With residence at altitude for more than a few days, there is a rapid reduction in serum levels of EPO and probably of production. With acclimatisation there is some reduction in the hypoxic stimulus, due to a small rise in PO_2 due to hyperventilation as respiratory acclimatisation takes place (see Chap. 3). There is also a rise in haematocrit due, initially, to a reduction in plasma volume and later to a rise in RCM. However, at the time of most rapid fall in serum EPO, at 3–7 days of altitude exposure, the effect of any reduction in the hypoxic stimulus, due to these mechanisms, is small. For instance, a subject taken very rapidly to 3,600 m will, after full respiratory acclimatisation, see a fall in PCO_2 of about 7 mmHg and a similar rise in PO_2 from values on arrival. This reduction in hypoxic stress is equivalent to a reduction in altitude of about 500–3,100 m, an altitude that would still result in a strong rise in serum EPO. The effect of a small rise in haematocrit from say, 44–48 % at most, over this time (3–7 days at altitude), would increase blood oxygen capacity by 10 % but oxygen delivery to tissues of less than this, since cardiac output would fall and flow to the kidneys would also reduce, so the effect again would be very small. Clearly, the regulation of the reduction in EPO is not a simple negative feedback with tissue hypoxia as the sensed stimulus.

In seeking for an explanation of the reduction in EPO after only 2–3 days at altitude, it seems likely that the mechanisms and feedback loops, which impinge on HIF production, are responsible. There are a number of possibilities, though which is likely to be of major importance is not clear, and it is also likely, in this rapidly changing field, that more will be discovered. Since the mechanism by which hypoxia induces the production and release of EPO via HIF is very complex, there are a variety of possible feedback mechanisms, positive as well as negative. Pugh et al. [21] discusses some seven possible mechanisms that could modulate the HIF response to hypoxia. These include:

1. HIF itself induces an antisense HIF transcript from the HIF-1-alpha gene, which in turn limits the amount of active HIF-1alpha mRNA. This does not seem to apply to HIF-2-alpha.
2. Hypoxia induces small non-protein-coding microRNAs, such as miR210, that negatively regulate expression of a variety of genes.
3. HIF also induces CITED2 (previously known as p35srj), and this protein competes with HIF for binding to p300/CBP thus inhibiting HIF transactivation.
4. PHD2 and 3, which hydroxylate HIF, thereby marking it for destruction, are themselves transcriptional targets for HIF and thus increase in HIF brings about its own degradation and will blunt the hypoxic response.
5. HIF is also inhibited by an enzyme called factor inhibiting HIF (FIH) which has a number of other substrates, including ankyrin proteins. These substrates compete with HIF for FIH, so changes in the abundance of these other substrates will affect HIF activity.
6. Hypoxia has complex effects on reactive oxygen species, the availability of iron and also the balance between Fe^{2+} and Fe^{3+} and the balance of oxoglutarate and succinate, all of which can affect the activity of the enzymes that normally inactivate HIF.
7. HIF-3-alpha may also modulate HIF activity. This isoform does not activate gene expression but can compete with HIF-1 and HIF-2-alpha, limiting formation of active transcriptional complexes.

This is a greatly simplified summary of some mechanisms, which may explain the downregulation of EPO in the situation of continuing hypoxia. For a fuller discussion and references to the original work on which it is based, see [21].

Continued Erythrocytosis with Near-Normal EPO Levels

Is there an alternative to the hypothesis suggested in an earlier review [1] that there is continued high EPO production and consumption which maintains erythrocytosis?

It is thought that, normally, the population of erythrocyte progenitor cells in the bone marrow is kept in check by programmed cell death, apoptosis. EPO acts by promoting their viability by repressing apoptosis [24]. This allows these progenitor cells to multiply. Once increased, this population causes an increase in the rate of red cell production. Perhaps it only requires a level of

EPO marginally above control values to maintain this enlarged population and hence support continued erythropoiesis.

Recent studies in mice [25] have suggested that there may be other mechanisms by which hypoxia causes erythrocytosis than via EPO. They found that hypoxia induced a change in the fate of haemopoietic stem cells towards an increase in megakaryocyte/erythroid progenitor, rather than granulocyte/monocyte progenitor cells. They found upregulation of the erythroid-specific GATA-1, the transcription factor for the EPO receptor. They also found that the supernatant from bone marrow cell cultures of hypoxic mice (but not the addition of EPO) could induce erythroid-priority differentiation and expansion of the haemopoietic stem cells. They suggested that this mechanism was activated by direct oxygen sensing via HIF-1α. They also found raised levels of IL3 and IL6, which, they suggest, may add to the effect of increasing differentiation towards erythroid progenitors. All these effects would increase the production of red cells.

Descent After Altitude Sojourn

When the prevailing hypoxia is relieved by descent to sea level, EPO levels drop rapidly. This reduction to below normal sea-level values may result in a rapid reduction of progenitor cell population by apoptosis. Neocytolysis is also triggered, resulting in an even more rapid reduction in RCM.

Hepcidin, Iron and Altitude Erythrocytosis

Hepcidin is a hormone produced by the liver. It has the effect of inhibiting the absorption of iron by gut enterocytes and is involved in the release of iron stored in macrophages. It is therefore important in maintaining iron homeostasis, itself important in the erythrocytosis response to hypoxia.

Robach et al. [26] studied the effect of low-dose EPO treatment in subjects at sea level. They showed that urinary hepcidin levels were reduced by the second day of treatment (the first time point studied). This finding confirms in humans, previous animal studies to which they refer. They also found, surprisingly, that muscle iron stores were not depleted despite the uptake of iron by the bone marrow for erythrocytosis.

Piperno et al. [27] recently found that altitude hypoxia inhibited the release of hepcidin in 47 human subjects. Serum levels were decreased within 40 h of reaching an altitude 3,400 m and at 5,400 m were reduced by 80 %. There was a strong correlation between ferritin and hepcidin levels at each stage of the study. Talbot et al. [28] studied 24 male subjects taken to 4,300 m with and without prior iron loading. They found, without iron loading, a rapid reduction in hepcidin levels which was complete by the second day at altitude, before change was seen in any index of iron availability. Prior iron loading delayed the fall in hepcidin until after the transferrin saturation, but not ferritin concentration, had normalised.

It has been known for some time, from animal studies, that iron absorption is increased by hypoxia before iron stores are reduced [29, 30]. This recent work begins to provide some explanation of this phenomenon. It seems likely that the rapid effect of hypoxia on iron traffic is mediated either by EPO itself or via some other gene product induced by HIF.

High-Altitude Resident Populations

Haemoglobin Concentration Levels

In an earlier review [1], as summarised in the introduction, Grover pointed out the difference between Andean and Tibetan high-altitude residents. The Andean (and USA) high-altitude populations have higher [Hb] than Tibetans at any given altitude. Recently Beall and her colleagues [31] studied Ethiopian high-altitude populations and found that they seemed to have a response to altitude hypoxia intermediate to that of Andean and Tibetan high-altitude residents. This is illustrated in Fig. 10.3a, b, drawn from their paper.

Considering [Hb], Fig. 10.3a shows the cumulative percentage [Hb] for the four populations (males), American at sea level, Ethiopian at

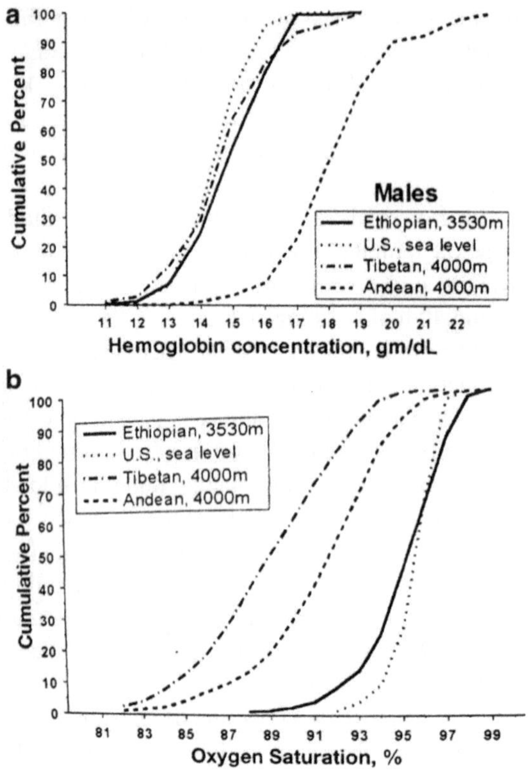

Fig. 10.3 Cumulative haemoglobin concentration (**a**) and oxygen saturation (**b**) in four populations. US subjects at sea level (*dotted line*), Ethiopian high-altitude residents at 3,530 m (*solid line*), Tibetan (*dot dash line*) and Andean (*dashed line*) high-altitude residents at 4,000 m (Composite from Beall et al. [31] Figs. 10.1 and 10.3, with permission)

3,530 m, Tibetan and Andean at 4,000 m. Clearly the Andean has significantly higher [Hb] than the other two high-altitude populations, which have [Hb] values similar to those of a sea-level population. But, of course, the Ethiopian subjects were at a rather lower altitude than the Andean and Tibetan subjects. Acclimatised lowlanders' [Hb] would fall close to the Andean line.

Figure 10.3b shows the cumulative percentage lines for arterial oxygen saturation (SpO_2) for the same populations. This shows three distinct responses to living at altitude by different high-altitude populations. Andeans show a moderate reduction in SpO_2 and Tibetan significantly lower values than Andean, whilst Ethiopians (at 3,530 m) show no desaturation compared with sea-level subjects. The mechanism, by which this latter population achieves a higher SpO_2 than other high-altitude populations, is unknown. It could result from higher ventilation, higher diffusing capacity or Hb with higher oxygen affinity.

Recent Genetic Studies

Beall et al. [32], using a genome-wide approach, compared Tibetans with Han Chinese and found eight single nucleotides polymorphisms (SNPs), out of 500,000 examined, to be significantly over represented in the Tibetans. All eight were on chromosome 2 close to the EPAS1 gene, which codes for HIF2α. This change in EPAS1 gene results in a reduced erythrocytic response and probably explains lower Hb levels and the reduced risk of CMS in Tibetans. There may well be other benefits conferred by this gene change to residence at high altitude.

Simonson et al. [33] reported a large, collaborative study between the Universities of Utah in Salt Lake City and Quinhai in Xining in Western China. They also compared Tibetans with Han Chinese but also with Japanese subjects. They chose to study candidate genes likely to be involved with adaptations to high altitude. These included genes whose products were involved with oxygen delivery, including the HIF pathway, and with nitric oxide metabolism. Genome-wide scans were carried out. They identified the genes EGLN1, PPARA and EPAS1 as being significantly associated with decreased haemoglobin phenotype.

Yi et al. [34] was a collaborative study with over 70 authors from the USA, China, Tibet and Denmark. They used yet another approach. They examined 50 exomes in three populations, comparing Tibetans from two villages at about 4,300 m with Han Chinese and Danes, from low altitude. Thirty-four of the fifty exomes studied were related to "response to hypoxia". The strongest signal of natural selection came from the EPAS1 gene. In this paper the authors also calculated the likely time since the two populations, Han and Tibetans, diverged and report that a bes-tfit model gave a result of only 2,750 years!, which, if true, would represent an even faster rate of frequency change than even the lactase persistence allele in northern European subjects.

However, archaeological evidence suggests much older settlements of the Tibetan plateau.

Three other studies using a variety of methods also found that the EPAS1 gene was overrepresented in high-altitude Tibetan populations (Bigham et al. [35], Peng et al. [36] and Xu et al. [37]). Bigham et al. [35] also found the EGLN1 gene to be selected in Andean as well as Tibetan populations. These studies did found no evidence that the HIF-1-A (the gene coding for HIF-1-α) was involved in altitude adaptation in high-altitude populations.

That these papers, published within a few months of each other, using quite different methods of approach, all find the EPAS1 gene to be importantly different in Tibetans compared with lowland populations is quite astonishing. No doubt other genes are also involved in adaptation to altitude, and it seems that different altitude populations have adapted in different ways. Also it is clear that we can expect more exciting results from further genetic studies in these and other populations.

For further discussion of this topic, the interested reader should also read Chap. 19, Human Evolution at High Altitude, in this book.

Summary and Questions Still to Be Addressed

In the last 10 years, the effect of intermittent hypoxia on erythrocytosis has been studied, usually in projects whose primary aim has been athletic performance. Many studies found no effect, because the level of hypoxia was probably of insufficient intensity and duration and did not result in increased RCM and [Hb]. In order to increase RCM, the equivalent altitude must be above 2,100 m, and the duration of hypoxia must be above 14 h a day for at least 12 days.

An increase in [Hb] and RCM improves performance at sea level and at altitudes up to about 4,000 m. Above this altitude an increase in these indices does not result in improvement in aerobic performance, though further work is desirable to confirm the result of the few studies that have addressed this.

Neocytolysis, triggered by a reduction in EPO levels to below normal values, is an important mechanism for rapidly reducing the number of red blood cells on descent after a period at high altitude. Thus, the performance-enhancing effect of altitude training for sea-level athletic events is limited.

The reduction in EPO levels after the first few days at altitude is discussed, and possible mechanisms underlying this via HIF regulation were suggested. The way EPO causes continued erythrocytosis at altitude after levels have fallen to values only marginally above normal is discussed and a mechanism is proposed. The evidence for non-EPO erythrocytosis comes only from animals so far. Studies in humans will not be easy but clearly are needed. There are obviously many questions to be answered in both these areas.

The hepcidin response to hypoxia has recently been shown, and its importance in the economy of iron and hence in the erythrocytic response to altitude is obvious. Iron levels are also important in the HVR (Chap. 3) and hypoxic pulmonary vasoconstriction (Chap. 5). Clearly there are many unanswered questions regarding control and significance of iron levels.

Recent work on the responses of different populations of high-altitude residents to altitude hypoxia seems to imply the evolution of different solutions to the same altitude hypoxia. In the last 3 years, there have been an exciting flush of papers on genetic adaptation, especially of Tibetan high-altitude populations, pinpointing variations in the EPAS1, EGLN1 and probably other genes involved with oxygen sensing. Again these findings open many questions for molecular and genetic studies to answer.

Blood Coagulation and Fibrinolysis

Acute Exposure

Platelets

In an earlier review [1], it was concluded that gradual ascent to altitudes of 4,500 m had no significant effect on platelet count, on most assays and markers of platelet activity in vitro or in vivo, whereas rapid decompression and/or more severe hypoxia and advanced high-altitude pulmonary oedema may enhance platelet activation or aggregation.

Contrary to these previous studies, more recent work found platelet count to be decreased and p-selectin, a nonspecific marker of platelet secretion, increased after rapid ascent to 4,559 m [38]. Enhanced ex vivo platelet aggregation suggested increased platelet consumption whilst in vivo thrombin formation was unchanged. The findings were not different between those with and without AMS or HAPE. Increased monocyte-platelet aggregates were found after 8 h at a simulated altitude of 2,400 m, but beta-thromboglobulin (BTG), a highly specific marker of platelet secretion, was decreased [39], whilst p-selectin was unchanged in this and another study after 3 h in normobaric hypoxia with an FIO_2 of 0.12 [40]. Prolonged exposure over several days to 1–2 weeks was consistently associated with increased platelet counts attributable to increased thrombopoietin [41] and possibly also EPO [42] plasma levels. Exposure of soldiers at 4,100–4,500 m leads, however, to a decrease of platelet count and platelet aggregation after 3 and 13 month by 13–31 % and 24–75 %, respectively [43].

In summary, platelet count and activity during the first few days of exposure to hypoxia are inconsistent between studies whereas platelet count with prolonged stay increases over 2 weeks and is decreased after 3–13 months. Discrepancies could be due to diurnal variation of platelet count and time of sampling, techniques of counting, confounding effects of exercise or differences between in vivo and in vitro activity.

Coagulation

An earlier review [1] concluded: data obtained in chamber and field studies indicate that blood coagulation after more or less gradual exposure (over 1 to several days) to high altitude is not altered at rest. Sudden exposure in a hypobaric chamber leads to a minor increase in factor VIII:C and a shortening of activated partial thromboplastin time (aPTT), which, however, do not give rise to increased thrombin or fibrin formation. The question as to whether exercise at high altitude can cause greater thrombin or fibrin generation has not been explored sufficiently. Enhanced coagulation or impaired fibrinolysis does not accompany the development of AMS or HAPE, whilst advanced HAPE gives rise to increased fibrin formation.

Because of an association of venous thromboembolism with long-haul flights [44, 45], much interest was given to the influence of moderate hypoxia, corresponding to the cabin pressure of 2,400 m, on coagulation. A placebo-controlled study simulating air cabin conditions in a hypobaric chamber [39] confirmed that prothrombotic alterations resulting in increased thrombin formation do not occur with this degree of hypobaric hypoxia. This is not surprising in view of earlier studies at 4,559 m cited in the first edition and of three more recent investigations that were all negative regarding markers of in vivo thrombin or fibrin formation [38, 40, 46]. Two uncontrolled studies, both using indwelling lines for blood sampling, reported increased in vivo thrombin formation during an 8 h exposure to a simulated altitude of 2,400 m [47] and after an 8 min exposure to normobaric hypoxia resulting in an SaO_2 of 80–85 % [48]. It is very likely that the observed activation of coagulation was caused by the indwelling lines [49].

These data demonstrate that activation of coagulation by hypoxia is not a contributing factor in thromboembolism associated with long-haul flights. However, the combination of factor V Leiden, which by itself increases the life time risk for venous thrombosis two-fold, with oral contraceptives may enhance thrombin and fibrin formation already at an altitude of 2,400 m, whilst oral contraceptives or factor V Leiden alone have no effect [50]. Based on this finding, it is conceivable that an interaction of hypoxia with risk factors of thrombosis also applies to other pre-thrombotic states, particularly at higher altitudes. There are no studies which have investigated these questions. However, several case reports of venous thromboses in individuals with protein C or S deficiency and anti-phospholipid syndrome have been reported [51–53]. Although exercise can enhance thrombin and fibrin formation [54], there is no interaction with hypoxia.

Two studies demonstrate that acute normobaric hypoxia (FIO_2 0.12) does not enhance exercise-induced activation of blood coagulation [55, 56].

Comprehensive studies of blood coagulation above 4,500 m are still lacking. The previously reported increase of D-dimer at 6,452 m [57] was confirmed during ascending Muztagh Ata between 5,533 and 6,865 m [58]. This study also showed indirect evidence of a minor activation of blood coagulation by assessing coagulation times, APC-resistance and activity of von Willebrand factor without the possibility to distinguish between the effects of exercise and hypoxia. Activation of blood coagulation obtained in a large group of 61 subjects hospitalised at 3,600 m with HAPE [59] confirmed previous findings in advanced HAPE [60, 61]. The severity of HAPE also correlated with changes in haemostasis as had been reported previously in a small sample [62].

Fibrinolysis

The previous conclusion that acute exposure to an altitude of 4,500 m or an FIO_2 of 0.12 has no effect on fibrinolysis is supported by further studies performed after 3–8 h of exposure in hypobaric hypoxia of 2,400 m [39] and normobaric hypoxia corresponding to 3,600 m [40]. Furthermore, exercise-induced activation of fibrinolysis is not altered in hypoxia (FIO_2 0.12) [56] and was enhanced in subjects hospitalised with HAPE [59].

Chronic Exposure

A prospective cohort study at 3,500 m followed 38 soldiers over 8 months and showed an increase of haemoglobin, platelet count, BTG, fibrinogen and plasminogen activator inhibitor (PAI)-1 whilst D-Dimer did not change [63]. Compatible with this activation of platelets, a further study calculated a 12-fold increase in the prevalence of stroke in soldiers stationed above 3,000 m (average altitude 4,000 m) vs. low altitude, which was not specified [64].

Furthermore, cases of portal system thrombosis in soldiers at low altitude were associated in 73 % with a pre-thrombotic condition vs. 19 % in 26 cases occurring in soldiers living between 4,200 and 6,500 m [65]. These data suggest that Indian soldiers stationed above 4,200 m over several months have an increased risk of thromboembolism in the arterial and venous system. A role for inherited thrombophilia was, however, not systematically excluded in this study. The results contrast with a reduced prevalence of thromboembolic disease associated with atherosclerosis in the native Andean population as discussed in an earlier review [1]. This difference may in part be explained by the role of lifestyle and possibly genetics on the risk factors for cardiovascular disease. Polymorphisms of the β-fibrinogen gene that are associated with lower plasma levels of fibrinogen were found in the Peruvian Quechua [66].

Mechanisms of Activation of Coagulation

It is well established that plasma concentration of the coagulation factor VIII (FVIII) is elevated at altitudes above 4,000 m and that it most likely accounts for the shortening of the aPTT. It also accounts for an enhanced ex vivo thrombin generation (TG), a test that may indicate a hypercoagulable state [67]. FVIII-dependent TG is enhanced in pronounced hypoxia (FIO_2 0.12) but not at more moderate levels (FIO_2 0.15 and 0.18), and it is abolished after oral intake of vitamin E which suggests that oxygen radicals mediate this response [68]. Considering that thrombin and fibrin generation do not occur with acute exposure to this degree of hypoxia, the clinical significance of this finding is not clear.

Activation of coagulation with fibrin formation in the lungs has been demonstrated in mice exposed to an FIO_2 of 0.06 (corresponding to 9,400 m) [68] or in mountaineers with HAPE and an arterial PO_2 of 20 mmHg or SaO_2 below 50 % [69]. Cultures of smooth muscle cells and hepatoma cells kept in severe hypoxia (FIO of 0.01–0.03) produce more tissue factor and PAI-1. Expression of PAI-1 returns to normal already at an FIO_2 of 0.035. It is mediated by the transcription factor HIF-1α [70], and the transcription factor early growth response gene (EGR-1) induces both PAI-1 and tissue factor. Thus, activation of coagulation by tissue factor and inhibition of fibrinolysis by PAI-1 may contribute to fibrin formation and deposition in very hypoxic tissue as it occurs in severe, advanced HAPE [69]. The increased incidence of thromboembolism in people living between 4,200 and 6,000 m is also compatible with this hypothesis [65]. It has been suggested, based on indirect evidence, that a hypoxia-induced pro-inflammatory upregulation of coagulation plays a major role of clotting during air travel with thrombophilia [71]. If true, such a mechanism might enhance blood coagulation in healthy individuals at extreme altitudes.

Summary and Recommendations for Further Research

Since most research has focused on the role of hypoxia for thromboembolism with long-haul flights and since it is very difficult for methodological reasons to address the open questions listed in the last edition, the suggestions for further studies remain the same:

1. Proper studies using markers of in vivo thrombin formation and in vivo fibrinolysis at altitudes above 5,000–6,000 m
2. Identifying individuals who are at greater risk for thromboembolism at high altitude, which means investigating the interaction of the various states of thrombophilia with hypoxia
3. Epidemiologic studies on the risk of thromboembolism in newcomers and in high-altitude populations
4. Blood coagulation and fibrinolysis in CMS

References

1. Grover RF, Bärtsch P. Blood. In: Hornbein TF, Schoene RB, editors. High altitude—an exploration of human adaptation. New York: Marcel Dekker; 2001. p. 493–523.
2. Chapman RF, Stickford JL, Levine BD. Altitude training considerations for the winter sport athlete. Exp Physiol. 2010;95:411–21.
3. Schumacker PT, Suggett AJ, Wagner PD, West JB. Role of hemoglobin P50 in O_2 transport during normoxic and hypoxic exercise in the dog. J Appl Physiol. 1985;59:749–57.
4. Wagner PD. A theoretical analysis of factors determining VO_2max at sea level and altitude. Respir Physiol. 1996;106:329–43.
5. Siebenmann C, Robach P, Jacobs RA, Rasmussen P, Nordsborg N, Diaz V, et al. "Live high-train low" using normobaric hypoxia: a double-blinded, placebo-controlled study. J Appl Physiol. 2012;112:106–17.
6. Schmidt W, Prommer N. Effects of various training modalities on blood volume. Scand J Med Sci Sports. 2008;18 Suppl 1:57–69.
7. Gore CJ, Rodriguez FA, Truijens MJ, Townsend NE, Stray-Gundersen J, Levine BD. Increased serum erythropoietin but not red cell production after 4 wk of intermittent hypobaric hypoxia (4,000-5,500 m). J Appl Physiol. 2006;101:1386–93.
8. Buick FJ, Gledhill N, Froese AB, Spriet L, Meyers EC. Effect of induced erythrocythemia on aerobic work capacity. J Appl Physiol. 1980;48(4):636–42.
9. Young AJ, Sawka MN, Muza SR, Boushel R, Lyons T, Rock PB, et al. Effects of erythrocyte infusion on VO_2max at high altitude. J Appl Physiol. 1996;81(1):252–9.
10. Calbet JAL, Radegran G, Boushel R, Sondergaard H, Saltin B, Wagner PD. Effect of blood haemoglobin concentration on VO_2, max and cardiovascular function in lowlanders acclimatised to 5260 m. J Physiol. 2002;545(2):715–28.
11. Robach P, Calbet JAL, Thomsen JJ, Boushel R, Mollard P, Rasmussen P, et al. The ergogenic effect of recombinant human erythropoietin on VO_2max depends on the severity of arterial hypoxemia. PLoS One. 2008;3(8):e2996.
12. Milledge JS, Cotes PM. Serum erythropoietin in humans at high altitude and its relation to plasma renin. J Appl Physiol. 1985;59:360–4.
13. Risso A, Turello M, Biffoni F, Antonutto G. Red blood cell senescence and neocytolysis in humans after high altitude acclimatization. Blood Cells Mol Dis. 2007;38(2):83–92.
14. Rice L, Ruiz W, Driscoll T, Whitley CE, Tapia R, Hachey DL. Neocytolysis on descent from altitude: a newly recognized mechanism for the control of red cell mass. Ann Intern Med. 2001;134:652–6.
15. Alfrey CP, Udden MM, Leach-Huntoon CS, Driscoll T, Pickett MH. Control of red blood cell mass in space flight. J Appl Physiol. 1996;81:98–104.
16. Rice L, Alfrey CP. The negative regulation of red cell mass by neocytolysis: physiologic and pathophysiologic manifestations. Cell Physiol Biochem. 2005;15:245–50.
17. Chang CC, Chen Y, Modi K, Awar OG, Alfrey CP, Rice L. Changes of red blood cell surface markers in a blood-doping model of neocytolysis. J Investig Med. 2009;57:650–4.
18. Risso A, Santamaria B, Pistarino E, Cosulich ME, Pompach P, Bezouska K, et al. Fragmentation of

human erythrocyte actin following exposure to hypoxia. Acta Haematol. 2010;123:6–13.
19. Trial J, Rice L, Alfrey CP. Erythropoietin withdrawal alters the interaction between young red blood cells, splenic endothelial cells and macrophages: an in vitro model of neocytolysis. J Investig Med. 2001;49:335–45.
20. Alfrey CP, Rice L, Udden MM, Driscoll T. Neocytolysis: a physiologic down-regulator of red blood cell mass. Lancet. 1997;349:1389–90.
21. Pugh CW. Modulation of the hypoxic response. In: Roach RC, Hackett PH, editors. Hypoxia and exercise (in press).
22. Rosenberger C, Mandriota S, Jürgensen JSJ, Wiesener MS. Hörstrup. JH, Frei U, et al. Expression of hypoxia-inducible factor-1alpha and -2alpha in hypoxic and ischemic rat kidneys. J Am Soc Nephrol. 2002;13:1721–32.
23. Tan CC, Eckardt K-U, Frith Ratcliffe PJ. Feedback modulation of renal and hepatic erythropoietin mRNA in response to graded anemia and hypoxia. Am J Physiol Renal Physiol. 1992;263:474–81.
24. Silva M, Grillot D, Benito A, Richard C, Nunez G, Fernåndez-Luna JL. Erythropoietin can promote erythroid progenitor survival by repressing apoptosis through Bcl-XL and Bcl-2. Blood. 1996;88:1576–82.
25. Li P, Huang J, Tian HJ, Huang QY, Jiang CH, Gao YQ. Regulation of bone marrow hematopoietic stem cell is involved in high-altitude erythrocytosis. Exp Hematol. 2011;39:37–46.
26. Robach P, Recalcati S, Girelli D, Gelfi C, Aachmann-Andersen NJ, Thomsen JJ, et al. Alterations of systemic and muscle iron metabolism in human subjects treated with low-dose recombinant erythropoietin. Blood. 2009;113:6707–15.
27. Piperno A, Galimberti S, Mariani R, Pelucchi S, Ravasi G, Lombardi C, et al. Modulation of hepcidin production during hypoxia-induced erythropoiesis in humans in vivo: data from the HIGHCARE project. Blood. 2011;117:2953–9.
28. Talbot NP, Lakhatl S, Smith TG, Privat C, Nickol AH, Rivera-Ch M, et al. Regulation of hepcidin expression at high altitude. Blood. 2012;119:857–60.
29. Hathorn MKS. The influence of hypoxia on iron absorption in the rat. Gastroenterology. 1971;60:76–81.
30. Raja KB, Pippard MJ, Simpson RJ, Peters TJ. Relationship between erythropoiesis and the enhanced intestinal uptake of ferric iron in hypoxia in the mouse. Br J Haematol. 1986;64:587–93.
31. Beall CM, Decker MJ, Brittenham GM, Kushner I, Gebremedhin A, Strohl KP. An Ethiopian pattern of human adaptation to high-altitude hypoxia. Proc Natl Acad Sci USA. 2002;99:17215–8.
32. Beall CM, Cavalleri GL, Deng L, Elston RC, Gao Y, Knight J, et al. Natural selection on *EPAS1 (HIF2α)* associated with low hemoglobin concentration in Tibetan highlanders. Proc Natl Acad Sci USA. 2010;107:11459–64.
33. Simonson TS, Yang Y, Huf CD, Yun H, Qin G, Witherspoon DJ, et al. Genetic evidence for high-altitude adaptation in Tibet. Science. 2010;329:72–5.
34. Yi X, Liang Y, Huerta-Sanchez E, Jin X, Cuo ZX, Pool JE, et al. Sequencing of 50 human exomes reveals adaptation to high altitude. Science. 2010;329:75–8.
35. Bigham A, Bauchet M, Pinto D, Mao X, Akey JM, Mei R, et al. Identifying signatures of natural selection in Tibetan and Andean populations using dense genome scan data. PLoS Genet. 2010;6:e1001116.
36. Peng Y, Yang Z, Zhang H, Cui C, Qi X, Luo X, et al. Genetic variations in Tibetan populations and high-altitude adaptation at the Himalayas. Mol Biol Evol. 2011;28:1075–81.
37. Xu S, Li S, Yang Y, Tan J, Lou H, Jin W, et al. A genome-wide search for signals of high-altitude adaptation in Tibetans. Mol Biol Evol. 2011;28:1003–11.
38. Lehmann T, Mairbäurl H, Pleisch B, Maggiorini M, Bärtsch P, Reinhart WH. Platelet count and function at high altitude and in high-altitude pulmonary edema. J Appl Physiol. 2006;100:690–4.
39. Toff WD, Jones CI, Ford I, Pearse RJ, Watson GG, Wattt SJ, et al. Effect of hypobaric hypoxia, simulating conditions during long-haul air travel, on coagulation, fibrinolysis, platelet function and endothelial activation. JAMA. 2006;295:2251–61.
40. Hodkinson PD, Hunt BJ, Parmar K, Ernsting J. Is mild normobaric hypoxia a risk factor for venous thromboembolism? J Thromb Haemost. 2003;1:2131–3.
41. Hartmann S, Krafft A, Huch R, Breymann C. Effect of altitude on thrombopoietin and the platelet count in healthy volunteers. Thromb Haemost. 2005;93:115–7.
42. Hudson JG, Bowen AL, Navia P, Rios-Dalenz J, Pollard A, Williams D, et al. The effect of high altitude on platelet counts, thrombopoietin and erythropoietin levels in young Bolivian airmen visitin the Andes. Int J Biometeorol. 1999;43:85–90.
43. Vij AG. Effect of prolonged stay at high altitude on platelet aggregation and fibrinogen levels. Platelets. 2009;20:421–7.
44. Lapostolle F, Surget V, Borron SW, Desmaizieres M, Sordelet D, Lapandry C, et al. Severe pulmonary embolism associated with air travel. N Engl J Med. 2001;345:779–83.
45. Scurr JH, Machin SJ, Bailey-King S, Mackie IJ, McDonald S, Coleridge Smith PD. Frequency and prevention of symptomless deep-vein thrombosis in long-haul flights: a randomised trial. Lancet. 2001;357:1485–9.
46. Crosby A, Talbot NP, Harrison P, Keeling D, Robbins PA. Relation between acute hypoxia and activation of coagulation in human beings. Lancet. 2003;361:2207–8.
47. Bendz B, Rostrup M, Sevre K, Andersen TO, Sandset PM. Association between acute hypobaric hypoxia and activation of coagulation in human beings. Lancet. 2000;356:1657–8.

48. van Känel R, Loredo JS, Powell FL, Adler KA, Dimsdale JE. Short-term isocapnic hypoxia and coagulation activation in patients with sleep apnea. Clin Hemorheol Microcirc. 2005;33:369–77.
49. Bärtsch P, Straub PW, Haeberli A. Hypobaric hypoxia. Lancet. 2001;357:955.
50. Schreijer AJM, Cannegieter SC, Meijers JCM, Middeldorp S, Buller HR, Rosendaal FR. Activation of coagulation system during air travel: a crossover study. Lancet. 2006;367(9513):832–8.
51. Boulos P, Kouroukis C, Blake G. Superior sagittal sinus thrombosis occurring at high altitude associated with protein C deficiency. Acta Haematol. 1999; 102:104–6.
52. Basnyat B, Graham L, Lee S-D, Lim Y. A language barrier, abdominal pain, and double vision. Lancet. 2001;357:2022.
53. Nair V, Mohapatro AK, Sreedhar M, Indrajeet IK, Tewari AK, Anand AC, et al. A case of hereditary protein S deficiency presenting with cerebral sinus venous thrombosis and deep vein thrombosis at high altitude. Acta Haematol. 2008;119:158–61.
54. Weiss C, Seitel G, Bärtsch P. Coagulation and fibrinolysis after moderate and very heavy exercise in healthy male subjects. Med Sci Sports Exerc. 1998;30:246–51.
55. Bärtsch P, Siedler K, Kreutzberger R, Menold E, Weiss C. Acute mormobaric hypoxia does not enhance exercise-induced thrombin formation (Abstract). Med Sci Sports Exerc. 2001;33:S99.
56. DeLoughery TG, Robertson DG, Smith CA, Sauer D. Moderate hypoxia suppresses exercise-induced procoagulant changes. Br J Haematol. 2004;125: 369–72.
57. Le Roux G, Larmignat P, Marchal M, Richalet J-P. Haemostasis at high altitude. Int J Sports Med. 1992;13:S49–51.
58. Pichler-Hefti J, Risch L, Hefti U, Scharrer I, Risch G, Merz TM, et al. Changes of coagulation parameters during high altitude expedition. Swiss Med Wkly. 2010;140:111–7.
59. Ren Y, Cui F, Lei Y, Fu Z, Wu Z, Cui B. High-altitude pulmonary edema is associated with coagulation and fibrinolytic abnormalities. Am J Med Sci. 2012; 344(3):186–9.
60. Bärtsch P, Haeberli A, Franciolli M, Kruithof EKO, Straub PW. Coagulation and fibrinolysis in acute mountain sickness and beginning pulmonary edema. J Appl Physiol. 1989;66:2136–44.
61. Bärtsch P, Waber U, Haeberli A, Maggiorini M, Kriemler S, Oelz O, et al. Enhanced fibrin formation in high-altitude pulmonary edema. J Appl Physiol. 1987;63:752–7.
62. Bärtsch P, Haeberli A, Nanzer A, Lämmle B, Vock P, Oelz O, et al. High altitude pulmonary edema: blood coagulation. In: Sutton JR, Houston CS, Coates G, editors. Hypoxia and molecular medicine. Burlington: Queen City Printers; 1993. p. 252–8.
63. Kotwal J, Apte CV, Kotwal A, Mukherjee B, Jayaram J. High altitude: a hypercoagulable state: results of a prospective cohort study. Thromb Res. 2007;120:391–7.
64. Jha SK, Anand AC, Sharma V, Kumar N, Adya CM. Stroke at high altitude: Indian experience. High Alt Med Biol. 2002;3:21–7.
65. Anand AC, Saha A, Seth AK, Chopra GS, Nair V, Sharma V. Symptomatic portal system thrombosis in soldiers due to extended stay at extreme altitude. J Gastroenterol Hepatol. 2005;20(5):777–83.
66. Rupert JL, Devine DV, Monsalve MV, Hochachka PW. Beta-fibrinogen allele frequencies in Peruvian Quechua, a high-altitude native population. Am J Phys Anthropol. 1999;109:181–6.
67. van Veen JJ, Makris M. Altitude and coagulation activation: does going high provoke thrombosis? Acta Haematol. 2008;119:156–7.
68. Wang J-S, Cheng M-L, Yen H-C, Lou B-S, Liu H-C. Vitamin E suppresses enhancement of factor VIII-dependent thrombin generation by systemic hypoxia. Stroke. 2009;40:656–9.
69. Schoene RB. Pulmonary edema at high altitude. Review, pathophysiology, and update. Clin Chest Med. 1985;6:491–507.
70. Fink T, Kazlauskas A, Poellinger L, Ebbesen P, Zachar V. Identification of a tightly regulated hypoxia-response element in the promoter of human plasminogen activator inhibitor-1. Blood. 2002;99:2077–83.
71. Schreijer AJM, Hoylaerts MF, Meijers JCM, Lijnen HR, Middeldorp S, Büller HR, et al. Explanations for coagulation activation after air travel. J Thromb Haemost. 2010;8:971–8.

Renal Function and Fluid Homeostasis

11

Erik R. Swenson and Niels V. Olsen

Abstract

Hypoxia can directly affect the kidneys, but more importantly, its effects on systemic acid-base balance, ventilation, neuroendocrine reflexes, and hemodynamics all play a far greater part in altering renal function and fluid balance. Acute and chronic effects of hypoxia and their magnitude may differ and these will be highlighted. These changes will be related to the common diseases of high altitude and to their impact on patients with chronic renal disease. Other features of high altitude separate from hypoxia, either alone or in combination, including hypobaria, exercise, and cold may also significantly perturb renal function.

Introduction

Hypoxia directly affects the kidneys, but, importantly, has effects on acid–base status, ventilation, neuroendocrine reflexes, and hemodynamics that also contribute to fluid balance. Exercise also can significantly alter renal function. Acute and chronic responses to mild and moderate hypoxia will be contrasted to those developing with more severe hypoxia, which are associated with acute high-altitude illness and in those with chronic renal disease (Chaps. 20–23).

E.R. Swenson, M.D. (✉)
VA Puget Sound Health Care System,
University of Washington, Seattle, WA, USA
e-mail: eswenson@u.washington.edu

N.V. Olsen, M.D., D.M.Sc.
University of Copenhagen, Copenhagen, Denmark
e-mail: nvolsen@sund.ku.dk

Effects of Hypoxia on Salt, Water, and Acid–Base Balance

Renal Function

Urinary Output

A dose–response relationship is found in humans with urine output and sodium excretion in acute hypoxia (Fig. 11.1); diuresis and natriuresis with mild hypoxia reaching a maximum between F_IO_2 of 0.12–0.14. Below 0.08–0.10 antidiuresis and sodium retention ensue. Variability between studies relates to differences in "ascent" rate, duration of hypoxia, barometric pressure (i.e., normobaric vs. hypobaric hypoxia), exercise, water and salt intake, fitness, hypoxic ventilatory responsiveness (HVR), arterial hypertension, and acute mountain sickness (AMS).

E.R. Swenson and P. Bärtsch (eds.), *High Altitude: Human Adaptation to Hypoxia*,
DOI 10.1007/978-1-4614-8772-2_11, © Springer Science+Business Media New York 2014

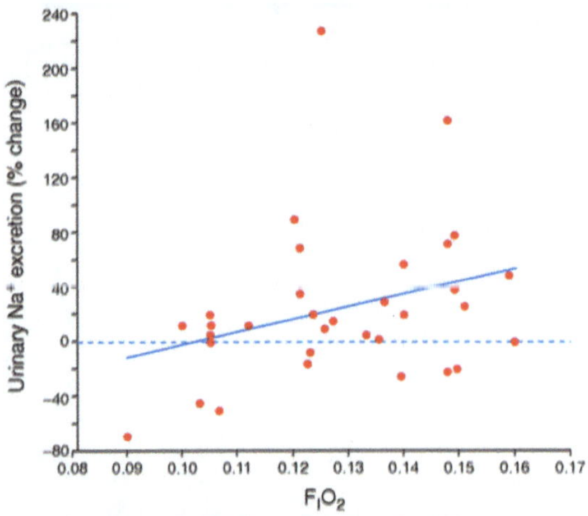

Fig. 11.1 Summary of 32 high-altitude and normobaric hypoxic studies in humans in which sodium excretion was measured between 1 and 24 h of indicated inspired O_2. Sodium excretion with hypoxia is given as the percent change above or below the preceding normoxic baseline period (From ref [206])

The diuresis is characterized by increased sodium, potassium, and bicarbonate excretion, increased urinary pH and a fall in ammonium and titratable acid [1]. Total volume and solute output rise in parallel so urinary osmolality is not greatly altered [2]. These responses occur within the first hour and persist for 1–2 days. Diuresis and natriuresis occur with subsequent new and greater hypoxic challenges [3]. Antidiuresis, particularly associated with ADH release, causes the expected urinary concentration. Subjects with borderline hypertension have considerably less hypoxic natriuresis and more AMS with simulated moderate altitude [4].

With chronic hypoxia, urinary composition and output revert to that dictated by dietary intake and extrarenal losses [1, 5]. In general, adaptation to chronic hypoxia results in persistence of the original net negative fluid balance, but with severe hypoxia and maladaptation fluid retention may persist (Chap. 20).

Renal Hemodynamics

Glomerular filtration rate (GFR) and renal blood flow (RBF) determine the amounts of water and solute entering the nephron. With systemic hypoxia of mild-moderate intensity (Fio_2 ~0.09–0.14) and short duration (hours), GFR changes minimally [6–9]. At >4,400 m there is a 0–10 % increase in GFR for several days [10–12] after which GFR may fall by about 20 % [12, 13]. GFR can fall almost 50 % at 24 h in AMS [14], but if AMS develops at very high altitude after some period of acclimatization, there is a slight rise in GFR that correlates modestly with AMS severity [13]. RBF rises slightly (8–20 %) with acute hypoxia in well-acclimatizing humans [6, 9], but is not sustained after several days [10, 11]. RBF and renal plasma flow (RPF) may fall as much as 20–30 % with longer term residence at high altitude [15].

In polycythemic high-altitude Andean natives RBF is only marginally reduced, but RPF is depressed by 30–40 % due to lower fractional plasma content. Despite a reduction in RPF, GFR is maintained and filtration fraction (GFR/RPF) rises proportionally to the RPF drop; renal O_2 delivery, a–v O_2 content gradients, and renal O_2 consumption are equal to sea level residents [16]. However, with greater hypoxia, RBF ultimately falls by about 30 % in healthy subjects newly residing at 6,000 m [17] and by roughly the same magnitude in chronic mountain sickness [18].

In conclusion, moderate hypoxia reduces renal vascular resistance and causes GFR and RBF to rise slightly or remain unchanged. The mechanism of direct hypoxic renal vasodilation is not established, but like many systemic vascular beds it may involve local release of adenosine, nitric oxide, and prostaglandins [19–21]. Hypoxia evoked neuroendocrine influences may attenuate direct hypoxic vasodilation. In addition, hypocapnia may limit vasoconstrictor input via the renal sympathetic nerves [22]. In contrast, it is well established if the systemic response to hypoxia causes hypotension, GFR and RBF fall with attendant antidiuresis and antinatriuresis.

Tubular Function

With its high blood flow and low overall O_2 extraction (renal SvO_2 is 80–90 %), the kidney is conventionally thought to be hypoxia resistant [16]. Yet, this picture neglects complexities of the renal vasculature and tubular geometry, metabolism and function that generate a marked heterogeneity of local PO_2 within the cortex and medulla [19, 23, 24]. The cortex and outer medulla are richly perfused to accommodate a high GFR and active salt reabsorption by the proximal tubule. Owing to a preglomerular O_2 diffusion [25] shunt from preglomerular arteries to adjacent veins, PO_2S in the cortex can vary between 20 and 50 mmHg [26]. The series arrangement of glomerular perfusion and cortical tubular blood flow means that a reduction of RBF generally reduces GFR, which decreases the dominant metabolic work of salt reabsorption and limits the fall in renal tissue PO_2. In contrast, reduced renal O_2 delivery by anemia or arterial hypoxemia has much less effect on GFR, and thus local PO_2 falls in the vicinity of the peritubular erythropoietin (EPO)-producing cells. This unique perfusion to metabolism (i.e., filtration-tubular reclamation) relation makes the kidney ideal for EPO release in response to anemia or hypoxemia but appropriately and largely unresponsive to changes in its own blood flow [27].

The medulla, by contrast, receives only about 10 % of RBF, and yet the medullary thick ascending limbs of the loop of Henle accomplish about one third of the salt and water reabsorption [28]. Its relative hypoperfusion is a result of the countercurrent arrangement of the medullary vasculature necessary for urinary concentration and dilution [29] and leaves this region surprisingly hypoxic (PO_2 8–10 mmHg) due to both lower blood flow and an oxygen diffusion shunt [30–32] (Fig. 11.2). These PO_2 values are close to those in vitro that cause biochemical dysfunction and anatomical derangement [33]. In vitro and in vivo there is stimulation of a variety of oxygen conserving and protective responses that are hypoxia inducible factor (HIF) dependent [34–36] and linked to AMP-activated protein kinase (AMPK), a ubiquitous metabolic sensor that phosphorylates a myriad of proteins to reduce and shift metabolic activity away from O_2 dependence [37] and toward greater non-aerobic glycolytic metabolism. Hypoxic up-regulation of several endogenous gasotransmitters; nitric oxide (NO) generation by neuronal and endothelial NO synthases, carbon monoxide (CO) by heme-oxygenase, hydrogen sulfide (H_2S) by cystathionine gamma lysase, cystathionine beta synthetase, and cysteine aminotransfersase [37], as well as prostaglandin E_2 (PGE_2) by cyclo-oxygenase, all afford hypoxia tolerance. With chronic hypoxia, rats (15 days at 5,500 m) there is also marked up-regulation of antioxidant enzymes, superoxide dismutase, and catalase with corresponding decreases in the oxidant generating enzymes, xanthine oxidase, and NAPDH oxidase [38]. The myriad of protective responses that mediate vasodilation, reduce sodium reabsorption and alter the antioxidant/oxidant ratio confer to healthy kidneys a remarkable resistance to acute and chronic arterial hypoxemia.

Although there is no evidence that the diuresis and natriuresis of mild to moderate hypoxia arises from ion pump dysfunction due to ATP depletion, hypoxia causes a down-regulation in vivo of several membrane Na^+-transporting proteins in the kidney [39]. This may be in part mediated by AMPK that decreases O_2 consumption by down-regulation and withdrawal of a number of membrane ion transporters [37] as well as by activation of the HIF pathways [40] that depress transporter gene expression. In the hypoxic kidney this can be considered a protective response to reduce O_2 consumption in the face of decreasing availability. In fact most means of inducing natriuresis and diuresis is oxygen sparing [31, 41, 42]. In studies of urinary output with global renal oxygenation assessment, natriuresis and diuresis with hypoxia occur before there is measurable reduction in total oxygen consumption, RBF, O_2 delivery, and GFR. Indeed, antinatriuresis and antidiuresis develop only when there is severe arterial hypoxemia ($SaO_2 < 50$ %) or reductions in systemic blood pressure, RBF, and GFR [43]. High-altitude natives have no deficits in tubular functions [44, 45]. When GFR falls with more extreme hypoxemia, O_2 consumption falls with the decreased reabsorptive work with reduced filtration.

Taken together tubular function studies demonstrate a dominating influence of RBF, systemic hemodynamics, and hormonal effects rather than arterial PO_2 in affecting tubular function at high altitude. It appears that acute arterial hypoxemia, which is not associated with reductions in RBF, is remarkably well tolerated

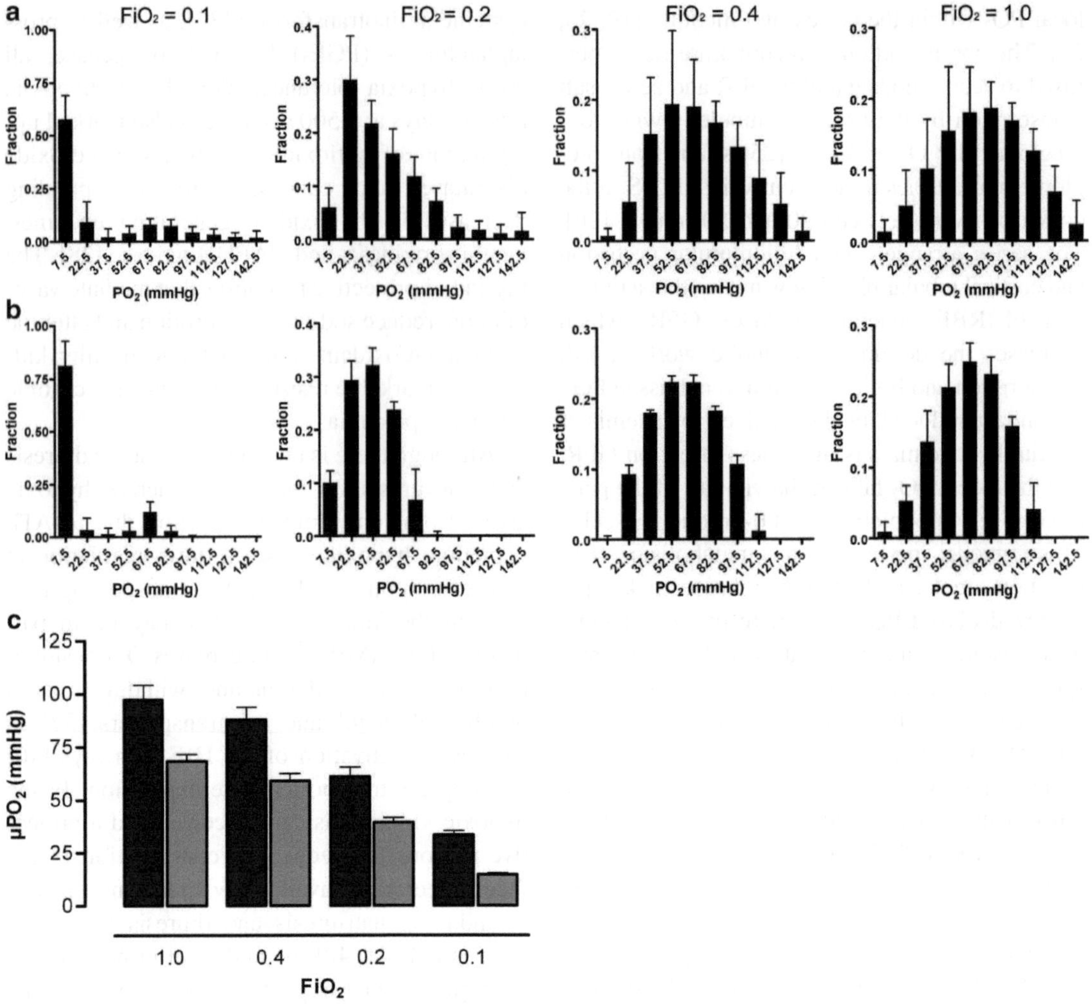

Fig. 11.2 Histograms showing oxygen distributions in the cortex (**a**) and outer medulla (**b**) of the rat kidney for four different FiO_2. A stepwise reduction in inspired oxygen fraction resulted in a decline in the average micrvascular PO_2 (μPO_2) and a leftward shift of the distribution (From ref [32])

and does not lead to significant renal function impairment despite the significant intrarenal PO_2 heterogeneity prevailing at sea level, which predictably like normobaric hypoxia [32] should worsen at high altitude.

Total Body Water and Volume Distribution

With hypoxia adaptation there is a 2–3 L loss of total body water in humans over several days [46] equally shared by plasma, extracellular, and intracellular spaces [46, 47]. The loss is a result of decreased intake, but also of increased respiratory, cutaneous, and renal losses [2]. Upon return to normoxia or lower altitudes, the water loss is quickly corrected [48]. In chronic hypoxia, hypohydration persists [49]. The magnitude of hypohydration is greatest in high-altitude populations, in whom total body water was only 43 % of total body mass compared to 53 % in sea level residents, possibly related to a relative insensitivity to ADH [12] and diminished expression and membrane targeting of aquaporin 2 in the collecting ducts.

Electrolytes and Acid–Base Status

The net result at altitude in extracellular fluid composition is one of little change in serum osmolality and sodium concentration acutely [14, 50]. However, after 2 days at 4,559 m there is a 5 mOsm increase in plasma osmolality concurrently with a decrease in circulating ADH [12]. With very high altitude (>5,400 m), serum sodium and osmolality rise 5 mM and 10 mOsm, respectively, in the face of a normal serum ADH [51]. These findings suggest a failure of normal osmoregulation and/or insensitivity of CNS osmoresponsive cells, which may develop as early as 3 days [52, 53].

The acute hypocapnia of hypoxia leads to only a slight reduction of serum potassium, both from intracellular uptake and kaliuresis [54–56]. Plasma chloride rises and bicarbonate falls as expected with acute and chronic hyperventilation [55]. In humans, the net acid–base effect of acute hypoxia is a mild respiratory alkalemia with incomplete renal compensation [1] in a dose-dependent manner with a time course of 1–2 days [57]. In high-altitude natives, respiratory alkalosis appears to be less pronounced owing to a combination of greater pulmonary gas exchange efficiency and less obligatory hypoxic ventilatory response [58].

Mechanisms of Changes in Salt and Water Balance at High Altitude

Sites of Hypoxic Sensation

The sensing of hypoxia relevant to renal function and fluid balance at high altitude may occur at several sites, including peripheral central chemoreceptors, brain, baroreceptors, and kidneys. The ultimate response will be a complex summation of differing hypoxic sensitivities of each site and their neurohumoral response.

Peripheral Chemoreceptors

In addition to their dominant role in the ventilatory response to hypoxia, the peripheral chemoreceptors have considerable influence on renal function and fluid balance, particularly on systemic cardiovascular regulation, renal vasculature, and sympathetic innervation [59, 60] and are the principal sensors initiating the diuresis and natriuresis of hypoxia. Peripheral chemoreceptor stimulation either by hypoxemia or almitrine increases absolute and fractional excretion of water and sodium [60, 61]. If the peripheral chemoreceptors are denervated, the response is ablated and sodium and water excretion decrease [60]. When peripheral chemoreceptors are left normoxic and only the brain and kidneys are rendered hypoxic, the response is one of antidiuresis and antinatriuresis [62]. Studies in man show a similar picture. Almitrine is a diuretic in normoxic subjects [63] and in subjects with a tenfold range of HVR, there is a very strong correlation between a subject's urinary sodium and water excretion over 6 h with 14 % O_2 breathing and HVR (a marker of peripheral chemosensitivity and afferent output to the brain and/or kidneys [64] (Fig. 11.3). The diuretic effect occurs within a few hours and may precede the natriuresis [8]. Recently an approximate surrogate of HVR (rise in ventilation from sea level to 5,050 m after 2 weeks) correlated with the magnitude of diuresis with an acute water load [65] providing further evidence of a persisting influence of the peripheral chemoreceptors.

How hypoxic or pharmacological stimulation of peripheral chemoreceptors stimulates diuresis and natriuresis and at what nephron site(s) these occur remain unknown. Lacking definitive tubular micropuncture studies, lithium clearance studies in humans point to a more distal site of suppressed sodium reabsorption with hypoxia [9, 66]. These renal responses to hypoxia are not initiated via renal innervation because denervation, in fact, augments the diuresis or natriuresis, by elimination a countervailing influence of peripheral chemoreceptor-mediated increase in RSNA [67, 68], which may be more dominant in hypertensives who have less hypoxic natriuresis than normotensives [4]. The response is not dependent upon changes in PCO_2, acid–base status, ventilation, and intrathoracic pressure variations [60]. Peripheral chemoreceptor-mediated natriuresis and

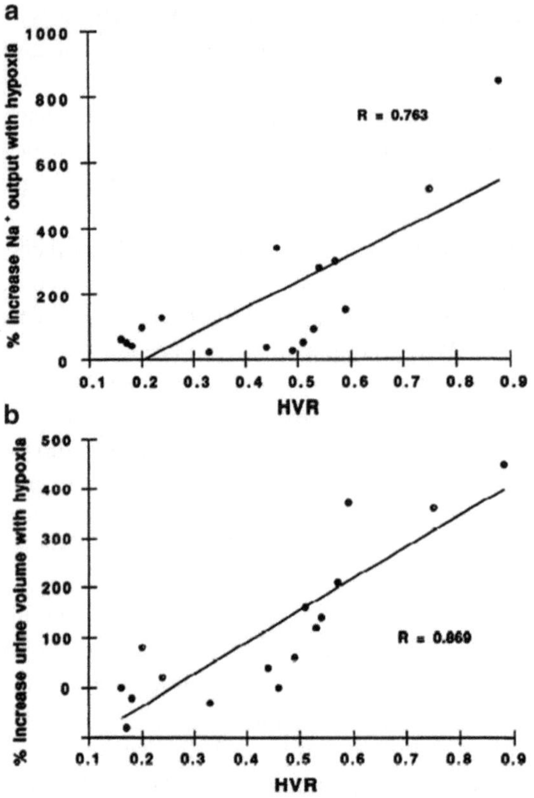

Fig. 11.3 Correlation between isocapnic hypoxic ventilatory response (HVR) and increase in urinary sodium excretion (**a**) and urinary volume (**b**) with 6 h of hypoxia (O_2=14 %) in 16 healthy human subjects. Correlation coefficients (*R*) are given (From ref [64])

diuresis, however, depend upon normal circulatory volume and low firing rate of cardiovascular stretch/pressure receptors [69]. Salt intake is important since there is an attenuation of hypoxic diuresis with salt restriction and augmentation with liberal intake [1, 55, 64]. The specific mechanism(s), by which peripheral chemoreceptors mediate increased sodium and water output with hypoxia points to an endogenous humoral factor(s) or intra RBF redistribution, to be discussed in the following section.

Brain and Sympathetic Nervous System

Severe CNS hypoxia excites vasomotor neurons and sympathetic tone, independent of peripheral chemoreceptor input [70, 71], and this itself will decrease sodium and water excretion. However, the impact of mild-moderate hypoxia in the brain on the CNS mediation of renal vascular and tubular responses is heavily dependent and modulated upon peripheral chemo- and baroreceptor input [68, 70]. However, with normal peripheral input (i.e., normoxia at the peripheral chemoreceptors), equivalent hypoxia isolated to the CNS by separate perfusion of the brain with hypoxic blood causes potent renal vasoconstriction, decreased RBF, GFR, and urinary sodium and water excretion [62], suggesting intense renal sympathetic nerve activity [72].

Kidneys

In the kidney, hypoxia both activates AMPK [73] and several HIF isoforms [74]. HIF-2 alpha is particularly important in that it initiates erythropoietin (EPO) production by proximal tubular fibroblasts and drives the hematopoietic response to hypoxia, but is of uncertain relevance in the monitoring of hypoxia relevant to fluid balance and renal function. There have been no studies examining whether HIF isoforms directly alter tubular function in hypoxia. Given the manifold effects of HIF systemically, the question would have to be studied in animals where HIF could be conditionally knocked out or suppressed by siRNA, and overexpressed solely in the kidneys. Although EPO has no known effects on tubular function [27], it is vasoconstricting in the renal vasculature [75]. Low doses of EPO in normal subjects reduce plasma concentrations of renin and aldosterone and decrease plasma volume [76]. Thus, its release into the renal interstitium with hypoxia could alter intrarenal hemodynamics and blood flow distribution and secondarily alter renal salt and water handling. Furthermore, because EPO has an array of cytoprotective effects in hypoxia and ischemia [77], its local release may contribute to defense of the kidney at high altitude. Whether EPO-producing cells relay neural input to the brain or feedback on efferent RSNA is not known, but it is clear that renal nerve afferents may be important in the response of the kidney to other changes in cardiovascular and volume status [78]. Since denervation of the kidneys markedly augments the diuretic and natriuretic response to hypoxia [59, 60], it is conceivable that afferent information from the kidney may diminish the final response.

Mechanisms of Hypoxic Salt and Water Regulation

Humoral Factors

Numerous hormones with salt- and water-regulating properties have been investigated in hypoxia to determine whether their plasma concentrations correlate with the urinary responses to hypoxia without conclusive results. Definitive assignment of a hormone or mediator role should include consistent findings across a number of studies, loss or blunting of effect with a specific antagonist, or genetic manipulation of a hormone and/or its receptor. No hormone has fully met these strict criteria. The lack of consistent findings in the classic salt and water hormones may relate to not measuring tissue levels at the sites of action, but point also to other compounds with diuretic and natriuretic action as possible mediators.

Antidiuretic Hormone

Hypoxic exposure has variable effects on antidiuretic hormone (ADH) release and action. ADH is either mildly suppressed or not affected by mild to moderate ($FiO_2 > 0.10$) normobaric hypoxia [6, 56, 64, 79]. ADH release with water deprivation and increased osmolality is preserved and augmented with prolonged stay at high altitude [80]. Hypobaric exposure for several days also does not alter ADH in subjects without illness [50, 81, 82]. Only with severe hypoxia ($FiO_2 < 0.10$) in humans (often associated with the nausea of AMS) [50, 81] does ADH rise.

Although isolated peripheral chemoreceptor stimulation may cause ADH release [83], why ADH is suppressed with mild hypoxia in vivo is not known. It may involve baroreceptor input as well as cortisol release [84], or hypocapnic suppression [50, 85]. That diuresis can occur without suppression of ADH, as with almitrine [63], implies that either other factors such as intrarenal osmolality changes promote water loss (washout of the hypertonic medullary interstitium with changes in medullary blood flow) or that hypoxia induces a functional state of renal ADH unresponsiveness. The latter is unlikely because ADH given to acutely hypoxic humans is antidiuretic [52].

Renin-Angiotensin-Aldosterone System

Data on renin activity at high altitude and with hypoxia are conflicting, but most find no change or a decline [64, 86, 87]. The differences may rest on whether the subjects had exercised (which raises renin) and whether there was any fall in effective circulating volume and thus renal perfusion pressure. Angiotensin II levels are not elevated with hypoxia [86]. Although earlier work suggested that hypoxia suppressed serum ACE activity, this has not been borne out in subsequent studies [88, 89].

A remarkable and common finding is a suppression of serum and urinary aldosterone with hypoxia [6, 54, 87, 89–91]. Aldosterone suppression occurs even with hypoxic exercise despite a rise in renin [90–92]. Stimulation of the peripheral chemoreceptors does not appear responsible because renin and aldosterone are unaltered in normoxic rats given almitrine [93]. With more chronic hypoxia, aldosterone levels remain normal [94] and rise only in those with overt illness and fluid retention [17].

Several mechanisms are responsible for hypoxic suppression of aldosterone, including direct inhibition by hypoxia of ACTH and angiotensin II-stimulated aldosterone secretion by adrenocortical cells [95], plasma potassium decreases [96], and possibly hypocapnia [97]. In vivo, hypoxia did not alter the aldosterone response to angiotensin II in humans [54] but did suppress the response to ACTH [98]. Because not all studies, in fact, find hypoxic aldosterone suppression, one explanation for natriuresis without aldosterone suppression may be down-regulation of aldosterone receptors in hypoxic cultured renal cells [99].

Natriuretic Peptides

Acute hypoxia stimulates production of many natriuretic peptides [100], including atrial natriuretic peptide (ANP) [98, 101], although it is not a uniform finding [44, 64, 102]. Possible factors involved in hypoxia-mediated ANF release include right atrial and ventricular stretch and a direct effect of low PO_2 on atrial tissue. The natriuretic actions of ANP include inhibition of angiotensin II-mediated aldosterone synthesis [103], direct inhibition of tubular sodium reabsorption, increased GFR, and inhibition of ADH release [104, 105]. A critical ANP role in hypoxic natriuresis remains uncertain because ANP is elevated in those with positive and negative fluid balance [102, 106]. These contradictory findings suggest that either ANP is not an important contributor to sodium excretion, its natriuretic effects are countered by other aspects of the hypoxic state, or to ANP's other actions, such as vasodilation and increased capillary permeability leading to extravascular escape of plasma water and loss of circulating volume [107, 108].

Brain natriuretic peptide (BNP), synthesized in the cardiac atria and ventricles, has many of the same actions as ANP [109] and is released in response to local or generalized hypoxia [110]. Evidence for BNP release at high altitude, however, is contradictory. In a hypobaric chamber study of several hours that caused oxygen saturations as low as 60 % this degree of hypoxemia did not increase BNP secretion [111]. Two studies of climbers ascending higher than 5050 m showed variable increases of up to fourfold over normal, but no relation to urinary sodium, urine volume or arterial saturation. In contrast, n-terminal proBNP was not elevated acutely in climbers ascending up to altitudes between 5,200 and 6,300 m and decreased slightly over several days [13, 111, 112]. AMS scores correlated with BNP concentrations [113–115] but it appears more likely that elevations in BNP may be a consequence of AMS once fluid retention sets in [114]. These data taken together, similar to ANP, do not point to any important contribution of BNP to early renal events at high altitude.

Urodilatin is another natriuretic peptide with sequence homology to ANP and BNP that is produced only in the kidney and appears to be much better correlated with urinary sodium excretion than ANP in a number of different conditions [116]. Urinary urodilatin excretion is increased sixfold in the hypoxic isolated perfused kidney [116]. However, with 1 h of hypoxia in humans there is no increase in urinary urodilantin excretion [8], or after 24 h at 4,559 m [12], although in a hypobaric simulation of 6,000 m there was a slight increase [117]. On the other hand, after 2 days at 4,559 m urinary urodilatin excretion was halved [12]. These data do not make a good case for a role of urodilatin as a hypoxic natriuretic peptide.

Endothelin

The endothelins are vasoactive peptides with effects beyond blood pressure control and are of potential significance in hypoxia. Acute normobaric and hypobaric hypoxia cause circulating endothelin to rise about 1.5- to 2-fold [118] and increase its urinary excretion [119]. Although endothelins are vasoconstricting, endogenous circulating and intrarenal levels of endothelin may be vasodilating in the kidney, because receptor blockade increases renal vascular resistance and depresses sodium excretion. Furthermore, when endothelin-1 is infused in low doses [120] or rises with hypoxia there is diuresis and natriuresis. In these situations, endothelin blocks sodium reabsorption by inhibiting tubular Na^+/K^+-ATPase [121] and increasing nitric oxide (NO) production via stimulation of NO synthase 1 (NOS) [122]. Endothelin also inhibits ADH action in the collecting duct to promote water excretion [119, 123], and renin secretion [124]. Small elevations in renal endothelin synthesis with hypoxia may promote a redistribution of blood flow to the medulla [125] and help to explain hypoxic diuresis and the tolerance of the poorly perfused medulla to arterial hypoxemia. Favoring this hypothesis, a mixed endothelin-1 receptor antagonist at 4,559 m blunted the normal increase in free water clearance and led to fluid retention despite a reduction in pulmonary artery pressure [126].

Cortisol and Other Steroid Hormones

Acute hypoxia in humans causes a 2- to 3-fold increase in circulating ACTH and cortisol within 1–2 h, with loss of the normal diurnal variation suggesting a stress response [53, 127], although elevated cortisol is not always observed especially in normobaric hypoxic studies [88]. In animals, cortisol and ACTH rise with hypoxia [85,

93], which requires intact carotid sinus innervation [128]. The peripheral chemoreceptors appear to be responsible, since cortisol and ACTH rise in normoxic rats given almitrine and the response is absent with carotid body denervation [93]. With more chronic hypoxia, there is a return to normal concentrations [44] except in subjects remaining above 6,000 m, who develop chronic mountain sickness [17]. Elevations in cortisol may be partly responsible for increased free water clearance at high altitude, independent of its ability to suppress ADH secretion [129] and for the lack of antidiuresis in some cases of ADH elevation [56]. Interesting in this regard is increased glucocorticoid receptor expression in hypoxic rat renal cortical epithelial cells [99].

Other adrenal-like steroids have been implicated because diuresis and natriuresis are attenuated with adrenalectomy despite cortisol replacement [130]. Endogenous cardiac glycosides, which have Na^+/K^+-ATPase-inhibiting effects are secreted by the adrenal glands and hypothalamus in response to hypoxia, raise blood pressure and promote natriuresis [131, 132]. These attributes make them attractive candidates in the peripheral chemoreceptor-mediated natriuretic response to hypoxia since there are afferent projections from the peripheral chemoreceptors to both these sites [133]. In hypoxemic patients with COPD, a plasma digoxin-like substance was elevated and correlated with urinary sodium excretion [134]. While it was found to be elevated in healthy trekking subjects [135], this was not found in normobaric hypoxic subjects undergoing diuresis and natriuresis [64]. As yet, no inhibitors have been tested in hypoxia to better assess their role.

Circulating Catecholamines

Although chronic hypoxic activation of the sympathetic nervous system leads to a raised circulating norepinephrine concentration, it is not elevated with acute hypoxic exposure [136, 137]. Muscle sympathetic nerve activity is increased. These findings suggest either heterogeneity of response in the sympathetic nervous system, or increased clearance of catecholamines and decreased synthesis by oxygen-dependent tyrosine hydroxylase. Epinephrine, in contrast, only transiently increases with acute hypoxia but returns to normal or below normal levels with chronic hypoxia [138]. Elevated circulating catecholamines as well as increased RSNA in hypoxia would be antidiuretic and antinatriuretic by renal vasoconstriction and direct stimulation of renal tubular sodium and water reabsorption [139]. The changes in the autonomic nervous system with hypoxia and high altitude are the subject of Chap. 8.

Other Natriuretic Factors

All of the established classic salt- and water-regulating hormones known to cause or be associated with edema such as ANP, BNP, renin, aldosterone, vasopressin, and angiotensin have been reported to be increased by hypoxia in those with salt and water retention. However, no single hormone or set of hormone changes predictive of natriuresis and diuresis have been consistently associated with hypoxic diuresis and natriuresis [60, 64]. This raises the possibility that other substances with natriuretic effect will prove more important.

Adrenomedullin is synthesized by many organs including adrenal medulla, kidney, lung, brain, and heart and is a potent natriuretic vasodilator [140]. Its plasma and renal tissue concentrations are elevated in acute hypoxia [140, 141]. Adrenomedullin suppresses angiotensin II-mediated aldosterone production [142], providing another possible control element in the phenomenon of aldosterone reduction. Three studies at high altitude (above 4,500 m) have reported on adrenomedullin [143–145]. With passive airlift plasma adrenomedullin concentrations did not change [145], but in climbers that ascended by foot, there was a doubling in concentrations. In a study of longer duration at higher altitude, plasma concentrations were only minimally elevated, but urinary adrenomedullin excretion was increased and showed a strong correlation with urinary sodium and volume output [143]. This study is first to show a strong correlation of a natriuretic peptide and fluid balance at high altitude and suggests the relevant concentrations may be at the tissue level and not in the circulation. Further work ideally coupled with possible antagonists of adrenomedullin will be necessary to confirm this interesting observation.

Nitric oxide (NO) is generated by a variety of cells in the kidney and has multiple vascular and tubular actions. Its release is stimulated directly by polycythemia [146], arterial hypoxemia [137], and local renal hypoxia [147] as well as secondarily to changes in renal vascular hemodynamics [148], neural tone [149], or other hormones [150]. Although reduction of NO production by NOS inhibition causes renal vasoconstriction, NOS inhibitors do not cause further renal vasoconstriction in hypoxia over that observed with hypoxia alone [151]. NO is vasodilatory [152], reduces renal mitochondrial O_2 consumption and increases O_2 utilization efficiency [153], blocks sodium reabsorption at multiple nephron sites [154, 155], and reduces renin secretion [156]. Another source of renal NO in hypoxia is nitrite (concentrated over 30-fold in urine) which is reduced to NO by various heme-containing proteins, such as deoxyhemoglobin, xanthine oxidoreductase, aldehyde dehydrogenase, and even NOS in hypoxic and acidic conditions [157]. Any of these effects may play an important role in hypoxic renal function and preserving renal function in the face of decreased O_2 supply.

Reactive oxygen species (ROS), such as superoxide, hydroxyl radical, peroxynitrite, and hydrogen peroxide are generated under hypoxic conditions and may play an important function as mediators of hypoxic signaling on gene transcription, membrane functions, and enzyme activity, although in excess they can be injurious. In the kidney, various forms of ROS have been shown to reduce tubular ion transport and increase medullary perfusion, but this complicated area is in its infancy and dissecting the physiological and pathophysiological roles of renal ROS will be critical but very difficult [158].

Local production in hypoxia of carbon monoxide and hydrogen sulfide in the kidney, especially in the medulla, may alter fluid and sodium reabsorption. CO, similar to NO, is a stimulator of soluble guanyl cyclase, and so acts by increasing cGMP. While some studies show CO promotes sodium excretion both by direct tubular effects and renovasodilation [159, 160], not all work is confirmatory [161–163]. While there are no direct studies of H_2S on tubular sodium reabsorption, H_2S activates AMPK, which in turn down-regulates Na^+/K^+ ATPase, ENaC, and the NaCl cotransporter; thus it should be natriuretic on this basis in addition to its known vasodilating action [161–163]. A recent finding in mice breathing 10 % O_2 that H_2S inhibits HIF-1 activation [164] adds further complexity to any role of H_2S in renal hypoxia.

Medullipin, a not fully characterized lipid moiety, is secreted by renomedullary cells and has natriuretic activity [165]. It may counter rennin–angiotensin–aldosterone system (RAAS) actions and be integral in the pressure-natriuresis response discussed below [28]. If hypoxic natriuresis is a pressure-natriuresis-like response, then acute hypoxic exposure should raise plasma and urinary medullipin levels, but this has not been tested. The synthesis of another lipid in the kidney, prostaglandin, PGE_2, rises with mild-moderate hypoxia; causing natriuresis by medullary vasodilation and reduction in O_2 consumption [42, 166]. Other natriuretic hormones and peptides exist but have not been tested in hypoxic states. Given that no definite hormonal mediator(s) of hypoxic diuresis and natriuresis has been identified following the dictates listed earlier, the search should continue.

Systemic and Intrarenal Hemodynamics

When hypoxia reduces RBF, with or without systemic hypotension, there is almost invariably sodium and water retention [17, 18], in part due to the fall in GFR. When RBF is maintained and systemic blood pressure rises, increased sodium and water excretion might be explained in part by a pressure natriuresis and diuresis that arise from the lack of renal medullary blood autoregulation [167]. As medullary blood flow increases with increasing blood pressure, there is a steep pressure-natriuresis relationship (~15 % increase in sodium excretion per mmHg rise in renal perfusion pressure). The basis for this response is a rising interstitial pressure [168] that suppresses sodium reabsorption at multiple sites by decreasing the hydrostatic pressure differences across the tubule. Furthermore, medullary solute washout from increased blood flow reduces osmotic and electrochemical gradients favoring luminal sodium and water exit [169]. Whether pressure natriuresis and diuresis play a role at high altitude has not been studied adequately, but in humans, diuresis and natriuresis occur [55, 56, 64, 116] most of whom who do not become hypertensive. Hypoxia appears to lead to a selective increase in medullary blood flow without changing systemic blood pressure [170], as is seen with low-dose endothelin infusion [125]. Thus an extreme sensitivity of the medullary pressure-natriuresis response could account for hypoxic natriuresis and diuresis. This does not appear to be mediated by the peripheral chemoreceptors because almitrine does not increase medullary blood flow [171].

Renal Innervation

The response of the autonomic system to hypoxia is that of generalized sympathetic and parasympathetic activation. The implications for the kidney are increased RSNA which when unopposed acts as antidiuretic and antinatriuretic influences. The antidiuretic and antinatriuretic influence of peripheral chemoreceptor-mediated RSNA is revealed by the 30–50 % greater diuresis and natriuresis observed in animals whose kidneys are denervated [172, 173]. Although there is recent evidence that renal sympathetic nerves release neuropeptide Y [174] and renal nerves can generate NO [175, 176], both of which are natriuretic when given exogenously, it appears that the catecholamine effects of sympathetic nerve stimulation are dominant.

Although the kidney is not thought to have any important non-sympathetic innervation, there is some biochemical and pharmacological evidence for opposing renal cholinergic innervation in the dog [159, 177], analogous to the hypoxic stimulation of parasympathetic activity to the heart [68]. However, an important parasympathetic-mediated diuresis and natriuresis would not be compatible with the greater hypoxic salt and water excretion following renal denervation, and the diuresis and natriuresis with acetylcholine infusion [159] could also be explained by generation of intrarenal NO, which is both vasodilatory and natriuretic.

Cardiopulmonary Innervation

The ventilatory response to hypoxia includes increased tidal volume and intrathoracic pressure swings, which alone stimulate intrathoracic stretch receptors in the venous, atrial, pulmonary vasculature [178, 179], and lung parenchyma [180]. In humans, voluntary normoxic isocapnic hyperventilation at rest (>30 L/min) causes diuresis and natriuresis [181]. However, hypoxic hyperventilation in normal resting subjects does not usually exceed 10–15 L/min at least above 10 % O_2 [64], and diuresis and natriuresis occur in animals with fixed ventilation [60]. Furthermore, lung denervation in spontaneously breathing rabbits did not significantly alter increased RSNA with inspired hypoxia [182], suggesting that stretch receptors do not attenuate RSNA in hypoxia.

The hemodynamic response to hypoxia includes an increase in cardiac output, pulmonary artery pressure, and increased venous return. These processes shift blood volume into the central circulation and stimulate intrathoracic low pressure baroreceptors, which increase urine output [183] and dampen RSNA with hypoxia [159, 183]. However, data on intrathoracic volume and pressure changes in the vena cavae and atria are not in agreement, and in many cases no elevations in central venous pressures or blood volume with acute hypoxia are found [82, 183, 184]. Cervical vagotomy studies would be useful to assess the role that these intrathoracic baroreceptors play in hypoxic diuresis [185].

Hypocapnia

Hypocapnia depresses proximal tubular bicarbonate reabsorption [186], with bicarbonaturia and water, Na^+, and K^+ losses [55]. Bicarbonate excretion does not appear essential because natriuresis occurs with hypoxia in mechanically ventilated animals whose ventilation and $PaCO_2$ are fixed [60]. In spontaneously breathing humans, there is a very weak correlation between sodium and bicarbonate excretion with acute hypoxia [64], and hypoxic natriuresis exceeds bicarbonate output 2–5 fold [1, 55]. Hypoxic natriuresis in man is not simply a hypocapnic natriuresis, because hypocapnic normoxia and hypoxia, but not isocapnic hypoxia, reduce proximal tubular fluid reabsorption, while the natriuretic response remains [9].

Hypocapnia may also alter the renal response to hypoxia indirectly by blunting sympathetic responses in the CNS [70, 187] and peripheral chemoreceptors [68]. A decrease in RSNA [70] may partly explain higher RBF in hypoxic spontaneously breathing (hypocapnic) rats than in those given CO_2 to prevent hypocapnia [22]. Hypocapnia may also alter the balance of salt- and water-regulating hormones and blunt renal sodium avidity, since hypercapnia activates the renin-angiotensin-aldosterone axis and RSNA [188, 189]. These findings could explain the adverse effects on AMS symptoms and prevention of negative fluid balance of preventing or failing to achieve a degree of hypocapnia when hypoxic [190, 191]. These studies reveal a complex interaction of CO_2 and O_2 on peripheral and central chemoreceptors on ventilation [192], hemodynamics (250), and renal hormones, but in general it appears that hypocapnia has moderating effects on those neurohumoral factors promoting salt and water retention in hypoxia.

Hypobaria

Lower barometric pressure itself does not greatly alter renal function and fluid balance unlike normobaric and hypobaric hypoxia [3, 56, 72]. In humans [2] there is an antidiuresis with rapid decompression from 630 to 430 mmHg but not with equivalent normobaric hypoxia. Breathing 100 % oxygen before decompression results in no antidiuresis, suggesting an element of decompression stress.

Chronic normoxic hypobaria (Pb = 258 mmHg) was associated with greater sodium and water retention, less weight loss and a 40 % fall in GFR with sodium restriction of sodium than with hyperoxic ($PiO_2 = 258$ mmHg) equivalent hypobaria [193]. A smaller fall in GFR was found in a 56-day sojourn at 258 mmHg ($PiO_2 = 175$ mmHg), suggesting that, analogous to the acute studies, there may be some mitigating effect of hyperoxia on the renal responses to hypobaria [194].

Hypobaria appears to possibly enhance a sodium- and water-retaining influence of hypoxia. One explanation for differences between hypobaric and normobaric hypoxia is the resulting PiO_2 and PaO_2 are less in hypobaric hypoxia than equivalent normobaric hypoxia due to a deleterious effect of hypobaria on pulmonary capillary permeability and gas exchange efficiency [195] and an increase in dead space ventilation [196]. Another possibility is greater activation of the sympathetic nervous system in hypobaric hypoxia vs. normobaric hypoxia [53, 136]. Lastly, the elusive question remains of whether possible nitrogen bubble formation with decompression may alter intrinsic renal function [197].

Exercise

Exercise has profound effects on systemic and renal hemodynamics, and renal humoral mediators. Acute normoxic exercise [198] reduces RBF in proportion to exercise intensity, although GFR falls only by 25 % with maximal exercise. Urinary water and sodium outputs decrease even in the absence of changes in GFR, ANP, and insensitivity to ADH. The antidiuretic and antinatriuretic states persist for 30–60 min following exertion. With normoxic exercise ANP, aldosterone, renin, serum potassium, catecholamines, and ACTH all rise [199]. ADH increases with short-term heavy exercise but rises less in well-trained individuals doing low level sustained work [200]. Exercise also increases RSNA in proportion to exercise intensity [201], and renal hemodynamic effects and urine output changes are blunted by renal denervation and loss of RSNA [198].

Exercise in hypoxia produces qualitatively many of the same neurohumoral responses as normoxic exercise [87]. There is a mild suppression of the normal aldosterone increase [90, 91]. Some of these same features are exhibited by exercising subjects at 4,350 m after 2–5 days of adaptation [11, 202, 203]. The differences between these studies and those performed with acute hypoxic exposures suggest that 2–5 days of acclimatization may blunt certain humoral responses to exercise considered to promote fluid retention and acute illnesses of high altitude [90, 91].

Less vigorous but sustained exercise affects fluid balance whether done at low (1,100 m) or high (3,600–4,400 m) altitudes [94, 204, 205]. Subjects maintaining a fixed salt intake while climbing exhibited marked sodium and water retention. Calculations of body fluid compartment changes show an expansion of the extracellular and plasma volumes at the expense of intracellular space. Associated with these changes were sustained elevations in aldosterone, plasma renin, and ANP, but not ADH. Thus, exercise promotes salt and water avidity and frank retention if the stimulus is maintained. The additional effect of hypoxia on exercise appears small because very little differences in salt and water retention between normoxic and hypoxic climbing.

Hypoxia and Renal Disease

The stress of renal and systemic hypoxia at high altitude in this chapter has focused on the normal healthy kidney. It is well established that early in course of renal disease, many areas of the kidney become hypoxic and the main defense mechanisms against hypoxia become impaired [206, 207]. A recent finding that ACE inhibition improves renal cortical and medullary oxygen content in health humans [208] needs to be tested formally in those with renal disease at high altitude, many of whom particularly with proteinuria may already be on these drugs. These considerations would suggest that patients with chronic kidney injury might have more rapid decline in renal function living at higher altitudes [206, 209] simply as a result of a reduction in blood oxygen content or driving gradient for O_2 diffusion. The only study to examine the question [210] drew the opposite conclusion from population-based data in the Canary Islands in finding a protective effect of higher altitude residence. However, the results showing an inverse relationship between altitude of residence and kidney disease was weighted largely on data for those living below 400 m, at which there is little evidence for any pathophysiological effect of such mild reductions in inspired PO_2 [211]. In fact the 10 % of the population residing above this altitude (500–3,300 m) had a higher incidence of end-stage renal disease prevalence. This important topic and the risks for patients with renal disease are discussed more fully in Chap. 23.

Conclusions

Negative fluid balance at high altitude reflects the dominance of peripheral chemoreceptor stimulation overriding the fluid retaining consequences of cerebral hypoxia and hypoxic systemic vasodilation and hypotension. Without this opposition, especially when hypobaria and exercise are involved, there is intense activation of the sympathetic nervous system and RSNA as well as increased ADH, aldosterone, and ANP. The complexity of the picture is further increased by the possibility of other non-peripheral chemoreceptor-mediated mechanisms of natriuresis, such as direct hypoxic down-regulation of renal tubular membrane Na transport proteins.

The several ways in which peripheral chemoreceptor afferent activity may oppose hypoxic fluid retention are shown in Fig. 11.3 and include ventilatory stimulation, possible natriuretic substance(s) release, and sufficient (but not excessive) sympathetic nervous system activation to counteract the vasodilating effect of hypoxia. Increased ventilation minimizes arterial hypoxemia and generates a respiratory alkalosis, which promotes sodium excretion with the renal loss of bicarbonate and limits RSNA, as does hyperpnea via stimulation of intrathoracic stretch receptors.

A strong argument can be made that diuresis and negative sodium balance may be preemptive responses to limit the edema associated with high-altitude illness, but whether there exists a causal link of AMS with fluid retention has not been convincingly established (see Chap. 20). Subjects ill at high altitude have relative hypoventilation and a preceding interval of salt and water retention [191], but these results raise the question of cause and effect. The earliest field study [212], found fluid retention followed rather than preceded the development of AMS. In a hypobaric chamber study, residents of high altitude (1,650 m) tolerated a rapid ascent to 6,000 m better than lowland residents despite the fact that the latter group had a much greater natriuresis and diuresis than did the highlanders [116]. Furthermore, it is not clear that peripheral chemoreceptor sensitivity plays an important role in fluid balance of climbers, because urine output in mountaineers with or without AMS and in subjects airlifted to high altitude (4,559 m) did not correlate with the hypoxic ventilatory response and that fluid retention was only evident after subjects became ill [213]. In subjects climbing to this same altitude there was fluid retention and only those without AMS showed a natriuresis while sedentary over the next 2 days [214]. These studies do not rule out the possibility that internal fluid balance differences may be important in the pathogenesis of AMS, as suggested by data in subjects developing AMS who appeared to differ from those not becoming ill in experiencing an increase in PV that correlates with AMS [215]. Additional prospective well-controlled field studies with rigorous renal and compartmental fluid balance measurements are clearly needed.

Fluid balance at high altitude is an integration of many (sometimes opposing) organ system responses. Future human investigation should explore further the peripheral chemoreceptor role in normobaric and hypobaric hypoxia and in exercise. Specific hormonal antagonists will be useful to determine the importance of a particular hormonal change with hypoxia. In any animal work, anesthesia should be avoided. Transgenic mice with knockout and/or overexpression of hormonal mediators and their receptors should be pursued, coupled with use of humoral antagonists and peripheral chemoreceptor manipulation. More focus on intrarenal hemodynamics, blood flow distribution and innervation, as well as measurements of newer humoral mediators affecting salt and water transport is needed. Lastly, effects of hypoxia in the kidney on hormone receptors and ion transporters, their binding characteristics, and membrane localization as well as gene expression, will likely yield important new insights into renal physiology in general and hypoxia adaptation in particular.

References

1. Krapf R, Beeler I, Hertner D, et al. Chronic respiratory alkalosis: the effect of sustained hyperventilation on renal regulation of acid–base regulation. N Engl J Med. 1991;324:1394–401.
2. Tucker A, Reeves J, Robertshaw D, et al. Cardiopulmonary response to acute altitude exposure: H_2O loading and denitrogenation. Respir Physiol. 1983;54:363–80.
3. Loeppky JA, Roach RC, Maes D, et al. Role of hypobaria in fluid balance response to hypoxia. High Alt Med Biol. 2004;6:60–71.
4. Ledderhos C, Pongratz H, Exner J, et al. Reduced tolerance of simulated altitude (4200 m) in young men with borderline hypertension. Aviat Space Environ Med. 2002;73:1063–6.
5. Neylon M, Marshall JM, Johns EJ. The effects of chronic hypoxia on renal function in the rat. J Physiol. 1997;501:243–50.
6. Ashack R, Farber MO, Weinberger MH, et al. Renal and hormonal responses to acute hypoxia in normal individuals. J Lab Clin Med. 1985;106:12–6.
7. Armstrong HG. Principles and practice of aviation medicine. Baltimore: Williams & Wilkins; 1939. p. 193–261.
8. Hildebrandt W, Offenbacher A, Schuster M, et al. Diuretic effect of hypoxia, hypocapnia, and hyperpnea in humans. Relation to hormones and O_2 chemosensitivity. J Appl Physiol. 2000;88:599–610.
9. Olsen NV, Christiansen H, Klausen T, et al. Effects of hyperventilation and hypocapnic/normocapnic hypoxemia on renal function and lithium clearance in humans. Anesthesiology. 1998;89:1389–400.
10. Olsen NV, Hansen JM, Kanstrup I-L, et al. Renal hemodynamics, tubular function, and response to low dose dopamine during acute hypoxia in humans. J Appl Physiol. 1993;74:2166–73.
11. Olsen NV. Kanstrup 1-L, Richalet J-P, et al. Effects of acute hypoxia on renal and endocrine function at rest and during graded exercise in hydrated humans. J Appl Physiol. 1992;73:2036–43.
12. Bestle MH, Olsen NV, Poulsen TD, et al. Prolonged hypobaric hypoxemia attenuates vasopressin secretion and the renal response to osmostimulation in men. J Appl Physiol. 2002;92:1911–22.
13. Pichler J, Risch L, Hefti U, et al. Glomerular filtration rate estimates decrease during high altitude expedition but increase with Lake Louise acute mountain sickness scores. Acta Physiol (Oxf). 2008;192:443–50.
14. Jankowska AH, Whitten BK, Shields JL, et al. Electrolyte patterns and regulation in man during acute exposure to high altitude. Fed Proc. 1969;28:1185–9.

15. Singh MV, Salhan AK, Rawal SB, et al. Blood gases, hematology and renal blood flow during prolonged mountain sojourns at 3500 and 5800 m. Aviat Space Environ Med. 2003;74:533–6.
16. Rennie D, Lozano R, Monge C, et al. Renal oxygenation in male Peruvian natives living at high altitude. J Appl Physiol. 1971;30:450–6.
17. Anand IS, Chandrashenkar Y, Rao SK, et al. Body fluid compartments, renal blood flow, and hormones at 6000 m in normal subjects. J Appl Physiol. 1993;74:1234–9.
18. Lozano R, Monge C. Renal function in high altitude natives and in high altitude natives with chronic mountain sickness. J Appl Physiol. 1965;20:1026–7.
19. Blantz RC, Deng A, Miracle CM, et al. Regulation of kidney function and metabolism: a question of supply and demand. Trans Am Clin Climatol Assoc. 2007;118:23–43.
20. Brezis M, Heyman SN, Dinour D, et al. Role of nitric oxide in renal medullary oxygenation; studies in isolated and intact rat kidneys. J Clin Invest. 1991;88:390–5.
21. Neylon M, Marshall JM. The role of adenosine in the respiratory and cardiovascular response to systemic hypoxia in the rat. J Physiol. 1991;440:529–45.
22. Marshall JM, Metcalfe JD. Influences of the cardiovascular response to graded levels of systemic hypoxia of the accompanying hypocapnia in the rat. J Physiol. 1989;410:381–94.
23. Brezis M, Rosen S. Hypoxia of the renal medulla—its implications for disease. N Engl J Med. 1995;332:647–55.
24. Evans RG, Gardiner BS, Smith DW, et al. Intrarenal oxygenation: unique challenges and the biophysical basis of homeostasis. Am J Physiol. 2008;295: F1259–70.
25. Schurek HJ, Jost U, Baumgartl H, et al. Evidence for a glomerular O_2 diffusion limitation shunt in rat renal cortex. Am J Physiol. 1990;259:F910–5.
26. Leichtweiss H-P, Lubbers DW, Weiss C, et al. The oxygen supply of the rat kidney: measurement of intrarenal PO_2. Pflugers Arch. 1969;309:328–49.
27. Erslev AJ, Caro J, Besarab A. Why the kidney? Nephron. 1985;41:213–6.
28. Cowley AW. Renal medullary oxidative stress, pressure-natriuresis, and hypertension. Hypertension. 2008;52:777–86.
29. Michel CC. Renal medullary microcirculation: architecture and exchange. Microcirculation. 1995;2: 125–39.
30. Baines AD, Adamson G, Wojciechowski P, et al. Effect of modifying O_2 diffusivity and delivery on glomerular and tubular function in hypoxic perfused kidney. Am J Physiol. 1998;274:F744–52.
31. Brezis M, Agmon Y, Epstein FH. Determinants of intrarenal oxygenation. Effects of diuretics. Am J Physiol. 1994;267:F1059–62.
32. Johannes T, Mik EG, Ince C. Dual-wavelength phosphorimetry for determination of cortical and subcortical microvascular oxygenation in rat kidney. J Appl Physiol. 2006;100:1301–10.
33. Ruegg CE, Mandel LJ. Bulk isolation of renal PCT and PST: differential responses to anoxia and hypoxia. Am J Physiol. 1990;259:F176–85.
34. Gunaratnam L, Bonventre JV. HIF in kidney disease and development. J Am Soc Nephrol. 2009;20: 1877–87.
35. Leonard MO, Cotell DC, Goodson C, et al. The role of HIF-1alpha in transcriptional regulation of the renal tubular epithelial cell response to hypoxia. J Biol Chem. 2003;278:40296–304.
36. Nangaku M, Rosenberger C, Heyman SN, Eckardt K-U. HIF regulation in kidney disease. Clin Exp Pharmacol Physiol. 2013;40(2):148–57.
37. Steinberg GR, Kemp BE. AMPK in health and disease. Physiol Rev. 2009;89:1025–78.
38. Yang CC, Ma MC, Chien CT, et al. Hypoxic preconditioning attenuates LPS-induced oxidative stress in rat kidneys. J Physiol. 2007;582:407–19.
39. Adachi S, Zelenin S, Matsuo Y, et al. Cellular response to renal hypoxia is different in adolescent and infant rats. Pediatr Res. 2004;55:485–91.
40. Ibla JC, Khoury J, Kong T, et al. Transcriptional repression of Na-K-2Cl cotransporter KKCC1 by hypoxia inducible factor-1. Am J Physiol. 2006;291:C282–90.
41. Cohen JJ. Relationship between energy requirements for Na+ reabsorption and other renal functions. Kidney Int. 1986;29:32–40.
42. Prasad PV, Epstein FH. Changes in renal medullary PO_2 during water diuresis as evaluated by blood oxygenation dependent magnetic resonance imaging: effects of aging and cycloxygenase inhibition. Kidney Int. 1999;55:294–8.
43. Gotshall RW, Miles DS, Sexson WR. Renal oxygen consumption during progressive hypoxemia in anesthetized dogs. Proc Soc Exp Biol Med. 1983; 174:363–7.
44. Ramirez G, Pineda DO, Bittle PA, et al. Salt excretory capacity in natives adapted to moderate high altitude living after acute mobilization to sea level. Aviat Space Environ Med. 1995;66:1063–70.
45. Winslow RM, Monge C. Renal Function in high-altitude polycythemia. In: Winslow RM, Monge C, editors. Hypoxia, Polycythemia, and Chronic Mountain Sickness. Baltimore: Johns Hopkins University Press; 1987. p. 119–41.
46. Claybaugh JR, Wade CE, Cucinelli SA. Fluid and electrolyte balance and hormonal response to the hypoxic environment. In: Claybaugh JR, Wade CE, editors. Hormonal regulation of fluid and electrolytes. New York: Plenum; 1989. p. 187–214.
47. Westerterp KR, Robach P, Wouters L, et al. Water balance and acute mountain sickness before and after arrival at high altitude of 4350 m. J Appl Physiol. 1996;80:1968–72.
48. Robach P, Lafforgue E, Olsen NV, et al. Recovery of plasma volume after one week of exposure to 4,350 m. Pflugers Arch. 2002;444:821–8.

49. Durkot MJ, Hoyt RW, Darigrand A, et al. Chronic hypobaric hypoxia decreases intracellular and total body water in microswine. Comp Biochem Physiol. 1996;114A:117–21.
50. Claybaugh JR, Wade CE, Sato AK, et al. Antidiuretic hormone responses to eucapnic and hypocapnic hypoxia in humans. J Appl Physiol. 1982;53: 815–23.
51. Blume FD, Boyer SJ, Braverman LE, et al. Impaired osmoregulation at high altitude: studies on Mt Everest. JAMA. 1984;252:524–6.
52. Bestle MH, Olsen NV, Roach RC, et al. Renal sensitivity to vasopressin in man in acute hypoxia. FASEB J. 1998;12:A722.
53. Ramirez G, Hammond M, Bittle PA, et al. Sodium excretion and hormonal changes during salt loading at moderately high altitude and acute hypoxemia at sea level. Aviat Space Environ Med. 1992;63: 891–8.
54. Colice G, Ramirez G. Aldosterone response to angiotensin II during hypoxemia. J Appl Physiol. 1986;61:150–4.
55. Gledhill N, Beirne GJ, Dempsey JA. Renal response to short-term hypocapnia in man. Kidney Int. 1975;8:376–86.
56. Heyes MP, Farber MO, Manfredi F, et al. Acute effects of hypoxia on renal and endocrine function in normal humans. Am J Physiol. 1982;243:R265–70.
57. Ge RL, Babb TG, Sivieri M, et al. Urine acid–base composition at simulated moderate high altitude. High Alt Med Biol. 2006;7:64–71.
58. Wagner PD, Araoz M, Calbert JL, et al. Pulmonary gas exchange and acid–base state at 5260 m in high-altitude Bolivians and acclimatized lowlanders. J Appl Physiol. 2002;93:1393–400.
59. Behm R, Gerber B, Habeck J-O, et al. Effect of hypobaric hypoxia and almitrine on voluntary salt and water intake in carotid body–denervated spontaneously hypertensive rats. Biomed Biochim Acta. 1989;48:689–95.
60. Honig A. Peripheral arterial chemoreceptors and reflex control of sodium and water homeostasis. Am J Physiol. 1989;257:R1282–302.
61. Bardsley PA, Johnson BF, Barer GR. Natriuresis secondary to carotid chemoreceptor stimulation with almitrine bismethylate in the rat: the effect on kidney function and the response to renal denervation and deficiency of antidiuretic hormone. Biomed Biochim Acta. 1991;50:175–82.
62. Gomori P, Kovacs AGB, Takacs L, et al. The control of the renal circulation in hypoxia. Acta Med Hung. 1960;16:43–60.
63. Koller EA, Schopen M, Keller M, et al. Ventilatory, circulatory, endocrine, and renal effects of almitrine infusion in man. A contribution to high altitude physiology. Eur J Appl Physiol. 1989;58:419–25.
64. Swenson ER, Duncan TB, Goldberg SV, et al. Diuretic effect of acute hypoxia in humans: relationship to hypoxic ventilatory responsiveness and renal hormones. J Appl Physiol. 1995;78:377–83.
65. Valli G, Bonardi D, Campigotto F, et al. Relationship between individual ventilatory response and acute renal water excretion at high altitude. Respir Physiol Neurobiol. 2008;162:103–8.
66. Brauer H, Gens H, Lederhos C, et al. Cardiorespiratory and renal responses to arterial chemoreceptor stimulation by hypoxia or almitrine in men. Clin Exp Pharmacol Physiol. 1996; 23:1021–7.
67. DiBona G, Kopp E. Neural control of renal function. Physiol Rev. 1997;77:75–197.
68. Marshall JM. Peripheral chemoreceptors and cardiovascular regulation. Physiol Rev. 1994;74:543–94.
69. Fox WC, Watson R, Lockette W. Acute hypoxemia increases cardiovascular baroreceptor sensitivity in humans. Am J Hypertens. 2006;19:958–63.
70. Fukoda Y, Sato A, Suzuki A, et al. Autonomic nerve and cardiovascular responses to changing oxygen and carbon dioxide levels in the rat. J Auton Nerv Syst. 1989;28:61–74.
71. Sun M-K, Reis DJ. Hypoxia selectively excites vasomotor neurons of rostral ventrolateral medulla in rats. Am J Physiol. 1994;266:R245–56.
72. Shibamato T, Uematsu H, Matsuda Y, et al. Acute effect of hypobaria and hypoxia on renal nerve activity in anaesthetized rabbits. Acta Physiol Scand. 1992;144:47–53.
73. Hallows KR, Mount PF, Pastor-Soler NM, Power DA. Role of the energy sensor AMP activated protein kinase in renal physiology and disease. Am J Physiol. 2010;298:F1067–77.
74. Haase VH. Hypoxia-inducible factors in the kidney. Am J Physiol. 2006;291:F271–81.
75. Heidenreich S, Rahn KH, Zideck W. Direct vasopressor effect of erythropoietin on renal resistance vessels. Kidney Int. 1991;39:259–65.
76. Lundby C, Thomsen JJ, Boushel R, et al. Erythropoietin treatment elevates haemoglobin concentration by increasing red cell volume and depressing plasma volume. J Physiol. 2007;578:309–14.
77. Joyeux-Faure M. Cellular protection by erythropoietin: new therapeutic implications? J Pharmacol Exp Ther. 2007;323:759–62.
78. Stella A, Zanchetti A. Functional role of renal afferents. Physiol Rev. 1991;71:659–99.
79. du Souich P, Saunier C, Hartemann D, et al. Effect of moderate hypoxia on atrial natriuretic factor and arginine vasopressin in normal man. Biochem Biophys Res Commun. 1987;148:906–12.
80. Maresh CM, Kraemer WJ, Judelson DA, et al. Effects of high altitude and water deprivation on arginine vasopression release in man. Am J Physiol. 2004;286:E20–4.
81. Hackett PH, Forsling ML, Milledge J, et al. Release of vasopressin in man at altitude. Horm Metab Res. 1978;10:571.
82. Koller EA, Bischoff M, Buhrer A, et al. Respiratory, circulatory, and neuropsychological responses to acute hypoxia in acclimatized and non acclimatized subjects. Eur J Appl Physiol. 1991;62:67–72.

83. Share L, Levy MN. Effect of carotid chemoreceptor stimulation on plasma antidiuretic hormone titer. Am J Physiol. 1966;210:157–61.
84. Raff H, Shinsako J, Keil LC, et al. Feedback inhibition of adrenocorticotropin and vasopressin responses to hypoxia by physiological increases in endogenous corticosteroids in dogs. Endocrinology. 1984;114:1245–9.
85. Raff H, Shinsako J, Keil LC, et al. Vasopressin, ACTH, and corticosteroids during hypercapnia and graded hypoxia in dogs. Am J Physiol. 1983;244:E453–8.
86. Millar EA, Angus RM, Nally JE, et al. Effect of hypoxia and beta agonists on the activity of the renin-angiotensin system in normal subjects. Clin Sci. 1995;89:273–6.
87. Zaccaria M, Rocco S, Noventa D, et al. Sodium regulating hormones at high altitude: basal and post exercise levels. J Clin Endocrinol Metab. 1999;83: 570–4.
88. Colice G, Ramirez G. Effects of hypoxemia on the renin-angiotensin-aldosterone system in humans. J Appl Physiol. 1985;58:724–30.
89. Milledge JS, Catley DM. Angiotensin converting enzyme activity and hypoxia. Clin Sci. 1987;72:149.
90. Milledge JS, Catley DM. Renin, aldosterone and converting enzyme during exercise and acute hypoxia in humans. J Appl Physiol. 1982;52:320–3.
91. Bärtsch P, Maggiorini M, Schobersberger W, et al. Enhanced exercise-induced rise of aldosterone and vasopressin preceding mountain sickness. J Appl Physiol. 1991;71:136–41.
92. Shigeoka JW, Colice GL, Ramirez G. Effect of normoxic and hypoxic exercise on renin and aldosterone. J Appl Physiol. 1985;59:142–8.
93. Honig A, Wedler B, Opperman H, et al. Effect of arterial chemoreceptor stimulation with almitrine bismethylate on plasma renin activity, aldosterone, ACTH and cortisol in anaesthetized, artificially ventilated cats. Clin Exp Pharmacol Physiol. 1996;23: 106–10.
94. Milledge JS, Catley DM, Williams ES, et al. Effect of prolonged exercise at altitude on the renin-aldosterone system. J Appl Physiol. 1983;55:413–8.
95. Raff H, Ball DL, Goodfriend TL. Low oxygen selectively inhibits aldosterone secretion from bovine adrenocortical cells in vitro. Am J Physiol. 1989; 256:E640–4.
96. Young D. Analysis of long term potassium regulation. Endocrinol Rev. 1985;6:24–45.
97. Raff H, Jankowska B. Effect of CO_2/pH on the aldosterone response to hypoxia in bovine adrenal cells in vitro. Am J Physiol. 1993;265:R820–5.
98. Ramirez G, Bittle PA, Hammond M, et al. Regulation of aldosterone secretion during hypoxemia at sea level and moderately high altitude. J Clin Endocrinol Metab. 1988;67:1162–5.
99. Jenq W, Rabb H, Wahe M, et al. Hypoxic effects on the expression of mineralo-corticoid and glucocorticoid receptors in human renal cortex epithelial cells. Biochem Biophys Res Commun. 1996;218:444–8.
100. Arjamaa O, Nikinmaa M. Natriuretic peptides in hormonal regulation of hypoxia responses. Am J Physiol. 2009;296:R257–64.
101. Baertschi AJ, Teague WG. Alveolar hypoxia is a powerful stimulus for ANF release in conscious lambs. Am J Physiol. 1989;256:H990–8.
102. Bärtsch P, Shaw S, Franciolli M, et al. Atrial natriuretic peptide in acute mountain sickness. J Appl Physiol. 1988;65:1929–37.
103. Metzler CH, Ramsey DJ. Physiological doses of atrial peptide inhibit angiotensin II stimulated aldosterone secretion. Am J Physiol. 1989;256: R1155–9.
104. Garvin JL. ANF inhibits norepinephrine-stimulated fluid absorption in rat proximal straight tubules. Am J Physiol. 1992;263:F581–5.
105. Iitake K, Share L, Crofton JT, et al. Central atrial natriuretic factor reduces vasopressin secretion in rats. Endocrinology. 1986;119:438–40.
106. Kawashima A, Kubo K, Matsuzawa Y, et al. Hypoxia-induced ANP secretion in subjects susceptible to high altitude pulmonary edema. Respir Physiol. 1992;89:309–17.
107. Ebert TJ, Groban L, Muzi M, et al. ANP-mediated volume depletion attenuates renal responses in humans. Am J Physiol. 1992;263:R1303–8.
108. Sarelius IH, Huxley VH. A direct effect of atrial peptide on arterioles of the terminal microvasculature. Am J Physiol. 1990;258:R1224–9.
109. La Villa G, Stefani L, Zurli C, et al. Acute effects of physiological increments of brain natriuretic peptide in humans. Hypertension. 1995;26:628–33.
110. Hill NS, Klinger JR, Warburton RR, et al. Brain natriuretic peptide: possible role in the modulation of hypoxic pulmonary hypertension. Am J Physiol. 1994;266:L308–15.
111. Woods D, Hooper T, Mellor A, et al. Brain natriuretic peptide and acute hypobaric hypoxia in humans. J Physiol Sci. 2011;61:217–20.
112. Toscher MR, Thompson AAR, Irving JB, et al. NT-proBNP does not rise on acute ascent to high altitude. High Alt Med Biol. 2008;9:307–10.
113. Feddersen B. AussererH, Haditsch B, et al. Brain natriuretic peptide at altitude: relationship to diuresis, natriuresis and mountain sickness. Aviat Space Environ Med. 2009;80:108–11.
114. Woods D, Begley J, Stacey M, et al. Severe acute mountain sickness, brain natriuretic peptide and NT-proBNP in humans. Acta Physiol (Oxf). 2012; 205:349–55.
115. Woods D, Hooper T, Hodkinson P, et al. Effects of altitude exposure on brain natriuretic peptide in humans. Eur J Appl Physiol. 2011;111:2687–93.
116. Goetz K, Drummer C, Ahu JL, et al. Evidence that urodilatin, rather than ANP, regulates renal sodium excretion. J Am Soc Nephrol. 1990;1:867–74.
117. Koller EA, Lesniewska B, Buhrer A, et al. The effects of acute altitude exposure in Swiss highlanders and lowlanders. Eur J Appl Physiol. 1993;66: 146–54.

118. Sartori C, Vollenweider L, Loffler B-M, et al. Exaggerated endothelin-1 release in high altitude pulmonary edema. Circulation. 1999;99:2665–8.
119. Modesti PA, Cecioni I, Miglioniri A, et al. Increased renal endothelin formation is associated with sodium retention and increased free water clearance. Am J Physiol. 1998;275:H1070–7.
120. Sandgaard NCF, Bie F. Natriuretic effect of non-pressor doses of endothelin-1 in conscious dogs. J Physiol. 1996;494:809–18.
121. Ziedel ML, Brady HR, Kone BC, et al. Endothelin, a peptide inhibitor of Na^+-K^+-ATPase in intact renal tubular epithelial cells. Am J Physiol. 1989;257: C1101–7.
122. Nakano D, Pollock JS, Pollack DM. Renal medullary ETB receptors produce diuresis and natriuresis via NOS1. Am J Physiol. 2008;294:F1205–11.
123. Schnermann I, Lorena IN, Briggs JP, et al. Induction of water diuresis by endothelin in rats. Am J Physiol. 1992;263:F516–26.
124. Scholz H, Kramer BK, Hamann M, et al. Effects of endothelins on renin secretion from rat kidneys. Acta Physiol Scand. 1995;155:173–82.
125. Rubinstein I, Gurbanov K, Hoffman A, et al. Differential effect of endothelin-l on renal regional blood flow: role of nitric oxide. J Cardiovasc Pharmacol. 1995;26 Suppl 3:S208–10.
126. Modesti PA, Vanni S, Morabito M, et al. Role of endothelin-1 in exposure to high altitude. Acute mountain sickness and endotheli-1 (ACME-1) study. Circulation. 2006;114:1410–6.
127. Richalet J-P, Rutgers V, Bouchet P. Diurnal variation of acute mountain sickness, color vision and plasma cortisol and ACTH at high altitude. Aviat Space Environ Med. 1989;60:105–11.
128. Marotta SF. Roles of aortic and carotid chemoreceptors in activating the hypothalamic-hypophysial-adrenocortical system during hypoxia. Proc Soc Exp Biol Med. 1972;141:915–27.
129. Raff H. Glucocorticoid inhibition of hypophysial vasopressin secretion. Am J Physiol. 1987;252: R635–44.
130. Lewis RA, Thorn GW, Koept GF, et al. The role of the adrenal cortex in acute anoxia. J Clin Invest. 1942;21:33–46.
131. DeAngelis C, Haupert GT. Hypoxia triggers release of an endogenous inhibitor of Na^+-K^+-ATPase from midbrain and adrenal. Am J Physiol. 1998;274: F182–8.
132. Murrell JR, Randall JD, Rosoff J, et al. Endogenous ouabain. Circulation. 2005;112:1301–8.
133. Cruz JC, Bonagamba LG, Machado BH, et al. Intermittent activation of peripheral chemoreceptors in awake rats induces Fos expression in rostral ventrolateral medulla-projecting neurons in the paraventricular nucleus of the hypothalamus. Neuroscience. 2008;157:463–72.
134. Ferri C, Bellini C, Coassin S, et al. Plasma endogenous-like substance levels are dependent on blood O_2 in man. Clin Sci. 1994;87:447–51.
135. DeAngelis C, Farrace S, Urbani L, et al. Effects of high altitude exposure on plasma and urinary digoxin-like immunoreactive substance. Am J Hypertens. 1992;5:600–7.
136. Mazzeo RS, Wolfel EE, Butterfield GE, et al. Sympathetic response during 21 days at high altitude (4300 m) as determined by urinary and arterial catecholamines. Metabolism. 1994;43:1226–32.
137. Rostrup M. Catecholamines, hypoxia and high altitude. Acta Physiol Scand. 1998;162:389–99.
138. Young PM, Rose MS, Sutton JR. Operation Everest II: plasma lipid and hormonal responses during a simulated ascent of Mt Everest. J Appl Physiol. 1989;66:1430–5.
139. Wang T, Chan YI. Neural control of distal tubular bicarbonate and fluid transport. Am J Physiol. 1989;257:F72–6.
140. Sandner P, Hofbaure KH, Tinel H, et al. Expression of adrenomedullin in hypoxic and ischemic rat kidneys and human kidneys with arterial stenosis. Am J Physiol. 2004;286:R942–51.
141. Hofbauer K-H, Jensen BL, Kurtz A, et al. Tissue hypoxygenation activities of the adrenomedullin system in vivo. Am J Physiol. 2000;278:R513–9.
142. Mazzocchi G, Refubbat P, Gottardo G, et al. Adrenomedullin and calcitonin related gene peptide inhibit aldosterone secretion in rats, acting via a common receptor. Life Sci. 1996;58:839–44.
143. Haditsch B, Roessler A, Hinghofer-Szalkay HG. Renal adrenomedullin and high altitude diuresis. Physiol Res. 2007;56:779–87.
144. Toepfer M, Hartman G, Schlosshauer M, et al. Adrenomedullin: a player at high altitude? Chest. 1998;113:1428.
145. Hasbek P, Lundby C, Olsen NV, et al. Calcitonin gene-related peptide and adrenomedullin release in humans: effects of exercise and hypoxia. Regul Pept. 2002;108:89–95.
146. Ruszchitzka FT, Wenger RH, Stallach T, et al. Nitric oxide prevents cardiovascular disease and determines survival in polyglobulic mice over expressing erythropoietin. Proc Natl Acad Sci. 2000;97: 11609–13.
147. Heyman SN, Goldfarb M, Darmon D, et al. Tissue oxygenation modifies nitric oxide bioavailability. Microcirculation. 1999;6:199–203.
148. Baumann JE, Persson PB, Ehmke H, et al. Role of endothelium-derived relaxing factor in renal autoregulation in conscious dogs. Am J Physiol. 1992; 263:F208–13.
149. Thomson SC, Vallon V. Alpha 2 adrenoreceptors determine the renal response to nitric oxide in rat glomerulus and proximal tubule. J Am Soc Nephrol. 1995;6:1482–90.
150. McLay JS, Chatterje PK, Mistry SK, et al. Atrial natriuretic factor and angiotensin II stimulate nitric oxide release from human proximal tubular cells. Clin Sci. 1995;89:527–31.
151. Perella MA, Edell ES, Krowka MJ, et al. Endothelium-derived relaxing factor in pulmonary

and renal circulations during hypoxia. Am J Physiol. 1992;263:R45–50.
152. Raij L, Baylis C. Glomerular actions of nitric oxide. Kidney Int. 1995;48:20–32.
153. Deng A, Miracle CM, Suarez JM, et al. Oxygen consumption in the kidney: effects of nitric oxide synthase isoforms and angiotensin II. Kidney Int. 2005;68:723–30.
154. Romero JC, Lahera V, Salom MG, et al. Role of the endothelium-dependent relaxing factor nitric oxide on renal function. J Am Soc Nephrol. 1992;2:1371–87.
155. Stoos BA, Garcia NH, Garvin JL. Nitric oxide inhibits sodium reabsorption in the isolated perfused cortical collecting duct. J Am Soc Nephrol. 1995;6:89–94.
156. He X-R, Greenberg SG, Briggs JP, et al. Effect of nitric oxide on renin secretion II: studies in the perfused juxtaglomerular apparatus. Am J Physiol. 1995;268:F953–9.
157. Tripatara P, Patel NS, Webb A, et al. Nitrite-derived nitric oxide protects the rat kidney against ischemia/reperfusion injury in vivo: role for xanthine oxidoreductase. J Am Soc Nephrol. 2007;18:570–80.
158. Nistala R, Whaley-Connell A, Sowers JR. Redox control of renal function. Antioxid Redox Signal. 2008;10:2047–89.
159. Nahmod VE, Lanari A. Abolition of autoregulation of renal blood flow by acetylcholine. Am J Physiol. 1964;207:123–7.
160. Jackson KE, Jackson DW, Quadi S, Reitzell MJ, Navar LG. Inhibition of heme oxygenase augments tubular sodium reabsorption. Am J Physiol. 2011;300(4):F941–6.
161. Wang T, Sterling H, Shao WA, Yan Q, Bailey MA, Giebisch G, et al. Inhibition of heme oxygenase decreases sodium and fluid absorption in the loop of Henle. Am J Physiol. 2003;285:F484–90.
162. Althaus M. Gasotransmitters: novel regulators of epithelial Na^+ transport. Front Physiol. 2012;3:1–10.
163. Beltowski J. Hypoxia in the renal medulla: implications for hydrogen sulfide signaling. J Pharmacol Exp Ther. 2010;334:358–63.
164. Kai S, Tanaka T, Daijo H, Harade H, Kishimoto S, Suzuki K, et al. Hydrogen sulfide inhibits hypoxia- but not anoxia-induced hypoxia inducible factor 1 activation in a von Hippel-Lindau- and mitochondria-dependent manner. Antioxid Redox Signal. 2012;16:203–16.
165. Folkow B. Secretory renal functions—Tigerstedt, renin and its neglected antagonist medullipin. Acta Physiol (Oxf). 2007;190:99–102.
166. Roszinski S, Jelkmann W. Effect of PO_2 on prostaglandin E2 production in renal cell cultures. Respir Physiol. 1987;70:131–41.
167. Roman RJ, Cowley AJ, Garcia-Estan J, Lombard H. Pressure-diuresis in volume-expanded rats: cortical and medullary hemodynamics. Hypertension. 1988;12:168–72.
168. Garcia-Estan J, Roman RJ. Role of renal interstitial hydrostatic pressure in the pressure diuresis response. Am J Physiol. 1989;256:F63–70.
169. Roman RJ, Zou A-P. Influence of the medullary circulation on the control of sodium excretion. Am J Physiol. 1993;265:R963–73.
170. Leonard BG, Malpas SC, Denton KM, et al. Differential control of intrarenal blood flow during reflex measures in sympathetic nerve activity. Am J Physiol. 2001;280:R62–8.
171. Ledderhos C, Gross V, Cowley AW. Pharmacological stimulation of arterial chemoreceptors in conscious rats produces different responses in renal cortical and medullary blood flow. Clin Exp Pharmacol Physiol. 1998;25:536–40.
172. Karim F, Poucher SM, Summerill RA. The effects of stimulating carotid chemoreceptors on the renal hemodynamics and function in dogs. J Physiol. 1987;392:451–62.
173. Ledderhos C, Queis W, Schuster R, et al. Renal hemodynamics and excretory function of healthy young men during stimulation of their peripheral arterial chemoreceptors by almitrine bismethylate. Biomed Biochim Acta. 1987;46:1035–42.
174. Bischoff A, Erdbrugger W, Smits J, et al. Neuropeptide Y-mediated diuresis and natriuresis in anesthetized rats is independent of renal blood flow reduction. J Physiol. 1996;495:525–34.
175. Liu GL, Liu L, Batrajas L. Development of NOS-containing neuronal somata in the rat kidney. J Auton Nerv Syst. 1996;58:81–8.
176. Mattson DL, Bellehumeur TG. Neural nitric oxide synthase in the renal medulla and blood pressure regulation. Hypertension. 1996;28:297–303.
177. Pirola JP, Alvarez AL, Balda MS, et al. Evidence for cholinergic innervation in dog renal tissue. Am J Physiol. 1989;257:F746–54.
178. Kirchheim HR. Systemic arterial baroreceptor reflexes. Physiol Rev. 1976;56:100–76.
179. Baekey DM, Molkov YI, Paton JF, et al. Effect of baroreceptor stimulation on the respiratory pattern: insights into respiratory-sympathetic interactions. Respir Physiol Neurobiol. 2010;174:135–45.
180. Kaufman MP, Cassidy SS. Reflex effects of lung inflation and other stimuli on the heart and circulation. In: Scharf SM, Cassidy SS, editors. Heart-Lung Interactions in Health and Disease. New York: Marcel Dekker; 1989. p. 339–63.
181. Currie JCM, Ullman E. Polyuria during experimental modification of breathing. J Physiol. 1961;155:438–55.
182. O'Hagan KP, Bell LB, Clifford PS. Effects of pulmonary deneruation on renal sympathetic and heart rate responses to hypoxia. Am J Physiol. 1995;269:R293–9.
183. Rutherford JD, Vatner SF. Integrated carotid chemoreceptor and pulmonary inflation reflex control of peripheral vasoactivity in conscious dogs. Circ Res. 1978;43:200–8.

184. Koller EA, Buhrer A, Felder L, et al. Altitude diuresis, endocrine and renal responses to acute hypoxia of acclimatized and non acclimatized subjects. Eur J Appl Physiol. 1991;62:228–34.
185. Vatner SF, Manders WT, Knight DR. Vagally mediated regulation of renal function in conscious primates. Am J Physiol. 1986;250:H546–9.
186. Gennari FJ, Goldstein MB, Cohen JJ. The nature of the renal adaptation to chronic hypocapnia. J Clin Invest. 1972;51:1722–30.
187. Hainsworth R, Rankin JA, Soladoye AO. Effect of cephalic carbon dioxide tension on the cardiac inotropic response to carotid chemoreceptor stimulation in dogs. J Physiol. 1985;358:405–16.
188. Anderson RJ, Henrich W, Gross PA, et al. Role of the renal nerves, angiotensin II and prostaglandins in the antinatriuretic response to acute hypercapnic acidosis in the dog. Circ Res. 1982;50:294–300.
189. Raff H, Roarty TP. Renin, ACTH, and aldosterone during acute hypercapnia and hypoxia in conscious rats. Am J Physiol. 1988;254:R431–5.
190. Bärtsch P, Baumgartner RW, Waber U, et al. Comparison of carbon dioxide enriched, oxygen enriched, and normal air in treatment of acute mountain sickness. Lancet. 1990;336:772–5.
191. Hackett PH, Rennie D, Grover RF, et al. Fluid retention and relative hypoventilation in acute mountain sickness. Respiration. 1982;43:321–9.
192. Moore LG, Huang SY, McCullough RE, et al. Variable inhibition by falling CO_2 of hypoxic ventilatory response in humans. J Appl Physiol. 1984;56:207–10.
193. Epstein M, Saruta T. Effects of a hyperoxic hypobaric environment on renin-aldosterone in normal man. J Appl Physiol. 1973;34:49–52.
194. Glatte HV, Giannetta CL. Study of man during a 56 day exposure to an oxygen-helium atmosphere at 258 mmHg total pressure: renal response. Aerosp Med. 1966;37:559–62.
195. Levine BD, Kubo K, Kobayashi T, et al. Role of barometric pressure in pulmonary fluid balance and oxygen transport. J Appl Physiol. 1988;64:419–28.
196. Loeppky JA, Icenogle M, Scotto P, et al. Ventilation during simulated altitude, normobaric hypoxia and normoxic hypobaria. Respir Physiol. 1997;107:231–9.
197. Conkin J, Wessel JH. Critique of the equivalent air altitude model. Aviat Space Environ Med. 2008;79:975–82.
198. Castenfors J. Renal function during exercise. Acta Physiol Scand. 1967;70 Suppl 293:1–40.
199. Schmidt W, Brabant G, Kroger G. Atrail natriuretic peptide during and after maximal and submaximal exercise under normoxic and hypoxic conditions. Eur J Appl Physiol. 1990;61:398–407.
200. Wade CE. Response, regulation, and actions of vasopressin during exercise: a review. Med Sci Sports Exerc. 1984;16:506–11.
201. O'Hagan KP, Bell LB, Mittelstadt SW, et al. Effect of dynamic exercise on renal sympathetic activity in conscious rabbits. J Appl Physiol. 1993;74:2099–104.
202. Bouissou P, Guezennec CY, Defer G, et al. Dissociated response of aldosterone from plasma renin activity during prolonged exercise under hypoxia. Horm Metab Res. 1988;20:517–21.
203. Bouissou P, Perronet F, Brisson G, et al. Metabolic and endocrine responses to graded exercise under hypoxia. Eur J Appl Physiol. 1986;555:290–4.
204. Williams ES, Ward MP, Milledge JS. Effect of exercise of seven consecutive days hill-walking on fluid homeostasis. Clin Sci. 1979;56:305–14.
205. Bocquerez O, Kohlmann N, Guigas B, et al. Fluid-regulatory hormone responses with cycling exercise in acute hypobaric hypoxia. Med Sci Sports Exerc. 2004;36:1730–6.
206. Luks AM, Johnson RJ, Swenson ER. Chronic kidney disease at high altitude. J Am Soc Nephrol. 2008;19:2262–71.
207. Evans RG, Goddard D, Eppel GA, et al. Factors that render the kidney susceptible in hypoxemia. Am J Physiol. 2011;300:R931–40.
208. Stein A, Goldmeier S, Voltolini S, Setogutti E, Feldman C, Figueiredo E, et al. Renal oxygen content is increased in healthy subjects after angiotensin converting enzyme inhibition. Clin Sci. 2012;67:761–5.
209. Arestegui AH, Fuquay R, Sirota J, Swenson ER, Schoene RB, Jefferson JA, et al. High altitude renal syndrome. J Am Soc Nephrol. 2011;22:1963–8.
210. Ghahramani N, Ahmed F, Al-Laham A, Lengerich EJ. The epidemiological association of altitude with chronic kidney disease: Evidence of protective effect. Nephrology. 2011;16:219–24.
211. Swenson ER. Hemodynamic and metabolism at low vs moderate altitudes. High Alt Med Biol. 2011;12:407–8.
212. Stampfli VR, Eberle A. Menge, spezifisches Gewicht und Leitfähigkeit des menschlichen Harns im Hochgebirge. Helv Physiol Acta 1944 (Suppl III):221–32.
213. Bärtsch P, Jülg B, Hohenhaus E. Urine volume in acute mountain sickness is not related to hypoxic ventilatory response. Eur Respir J. 1995;8:625.
214. Bärtsch P, Swenson ER, Paul A, et al. Hypoxic ventilatory response, ventilation, gas exchange and fluid balance in acute mountain sickness. High Alt Med Biol. 2002;4:361–76.
215. Roach RC, Maes D, Riboni K, et al. Increased plasma volume at simulated altitude and the onset of acute mountain sickness. FASEB J. 1998;12:A57.

Endocrine Function

12

Jean-Paul Richalet

Abstract

Adaptation to a stressful environment such as altitude hypoxia will affect most hormonal systems. Stress hormones (cortisol, epinephrine, norepinephrine, prolactin) are stimulated in acute hypoxia but adrenergic system is down-regulated in prolonged hypoxia. Thyroid hormone is increased in hypoxia, with no change in TSH. Parathromone is stimulated and parathromone receptors desensitized with prolonged exposure. Leptin is increased and ghrelin decreased with exposure to hypoxia. Growth hormone is poorly affected by acute hypoxia and exercise-induced GH response is potentiated by acute hypoxia. LH and FSH decrease with acute hypoxia. The hypophyseal response to hypothalamic factors is not substantially modified in altitude hypoxia. The possible role of hypoxia-inducible factors in the regulation of hormones at high altitude remains to be explored.

Introduction

Adaptation to a new environment needs an information system which first informs the body of the characteristics of the environment and second triggers biological responses which may be more or less "adaptive" to new environmental conditions. The nervous system and the endocrine (or neuro-endocrine) system are where the perception and the adaptive processes occur.

J.-P. Richalet, M.D., Ph.D. (✉)
Université Paris 13, Sorbonne Paris Cité, 74 rue Marcel Cachin, 93017 Bobigny, France
e-mail: richalet@univ-paris13.fr

The effect of acute and sustained hypoxia on some endocrine systems has been extensively studied, as for example hormones involved in water and electrolyte balance or the adrenergic system. Much less is known about others, such as thyroid or sex hormones.

Hormonal receptors are classified as membrane receptors (catecholamines, angiotensin, anti-diuretic hormone (ADH), growth hormone (GH), parathormone (PTH), prolactin) and intracellular receptors, nuclear or cytoplasmic (cortisol, aldosterone, estrogens, progesterone, T3, T4). Little is known about the effect of hypoxia on nuclear receptors. However, the expression of these receptors can be modulated by hypoxia [1]. More is known about some membrane receptors, e.g., G-protein-mediated receptors such as

E.R. Swenson and P. Bärtsch (eds.), *High Altitude: Human Adaptation to Hypoxia*,
DOI 10.1007/978-1-4614-8772-2_12, © Springer Science+Business Media New York 2014

β-adrenoceptors, muscarinic receptors, or adenosinergic receptors [2].

Exposure to hypoxia has been shown to produce both down- and upregulation. For example, release of cortisol is activated and release of aldosterone is blunted by hypoxia, although these two hormones are secreted by adjoining zones in the same organ. Some endocrine or neuroendocrine pathways are activated in response to the hypoxic stressor, other pathways are blunted. A resistance appears to certain stimuli, limiting the effects of the activated pathways. Finally, an optimal adaptation to the stressful environment would depend on an adequate balance between activation and resistance [3]. One can speculate about a common pathway involving hypoxia-inducible factors (HIF) in the modulation (activation or inhibition) of the secretion of some hormones in hypoxic conditions. Extensive studies are in progress on the mechanisms of oxygen sensing at the cellular and molecular levels (see Chaps. 1 and 2).

Stress Hormones

Stress is a generic term for any circumstance that tends to disturb an individual or system from its "normal" resting equilibrium. In that sense, hypoxia is a stressor that induces an endocrine reaction from adrenal glands. Whether the physiological response of these glands to hypoxia is specific or similar to the response to other stressors remains to be clarified.

Cortisol

Both acute and sustained altitude hypoxia stimulate the adrenal cortex. Plasma concentrations of ACTH and cortisol are increased, often transiently [4–13]. However, not all studies report a change in plasma cortisol [14–18]. When altitude exposure is associated with another stress such as exercise, cold or danger, increase in cortisol levels is more likely [8, 10, 19].

Acute mountain sickness: Clinical status influences the release of corticosteroids. In general cortisol increases more when the ascent is associated with AMS [20]. The rise in plasma cortisol may precede the onset of AMS symptoms [8]. The diurnal rhythm of cortisol is maintained at high altitude and is accompanied by a parallel variation in AMS score, with maximal values at 8 a.m. [21]. After 1 week at 6,542 m plasma cortisol is elevated, then decreases with subsequent acclimatization, further supporting an association with AMS [22, 23].

ACTH is not the sole controller of cortisol secretion during hypoxic stress. Other factors can also interfere, such as adrenal blood flow (higher in spontaneous ventilation), stimulation of pulmonary stretch receptors (which can inhibit ACTH response), and low adrenal tissue PO_2 [24]. A possible relation between hypoxic ventilatory response (HVR) and plasma cortisol is evoked [25].

Whatever the mechanism, the stimulation of adrenal cortex seems to be a common feature at high altitude. Whether it is a specific response to hypoxia or a non specific response to stress remains to be established.

Adrenergic System

Altitude hypoxia activates the adrenergic system both at rest and during submaximal exercise (see also Chaps. 6, 8 and 16 for more details). With sustained exposure to hypoxia, hyperplasia of the adrenal medulla occurs. Plasma or urine norepinephrine (NE) constantly increases [12, 13, 27–30] while epinephrine shows more contrasted changes, either a transient increase or no change [13, 30, 31]. Dopamine has also been shown to increase in hypoxia [32]. The elevation of plasma arterial NE concentration is probably caused by a higher whole-body NE release since whole-body NE clearance is unchanged. These changes occur despite a greater systemic O_2 delivery after altitude acclimatization [33]. Permanent inhabitants of high altitude show higher plasma NE levels than sea level natives at sea level, but similar epinephrine and dopamine levels [34].

A major characteristic of the adrenergic system in response to prolonged hypoxia is the desensitization of β-adrenergic receptors [35, 36]. A number of human and animal studies show a decrease in maximal heart rate during exercise [35–37], a lower heart rate response to adrenergic activation (exercise or isoproterenol infusion) [38, 39], a downregulation of β-receptors [28, 39–41] and blood cells α2-adrenoceptors receptors [31], and a modulation of G proteins gene expression and function [2].

This blunting of β-adrenergic response is balanced by the upregulation of muscarinic receptors, the downregulation of receptors to adenosine [42–44]. The decreased maximal heart rate in chronic hypoxia prevents myocardial oxygen need from exceeding supply and protects the myocardium against the risk of tissue anoxia. This blunting is a clear mechanism of autoregulation of cardiac function and oxygen balance within the myocardium [35, 36, 45].

Adrenergic receptors are found not only in the heart and blood vessels but also in many other organs (body fat, kidney, etc.), involved in the control of numerous functions and endocrine systems, e.g., control of blood glucose and substrate utilization, renin release by the kidney, thyroid hormones. Thus, activation of the adrenergic system and subsequent receptor desensitization induced by hypoxia may influence various hormonal responses. During a simulated ascent to Mount Everest, the in vitro lipolytic response to epinephrine, isoproterenol, GH, and PTH decreased. Prolonged exposure to hypobaric hypoxia led to a potent reduction in lipid mobilization, through a decrease in the efficiency of β-adrenergic, GH and PTH lipolytic pathways [46] as will be discussed further in the chapter.

Thyroid Hormones

Thyroid hormones play an important role in regulating the overall rate of body metabolism and are critical for normal growth and development. Thyroid hormones increase oxygen consumption and heat production. They also increase oxygen availability by increasing ventilation and cardiac output as well as red blood cell mass. In that sense, they could play a role in long-term adaptation to altitude hypoxia.

High altitude stimulates thyroid hormone release. An increase in protein-bound iodine, first reported in 1966 [47], has been confirmed by observations of increased uptake and release of iodine from the thyroid gland [48]. Most authors agree that high altitude induces an elevation in plasma concentration of both free and total T4 fractions of thyroxin [30, 47, 49–57]. T3 also increases [30, 49, 50, 55], although to a lesser extent than T4. A few authors report no increase in T3, but rather an increase in reverse T3 (rT3) [51, 53, 58], perhaps because of an inhibition of the conversion of T4–T3 due to hypoxia, with a concomitant rise in rT3 concentration and an elevated T4:T3 ratio [53, 58]. Such changes are also seen in other stressful states and are possibly caused by an increased release of corticosteroids. Exhaustion may play some role, for depressed levels of thyroid hormones were observed after an expedition to 4,360–6,194 m under strenuous and difficult conditions [58]. Cold may also influence thyroid hormone levels. T3, likely to be the most important thyroid hormone for cold habituation, decreases with cold exposure, while T4 and TSH remain unaltered [59]. A concomitant elevation of basal metabolic rate and thyroid hormones at high altitude suggests that these changes are causally related [57].

Most studies show unaltered baseline TSH concentration at high altitude despite elevated levels of T3 and T4 [18, 30, 50–52, 54, 55]. This "dissociation" is remarkable as TSH is considered to be the main regulator of thyroid activity, and high levels of thyroid hormones either suppress this pituitary stimulus or reflect the increased TSH concentration.

Several explanations for the TSH-independent thyroxin rise have been proposed. In men trekking at altitudes of 3,505–3,658 m, thyroxin binding globulin (TBG) was found sufficiently elevated to explain the T4 rise [50, 57]. The T3 and T4 increase at high altitude is often accompanied by an increase in noradrenaline or epinephrine plasma concentrations [30, 56]. The thyroid gland is innervated by sympathetic branches and adrenaline can stimulate thyroid hormone secretion via β-adrenergic receptors. The unchanged level of

TSH in the presence of increased levels of T4 could reflect an altered feedback regulation or disturbed hypophyseal function. But most studies do not show any evidence of a hypophyseal malfunction, as a completely normal TSH response is seen after TRH administration [51, 53–55].

To compare altitude-induced neuroendocrine changes with or without energy imbalance and to explore how energy sufficiency alters the endocrine acclimatization process, 26 young men were studied for 3 weeks at 4,300 m, with one group adequately fed to maintain body weight and the other with calories restricted. Free T4, TSH, and NE showed similar patterns between the two groups (T4 increased but TSH did not change), suggesting that the hypoxic stimulus is capable of overriding the fall in T4 induced by caloric restriction [60].

High levels of thyroid hormones in prolonged and severe hypoxia suggest hyperthyroidism as a possible protective adaptation to this extreme environment [49]. Increased thyroid hormones release would be beneficial in hypoxia since these hormones increase 2,3-diphophoglycerate in erythrocytes, facilitating oxygen release to the tissues, by shifting the oxyhemoglobin dissociation curve to the right. Moreover, hypoxia-induced increase in 2,3-diphophoglycerate would decrease the response of erythropoietin to hypoxia [61]. In contrast, no difference in thyroid response was observed between subjects susceptible to AMS and resistant subjects [19, 58], suggesting that thyroid hormones play a minor role in the early phase of acclimatization. However, it has been shown that thyroid hormones increase alveolar fluid clearance through an increased Na, K ATPase activity [62], which could play a protective role in the development of high-altitude pulmonary edema.

In summary, plasma concentrations of thyroid hormones, in particular T4, increase with acute and sustained hypoxia. This increase is not accompanied by any consistent change in TSH level but is probably due to the increased adrenergic activity. In many studies done in the mountains, cold exposure may have contributed to the stimulation of the thyroid gland.

Hormones of Calcium and Phosphate Balance

A complex interaction of three hormones (PTH, calcitonin, and vitamin D) on various tissues (intestine, bone, and kidney) is responsible for maintenance of normal concentrations of calcium and phosphate in body fluids. In four men exposed for 6 days to 3,450 m, plasma ionized calcium decreased, excretion of nephrogenous cAMP decreased, but serum intact PTH did not change, suggesting an impaired renal responsiveness to PTH [63]. Altitude-induced respiratory alkalosis may interfere with calcium homeostasis since alkalosis has been shown to reduce ionized calcium and therefore stimulate PTH secretion [64].

In ten lowlanders living for 3 weeks at 6,542 m, total serum calcium and phosphate were unchanged while PTH and 25-OH D3 were elevated and 1,25-$(OH)_2$ D3 was decreased. These observations could be explained either by a hypoxia-induced inhibition of renal 1α-hydroxylase resulting in a tendency to hypocalcemia and a secondary hyperparathyroidism, or by a resistance to PTH [22, 23]. The conversion of 25-OH D3 into 1,25-$(OH)_2$ D3 within the kidney by the enzyme 1α-hydroxylase might be O_2-dependent. In response to this decreased production of active vitamin D, PTH would be released in excess, maintaining a normal level of plasma calcium, in spite of a low calcium intake. High PTH and low 1,25-$(OH)_2$ D3 levels could be due to a desensitization of PTH receptors. Observations performed at 4,350 m showing a blunted urinary cyclic AMP response to PTH favor the resistance to PTH hypothesis [65].

In summary, altitude seems to stimulate PTH release and inhibit active vitamin D, resulting in an unaltered calcium and phosphate balance. The desensitization of PTH receptors, which are G protein-coupled receptors like ß-adrenoceptors, should be examined as a protective mechanism similar to what occurs within the heart or fat tissue with desensitization of adrenergic system.

Hormones Controlling Blood Glucose and Appetite

Glucose, Insulin, and Glucagon

Resting blood glucose has been reported to be reduced [18, 56], unchanged [66], or transiently increased [9] at high altitude. Acute (1 h) and, in a more marked manner, prolonged (3 weeks) exposure to 4,300 m increased utilization of blood glucose during exercise [67]. Altitude and beta-blockade augment glucose utilization during submaximal exercise [68]. Plasma insulin increases after prolonged exposure to high altitude such as in Operation Everest II [69] and the Sajama expedition [70], but was found unchanged after a 7-week expedition to Mount Everest [18]. Serum insulin and C-peptide were found increased in acute hypoxia during intense exercise by sedentary subjects [71]. Tolerance to glucose decreases with acute hypoxia, suggesting that the decrease in peripheral use of glucose during hypoxic exercise could be an additional factor accounting for hypoxia-induced hyperglycemia [72]. Hypoxia-induced activation of the adrenergic system could suppress insulin release [73]. This mechanism could facilitate hepatic glycogenolysis and increase blood glucose. Fasting glucagon was found unchanged at high altitude [69]. Glucose and insulin concentrations on an oral glucose tolerance test were significantly lowered by a 25-day mountaineering activity only in subjects with high dehydroepiandrosterone-sulfate (DHEA-S), suggesting that DHEA-S is essential for physiologic acclimatization to mountaineering challenge [74]. To explore how energy sufficiency alters the endocrine acclimatization process, 26 men were studied for 3 weeks at 4,300 m, with one group adequately fed to maintain body weight (ADQ) and the other with calories restricted (DEF). Fasting glucose and insulin exhibited a blunted rise in DEF compared with ADQ [60] (Fig. 12.1a, b).

In summary, no clear modification of glucose homeostasis and related hormones was demonstrated at high altitude, probably because conditions of food intake were not controlled in most studies.

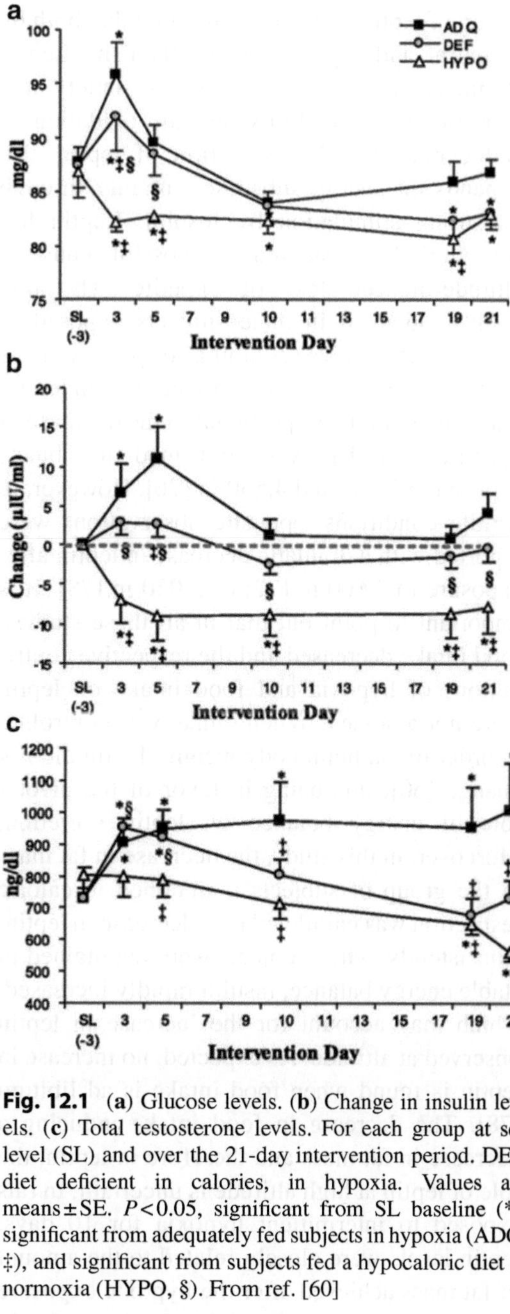

Fig. 12.1 (**a**) Glucose levels. (**b**) Change in insulin levels. (**c**) Total testosterone levels. For each group at sea level (SL) and over the 21-day intervention period. DEF, diet deficient in calories, in hypoxia. Values are means ± SE. $P<0.05$, significant from SL baseline (*), significant from adequately fed subjects in hypoxia (ADQ, ‡), and significant from subjects fed a hypocaloric diet in normoxia (HYPO, §). From ref. [60]

Leptin and Ghrelin[1]

Leptin is mainly secreted by the adipose tissue proportionally to its mass. The most emblematic

[1] This paragraph was written in collaboration with Didier Chapelot.

action of leptin is to reduce food intake, both by a central and a peripheral mechanism. Leptin stimulates the neuromediators of satiety and increases the availability and the oxidation of endogenous fat. The secretion of leptin also depends on energy substrates, mainly through its strong stimulation by insulin. Leptin has been initially considered as a possible cause of altitude-induced loss of appetite. The first results, obtained in male subjects exposed to 4,559 m after active (climbing) or passive (helicopter) ascent, showed a rapid increase in leptin, with a magnitude proportional to the decrease in appetite [75]. This was confirmed in subjects exposed to 3,600 and 4,550 m [76]. However, in similar conditions, opposite observations were reported with a dramatic decrease in leptin after exposure to 3,600 m [77] or 5,050 m [78]. It is important to point out that in all these studies, food intake decreased and the respective contributions of hypoxia and food intake on leptin were not assessed. When intake was controlled in order to maintain body weight, leptin did not change [60], this being in favor of the pivotal role of energy balance on leptin secretion. Moreover, in this study, the decrease in fat mass in the group of subjects maintained in caloric restriction was paralleled by a decrease in leptin. Consistently, when subjects were maintained in stable energy balance, insulin rapidly increased, which may account for the increase in leptin observed at altitude. As expected, no increase in leptin is found when food intake is ad libitum [78]. The decrease in food intake inducing a decrease in fat mass and therefore in leptin, the role of leptin at high altitude is uncertain. In rats exposed to intermittent hypoxia for 10 days, leptin levels were closely related to the amount of fat mass achieved after the hypoxic exposure [79]. To date, the effect of hypoxia on leptin secretion is highly controversial (see debate in the *Journal of Applied Physiology*, [80, 81]). Altogether, it seems that acute exposure to hypoxia (24 h) induces an increase in leptin release, independently of the food intake condition [82]. With acclimatization to altitude, leptin is higher than expected from the changes in body weight, although the measurement of fat mass is often lacking [18]. These observations led some authors to suggest that hypoxic exposure (intermittent?) could be proposed for the treatment of obesity [83]. In high-altitude natives, leptin is found consistently lower than in sea level natives, with an inverse relationship between altitude and leptin concentration [84, 85]. However, it is still not clear if a lower fat mass in high-altitude natives could contribute to this effect. Finally, the strongest argument against a role for leptin in altitude-induced body weight loss was given by a study showing that leptin-resistant rats reduce their food intake and decrease their body weight when exposed to hypoxia [86].

Ghrelin, an endogenous GH secretagogue secreted by gastric cells, is a polypeptide with orexigenic properties. It has been considered a candidate for the role of meal initiation since an increase of its blood concentration was reported prior to meals taken at fixed hours [87] or spontaneously requested [88]. However, this putative role has been challenged by results showing that such an increase may primarily depend on conditioning processes [89]. A possible involvement of ghrelin in the hypoxia-induced anorexia has received no definitive experimental demonstration but after 7 days at 4,300 m, ghrelin is actually reduced [76] and these lower levels are found again after 7 weeks of exposure to 5,200 m [18]. An inhibition of ghrelin by hypoxia is supported by results in rats bred in simulated altitude [90]. However, this effect may, once again, be mediated by insulin, which was found to be elevated in this animal model. Moreover, a strong negative correlation was reported in humans between ghrelin and insulin [88], an insulin threshold being necessary to reach for ghrelin to increase. Interestingly, ghrelin has vasodilator properties and was recently found to reduce the development of pulmonary arterial hypertension, pulmonary vascular remodeling, and right ventricular hypertrophy in decreasing overexpression of both endothelial nitric oxide synthase and endothelin-1 [91]. Thus, this hypoxia-induced decrease in ghrelin may represent a deleterious mechanism in terms of cardiovascular adaptation to hypoxia.

Other Hormones

Prolactin

Prolactin release is stimulated by TRH and inhibited by dopamine. Prolactin release increases in response to various stressors, including exercise, hyperthermia, fasting, surgery. The reports on prolactin response to high altitude are few and inconclusive: elevated [92] or reduced [30, 93] resting levels of prolactin have been described.

Studies on acute hypoxic exposure have mostly been performed in male lowlanders, while chronic hypoxia has been more frequently studied in female high-altitude natives. Prolactin concentration in women born and living at 4,340 m, compared to that of women at sea level, is diminished in chronic hypoxia [94]. Increased fertility and decreased frequency of menstrual disturbances have also been observed among high-altitude women compared with women living at sea level [94]. Dopamine and possibly noradrenaline, inhibitors of prolactin, have been suggested as a cause of this alteration as dopamine concentration increases at high altitude (see above). The exercise-induced increase in prolactin is well documented [93, 95–97], though the mechanism is unclear. During exercise in normobaric hypoxia, plasma prolactin was found increased [97], unchanged [93] or decreased [96], when compared to normoxia. A blunted prolactin response was observed during exercise in normoxia with concomitant face-cooling, suggesting that cold exposure may diminish the exercise-induced prolactine release [95]. After 7 weeks at high altitude, a prolactin increase coupled with testosterone decrease and progesterone increase, without any change in estradiol, probably due to chronic stressful conditions [18].

Prolactin secretion is under the influence of numerous regulators that could be altered by hypoxia. High work intensity, blood lactate level, a fat-rich diet induce a prolactin increase and could explain the varying results at altitude. Whatever the mechanism that alters prolactin baseline levels, it seems to be overridden by hypothalamic influence since the prolactin response to TRH administration is not altered in hypoxia [30, 53, 54].

In conclusion, prolactin secretion seems to be altered only on acute exposure as a non specific stress. In women exposed to chronic hypoxia prolactin level is decreased when compared to women at sea level.

Endothelin

The endothelium has potent actions on hemostasis, vasomotor tone, and vascular permeability. The endothelial cell produces vasodilators, PGI2 (with cAMP as second messenger) and nitrogen monoxide (NO) (with cGMP as second messenger). The endothelial cell produces also vasoconstrictor agents such as endothelin, endothelium derived growth factor (EDGF) and endothelium derived contracting factor (EDCF2). Endothelin increases catecholamine, renin, and ADH actions and modulates smooth muscle growth. Endothelin increases the synthesis and secretion of ANP [98]. Endothelin inhibits renin release from isolated rat glomeruli, via a calcium entry mechanism, and the isoproterenol-induced increase in renin release is inhibited by endothelin [99].

Hypoxia increases the release of endothelin from rat mesenteric arteries in vivo [100], and induces endothelin gene expression and secretion in cultured human endothelium [101]. Endothelin-1 and endothelin receptor gene expression is enhanced in rat lungs exposed to chronic hypoxia [102].

In humans, no change in endothelin response to maximal exercise in acute hypobaric hypoxia (70 min at 3,000 m) was noted when compared to normoxia [17]. In subjects acutely exposed to 4,559 m, plasma endothelin increased twofold, correlated with pulmonary arterial pressure; nifedipine lowered pulmonary pressure but had no effect on plasma endothelin [103]. The administration of an endothelin receptor blocker (bosentan) reduced the pulmonary artery pressure in subjects acutely exposed to 4,559 m [104]. In 34 subjects exposed to 4,559 m, the increase in pulmonary artery pressure was correlated to the gain

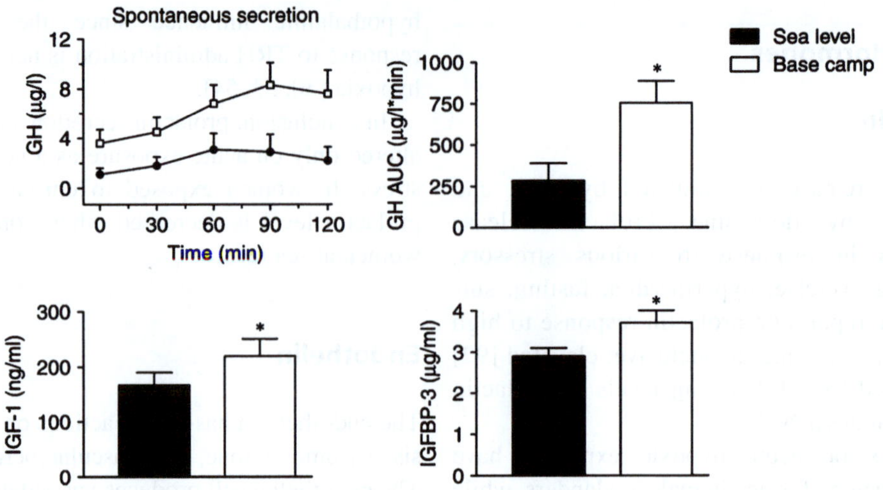

Fig. 12.2 Resting plasma concentrations of GH, IGF-I, and IGFBP-3 in nine male elite climbers at sea level and after high-altitude chronic hypoxia exposure. Mean (±SEM) (*$P<0.05$ vs. sea level). From ref. [18]

in plasma endothelin from venous to arterial blood through the lungs; however, it was not determined if this gain was a cause or a consequence of the increase in pulmonary pressure [105]. The Sajama expedition showed the first data on plasma endothelin during exercise and prolonged hypoxia. Although plasma endothelin may not reflect local concentrations of released endothelin, the observation that plasma endothelin did increase with hypoxic exposure, even at rest, but not with severe exercise suggesting that shear stress is not a major stimulus for endothelin release at high altitude and that metabolic changes in the endothelial cell, induced by hypoxia, may play a central role [23, 104].

Growth Hormone

GH is synthesized and released from the hypophysis under the hypothalamic control of GH-releasing hormone (stimulating) and somatostatin (inhibiting). GH has an important anabolic action on various tissues. The pulsatile pattern of GH secretion complicates ascertaining changes in baseline levels.

In high-altitude residents, several studies have documented elevated levels of GH [9, 106, 107] with maintained circadian rhythm [9]. Sea level residents acutely exposed to high altitude did not show any change in the resting level of GH [106, 107]. However, greatly increased values of GH were found in men after ten weeks of exposure to 6,000 m, possibly due to exertion of the subjects and the prolonged exposure to a stressful and cold environment [10].

During intense exercise in normoxia, GH concentration increases [108]. One of the mediators of GH action, IGF-1 (insulin-like growth factor), also increases during heavy exercise, though the increase is small, brief and shows a GH-independent pattern [109] (Fig. 12.2). The exercise-induced GH increase is strongly potentiated in hypoxia, producing a much earlier and faster rise than in normoxia [8, 18, 97, 106, 110]. However, acute exposure to moderate altitude (2,325 m) blunts the GH response to submaximal physical exercise in untrained individuals [111]. It has been suggested that the influence of hypoxia on GH levels during exercise is mainly due to changes in relative workload, rather than to an effect of hypoxia per se [112].

The GH response to hypothalamic stimulation (by GH-releasing hormone) was potentiated at 2,600 m when compared to sea level [54]. It was found very much variable among eight men exposed to 4,350 m and a blunted response to GHRH was found in some subjects [30]. The concentration of IGF-1 was elevated [54] or unchanged [30] at high altitude while the binding protein IGFBP-3 was unchanged [30, 54]. However, in subjects exposed for 3 weeks at 6,542 m, IGF-1 was normal [70], as well as in marathon runners at 4,000 m [113].

In summary, resting GH is poorly affected by acute hypoxia, whereas the exercise-induced GH response is potentiated by acute hypoxic exposure. With sustained hypoxia and in high-altitude residents, GH concentration is generally increased. This increase could play a role in modifying the metabolism to satisfy increased needs at altitude.

Sex Hormones

The male sex hormones, including testosterone, luteinizing hormone (LH) and follicle-stimulating hormone (FSH), are integrated in a number of regulatory systems with complex feedback mechanisms. Hypothalamic gonadotropin releasing hormone (GnRH) stimulates the hypophysis to secrete LH, an important stimulus for testosterone production, and FSH, influencing testosterone production indirectly by sensitizing LH-receptors. Testosterone increases quickly and transiently during short-term exercise [11, 108, 114, 115]; in contrast, prolonged exercise decreases the testosterone level [11, 108].

Acute exposure to altitudes (from 1,650 to 3,048 m) has been reported to augment resting levels of testosterone within a few days [7, 116], an effect possibly related to increased adrenal activity. Others have recorded a decrease in testosterone at 3,048–4,250 m [92, 117]. An influence from a concomitant high prolactinemia [92], as well as an increased level of estradiol [117], could account for a decrease in testosterone. Consistently, chronic hypoxia does not induce any change in testosterone levels of adult high-altitude residents [118] although an earlier increase in testosterone and onset of puberty has been recorded in young high-altitude males [119]. In elderly high-altitude men, high levels of testosterone were associated with increased hemoglobin and more hypoxemia during sleep, potentially compromising adaptation to high altitude [120]. Conversely, progesterone is likely to increase the HVR in polycythemic high-altitude residents [121]. However, no evidence has been found that progesterone might mediate the increased HVR during pregnancy [122].

During intense anaerobic exercise the LH–testosterone relationship is modified, i.e., testosterone increases without any significant elevation of LH [115]. At high altitude on the other hand (without intense exercise), LH and FSH have a tendency to decrease [7, 92, 117, 123]. Interestingly, as with intense exercise, the LH change did not parallel that of testosterone, perhaps reflecting influences of other mediators such as catecholamines [92].

After 7 weeks at high altitude a testosterone decrease and progesterone increase was found without any change in estradiol levels [18]. When intake was controlled in order to maintain body weight at 4,300 m, testosterone acutely increased at altitude but did not decline over time (3 weeks), as did the control group with deficient food intake [60] (Fig. 12.1c). When lowlanders acclimatized to 3,542 m trekked to an extreme altitude of 5,080 m, plasma testosterone decreased but had returned to SL values when measured after 6 months' continuous stay at 6,300 m. Plasma LH after trekking to 5,080 m was higher than at an altitude of 3,542 m, but decreased to levels found at 3,542 m or SL after prolonged residence at extreme altitude. Plasma progesterone was increased after a 6-month stay at extreme altitude. Prolonged residence at lower as well as at extreme altitude does not appreciably alter blood levels of pituitary, gonadal, or adrenal hormones except for progesterone. This rise in progesterone may be important in increasing the sensitivity of the HVR and activating the hemoglobin synthesis [124].

The male reproductive functions of the members of an expedition (Masherbrum, 7,821 m) were examined via semen analyses and endocrine tests. Sperm counts decreased after 1 and 3 months but had recovered after 2 years in all subjects. An increase in abnormally shaped sperm was also observed after 1 month, but had nearly recovered to the pre-expedition state after 3 months. Endocrine tests revealed slightly decreased testosterone in the blood after 1 and 3 months. The tests were completely normal after 2 years, suggesting that a high-altitude sojourn may induce reversible spermatogenic and Leydig cell dysfunction [125].

Sex hormones in nonpregnant women were mainly studied in relation to their influence on the control of ventilation at high altitude. There is a better blood oxygenation and a lower hemoglobin level before than after menopause in women long residing at altitude, and this is associated with higher levels of progesterone in the luteal phase of the cycle [126]. In a comprehensive study comparing in Peru ten women living at sea level and ten women living at 4,340 m, ovulation after LH peak occurred earlier at high altitude than at sea level [127]. Serum FSH levels were higher at late luteal phase and early follicular phase at high altitude than at sea level. During the early follicular phase serum estradiol levels were significantly higher at high altitude than at sea level. At ovulation, the serum estradiol levels in women at sea level were 55.1 % of the peak, but remained at high levels (80 % of the peak) in women at high altitude. The second increase of serum estradiol occurred earlier at sea level than at high altitude. From days +12 to +15, there was a significant decline in serum estradiol levels in women at sea level but not in those from high altitude. Serum progesterone levels at days +5, and +8 to +12 were significantly higher at sea level than at high altitude [127]. Among women older than 50 years of age, a greater decline in serum concentrations of dehydroepiandrosterone (DHEA) was observed in those living at high altitude than in those living at sea level, suggesting that adrenopause is attained earlier and is of greater magnitude at high altitude than at sea level [128]. The ventilatory response to hypoxia was recently found decreased in a large series of menopausal women compared to non-menopausal women, confirming the probable influence of female sex hormones. Interestingly, when menopausal women maintain a regular physical activity, their decline in ventilatory response to hypoxia at exercise is blunted [129].

In summary, LH and FSH decrease with acute hypoxia, while testosterone response is variably described, increasing at moderate altitudes and decreasing at higher altitudes, while progesterone increases. The hypophyseal axis seems disturbed, as LH and testosterone show different patterns. However, high-altitude natives have normal hormone levels and the modifications are probably only transient, normalizing on chronic exposure. The female hormonal cycle is slightly modified in women residing at high altitude and these changes may have consequences on health status after menopause.

Hormonal Response to Hypothalamic Factors

Hypothalamic factors play a major role in the regulation of all the hypophyseal hormones and a malfunction of this axis could possibly explain the observed altitude-induced alterations [30]. In a recent study, the responses of TSH, thyroid hormones, prolactin, sex hormones, and GH to injection of TRH, GnFlH, and GRH were studied in eight men in normoxia and on acute exposure to an altitude of 4,350 m (Table 12.1). Elevated concentrations of thyroid hormones were found at high altitude, but TSH levels were unchanged in hypoxia both before and alter hormonal injection. Prolactin baseline level decreased at altitude, but the response after hypophyseal stimulation was similar in both environments. The decreasing tendency observed in LH and FSH concentrations at altitude was not significant. GH baseline level was unaltered, but the response after hormone administration varied among subjects in hypoxia. Many factors have been proposed to modify the hormone levels in hypoxia, but most theories are still speculative. We conclude that exposure to acute hypoxia induces various changes in hormonal levels, but the hypophyseal response to hypothalamic factors does not appear to be blunted [30].

Table 12.1 Concentration of hormones at sea level and after 3–4 days at high altitude (4,350 m) in eight male sea level natives (From ref [30])

	Sea level	4,350 m	*P*
Total T4, nmol/mL	85.1 ± 11.8	103.1 ± 18.5	<0.001
Free T4, pmol/L	13.9 ± 2.4	16.8 ± 3.4	<0.001
Total T3, nmol/L	1.72 ± 0.20	1.99 ± 0.29	0.006
Free T3, pmol/L	6.51 ± 0.41	7.54 ± 0.69	<0.001
T4/T3	49.7 ± 5.1	51.9 ± 5.1	0.04
TSH, mU/mL	1.51 ± 0.80	1.37 ± 0.93	ns
Prolactin, ng/mL	14.5 ± 7.0	7.0 ± 4.7	0.006
FSH, m/mL	4.15 ± 2.44	3.44 ± 1.73	0.02
LH, mU/mL	2.79 ± 0.94	2.33 ± 0.99	ns
Norepinephrine, nmol/L	2.98 ± 0.98	4.96 ± 1.13	0.005
Epinephrine, nmol/L	1.22 ± 0.86	1.06 ± 0.51	ns
Dopamine, nmol/L	1.90 ± 1.05	2.22 ± 1.02	ns
Cortisol, ng/mL	94.5 ± 38.9	119.5 ± 49.6	0.07
GH, ng/mL	1.40 ± 1.86	0.73 ± 1.05	ns
IGF-1, µg/L	168 ± 55	184 ± 53	ns
IGF-BP3, mg/L	2.17 ± 0.28	2.16 ± 0.28	ns

Table 12.2 Direction of hormonal changes in response to hypoxia, at rest, with exercise and in the case of AMS or HAPE

	Normal acclimatization rest	Normal acclimatization exercise	AMS or HAPE
Norepinephrine	↗	↗	↗
Epinephrine	→	↗	
Dopamine	↗	↗	
Cortisol	↗	↗	↗↗
Prostaglandins	↗		↗↗
T3, T4	↗	↗	
Parathormone	↗		
Insulin	↗	↗	
Glucagon	→		
Leptin	↗		
Ghrelin	↘		
Prolactin	↘		
Endothelin	↗	↗	
Growth hormone	↗	↗↗	
Testosterone	↗ ↘		
LH, FSH	↘		

These variations may be controversial in some cases but the direction shown is the most frequent found in the literature

Summary and Future Directions

The direction of hormonal changes at rest and exercise in response to acute and chronic hypoxia of high altitude, to the extent it is known, is summarized in Table 12.2. Among future research challenges in studying responses of human hormones to high altitude is to design clinical outcome studies which can distinguish the effect of hypoxia from that of many other stressors altering hormone levels. Such studies might better be done under simulated conditions rather than in the field. The perspectives of fundamental research in the domain of endocrines at high altitude is clearly directed towards the study of receptors, gene expression in hormonal systems and the possible role of hypoxia-inducible factors.

References

1. Jenq W, Rabb H, Wahe M, Ramirez G. Hypoxic effects on the expression of mineralocorticoid and glucocorticoid receptors in human renal cortex epithelial cells. Biochem Biophys Res Commun. 1996;218:444–8.
2. Kacimi R, Moalic JM, Aldashev A, et al. Differential regulation of G protein expression in rat hearts exposed to chronic hypoxia. Am J Physiol. 1995;38:H1865–73.
3. Richalet JP. Oxygen sensors in the organism. Examples of regulation under altitude hypoxia in mammals. Comp Biochem Physiol. 1997;118A:9–14.
4. Ayres PJ, Hurter WG, Williams ES. Aldosterone excretion and potassium retention in subjects living at high altitude. Nature. 1961;191:78–80.
5. Frayser R, Rennie ID, Gray GW, Houston CS. Hormonal and electrolyte response to exposure to 17,500ft. J Appl Physiol. 1975;38:636–42.
6. Heyes MP, Farber MO, Manfredi F, et al. Acute effects of hypoxia on renal and endocrine function in normal humans. Am J Physiol 1982; 243 (Reg Int Comp Physiol 12): R265-70.
7. Humpeler E, Skrabal F, Bartsch G. Influence of exposure to moderate altitude on the plasma concentration of cortisol, aldosterone, renin, testosterone, and gonadotropins. Eur J Appl Physiol. 1980;45:167–76.
8. Sutton JR, Viol GW, Gray GW, et al. Renin, aldosterone, electrolyte, and cortisol responses to hypoxic decompression. J Appl Physiol. 1977;43:421–4.
9. Sawhney RC, Malhotra AS, Singh T. Glucoregulatory hormones in man at high altitude. Eur J Appl Physiol. 1991;62:286–91.
10. Anand IS, Chandrashekhar Y, Rao SK, et al. Body fluid compartments, renal blood flow, and hormones at 6,000m in normal subjects. J Appl Physiol. 1993;74:1234–9.
11. Marinelli M, Roi GS, Giacometti M, et al. Cortisol, testosterone and free testosterone in athletes performing a marathon at 4000m altitude. Horm Res. 1994;41:225–9.
12. Panjwani U, Thakur L, Anand JP, et al. Effect of simulated ascent to 3500 meter on neuro-endocrine functions. Indian J Physiol Pharmacol. 2006; 50:250–6.
13. Larsen JJ, Hansen JM, Olsen NV, et al. The effect of altitude hypoxia on glucose homeostasis in men. J Physiol. 1997;504:241–91.
14. Colice GL, Ramirez G. Effect of hypoxemia on the renin-angiotensin-aldosterone system in humans. J Appl Physiol. 1985;58:724–30.
15. Bouissou P, Fiet J, Guezennec CY, Pesquies PC. Plasma adrenocorticotrophin and cortisol responses to acute hypoxia at rest and during exercise. Eur J Appl Physiol. 1988;57:110–3.
16. Tunny TJ, van Gelder J, Gordon RD, et al. Effects of altitude on atrial natriuretic peptide, the bicentennial Mount Everest Expedition. Clin Exp Pharmacol Physiol. 1989;16:287–91.
17. Vuolteenaho O, Koistinen P, Martikkala V, et al. Effect of physical exercise in hypobaric conditions on atrial natriuretic peptide secretion. Am. J. Physiol. 1992; 263 (Reg. Int. Comp. Physiol. 32): 647–52.
18. Benso A, Broglio F, Aimaretti G, et al. Endocrine and metabolic responses to extreme altitude and physical exercise in climbers. Eur J Endocrinol. 2007;157:733–40.
19. Imoberdorf R, Garlick PJ, McNurlan MA, et al. Skeletal muscle protein synthesis after active or passive ascent to high altitude. Med Sci Sports Exerc. 2006;38:1082–7.
20. Bärtsch P, Shaw S, Francioli M, et al. Atrial natriuretic peptide in acute mountain sickness. J Appl Physiol. 1988;65:1929–37.
21. Richalet JP, Rutgers V, Bouchet P, et al. Diurnal variations of acute mountain sickness, colour vision, and plasma cortisol and ACTH at high altitude. Aviat Space Environ Med. 1989;60:105–11.
22. Richalet JP, Antezana AM, Bienvenu A, et al. Physiological factors in survival at extreme altitude. In: Sutton JR, Houston CS, Coates G, editors. Hypoxia and molecular medicine. Burlington: Quenn City Printers; 1993. p. 235–51.
23. Richalet JP, Déchaux M, Bienvenu A, et al. Erythropoiesis and renal function at the altitude of 6,542 m. Jpn J Mount Med. 1995;15:135–50.
24. Raff H, Tzankoff SP, Fitzgerald RS. ACTH and cortisol responses to hypoxia in dogs. J Appl Physiol. 1981;51:1257–60.
25. Honig A. Peripheral arterial chemoreceptors and reflex control of sodium and water homeostasis. Am J Physiol. 1989;257:R1282–302.
26. Richalet JP, Mehdioui H, Rathat C, et al. Acute hypoxia decreases cardiac response to catecholamines in exercising humans. Int J Sports Med. 1988;9:157–62.
27. Antezana AM, Kacimi R, Le Trong JL, et al. Adrenergic status of humans during prolonged exposure to the altitude of 6542 m. J Appl Physiol. 1994;76:1055–9.
28. Mazzeo RS, Wolfel EE, Butterfield GE, Reeves JT. Sympathetic response during 21 days at high altitude (4,300 m) as determined by urinary and arterial catecholamines. Metabolism. 1994;43:1226–32.

29. Bestle MH, Olsen NV, Poulsen TD, et al. Prolonged hypobaric hypoxemia attenuates vasopressin secretion and renal response to osmostimulation in men. J Appl Physiol. 2002;92:1911–22.
30. Richalet JP, Letournel M, Souberbielle JC. Effects of high altitude hypoxia on the hormonal response to hypothalamic factors. Am J Physiol Regul Integr Comp Physiol. 2010;299:R1685–92.
31. Fischetti F, Fabris B, Zaccaria M, et al. Effects of prolonged high-altitude exposure on peripheral adrenergic receptors in young healthy volunteers. Eur J Appl Physiol. 2000;82:439–45.
32. Olsen NV, Hansen JM, Kanstrup IL, et al. Renal hemodynamics, tubular function, and the response to low-dose dopamine during acute hypoxia in humans. J Appl Physiol. 1993;74:2166–73.
33. Calbet JA. Chronic hypoxia increases blood pressure and noradrenaline spillover in healthy humans. J Physiol. 2003;551:379–86.
34. Antezana AM, Richalet JP, Noriega I, et al. Hormonal changes in normal and polycythemic high altitude natives. J Appl Physiol. 1995;79:795–800.
35. Richalet JP. The heart and adrenergic system in hypoxia. In: Sutton JR, Coates G, Remmers JE, editors. Hypoxia. The adaptations. Toronto: B.C. Dekker; 1990. p. 231–40.
36. Favret F, Richalet JP. Exercise and hypoxia, the role of the autonomic nervous system. Respir Physiol Neurobiol. 2007;158:280–6.
37. Lundby C, Araoz M, van Hall G. Peak heart rate decreases with increasing severity of acute hypoxia. High Alt Med Biol. 2001;2:369–76.
38. Richalet JP, Le Trong JL, Rathat C, et al. Reversal of hypoxia-induced decrease in human cardiac response to isoproterenol infusion. J Appl Physiol. 1989;67:523–7.
39. León-Velarde F, Richalet JP, Chavez JC, et al. Hypoxia- and normoxia-induced reversibility of autonomic control in Andean guinea pig heart. J Appl Physiol. 1996;81:2229–34.
40. Kacimi R, Richalet JP, Corsin A, et al. Hypoxia-induced downregulation of b-adrenergic receptors in rat heart. J Appl Physiol. 1992;73:1377–82.
41. Voelkel NF, Hegstrand L, Reeves JT, et al. Effects of hypoxia on density of β-adrenergic receptors. J Appl Physiol. 1981;50:363–6.
42. Kacimi R, Richalet JP, Crozatier B. Hypoxia-induced differential modulation of adenosinergic and muscarinic receptors in rat heart. J Appl Physiol. 1993;75:1123–8.
43. León-Velarde F, Richalet JP, Molinatti G, Crozatier B. Up-regulation of cardiac a-adrenergic receptors (a-AR) in rats exposed to prolonged hypoxia. Tenth hypoxia symposium. Lake Louise, Canada, Feb 18–22, 1997.
44. Bao X, Kennedy BP, Hopkins SR, et al. Human autonomic activity and its response to acute oxygen supplement after high altitude acclimatization. Auton Neurosci. 2002;102:54–9.
45. Richalet JP, Kacimi R, Antezana AM. The control of chronotropic function in hypobaric hypoxia. Int J Sports Med. 1992;13:S22–4.
46. de Glisezinski I, Crampes F, Harant I, et al. Decrease of subcutaneous adipose tissue lipolysis after exposure to hypoxia during a simulated ascent of Mt Everest. Pflugers Arch. 1999;439:134–40.
47. Surks MI. Elevated PBI, free thyroxine and plasma protein concentration in man at high altitude. J Appl Physiol. 1966;21:1185–90.
48. Rawal SB, Singh MV, Tyagi AK, Chaudhuri BN. Thyroidal handling of radioiodine in sea level residents exposed to hypobaric hypoxia. Eur J Nucl Med. 1993;20:16–9.
49. Basu M, Pal K, Malhotra AS, et al. Free and total thyroid hormones in humans at extreme altitude. Int J Biometeorol. 1995;39:17–21.
50. Chakraborty S, Samaddar J, Batabyal SK. Thyroid status of humans at high altitude. Clin Chim Acta. 1987;166:111–3.
51. Férézou J, Richalet JP, Sérougue C, et al. Reduction of postprondial lipemia after acute exposure to high altitude hypoxia. Int J Sports Med. 1993;14:78–85.
52. Kotchen TA, Mougey EH, Hogan RP, et al. Thyroid responses to simulated altitude. J Appl Physiol. 1973;34:165–8.
53. Mordes JP, Blume FD, Boyer S, et al. High-altitude pituitary-thyroid dysfunction on Mount Everest. N Engl J Med. 1983;308:1135–8.
54. Ramirez G, Herrera R, Pineda D, et al. The effects of high altitude on hypothalamic-pituitary secretory dynamics in men. Clin Endocrinol (Oxf). 1995;43:11–8.
55. Sawhney RC, Malhotra AS. Thyroid function in sojourners and acclimatised low landers at high altitude in man. Horm Metab Res. 1991;23:81–4.
56. Surks MI, Beckwitt HJ, Chidsey CA. Changes in plasma thyroxine concentration and metabolism, catecholamine excretion and basal oxygen consumption in man during acute exposure to high altitude. J Clin Endocrinol Metab. 1967;27:789–99.
57. Wright AD. Birmingham Medical Research Expeditionary Society 1977 Expedition, Thyroid function and acute mountain sickness. Postgrad Med J. 1979;55:483–6.
58. Hackney AC, Feith S. Pozos, Seale J. Effects of altitude and cold exposure on resting thyroid hormone concentrations. Aviat Space Environ Med. 1995;66:325–9.
59. Savourney G, Caravel JP, Barnavol B, Bittel JHM. Thyroid hormone changes in a cold air environment after local cold acclimation. J Appl Physiol. 1994;76:1963–7.
60. Barnholt KE, Hoffman AR, Rock PB, et al. Endocrine responses to acute and chronic high-altitude exposure

(4,300 meters), modulating effects of caloric restriction. Am J Physiol Endocrinol Metab. 2006;290:E1078–88.
61. Savourey G, Launay JC, Besnard Y, et al. Control of erythropoiesis after high altitude acclimatization. Eur J Appl Physiol. 2004;93:47–56.
62. Bhargava M, Runyon MR, Smirnov D, et al. Triiodo-L-thyronine rapidly stimulates alveolar fluid clearance in normal and hyperoxia-injured lungs. Am J Respir Crit Care Med. 2008;178:506–12.
63. Krapf R, Jaeger P, Hulter HN, et al. Chronic respiratory alkalosis induces renal PTH-resistance, hyperphosphatemia and hypocalcemia in humans. Kidney Int. 1992;42:727–34.
64. Lopez I, Rodriguez M, Felsenfeld AJ, et al. Direct suppressive effect of acute metabolic and respiratory alkalosis on parathyroid hormone secretion in the dog. J Bone Miner Res. 2003;18:1478–85.
65. Souberbielle JC, Richalet JP, Garabedian M, et al. Effect of high altitude hypoxia on calcium metabolism and bone markers. XII Int. Conf. on calcium regulating hormones. Bone. 1995;16:210S.
66. Sawhney RC, Malhotra AS, Singh T, et al. Insulin secretion at high altitude in man. Int J Biometeorol. 1986;30:231–8.
67. Brooks GA, Butterfield GE, Wolfe RR, et al. Increased dependence on blood glucose after acclimatization to 4,300 m. J Appl Physiol. 1991; 70:919–27.
68. Roberts AC, Reeves JT, Butterfield GE, et al. Altitude and β-blockade augment glucose utilization during submaximal exercise. J Appl Physiol. 1996; 80:605–15.
69. Young PM, Rose MS, Sutton JR, et al. Operation Everest II, plasma lipid and hormonal responses during a simulated ascent of Mont Everest. J Appl Physiol. 1989;66:1430–5.
70. Richalet JP, Souberbielle JC, Antezana AM, et al. Control of erythropoiesis in humans during prolonged exposure to the altitude of 6542 m. Am J Physiol 1994;266 (Reg Int Comp Phys 35) R756–64.
71. Kullmer T, Gabriel H, Jungmann E, et al. Increase of serum insulin and stable C-peptide concentrations with exhaustive incremental graded exercise during acute hypoxia in sedentary subjects. Exp Clin Endocrinol. 1995;103:156–61.
72. Sutton JR. Effects of acute hypoxia on the hormonal response to exercise. J Appl Physiol. 1977; 42:587–92.
73. Baum D. Stress hyperglycemia and the adrenergic regulation of pancreatic hormones in hypoxia. Metabolism. 1980;29:1176–85.
74. Lee WC, Chen SM, Wu MC, et al. The role of dehydroepiandrosterone levels on physiologic acclimatization to chronic mountaineering activity. High Alt Med Biol. 2006;7:228–36.
75. Tschöp M, Strasburger CJ, Hartmann G, et al. Raised leptin concentrations at high altitude associated with loss of appetite. Lancet. 1998;352:1119–20.
76. Shukla V, Singh SN, Vats P, et al. Ghrelin and leptin levels of sojourners and acclimatized lowlanders at high altitude. Nutr Neurosci. 2005;8:161–5.
77. Vats P, Singh VK, Singh SN, Singh SB. High altitude induced anorexia, effect of changes in leptin and oxidative stress levels. Nutr Neurosci. 2007;10:243–9.
78. Zaccaria M, Ermolao A, Bonvicini P, et al. Decreased serum leptin levels during prolonged high altitude exposure. Eur J Appl Physiol. 2004;92:249–53.
79. Morel OE, Aubert R, Richalet JP, Chapelot D. Simulated high altitude selectively decreases protein intake and lean mass gain in rats. Physiol Behav. 2005;86:145–53.
80. Sierra-Johnson J, Romero-Corral A, Somers VK, Johnson BD. Effect of altitude on leptin levels, does it go up or down? J Appl Physiol. 2008;105:1684–5.
81. Ye J. Regulation of leptin by hypoxia. J Appl Physiol. 2008;105:1687–90.
82. Snyder EM, Carr RD, Deacon CF, Johnson BD. Overnight hypoxic exposure and glucagon-like peptide-1 and leptin levels in humans. Appl Physiol Nutr Metab. 2008;33:929–35.
83. Yingzhong Y, Droma Y, Rili G, Kubo K. Regulation of body weight by leptin, with special reference to hypoxia-induced regulation. Intern Med. 2006;45:941–6.
84. Santos JL, Perez-Bravo F, Albala C, et al. Plasma leptin and insulin levels in Aymara natives from Chile. Ann Hum Biol. 2000;27:271–9.
85. Cabrera de León A, González DA, Méndez LI, et al. Leptin and altitude in the cardiovascular diseases. Obes Res 2004;12: 1492–8.
86. Simler N, Grosfeld A, Peinnequin A, et al. Leptin receptor-deficient obese Zucker rats reduce their food intake in response to hypobaric hypoxia. Am J Physiol Endocrinol Metab. 2006;290:E591–7.
87. Cummings DE, Purnell JQ, Frayo RS, et al. A preprandial rise in plasma ghrelin levels suggests a role in meal initiation in humans. Diabetes. 2001; 50:1714–9.
88. Cummings DE, Frayo RS, Marmonier C, et al. Plasma ghrelin levels and hunger scores in humans initiating meals voluntarily without time- and food-related cues. Am J Physiol Endocrinol Metab. 2004;287:E297–304.
89. Drazen DL, Vahl TP, D'Alessio DA, et al. Effects of a fixed meal pattern on ghrelin secretion, evidence for a learned response independent of nutrient status. Endocrinology. 2006;147:23–30.
90. Chaiban JT, Bitar FF, Azar ST. Effect of chronic hypoxia on leptin, insulin, adiponectin, and ghrelin. Metabolism. 2008;57:1019–22.
91. Schwenke DO, Tokudome T, Shirai M, et al. Exogenous ghrelin attenuates the progression of chronic hypoxia-induced pulmonary hypertension in conscious rats. Endocrinology. 2008;149:237–44.
92. Sawhney RC, Chhabra PC, Malhotra AS, et al. Hormone profiles at high altitude in man. Andrologia. 1985;17:178–84.

93. Knudtzon J, Bogsnes A, Norman N. Changes in prolactin and growth hormone levels during hypoxia and exercise. Horm Metab Res. 1989;21:453–4.
94. Gonzales GF, Carrillo CE. Low serum prolactin levels in native women at high altitude. Int J Gynecol Obstet. 1993;43:169–75.
95. Brisson GR, Boisvert P, Péronnet F, et al. Face cooling-induced reduction of plasma prolactin response to exercise as part of an integrated response to thermal stress. Eur J Appl Physiol. 1989; 58:816–20.
96. Bouissou P, Brisson GR, Péronnet F, et al. Inhibition of exercise-induced blood prolactin response by acute hypoxia. Can J Sport Sci. 1987;12:I49–50.
97. Strüder HK, Hollmann W, Platen P. Increased prolactin response to hyperoxia at rest and during endurance exercise. Int J Sports Med. 1996; 17:390–2.
98. Gardner DG, Newman ED, Nakamura KK, Nguyen PT. Endothelin increases the synthesis and secretion of atrial natriuretic peptide in neonatal rat cardiocytes. Am J Physiol. 1991;261:E177–82.
99. Moe O, Tejedor A, Campbell WB, et al. Effects of endothelin on in vitro renin secretion. Am J Physiol. 1991;260:E521–5.
100. Rakugi H, Tabuchi Y, Nakamaru M, et al. Evidence for Endothelin-1 release from resistance vessels of rats in response to hypoxia. Biochem Biophys Res Commun. 1990;169:973–7.
101. Kourembanas S, Marsden PA, McQuilan LP, Faller DV. Hypoxia induces endothelin gene expression and secretion in cultured human endothelium. J Clin Invest. 1991;88:1054–7.
102. Li H, Chen SJ, Chen YF, et al. Enhanced endothelin-1 and endothelin receptor gene expression in chronic hypoxia. J Appl Physiol. 1995;77:1451–9.
103. Goerre S, Wenk M, Bärtsch P, et al. Endothelin-1 in pulmonary hypertension associated with high-altitude exposure. Circulation. 1995;91:359–64.
104. Modesti PA, Vanni S, Morabito M, et al. Role of endothelin-1 in exposure to high altitude, Acute Mountain Sickness and Endothelin-1 (ACME-1) study. Circulation. 2006;114:1410–6.
105. Berger MM, Dehnert C, Bailey DM, et al. Transpulmonary plasma ET-1 and nitrite differences in high altitude pulmonary hypertension. High Alt Med Biol. 2009;10:17–24.
106. Raynaud J, Drouet L, Martineaud JP, et al. Time course of plasma growth hormone during exercise in humans at altitude. J Appl Physiol. 1981; 50:229–33.
107. Sawhney RC, Malhotra AS. Circadian rhythmicity of growth hormone at high altitude in man. Indian J Physiol Pharmacol. 1991;35:55–7.
108. Viru A. Plasma hormones and physical exercise. A review. Int J Sports Med. 1992;13:201–9.
109. Cappon J, Brasel JA, Mohan S, Cooper DM. Effects of brief exercise on circulating insulin-like growth factor I. J Appl Physiol. 1994;76:2490–6.
110. Schmidt W, Doré S, Hilgendorf A, et al. Effects of exercise during normoxia and hypoxia on the growth hormone-insulin-like growth factor I axis. Eur J Appl Physiol. 1995;71:424–30.
111. Gutiérrez A, Gonzalez-Gross M, Ruiz JR, et al. Acute exposure to moderate high altitude decreases growth hormone response to physical exercise in untrained subjects. J Sports Med Phys Fitness. 2003;43:554–8.
112. Kjær M, Banhsbo J, Lortie G, Galbo H. Hormonal response to exercise in humans, influence of hypoxia and physical training. Am J Physiol. 1988; 254:R197–203.
113. Banfi G, Marinelli M, Roi GS, et al. Growth hormone and insulin-like growth factor I in athletes performing a marathon at 4000 m of altitude. Growth Regul. 1994;4:82–6.
114. Bouissou P, Péronnet F, Brisson G, et al. Metabolic and endocrine responses to graded exercise under acute hypoxia. Eur J Appl Physiol. 1986;55:290–4.
115. Hackney AC, Premo MC, McMurray RG. Influence of aerobic versus anaerobic exercise on the relationship between reproductive hormones in men. J Sports Sci. 1995;13:305–11.
116. Vasankari TJ, Rusko H, Kujala UM, Huhtaniemi IT. The effects of ski training at altitude and racing on pituitary, adrenal and testicular function in men. Eur J Appl Physiol. 1993;66:221–5.
117. Friedl KE, Plymate SR, Bernhard WN, Mohr LC. Elevation of plasma estradiol in healthy men during a mountaineering expedition. Horm Metab Res. 1988;20:239–42.
118. Garmendia F, Valdivia H, Castillo O, et al. Hypothalamo-hypophyso-gonadal response to clomiphene citrate at median high altitude. Horm Metab Res. 1982;14:679–80.
119. Fellmann N, Bedu M, Spielvogel H, et al. Anaerobic metabolism during pubertal development at high altitude. J Appl Physiol. 1988;64:1382–6.
120. Beall CM, Worthman CM, Stallings J, et al. Salivary testosterone concentration of Aymara men native to 3600m. Ann Hum Biol. 1992;19:67–78.
121. Kryger M, Glas R, Jackson D, et al. Impaired oxygenation during sleep in excessive polycythemia of high altitude, improvement with respiratory stimulation. Sleep. 1978;1:3–17.
122. Moore LG, McCullough RE, Weil JV. Increased HVR in pregnancy, relationship to hormonal and metabolic changes. J Appl Physiol. 1987;62:158–63.
123. Bangham CRM, Hackett PH. Effects of high altitude on endocrine function in the sherpas of Nepal. J Endocrinol. 1978;79:147–8.
124. Basu M, Pal K, Prasad R, et al. Pituitary, gonadal and adrenal hormones after prolonged residence at extreme altitude in man. Int J Androl. 1997; 20:153–8.
125. Okumura A, Fuse H, Kawauchi Y, et al. Changes in male reproductive function after high altitude mountaineering. High Alt Med Biol. 2003;4:349–53.

126. León-Velarde F, Rivera-Chira M, et al. Relationship of ovarian hormones to hypoxemia in women residents of 4,300 m. Am J Physiol Regul Integr Comp Physiol. 2001;280:R488–93.
127. Escudero F, Gonzales GF, Góñez C. Hormone profile during the menstrual cycle at high altitude. Int J Gynaecol Obstet. 1996;55:49–58.
128. Gonzales GF, Góñez C, Villena A. Adrenopause or decline of serum adrenal androgens with age in women living at sea level or at high altitude. J Endocrinol. 2002;173:95–101.
129. Lhuissier F, Canouï-Poitrine F, Richalet JP. Ageing and cardiorespiratory response to hypoxia. J Physiol (London). 2012;590:5461–74.

Gastrointestinal Function 13

Noor Hamad and Simon Travis

Abstract

Climbers lose weight above 5,000 m, which impairs physical performance and reduces safety margins. Although widely assumed to be due to energy imbalance, with expenditure exceeding nutritional intake, weight loss has been observed in mountaineers at rest at high altitude. Basal metabolic rate is increased and some evidence points to carbohydrate malabsorption. This chapter examines how the normal physiological processes of carbohydrate, fat, and protein absorption change at altitudes above 5,000 m, in a standard format briefly describing normal physiology, experimental models, and then field studies at altitude. Other aspects of gut function, from gastric acid secretion to mucosal morphology, mesenteric blood flow, motility, liver function, and the effect of hypoxia inducible factor on gut function are then described before gastrointestinal diseases in short-term visitors and residents at high altitude are addressed.

Abbreviations

3mG	3-*O*-methyl-D-glucose
AUGIB	Acute upper gastrointestinal bleeding
CHO	Carbohydrate
COPD	Chronic obstructive pulmonary disease
CRP	C-reactive protein
CYP	Cytochrome P-450 enzyme (variants)
DcytB	Duodenal cytochrome b
DMT1	Divalent metal transporter 1
FA	Fatty acid
GLUT2	Glucose transporter 2
GLUT5	Glucose transporter 5
HIF	Hypoxia inducible factor
HPV	Hepatic portal vein
L/R ratio	Lactulose–rhamnose ratio for intestinal permeability
LPS	Lipopolysaccharide
MG	Monoglyceride(s)
MJ	Megajoules
PGA/PGC	Pepsinogen A/pepsinogen C ratio

N. Hamad, B.M. B.Ch. (Oxon) M.A. (Cantab)
Perranporth Surgery, Perranporth, Cornwall
TR6 0PS, UK

S. Travis, D.Phil. F.R.C.P. (✉)
Translational Gastroenterology Unit, John Radcliffe Hospital, Oxford, Oxon OX3 9DU, UK
e-mail: simon.travis@ndm.ox.ac.uk

E.R. Swenson and P. Bärtsch (eds.), *High Altitude: Human Adaptation to Hypoxia*,
DOI 10.1007/978-1-4614-8772-2_13, © Springer Science+Business Media New York 2014

SGLT1	Sodium-dependent glucose transporter 1
SMA	Superior mesenteric artery
TG	Triglyceride(s)
TNF	Tumor necrosis factor α

Introduction

Weight loss has long been documented at high altitude and this potentially impairs performance. A variety of potential causes have been proposed (Table 13.1), some supported by experimental evidence, although others remain speculative [1]. In order for weight and body composition to remain stable over a period of time, energy intake must equal energy expenditure. Energy intake is controlled by a feedback loop involving a "set-point" that maintains stability of total body energy stores. The hypothalamus is a key modulator. Signals that reflect nutritional state are detected by hypothalamic neurons and consequently food intake and energy expenditure are modified via endocrine, behavioral, and autonomic effector mechanisms [2]. The brainstem, cerebral cortex, and olfactory areas are involved in the central control. However, environmental, cultural, and genetic factors also influence energy balance. Under hypobaric hypoxia several factors in the feedback loop are disturbed, resulting in weight loss.

Some studies have demonstrated continuing weight loss at altitudes above 5,000 m, despite an energy intake that would be adequate at sea level and reduced energy expenditure during study at altitude. The possibilities of malabsorption or malutilization of nutrients as a cause of weight loss have been explored [3, 4]. Work on gut function at high altitude best uses currently accepted definitions (Table 13.2), although terms may be used loosely in older literature. Most evidence for a change in gut function has only been detectable above 5,000 m. Different cohorts of people will therefore be exposed. Residents generally live below 3,500 m, while trekkers generally limit themselves to <5,000 m, while only mountaineers tend to exceed 5,000 m, and are therefore exposed to altitude-induced changes in gut function.

Table 13.1 Possible causes of weight loss at high altitude (>5,000 m)

Potential causes of altitude-induced weight loss
1. Decreased food intake, due to loss of appetite
2. Imbalance of energy intake and expenditure, due to increased basal metabolic rate and high activity levels that are not matched by food intake
3. Impaired intestinal function
4. Changes in endocrine parameters controlling homeostasis
5. Decrease in body water, through increased ventilation, decreased water intake, or altered metabolism

Table 13.2 Definitions of altitude for gastrointestinal function studies [5]

Definitions of altitude	
• Near sea level	<500 m
• Intermediate altitude	500–2,000 m
• Moderate altitude	2,000–3,000 m
• High altitude	3,000–5,500 m
• Extreme altitude	>5,500 m

The practical importance of the topic relates to high altitude deterioration, where weight loss has the potential to alter thermoregulation, muscle function, and hence safety margins at extreme altitude. Changes in gut physiology also have the potential to influence nutritional requirements and choice of food for expeditions [6] (see Chap. 15) and the risk of gastrointestinal illness, such as diarrhea or intestinal bleeding. This chapter will review the effects of hypoxia on gastrointestinal physiology, looking individually at the absorption and transport of carbohydrates, protein, and fat, before considering the clinical effects of high altitude on human gastrointestinal disease.

Effect of Hypoxia on Carbohydrate Absorption

Normal CHO Digestion and Absorption

Carbohydrates are ingested as a mixture of complex carbohydrates (starch or polysaccharides) and simple sugars (monosaccharides such as glucose, fructose, and galactose, or disaccharides such as sucrose and lactose). Food also contains non-starch polysaccharides (better known as fiber), which are indigestible by the human gut although subject to fermentation by colonic bacteria to release methane or hydrogen. Carbohydrates that are undigested or escape absorption in the small intestine are also utilized by colonic bacteria to produce short chain fatty acids (FAs; such as butyrate, the principal fuel for the colonic epithelium, acetate, and propionate), in a process sometimes termed "colonic carbohydrate salvage" [7].

The normal process of digestion of carbohydrate starts with the hydrolysis of starch, initially by salivary amylase and then, principally, by pancreatic amylase. This converts starch to maltose, maltotriose, and α-dextrins, although some glucose is also produced. These products of α-amylase digestion are further hydrolyzed into their component monosaccharides by enzymes (disaccharidases) expressed on the brush border of the small intestinal epithelium, the most important of which are maltase, sucrase, isomaltase, and lactase. Intestinal brush border disaccharidases tend to be inducible. For example, a high sucrose intake induces intestinal sucrase activity, resulting in an increased rate of glucose absorption which in turn increases the secretion of insulin and gastric inhibitory polypeptide, a paracrine hormone that influences gastric emptying. Lactase, and possibly other disaccharidases, is sensitive to hypoxia inducible factor (HIF, below). Lack of brush border disaccharidases results in an inability to absorb specific carbohydrates. Lactase deficiency is common in non-Caucasian populations. After digestion of complex carbohydrates, monosaccharides are then absorbed across the intestinal epithelium [8].

Carbohydrate absorption involves mediated transport across the gastrointestinal epithelium (in other words, involving a specific membrane transport protein), which may involve active transport, or passive, facilitated diffusion. Passive, unmediated diffusion is unimportant in terms of carbohydrate absorption. Nevertheless, this process provides a useful measure of intestinal permeability when assessed by non-metabolizable carbohydrates. Only D-glucose and D-galactose are actively absorbed in the human small intestine. The sodium-dependent glucose transporter, SGLT1, is responsible for the active transport of glucose or galactose with an equimolar amount of sodium against a concentration gradient into the cytoplasm of the enterocyte, utilizing the favorable gradient established by active extrusion of sodium by Na+/K+ ATPase. Glucose exits the enterocyte into the intracellular space by the glucose transporter 2 (GLUT2) [9]. GLUT2 expression appears sensitive to HIF (see below). D-Fructose is not actively absorbed, but has a rate of diffusion greater than would be expected by passive diffusion. Fructose is taken up by facilitated transport by the glucose transporter 5 (GLUT5) and exits through the same pathway as glucose (GLUT2).

The molecular mechanisms are still incompletely understood. Disaccharidase activity (lactase, at least) is HIF-responsive, which may protect carbohydrate absorption at moderate altitudes (below). In vitro studies on epithelial cell membrane transporters (specifically the organic cation transporter, OCTN2) have shown impaired function, upregulation of gene transcription, but no increase in transporter protein, in response to cell culture under profound hypoxia (2 % O_2) [10]. The same is likely to apply to other membrane transporters, although such profound hypoxia is not physiologically relevant. In vivo studies in rats have shown that either acute or chronic hypoxia elicit changes in transcript abundance for genes involved in carbohydrate metabolism and glycolysis, as well as lipid metabolism and protein amino acid phosphorylation [11].

Any of these processes might conceivably be affected by hypobaric hypoxia, but only some have been studied in any detail. It is easier to investigate carbohydrate digestion and absorption than other macronutrients, especially in the field, so publications are biased in their favor. This is because non-metabolizable sugars can be used to investigate carbohydrate transport, urine samples are readily collected (and easier to preserve than blood samples), and there is no need for radioisotopes. Studies on the effect of hypoxia on carbohydrate absorption have been performed in patients with chronic hypoxia at sea level, in the field at high altitude and in hypobaric chambers.

CHO Absorption During Hypoxia at Sea Level

During early investigations into the effect of hypoxia on gut function, it was conceived that part of the malnutrition associated with hypoxic states (chronic pulmonary disease) might be related to malabsorption. Measuring intestinal functional capacity has never been part of routine clinical practice in the way that measurement of creatinine clearance has been in renal medicine. Nevertheless, single sugar tests of intestinal absorption (using xylose) were developed. In 1972, Milledge studied patients with chronic hypoxia at sea level due to either chronic obstructive pulmonary disease or congenital heart disease [12]. Decreasing arterial oxygen saturations (below 70 %) corresponded to a linear decrease in xylose absorption. Hypoxia was subsequently relieved by corrective cardiac surgery or at least 13 h of supplementary oxygen. All cases demonstrated an improvement in xylose absorption. Hypoxia has also been shown to suppress amylase production by rat parotid glands [13] and may conceivably have a similar effect on pancreatic amylase. In a study of 26 patients with chronic obstructive pulmonary disease (COPD) and oxygen saturations above 78 %, no malabsorption of D-xylose was detectable [14], in contrast to the study in patients with an SaO_2 <70 % [12]. This suggests that carbohydrate absorption is preserved until hypoxia is severe.

The trouble with single sugar tests is that absorption of any non-metabolizable sugar is contingent on so many factors within and between individuals that there is wide variation in normal people. The concept of sugar absorption testing is simple enough (take a non-metabolizable sugar absorbed by the same process as glucose and measure excretion in the urine or plasma as a function of absorptive capacity), but the test–re-test validity is low. This reduces the sensitivity. Such factors include variation in gastric emptying, the dilutional effect of intestinal secretion, mucosal surface area, osmolality of the test solution, intestinal transit, renal function, and the completeness of urine collection [15]. This is the main reason that xylose absorption is no longer used in clinical practice. To circumvent this problem, the technique of dual sugar absorption was developed, whereby two sugars subject to the same influences bar the one under study (the process of absorption), are ingested at the same time, and the results expressed as a ratio of the absorption of one sugar to another, to give a measure of intestinal absorptive capacity or intestinal permeability [15, 16].

Field Studies on CHO Absorption

It was an observation by Pugh on the Silver Hut expedition in 1960–1961, who noted that stools became greasy and that people lost weight at extreme altitude, although this returned to normal during spells at base camp [17]. This lead to the idea that malabsorption might be a feature of exposure to extreme altitude and contribute to the weight loss. Field studies have demonstrated that carbohydrate absorption is only affected at very high or extreme altitude (>5,000 m), when hypoxia becomes severe. Preliminary studies, using single sugar (xylose) absorption in 11 subjects studied at 4,846 m, showed xylose levels within the normal range (0.6–1.3 mmol/L) [18]. These levels were unchanged from those observed at 3,100 m. In those who ascended to 5,600 m, however, the 60-min plasma xylose concentrations were reduced to 0.1–0.5 mmol/L, which could indicate a critical threshold above which significant malabsorption may occur. Furthermore, in 1984 Boyer and Blume found a mean 24 % reduction in xylose absorption in six out of seven subjects at 6,300 m [19].

The more accurate dual (or combined) sugar test was used in 1994 to measure intestinal carbohydrate absorptive capacity and intestinal permeability in 14 volunteers [20] at an altitude of 5,730 m. Subjects drank a solution of lactulose, L-rhamnose, D-xylose, and 3-*O*-methyl-D-glucose (3mG). Absorption of carbohydrates by mediated transport in the small intestine was measured by the ratio of urinary xylose to 3-*O*-methyl-D-glucose (3mG) excretion. Intestinal permeability (a measure of barrier function) was measured as the ratio of lactulose to L-rhamnose excretion (L/R ratio) at sea level, 1,050 m in Nepal, and at

5,730 m. Body weight decreased by 5.7 ± 1.2 kg from that at sea level, even though there was no change in energy intake. The xylose/3mG ratio decreased by 22 %, with a 34 % decrease in mediated absorption of xylose and 15 % decrease in 3-*O*-methyl-D-glucose compared to sea level. Intestinal permeability increased from an L/R ratio of 0.036 ± 0.014 at sea level to 0.084 ± 0.042 at 1,050 m ($p = 0.006$), possibly due to infective enteropathy after arrival in Nepal, but reverted to normal (0.045 ± 0.013, $p = 0.062$) at 5,730 m. Absorption of all carbohydrates returned to normal after return to the UK. It confirms a decrease in mediated (D-xylose or 3mG) and diffusional (L-rhamnose) monosaccharide absorption occurs at high altitude, but shows that intestinal permeability (gut "porosity") at 5,730 m is unchanged. It indicates that there are changes in epithelial membrane function, possibly due to a change in membrane fluidity [13].

The change in carbohydrate absorptive capacity is notable. This could represent a real decrease in nutrient carbohydrate absorption, since glucose and galactose are absorbed in a similar manner to the test monosaccharides, but since energy balance was not measured in this study, this is not necessarily the case [1]. Consequently a further study examining energy intake, urine, and fecal energy wastage was conducted by the original investigators [21]. "Wastage" in this context means the energy content of urine and stool collections measured over 3 days at 5,650 m eating a standard diet compared to the values at sea level in the same subjects on the same diet. In this study, 19 subjects ingested identical rations at sea level and at 5,650 m. In addition to energy wastage, carbohydrate absorptive capacity (xylose/3mG ratio), intestinal permeability (L/R ratio), disaccharidase activity (lactase, sucrase, and isomaltase, measured as a ratio of lactose, sucrose, or palatinose permeation to that of lactulose, with a normal ratio of <0.3 and 1.0 representing absent enzyme activity), and food preference studies were performed. The results showed that fecal and urinary energy wastage was no different at high altitude than at sea level (median wastage 0.83 (range 0.28–1.05) MJ/day at sea level and 0.77 (0.40–1.13) MJ/day at 5,650 m, $p = 0.78$, where 1 MJ = 239 kcal; http://www.unitconversion.org/energy/megajoules-to-kilocalories-itconversion.html). It confirmed a decrease in carbohydrate absorptive capacity (xylose/3mG ratio) and no change in intestinal permeability. But interestingly, it demonstrated a reduction in disaccharidase activity by 41 % for sucrase ($p = 0.09$), 50 % for isomaltase ($p = 0.02$), and 62 % for lactase ($p < 0.01$) [21, 22]. There was a corresponding significant, although small (2 %), increase in carbohydrate intake at altitude [6]. Studies in rats exposed to a simulated altitude equivalent to 6,960 m for 18 days after 5 days treadmill training showed that hypobaric hypoxia resulted in a significant decrease in food intake, body weight, and reduced endurance exercise capacity. The effect of a carbohydrate supplement, offered as a diet option, did not ameliorate the hypoxia-induced weight loss, but significantly delayed the onset of fatigue during exercise in the supplemented rats compared to a hypoxic control group [23].

Overall the results suggest that although disaccharidase activity may be significantly reduced at high altitude, any unabsorbed carbohydrate is not wasted, but probably changed into short chain fatty acids that are then absorbed in the colon. Energy intake when consuming identical rations did not differ (9.7 ± 2.2 vs. 9.8 ± 2.4 MJ/day (mean ± SD) at sea level and 5,650 m, respectively), suggesting that an energy deficit accounted for the 4.4 ± 1.6 kg weight loss at altitude [21]. This agrees with other studies on energy balance at extreme altitude (see below).

Effect of Hypoxia on Fat Absorption

Normal Fat Digestion and Absorption

Dietary fat is the main fuel for endurance exercise at sea level. It has an energy density of twice that of carbohydrate or protein, releasing 37 kJ/g (9 kcal/g), compared to 17 kJ/g (4 kcal/g) for carbohydrate. This dependence on fat for endurance exercise appears to change at high altitude, with greater dependence on carbohydrate, possibly because for the same oxygen utilization, glucose yields more ATP than fat or protein [24]. The process of digestion and absorption of fat has four distinct phases:

(a) Lipolysis of dietary triglyceride (TG) by pancreatic lipase to fatty acid (FA) and β-monoglyceride (MG)
(b) Micellar suspension with bile acid
(c) Uptake into the mucosal cell, with reesterification of the MG with FA to form TG, and chylomicron formation in the presence of cholesterol, cholesterol esters, phospholipids, and protein
(d) Delivery of chylomicrons in lymphatics to the body for utilization of fat

Before lipolysis is initiated, however, emulsification of fat begins in the stomach, through gastric contractions and mixing with gastric acid [25]. This makes it relevant to consider gastric acid secretion and motility at altitude (below), as well as fat absorption. Fat malabsorption can occur due to rapid gastric emptying and improper mixing, altered duodenal pH that affects the function of lipase, altered mucosal function, pancreatic insufficiency, interruption of the enterohepatic circulation of bile acids, or altered processing of chylomicrons or lymphatic flow.

Fat Absorption During Hypoxia at Sea Level

Remarkably, there have been no studies on the impact of hypoxia on fat digestion and absorption at sea level, whether in patients with chronic pulmonary disease reversed by oxygen supplementation, or cyanotic congenital heart disease that is then reversed by surgery. In an animal model of mice exposed to hypoxia equivalent to 4,500 m for up to 32 days, leptin receptor genes in particular and others involved in lipid metabolism were upregulated, suggesting that the leptin signaling pathway is an important feature of acclimatization to hypoxia [11].

Field Studies on Fat Absorption

Studies at high altitude into the effect of hypobaric hypoxia on fat absorption are conflicting. In 1975 Rai et al. found no disturbance in digestibility and utilization of fat up to 4,700 m [26]. The investigators measured fecal fat excretion at 3,800 m on a fat intake of 364 g/day and at 4,700 m on an intake of 232 g/day. Fecal fat content was 11.5 g at 3,800 m, indicating 97 % digestibility, which was similar at 4,700 m (98 %). This concurs with a subsequent study in 1987 of 11 subjects who ingested 100 g fat, followed by a plasma chylomicron assay with a nephelometer [18]. The results showed an increase in chylomicrons in 9/12 subjects at 4,846 m compared to 3,100 m, indicating increased fat absorption (or decreased clearance of chylomicrons). Other techniques to measure fat absorption, including the ^{14}C-triolein breath test and fecal excretion of volatile fatty acids at 5,500 m on Aconcagua [27] and 4,300 m on Pike's Peak [28], respectively, also found no evidence of fat malabsorption.

The exception is a study performed by Boyer and Blume in 1984 during the American Medical Research Expedition to Everest [19]. They demonstrated a 49 % decrease in fat absorption at 6,300 m compared to sea level in three acclimatized subjects, evaluated by an increase in 3-day stool fat excretion. They postulated that this might be the result of hypoxia affecting pancreatic function, perhaps by affecting the energy-dependent process of enzyme secretion, or a nonspecific effect of relative ischemia on bowel function. This study has not been replicated, but mesenteric blood flow has been studied more recently (below). More work is required to clarify the effect of very high or extreme altitude on fat absorption, but it suggests that as for carbohydrates, there may be an altitude above which all absorption is impaired, whether by insufficient mucosal blood flow, general mucosal dysfunction, or the influence of HIF on digestive enzyme function.

Effect of Hypoxia on Protein Absorption

Normal Protein Digestion and Absorption

Protein digestion is initiated by acid breakdown of dietary polypeptides in the stomach, through the action of pepsinogen secreted by the stomach, which is converted to the active protease pepsin by the action of acid. It principally occurs in the lumen of the small intestine through the action of peptidases (trypsin, chymotrypsin, and carboxypeptidases) which hydrolyze proteins into oligopeptides. The brush border of the small intestine contains a family of endopeptidases, which are integral membrane proteins rather than soluble enzymes. They further hydrolyze luminal peptides, converting them to free amino acids and very small peptides formed on the surface of the enterocyte, which are then ready for absorption [29].

Amino acids are absorbed in a similar fashion to monosaccharides. The enterocyte membrane contains at least four sodium-dependent amino acid transporters, one each for acidic, basic, and neutral amino acids. These transporters bind amino acids only after binding sodium. The transporter then undergoes a conformational change that dumps sodium and the amino acid into the cytoplasm, followed by its reorientation back to the original form. Consequently absorption of amino acids is not only dependent on the electrochemical gradient of sodium across the epithelium but also helps generate the osmotic gradient that drives water absorption.

There is abundant absorption of di- and tripeptides by cotransport with H^+ ions by a transporter called PepT1, but almost none of any larger peptides [29]. Once inside the enterocyte, the absorbed di- and tripeptides are digested into amino acids by cytoplasmic peptidases and exported from the cell into blood. Only a tiny proportion of these small peptides enter blood intact. There are further transporters in the basolateral membrane that export amino acid from the cell into blood that are not dependent on sodium gradients. Genes for protein amino acid phosphorylation are generally upregulated in response to hypoxia as are those for lipid metabolism and carbohydrate absorption [11]. It seems likely that transporter function is also affected by hypoxia [10], but the hypoxic threshold (and altitude) for protein metabolism has yet to be defined. In an animal model where rats were exposed to intermittent extreme hypobaric hypoxia (equivalent to 7,260 m) sufficient to induce 28–30 % weight loss, glutamine synthetase activity in muscle was significantly higher after 14 days' hypobaric hypoxia (4.32 μmol γ-glutamyl hydroxamate formed/g protein/min) compared to normal (1.53) [30]. The same changes occurred in liver enzyme activity, but were not sustained after 21 days' exposure. This suggests that since no dramatic changes in the levels of protein were observed in the muscle or liver, alterations in glutaminase and glutamine synthetase activity simply maintain nitrogen metabolism in the initial phase of hypoxic exposure.

The amino acid glutamine is the main fuel for small intestinal enterocytes. Although oral glutamine supplementation halved bacterial translocation to mesenteric lymph nodes (a measure of enterocyte barrier dysfunction) in starved rats exposed to a simulated altitude of 7,000 m for 72 h [31], the dose was 0.5/100 g body weight, equivalent to 350 g glutamine/day for a 70 kg human exposed to these artificial conditions. Despite speculation that glutamine supplementation might have beneficial effects on muscle function [32], we have seen that glutamine synthetase actually increases on short-term exposure to high altitude [30] and there is no objective evidence that glutamine supplementation enhances or protects small intestinal function in normal individuals at high altitude.

Protein Absorption During Hypoxia at Sea Level

As with fat absorption, there have been no formal studies of the effect of hypoxia on human intestinal protein absorption. Neonatal hypoxia has been associated with delayed intestinal enzyme development in animal models [33] and the cat small intestine has been used as a model to examine intestinal lymphatic protein clearance, with a reduction in arterial oxygen tension from 108 to 35 mmHg having no effect [34].

Field Studies on Protein Absorption

Studies on protein absorption have not been conducted at an altitude at which weight loss is greatest (above 5,500 m). This applies to almost all studies on nutrient absorption, so generic comments on a threshold for hypoxia- and hypobaric-induced changes in mucosal function apply, as for carbohydrate or fat absorption, or mucosal blood flow. Evolutionary pressures seem likely to have allowed protective mechanisms to develop for nutrient absorption during hypoxia up to 5,000 m. Above this altitude there would be no selection pressure in human evolution, but it would be interesting to compare gene transcription and function between higher altitude species and their lowland counterparts.

Climbers have reported subjective improvement in digestion and strength when taking digestive enzyme tablets [19], but an objective benefit must be treated with scepticism. Protein absorption after ingestion of ^{15}N-labelled soya protein and injection of ^{15}N-glycine has been studied by measuring urinary and fecal ^{15}N-excretion by mass spectrometry [35]. In six subjects studied at 122 m and then after 3 weeks at 5,000 m, protein absorption (calculated as 100 – [fecal excretion of ^{15}N after ingestion of ^{15}N soya protein (% of dose given) – fecal excretion of ^{15}N after injection of ^{15}N glycine (% of dose given)]), was not significantly impaired at altitude compared with sea level (96 % vs. 97 %, respectively). The overall digestible energy at altitude, calculated as 100 – [percent undigested gross energy in the feces], amounted to 96 %, indicating that malabsorption of protein or other nutrients did not contribute to the 3 % weight loss observed in these subjects at 5,000 m. At extreme altitude (up to 8,000 m equivalent), Westerterp reported up to 15 % fecal loss of energy [36].

Other Effects of High Altitude on Gastrointestinal Physiology

The primary function of the gut is nutrient digestion and absorption, although this ignores its role in immune function (being the largest repository of lymphoid tissue in the body), metabolic contribution from the gut microbiota (which constitute 90 % of all cells in the human body), ancillary neuroenteric function, and splanchnic blood flow. Nevertheless, there has been little work on the effect of high altitude on gastrointestinal physiology other than nutrient absorption. Some studies have been carried out in hypobaric chambers in which high altitude is simulated in order to minimize confounding factors. Field studies are generally difficult to coordinate and perform and those that have been done tend to involve small subject numbers. This may be because few altitude physiologists have a background in gastroenterology. The principal consequence of impaired nutrient absorption is on energy balance, leading to weight loss.

Energy Balance and Weight Loss

That there may be an energy deficit to account for high altitude weight loss seems self-evident, although this cannot be assumed to be the only cause of high altitude weight loss. When seven men stayed at 4,300 m for 21 days basal metabolic rate (BMR) increased by 27 % on day 2 and reached a plateau of 17 % by day 10 for the remainder of the stay at altitude [26]. There was a significant rate of weight loss in the first 7 days at altitude when energy requirement was elevated and energy intake equalled that at sea level. When the increased BMR was balanced by increased food intake, weight loss was minimized. BMR accounts for 50 % of our daily energy expenditure, while diet-induced energy expenditure only accounts for 10 %. The mechanism for a raised BMR at high altitude is not fully understood, but thermogenesis driven by the thyroid to compensate for cold is one hypothesis, as well as the heightened sympathetic tone that accompanies acute or chronic hypoxia (Chap. 12).

An energy deficit can occur through lack of intake, or increased expenditure. Loss of appetite is a common symptom of acute mountain sickness. In addition, discomfort and lack of palatable food on climbs result in a reduced energy intake. Below 4,500 m appetites and food intake return to normal once acclimatized, while above 6,000 m anorexia becomes more pronounced as duration at this altitude increases. Whether impaired mesenteric blood flow contributes to anorexia has been investigated (see section "Mesenteric Blood Flow"), but it appears not up to 4,400 m [37]. To determine the extent that reduced energy intake contributes to weight loss, the effect of chronic hypobaric hypoxia on eight men has been studied in a hypobaric chamber, simulating several altitudes up to the summit of Mount Everest (8,848 m [36]). A hypobaric chamber eliminates confounding factors such as exertion, cold, stress, and limited food supply. Weight loss of 5.0 ± 2.0 kg occurred over 31 days. Energy intake at normoxia was 13.6 ± 1.8 MJ/day. Energy intake decreased from 10.4 ± 2.1 to 8.3 ± 1.9 MJ/day ($p < 0.001$) and energy expenditure from 13.3 ± 1.6 to 12.1 ± 1.8 MJ/day ($p < 0.001$) over the first and second 15-day intervals of progressive hypoxia, which accounted for the weight loss [36]. Approaches to stimulating appetite, facilitating food intake with liquid carbohydrate drinks, or nibbling throughout the day are important considerations when trying to maintain energy intake (see Chap. 15). There is also some evidence that hypoxia stimulates leptin production through the action of HIF, that in turn decreases appetite [2, 38, 39]. As a consequence, the translational application of hypobaric hypoxia on body weight for bariatric medicine has not escaped notice [40].

Energy expenditure when climbing at extreme altitude has also been measured and the energy deficit calculated in the field. Stable isotope techniques using doubly labelled water (deuterium and oxygen-18) allow measurement of energy expenditure over a period of time. Deuterium (^2H) is eliminated as water, while ^{18}O is eliminated as both water and carbon dioxide. Carbon dioxide production can therefore be calculated from the different elimination rates [41]. Using this technique in four subjects climbing Mt Everest in a 7- to 10-day interval up to the summit (5,300–8,872 m), average daily energy expenditure was 13.6 MJ, which was similar to the preparation training in the Alps (14.7 MJ/day at 2,500–4,800 m). The negative energy balance was 5.7 ± 1.9 MJ/day and weight loss correlated with the energy deficit ($r = 0.84$, $p < 0.05$) [42]. Consequently the energy deficit, largely through decreased intake, generally overrides any other mechanism of weight loss at extreme altitude.

Gastric Acid Secretion

Gastric acid contributes to the absorption of fat and protein (above). Gastric secretion, acid concentration, and peptic activity have been studied before and after pentagastrin stimulation in ten sea level residents before and during 22 days at 3,500 m [43]. Results were compared to two other groups of ten men, either natives of high altitude or lowlanders acclimatized to high altitude. There was no change in peptic activity or acid output at 3,500 m and (consistent with other work) D-xylose absorption was also unchanged. On the other hand, work on an animal model (rats exposed to profound hypoxia, 7.6 or 10.5 % O_2) has shown that profound hypoxia equivalent to extreme altitude may reduce gastric acid secretion and confirms that hypoxia impairs gastric emptying [44].

Gut Morphology

A pertinent question is whether gut morphology changes at high altitude, but this remains unknown in humans, despite an attempt by one of the authors to take small intestinal biopsies at sea level and then again at 5,650 m. Ethical considerations prevented exposure of healthy individuals to a Crosby capsule biopsy, which carries a small risk of intestinal perforation, so the author subjected himself to the procedure. Unfortunately the capsule at altitude obtained a sample of healthy gastric and not small intestinal mucosa. Endoscopic biopsy is impracticable at extreme altitude, although endoscopic small intestinal biopsies have been taken at 4,559 m in the Margherita Hut. Detailed morphometry was not reported, since the goal was to examine homogenates [45]. Nevertheless, studies have been performed on chickens exposed to moderate hypobaric hypoxia (2,900 m equivalent altitude) in an attempt to select a strain resistant to ascites. Under hypobaric hypoxic conditions, duodenal villous surface area increased in one strain of chicken from 181.3 ± 16.8 to 219 ± 10.9 μm^2 compared to 130.1 ± 10.5 and 134.3 ± 9.3 μm^2 in another. No differences in ileal villous morphology were observed [46].

This suggests that intestinal mucosal proliferation might occur during hypobaric hypoxia and intuitively this makes sense as a compensatory mechanism if hypobaric hypoxia impairs nutrient absorption.

Also of note is further work on chickens from the same authors. It was hypothesized that promoting neonatal gut development with a prebiotic, such as Aspergillus meal (Prebiotic-AM), would enhance gut efficiency and decrease the oxygen demand of the gut. Prebiotic-treated birds reared in hypobaric hypoxic conditions (2,900 m equivalent) or sea level had similar duodenal and villous morphology compared to control birds reared at equivalent altitudes [47]. Nevertheless, the concept that the intestinal microbiota might change at altitude and influence gut morphology is intriguing. The effect of altitude on fecal or mucosa-associated microbiota in humans has been little studied (below).

Gastrointestinal Motility

In 2004 Yoshimoto et al. examined the effect of acute exposure to hypobaric hypoxia on gastric and colonic motility in Wistar rats [48]. Rats were instrumented with transducers to measure gastric and colonic motility. In a separate group gastric vagotomy was performed. Acute exposure to hypobaric hypoxia resulted in decreased frequency and area of the gastric contraction waves. Gastric vagotomy abolished this suppression. Colonic motility, however, increased. This study concluded that acute suppression of gastric motility occurs on exposure to hypobaric hypoxia and is likely to be mediated by the vagus nerve. A separate group of Chinese investigators examined gastric emptying and intestinal transit in 80 rats subject to hypobaric hypoxia [49]. Using crude, but direct methods of measurement (excision of the stomach to measure contents and enterectomy to measure ileocecal transit of charcoal at defined time periods), they showed that gastric emptying in the 5,000 m-simulated altitude group was only 41 % (±SD 10 %), compared to a 62 % (±12 %) in the sea level group ($p<0.01$). Charcoal transit rate in the 5,000 m-simulated altitude group was also slower than at sea level (37±8 % vs. 61±13 %, $p<0.01$). Delayed gastric emptying may decrease the absorption rate of nutrients and fluid which, with slow intestinal transit, may contribute to the distension, nausea, or vomiting that some climbers experience.

This has pragmatic implications for common gastrointestinal symptoms experienced by long haul aircrew and airline passengers, as well, perhaps, by mountaineers. A questionnaire of long-distance aircrew reported significantly more dyspeptic symptoms than short-haul crewmember and ground personnel (belching: 57 % vs. 37 %, bloating: 51 % vs. 36 %). Consequently 16 healthy men were subjected to a high fiber (20 g) or low fiber (2 g) meals on separate days in a hypobaric chamber, assigned to a flight altitude of either 2,500 or 1,000 m [50]. Gastric emptying was assessed by 13C-octanoic acid breath test; although the technique may overestimate the rate at altitude [45], emptying was significantly delayed at 2,500 m when a high-fiber meal was given ($T_{1/2}=146\pm58$ min low fiber vs. 194 ± 54 min high fiber). The symptom score for gastric distension and bloating was also significantly increased at 2,500 m for the high-fiber meal compared to the low-fiber meal. Intestinal gas production from fiber and the natural consequences of Boyle's law probably explain this.

Mesenteric Blood Flow

Mesenteric blood flow is a challenge to measure even in laboratories near sea level: variables include the size of vessel evaluated (mucosal vs. small- or medium-sized artery); postprandial state; hydration; rest or physical exercise; sympathetic stimulation; core temperature and medication, to name just some. Differences between rest and exercise are particularly relevant to field conditions. Gastric mucosal blood flow at rest may decrease at high altitude, as measured by automated air gastric tonometry at 5,000 m [51]. It is mucosal blood flow that is most likely to matter most as far as intestinal absorption is concerned and, if mucosal blood flow is decreased, this may contribute to susceptibility to peptic (gastroduodenal) ulceration (below). There is some evidence that carotid chemoreceptors activated by hypoxia might lead to reduced gastric blood flow [52].

During exercise, it has been suggested that the origin of the gastrointestinal distress experienced by long-distance runners or mountaineers is due to transient ischemia of the gastric mucosa. Whether mountaineering at high or extreme altitude constitutes the same hemodynamic stress and leads to similar redistribution of blood flow as does marathon running may be debated. To evaluate gastric mucosal function during endurance exercise at altitude, serum pepsinogen measurements have been used, since in many gastrointestinal pathologies the ratio between pepsinogen A/pepsinogen C (PGA/PGC) decreases [53]. In 13 athletes who performed a marathon at 4,300 m [53], 40 % experienced gastrointestinal symptoms. After the race the ratio of PGA/PGC decreased, but no relationship was observed with gastrointestinal symptoms. A control group of five subjects, who had been exposed to the same environmental conditions, showed no gastrointestinal or hormonal alteration.

In larger vessels at rest, mesenteric and hepatic venous blood flow might be expected to be impaired at high altitude, but in contrast it appears to increase, at least up to 4,400 m [37]. Blood flow in the superior mesenteric artery (SMA) and hepatic portal vein (HPV) was studied by Doppler ultrasonography in 12 subjects following an overnight fast and a standard meal, to investigate the hypothesis that relative mesenteric ischemia might contribute to high altitude anorexia [37]. All subjects experienced reduced appetite at 4,392 m. Preprandial flow was significantly higher in the SMA at 4,392 m than at sea level (2,020 vs. 1,024 cm^3/min, $p<0.01$) and so too was HPV flow (708 vs. 505 cm^3/min, respectively). The changes were due to increased vessel diameter and increased flow velocity. There was no difference in postprandial flow between sea level and 4,392 m in the HPV, although the increase in postprandial flow was greater at sea level than high altitude (254 % increase vs. 144 %).

The effect of exercise on blood flow in larger vessels is unclear, because it is not readily measured in these conditions. It is, however, notable that there is a high prevalence of portal vein thrombosis reported in 26 young military personnel after residence at high altitude (>3,000 m) [54]. Consequently it is conceivable that exercise (or dehydration) might have an impact, since it is difficult to see how an increase in flow rates in the HPV measured at rest might also be associated with portal vein thrombosis in the field. The effect of chronic hypobaric hypoxia (long-term residence at high altitude) on mesenteric blood flow has not been studied. The relationship between gastric mucosal and mesenteric blood flow is unclear and the techniques of gastric mucosal tonometry and ultrasound are not only quite different, but measure flow in different sized vessels.

Hepatic Function

The liver is a major metabolic organ with a multiplicity of enzyme-dependent functions, so it would be surprising if hypoxia (let alone hypobaric hypoxia) did not have some effect. This would have implications for drug metabolism, hepatic blood flow, exacerbation of underlying liver disease, and clinical consequences. There has, however, been little work at altitude even though it was of early interest [55].

So it is rather surprising that the pivotal metabolic enzyme complex, cytochrome P-450 (CYP), appears unaffected by hypoxia, until one considers the generic protective mechanisms against hypoxia that appear effective up to 5,000 m. Twelve healthy subjects living at sea level were exposed to altitude-induced hypoxia for 7 days at 4,559 m [56]. Hepatic CYP enzyme activity was measured before departure, at 24 and 96 h after arrival at high altitude, and at 1 month after return to sea level. The metabolic ratio of sparteine, a measure of CYP2D activity, increased after 24 h at high altitude (median difference 0.15; 95 % CI 0.05–0.28) and remained increased after 96 h. The ratio decreased after return to sea level (median difference −0.15; 95 % CI, −0.29 to −0.03; $p=0.016$). The metabolic ratio of endogenous cortisol (CYP3A4) decreased after 24 h (median difference −2.0; 95 % CI −3.5 to −0.5), but returned to sea level values after 96 h at high altitude (median difference 1.6; 95 % CI 1.0–4.2, $p=0.047$). These small changes in the activity of CYP2D6 and CYP3A4 suggest that acute hypoxia has no clinically significant effects on CYP enzymes in humans. Another study confirms this, examining the effect of acute hypoxia on CYP1A2 (theophylline metabolism) and CYP3A4 (verapamil) in 20 subjects in a normobaric hypoxic chamber [57]. Subjects were randomly allocated to receive intravenous infusions of theophylline (6 mg/kg) or verapamil (5 mg) in a crossover design, with exposure to normoxia and hypoxia (12 % oxygen) corresponding to the ambient PaO_2 at 4,500 m. Acute hypoxia did not alter the pharmacokinetics of theophylline ($T_{1/2}\pm$SD: 9.29 ± 1.77 vs. 9.39 ± 1.40 (hypoxia)), verapamil (2.00 ± 0.98 vs. 1.79 + 0.58), or their principal metabolites. Unlike the study at 4,392 m [37], hypoxia did not alter portal venous blood flow, also assessed by Doppler ultrasound, at this normobaric simulated altitude. In practical terms it suggests that drugs metabolized by CYP1A2, CYP2D6, and CYP3A4 do not need dose-adjustment for people during a short stay below 5,000 m.

Only animal models have been used to assess the effects on hepatic function of longer term exposure to high altitude. Exposure of rats for 6 h/day to a simulated extreme altitude of 7,260 m in a hypobaric chamber for up to 21 days was sufficient to decrease body weight by 28–30 % compared to controls [30]. Liver glycogen increased over threefold following 1 day of hypoxic exposure (4.8 ± 0.78 mg/g wet tissue in control rats vs. 15.8 ± 2.30 mg/g in rats exposed to hypoxia). This returned to normal in the later stages of exposure. There was, however, no change in glycogen synthetase activity except for a decrease in the 21-day hypoxia-exposed group. Both glutamine synthetase and glutaminase activity increased over 14 days, but only glutaminase activity was sustained over 21 days, suggesting adaptation to maintain hepatic nitrogen metabolism in the initial phase of hypoxic exposure. During chronic hypoxic exposure of mice to a simulated altitude of 4,500 m for up to 32 days in another animal model, the hematocrit increased up to a remarkable 80 % [11]. Upregulation of genes peculiar for liver tissue involved in hematopoiesis and oxygen transport raises the possibility that the liver may engage in extramedullary hematopoiesis in response to severe, chronic hypoxic stress.

From a clinical point of view, the most substantial risk to hepatic function is the high prevalence of portal vein thrombosis reported in Indian soldiers living above 3,000 m (below [54]).

HIF and the Gut

HIF affects physiological processes in the gut and liver as it does elsewhere in the body (for review, see [58] and Chap. 2). The oxygen gradient between epithelial cells exposed to luminal anaerobic bacteria on one surface and the rich intestinal mucosal blood supply on the other are steeper than anywhere else in the body. It means that intestinal HIF expression may protect barrier function against intestinal inflammatory disease, with evidence that HIF1α-knockout mice are more susceptible to colitis [59]. Of greater interest is the concept that gut inflammation and loss of gut barrier function are implicated as the triggering events that contribute to the development of the systemic inflammatory response, with HIF-1 as a key regulator of the physiological and pathophysiological response to hypoxia [60, 61]. Although this has been related to trauma, reperfusion injury, and systemic inflammatory response, it is provocative to speculate, but quite conceivable, that the profound hypoxic stress in the gut at extreme altitude may similarly provoke acute lung injury characterized by pulmonary oedema. This merits investigation, perhaps starting with studies on whether gut epithelial cells are a primary source of HIF at extreme altitude or equivalent hypoxia.

HIF regulates intestinal mucosal function through several pathways. Iron absorption is among the best studied—and given the stimulation of erythropoiesis by hypoxia, this is intuitively appropriate. Duodenal cytochrome b (DcytB) and divalent metal transporter 1 (DMT1) regulate iron absorption and their expression is increased during high systemic requirements for iron [62]. The molecular mechanisms that regulate DcytB and DMT1 expression are HIF-dependent. Duodenal HIF signaling is induced by acute iron deficiency, resulting in activation of DcytB and DMT1 expression and an increase in iron uptake [63]. Genetic disruption of HIF signaling (HIF-2α) in the intestine abolishes the adaptive induction of iron absorption following iron deficiency, resulting in low systemic iron and anemia. Intestinal HIF signaling is a critical regulator of systemic iron homeostasis, through the control of ferroportin expression [64]. Intestinal mucosal disaccharidase activity is also susceptible to HIF with induction of lactase gene transcription that may be an adaptive response to gut hypoxia [65]. Although this is apparently at odds with the reduced disaccharidase activity at extreme altitude measured in man [21], it is conceivable that the upregulation is insufficient to meet physiological demand at altitudes above 5,500 m. Other key processes in absorption are also likely to be susceptible to HIF, but have been little studied. GLUT1 expression is HIF-dependent, playing a pivotal role in the hyperglycolysis of malignancy, including intestinal epithelial cancers [66], but the function of SGLT1, GLUT2, and GLUT5 has been little studied. In a Von Hippel Lindau factor knockout model, which renders HIF expression unresponsive, hepatic GLUT2 expression was downregulated, in contrast to upregulation of GLUT1 [67] and induction of HIF1α upregulated GLUT2 over threefold in another animal model [68].

It is interesting to speculate that HIF-induced regulation of physiological processes, including nutrient absorption, is a consequence of evolutionary pressures to maintain function during the low oxygen tension of ancient atmospheres [69]. These fluctuated, but were around 15 % at their lowest, equivalent to the partial pressure of oxygen at about 5,000 m. This would allow function to be preserved up to this altitude and residence at high altitude, but there would be no selection pressure to preserve function at extreme altitude, where mountaineers choose to climb.

Gastrointestinal Disorders at High Altitude

Gastrointestinal disorders are commonly reported at high altitude, but are also common at sea level. It helps to discriminate between short-term visitors, long-term residents, and native highlanders [70].

Short-Term Visitors

Anorexia, nausea, and occasionally vomiting are early features of acute extreme altitude (above) and are also nonspecific features of gastrointestinal upset, such as that caused by travellers' diarrhea, or any intercurrent illness. Other common, non-specific symptoms include dyspepsia and flatulence. The former is so common that medical kits for expeditions had best contain antacids or proton pump inhibitors. The latter may be a feature of altered bacterial flora in the gut, especially after a diarrheal illness at any altitude, but the simple effect of Boyle's law (that the volume of gas is inversely proportional to the pressure, if the temperature is kept constant within a closed system) means that the volume of intestinal gas increases at high altitude. This is observed even in aircrew on short-term exposure to moderate altitude (up to 2,500 m, above [50]).

Travellers' diarrhea is common and troublesome, but there is little evidence that this is any more prevalent at high altitude, other than as a consequence of the remote places in which mountains are found and the relative lack of sanitation. The symptoms are well known and widely experienced, since half of travellers to the Himalaya or the Andes are likely to be affected: sudden onset of diarrhea, with abdominal cramps and sometimes vomiting or (in severe cases) dehydration. It is self-limiting, but prevention and treatment are important considerations for short-term visitors to altitude. A systematic review found no evidence for the widespread advice to take care in what you eat or drink [71]. Common sense, however, is not commonly subject to science.

Chemoprophylaxis (preventative treatment with antibiotics) should, however, be considered if the impact of an episode will affect the goals of the expedition. Rifaximin (a nonabsorbable antibiotic) is effective, as is bismuth subsalicylate or a fluoroquinolone (such as ciprofloxacin) [72]. Treatment consists of a single dose of ciprofloxacin (750 mg), or a 3-day course of rifaximin, or other appropriate antibiotic [73]. Of greatest interest, vaccines against enterotoxigenic *Escherichia coli* (the principal cause of travellers' diarrhea) are being developed [74].

Aside from pathogenic infections, the bacterial flora may change at altitude. When fecal bacterial population groups and serum responses in seven mountaineers on a 47-day expedition to the Himalaya were examined, there was distinct alteration in the microflora above 5,000 m [75]. Bifidobacteria and species belonging to the *Atopobium*, *Coriobacterium*, and *Eggerthella lenta* group decreased, whereas potentially pathogenic bacteria of the γ-subdivision of *Proteobacteria* and specific Enterobacteriaceae such as *Escherichia coli* increased. There was a reduction in serum levels of IgM- and/or IgA anti-LPS and elevation in C-reactive protein (CRP). Whether this was a consequence of hypoxia or travel in a remote area is unclear.

Preexisting gastrointestinal conditions (such as Crohn's disease, ulcerative colitis, or celiac disease, see also Chap. 23) are unlikely to be affected by altitude, but the consequences of a gastrointestinal infection superimposed on such a condition have to be considered, as well as the implications for treatment. Dietary adherence to a gluten-free diet, for instance, may be difficult and anti-TNF therapy for inflammatory bowel disease is a relative contraindication to travel in remote areas. Expert advice is appropriate.

Residents and Native Highlanders

Peptic ulceration and gastrointestinal bleeding are common in native highlanders in the Peruvian Andes (9.6/10,000 per year [70]). Although this may reflect a high prevalence of *Helicobacter pylori* in the Sierra compared to the coast [76, 77], high altitude (3,150 m) was not associated with a higher prevalence of duodenal ulceration or frequency of *Helicobacter pylori* compared to low altitude areas in the Kingdom of Saudi Arabia [78].

Nevertheless, acute upper gastrointestinal bleeding (AUGIB) is a real risk at high altitude, with one of the authors having to deal with three cases during a single expedition to Mount Everest. During the construction of the Beijing-Lhasa railroad (up to 4,905 m) a high prevalence of AUGIB was noted in a young cohort ($n = 13{,}502$; mean age ± SD 29.5 + 7.4 year; 99 % male), risk factors being alcohol, aspirin, dexamethasone, a previous history of AUGIB, or high altitude polycythemia [79, 80]. The overall incidence of AUGIB was 0.49 %. Endoscopy in those evacuated to Golmud (2,808 m) or Xining (2,261 m) showed hemorrhagic gastritis, gastric ulcer, duodenal ulcer, or gastric erosions. The reason for increased susceptibility is likely to be the susceptibility of the mucosal barrier to hypobaric hypoxia, with disruption of mucosal integrity, with evidence from an animal model [81]. It has practical implications for medical planning for mountaineering expeditions: people with a history of peptic ulceration should continue maintenance of acid suppression at high altitude [82].

Apart from peptic ulceration, another cause of abdominal pain (with or without gastrointestinal bleeding) that is more common at high altitude than sea level appears to be portal vein thrombosis. At an Indian Army Hospital between 1998 and 2003, 37 cases of portal vein thrombosis were seen, of which 26 (mean age 27 ± 4 years) had stayed at >3,000 m for a year (mean 11.7 ± 6.2 months). The presenting symptom was abdominal pain, followed by gastrointestinal bleeding (38 %) or fever (38 %), and vomiting in 51 %. A prothrombotic state was detected in 5/26 cases from high altitude and 8/11 cases from low altitude ($p \leq 0.01$) [54]. This is notable, although difficult to square with increased flow rates in the portal vein after short-term exposure to 4,400 m (above, [37]). It means that portal vein thrombosis should be given a higher profile in the differential diagnosis of acute abdominal pain at high altitude than at low altitude.

Sigmoid volvulus and megacolon may also be more common at high altitude with volvulus accounting for 79 % of all intestinal obstruction in Bolivian hospitals [70]. Whether this is a consequence of altitude and gas expansion, diet, or the length of the mesocolon in native Bolivians is unclear. There is plenty of scope for research.

Residents for 1–3 years at high altitude are, however, not necessarily subject to a higher prevalence of gastrointestinal disorders. In a discursive report comparing illness reported to military medical centers by 130,700 soldiers stationed in the plains (up to 760 m) with those from 20,000 men stationed between 3,692 and 5,538 m in 1965–1972 [83], annual differences in the incidence of bacterial, protozoal, or viral infections (many of which will have been gastrointestinal) were significantly lower at high altitude than at sea level. The incidence of amoebic hepatitis, goiter, and lobar pneumonia were higher, but the study took no account of the fact that sick soldiers were more likely to be kept at base and not sent to high attitude.

Conclusions

The effect of hypobaric hypoxia on gastrointestinal physiology is incompletely understood. For practical purposes there is no discernible effect below 5,000 m. Weight loss at extreme altitude (>5,800 m) is consistently documented in field and hypobaric chamber studies. There appears to be a reduction in carbohydrate absorptive capacity and disaccharidase activity above 5,000 m, but the principal cause of weight loss is an energy deficit caused by inadequate food intake, not nutrient malabsorption. Gastrointestinal disorders are likely to be at least as common at altitude as at sea level, but the main cause relates to hygiene in remote environments. There are practical measures that can be taken to reduce the chance of acquiring travellers' diarrhea. Specialist advice on this and for people with a preexisting condition travelling to high altitude is appropriate.

Most work on gut function to date has examined nutrient absorption for reasons of practicality. Areas for further research (in no particular order) include:

- Comparing small intestinal mucosal biopsies performed at endoscopy in lowland and high altitude residents, to answer the question whether villous surface area increases at high altitude, although allowance would need to be made for comorbidity, drug ingestion, and intestinal pathogens
- Examining mechanisms and causes of anorexia at both high and extreme altitudes
- Examining intestinal blood flow at extreme altitude in humans
- Investigating the effect of high or extreme altitude on the luminal or mucosal microbiota
- Evaluating hepatic function in residents at high altitude
- Investigating whether hypoxic stress in the gut may provoke acute lung injury through the release of HIF, cytokine activation, or translocation of bacterial peptides, starting with studies on whether gut epithelial cells are a primary source of HIF at extreme altitude or equivalent hypoxia

The new disciplines of metagenomics, metabolomics, and others lend themselves to examining metabolism in human subjects at high altitude, through simple sample collection in the field and analysis in appropriate laboratories. New frontiers and new horizons beckon.

References

1. Kayser B. Nutrition and energetics of exercise at altitude: theory and possible practical implications. Sports Med. 1994;17:309–21.
2. Tschöp M, Strasburger CJ, Hartmann G, et al. Raised leptin concentrations at high altitude associated with loss of appetite. Lancet. 1998;352:1119–20.
3. Hamad N, Travis SPL. Weight loss at high altitude: pathophysiology and practical implications. Eur J Gastroenterol Hepatol. 2006;18:5–10.
4. Milledge JS, Travis SPL. Intestinal function and nutrition. In: Ward M, Milledge JS, West J, editors. High altitude medicine and physiology. 2nd ed. London: Chapman Hall; 1995. p. 285–300.
5. Gatterer H, Faulhaber M, Netzer N. Hypoxic training for football players. Scand J Med Sci Sports. 2009; 19:607.

6. Edwards JSA, Dinmore AL, Travis SPL. Food and nutritional intake at high altitude. J Nutr Food Sci. 1998;1:5–8.
7. Read NW. Diarrhoea: the failure of colonic salvage. Lancet. 1982;2(8296):481–3.
8. Drozdowski LA, Thomson AB. Intestinal sugar transport. World J Gastroenterol. 2006;12:1657–70.
9. Kellett GL, Brot-Laroche E, Mace OJ, et al. Sugar absorption in the intestine: the role of GLUT2. Annu Rev Nutr. 2008;28:35–54.
10. Rytting E, Audus KL. Effects of low oxygen levels on the expression and function of transporter OCTN2 in BeWo cells. J Pharm Pharmacol. 2007;59:1095–102.
11. Baze MM, Schlauch K, Hayes JP. Gene expression of the liver in response to chronic hypoxia. Physiol Genomics. 2010;41:275–88.
12. Milledge JS. Arterial oxygen desaturation and intestinal absorption of xylose. Br Med J. 1972;3(5826): 557–8.
13. Morawa AP, Han SS. Studies on hypoxia. 8. Ultrastructural and biochemical effects of prolonged exposure on rat parotid glands. Exp Mol Pathol. 1974;21:268–87.
14. Pritchard JS, Lane DJ. Intestinal absorption studied in patients with chronic obstructive airways disease. Thorax. 1976;29:609.
15. Travis SPL, Menzies IS. Intestinal permeability: functional assessment and significance. Clin Sci. 1992; 82:471–88.
16. Debnam ES, Grimble GK. Methods for assessing intestinal absorptive function in relation to enteral nutrition. Curr Opin Clin Nutr Metab Care. 2001; 4:355–67.
17. Pugh LGCE. Physiological and medical aspects of the Himalayan scientific and mountaineering expedition, 1960-61. Br Med J. 1962;2(5305):621–7.
18. Chesner IM, Small NA, Dykes PW. Intestinal absorption at high altitude. Postgrad Med J. 1987;63:173–5.
19. Boyer SJ, Blume FD. Weight loss and changes in body composition at high altitude. J Appl Physiol. 1984;57:1580–5.
20. Dinmore AJ, Edwards JS, Menzies IS, et al. Intestinal carbohydrate absorption and permeability at high altitude (5,730 m). J Appl Physiol. 1994;76:1903–7.
21. Travis SPL, Edwards JSA, Westerterp K, et al. Carbohydrate intake, hydrolysis and permeation at extreme altitude (5650m). Gastroenterology. 1996; 110:A845.
22. Bjarnason I, Batt R, Catt S, et al. Evaluation of differential disaccharide excretion in urine for non-invasive investigation of altered intestinal disaccharidase activity caused by alpha-glucosidase inhibition, primary hypolactasia, and coeliac disease. Gut. 1996;39:374–81.
23. Sharma A, Singh SB, Panjwani U, et al. Effect of a carbohydrate supplement on feeding behaviour and exercise in rats exposed to hypobaric hypoxia. Appetite. 2002;39:127–35.
24. Brooks GA, Butterfield GE, Wolfe RR, et al. Increased dependence on blood glucose after acclimatization to 4300m. J Appl Physiol. 1991;70:919–27.
25. Mansbach CM, Tso P, Kuksis A. Intestinal lipid metabolism. New York: Kluwer Academic/Plenum; 2001.
26. Rai RM, Malhotra MS, Dimri GP, et al. Utilization of different quantities of fat at high altitude. Am J Clin Nutr. 1975;28:242–7.
27. Imray CH, Chesner I, Winterbourn M, et al. Fat absorption at altitude: a reappraisal (abstract). Int J Sports Med. 1992;13:87.
28. Butterfield GE, Gates J, Fleming S, et al. Increased energy intake minimizes weight loss in men at high altitude. J Appl Physiol. 1992;72:1741–8.
29. Gilbert ER, Wong EA, Webb Jr KE. Board-invited review: peptide absorption and utilization: implications for animal nutrition and health. J Anim Sci. 2008;86:2135–55.
30. Vats P, Mukherjee AK, Kumria MM, et al. Changes in the activity levels of glutamine synthetase, glutaminase and glycogen synthetase in rats subjected to hypoxic stress. Int J Biometeorol. 1999;42:205–9.
31. Zhou QQ, Yang DZ, Luo YJ, et al. Over-starvation aggravates intestinal injury and promotes bacterial and endotoxin translocation under high-altitude hypoxic environment. World J Gastroenterol. 2011; 17:1584–93.
32. Wagenmakers AJ. Amino acid metabolism, muscular fatigue and muscle wasting. Speculations on adaptations at high altitude. Int J Sports Med. 1992;13 Suppl 1:S110–3.
33. Lee PC, Struve M, Ruff H. Effects of hypoxia on the development of intestinal enzymes in neonatal and juvenile rats. Exp Biol Med. 2003;228:717–23.
34. Perry MA, Shepherd AP, Kvietys PR, et al. Effect of hypoxia on feline intestinal capillary permeability. Am J Physiol. 1985;248:G272–6.
35. Kayser B, Acheson K, Decombaz J, et al. Protein absorption and energy digestibility at high altitude. J Appl Physiol. 1992;73:2425–31.
36. Westerterp-Plantenga MS, Westerterp KR, Rubbens M, et al. Appetite at "high altitude" [Operation Everest III (Comex-'97)]: a simulated ascent of Mount Everest. J Appl Physiol. 1999;87:391–9.
37. Kalson NS, Hext F, Davies AJ, et al. Do changes in gastro-intestinal blood flow explain high-altitude anorexia? Eur J Clin Invest. 2010;40:735–41.
38. Westerterp KR, Meijer EP, Rubbens M, et al. Operation Everest III: energy and water balance. Pflugers Arch. 2000;439:483–8.
39. Yingzhong Y, Droma Y, Rili G, et al. Regulation of body weight by leptin, with special reference to hypoxia-induced regulation. Intern Med. 2006;45:941–6.
40. Quintero P, Milagro FI, Campión J, et al. Impact of oxygen availability on body weight management. Med Hypotheses. 2010;74:901–7.
41. Ainslie P, Reilly T, Westerterp K. Estimating human energy expenditure: a review of techniques with particular reference to doubly labelled water. Sports Med. 2003;33:683–98.
42. Westerterp KR, Kayser B. Body mass regulation at altitude. Eur J Gastroenterol Hepatol. 2006;18:1–3.

43. Sridharan K, Malhotra MS, Upadhayay TN, et al. Changes in gastrointestinal function in humans at an altitude of 3500 m. Eur J Appl Physiol. 1982; 50:145–54.
44. Yamaji R, Sakamoto M, Miyatake K, et al. Hypoxia inhibits gastric emptying and gastric acid secretion in conscious rats. J Nutr. 1996;126:673–80.
45. Vuille-ditBrill C, Meier D, Kymmer EE, et al. Hypoxia induces intestinal isocitrate dehydrogenase expression and enhances 13C-octanoate metabolism in healthy mountaineers after rapid ascent to 4559m (abstract). Gastroenterology. 2011;138:1708.
46. de Los SF, Tellez G, Farnell MB, et al. Hypobaric hypoxia in ascites resistant and susceptible broiler genetic lines influences gut morphology. Poult Sci. 2005;84:1495–8.
47. Solis de los SF, Farnell MB, Téllez G, et al. Effect of prebiotic on gut development and ascites incidence of broilers reared in a hypoxic environment. Poult Sci. 2005;84(7):1092–100.
48. Yoshimoto M, Sasaki M, Naraki N, et al. Regulation of gastric motility at simulated high altitude in conscious rats. J Appl Physiol. 2004;97:599–604.
49. Yang CM, Chen Y, Mao GP, et al. Effects of acute hypobaric hypoxia on gastric emptying and intestinal propulsion: experiment with rats. Zhonghua Yi Xue Za Zhi. 2006;86:2391–4 (Eng abstract).
50. Hinninghofen H, Musial F, Kowalski A, et al. Gastric emptying effects of dietary fiber during 8 hours at two simulated cabin altitudes. Aviat Space Environ Med. 2006;77:121–3.
51. Martin D, McCorkell S, Vercueil A, et al. Increased gastric-end tidal P(CO2) gap during exercise at high altitude measured by gastric tonometry. High Alt Med Biol. 2007;8:50–5.
52. Siński M, Kowalczyk P, Stolarczyk A, et al. Influence of the stimulation of carotid body chemoreceptors on the gastric mucosal blood flow in artificially ventilated and spontaneously breathing rats. J Physiol Pharmacol. 2002;53:359–69.
53. Banfi G, Marinelli M, Bonini P, et al. Pepsinogens and gastrointestinal symptoms in mountain marathon runners. Int J Sports Med. 1996;17:554–8.
54. Anand AC, Saha A, Seth AK, et al. Symptomatic portal system thrombosis in soldiers due to extended stay at extreme altitude. J Gastroenterol Hepatol. 2005; 20:777–83.
55. Berendsohn S. Hepatic function at high altitudes. Arch Intern Med. 1962;109:56–64.
56. Jürgens G, Christensen HR, Brøsen K, et al. Acute hypoxia and cytochrome P450-mediated hepatic drug metabolism in humans. Clin Pharmacol Ther. 2002;71:214–20.
57. Streit M, Göggelmann C, Dehnert C, et al. Cytochrome P450 enzyme-mediated drug metabolism at exposure to acute hypoxia (corresponding to an altitude of 4,500 m). Eur J Clin Pharmacol. 2005;6:39–46.
58. Majmundar AJ, Wong WJ, Simon MC. Hypoxia-inducible factors and the response to hypoxic stress. Mol Cell. 2010;40:294–309.
59. Glover LE, Colgan SP. Hypoxia and metabolic factors that influence inflammatory bowel disease pathogenesis. Gastroenterology. 2011;140:1748–55.
60. Feinman R, Deitch EA, Watkins AC, et al. HIF-1 mediates pathogenic inflammatory responses to intestinal ischemia-reperfusion injury. Am J Physiol Gastrointest Liver Physiol. 2010;299:G833–43.
61. Kannan KB, Colorado I, Reino D, et al. Hypoxia-inducible factor plays a gut-injurious role in intestinal ischemia reperfusion injury. Am J Physiol Gastrointest Liver Physiol. 2011;300:G853–61.
62. Evstatiev R, Gasche C. Iron sensing and signalling. Gut. 2012;61(6):933–52.
63. Shah YM, Matsubara T, Ito S, et al. Intestinal hypoxia-inducible transcription factors are essential for iron absorption following iron deficiency. Cell Metab. 2009;9:152–64.
64. Taylor M, Qu A, Anderson ER, et al. Hypoxia-inducible factor-2α mediates the adaptive increase of intestinal ferroportin during iron deficiency in mice. Gastroenterology. 2011;140:2044–55.
65. Lee SY, Madan A, Furuta GT, et al. Lactase gene transcription is activated in response to hypoxia in intestinal epithelial cells. Mol Genet Metab. 2002; 75:65–9.
66. Griffiths EA, Pritchard SA, McGrath SM, et al. Increasing expression of hypoxia inducible proteins in the Barrett's metaplasia-dysplasia-adenocarcinoma sequence. Br J Cancer. 2007;96:1377–83.
67. Park SK, Haase VH, Johnson RS. von Hippel Lindau tumor suppressor regulates hepatic glucose metabolism by controlling expression of glucose transporter 2 and glucose 6-phosphatase. Int J Oncol. 2007; 30:341–8.
68. Wan J, Chai H, Yu Z, et al. HIF-1α effects on angiogenic potential in human small cell lung carcinoma. J Exp Clin Cancer Res. 2011;30:77–91.
69. Taylor CT, McElwain JC. Ancient atmospheres and the evolution of oxygen sensing via the hypoxia-inducible factor in metazoans. Physiology (Bethesda). 2010;25:272–9.
70. Anand AC, Sashindran VK, Mohan L. Gastrointestinal problems at high altitude. Trop Gastroenterol. 2006;27:147–53.
71. DuPont HL. Systematic review: prevention of travellers' diarrhoea. Aliment Pharmacol Ther. 2008;27: 741–51.
72. Parry H, Howard AJ, Galpin OP, et al. The prophylaxis of travellers' diarrhoea; a double blind placebo controlled trial of ciprofloxacin during a Himalayan expedition. J Infect. 1994;28:337–8.
73. Hill DR, Ryan ET. Management of travellers' diarrhoea. Br Med J. 2008;337:1746.
74. Frech SA, Dupont HL, Bourgeois AL, et al. Use of a patch containing heat-labile toxin from Escherichia coli against travellers' diarrhoea: a phase II, randomised, double-blind, placebo-controlled field trial. Lancet. 2008;371:2019–25.
75. Kleessen B, Schroedl W, Stueck M, et al. Microbial and immunological responses relative to high-altitude

exposure in mountaineers. Med Sci Sports Exerc. 2005;37:1313–8.

76. Ecology of Helicobacter pylori in Peru: infection rates in coastal, high altitude, and jungle communities. The Gastrointestinal Physiology Working Group of the Cayetano Heredia and the Johns Hopkins University. *Gut* 1992;33:604–5.
77. Recavarren-Arce S, Ramirez-Ramos A, Gilman RH, et al. Severe gastritis in the Peruvian Andes. Histopathology. 2005;46:374–9.
78. Ahmed ME, al-Knawy BA, al-Wabel AH, et al. Duodenal ulcer and *Helicobacter pylori* infection at high altitude: experience from southern Saudi Arabia. Can J Gastroenterol. 1997;11:313–6.
79. Wu TY, Ding SQ, Liu JL, et al. Who should not go high: chronic disease and work at altitude during construction of the Qinghai-Tibet railroad. High Alt Med Biol. 2007;8:88–107.
80. Wu TY, Ding SQ, Liu JL, et al. High-altitude gastrointestinal bleeding: an observation in Qinghai-Tibetan railroad construction workers on Mountain Tanggula. World J Gastroenterol. 2007;13:774–80.
81. Dong Y, Zheng J, Wang XY, et al. Effect of oxygen therapy for injury of intestinal mucosal barrier of rabbits in high altitude hemorrhagic shock. Zhongguo Wei Zhong Bing Ji Jiu Yi Xue. 2005;17:32–5 [Eng abstract].
82. A'Court C, Stables R, Travis SPL. How to do it: be a doctor to a high altitude mountaineering expedition. Br Med J. 1995;310:1248–52.
83. Singh I, Chohan IS, Lal M, et al. Effects of high altitude stay on the incidence of common diseases in man. Int J Biometeorol. 1977;21:93–122.

Immune System

14

Robert S. Mazzeo and Erik R. Swenson

Abstract

The human immune system is a complex network of molecular, cellular, and genetic components designed to provide defense against foreign organisms and substances. It is a highly regulated system that is sensitive to a number of extrinsic factors including environmental stress (e.g., heat, cold, microgravity, and hypoxia). The impact that any stressor is likely to have on immune function is dependent on whether the stress is acute or chronic, the severity of the stress, and the individual's current physiological and emotional state. While the effects of hypoxia on the immune system are evident at almost all levels and so will have potential impact on infectious risks and outcomes, high altitude and hypoxia also affect certain pathogens and their vectors to alter human health at high altitude. We will review how hypoxia affects the innate and adaptive immune systems both positively and negatively, and the central role of hypoxia inducible factor(s) (HIF) in these changes, how the central nervous system (CNS) contributes particularly to the adaptive immune responses at high altitude, and how these responses affect aspects of defense against pathogens, vaccine effectiveness, and immunosurveillance of cancer.

Introduction

The human immune system is a complex network of molecular, cellular, and genetic components designed to provide defense against foreign organisms and substances. It is a highly regulated system that is sensitive to a number of extrinsic factors including environmental stress (e.g., heat, cold, microgravity, and hypoxia). The impact that any stressor is likely to have on immune function is dependent on whether the stress is acute or chronic, the severity of the stress, and the individual's current physiological and emotional state. While the effects of hypoxia on the immune system are evident at almost all levels and so will have potential impact on infectious risks and outcomes, high altitude and hypoxia also affect certain pathogens and their vectors to alter human

R.S. Mazzeo, Ph.D.
Department of Integrative Physiology, University of Colorado, 354 UCB, Boulder, CO 80309, USA
e-mail: mazzeo@colorado.edu

E.R. Swenson, M.D. (✉)
VA Puget Sound Health Care System,
University of Washington, Seattle, WA, USA
e-mail: eswenson@u.washington.edu

E.R. Swenson and P. Bärtsch (eds.), *High Altitude: Human Adaptation to Hypoxia*,
DOI 10.1007/978-1-4614-8772-2_14, © Springer Science+Business Media New York 2014

health at high altitude. We will review how hypoxia affects the innate and adaptive immune systems both positively and negatively, and the central role of hypoxia inducible factor(s) (HIF) in these changes, how the central nervous system (CNS) contributes particularly to the adaptive immune responses at high altitude, and how these responses affect aspects of defense against pathogens, vaccine effectiveness, and immunosurveillance of cancer.

The Stress of Altitude and the CNS

Before discussing direct effects of hypoxia on cells involved in immune responses it must be recognized that ascent to high altitude is a stress known to alter a number of physiologic and metabolic functions. These adjustments to high altitude are a necessary attempt to maintain homeostasis in the presence of a reduction in arterial oxygen pressure (P_aO_2). Hypoxemia represents a major disruption in normal homeostasis and thereby induces both systemic and local effects to compensate for the reduced arterial oxygen content (C_aO_2) and O_2 delivery. As with many other stressors, hypoxia elicits systemic responses of which those controlled by the CNS are most highly relevant for the immune system. The critical elements of the CNS response to high altitude on the immune system are the sympathetic nervous system (SNS) and the hypothalamic–pituitary–adrenal (HPA) axis. A typical neuroendocrine response to stress involves the activation of both the HPA axis and the SNS.

The HPA axis is responsible for the release of cortisol from the adrenal cortex. Cortisol, as with other glucocorticoids, are well documented to be immunosuppressive at the high circulating levels associated with stress [1]. With initial exposure to high altitude, free and total cortisol levels increase when compared to sea level values [2, 3]. However, as an individual acclimatizes over time, cortisol levels generally return to sea level values (Fig. 14.1). Thus, it would appear that the initial hypoxic stress activates the HPA axis and subsequent cortisol release but as C_aO_2 improves with acclimatization, HPA activity and cortisol levels return to baseline.

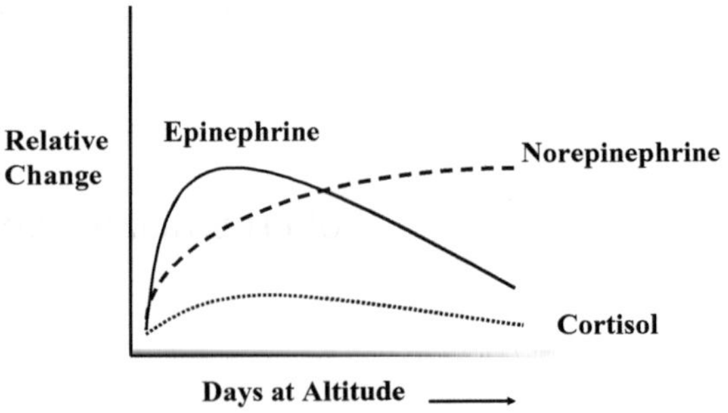

Fig. 14.1 In response to stress, such as hypoxia, the central nervous system activates both the hypothalamic–pituitary–adrenal (HPA) axis and the sympathetic nervous system (SNS). The resulting production of circulating glucocorticoids and catecholamines can influence leukocyte number and function. Also, norepinephrine produced at SNS nerve endings can modulate immune-cell function via direct innervation of lymphoid organs. Cytokines produced by immune cells can modulate the activity of the hypothalamus. *APC* antigen-presenting cell, *IL-1* interleukin-1, *NK* natural killer (From: Glaser R., and Kiecolt-Glaser J.K., Nat. Rev. Immunol. 5:243–51, 2005—with permission)

The SNS directly innervates the adrenal medulla and thus, upon activation, causes the release of epinephrine into the circulation. Additionally, sympathetic nerve fibers innervate both primary and secondary lymphoid organs including the spleen, thymus, and lymph nodes with obvious immunomodulatory ramifications during periods of stress as norepinephrine is released from excited SNS terminals. Almost all immune cells express the beta-2-adrenergic receptor (β_2AR) that binds both norepinephrine and epinephrine activating intracellular signaling [4]. It is important to note that the increase in SNS activity and norepinephrine levels associated with high altitude exposure [5] can exert their effect via both the α- and β-adrenergic receptors while epinephrine primarily acts upon β-adrenergic receptors. With exposure to hypoxia, elevations in SNS and adrenal medullary activity are associated with impairment of immunocompetent cells, mobilization of T-cells and natural killer (NK) cells, and regulation of cytokine production and release [6–9].

There is a clear direct effect of hypoxia on stimulating adrenal medullary release of epinephrine as well as increasing SNS activity [10–12]. A number of studies (summit of Pikes Peak, 4,300 m,

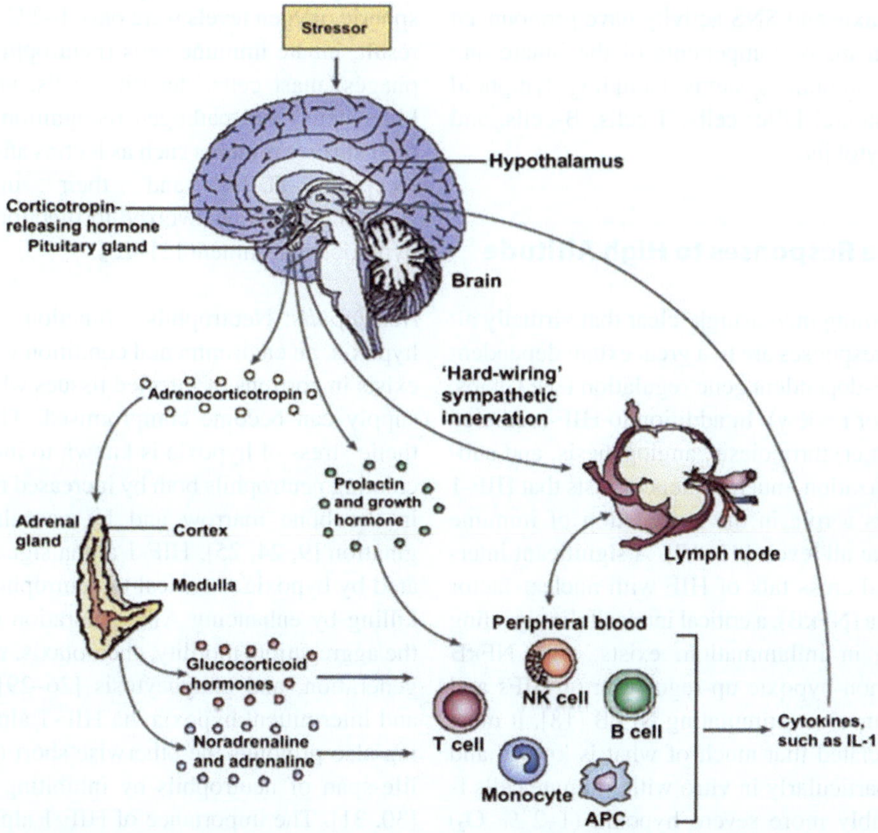

Fig. 14.2 The response of key stress hormones to both acute and chronic high altitude exposure. With acute exposure, hypoxia acts directly to stimulate adrenal medullary activity, thereby increasing circulating epinephrine levels dramatically. With acclimatization, oxygen content improves reducing the stimulus on adrenal medullary activity, and epinephrine levels return toward sea level values. However, sympathetic nerve activity increases steadily over time at altitude influencing various systems, including the immune system. Finally, there is a transient increase in cortisol levels with initial exposure to altitude

~462 mmHg) have extensively documented the sympathoadrenal responses during both acute and chronic exposure to high altitude conditions. It has been consistently demonstrated that upon acute exposure (within 4 h of arrival) arterial concentrations of epinephrine are elevated at rest compared to sea level (Fig. 14.1). This response is dependent upon the degree and severity of hypoxia as it is causally related to the extent that hypoxia stimulates adrenal medullary epinephrine release. With acclimatization, improvements in oxygen saturation reduce the hypoxic stress and epinephrine levels decline toward sea level values. Thus, an inverse relationship between C_aO_2 and arterial epinephrine concentration exists.

Interestingly, unlike the adrenal medullary release of epinephrine, SNS activity and the subsequent release of norepinephrine remain at elevated levels despite improvements in oxygen saturation. Arterial levels of norepinephrine are a function of the rate of spillover into the circulation released by SNS, and its eventual clearance. Resting levels of arterial norepinephrine during acute exposure are generally found to be similar to those observed at sea level ([10–14]; see also Chaps. 8 and 12). However, norepinephrine levels rise significantly with time during more prolonged stay at high altitude reaching a peak after 1 week.

In summary, an elevation in both HPA axis and SNS activity is well documented to occur in response to both acute and chronic high altitude exposure [2, 3, 10–12]. As shown schematically in Fig. 14.2, in response to stress, activation of the HPA axis and SNS activity have pronounced effects on many components of the innate and adaptive immune systems including lymphoid organs, natural killer cells, T-cells, B-cells, and various cytokines.

Immune Responses to High Altitude

It is becoming increasingly clear that virtually all hypoxic responses are to a great extent dependent upon HIF-dependent gene regulation (see Chaps. 1 and 2 for review). In addition to HIF-mediated actions in erythropoiesis, angiogenesis, and substrate utilization, much evidence exists that HIF-1 also plays a role in the modulation of immune function at all levels [15–17]. A significant interaction and cross talk of HIF with nuclear factor kappa beta (NFκB), a critical intracellular signaling molecule in inflammation, exists, with NFκB causing non-hypoxic up-regulation of HIFs and HIFs themselves stimulating NFκB [18]. It must be appreciated that much of what is known and studied particularly in vitro with immune cells is considerably more severe hypoxia (1–2 % O_2) typical of injured and inflamed tissues than that experienced in otherwise healthy humans at habitable altitudes. In addition, other factors relevant to injured tissues including acidosis, depletion of glucose, and accumulation of lactate and other metabolites, cytokines, and microbial products are not usually present. Nonetheless, even modest altitudes at around 4,000 m (equivalent to 12 % O_2) are sufficient to elicit HIF-1 alpha increases in human leukocytes and HIF-1 alpha DNA binding in vivo [19]. More complete dose response studies of hypoxia levels to immune system responses are needed to better understand what occurs at high altitude.

Innate Immune System

Innate immune responses represent the earliest evolution of defense against pathogens. Elements of the innate immune system arose in single cell and early multicellular organisms at a time (1–2 billion years ago) before the proliferation of O_2-generating, photosynthesizing plants when atmospheric oxygen levels were only 1–2 % [20]. As a result, innate immune cells (neutrophils, macrophages, mast cells, dendritic cells, and natural killer cells) and pathogen recognition receptors (cell surface proteins such as lectins and Toll-like receptors—TLRs) and their intracellular signaling pathways evolved to function well in a hypoxic environment [21–23].

Neutrophils: Neutrophils function well in hypoxia, an environmental condition which often exists in portions of infected tissues whose blood supply can become compromised. The sympathetic stress of hypoxia is known to increase circulating neutrophils both by increased maturation in the bone marrow and by vascular demargination [9, 24, 25]. HIF-1 alpha signaling initiated by hypoxia is critical to neutrophil bacterial killing by enhancing ATP generation to support the aggregation, motility, chemotaxis, superoxide generation, and phagocytosis [26–29]. Chronic and intermittent hypoxia via HIF-1 alpha signaling also prolongs the otherwise short (3–5 days) life-span of neutrophils by inhibiting apoptosis [30, 31]. The importance of HIF-1 alpha in neutrophils is manifested by the impaired ability of HIF-1 alpha null phagocytes to effectively eliminate bacterial growth [27, 28] and by the enhanced life-span and phagocytic capacity of neutrophils in von Hippel–Lindau disease, in which degradation of HIF-1 alpha is impaired [32], or the augmented phagocytic ability of neutrophils in mice given the HIF-1 agonist, mimosine [33]. In opposition to HIF-1 alpha, adenosine, an intermediate metabolite of ATP, which is increased with hypoxia, via binding to the adenosine A-2a receptor on neutrophils, as well as to natural killer cells and lymphocytes down-regulates pro-inflammatory signaling [34, 35].

Monocytes and macrophages: These cells are similarly responsive to hypoxia with up-regulation of HIF-1 and HIF-2 leading to an array of both up- and down-regulated genes [36, 37]. In general these gene expression changes allow macrophages to better function in hypoxic areas in their phagocytic and killing abilities (generation of large amounts of NO, oxidants, proteases, and defensins) and in their immune activating capacity. An important pro-inflammatory cytokine generated by a variety of hypoxic cells including macrophages is macrophage migration inhibitory protein (MIF), which acts to limit random macrophage movement in order to concentrate macrophages at the sites of injury and infection [38]. When made hypoxic-resident alveolar macrophages in the lung release into the circulation monocyte chemoattractant protein (MCP-1), a chemokine which activates systemic perivascular mast cells to initiate microvascular inflammatory cascade and may account for some of the systemic effects of alveolar hypoxia [39]. How relevant these acute studies in mice at 10 % O_2 are to humans is uncertain since this more severe hypoxia, equivalent to 6,000 m, is rarely experienced and small rodents with their greater metabolism may be more sensitive to any level of hypoxia than humans. Alveolar macrophages harvested from humans at 4,559 m showed no evidence of activation compared to those harvested at near sea level [40].

Dendritic cells: Dendritic cells (DC) are specialized monocytes that migrate to sites of infection and then travel to regional lymph nodes as antigen-presenting cells. Hypoxia leads to a more pro-inflammatory profile of these cells [41–44]. HIF-1 is expressed in high levels in these cells [45] and is critical to these functions. Hypoxic exposure of dendritic cells with a conditional deletion of HIF-1 leads to less migratory capacity, decreased production of interleukin-22, a cytokine important in the production of antimicrobial peptides, and greater apoptosis [45, 46].

Natural killer cells: Natural killer (NK) cells are part of the innate immune system and do not express markers of either T- or B-cell lineage. NK cells primarily kill virus-infected cells and some tumor cells and are therefore important in fighting infection and cancer. Several studies have demonstrated that acute hypoxia results in increases in natural killer cell numbers and activity [6, 7, 9]. The role of HIF in NK cell function studied either by gene deletion or by siRNA knockdown has not yet been explored in any depth, although one study unexpectedly found when HIF-1 was repressed by siRNA, the population of NK cells was not increased as compared with control cells under hypoxia. These results indicate that the inhibition of NK cell differentiation is regulated by HIF-1-independent mechanism under hypoxia [47]. The finding of enhanced NK cell number and activity with acute hypoxia in vivo appears to be transient as levels return to normal with more prolonged exposure to hypoxia [6, 7, 48]. A potential mechanistic role involving the sympathoadrenal pathways has been suggested. In support of this, epinephrine infusion can mimic the effect of hypoxia on NK cells while β-adrenergic blockade abolishes the increase in NK cell number [9].

Pathogen-associated molecular pattern (PAMP) responses: In addition to the cellular arms of the innate immune system, most cells of the body have membrane surface proteins that upon binding to pathogens or pathogen-associated molecules trigger intracellular signaling pathways to up-regulate pro-inflammatory responses and cytokine expression and alter apoptosis. This is a highly active area of research and multiple PAMPs and related damage-associated molecular pattern (DAMP) molecules are being identified and their roles explored. Most studied of these include the many TLRs, which evolved early in evolution in single cell organism host defense. Several of the TLRs are HIF-1 alpha-sensitive [49–51] and themselves can up-regulate HIF-1 alpha to act in a synergistic fashion [52]. In macrophages hypoxia increases TLR-4 expression [53], while in dendritic and monocytic cells, hypoxia causes induction of TLR-2 and TLR-6 via HIF-1 alpha [54]. HIF-1 alpha stimulates inflammatory gene up-regulation via NFκB and reactive oxygen species-dependent pathways as do the TLRs [55, 56].

Cytokines and other signaling molecules: Cytokines are small proteins with hormone-like characteristics involved in cell-to-cell signaling. They help regulate the intensity and duration of the immune response. Recently, the influence of high altitude exposure and hypoxia on inflammatory cytokines has received more attention with a special interest in IL-6. It appears that the environmental stress of altitude/hypoxia alone is sufficient to cause an elevation in circulating IL-6. In

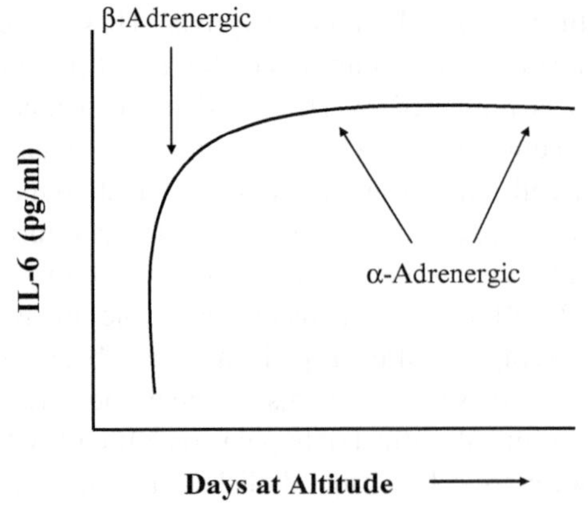

Fig. 14.3 The response of circulating interleukin-6 following both acute and chronic exposure to the summit of Pikes Peak (4,300 m). Initially, the large increases in IL-6 levels with acute hypoxia are likely mediated via the epinephrine-stimulated β-adrenergic pathway. During acclimatization, epinephrine levels decline; however, IL-6 levels remain elevated. This is likely related to the continued increase in SNS activity with acclimatization and the direct norepinephrine stimulation of the α-adrenergic pathway, as α-blockade completely abolishes this IL-6 response observed over time at altitude

humans, acute hypoxia elevates serum IL-6 levels significantly while other pro-inflammatory cytokines (IL-1 beta, TNF-alpha) remained unchanged [57]. Animal studies extend these findings as cultured endothelial cells [58] acutely exposed to hypoxia demonstrate an increase in the expression of IL-6 as do neonatal cardiac myocytes after 4 h of hypoxia [59]. Results from investigations conducted at the summit of Pikes Peak indicate that resting IL-6 levels increase immediately upon arrival to altitude and remain elevated for several weeks with continued exposure [8]. Therefore, both acute and chronic altitude exposure result in elevated resting IL-6 levels in humans.

A sympathoadrenal mechanism responsible for this observation has been proposed [8]. It is known that the catecholamines can act as a potent stimulator for IL-6 production via activation of the β-adrenergic pathways [60–63]. β-Adrenergic activation upon initial exposure to high altitude is consistent with the rapid increase in both epinephrine and IL-6 levels associated with acute altitude exposure (Fig. 14.3).

However, resting IL-6 levels remain elevated over time at altitude while epinephrine levels return toward sea level values. This would suggest that other mechanisms are responsible to the sustained increase in IL-6 at altitude. Increase in SNS activity over time at altitude likely contributes to this observation [8]. Specifically, the release of norepinephrine from SNS nerves can act directly on α-adrenergic receptors. In the Pikes Peak studies, the administration of the α-adrenergic blocker prazocin completely abolished the sustained increase in resting IL-6 levels [8]. Thus, while IL-6 levels remained elevated throughout 12 days duration at altitude for the group receiving the placebo, IL-6 levels returned to sea level values by day 3 and remained there throughout the remainder of stay at altitude in subjects receiving α-adrenergic blockade. This is further supported by a significant correlation ($p = 0.004$) between IL-6 levels and urinary norepinephrine excretion rates (a marker of overall sympathetic nerve activity) for subjects over the course of time while at altitude. Other studies have reported a similar relationship between peak plasma norepinephrine and IL-6 levels during high-intensity exercise in humans [63, 64]. The significance of the elevated IL-6 response to altitude on immune function remains unknown. It is classically thought of as a pro-inflammatory cytokine, but it may have anti-inflammatory properties as well, so its overall impact may vary in different circumstances [65]. Furthermore it may promote angiogenesis [66], induce vascular endothelial growth factor (VEGF, [67]), and modulate erythropoietin production [57, 68].

Adaptive Immune System

T lymphocytes: T lymphocytes are developed in the thymus and are integral components of cell-mediated immunity. T-cell subsets carry out a variety of immunological functions including helper, suppressor, and cytotoxic (killing cells already infected with viruses and other intracellular pathogens). The literature clearly indicates that cell-mediated immunity and T lymphocyte function is the branch of the immune system most adversely affected by acute exposure to high altitude [48]. Several investigations involving human subjects suggest that acute exposure to high altitude can suppress T-cell function (proliferation) as well as alter T-cell subset ratios [69, 70]. In a well controlled, multifaceted study now referred to as Operation Everest II, seven male subjects were progressively brought to a simulated altitude of 7,620 m (25,000 ft) in a hypobaric chamber over the course of 4 weeks [48]. Results indicated that T-cell function was significantly impaired during hypoxia as indicated by significant reductions in both phytohaemagglutinin (PHA)-stimulated lymphocyte proliferation and protein synthesis. It was postulated that alterations in circulating stress hormones associated with acute hypoxia may be, in part, responsible for this finding. This supports two earlier studies examining PHA-stimulated lymphocyte proliferation in human subjects exposed to altitudes of 3,200–3,800 m [71, 72]. In these studies a significant reduction was found in T-cell-specific blastogenesis after 5 days at high altitude with the added observation that values returned to baseline after 25 days at altitude.

It has been suggested [69] that the decrease in PHA-stimulated lymphocyte proliferation associated with high altitude (5,050 m) may be related to the decline in the specific subset of CD4+ T-cells (Th1). Possible alterations in IFN-γ and IL-2 production related to lower numbers of Th1 cells are likely responsible. More recently, it has been suggested that this observed impairment of Th1/Th2 immune balance is, in part, regulated by the alterations in the sympathoadrenal axis during exposure to high altitude [73].

Specifically, T-cell receptor signal transduction has been shown to be inhibited by HIF-1 via a calcium signaling pathway [74]. The HIF-1 response favors T helper2 (Th2)-regulated pathways while down-regulating the T helper1 (Th1) circuit [75]. Ben-Shoshan et al. [15] have reported an HIF-1α-mediated association between hypoxia and regulatory T-cells resulting in favorable anti-inflammatory conditions. In contrast targeted deletion of HIF-1 alpha in lymphocytes improves antibacterial response in septic mice [76]. Others [77] have demonstrated that in response to acute hypoxia, both HIF-1 and NFκB play a key role in this anti-inflammatory response. In part this may be related to HIF-1 alpha-induced differentiation of T regulatory cells (Treg), whose role is generally to blunt inflammation [78]. Redistribution of T lymphocytes and a reduction in CD4+ T-cell numbers as well as a suppression in T-cell activation and proliferation have been consistent observations [6, 7, 48, 67].

B lymphocytes: B lymphocytes are developed in the bone marrow of most mammals and make up an integral component of the humoral immune response. The principal functions of B-cells are to make antibodies (immunoglobulins—Ig) against specific antigens. B-cells also play a role in immunological memory after activation by antigen interaction. It is generally observed that this branch of immunity is only marginally affected by high altitude exposure [48, 69, 79–81], although lymphopenia has been reported to occur with acute exposure to hypoxia [9, 24] possibly as a result of the associated increase in cortisol. In the Operation Everest II study cited above, it was reported that no differences were found at any altitude in polyclonal B-cell function [48]. After 9 days at altitude, no differences in spontaneous or pokeweed mitogen (PWM)-stimulated IgA, IgG, or IgM production were found. Normal and increased antibody responses to both bacterial and viral vaccinations have been consistently reported in other studies at altitudes ranging from 2,300 to 4,200 m [79, 81, 82]. Consequently, B-cell function appears to be less affected by high altitude exposure.

Intermittent Hypoxia/Altitude Exposure

Intermittent hypoxia exposure has been employed in attempts to pre-acclimatize individuals before ascent to high altitude (military personnel, mountaineers, athletes). This can be achieved by either breathing hypoxic gas mixtures or exposure to hypobaric hypoxia via an altitude chamber. A few studies have assessed the impact of intermittent hypoxia on components of immune function [83, 84]. Recently, it was demonstrated that in young, healthy men, 14 days of intermittent hypoxia (four 5-min bouts/day breathing 10 % oxygen) resulted in enhanced innate immunity by mobilizing hematopoietic stem and progenitor cells thereby activating neutrophils and increasing circulating complement and immunoglobulins [84].

Using hypobaric hypoxia to simulate an altitude of 4,300 m (barometric pressure = 446 mmHg) it was observed that white blood cell and several leukocyte counts were elevated during acute altitude exposure but returned to baseline values after 3 weeks of intermittent altitude exposure [83]. Stress hormones (cortisol, epinephrine, and norepinephrine) were not affected by either acute or chronic exposure to intermittent altitude exposure. It was concluded that this method of pre-acclimatization can safely be employed for inducing altitude acclimatization without negatively impacting immune function.

Exercise at Altitude: An Added Stressor

Importantly, individuals are generally physically active while at altitude (hiking, climbing, skiing, etc.). It has been demonstrated that exercise represents an added stress that, when combined with hypoxia, elicits a more pronounced neuroendocrine response when compared to the same absolute intensity at sea level [8, 10–12]. Consequently, exercise at high altitude is likely to have a greater impact on immune function then the hypoxic exposure alone.

There have been numerous studies documenting the effect of a single bout of exercise on immune function at sea level (for reviews, see [85, 86]). It is clear that an acute bout of exercise is a physical stressor that can transiently affect immune function. A key factor dictating the impact of the exercise bout on suppressing immune function is related to the exercise intensity or relative stress. The greater the relative exercise intensity, the greater the disruption in homeostasis and subsequent neuroendocrine responses. As the same absolute workload represents a greater relative intensity at high altitude when compared to sea level (as maximal oxygen consumption, VO_2max, declines with ascending altitude) the greater relative stress will have a more pronounced effect on immune function. Unfortunately, there have been only a few studies that have examined the effect of exercise on immune function during altitude/hypoxic exposure. However, it appears that even when controlling for the relative exercise intensity, hypoxia represents an added stressor to that imposed by exercise alone. When subjects exercised for 20 min under both normoxic and hypoxic (11.5 % O_2) conditions, it was demonstrated that exercise during hypoxia resulted in a significantly greater NK cell response when compared to exercise in normoxic conditions [7], but an inhibition of neutrophil cytotoxic activity [25]. A training program consisting of daily exercise with 15 % O_2 for 4 weeks also found enhanced NK cell function [87]. Hypoxic exercise appears to induce a more pronounced immunological stress response than exercise under sea level conditions.

Further, it has been shown that when exercise is performed at the same relative workload (indicated by similar blood lactate levels), sympathetic nerve activity is elevated to a greater extent at altitude (1,800 m) when compared to sea level [63]. More importantly, the increase in circulating epinephrine and norepinephrine correlated with the greater exercise-induced increase in IL-6 levels at altitude. These findings are consistent with the concept of an exacerbated immunological stress response when exercise and hypoxia are combined. Several studies conducted at sea level have identified skeletal muscle as a potential source for the elevated circulating IL-6 during exercise and this likely holds true with exercise at altitude [9, 88, 89].

Similar findings have been reported during submaximal exercise (50 % of VO_2max for 50 min) under conditions of both acute (day 1) and chronic (day 12) altitude exposure at 4,300 m. With acute altitude exposure, IL-6 levels were significantly elevated during exercise and correlated with increasing epinephrine levels [8]. As described above, this appears to be mediated primarily via β-adrenergic stimulation as α-adrenergic blockade had no effect in reducing the exercise-induced increase in IL-6 during acute hypoxia. After 12 days of acclimatization, however, α-adrenergic blockade significantly lowered IL-6 levels during exercise to values found at sea level. The placebo group still demonstrated elevated IL-6 levels during exercise suggesting a strong α-adrenergic component with more prolonged residence at high altitude. Taken together, these studies suggest an additive effect of the physical stress imposed by exercise with that of hypoxic stress. This results in a more pronounced sympathoadrenal response that has implications for immune function.

Exercise training: Currently, the "Live High-Train Low" paradigm is popular among endurance athletes for the purported benefits leading to an improvement in performance at sea level [90–92]. This design calls for altitude exposure of at least 12–16 h/day for 3–4 weeks to elicit the desired adaptations (erythropoietin, red blood cell volume, etc.) while returning to a lower altitude for training.

To date, only one study has examined the effect of "Live High-Train Low" on immune function [93]. In a study directed at mucosal immunity, trained cross-country skiers after an 18-day period of training low at 1,200 m and living high at 3,500 m demonstrated a significant decrease in secretory immunoglobulin A (sIgA). The control group (lived and trained at 1,200 m) had no changes in sIgA levels. It was suggested that a cumulative adverse effect of exercise training and hypoxia on mucosal immunity occurred over time at altitude.

When athletes both live and train at high altitude, the effect of hypoxia on suppressing some immune function is evident. When members of the Australian Olympic swimming team were exposed to a 21-day training camp at 2,100 m, leukocyte numbers and concanavalin-A (ConA)-induced blastogenesis were reduced 38 % and 32 %, respectively, when compared to sea level values [70].

Clearly, more studies need to be conducted to determine the extent to which these various training models involving altitude/hypoxia influence immune function and whether the actual susceptibility and incidence of illness is affected.

Relation to Infections and Cancer at High Altitude

The many and variable responses of the neuroimmune system to acute and chronic hypoxia are predicted to have possible impact on infectious risks and severity of infections [94]. However, given the myriad associated aspects of high altitude for both residents and visitors separate from hypoxia that may influence immune status or transmission of infection, the ability to study the separate impact of hypoxia is daunting. Other relevant climatic and socioeconomic differences of high altitude include UV radiation, cold, hazardous weather, poverty, personal and community hygiene, home and outdoor air pollution, nutritional status, population density, and access to basic medical prevention and care. In addition to their influence on host defenses, UV radiation, humidity, and cold will affect the viability of infectious agents and/or their insect vectors, such as mosquitoes [95, 96].

Notwithstanding the inherent limitations of cross-sectional and observational investigations owing from lack of control of confounding factors, surveys have found both positive and inverse associations with certain infections at high altitude. Historically tuberculosis (TB) has been of greatest interest. Numerous studies have found a lower mortality from incidence/prevalence of TB with altitudes above 2,000–3,000 m [97–104]. The virulence and transmissibility of *Mycobacterium tuberculosis* is well known to be reduced under hypoxic conditions [105, 105a], low humidity [106], and UV radiation [107]. These environmental factors, in addition to the rest, better nutrition, and escape from crowded and polluted lowland cities afforded by the mountain environment in the decades before availability of antituberculous drug therapy, have been considered instrumental to the therapeutic success of the alpine sanatorium.

In contrast, the general view is that most infections are worse at high altitude. This seems to be the case for childhood pneumonia [108] and respiratory syncytial virus infection [109]. Because the majority of bacteria, fungi, and viruses are either anaerobic or can grow in anaerobic conditions either by fermentation or using nitrate and nitrite as terminal electron acceptors, it appears that hypoxia is not a deterrent to their growth, and may in fact enhance growth by depression of lymphocyte function and adaptive immunity [94]. Interestingly, a 14-day hypobaric exposure to 18,000 ft in mice likely to upregulate HIF-1 alpha in all leukocytes reduced mortality from pneumococcal infection when the animals were infected under normoxic conditions when compared to mice not given the hypoxic preexposure [110].

There has been very little work on the effect of altitude on vaccine efficacy, but one study in healthy persons living at 16,000 ft showed equivalent specific antibody response to polysaccharide C of *N. meningitides* compared to a group vaccinated at sea level. The same findings were observed in normoxic mice and those maintained in a hypobaric chamber at a barometric pressure of 360 mmHg [111].

The immune system can recognize neoantigens on malignantly transformed cells as foreign and act to repress tumor growth and metastatic spread. It is reasonable to ask whether the immune-modulating effects of hypoxia might also extend to the immunosurveillance of cancer. Dendritic cells and lymphocytes are critical elements in host defense against malignancy and the functions of both are altered with hypoxia. When mice were studied at 4,500 m, spontaneous lung tumors [112] and metastatic spread of experimental-injected cancer cells [113, 114] were greater than in mice at sea level. In contrast, epidemiological surveys of cancer incidence and altitude find in general that more modest altitudes (2,100–3,500) may be associated with decreased cancer rates than in low altitude (<200 m) populations. For all cancers, breast cancer, respiratory tract cancers, and non-Hodgkin lymphoma in the United States, there appears to be about a 10–20 % lower standard mortality rate (SMR) in people living above 2,100 m [115]. These recent data confirm earlier work, which also found lower SMRs for leukemia, oropharyngeal cancer, and gastrointestinal and multiple myeloma [116–118]. The only cancers to show a greater rate at higher altitudes are melanoma likely due to greater UV radiation exposure [119], carotid body tumors or chemodectomas [120], and placental chorioangioma [121]. As with infections at high altitude, other environmental, socioeconomic, cultural, and ethnic aspects of life are difficult to control as possible confounding factors. Although hypoxia in general may depress lymphocyte-mediated functions involved in cancer recognition, hypoxia and HIF-1 alpha stimulation of NK cell tumor surveillance [122, 123] may be more important and underlie the epidemiological findings of reduced cancer at high altitude.

Summary

Exposure to high altitude is an environmental stress that elicits a variety of changes in the immune system that are additionally modified by neuroendocrine response to hypoxia. The ultimate elevation achieved and/or degree of the hypoxia is a primary factor that influences the extent of physiological stress, neuroendocrine adjustments, and immunological changes associated with exposure. Additionally, how physically active an individual is while at altitude may also play a role in the immunological responses. Virtually all components of the immune system are affected. In general, elements of the innate immune arm appear to be enhanced with hypoxia in contrast to impairment in T-cell-mediated immunity. Despite what appears to more robust innate immunity with hypoxia, it seems that the risk of most infections is greater during the initial days of exposure to high altitude. With acclimatization, hypoxic stress is lessened as arterial oxygen content increases over time and adrenal medullary release of epinephrine returns toward sea level values. Consequently, T-cell function returns toward normal approaching sea level values and the risk of infection may decline. It is important to note that any suppression, as for example, of T-cell function, that depends on oxygen availability will improve with prolonged exposure because oxygen supply of the tissues increases with acclimatization. There is, however, a great interindividual variability regarding responses and sensitivity to acute and chronic high altitude exposure (e.g., neuroendocrine, acute mountain sickness, red blood cell production). Thus, some individuals may be more susceptible to alterations in immune function and infection than others, possibly at genetic level. Future studies should focus on elucidating the precise mechanisms responsible for alterations in immune function during high altitude exposure including assessment of the time course of these immunological changes. Furthermore, well designed studies are needed to determine what contribution the changes in immune function with hypoxia have on the incidence and severity of infections at high altitude, as well as cancer. It would be useful to know if preacclimatization strategies used to enhance physical performance at high altitude might also lower infectious risks for those going to high altitude.

References

1. Webster JI, Tonelli L, Sternberg EM. Neuroendocrine regulation of immunity. Annu Rev Immunol. 2002; 20:125–63.
2. Humpeler E, Skrabal F, Bartsch G. Influence of exposure to moderate altitude on the plasma concentration of cortisol aldosterone renin testosterone and gonadotropins. Eur J Appl Physiol. 1980;17: 167–76.
3. Sawhney RC, Malhotra AS, Singh T. Glucoregulatory hormones in man at high altitude. Eur J Appl Physiol. 1991;62:286–91.
4. Sanders VM, Kavelaars A. Adrenergic regulation of immunity. Psychoneuroimmunology. 2007;1:63–83
5. Duplain H, Vollenweider L, Delabays A, et al. Augmented sympathetic activation during short-term hypoxia and high-altitude exposure in subjects susceptible to high-altitude pulmonary edema. Circulation. 1999;99:1713–8.
6. Klokker M, Kharazmi A, Galbo H, et al. Influence of in vivo hypobaric hypoxia on function of lymphocytes, neutrocytes, natural killer cells, and cytokines. J Appl Physiol. 1993;74:1100–6.
7. Klokker M, Kjaer M, Secher NH, et al. Natural killer cell response to exercise in humans: effect of hypoxia and epidural anesthesia. J Appl Physiol. 1995;78: 709–16.
8. Mazzeo RS, Donovan D, Fleshner M, et al. Interleukin-6 response to exercise and high-altitude exposure: influence of α-adrenergic blockade. J Appl Physiol. 2001;91:2143–9.
9. Pedersen BK, Steensberg A. Exercise and hypoxia: effects on leukocytes and interleukin-6-shared mechanisms? Med Sci Sports Exerc. 2002;34: 2004–12.
10. Mazzeo RS, Wolfel EE, Butterfield GE, et al. Sympathetic responses during 21 days at high altitude (4,300 m) as determined by urinary and arterial catecholamines. Metabolism. 1994;43:1226–32.
11. Mazzeo RS, Brooks GA, Butterfield GE, et al. Acclimatization to high altitude increases muscle sympathetic activity both at rest and during exercise. Am J Physiol. 1995;269:R201–7.
12. Mazzeo RS, Child A, Butterfield GE, et al. Sympathoadrenal responses to submaximal exercise in women after acclimatization to 4,000 m. Metabolism. 2000;49:1036–42.
13. Berger MM, Luks AM, Bailey DM, et al. Transpulmonary plasma catecholamines in acute high-altitude pulmonary hypertension. Wilderness Environ Med. 2011;22:37–45.
14. Richalet JP, Letournel M, Souberbielle JC. Effects of high-altitude hypoxia on the hormonal response to hypothalamic factors. Am J Physiol. 2010;299: R1685–92.
15. Ben-Shoshan J, Maysel-Auslender S, Mor A, et al. Hypoxia controls CD4+CD25+ regulatory T-cell homeostasis via hypoxia-inducible factor-1α. Eur J Immunol. 2008;38:2412–8.
16. Gale DP, Maxwell PH. The role of HIF in immunity. Int J Biochem Cell Biol. 2010;42:486–94.
17. Imtiyaz HZ, Simon MC. Hypoxia-inducible factors as essential regulators of inflammation. Curr Top Microbiol Immunol. 2010;345:105–20.
18. Taylor CT, Cummins EP. The role of NFκB in hypoxia-induced gene expression. Ann N Y Acad Sci. 2009;1177:178–84.
19. Tissot van Patot MC, Serkova NJ, Haschke M, et al. Enhanced leukocyte HIF-1 alpha and HIF-1 alpha DNA binding in humans after rapid ascent to 4300 m. Free Radic Biol Med. 2009;46:1551–7.
20. Martin W, Rotte C, Hoffmeister M, et al. Early cell evolution, eukarocytes, anoxia, sulfide, oxygen fungi first (?) and a tree of genomes revisited. IUBMB Life. 2003;55:193–204.
21. Sica A, Melillo G, Varesio L. Hypoxia: a double edged sword of immunity. J Mol Med. 2011;89:657–65.
22. Eltzschig KK, Carmeliet P. Hypoxia and inflammation. N Engl J Med. 2011;364:656–65.
23. Nizet V, Johnson RS. Interdependence of hypoxic and innate immune response. Nat Rev Immunol. 2009;9:609–17.
24. Thake CD, Mian T, Garnham AW, et al. Leukocyte counts and neutrophil activity during 4 h of hypocapnic hypoxia equivalent to 4000 m. Aviat Space Environ Med. 2004;75:811–7.
25. Chouker A, Demetz F, Martignoni A, et al. Strenuous exercise inhibits granulocyte activation induced by high altitude. J Appl Physiol. 2005;98:640–7.
26. Tamura DV, Moore EE, Pactrick DA, et al. Acute hypoxemia in humans enhances neutrophil inflammatory response. Shock. 2002;17:269–73.
27. Cramer T, Yamanishi Y, Clausen BE, et al. HIF-1alpha is essential for myeloid cell mediated inflammation. Cell. 2003;112:645–57.
28. Peyssonnaux C, Datta V, Cramer T, et al. HIF-1alpha expression regulates the bactericidal capacity of phagocytes. J Clin Invest. 2005;115:1806–15.
29. Hitomi Y, Miyamura M, Mori S, Suzuki K, Kizaki T, Itoh C, Murakami K, Haga S, Ohno H. Intermittent hypobaric hypoxia increases the ability of neutrophils to generate superoxide anion in humans. Clin Exp Pharmacol Physiol. 2003;30:659–64.
30. Walmsley SR, Print C, Farahi N, et al. Hypoxia-induced neutrophil survival is mediated by HIF-1alpha dependent NF-kappa beta activity. J Exp Med. 2005;201:105–11.

31. Dyugovskaya L, Polyakov A, Lavie P, Lavie L. Delayed neutrophil apoptosis in patients with sleep apnea. Am J Respir Crit Care Med. 2008;177: 544–54.
32. Walmsley SR, Cowburn AS, Clatworthy MR, et al. Neutrophils from patients with heterozygous germline mutations in the von Hippel-Lindau protein (pHVL) display delayed apoptosis and enhanced bacterial phagocytosis. Blood. 2006;108: 3176–8.
33. Zinkernagel AS, Peyssonnaux C, Johnson RS, Nizet V. Pharmacologic augmentation of HIF-1alpha with mimosine boosts the bactericidal capacity of phagocytes. J Infect Dis. 2008;197:214–7.
34. Linden J. Regulation of leukocyte function by adenosine receptors. Adv Pharmacol. 2011;61:95–114.
35. Eltzschig H, Sitkoivsky MV, Robson SC. Purinergic signaling during inflammation. N Engl J Med. 2012;367:2322–33.
36. Staples KJ, Pearson H, Frankenberger M, et al. Monocyte-derived macrophages matured under prolonged hypoxia transcriptionally upregulate HIF-1 alpha m RNA. Immunobiology. 2011;216: 832–9.
37. Fang HY, Hughes R, Murdoch C, et al. Hypoxia-inducible factors 1 and 2 are important transcriptional effectors in primary macrophages experiencing hypoxia. Blood. 2009;114:844–59.
38. Ietta F, Wy Y, Romagnoli R, et al. Oxygen regulation of macrophage migration inhibitory factor in human placenta. Am J Physiol. 2007;292:E272–80.
39. Chao J, Wood JG, Gonzalez NC. Alveolar macrophages initiate the systemic microvascular response to alveolar hypoxia. Respir Physiol Neurobiol. 2011;178:439–48.
40. Swenson ER, Mongovin S, Maggiorini M, et al. Alveolar macrophage interleukin 6 response to hypoxia and lipopolysacchride in HAPE-Susceptible and -resistant mountaineers. AMJ Respir Crit Care Med. 2001;163:A618.
41. Bosco MC, Varesio L. Dendritic cell reprogramming by the hypoxic environment. Immunobiology. 2012; 217:1241–9.
42. Blegio F, Raggi F, Pierobon D, et al. The hypoxic environment reprograms the cytokine/chemokine expression profile of human dendritic cells. Immunobiology. 2013;218:76–89.
43. Bosco MC, Pierobon D, Blengio F, et al. Hypoxia modulates the gene expression profile of immunoregulatory receptors in human mature dendritic cells: identification of TREM-1 as a novel hypoxic marker in vitro and in vivo. Blood. 2011;117: 2626–39.
44. Mancino A, Schioppa T, Larghi P. Divergent effects of hypoxia on dendritic cell functions. Blood. 2008;112:3723–34.
45. Naldini A, Morena E, Pucci A, et al. Hypoxia affects dendritic cell survival: role of hypoxia inducible factor-1alpha and lipopolysaccharide. J Cell Physiol. 2012;227:587–95.
46. Kohler T, Reizis B, Johnson RS, et al. Influence of hypoxia-inducible factor 1alpha on dendritic cell differentiation and migration. Eur J Immunol. 2012; 42:1226–36.
47. Yun S, Lee SH, Yoon SR, et al. Oxygen tension regulates NK cell differentiation from hematopoietic stem cells in vitro. Immunol Lett. 2011;137:70–7.
48. Meehan R, Duncan U, Neale L, et al. Operation Everest II: alterations in the immune system at high altitudes. J Clin Immunol. 1988;8:397–406.
49. Paone A, Galli R, Gabellini C, et al. Toll-like receptor 3 regulates angiogenesis and apoptosis in prostate cancer cell lines through hypoxia-inducible factor 1 alpha. Neoplasia. 2010;12:539–49.
50. Fan P, Zhang FP, Wang B, et al. Hypoxia-inducible factor-1 up-regulates the expression of Toll-like receptor 4 in pancreatic cancer cells under hypoxic conditions. Pancreatology. 2012;12:170–8.
51. Kim SY, Jeong E, Joung SM, Lee JY. PI3K/Akt contributes to increased expression of Toll-like receptor 4 in macrophages exposed to hypoxic stress. Biochem Biophys Res Commun. 2012;419: 466–71.
52. Jantsch J, Wiese M, Schodel J, et al. Toll-like receptor activation and hypoxia use distinct signaling pathways to stabilize hypoxia-inducible factor 1 alpha (HIF1A) and result in differential HIF-1A-dependent gene expression. J Leukoc Biol. 2011; 90:551–60.
53. Kim SY, Choi YJ, Juong SM, et al. Hypoxic stress up-regulates the expression of Toll-like receptor4 in macrophages via hypoxia inducible factor. Immunology. 2009;129:516–24.
54. Kuhicke J, Frick JS, Morote-Garcia JC, et al. Hypoxia inducible factor (HIF-1) coordinates induction of Toll-like receptors TLR2 and TLR-6 during hypoxia. PLoS One. 2007;12:e1364.
55. Mkaddem SB, Bens M, Vanderwalle A. Differential activation of Toll-like receptor-mediated apoptosis induced by hypoxia. Oncotarget. 2010;1:741–50.
56. Hallam S, Escorcio-Correia M, Spoer R, et al. Activated macrophages in the tumor environment-dancing to the tune of TLR and NFκB. J Pathol. 2009;219:143–52.
57. Klausen T, Olsen NV, Poulsen TD, et al. Hypoxemia increases serum interleukin-6 in humans. Eur J Appl Physiol Occup Physiol. 1997;76:480–2.
58. Yan SF, Tritto I, Pinsky D, et al. Induction of interleukin 6 (IL-6) by hypoxia in vascular cells. Central role of the binding site for nuclear factor-IL-6. J Biol Chem. 1995;270:11463–71.
59. Yamauchi-Takihara K, Ihara Y, Ogata A, et al. Hypoxic stress induces cardiac myocyte-derived interleukin-6. Circulation. 1995;91:1520–4.
60. DeRijk RH, Boelen A, Tilders FJ, et al. Induction of plasma interleukin-6 by circulating adrenaline in the rat. Psychoneuroendocrinology. 1994;19: 155–63.
61. Soszynski D, Kozak W, Conn CA, et al. Beta-adrenoceptor antagonists suppress elevation in body

temperature and increase in plasma IL-6 in rats exposed to open field. Neuroendocrinology. 1996; 63:459–67.
62. van Gool J, van Vugt H, Helle M, et al. The relation among stress, adrenalin, interleukin 6 and acute phase proteins in the rat. Clin Immunol Immunopathol. 1990;57:200–10.
63. Niess AM, Fehrenbach E, Strobel G, et al. Evaluation of stress responses to interval training at low and moderate altitudes. Med Sci Sports Exerc. 2003;35:263–9.
64. Papanicolaou DA, Petrides JS, Tsigos C, et al. Exercise stimulates interleukin-6 secretion: inhibition by glucocorticoids and correlation with catecholamines. Am J Physiol. 1996;271:E601–5.
65. Pedersen BK. The anti-inflammatory effect of exercise: its role in diabetes and cardiovascular disease control. Essays Biochem. 2006;42:105–17.
66. Motro B, Itin A, Sachs L, et al. Pattern of interleukin 6 gene expression in vivo suggests a role for this cytokine in angiogenesis. Proc Natl Acad Sci. 1990;87:3092–6.
67. Caldwell C, Kojima H, Lukashev D. Differential effects of physiologically relevant hypoxic conditions on T lymphocyte development and effector functions. J Immunol. 2001;167:6140–9.
68. Faquin WC, Schneider TJ, Goldberg MA. Effect of inflammatory cytokines on hypoxia-induced erythropoietin production. Blood. 1992;79:1987–94.
69. Facco M, Zilli C, Siviero M, et al. Modulation of immune response by the acute and chronic exposure to high altitude. Med Sci Sports Exerc. 2005;37:768–74.
70. Pyne DV, McDonald WA, Morton DS, et al. Inhibition of interferon, cytokine, and lymphocyte proliferative responses in elite swimmers with altitude exposure. J Interferon Cytokine Res. 2000;20:411–8.
71. Kitayev MI, Tokhtabayev AG. T and B lymphocytes as related to adaptation to high altitudes. Kosm Biol Avia Kosm Med. 1981;15:87–9.
72. Mirrakhimov MM, Kitayev MI. Problems and prospects of high-altitude immunology. Vestn Akad Med Nauk SSSR. 1979;4:64–9.
73. Ermolao A, Travain G, Facco M, et al. Relationship between stress hormones and immune response during high-altitude exposure in women. J Endocrinol Invest. 2009;32:889–94.
74. Neumann AK, Yang J, Biju MP, et al. Hypoxia inducible factor 1(alpha) regulates T cell receptor signal transduction. Proc Natl Acad Sci. 2005; 102:17071–6.
75. Pan F, Barbi J, Pardoll DM. Hypoxia-inducible factor 1: a link between metabolism and T cell differentiation and a potential therapeutic target. Oncoimmunology. 2012;1:510–5.
76. Theil M, Caldwell CC, Kreth S, et al. Targeted deletion of HIF-1alpha gene in T cells prevents their inhibition in hypoxic inflamed tissues and improves septic mice survival. PLoS One. 2009;9:3853.
77. Taylor CT. Independent roles for hypoxia inducible factor and nuclear factor-kappa B in hypoxic inflammation. J Physiol. 2008;586:4055–9.
78. Dang EV, Barbi J, Yang HY, et al. Control of TH17/Treg balance by hypoxia inducible factor 1. Cell. 2011;146:772–84.
79. Chohan IS, Sing I, Balakrishnan K. Immune responses in human subjects at high altitude. Int J Biometeorol. 1975;19:137–43.
80. Cohen T, Nahari D, Cerem LW, et al. Interleukin 6 induces the expression of vascular endothelial growth factor. J Biol Chem. 1996;271:736–41.
81. Krupina TN, Korotaev MM, Pukhova YI. Comparative evaluation of studies of different levels of hypoxia on the human immunological status. Kosm Biol Aviakosm. 1975;11:38–43.
82. Krupina TN, Korotaev MM, Pukhova YI. Characteristics of the human immunologic state during hypoxic hypoxia. Kosm Biol Avia Kosm Med. 1974;8:56–60.
83. Beidleman B, Muza SR, Fulco CF, et al. White blood cell and hormonal responses to 4300 m altitude before and after intermittent altitude exposure. Clin Sci. 2006;111:163–9.
84. Serebrovskaya TV, Nikolsky IS, Nikolska VV, et al. Intermittent hypoxia mobilizes hematopoietic progenitors and augments cellular and humoral elements of innate immunity in adult men. High Alt Med Biol. 2011;12:243–52.
85. Gleeson M, Nieman DC, Pedersen BK. Exercise, nutrition and immune function. J Sports Sci. 2004;22:115–25.
86. Pedersen BK, Hoffman-Goetz L. Exercise and the immune system: regulation, integration, and adaptation. Physiol Rev. 2000;80:1055–81.
87. Wang J-S, Weng T-P. Hypoxic exposure training promotes antitumor cytotoxicity of natural killer cells in young men. Clin Sci. 2011;121:343–53.
88. Ostrowski K, Hermann C, Bangash A, et al. A trauma-like elevation of plasma cytokines in humans in response to treadmill running. J Physiol. 1998;513:889–94.
89. Ullum H, Haahr PM, Diamant M, et al. Bicycle exercise enhances plasma IL-6 but does not change IL-1α, IL-1β, IL-6, or TNF-α pre-mRNA in BMNC. J Appl Physiol. 1994;77:93–7.
90. Hoppeler H, Vogt M. Hypoxia training for sea level performance. Training high-living low. Adv Exp Med Biol. 2001;502:61–73.
91. Levine BD, Stray-Gundersen J. "Living high-training low": effect of moderate-altitude acclimatization with low-altitude training on performance. J Appl Physiol. 1997;83:102–12.
92. Stray-Gundersen J, Chapman RF, Levine BD. "Living high-training low" altitude training improves sea level performance in male and female elite runners. J Appl Physiol. 2001;91:1113–20.
93. Tiollier E, Schmitt L, Burnat P, et al. Living high-training low altitude training: effects on mucosal immunity. Eur J Appl Physiol. 2005;94:298–304.
94. Schaible B, Schaffer K, Taylor CT. Hypoxia, innate immunity and infection in the lung. Respir Physiol Neurobiol. 2010;174:235–43.

95. Bodker R, Msangeni HA, Kisinza W, Lindsay SW. Relationship between the intensity of exposure to malaria parasites and infection in the Usambara mountains, Tanzania. Am J Trop Med Hyg. 2006; 14:716–23.
96. Achidi EA, Apinjoh TO, Mbunwe E, et al. Febrile status, malarial parasitemia and gastro-intestinal helmiththiasis in school children resident at different altitudes, in southwestern Cameroon. Ann Trop Med Parasitol. 2008;102:103–88.
97. Mansoer JR, Kibuga DK, Borgdorff MW. Altitude: a determinant for tuberculosis in Kenya. Int J Tuberc Lung Dis. 1999;3:156–61.
98. Tanrikulu AC, Acemoglu H, Palanci Y, Dagli CE. Tuberculosis in Turkey: high altitude and other socio-economic factors. Public Health. 2008;122:613–9.
99. Vree M, Hoa NB, Sy DN, et al. Low tuberculosis notification in mountainous Vietnam is not due to low case detection: a cross-sectional survey. BMC Infect Dis. 2007;7:109.
100. Saito M, Pan WK, Gilman RH, et al. Comparison of altitude effect on Mycobacterium tuberculosis infection between rural and urban communities in Peru. Am J Trop Med Hyg. 2006;75:49–54.
101. Gardiner CF, Webb GB, Ryder CT. Tuberculosis mortality in relation to altitude, humidity and population density. Trans Am Climatol Clin Assoc. 1923;39:197–208.
102. Pérez-Padilla R, Franco-Marina F. The impact of altitude on mortality from tuberculosis and pneumonia. Int J Tuberc Lung Dis. 2004;8:1315–20.
103. Vargas MH, Furuya ME, Pérez-Guzmán C. Effect of altitude on the frequency of pulmonary tuberculosis. Int J Tuberc Lung Dis. 2004;8:1321–4.
104. Singh I, Chohan IS, Lal M, et al. Effects of high altitude stay on the incidence of common diseases in man. Int J Biometeorol. 1977;21:93–122.
105. Sever JL, Youmans GP. The relation of oxygen tension to virulence of tubercle bacilli and to acquired resistance in tuberculosis. J Infect Dis. 1957;101: 193–202.
105a. Eisen S, Pealing L, Aldridge RW, et al. Effects of ascent to high altitude on human anti mycobacterial immunity. PLoS One. 2013;8:e74220.
106. Ko G, First MW, Burge HA. Influence of relative humidity on particle size and UV sensitivity of Serratia marcescens and Mycobacterium bovis BCG aerosols. Tuber Lung Dis. 2000;80: 217–28.
107. Riley RL, Knight M, Middlebrook G. Ultraviolet susceptibility of BCG and virulent tubercle bacilli. Am Rev Respir Dis. 1976;113:413–8.
108. Kan AJ, Hussain H, Omer SB, et al. High incidence of childhood pneumonia at high altitudes in Pakistan: a longitudinal cohort study. Bull World Health Organ. 2009;87:193–9.
109. Choudhuri JA, Ogden LG, Ruttenber J, et al. Effect of altitude on hospitalizations for respiratory syncytial virus infection. Pediatrics. 2006; 117:349–56.
110. Schmidt JP, Ball RJ. Atmospheric oxygen: effect on resistance in mice to pneumococcal pneumonia. Aerosp Med. 1970;41:1283–6.
111. Biselli R, Le Moli S, Matricardi PM, et al. The effects of hypobaric hypoxia on specific B cell responses following immunization in mice and humans. Aviat Space Environ Med. 1991;62:870–4.
112. Mori-Chavez P. Development of spontaneous pulmonary tumors at high altitude in strain A mice. J Natl Cancer Inst. 1962;28:55–65.
113. Mori-Chavez P, Salazar M. Study of metastasis after the intravenous injection of ascites carcinoma # 678 in to C3H mice at high altitude. J Natl Cancer Inst. 1965;35:193–9.
114. Mori-Chavez P, Salazar M. Growth of experimental tumors at high altitude. Natl Cancer Inst Monogr. 1964;14:309–19.
115. Youk AO, Buchanich JM, Fryzek J, et al. An ecological study of cancer mortality rates in high altitude counties of the United States. High Alt Med Biol. 2012;13:98–104.
116. Amsel J, Waterbor JW, Oler J, et al. Relationship of site-specific cancer mortality rates to altitude. Carcinogenesis. 1982;3:461–5.
117. Mason TJ, Miller RW. Cosmic radiation at high altitudes and US cancer mortality rate, 1950–1969. Radiat Res. 1974;60:301–6.
118. Weinberg CR, Brown KG, Hoel DG. Altitude, radiation and mortality from cancer and heart disease. Radiat Res. 1987;112:381–90.
119. Aceituno-Madera P, Buendia-Eisman A, Olmo FJ, et al. Melanoma, altitude and UV-B radiation. Acta Dermosifilogr. 2011;102:199–205.
120. Cuevas-Rodriguez S, Lopez-Garza J, Labastida-Almendaro S. Carotid body tumors in inhabitants of altitudes higher than 2000 m above sea level. Head Neck. 1998;20:374–8.
121. Rshetnikova OS, Burton GJ, Milovanov AP, Fokin EI. Increased incidence of placental choriocarcinoma in high altitude pregnancies: hypobaric hypoxia as a possible etiologic factor. Am J Obstet Gynecol. 1996;174:557–61.
122. Groth A, Klöss S, von Strandmann EP, et al. Mechanisms of tumor and viral immune escape from natural killer cell-mediated surveillance. J Immunol. 2011;3:344–51.
123. Wan J-S, Chia-Kuan C. Systemic hypoxia affects exercise-induced antitumor cytotoxicity of natural killer cells. J Appl Physiol. 2009;107: 1817–24.

Nutrition and Metabolism

15

George A. Brooks

Abstract

Like studies on isolated cells and tissues studied under hypoxic conditions, studies on men and women at altitude show a preference for carbohydrate-derived fuel energy sources (glycogen, glucose, and lactate). When dietary energy is adequate to cover need, at altitude working muscle utilizes little lipid. Classical concepts of a "Lactate Paradox" invoking a "Pasteur Effect" of presumed anaerobic metabolism are not supported by studies utilizing contemporary techniques. If anything, lactate shuttling is prominent at altitude, with lactate being produced in diverse tissue beds and serving at least two functions at altitude: lactate is a preferred fuel in working muscle and lactate is the major gluconeogenic precursor in support of glycemia at altitude. A third role of lactate, i.e., that of promoting cellular adaptations at altitude by increasing HIF-1 expression, is also suggested in the literature. While controlled laboratory studies show clear preference for carbohydrate-derived fuels under hypoxic conditions, cachexia is common among mountaineers. Under the stresses of altitude, loss of appetite and dietary energy insufficiency result in body wasting. Relative to energy content, carbohydrate foods are less efficient to carry or transport and CHO foods typically require water for cooking. Consequently, the tendency is to carry energy-dense, high-fat and protein foods at altitude. Because of the disparity between needs of the CNS and peripheral nerves and working muscles for CHO energy sources and dietary CHO supply, at altitude lean tissue is mobilized to supply gluconeogenic precursors and adipose is mobilized for glycerol (a lesser gluconeogenic precursor) and fatty acids (a less-preferred, but available energy substrate). The accomplishments of mountaineers are remarkable in many ways. That mountaineers accept and manage the challenge of working under multiple stresses, while having to rely on less-preferred fuel energy substrates is one more example of their extraordinary accomplishments.

G.A. Brooks, Ph.D. (✉)
Department of Integrative Biology,
University of California, 5101 VLSB, Berkeley,
CA 94720-3140, USA
e-mail: gbrooks@berkeley.edu

E.R. Swenson and P. Bärtsch (eds.), *High Altitude: Human Adaptation to Hypoxia*,
DOI 10.1007/978-1-4614-8772-2_15,

Introduction

In the previous volume of the series, Gail Butterfield and I reviewed the literature on the acute physiological response to hypobaric hypoxia encountered at elevations above 10,000 ft [1] following our extensive metabolic studies on men [1–8] and her similar studies on women on Pikes Peak [9–12]. In this update, I will draw seemingly disparate threads together from the efforts on Pikes Peak and the results of others to arrive at a new and perhaps better understanding of the dietary needs at altitude. In brief, it comes down to a matter of an extra 500 kcal/day for men at 4,300 m and perhaps even more at altitudes above 5,500 m. Consistent with the ability of women to maintain metabolic homeostasis when faced with an exercise challenge at sea level [13, 14], the energy deficit for women at 4,300 m was less than half that of men [15]. Although little explored, persistence at altitude and the ability to maintain glycemia are likely attributable in good part to the ability of the liver and kidneys to maintain glycemia by means of gluconeogenesis. Comparing data from various sources shows that so long as daily dietary energy intake is adequate at altitude, working muscle runs on CHO-derived fuels [6, 16–18], while the rest of the body largely utilizes fat for energy. And, during the remainder of the non-exercise, non-climbing parts of the day, the whole the body depends on the catabolism of lipid and protein stores to support energy substrate supply and gluconeogenesis. Hence, profound losses of total and lean body mass (LBM) can occur in lowland natives at altitude. Several factors contribute to energy deficits and body wasting at altitude. First, altitude exposure is associated with loss of appetite and undernutrition. Second, basal energy expenditure is elevated, likely due to sympathetic activation [5, 19]. And third, the energy expenditure of exercise [14] during climbing can be high, even if VO_2max is half or one third of that at sea level. Again, because every given effort is accomplished at a higher relative effort as given by $\%VO_2max$, the greater the ascent, the greater the demands on muscle glycogen, blood glucose, gluconeogenesis, and CHO nutrition. Inadequate energy intake and increased need force a shift toward catabolism of body protein and fat stores, resulting in body wasting and use of less oxygen-efficient fuels. Accordingly many of the "chronic" physiological responses in energy substrate utilization attributed to altitude exposure represent the consequences of malnutrition plus hypoxia, and not hypoxia alone. That mountaineers are capable of incredible metabolic adaptability by simultaneously managing high and extreme altitudes, while malnourished is a tribute to both their remarkable physiology and fortitude. Still, loss of strength and fitness at altitude is less than desirable and perhaps inadvisable if not dangerous.

There were several goals in developing the following narrative; first as a convenience for the readers, key elements of the previous chapter on men at altitude are reviewed; second more recent results including data on women at altitude are presented; third an attempt is made to distinguish the effects of hypoxia from cachexia on the metabolic responses to altitude, and fourth dietary recommendations for physical activity at altitude are reviewed. And finally, because of emerging data on the Tibetan genotype [20, 21], where possible attempts are made to distinguish between the effects of short-term acclimatization to high altitude that reflect phenotypic expression and long-term, generation-spanning adaptations in genotype to altitudes of 4,000–5,000 m.

Studies Revealing the Complication of Malnutrition

In preparations for the 1988 study on Pikes Peak, "Men at Altitude," our review of the extant literature and consultation with altitude sojourners and investigators revealed that dietary controls were seldom in place to support the needs of climbers or participants in chamber and laboratory studies. It was not that case that the issue of diet was neglected as, in some cases, investigators labored to provide luxurious foods. Rather, it was more an issue of climbers and study participants not consuming sufficient quantities of foods to cover energy need at altitude. From sea-level experience we know that energy insufficiency resulted in negative nitrogen balance as well as body weight loss. Hence, in an attempt to distinguish between the effects of hypoxia and malnutrition, planning for "Men at Altitude" and subsequently for "Women at Altitude" studies involved planning for dietary energy and fluid and electrolyte balances.

Lessons from Pre-1988 Efforts

Acute exposure (0–24 h) to altitudes greater than 10,000 ft has long been associated with reports of anorexia [1]. As well, diuresis, increased metabolic energy need, and shifts in circulating levels of fuel metabolites in sojourners suggested that altitude exposure caused changes in energy substrate utilization from normal sea-level metabolism [22–25]. Accompanying altitude anorexia may be acute mountain sickness [26], including nausea and vomiting. The resultant decrease in energy intake contributes to weight loss as was documented [1, 11, 12, 19]. Careful assessment of studies in which food intake has been documented both at sea level and at altitude showed a fairly consistent decrement in energy intake of approximately 200 kcal/day in men during exposure at an altitude of 4,300 m. Such a deficiency in energy intake carries with it an inadequate intake of most other essential nutrients, complicating the interpretation of research on the need for those nutrients in response to hypobaric hypoxia, as well.

Accompanying anorexia, diuresis at altitude is frequently reported [25, 27]. The diuresis serves to concentrate hemoglobin and thus improve the peripheral delivery of oxygen during the initial hours at altitude [28]. Although not rigorously evaluated, hemoconcentration is typical at altitude and the response is considered to be a mandatory part of the initial successful acclimatization to altitude exposure [29]. However, more water loss than is necessary for appropriate adaptation to acute exposure may adversely affect performance.

Lessons from the Pikes Peak Studies: "Men" and "Women" at Altitude

The effect of the anorexia in limiting energy and macronutrient intake on exposure to hypobaric hypoxia is aggravated by the effect of raised basal energy expenditure in both men and women [15, 19]. In the first 2 days of exposure to 4,300 m basal energy, needs of men were elevated by as much as 40 % above sea-level values and declined over a period of 3–4 days to stabilize at about 17 % above sea-level values [19]. The spike in basal energy consumption seen upon acute exposure was eliminated when energy intake was matched to energy requirement from the first day of exposure [1]. Hence, disregarding any increase in EEE due to climbing activity, in combination, at 4,300 m decreased energy intake and increased BEE can result in a 500 kcal/day energy deficit in men [12] but less than half that [19] in women [15]. Over time, such nutrient deficits have the potential to result in significant changes in body mass and composition, outcomes seen regularly in mountaineers and uncontrolled experiments.

Eating to maintain body mass and strength at altitude is easier said than done and requires not only planning but also logistical support and education of investigators and study participants alike. Nevertheless, in the 1988 and 1991 Pikes Peak studies on men [19] and subsequent studies on women [15], it was possible to maintain energy and nutrient balance in subjects at altitude. This assertion is supported by the preservation of body weight [1] and nitrogen balance [1]. Because the measures of body composition are complicated by shifts in body water at altitude, Nitrogen balance is the most sensitive measure of LBM and energy balance [30]. Indeed, the data provided by Butterfield and associates [15, 19] are unique in the field and of unrivaled importance as nitrogen balance measurements involving the collection of urine and feces, and their subsequent analysis, are seldom, if ever, attempted.

Within the context of appropriate nutritional controls, we utilized indirect calorimetry, stable isotope tracers, and mass balance measurements for metabolites and tracers to determine the effects of acute and chronic hypoxia on muscle substrate utilization at altitude. Our results obtained on men showed a shift toward carbohydrate [2, 3, 17, 18, 31] and away from lipid (adipose and intramuscular triglycerides [IMTG]) [6, 7]. Those results are strikingly similar to those reported on dogs made to breathe hypoxic gas mixtures during treadmill running [32]. As well, with cells studied in culture [33], and perfused rat hindlimb muscles hypoxia is a major stimulus to glycolysis [34, 35]. The controlled experiment, laboratory-based Pasteur-like effect of hypoxia on the balance of substrate utilization in controlled experiments stands in sharp contrast to "conventional wisdom" which interprets hypoglycemia and hyperlipidemia in mountaineers to reflect decreased glucose and increased lipid metabolism during exercise.

In neither the experiments we conducted nor those of others does there appear to be a problem at altitude with regard to glucose or free fatty acid (FAA) delivery at altitude. In response to hypoxemia, cardiac output and tissue perfusion increase [8]. As blood flow increases to maintain arterial O_2 delivery, it also increases arterial delivery of all substrates, including glucose, during rest and exercise. Thus, within the first several hours of altitude exposure, arterial glucose level is unaffected and decreases only slightly (5–7 %) after chronic exposure [2, 7].

In our experiments we studied glucose kinetics by using D_2- and [1-^{13}C]glucose [6, 7, 17, 18] and [3-^{13}C]lactate tracers. As well, we determined muscle uptake by measuring limb blood flow and arterial–venous (a–v) differences for metabolites and tracer isotopomers during rest and a standard exercise task (leg ergometer cycling at 100 W which elicited approximately 50 % of sea-level VO_2max and 65 % of altitude VO_2max). This power output was selected because it affords steady-state conditions eliciting constant arterial metabolite concentrations and isotopic enrichments and allows whole body as well as working limb (leg) oxygen consumption [2] to be maintained equivalent to values at sea level. Important also from the standpoint of the Crossover Concept [16], and the purported effect of hypoxia on sifting energy substrate use to CHO-derived fuels, is that a task requiring 50 % of VO_2max at sea level requires 65 % of VO_2max at altitude [2], the effect of which is to increase CHO dependence whether measured by pulmonary gas exchange or isotope tracers. Our results [2, 7] indicate that both tracer measured glucose disposal and limb net glucose uptake are increased during rest and exercise at altitude. Upon acute altitude exposure, limb blood flow increases over sea level, and it is not possible to detect any immediate increase in the (a–v) difference for glucose which is small at rest and decreases during exercise as the relative gain in muscle blood flow exceeds the gain in glucose uptake. Therefore, during exercise at altitude, glycemia and tissue glucose delivery is maintained. Hence, muscle glucose uptake is not limited by delivery during rest or exercise. That the Crossover Concept model describes energy substrate partitioning in individuals exercising at altitude as well as sea level is also to be found in the results of others [36].

The Lactate Shuttle at Altitude

Our understanding on the subject of lactate responses to altitude benefits from the perspective of the "Lactate Shuttle" hypothesis [37] positing that lactate is the metabolic intermediate that links glycolysis to mitochondrial substrate utilization. As well, because lactate–pyruvate interactions affect cell redox status and because lactate can serve as a reactive oxygen species (ROS) generator [38], lactate can act a signaling molecule, a "lactormone." From our experiments conducted on subjects exercising at high altitude, we concluded that an elevation in circulating lactate is one of the compensatory adjustments to the stresses of exercise and environment. And, though seemingly at odds with previous views on the causes and consequences of lactacidemia at altitude, our results and conclusions are similar to those of Gutierrez et al. [39], who showed hypoxia increased lactate uptake by rabbit muscle in situ, and Zinker et al. [32] who showed that hypoxia increased muscle lactate uptake in working dog muscle.

To know lactate production during rest and exercise at sea level and at high altitude, we used [3-^{13}C]lactate tracer in combination with D_2-glucose to evaluate the influences of altitude and exercise on carbohydrate utilization in men. Although we were unable to establish that working muscle was singularly responsible for blood lactate appearance during exercise, it is clear that the elevated circulating lactate level during exercise at constant power and VO_2 was attributable to increased lactate production [1] in working muscles and other tissues [3, 17]. The obvious candidate of an extra-muscular tissue supporting systemic lactate Ra during exercise is the liver. Enhanced hepatic lactate release and muscle uptake has previously been demonstrated in hypoxic running dogs [32], but hepatic lactate release has not been observed in exercising humans [31]. Possibly, adipose [40] and skin [41] are involved. That diverse tissues contribute to the circulating lactate level, thereby providing and energy fuel substrate to working muscle and other tissues at altitude, is a departure from conventional belief.

During the continuous, submaximal exercises we have studied, net lactate release from working muscle is responsible for the initial rise in circulating lactate at exercise onset. However, with continued work at sea level, the (v–a) difference for lactate across muscle changes from positive [31] to 0 (no release) [1, 17]. During exercise at altitude, net lactate release from working muscle is increased over sea level, but as at sea level, the larger lactate release from muscle working at altitude is attributable mostly to exercise onset. Therefore, the maintenance of stable and elevated circulating lactate [1] at altitude cannot be attributed solely to lactate production in working muscle [17]. The extra-muscular source of lactate is unidentified and likely will not be identified with current methodologies unless the liver or some other major tissue site is involved. This is because as the (v–a) difference for lactate across other low-flow tissues such as skin and adipose is likely small and difficult to determine.

Above we emphasized the role of vascular conductance [= (limb blood flow) (arterial concentration)] in the delivery and utilization of substrates during exercise and at altitude. However, in the 1988 Pikes Peak experiment in which we utilized, pulmonary gas exchange, [3-^{13}C] lactate tracer along with working muscle net exchange measurements and [3, 17] showed that lactate uptake in working muscle is directly related to arterial lactate concentration. Further, oxidation accounts for essentially all the lactate taken up by muscle working at altitude. We interpret these results to mean that, contrary to previous thought, during sustained exercise upon acute altitude exposure, when arterial lactate concentration is greatest, significant consumption and oxidation of lactate occur in active skeletal muscle. Thus, as shown previously [42] the use of lactate as fuel source at altitude depends on availability, a result consistent with presence of a sarcolemmal lactate transporter expression [43, 44].

Dual-isotope technology has allowed simultaneous measurements of glucose and lactate fluxes. At rest lactate appearance (Ra) approximates 30–50 % of glucose disappearance (Rd). However, during even mild exercise (e.g., 40–50 % VO_2max at sea level) [2, 3, 42], lactate Ra equals or exceeds glucose Rd. As previously noted, altitude exposure increases glucose flux, but the increase in lactate flux during exercise is far greater than the glucose flux [18]. Upon acute exposure to high altitude, lactate flux exceeds the glucose flux by 4–5-fold. These data indicate a role for glycogen in supplying substrate during exercise upon acute altitude exposure in adequately nourished subjects. As with the increase in glucose metabolism, the increase in rate of lactate utilization supports the conclusion that at altitude, there is increased use of carbohydrate energy sources.

Upon initial exposure to altitude, resting circulating FFA and glycerol levels are similar to those observed at sea level. However, after residency at 4,300 m, circulating FFA and glycerol levels rise [6]. The rise in circulating FFAs with chronic exposure has been interpreted as a switch to increased lipid utilization. Despite the elevation in circulating lipids, after a 3-week sojourn at 4,300 m, we observed minimal (a–v) differences for glycerol and FFA across working leg muslces [1] during rest and exercise. Therefore, an important distinction between muscle consumption of FFAs and glucose and their respective circulating levels is realized. Elevated FFA and glycerol levels after acclimatization are largely due to reduced tissue uptake, an extraordinary finding, given the usual assumption that glycerol and FFA levels are predictive of the rate of lipid mobilization and oxidation.

Given the data available indicating suppression of glycerol release and FFA uptake from working limbs at altitude, it is probably appropriate to conclude that intramuscular lipolysis as well as FFA uptake is suppressed by acclimatization. The basis for this conclusion is the absence of glycerol release [6]. Because skeletal muscle and adipose tissues lack glycerol kinase (the enzyme necessary to recycle to triacylglycerol), we can interpret the absence of glycerol release from muscle after acclimatization to mean suppression of lipolysis within muscle.

The result of suppressed glycerol release also assists with interpretation of the acclimatization-imposed limits on muscle FFA uptake. Glycerol release in the absence of significant FFA uptake could be interpreted to indicate utilization of intramuscular lipids and an acclimatization-imposed limitation in muscle FFA uptake such as could be imposed by downregulation of the sarcolemmal FABP. However, because after acclimatization limb muscle glycerol release is nil, for the present we can conclude that feed-forward regulation of glycolysis results in mitochondrial acetyl-CoA formation and downregulation of mitochondrial FFA uptake. For working cardiac muscle, it has been demonstrated that increased glycolytic flux results in activation of acetyl-CoA carboxylase (ACC) and formation of malonyl-CoA, a potent inhibitor of mitochondrial carnitine palmitoyl carboxylase 1 (CPT1 1) [16, 45]. Thus, we interpret our results to indicate that chronic altitude exposure both decreases intramuscular lipolysis and suppresses mitochondrial uptake of activated FFAs in adequately nourished subjects.

Altitude exposure and increments in exercise intensity both have the effect of shifting the balance of substrate toward carbohydrates and away from lipids [9]. In our experiments on Pikes Peak, we controlled the dietary status of subjects and we required them to perform the same absolute exercise tasks at sea level and altitude. We believe these protocols demonstrate the effects of hypoxia, independent of cachexia, on the balance of substrate utilization.

For the sojourner at altitude, it remains that altitude exposure may depress appetite and food availability or palatability may exacerbate this situation, leading to significant energy deficit. Therefore, the ill-prepared and ill-informed mountaineer, with inadequate dietary energy and carbohydrate, can be expected to have difficulty with prolonged exercise tasks. Each task is relatively more difficult at altitude, but the normal shift toward muscle glycogen and blood glucose with intense exercise at sea level could be limited by availability of fuel sources at altitude. The forced catabolism of lipid and lean tissue not only weakens the individual but also yields fuel sources with lesser enthalpies per unit O_2 consumed and, in any case, less preferred by working muscle. Therefore, whether one's perspective is from that of nutrition or exercise physiology, whether one considers the perceived effort or the actual effect of hypoxia on muscle metabolism, the primary concern should be for providing carbohydrate-rich foods for the altitude sojourner.

Studies on Women Uncomplicated by Malnutrition

As described by Mawson et al. [15] and Braun et al. [9], altitude increased basal oxygen consumption in women (Fig. 15.1) [15], but RER was depressed on exposure (Fig. 15.2) [46] such that the acute altitude effect approximated a 7 % increase that declined to a persistent 6 % increase in BMR over the SL baseline after day 6 at altitude. Accordingly, the altitude-induced perturbation in BEE in women approximates half that in men. Glucose tracer studies were conducted on women day 10 at 4,300 m.

As in men, at altitude blood glucose concentration rose during exercise at altitude, as did blood [lactate] RER. However, the increases in

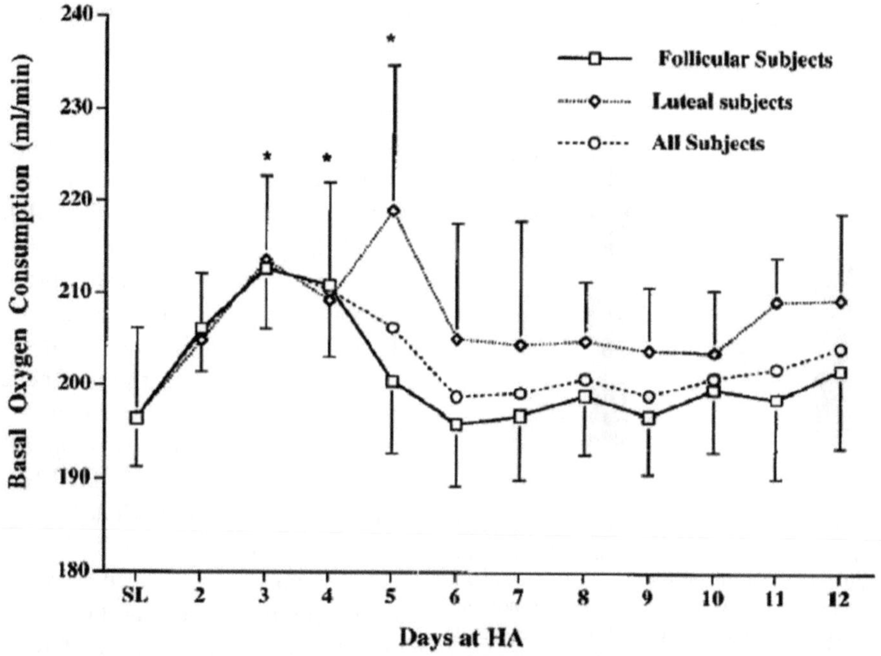

Fig. 15.1 Basal oxygen consumption for subjects studied in luteal and follicular phases at sea level (SL) and for *days 2–12* at high altitude (HA). Values are means ± SE. *Significantly different from SL ($P<0.05$). From [15] with permission

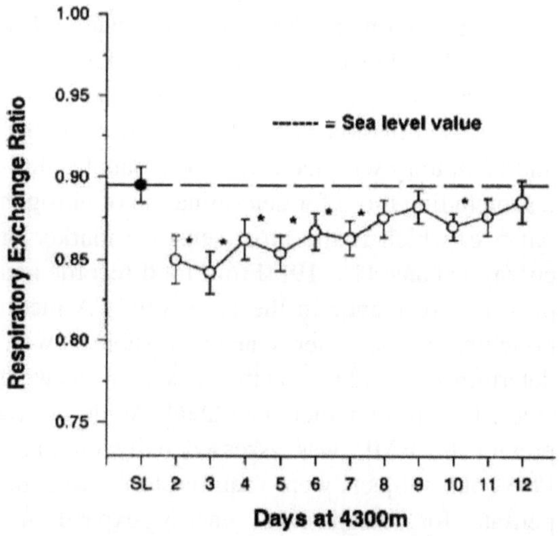

Fig. 15.2 Respiratory exchange ratio during measurements of basal metabolic rate at sea level and on each day at 4,300 m. Values are means ± SE. *Significantly different from sea-level value, $p<0.05$. From [10] with permission

glucose flux were less impressive in women than men, not really being different from the changes they experienced during exercise at sea level, especially if relative exercise intensity was considered (Fig. 15.3) [9]. The previous gender difference of a lower total CHO oxidation rate, but similar glucose oxidation rate, and therefore, lesser glycogen utilization rate, was observed in women at altitude. In general, compared to men at altitude, women behaved more like trained athletes at sea level.

Insulin Resistance at Altitude

Greater glucose disposal rates, but unchanged insulin levels, imply increased insulin sensitivity at altitude [5]. However, by stimulating epinephrine, altitude exposure has the possibility to elicit insulin resistance. Not surprisingly, in euglycemic-hyperinsulinemic glucose clamp studies on Monte Rosa (4,559 m), Larsen et al. [46] described an apparent insulin resistance in subjects studied 2 days after arrival at altitude. Similarly, in studies using food-tolerance tests on diet-controlled subjects, Braun [9, 47] demonstrated reduced insulin

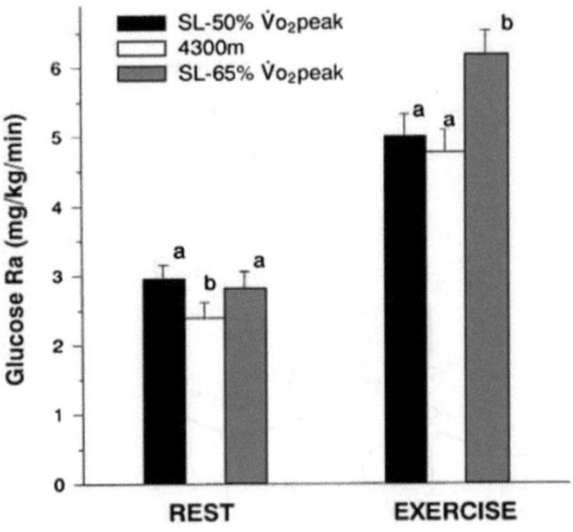

Fig. 15.3 Glucose rates of disappearance (Rd) at rest and during submaximal exercise. Values are means ± SE. SL-50 and SL-65, sea-level exercise trials performed at 50 and 65 % VO_2peak, respectively. Resting values represent mean values from two pre-exercise time points; exercise values represent mean values between 30 and 45 min. Conditions that do not share a superscripted letter (*a*, *b*) are significantly different from each other. From [10] with permission

sensitivity in subjects 10 days after arrival on Pikes Peak. Given that tracer studies to date [2, 7] demonstrate increased glucose use during exercise at altitude, what do observations of apparent insulin resistance during clamp and food-tolerance tests on resting subjects really mean? As described below, the global insulin resistance is a glucose-sparing mechanism that likely benefits the use of CHO-derived energy substrates in working muscle during exercise at altitude.

The Power of Hypoxia-Induced Cachexia: A Comparison of Results of Studies with and Without Dietary Controls

A comparison of results of studies with and without dietary controls may represent one method to distinguish between the separate effects of hypoxia and malnutrition on metabolism at altitude. To accept any outcome of such a comparison, wide latitude needs to be accepted in comparing data from studies on different subjects tested in different places and for different durations. The studies compared were extremely difficult to conduct, not only because of their invasiveness and technical sophistication but also because of the logistics necessitating the need to collect, preserve, and transport samples for subsequent analysis. Where similarities in the results exist, the results are taken to be robust. Indeed, this is evident in a side-by-side comparison of the results of four studies, two on Pikes Peak, Colorado (4,300 m) [3, 6, 7, 17, 18], and two in Bolivia, one centered at Chacaltaya (5,600 m) [48], and the other El Alto (4,100 m) [49].

Studies on Pikes Peak represented collaboration among researchers at the Universities of Colorado and California, Stanford University, and the US Army Research Institute of Environment Medicine (USARIEM). For studies on Pikes Peak, the sea-level control studies took place at Stanford and the Palo Alto, CA Veterans' Administration Medical Center, where a metabolic unit with kitchen was in place. The USARIEM Maher Laboratory on Pikes Peak, where subjects were studied acutely and after a 3-week acclimation period, also had a kitchen and laboratory with facilities for metabolic studies, including those for determination of nitrogen balance, which is the most sensitive marker of energy balance [15, 19]. From food records and pre-study residency in the Palo Alto, CA metabolic unit, dietary energy needs of subjects were determined. There and on the peak, subjects were studied 12 h after their last meal. As described previously, BMR was assessed daily on Pikes Peak, and subjects were required to eat to compensate for raised basal energy expenditure. Subjects were weight stable and in positive nitrogen balance.

Most investigators in the two studies in Bolivia were members of the Copenhagen Muscle Research Center (CMRC). The acclimatization period was longer (2 or 9 weeks), but there were no pre-altitude, acute exposure studies, or dietary controls in place. In the first report, subjects lost 7.3 kg (16 lb) after 5 weeks at 5,600 m, and after checking with the first author (G. Van Hall, personal communication), subjects were weight stable in the subsequent study at 4,100 m [49]. The first Pikes Peak study focused on glucose–lactate interactions, whereas the second focused on glucose–lipid interactions. The first CMRC study at Chacaltaya also focused on glucose–lactate interactions, whereas the second was focused solely on the lactate concentration response. To reiterate, ordinarily no one would attempt to contrast metabolic studies when body mass and energy and nitrogen balances were uncontrolled. Nonetheless, the studies on Pikes Peak and Chacaltaya contain data from isotope tracers, limb blood flow, arterial–venous difference measurements for glucose and lactate, and exercise power outputs approximated 46–50 % sea-level VO_2peak, representing 62–62 % altitude VO_2peak, the point at which energy supplied by carbohydrate oxidation predominates over that from lipid oxidation, and that lactate production and oxidation are elevated at altitude as compared to sea level. Hence, *the clear message is for the predominance of CHO-derived fuel sources including lactate in working muscles of subjects exercising at altitude, even in subjects who had lost weight and had been in negative energy and nitrogen balances.*

Glucose

Blood glucose concentration was well supported in all trials. However, blood glucose in resting individuals (mostly men) fell progressively with duration of stay at altitude. The exercise response varied, but blood [glucose] generally rose during exercise, even as the pre-exercise baseline value was depressed at altitude. In all studies, the maintenance of blood [glucose] during exercise at altitude was supported by increased glucose production rates (tracer-measured rate of appearance, Ra).

Lactate

Lactate turnover (simultaneous production and removal) was prevalent during exercise, whether at SL or altitude. As predicted by Lactate Shuttle Theory, regardless of study (California, Colorado, or Bolivia), most lactate disposal was accomplished by oxidation. In the study on Pikes Peak, lactate oxidation was greatest on acute exposure, oxidation on acute exposure was not determined in Bolivia. Seemingly, however, the greatest muscle lactate oxidation was determined in well-acclimatized subjects in Bolivia. Perhaps this was due to the use of [1-^{13}C]- as opposed to [3-^{13}C]lactate tracer or in some way related to body energy storage depletion. Still, it is clear that working muscles simultaneously produced, released, took up, and oxidized lactate, with lactate delivery to working muscle providing a significant portion of the CHO energy, and therefore, total energy liberated in working muscle at altitude. As noted previously, the same effect of hypoxia increasing muscle lactate oxidation was observed by in dogs during treadmill running [32], and rabbit [34], or rat muscle preparations in situ [35].

Gluconeogenesis at altitude: In studies on Pikes Peak, gluconeogenesis (GNG) accounted for most of the non-oxidative lactate disposal [2]. As such, while most lactate was removed by oxidation, GNG from lactate was a major means to support glycemia during exercise at altitude. The data available from studies on both Pikes Peak and in Bolivia show that glycemia is well maintained and that the expected shift to CHO oxidation during exercise occurs irrespective of the adequacy of dietary energy intake. Carbohydrate dependence at altitude is supported by measurements of pulmonary respiratory exchange [20] or working muscle respiratory quotient (RQ) and both blood glucose and lactate fluxes at altitude compared to sea level. Admittedly the studies in Bolivia lacked pre-altitude exposure and acute altitude exposure trials, but the findings of increased CHO (glucose, glycogen, lactate) use at altitude were sufficiently robust to overcome differences in subject treatment. The ability to maintain glycemia and total body and working muscle CHO fuel oxidation even in the face of dietary deprivation is truly an amazing phenomenon that is dependent on hepatic and renal gluconeogenesis.

The liver at altitude: The persistence of CHO oxidation in working muscle is supported by the increases in glucose and lactate fluxes as well as the increase in hepatic gluconeogenesis, largely from lactate. Working muscles and other tissues [17, 31] release lactate for consumption by the gluconeogenic organs (liver and kidneys) as well as the working muscles. Systemic lactate is converted to glucose supplying 22–28 % of the glucose Ra during rest or exercise. Elevated precursor (i.e., lactate) supply during exercise at altitude helps support hepatic GNG and hence glycemia. As well, the catabolism of lean tissue under conditions of dietary inadequacy at altitude serves to supply gluconeogenic amino acids. Thus, these data unmask the liver as a major organ supporting life and exercise at altitude. There are both temporal and tissue considerations in considering the role of liver in energy substrate partitioning at altitude.

Whether the individuals exposed to moderate or extremely high altitudes are participants in research studies, whether the individuals are climbers exposed because of other reasons, and whether or not dietary controls are in place, the liver needs to perform. As shown by the high muscle blood flows, glucose and lactate uptake rates, and RQs during exercise at altitude, during exercise, CHO energy reserves stream to and center in working muscle. This means that during exercise, the remainder of the body needs to become glucose and insulin resistant and persist on lipid energy sources. Indeed the glucose clamp studies of Larsen et al. [46] and the food-tolerance test results of Braun et al. [9] support the notion of restricted glucose use by nonworking (muscle and other) body tissues at altitude. This also means that during the non-exercise (non-climbing) parts of the day, CHO oxidation needs to be minimized and alternative fuels such as FFAs and amino acids need to be provided from the catabolism of body adipose and protein reserves. Results for the CMRC investigators in Bolivia are taken to mean just this. They show robust lactate production and lactate and total CHO oxidation in sojourners even after significant weight loss [48, 49]. The complex physiological interactions brought on by altitude exposure involving cachexia and hypoxia have yet to be understood. As a first step, Friedlander and Fulco undertook a study in which one cohort of subjects were underfed at sea level, while other cohorts were either adequately or inadequately underfed at 4,300 m on Pikes Peak [50–53]. At sea level, subjects were underfed 40 % of predicted daily energy expenditure and over a 3-week period lost 3.9 kg body weight and 2 kg LBM. Over the course of the 3-week period, subjects maintained VO_2max and the ability to perform on a standardized leg cycle ergometer and other tests [50]. However, there was great variability in individual subject endurance capacities. At 4,300 m altitude, another cohort of men lost 5.8 kg body weight over a 3-week period [51]. Again, the investigators failed to find significant differences in VO_2peak or other measures of physical performance [52]. In sum, the results show great adaptability and resilience in physical performance in the face of significant weight loss at sea level or altitude. But, while impressive, the investigators were puzzled because CHO nutrition is known to enhance exercise performance at sea level. As a consequence, in another study on Pikes Peak, men lost 2.3 kg body weight over a 10-day period [53]. Then, in a double-blind fashion, half the subjects drank either a 10 % CHO solution or a flavor-matched no CHO placebo every 15 min during a cycle ergometer time trial test. Those supplemented with CHO during the test completed in less time and at higher relative exercise intensities when taking the CHO solution at altitude. Apparently, CHO supplementation during exercise at altitude has the same ergogenic effect at altitude as it does at sea level.

Lipid Metabolism

Data on whole-body and working lipid metabolism are least abundant. Nonetheless, studies at altitude in Colorado and Bolivia lead to a similar conclusion. Pulmonary muscle combustion quotients are high during exercise (*vide Sapra*). In only one study [6, 7] did investigators measure muscle net FFA and glycerol exchange rates. Working muscle glycerol release is a surrogate for lipolysis of IMTG as glycerol is neither oxidized nor used for reesterification in myocytes or adipocytes. Unique data on working muscle FFA uptake at SL show small, but significant, working muscle FFA uptake. However, this uptake is diminished at altitude and with time at altitude.

Pulmonary respiratory exchange ratios, muscle respiratory quotients, and energy substrate partitioning during rest and exercise at sea level and altitude: Both combustion coefficient measurements decline over time at altitude, whether obtained during rest or exercise. Resting muscle RQs indicate a rough equivalence of CHO and lipid oxidation at rest. However, during exercise at altitude working muscle RQs exceeded pulmonary RERs and RQs were close to unity. Accordingly, the data can be interpreted to indicate the presence of a shunt of CHO-related substrates to working muscle at SL and altitude and predominance of muscle CHO oxidation during exercise at SL and altitude. RQ measurements are admittedly difficult to make, but the data are more numerous than measurements of muscle FFA uptake and oxidation. Nevertheless, like FFA uptake and glycerol release data, working muscle RQ data show little lipid oxidation in working muscle at altitude. This is apparently true for subjects in energy and nitrogen balance as well as those who lost significant body mass at altitude.

Hypoxia-Inducible Factors Signaling and Lactate Shuttling at Altitude

Chapters 1 and 2 of this volume describe cellular and molecular mechanisms of O_2 sensing, and readers are referred to these chapters because oxygen sensing via hypoxia-inducible transcription factors (HIF) has multiple effects, including long- and short-term effects on metabolic enzyme regulation and expression [54–59]. As a whole, considerable evidence exists to support the conclusion that acclimatization to the hypoxia of high-altitude exposure involves HIF-signaling and the expression of a more glycolytic phenotype [60]. Indeed, generations of exposure to the hypoxia of high altitude may well have affected human adaptation to altitude in Tibet [21]. Hence, hypoxia-inducible factors have been termed "master regulators of hypoxic signaling." Because there are several arms of HIF-signaling, one of stimulating glycolytic adaptation, another of stimulating mitochondrial adaptation, and another stimulating circulatory responses via changes in erythropoietin (Epo) and vascular endothelial growth factor (VEGF), reiteration of some of what is known is appropriate here as it fits with what is know from studies of lactate shuttling [3, 17] and its interactions with glucose [2, 18] and fatty acid metabolism at altitude [6, 7].

As described by Semenza and associates [54–56], and reiterated by others [60], HIF-1 is a heterodimeric protein that is composed of an O_2-regulated HIF-1α subunit and a constitutively expressed HIF-1β subunit [61]. The levels of HIF-1α protein increase as O_2 concentration declines, with changes in HIF-1α protein being the result of O_2-dependent ubiquitination and proteasomal degradation of HIF-1α.

At the cellular level, HIF-signaling promoted development of a glycolytic phenotype by at least two mechanisms. First, HIF-1 contributes to increased glucose use in most tissues and increased glucose uptake and use as well as glycogen mobilization via increased expression of pyruvate dehydrogenase (PDH) kinase 1 (PDK1) [62]. As well, HIF has the effect of downregulating mitochondrial respiration [63]. Hence, at the mitochondrial level HIF-signaling has the effect of moderating glycolytic flux to the TCA cycle, thereby giving rise to lactate production and affecting cytosolic redox status, further promoting glycolytic flux and lactate production, all of which are seen at altitude [2, 3, 16–18].

Additionally, a second way that HIF-signaling acts is to affect the expression of muscle-type lactate dehydrogenase isoforms, specifically LDH-A that has a high affinity (low KM) for its substrate, pyruvate, high V_{max} and enormous K_{eq}. Hence, on exposure to hypoxia the predisposition to direct glycolytic flux to lactate production rather than acetyl-Co A formation and disposal via the TCA is encouraged. Seen for the perspective of lactate shuttling to provide energy substrate and gluconeogenic precursor at altitude, the HIF-supported increase in lactate production is an advantage because lactate is a preferred fuel during hypoxia and the main gluconeogenic precursor [3, 18, 32]. Hence, lactate produced in one cell domain is useful substrate in an adjacent or anatomically distant cell domain.

In addition to affecting mitochondrial respiratory rate and directing glycolytic flux to lactate production [63], it has also been proposed that HIF-1 also regulates respiration by directing a subunit switch in cytochrome c oxidase that appears to increase the efficiency of electron transfer at complex IV under hypoxic conditions [64]. It has been proposed that this response increases the efficiency of oxidative phosphorylation but there is contrary evidence that changes in muscle oxidative capacity occur in humans as the result of acclimatization to altitude [4]. One approach to assess the effect of adaptation to altitude has been to study the economy of walking of several ethnic groups, including high-altitude natives. However, those efforts were constrained by my methodology and did not directly address the issue of mitochondrial phosphorylative coupling efficiency [60].

Despite the promise of HIF-signaling mechanisms for explaining metabolic responses to exercise [65], at present data on humans are missing explaining the effects of acclimatization on moderating lactate production and stimulating hepatic and renal gluconeogenesis from lactate. As well, enthusiasm for one signaling factor (e.g., HIF) need not be interpreted to mean that alternative, and parallel signaling pathways (e.g., AMPK) [66], lactate-generated ROS [38], and changes in redox status [67, 68] are not necessary for full expression the altitude adapted phenotype.

And, finally before leaving the issue of HIF-1 signaling and human acclimatization and adaptation to high altitude, it needs to be acknowledged that the origin of modern humans is traceable to life in a high and dry environment as in Ethiopia and a significant role of HIF-signaling [65]. Given that background, a large genomic screening of Tibetans with contrasts of Han Chinese and Danes [21] shows the major adaption in Tibetans to be in the expression of the gene encoding for endothelial PAS domain protein 1 (EPAS1, also known as HIF-2αI. Although promising and intriguing, it is unclear how increased expression of HIF-2α promotes metabolic, ventilatory, or hematological or other adaptations in families resident to high altitude [65]. Fortunately, such delineation of function is now technically possible and, hopefully, achievable in the near term. As well, the rapidly developing field of hypoxamirs, that is, the effects of hypoxia and derivatives of hypoxia on microRNA expression, may provide missing insights on mechanisms of acclimatization and adaptation to altitude, for example, as some hypoxamirs (e.g., miR210 and mi273) may be induced by HIF, whereas others (e.g., miR20b and miR199a) may affect HIF). As well, there appear to be microRNAs that affect HIF expression independent of hypoxia [69].

Cachexia, Dietary Composition, and Malnutrition at Altitude

Cachexia (appetite suppression and loss of body mass wasting syndrome subsequent to loss of appetite) is common at altitude, and as predicted by the Crossover Concept, hypoxia shifts energy substrate partitioning toward increased used of CHO-derived fuels. But, is malnutrition and wasting syndrome inevitable if dietary CHO is emphasized?

Contrasting results of our efforts [11, 12, 15, 19] alongside past and more recent efforts of others [26, 48], we are left at an impasse regarding the effects of undernutrition on the physical performance of sojourners at high and extreme altitude [52]. It may be that a talented scientist and accomplished dietitian can create an environment in which body wasting syndrome is prevented at 4,300 m altitude. Additionally, it may be that in such an environment, physical performance can be promoted by providing a high carbohydrate diet [53]. However, looking at the literature as a whole, such achievements can be characterized as extraordinary, as opposed to typical. For instance, an important lesson may be found in a recent report describing their efforts to raise dietary CHO composition during a 21-day expedition in the Himalayas [26]. In their report Macdonald et al. concluded that dietary CHO supplementation did not prevent weight loss at extreme altitudes. In evaluating the data from that expedition, it is appropriate to judge that because climbers lost body mass they were malnourished, and not supplemented at all. Still, the alternative explanation may be that appetite suppression at extreme altitude is subject to powerful endocrine signaling mechanisms [51, 70] that are so pervasive with powerful effects on CNS function [71] that cachexia is inevitable at extreme altitude.

Nutrient Recommendations for High Altitude

Based on our previous results and more recent data of others, the previous recommendation of increasing daily energy intake above appetite by emphasizing carbohydrate-rich foods [11, 12] remains the recommendation. If cachexia prevents adequate energy intake from sojourners on a high-CHO diet, then the addition of energy-dense foods needs to be considered. The recommendation is based on part by the decrease in energy intake resulting from AMS and cachexia, in part by increase basal energy expenditure, and in part by the need for exercise energy expenditure at altitude. Responsibility for adequately provisioning food and water to meet needs of altitude sojourners is to be placed on organizers of high-altitude expeditions who need to arrange logistics and education of climbers who will be physically active at altitude, experience elevation in BMR and become cachexic, and need to understand the need for adequate nutrition and hydration at altitude.

The present analysis of data from the two research groups, one working on Pikes Peak, and the other in Bolivia, seemingly at odds over relevance of archaic terms (e.g., the Lactate Paradox), once again shows the importance of CHO energy for working under stress. That CHO-derived energy sources (glycogen, glucose, and lactate) remain as prime energy sources in working muscles of humans at altitude, even in undernourished sojourners, illustrates adaptability of diverse organ system, including the liver and perhaps also kidneys. As well, the ability of lowland natives to succeed at altitude physically speaks to their courage and dedication. However, body wasting cannot be considered a goal. Given what has been demonstrated in research in a few research studies and repeatedly in mountaineering as well as sea-level athletics, loss of some body mass to fuel the working muscles is tolerable and perhaps advantageous. Over time, however, major depletion of body reserves of energy and tissue mass could put altitude sojourners at risk. Hence, in an attempt to minimize the accrual of an energy deficit at altitude, "an extra chocolate bar, or two or three a day (adding at 500 kcal at moderate 4,300 m., and more at altitude extremes) could keep the rescuer away." That said, questions arise concerning the high-fat and solid nature of chocolate in comparison to the macronutrient compositions of other foods, and further, the recommendation unmasks the presence of numerous decisions that are made in provisioning sojourners at altitude. The sugar and fat contents of chocolate are high, and the food is energy dense. On one hand, palatability of chocolate at altitude may assist in resisting the effects of cachexia. But, on the other hand, while the energy density of chocolate is good in terms of providing calories, the organoleptic effects of a solid, as opposed to a liquid food, food may serve to displace interest in other solid foods, such as occurs at sea level [72, 73]. Perhaps a more viable idea would be the addition of liquid meals (e.g., Ensue or Muscle Milk) to truly add to (i.e., supplement) the diet at altitude. However, it remains to be seen if difficulties of provisioning liquid meal supplements outweigh the benefits.

References

1. Brooks GA, Butterfield GE. Metabolic responses of lowlanders to high-altitude exposure: Malnutrition versus the effect of hypoxia. In: Hornbein T, Schoene R, editors. High altitude: an exploration of human adaptation. New York: Marcel Dekker; 2001. p. 569–99.
2. Brooks GA, Butterfield GE, Wolfe RR, et al. Increased dependence on blood glucose after acclimatization to 4,300 m. J Appl Physiol. 1991;70:919–27.

3. Brooks GA, Butterfield GE, Wolfe RR, et al. Decreased reliance on lactate during exercise after acclimatization to 4,300 m. J Appl Physiol. 1991;71: 333–41.
4. Green HJ, Sutton JR, Wolfel EE, et al. Altitude acclimatization and energy metabolic adaptations in skeletal muscle during exercise. J Appl Physiol. 1992; 73:2701–8.
5. Mazzeo RS, Bender PR, Brooks GA, et al. Arterial catecholamine responses during exercise with acute and chronic high-altitude exposure. Am J Physiol. 1991;261:E419–24.
6. Roberts AC, Butterfield GE, Cymerman A, et al. Acclimatization to 4,300-m altitude decreases reliance on fat as a substrate. J Appl Physiol. 1996;81: 1762–71.
7. Roberts AC, Reeves JT, Butterfield GE, et al. Altitude and beta-blockade augment glucose utilization during submaximal exercise. J Appl Physiol. 1996;80: 605–15.
8. Wolfel EE, Groves BM, Brooks GA, et al. Oxygen transport during steady-state submaximal exercise in chronic hypoxia. J Appl Physiol. 1991;70:1129–36.
9. Braun B, Butterfield GE, Dominick SB, et al. Women at altitude: changes in carbohydrate metabolism at 4,300-m elevation and across the menstrual cycle. J Appl Physiol. 1998;85:1966–73.
10. Braun B, Mawson JT, Muza SR, et al. Women at altitude: carbohydrate utilization during exercise at 4,300 m. J Appl Physiol. 2000;88:246–56.
11. Butterfield GE. Elements of energy balance at altitude. In: Sutton JRCG, Remmers JE, editors. Hypoxia, the adaptations. Philadelphia: B.C. Decker; 1990. p. 88–93.
12. Butterfield GE. Maintenance of body weight at altitude: in search of 500 kcal/day. Nutritional needs in cold and high altitude environments. Washington, DC: National Academy Press; 1996.
13. Henderson GC, Fattor JA, Horning MA, et al. Glucoregulation is more precise in women than in men during postexercise recovery. Am J Clin Nutr. 2008;87:1686–94.
14. Henderson GC, Fattor JA, Horning MA, et al. Lipolysis and fatty acid metabolism in men and women during the postexercise recovery period. J Physiol. 2007;584:963–81.
15. Mawson JT, Braun B, Rock PB, et al. Women at altitude: energy requirement at 4,300 m. J Appl Physiol. 2000;88:272–81.
16. Brooks GA, Mercier J. Balance of carbohydrate and lipid utilization during exercise: the "crossover" concept. J Appl Physiol. 1994;76:2253–61.
17. Brooks GA, Wolfel EE, Butterfield GE, et al. Poor relationship between arterial [lactate] and leg net release during exercise at 4,300 m altitude. Am J Physiol. 1998;275:R1192–201.
18. Brooks GA, Wolfel EE, Groves BM, et al. Muscle accounts for glucose disposal but not blood lactate appearance during exercise after acclimatization to 4,300 m. J Appl Physiol. 1992;72:2435–45.
19. Butterfield GE, Gates J, Fleming S, et al. Increased energy intake minimizes weight loss in men at high altitude. J Appl Physiol. 1992;72:1741–8.
20. Bigham A, Bauchet M, Pinto D, et al. Identifying signatures of natural selection in Tibetan and Andean populations using dense genome scan data. PLoS Genet. 2010;6:e1001116.
21. Yi X, Liang Y, Huerta-Sanchez E, et al. Sequencing of 50 human exomes reveals adaptation to high altitude. Science. 2010;329:75–8.
22. Consolazio CF, Johnson HL, Krzywicki HJ, et al. Metabolic aspects of acute altitude exposure (4,300 meters) in adequately nourished humans. Am J Clin Nutr. 1972;25:23–9.
23. Consolazio CF, Matoush LR, Johnson H, et al. Energy, nitrogen and water requirements of normal adults residing at 4300 meters for 28 days. Lab Rep No 308. Report. United States. Army Medical Research and Nutrition Laboratory, Denver; 1967. p. 1–19.
24. Hannon JP, Klain GJ, Sudman DM, et al. Nutritional aspects of high-altitude exposure in women. Am J Clin Nutr. 1976;29:604–13.
25. Boyer SJ, Blume FD. Weight loss and changes in body composition at high altitude. J Appl Physiol. 1984;57:1580–5.
26. Macdonald JH, Oliver SJ, Hillyer K, et al. Body composition at high altitude: a randomized placebo-controlled trial of dietary carbohydrate supplementation. Am J Clin Nutr. 2009;90:1193–202.
27. Krzywicki HJ, Consolazio CF, Johnson HL, et al. Metabolic aspects of caloric restriction (420 kcal): body composition changes. Am J Clin Nutr. 1972;25:67–70.
28. Honig A. Peripheral arterial chemoreceptors and reflex control of sodium and water homeostasis. Am J Physiol. 1989;257:R1282–302.
29. Hackett PH, Rennie D, Grover RF, et al. Acute mountain sickness and the edemas of high altitude: a common pathogenesis? Respir Physiol. 1981;46:383–90.
30. Calloway DH. Nitrogen balance of men with marginal intakes of protein and energy. J Nutr. 1975;105: 914–23.
31. Ahlborg G, Felig P. Lactate and glucose exchange across the forearm, legs, and splanchnic bed during and after prolonged leg exercise. J Clin Invest. 1982;69:45–54.
32. Zinker BA, Wilson RD, Wasserman DH. Interaction of decreased arterial PO2 and exercise on carbohydrate metabolism in the dog. Am J Physiol. 1995;269: E409–17.
33. Cushman SW, Wardzala LJ. Potential mechanism of insulin action on glucose transport in the isolated rat adipose cell. Apparent translocation of intracellular transport systems to the plasma membrane. J Biol Chem. 1980;255:4758–62.
34. Cartee GD, Douen AG, Ramlal T, et al. Stimulation of glucose transport in skeletal muscle by hypoxia. J Appl Physiol. 1991;70:1593–600.
35. Gamboa JL, Garcia-Cazarin ML, Andrade FH. Chronic hypoxia increases insulin-stimulated glucose

uptake in mouse soleus muscle. Am J Physiol Regul Integr Comp Physiol. 2011;300:R85–91.
36. Lundby C, Van Hall G. Substrate utilization in sea level residents during exercise in acute hypoxia and after 4 weeks of acclimatization to 4100 m. Acta Physiol Scand. 2002;176:195–201.
37. Brooks GA. Glycolytic end product and oxidative substrate during sustained exercise in mammals—the "lactate shuttle". In: Gilles R, editor. Comparative physiology and biochemistry—current topics and trends, Respiration—metabolism—circulation, vol. A. Berlin: Springer; 1984. p. 208–18.
38. Hashimoto T, Hussien R, Oommen S, et al. Lactate sensitive transcription factor network in L6 cells: activation of MCT1 and mitochondrial biogenesis. FASEB J. 2007;21:2602–12.
39. Gutierrez G, Fernandez E, Hurtado FJ, et al. Hydroxymalonate inhibits lactate uptake by the rabbit hindlimb. J Appl Physiol. 1994;76:2735–41.
40. Liu C, Wu J, Zhu J, et al. Lactate inhibits lipolysis in fat cells through activation of an orphan G-protein-coupled receptor, GPR81. J Biol Chem. 2009;284:2811–22.
41. Johnson JA, Fusaro RM. The role of the skin in carbohydrate metabolism. Adv Metab Disord. 1972;60:1–55.
42. Stanley WC, Wisneski JA, Gertz EW, et al. Glucose and lactate interrelations during moderate intensity exercise in humans. Metabolism. 1988;37:850–8.
43. Roth DA, Brooks GA. Lactate transport is mediated by a membrane-bound carrier in rat skeletal muscle sarcolemmal vesicles. Arch Biochem Biophys. 1990;279:377–85.
44. Roth DA, Brooks GA. Lactate and pyruvate transport is dominated by a pH gradient-sensitive carrier in rat skeletal muscle sarcolemmal vesicles. Arch Biochem Biophys. 1990;279:386–94.
45. Saddik M, Gamble J, Witters LA, et al. Acetyl-CoA carboxylase regulation of fatty acid oxidation in the heart. J Biol Chem. 1993;268:25836–45.
46. Larsen JJ, Hansen JM, Olsen NV, et al. The effect of altitude hypoxia on glucose homeostasis in men. J Physiol. 1997;504(Pt 1):241–9.
47. Braun B, Rock PB, Zamudio S, et al. Women at altitude: short-term exposure to hypoxia and/or alpha(1)-adrenergic blockade reduces insulin sensitivity. J Appl Physiol. 2001;91:623–31.
48. Van Hall G, Calbet JA, Sondergaard H, et al. Similar carbohydrate but enhanced lactate utilization during exercise after 9 wk of acclimatization to 5,620 m. Am J Physiol Endocrinol Metab. 2002;283:E1203–13.
49. van Hall G, Lundby C, Araoz M, et al. The lactate paradox revisited in lowlanders during acclimatization to 4100 m and in high-altitude natives. J Physiol. 2009;587:1117–29.
50. Friedlander AL, Braun B, Pollack M, et al. Three weeks of caloric restriction alters protein metabolism in normal-weight, young men. Am J Physiol Endocrinol Metab. 2005;289:E446–55.
51. Barnholt KE, Hoffman AR, Rock PB, et al. Endocrine responses to acute and chronic high-altitude exposure (4,300 meters): modulating effects of caloric restriction. Am J Physiol Endocrinol Metab. 2006; 290:E1078–88.
52. Fulco CS, Friedlander AL, Muza SR, et al. Energy intake deficit and physical performance at altitude. Aviat Space Environ Med. 2002;73:758–65.
53. Fulco CS, Kambis KW, Friedlander AL, et al. Carbohydrate supplementation improves time-trial cycle performance during energy deficit at 4,300-m altitude. J Appl Physiol. 2005;99:867–76.
54. Semenza GL. Regulation of oxygen homeostasis by hypoxia-inducible factor 1. Physiology. 2009;24:97–106 Cold Spring Harb Symp Quant Biol. 2011;76:347-53.
55. Semenza GL. Regulation of metabolism by hypoxia-inducible factor 1. Cold Spring Harbor symposia on quantitative biology; 2011
56. Semenza GL, Wang GL. A nuclear factor induced by hypoxia via de novo protein synthesis binds to the human erythropoietin gene enhancer at a site required for transcriptional activation. Mol Cell Biol. 1992;12:5447–54.
57. Ugocsai P, Hohenstatt A, Paragh G, et al. HIF-1beta determines ABCA1 expression under hypoxia in human macrophages. Int J Biochem Cell Biol. 2010;42:241–52.
58. van Patot MC, Gassmann M. Hypoxia: adapting to high altitude by mutating EPAS-1, the gene encoding HIF-2alpha. High Alt Med Biol. 2011;12:157–67.
59. van Uden P, Kenneth NS, Webster R, et al. Evolutionary conserved regulation of HIF-1beta by NF-kappaB. PLoS Genet. 2011;7:e1001285.
60. Cerretelli P, Gelfi C. Energy metabolism in hypoxia: reinterpreting some features of muscle physiology on molecular grounds. Eur J Appl Physiol. 2011; 111:421–32.
61. Wang GL, Semenza GL. Purification and characterization of hypoxia-inducible factor 1. J Biol Chem. 1995;270:1230–7.
62. Kim JW, Tchernyshyov I, Semenza GL, et al. HIF-1-mediated expression of pyruvate dehydrogenase kinase: a metabolic switch required for cellular adaptation to hypoxia. Cell Metab. 2006;3:177–85.
63. Papandreou I, Cairns RA, Fontana L, et al. HIF-1 mediates adaptation to hypoxia by actively downregulating mitochondrial oxygen consumption. Cell Metab. 2006;3:187–97.
64. Fukuda R, Zhang H, Kim JW, et al. HIF-1 regulates cytochrome oxidase subunits to optimize efficiency of respiration in hypoxic cells. Cell. 2007;129:111–22.
65. Hochachka PW, Rupert JL, Monge C. Adaptation and conservation of physiological systems in the evolution of human hypoxia tolerance. Comp Biochem Physiol A. 1999;124:1–7.
66. Wadley GD. Effect of exercise intensity and hypoxia on skeletal muscle AMPK signaling and substrate metabolism in humans. Am J Physiol Endocrinol Metab. 2006;290:E694–702.
67. Brooks GA. Lactate shuttles in nature. Biochem Soc Trans. 2002;30:258–64.
68. Brooks GA. Cell–cell and intracellular lactate shuttles. J Physiol. 2009;587:5591–600.

69. Loscalzo J. The cellular response to hypoxia: tuning the system with microRNAs. J Clin Invest. 2010; 120:3815–7.
70. Snyder EM, Carr RD, Deacon CF, et al. Overnight hypoxic exposure and glucagon-like peptide-1 and leptin levels in humans. Appl Physiol Nutr Metab. 2008;33:929–35.
71. Yan X, Zhang J, Gong Q, et al. Appetite at high altitude: an fMRI study on the impact of prolonged high-altitude residence on gustatory neural processing. Exp Brain Res. 2011;209:495–9.
72. Campbell WW, Leidy HJ. Dietary protein and resistance training effects on muscle and body composition in older persons. J Am Coll Nutr. 2007;26: 696S–703.
73. Campbell WW. Synergistic use of higher-protein diets or nutritional supplements with resistance training to counter sarcopenia. Nutr Rev. 2007;65:416–22.

Exercise

16

Carsten Lundby

Abstract

Altitude has substantial effects on sport and exercise performance. This is the consequence of two domains; (1) physiological and (2) physical. A reduction in barometric pressure decreases partial pressure of inspired oxygen and thereby partial pressures and contents at every step along the oxygen cascade (Fig. 16.1). Thus the availability of oxygen at the mitochondrial level to produce ATP via oxidative phosphorylation is also reduced. This requires adjustments within the oxygen transport cascade for a given workrate and affects the maximal attainable oxygen uptake (VO_2max) and hence compromises endurance. In contrast, the decrease in air density reduces air resistance which will facilitate high-velocity performances such as sprint running and throwing events. Various forms of altitude training have been proposed to increase sea level exercise endurance capacity and these will be discussed.

Aerobic Power, Submaximal Exercise Performance, and O_2 Transport at High Altitude

The Response of VO_2max and Submaximal Performance to Altitude Exposure

Maximal oxygen uptake (VO_2max) is the product of the total quantity of blood transported to the tissues by the heart (i.e., cardiac output, Q) and the difference in systemic arterial and venous O_2 content (CaO_2 and C_vO_2, respectively). This is referred to as the Fick equation: $VO_2max = Q \times (CaO_2 - C_vO_2)$. Altitude-related adjustments in any of these parameters will affect VO_2max, and as described in the following sections both Q and CaO_2 are reduced at altitude, whereas oxygen extraction is maintained. First the response of VO_2max and submaximal exercise performance to altitude is described, and subsequently the underlying mechanisms are discussed.

The solid line in Fig. 16.2 illustrates the curvilinear relationship of the measured decrease (%) in VO_2max with increasing degree of altitude. Data were obtained from 67 research studies carried out in acute hypoxia (laboratory or at altitude)

C. Lundby, Ph.D. (✉)
Center for Integrative Human Physiology (ZIHP), University of Zürich, Winterthurerstrasse 190, Zürich, Switzerland
e-mail: carsten.lundby@uzh.ch

E.R. Swenson and P. Bärtsch (eds.), *High Altitude: Human Adaptation to Hypoxia*,
DOI 10.1007/978-1-4614-8772-2_16, © Springer Science+Business Media New York 2014

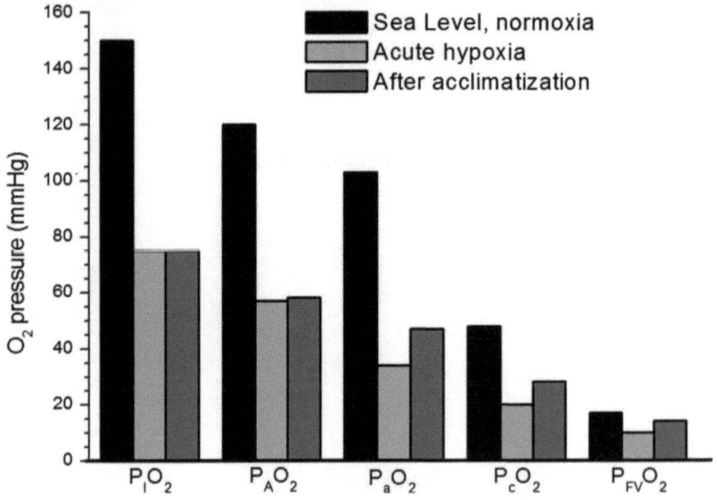

Fig. 16.1 Oxygen cascade at rest from the atmosphere to the femoral vein. The values of inspiratory O_2 pressure (P_IO_2), alveolar PO_2 (P_AO_2), arterial PO_2 (P_aO_2), estimated mean capillary PO_2 (P_cO_2), and femoral vein PO_2 ($P_{FV}O_2$) are represented during exercise on the cycle-ergometer during exercise in normoxia (*black bars*), acute hypoxia (*light grey bars*), and chronic hypoxia (*grey bars*). Adapted from [120]

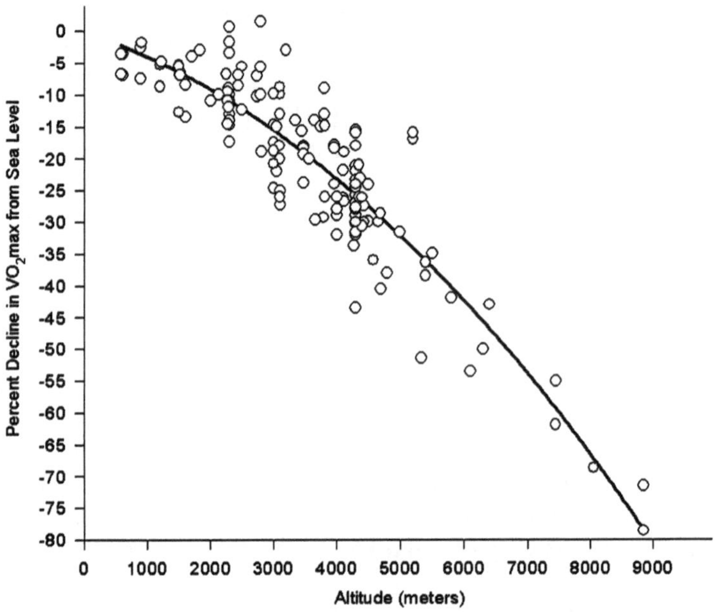

Fig. 16.2 Changes in VO_2max with changes in altitude. Each of the 146 point represents the average VO_2max decrement of a group of test subjects participating in one of 67 research studies conducted at actual or simulated altitudes from 580 to 8,848 m. Since each data point is a mean value of many intrastudy individual determinations of VO_2max, the drawn regression line represent thousands of VO_2max test values. From [18]. The figure is printed with permission

or after exposure to hypoxia (laboratory or altitude) of varying duration. In general a decrease in VO_2max is observed from approximately 1,500 m upwards, and thereafter by a ≈1 % reduction for every 100 m of altitude gained [1], but this may vary greatly, and should only be seen as a rough estimate. For example, on average, VO_2max declines by about 9 %, 14 %, 24 %, and 32 % at 2,000, 3,000, 4,000, and 5,000 m, respectively. At the summit of Mount Everest at 8,850 m the reduction in VO_2max is approximately 80 %, and thus leaves not much exercise capacity left for climbing. When men and women are matched to sea level aerobic fitness level, there is generally no difference in the altitude-induced decline in VO_2max. In elite athletes a decrease in VO_2max has been observed at altitudes starting as low as 580 m [2]. As mentioned above a decrease in

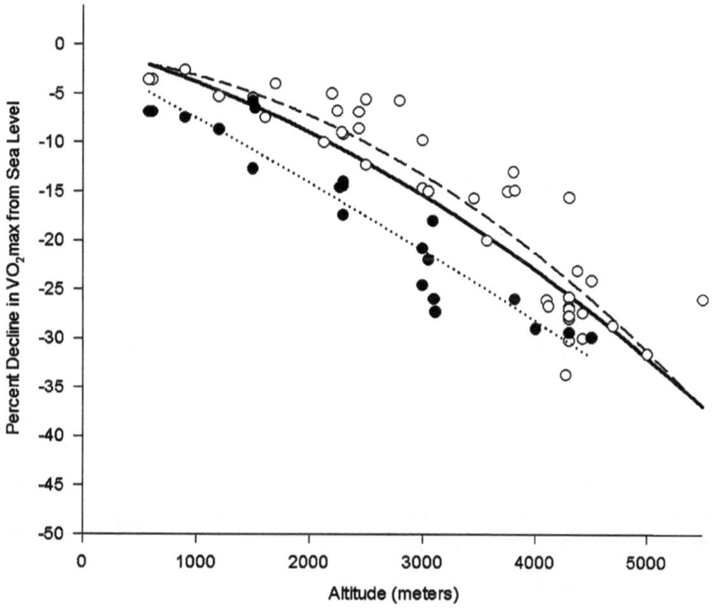

Fig. 16.3 The effect of sea level physical fitness on VO_2max decrement at high altitude. *White circles* and *dotted line* = Trained individuals (VO_2max > 63 mL kg^{-1} min^{-1}), and *black circles* and *broken line* = untrained individuals. From [18]. The figure is printed with permission

VO_2max at this elevation is normally not the case for the majority of human beings, and one of the biggest contributors to the variation in the decline in VO_2max at altitude is aerobic fitness level before exposure. Figure 16.3 illustrates the difference in the decline of VO_2max in trained vs. untrained subjects. The difference *may* become substantial; for example at 4,000 m a trained individual's VO_2max may be decreased by 22 % whereas an untrained individuals VO_2max may only be decreased by 13 %. However, also the opposite has been observed. An individual with a VO_2max of 82 mL min^{-1} kg^{-1} at sea level was reduced with only 6 % with acute exposure to 2,500 m, whereas another individual with a VO_2max of 63 mL min^{-1} kg^{-1} was reduced by 14 % at the same level of hypoxia in the very same study [3]. Hence, although it is generally accepted that athletes are relatively more limited at altitude as compared to less-trained individuals, the response is certainly not uniform. In line with this, Wehrlin and Hallen [4] have extended this body of knowledge by analyzing studies which included male, unacclimatized endurance-trained athletes with a mean VO_2max > 60 mL min^{-1} kg^{-1}. In this population VO_2max was reduced by 7.7 % per 1,000 m altitude. Furthermore, in an attempt to find a threshold altitude for reduced VO_2max in individual athletes, they observed a quite uniform and highly linear decline in VO_2max at altitude, beginning as low as 800 m altitude and extending through 2,800 m with a decline of 6.3 % per 1,000 m altitude. The greater decrease in VO_2max of trained vs. untrained humans seems to be dependent on a higher degree of arterial desaturation occurring in the more trained male and female subjects [2, 5], and this is the consequence of greater capillary diffusion limitation and ventilation-perfusion mismatch [6].

VO_2max and Submaximal Exercise Performance Following Acclimatization

Whether VO_2max is increased significantly with acclimatization has been debated for decades. Various factors such as the ability to maintain training intensity/volume and the amount of altitude-dependent muscle wasting are potential confounders. Also a different fatigue pattern at different altitudes may explain some of the controversies. Recent studies have revealed that from a certain threshold altitude, the origin of fatigue during exercise is shifted from being of predominantly peripheral to a centrally mediated origin (see more below) [7]. With degrees of hypoxia where the O_2 transport system may not be the main limiting factor for VO_2max, it makes sense that restoring, or even surpassing, sea level CaO_2 does not increase VO_2max. In an attempt to reduce any potential negative influence of altitude acclimatization on VO_2max, recent studies have been performed with acute hypoxic exposure before and after 1–2 months of erythropoietin (Epo) treatment. Following Epo treatment CaO_2 was significantly increased at maximum exercise at sea level, 1,500, 2,500, 3,500, 4,100, and 4,500 m [8, 9], but VO_2max only increased from sea level up to 3,500 m, but not at 4,500 m. These experimental laboratory data support the observation that VO_2max may increase with acclimatization to altitudes below 3,500–4,000 m [10, 11], but not above [12–14], although exceptions have occurred [15]. It should be recognized however that most of these studies include low subject numbers, and that clear-cut conclusions are difficult to draw. In some studies, acute exposure to altitudes above 5,000 m decreases VO_2max more than in chronic hypoxia, and thus a small increase may be observed with acclimatization. One reason here for *could* be an underestimated VO_2max obtained in acute hypoxia because most subjects don't tolerate this degree of stress well, i.e., the increase in VO_2max with acclimatization to such altitudes *may* merely just be a case of tolerating the stress better, or because the degree of central fatigue has become less with acclimatization (see also below).

Unlike VO_2max, submaximal exercise performance is increased in several investigated altitudes. Endurance time to exhaustion performed at 4,300 m at a given power output increased by 45 % on day 10 [16] and by 59 % on day 16 [15] compared to day 2 in individuals with VO_2max of approximately 49 mL min^{-1} kg^{-1}. At 2,340 m altitude time to exhaustion performed at 80 % of sea level maximal attainable workload during a VO_2max test was decreased by 25 % as compared to sea level on day 1 at altitude, but increased with acclimatization by 6 % on day 7, and by a further 5.7 % and 1.4 % after 14 and 21 days at altitude, respectively [11]. In the experiments conducted at 4,300 m altitude the increase in submaximal exercise performance occurred without a concomitant increase in VO_2max, whereas the augmented submaximal exercise capacity reported to occur at 2,340 m correlated well with a similar time-dependent increase in VO_2max. Therefore the increased submaximal exercise capacity at the lower altitude could be associated with a gradual reduction in relative exercise intensity, whereas this cannot be the case at the higher altitude. This is supported by the finding that submaximal exercise to exhaustion time was not altered with acute or chronic (16 days) exposure to 4,300 m as compared to sea level performance, if performed at the same relative exercise intensity [17]. One of the most likely causes for increasing submaximal exercise performance without a concomitant increase in VO_2max could be related to the restored CaO_2 resulting from hemoconcentration and ventilatory adaptation that allow a given VO_2 to be achieved with a lower *Q*, and thereby reducing cardiac work.

The altitude-induced changes in VO_2max and submaximal exercise performance obviously affect sport performance. In events lasting approximately 2 min (i.e., 800 m running) and longer the reduction in performances becomes greater the longer the duration of the event [18]. In general, acclimatization to altitude increases event performance by approximately 2–3 % within 2–3 weeks, but will never reach sea level values. On the other hand, events lasting less than approximately 2 min (or events performed at very high speeds, see below) may improve at altitude. The reason here for is the decrease in air density, which parallels the reduction in barometric pressure, and hence reduces air resistance. This is especially important in contests where the main obstacle to surpass is air resistance such as in throwing events, or in events performed at high speeds when air resistance quadruples when speed is doubled such as track cycling. Theoretical work suggests that although VO_2max and submaximal exercise performance are drastically reduced at 2,500 m, this may be the optimal altitude to pursue the 1 h solo cycling record because

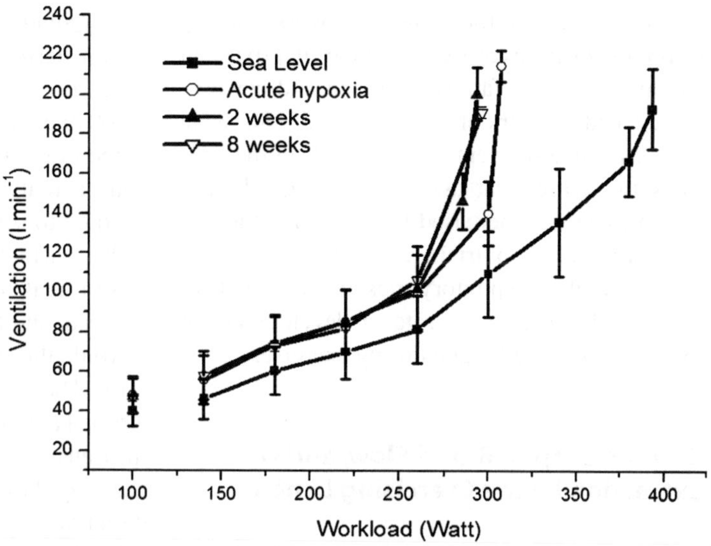

Fig. 16.4 The effect of increasing altitude on ventilation during incremental intensity exercise until exhaustion in lowlanders at sea level, with acute hypoxic exposure, and after 2 and 8 weeks acclimatization to 4,100 m. Note that the ventilatory response is increased more the higher the workload, and that the maximal attainable ventilation is not changed. Also note that exercise ventilation is not decreased with acclimatization. Adapted from [61]

the advantage of the reduced air resistance surpasses the negative effect of reduced exercise capacity [19].

Sports that use mainly anaerobic capacity such as weight lifting will be unaffected by altitude exposure, whereas throwing events will show increased performance because of the decrease in air density. At 2,300 m, the 24 % reduction in air density will increase the shot put, hammer throw, javelin, and discus distances by approximately 6, 53, 69, and 162 cm, respectively [20], and also sprint running (60, 100, 200, and 400 m) times may be decreased with increasing altitude because of a reduced air resistance and reliance on (mainly unaffected) anaerobic metabolism [21] although acclimatization improvements in skeletal muscle acid–base and buffering mechanisms [22] could lead to increased sprint performance also. Another factor to be considered is that the reduced air density also changes aerodynamics which for example affects the trajectory of a flying ball [23]. For example, a football kicked to spin in lateral direction for 4 m at sea level, will deviate about 0.4 m less at an altitude of 1,000 m, 0.8 m less at 2,000 m, and 1.2 m less at 3,000 m, and thus based on prior experience at sea level, it may become difficult for the players to judge exactly where the ball will go.

Ventilation and Exercise in Hypoxia

At altitude ventilation is increased at rest and during exercise. The hypoxic ventilatory response results in an increase in the partial pressure for oxygen (P_AO_2), and hence also increases P_aO_2, S_aO_2, and ultimately also CaO_2. Exercise at altitude is accompanied by an increase in submaximal ventilation, whereas maximal VE is generally not affected due to a lower VO_2max in spite of higher VE/VO_2 (Fig. 16.4). The higher the altitude, the higher the hypoxic ventilatory response at a given metabolic rate. As compared to resting conditions, exercise ventilation is increased much more. Increased ventilation during exercise is not caused by the increase in relative exercise intensity (due to a reduction in VO_2max, see later) but due to chemoreceptor stimulation. During exercise the ventilatory response to acute hypoxia is sensitized by a resetting of peripheral chemoreceptors [24]. At maximum effort, ventilation (in L min^{-1}) will be similar to those obtained at sea level, even at extreme elevations, it should be noted however, that the degree of hyperventilation $\left(V_A / V_{CO_2}\right)$ is of much greater magnitude. Since the last edition of the book, knowledge has emerged regarding the effects of ventilation upon locomotor fatigue in acute hypoxia. In an attempt to isolate the independent effects of inspiratory muscle work during heavy-intensity exercise in acute hypoxia on locomotor muscle fatigue, Amman and coworkers reduced the work of breathing by 35–80 % via proportional assist ventilation (PAV). In the PAV trials the subsequent assessment of quadriceps twitch force was increased by about one-third as compared to control trials [25]. This suggest that in hypoxia that inspiratory muscle work contributes significantly to the rate of development of locomotor muscle fatigue during exercise.

Cardiac Output, Blood Flow, and O_2 Extraction by the Exercising Limbs

Systemic O_2 transport is the product of CaO_2 and cardiac output (*Q*). With acute hypoxic exposure when CaO_2 is reduced, systemic oxygen transport is maintained at rest and during submaximal exercise by a higher cardiac output, skeletal muscle blood flow, and O_2 extraction [26, 27]. At maximal exercise *Q* is reduced as compared to at sea level, and as a result also maximal blood flow to the exercising legs is reduced. Since the maximal fractional O_2 extraction of the exercising muscles cannot precede the 90–91 % achieved already at sea level [13], ultimately the decrease in O_2 delivery to the exercising muscles is the cause for VO_2max to be reduced.

The initial increase in resting cardiac output with an acute exposure to hypoxia is, for the most part, a function of an increased heart rate (chronotropic response). This change is sustained by hypoxia-dependent sympathetic activation, directly via stimulation of the peripheral chemoreceptors [28], and vagal withdrawal, indirectly via an increase in ventilation. With acclimatization to altitudes >3,000 m (which corresponds to the commencement of the relatively steep part of the oxygen dissociation curve), the elevation in resting heart rate persists, although slightly decreasing toward sea level after 2–3 weeks. With acute hypoxic exposure, β-receptor blockade diminishes most of the initial increase in resting heart rate. However, this is not the case with acclimatization. In the course of such adaptation, heart rate remains somewhat elevated, even when administering supplemental oxygen or *beta* blockers. At altitudes above approximately 3,000 m maximum heart rate is reduced [29]. This is the effect of an elevated parasympathetic nervous activity, and maximum heart rate can be restored to sea level values by blocking parasympathetic activity [30]. Interestingly, when maximum heart rate is restored, maximum stroke volume is reduced, and thus leaving maximum *Q* unaltered. It would thus seem that the regulated variable is *Q* [30]. The reason for reducing *Q* at maximum effort at altitude could be to minimize the decrease the transit time of the red blood cell in the pulmonary capillaries in order to prevent further arterial desaturation, but this remains speculative [31]. Also the exact mechanisms causing the reduction in maximal *Q* remain uncertain.

As mentioned above the initial increase in resting cardiac output is not a function of an augmented stroke volume. However, changes in SV become important with acclimatization. Since heart rate is consistently elevated at altitude (Fig. 16.5), the reduction in cardiac output noted above must be a consequence of a decreased stroke volume, and this has been confirmed in various studies. This reduction is mainly caused by a decrease in plasma volume. Since stroke volume is the difference between end-diastolic and end-systolic volumes, it can only be reduced if end-systolic volume is increased due to impaired contractile function, or if end-diastolic volume is reduced due to impaired left ventricular filling. Laboratory studies conducted as part of Operation Everest II showed that myocardial contractility is well preserved, even at extreme altitudes [32], and stroke volume is therefore reduced in parallel with changes in left ventricular end-diastolic volume. Such changes are secondary to a hypoxia-induced plasma (blood) volume reduction [33].

The reduced maximum cardiac output at high altitude ultimately reduces maximal blood flow to the exercising legs during whole body exercise [13]. With acclimatization to high altitude there is a gradual decrease in maximal leg blood flow, but with a concomitant gradual increase in CaO_2,

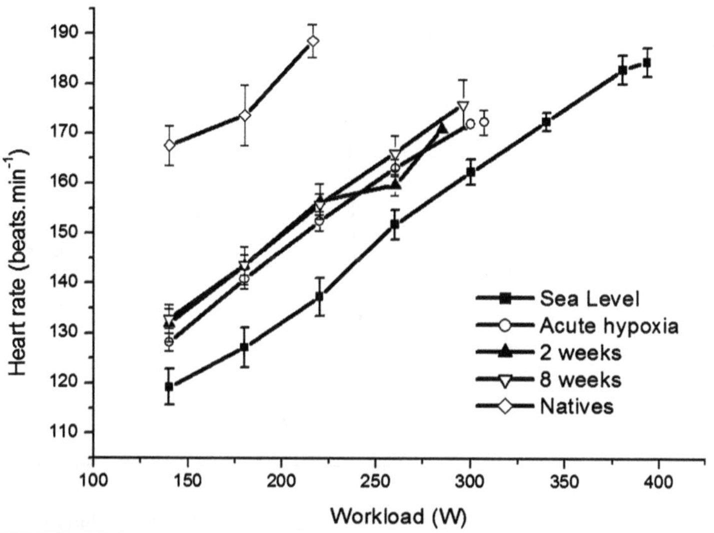

Fig. 16.5 Hear rate response to graded exercise in lowlanders at sea level, with acute hypoxia exposure, and after 2 and 8 weeks of acclimatization to 4,100 m altitude. Data are also shown for high altitude native Aymaras. Note that (1) in the lowlanders, that maximal heart rate is decreased with acute hypoxic exposure and that this remains unchanged throughout acclimatization (2) submaximal exercise heart rate is increased with acute hypoxic exposure and is not decreased with acclimatization despite normalization of arterial oxygen content. Adapted from [13]

O_2 delivery to the exercising legs during exercise is similar throughout acclimatization (at least up to 8 weeks), and therefore also VO_2max does not change from acute to chronic exposure at altitudes above approximately 4,000 m [34].

Substrate Utilization During Exercise at High Altitude

Resting metabolism is increased slightly with exposure to altitude (Fig. 16.6) and is most likely the consequence of an elevated energy expenditure caused by an elevated heart rate, ventilation, and other factors (see Chaps. 4, 6, 13 and 15). If this is not compensated for by an increase in caloric intake, loss of body mass will be unavoidable [35]. In a variety of experimental models, hypoxia causes a shift in substrate use in favor of glucose. One explanation for this phenomenon could be a selective advantage derived from the increased metabolic economy (more ATP derived per unit oxygen consumed; ATP/O_2 ratio for carbohydrate = 6.0–6.3 and 5.6 for lipids) than results when glucose is oxidized rather than lipid or protein. In support of this hypothesis, after acclimatization to hypoxia it has been demonstrated that when exercising at the same absolute exercise intensity as at sea level, and hence at a higher relative exercise intensity due to the altitude depended reduction in VO_2max, that there is a high dependence on blood glucose as a substrate [36]. However since exercise intensity is of major importance for substrate preference during exercise, enhanced glucose oxidation during exercise at altitude could be the consequence of relatively higher exercise intensity. To examine the potential contribution of exercise intensity on the increase in glucose oxidation at altitude, we investigated lowlanders during 60 min cycle exercise at 50 % of sea level VO_2max. The protocol was repeated in acute hypoxia and also after 4 weeks acclimatization to 4,100 m. In the hypoxic conditions the 60 min exercise trial was performed at the same absolute exercise intensity as at sea level (now corresponding to approximately 65 % of hypoxic VO_2max) and at the same relative exercise intensity, i.e., 50 % of the hypoxic VO_2max. The results obtained from this study confirmed previous studies in showing an increased

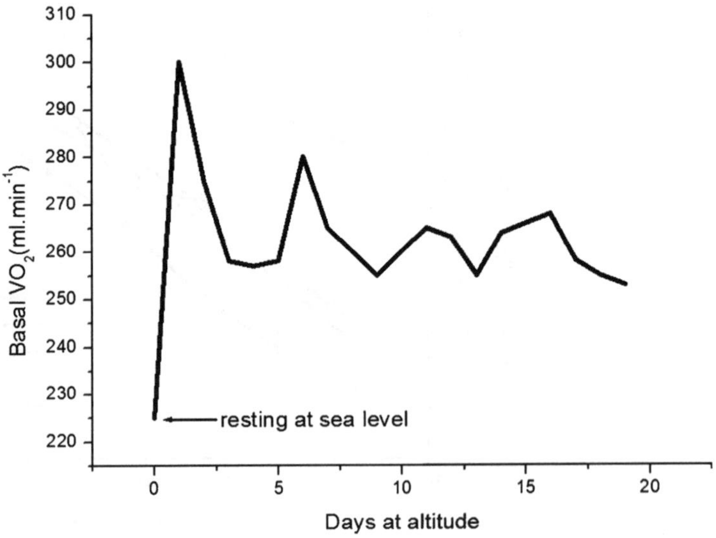

Fig. 16.6 Basal oxygen uptake at sea level and during acclimatization to 4,300 m altitude. Adapted from [35]. Note that basal oxygen uptake is increased with exposure to high altitude. If this is not compensated for by increased caloric intake a reduction in body weight is unavoidable

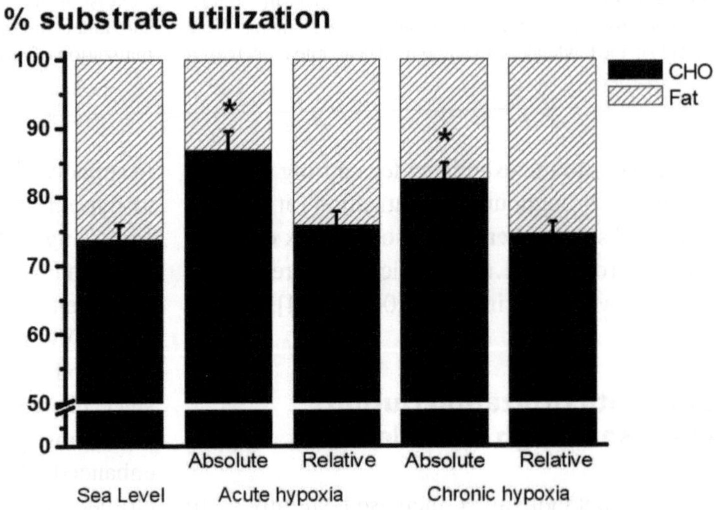

Fig. 16.7 Substrate utilization during submaximal exercise at sea level, and again at the same relative and same absolute intensity in acute and chronic hypoxia. Adapted from [37]. Note that substrate utilization is only altered with hypoxic exercise if performed at the same absolute exercise intensity, but not if performed at the same relative exercise intensity

glucose oxidation at the same absolute exercise intensities, but when the intensity was reduced to elicit the same relative intensity as at sea level, no differences in substrate utilization as compared to sea level were observed (Fig. 16.7) [37]. This suggests that at least during exercise the previously observed increase in glucose oxidation may have been a result of the increased exercise intensity, and not directly caused by hypoxia.

Lactate metabolism at submaximal and maximal exercise intensities with acclimatization have been the origin for a lively debate since the 1930s [38, 39]. Based on early work the term "lactate paradox" was later coined and refers to a decrease in both the maximal attainable lactate concentration following maximal exercise, and also to a gradual decrease in submaximal blood lactate concentrations following acclimatization. These findings were considered paradoxical since one would assume maximal lactate concentrations, if anything, to be higher at altitude, and since at submaximal exercise O_2 delivery to the exercising muscles remains constant with altitude exposure by regulating leg blood flow and O_2 extraction

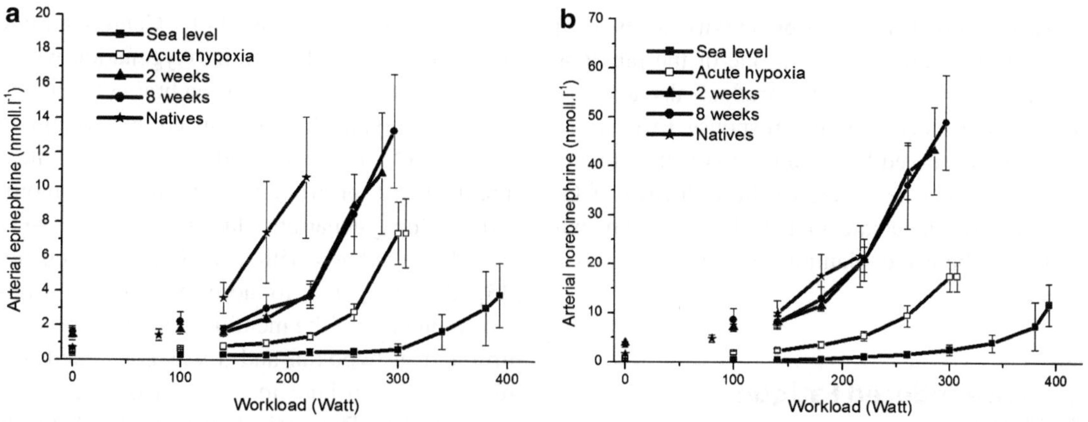

Fig. 16.8 Arterial epinephrine (**a**) and norepinephrine (**b**) concentrations during incremental exercise at sea level, in acute hypoxia, and after 2 and 8 weeks of acclimatization. Values are also presented for high altitude native Aymaras. Adapted from [13, 40]

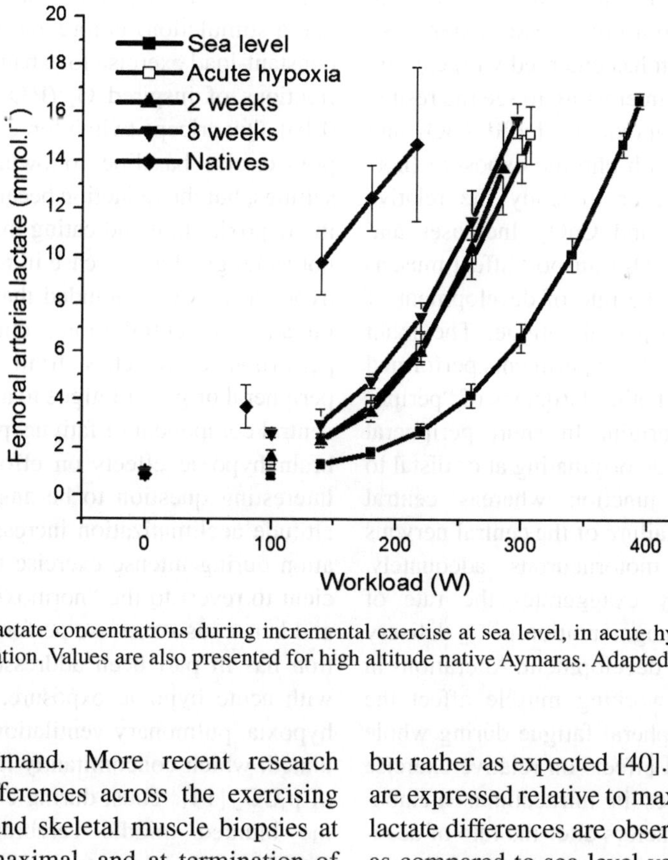

Fig. 16.9 Arterial lactate concentrations during incremental exercise at sea level, in acute hypoxia, and after 2 and 8 weeks of acclimatization. Values are also presented for high altitude native Aymaras. Adapted from [40]

according to demand. More recent research applying a–v differences across the exercising skeletal muscle and skeletal muscle biopsies at rest, during submaximal, and at termination of exercise clearly demonstrate that norepinephrine (Fig. 16.8), the activator of glycolysis and the blood lactate (Fig. 16.9) response to altitude acclimatization does not behave paradoxically, but rather as expected [40]. In fact, if workloads are expressed relative to maximum effort, then no lactate differences are observed at any time point as compared to sea level values. Whether this is also the case at very high altitudes remains unknown, but seems likely. The very elevated norepinephrine levels in Fig. 16.8 are supported by a tremendous hypoxia-induced increase in muscle sympathetic nervous activity at rest [41] and during exercise [42]. In one of the previous studies reporting a "lactate paradox" also a reduction in the catecholamine response to exercise has been observed [43], and perhaps the controversy may not be linked to the influence of O_2 availability on lactate metabolism as such, but rather to the catecholamine response to hypoxic exercise.

Exercise-Induced Fatigue in Hypoxia

Central Fatigue

In recent years an increasing number of studies have investigated the mechanisms for fatigue in acute hypoxia, and to a much lesser extent also important information has emerged with chronic hypoxia. It would be interesting to see the results obtained in acute hypoxia (reduced PaO_2 and CaO_2) being tested with chronic exposure (normalized CaO_2) in order to study the relative importance of PaO_2 and CaO_2. Increases and decreases in systemic O_2 transport affect muscular performance and the rate of development of both central and peripheral fatigue. The main question addressed in the experiments performed in acute hypoxia is whether fatigue is of "peripheral" or "central" origin. In short peripheral fatigue refers to fatigue originating at or distal to the neuromuscular junction whereas central fatigue includes the failure of the central nervous system to excite motorneurons adequately. Blunted O_2 delivery exaggerates the rate of fatigue, whereas an augmentation in O_2 delivery attenuates the rate of development. Alterations in O_2 delivery to the working muscle affect the development of peripheral fatigue during whole body exercise via its effects on relative exercise intensity and changes in intracellular metabolism. Both of these factors alter the rate of accumulation of metabolites known to cause failure of excitation-contraction coupling within the muscle fiber, which has been identified as the main factor that evokes loss of tension development during the fatigue process occurring under conditions of high-intensity exercise [44]. Central fatigue may be elicited by low brain oxygenation which may cause a mismatch between brain O_2 demand and O_2 supply, leading to reduced interstitial and cellular PO_2. Near infrared spectroscopy measurements of prefrontal, premotor, and motor cortex deoxygenation during exercise were recently completed with acute hypoxic exposure [45]. As compared to normoxic control trials deoxygenation of the mentioned areas were more profound in hypoxia, and it was concluded that this could contribute to a hastened decision to stop exercise. The critical interstitial PO_2 levels leading to central fatigue are however unknown, but it has been proven that an "altitude threshold" exists. The group of Dempsey and colleagues [7] assessed neuromuscular function (quadriceps twitch force by supramaximal magnetic femoral nerve stimulation) before and immediately after constant-load exercise to exhaustion with various fractions of inspired O_2 (F_IO_2=0.21, 0.15, and 0.10). Quadriceps twitch force was reduced from pre-exercise baseline following exercise in all settings, but the reduction became less the greater the hypoxia; thus indicating that the muscle was not as fatigued as the case in normoxia and mild hypoxia. It was concluded that the major determinants of central motor output and exercise performance switches from a predominantly peripheral origin of fatigue to a hypoxia-sensitive central component of fatigue, probably involving brain hypoxic effects on effort perception. An interesting question to be answered is whether altitude acclimatization increases brain oxygenation during intense exercise to an extent sufficient to revert to the "normoxic fatigue pattern" and hence also exercise performance. This question has in part been addressed experimentally with acute hypoxic exposure. With exercise in hypoxia, pulmonary ventilation is strongly stimulated, which concomitantly leads to a lowering of p_aCO_2 [14]. Thus, during exercise in hypoxia the blunted middle cerebral artery velocity (MCAv), secondary to hypocapnia [46] combined with the low CaO_2 causes brain deoxygenation and in turn this may facilitate centrally originating fatigue [45, 47, 48]. To test if an increased cerebral oxygenation during maximal exercise in acute hypoxia would also increase exercise performance Subudhi and coworkers [46] "clamped" end tidal CO_2 ($P_{ET}CO_2$) to 50 mmHg during exercise in normoxia and in hypoxia equivalent to 4,875 m altitude. Although the intervention increased cerebral oxygenation exercise performance became reduced. The authors speculated that this could be the cause of a too severe respiratory acidosis. In a set of additional experiments in which $P_{ET}CO_2$ was maintained at 40 mmHg from about 75 % of VO_2max exercise performance was not decreased, but tended to be ($P<0.10$). Siebenmann [49] obtained similar results when clamping $P_{ET}CO_2$ to 40 mmHg during exercise at 3,450 m altitude. Despite their limitations in regards to respiratory acidosis, the study questions the role of cerebral oxygenation and its potential role in limiting exercise performance in hypoxia. More studies are needed in this regard.

Peripheral Fatigue

A recently published study performed by van Hall and coworkers [40] clearly demonstrates that peripheral indicators of fatigue are unchanged throughout the process of acclimatization to 4,100 m. In muscle biopsies obtained a few seconds after the termination of a graded exercise test until exhaustion, metabolites classically associated with fatigue were not altered after 2 and 8 weeks of acclimatization as compared to acute exposure or to values obtained in normoxia. Analysis of the biopsies for ATP, ADP, AMP, IMP, creatine phosphate, creatine content, inorganic phosphate (P_i,), lactate and glycogen content all showed no signs of change in peripheral fatigue pattern. As mentioned before a recent study has revealed that from a certain threshold altitude, the origin of fatigue during exercise is moved from being of predominantly peripheral in origin to predominantly centrally mediated [7]. When central fatigue becomes prevailing, one can speculate peripheral signs of fatigue will be reduced. This however, does not seem to be the case as at least plasma lactate is concerned because it is equally high in both situations.

Pulmonary Vascular Limitations to Hypoxic Exercise

Whether hypoxic pulmonary vasoconstriction (HPV) is a limiting factor of exercise at high altitude or not is debated [50, 51], but at least some of the limitation to aerobic exercise capacity at altitude may be related to an elevation in pulmonary arterial pressure (PAP). Acute hypoxic exposure induces HPV, and depending on an individual's predisposition and on external factors such as the degree of hypoxia and rate of ascent to altitude, HPV may increase PAP to levels which are observed in patients with pulmonary arterial hypertension (PAH) at sea level [34]. PAH is a severe disease which ultimately decreases arterial O_2 saturation by inducing a modest mismatch between ventilation and perfusion, but more importantly also limits cardiac output by increasing right ventricular afterload. Both of these impair convective O_2 transport and limited exercise tolerance is thus a cardinal symptom for the diagnosis of PAH. It seems plausible to hypothesize that an altitude-dependent increase in PAP may contribute to the reduction in exercise tolerance by reducing arterial oxygenation or limiting cardiac output, or both. Unfortunately the majority of studies addressing these questions have applied noninvasive techniques with concomitant measurement uncertainties. In one of the few studies actually placing an arterial catheter both sildenafil and bosentan equally improved arterial O_2 saturation at the investigated exercise intensities (up to 90 % of VO_2max), and it was concluded that this could lead to an improved physical performance at altitude [52]. In most other studies administrating sildenafil it was observed that sildenafil diminished PAP and concomitantly also improved exercise capacity in hypoxia, while the treatment showed no effect on sea level exercise capacities [53–55]. Similar results have been obtained in studies in which pulmonary vasodilation was induced by the glucocorticoid dexamethasone [56, 57]. Lastly, bosentan has been shown to enhance exercise capacity without affecting arterial O_2 saturation, and it was concluded that PAP may limit hypoxic exercise capacity most likely by increasing right ventricular afterload [58]. In summary, the findings of several studies, but not all, have demonstrated that a hypoxia-dependent increase in PAH can be contributor to the reduced exercise capacity in hypoxic conditions. However, the underlying mechanisms are not entirely unravelled, but are most likely related to a reduced right ventricular afterload and/or improved arterial oxygenation. Before drawing any clear conclusions however it needs to be recognized that the assessment of cardiac output and arterial oxygenation in the above studies often relied on noninvasive methodologies which clearly leaves room for improvement. Besides conducting future studies in a randomized, double blinded crossover study design, investigators should also apply the appropriate invasive research tools.

Effects on Performance of Lifelong Acclimatization (High Altitude Natives)

A number of interesting findings have been made in regard to the exercise capacity of high altitude natives. Here special attention will be given to pulmonary diffusion capacity, the ability to recover a high percentage of sea level VO_2max with acclimatization and to exercise economy.

For a given P_AO_2 it is well documented that high altitude natives have a higher pulmonary diffusing capacity [59–63], and the smaller A-aPO_2 helps to preserve PaO_2 for any given level of ventilation. The likely basis of the low A-aPO_2 in Aymaras is the higher diffusion capacity associated with larger lung volumes [62, 64, 65]. The larger lung volumes may be the result of multiple factors, and from animal studies it is known that hypoxic exposure is associated with larger alveolar septal tissue volume and surface areas [66], which are further augmented by smaller alveolar duct volume and smaller mean harmonic diffusion-barrier thickness. In these animals the structural changes induced a higher oxygen diffusing capacity during exercise [67]. For lung adaptations to occur (in animal) high altitude exposure has to be initiated during lung maturation and altitude exposure in adult animal life does not induce structural or volume changes [68]. Once obtained however, even 2 years of reexposure to normoxia does not reverse the adaptations [67]. From the above studies it may be concluded that increases in gas-exchange efficiency arise largely from developmental exposure to hypoxia, and is further strengthened by the similar diffusion capacities of natives from Colorado, the Andes, and Tibet. It seems unlikely that the superior diffusion capacities of high altitude natives, which depend on structural adaptations in the lungs, may become available to acclimatizing low landers. Other factors however may also affect diffusion capacity and include polycythemia, cardiac output, PAH, and pulmonary vascular reactivity.

The higher diffusion capacity of highlanders becomes especially advantageous with exercise. In the transition from resting conditions to maximal exercise intensity, the exercise-induced degree of arterial desaturation is much lower in high landers. In lowlanders, for example, in the transition from rest to maximal exercise saturation decreased by 12.8 % (89.7–76.9 %) after 2 weeks acclimatization and 11.6 % (90.9–79.3 %) after 8 weeks of additional exposure to 4,100 m (Fig. 16.10). In Aymaras however, the exercise-induced desaturation was only 4.6 % (from 87.3 to 82.7 %), and thereby CaO_2 was also maintained at a higher level than in the lowlanders. In the Aymaras CaO_2 was 203 ± 6 mL L^{-1} at rest and 203 ± 2 mL L^{-1} at maximum exercise (Fig. 16.10). In contrast CaO_2 was decreased in the lowlanders from 204 ± 4 to 184 ± 2 mL L^{-1} and from 201 ± 4 to 182 ± 2 mL L^{-1} within the transition from rest to maximum effort after 2 and 8 weeks of acclimatization, respectively [69]. Hence, one clear advantage of Aymara during exercise at altitude is that exercise is not associated with the same degree of arterial desaturation as it is known to occur in lowlanders. This advantage is, however, partially offset by a lower arterial O_2 extraction at the level of the exercising muscle, and seems to be the consequence of a lower skeletal muscle diffusion capacity of high altitude natives [13]. It may also be that Aymaras with higher exercise training levels may desaturate more like the European lowlanders, but this remains to be investigated.

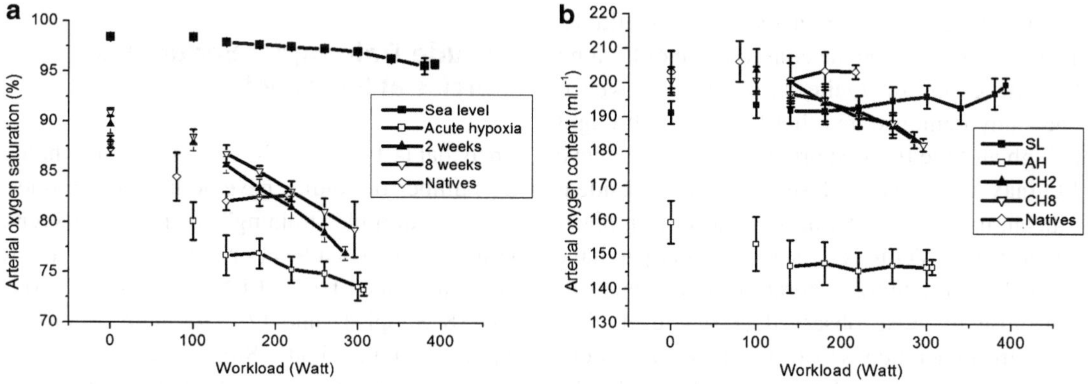

Fig. 16.10 Arterial (**a**) oxygen saturation and (**b**) content during incremental exercise at sea level, in acute hypoxia, and after 2 and 8 weeks of acclimatization in lowlanders and in high altitude Aymaras. Adapted from [13, 61]. Note that the lowlanders desaturate substantially in the transition from rest to maximal exercise in hypoxia whereas the high altitude Aymaras nearly don't desaturate. This affects arterial oxygen content tremendously

VO_2max values obtained from high altitude natives residing at various altitudes have been commented on as being relatively high as compared to lowlanders. Brutsaert [70] recently combined data from most (but not all) studies reporting VO_2max in Tibetans, Sherpas, and Andean groups to increase statistical power and tested the hypothesis that high altitude natives have higher values for VO_2max as compared to sea level residents, and indeed found that these groups have significantly higher mean values for VO_2max. Obviously this analysis assumes that confounding factors are randomly distributed between the studies and populations. While large data bases are available for VO_2max values over a wide range of ages, genders, and other factors in Western populations, data from high altitude natives are scarce. Although care is taken in most studies to match activity level, age, and gender of the investigated high altitude natives when comparing these to lowlanders, caution should be taken in extrapolating such data to the general population. One study that seems to have overcome at least some of these limitations was completed by Brutsaert and coworkers [71] when studying 150 adult males in Bolivia. High altitude natives ($n=75$) and low altitude natives ($n=75$) were studied at high altitude (3,600–3,850 m) and near sea level. A trend for increased VO_2max with increasing developmental high altitude residence did not reach statistical significance, and it was concluded that the results did not support the hypothesis that Andean high altitude natives have been selected to express a greater physical work capacity in hypoxia. To further test this hypothesis however, 30 men of mixed Spanish and Quechua origins with individual genetic admixture level (%Spanish ancestry ranging from 1 to 64 %) were studied at sea level and with acute exposure to 4,338 m. Admixture predicted the decrease in VO_2max with altitude exposure, and it was concluded that Quechuas possess a better ability to perform exercise at altitude due to an optimized gas-exchange system [72]. The diminished decrease in VO_2max of high altitude natives when exposed to higher than residing altitudes was recently also demonstrated in Tibetans. It was shown (albeit with a small sample power) that second generation Tibetan lowlanders born at 1,300 m and who had never been exposed to high altitude recovered 92 % of their sea level VO_2max after 30 days at 5,050 m. In comparison untrained and trained lowlanders were able to recover only 70 % and 55 % of their sea level VO_2max, respectively [73]. It would thus seem that the VO_2max of high altitude natives does not decrease as much as that of Caucasians when exposed to altitude, or to altitudes higher than their actual resident altitude. This feature could be the consequence of lifelong hypoxic exposure. Quite surprisingly however, one study demonstrated that lowlanders brought to high altitude for as short as 27 months achieve VO_2max values very close to those of resident Tibetan natives suggesting that, at least in absolute terms for VO_2max, it is not necessary to have resided at altitude for generations to achieve values close to those observed at sea level [74]. Since the general consensus is that VO_2max does not increase with acclimatization [13, 14], the ability of high altitude natives to achieve over 90 % of sea level VO_2max when exposed to high altitude is remarkable.

Besides the importance of a high pulmonary VO_2 in order to achieve a high level of aerobic performance, muscular efficiency is also important. Muscular efficiency is calculated as external + internal work/energy expenditure but is often measured simply as workload on the cycle-ergometer/pulmonary VO_2 because of the relative difficulties associated with assessing the contribution of glycolytic and phosphagen energy production and internal work. The potential effects of hypoxia on work efficiency were recently reviewed by Brutsaert [70]. In brief, one of the better studies addressing the question investigated 186 men and women that were distinguished by ethnicity, place of birth, and testing environment. In this study there was no evidence of higher work efficiencies in Aymara [75]. Also when analyzing multiple data obtained from independent investigations no trend was observed toward higher work efficiencies in high altitude natives from around the world [70]. This is in contrast to recent work performed by Marconi and coworkers [76]. They reported that Tibetan migrants born and living between 3,500 and 4,500 m had a better treadmill assessed locomotor economy (assessed at 1,300 m) as compared to Nepalese born and living in the Kathmandu valley [76]. It was concluded that the increased economy in Tibetans was due to chronic hypoxia-induced metabolic adaptations, although these still need to be elucidated. It should be mentioned however, that at least some of the differences may be attributed to differences in body mass between subject groups.

Altitude Training for Endurance Exercise at Sea Level

Since that last edition of this book an overwhelming number of studies have addressed the question if "altitude training" increases exercise performance at sea level or not (see for example the entire issue of Scand J Med Sci Sports. 2008 Aug;18 Suppl 1, and [77] for a more critical appraisal of the field). Studies have especially focused on the more recently developed altitude training methods, i.e., either "Live high–train low," "Live low–train high," or "Intermittent hypoxic exposure," whereas the "original" type of altitude training "Live high–train high" seems to have received less attention. In the below section the different studies are discussed. In general most altitude training studies have unfortunately not made use of double-blinded and placebo-controlled study designs (which is nearly impossible when using real altitude), and a placebo effect can never be ruled out in such studies. Furthermore many altitude training studies make use of *trained subjects* rather than the elite athletes for which altitude training is designed to increase performance. This may prove a mayor problem since a recent analysis questions the added for value for truly elite athletes to engage into altitude training. Figure 16.11a illustrates the relationship between baseline total hemoglobin mass measured prior to a live high–train low (LHTL) intervention and the relative increases in Hb_{mass} associated to LHTL, and may suggest that individuals with already high hemoglobin mass may not experience a further increase following LHTL.

Live High–Train High

The overall rationale for living and training at altitude is to increase red cell mass and to superimpose an additional training stimulus. Usually altitudes between 1,500 and 3,000 m are used. There exist numerous anecdotal reports on world class athletes who incorporate this type of altitude training into their preparations but well-controlled

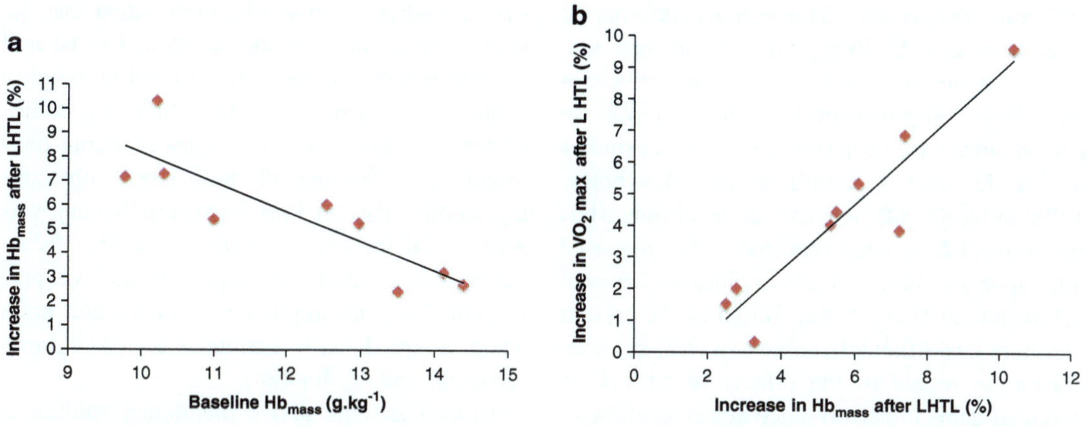

Fig. 16.11 (**a**) The correlation between the relative gain in Hb_{mass} following LHTL and the corresponding increase in VO_2max and (**b**) The correlation between baseline total hemoglobin mass (Hb_{mass}, body-weight adjusted) measured prior to Live high–train low (LHTL) intervention, and the relative increases in Hb_{mass} following LHTL. The present analysis is based on nine previously published LHTL studies conforming to an appropriate "dose" of hypoxia, i.e., an altitude >2,000 m and a daily exposure to hypoxia >12 h. Each point corresponds, for a given LHTL study, to the mean value (baseline or LHTL-induced change) reported by the authors for the LHTL group. Body-weight adjusted Hb_{mass} was either reported directly from the published data, or calculated by using the available mean body-weight values. Reproduced form [121] with permission

studies investigating the effects of live high–train high (LHTH) on sea level performance are scarce. One of the best controlled early studies was published by Mellerowicz in 1970 [78]. Before exposing 22 East German police officers (with low VO_2max's of ≈50 mL kg^{-1}) to either a 4 week altitude (2,020 m) or sea level training intervention with a rigorously controlled exercise training program, all volunteers were subjected to a 6 week long lead-in trial at sea level. In that study 3,000 m running performance and VO_2max were greatly increased in the altitude group as compared to the normoxic control group for up to 2 weeks after termination of the intervention. A few years later Adams and colleagues [79] enrolled 12 competitive track runners (2 miles in little more than 9 min) to a 3 week long altitude (2,300 m) or sea level training program in a crossover study design with controlled training. Altitude training decreased 2 mile running time by 7 s, but no statistical differences could be obtained for this or for VO_2max, and hence the results were not as promising as those previously reported by Mellerowicz. Although the applied crossover design is unmatched as of today, the conclusions may be somewhat hampered since training at altitude was performed at the same relative exercise intensity as at sea level, and hence at a lower absolute intensity. Whether a decrease in absolute workrate at a time when relative intensity is maintained affects performance remains however speculative. Many years later the group of Levine [80] subjected 39 college runners (VO_2max's of little less than 65 mL kg^{-1}) to 2 weeks of lead-in training, followed by 4 weeks of supervised sea level training. Subjects were then randomly assigned to 4 weeks of either living at 2,500 m and training at 2,500–2,700 m (LHTH), living and training at sea level (LLTL or Control), or living at 2,500 m and training at 1,200–1,400 m (LHTL). Following the various training camps, VO_2max was increased in LHTH and LHTL, but 5,000 m running performance was only increased in the LHTL group. The authors speculated that the reason for running performance not to be increased with LHTH was that running speeds during training could not be maintained at altitude due to the reduction in VO_2max at altitude, although as stated above this remains unknown. Others have not found an increase in sea level VO_2max after 4 weeks at altitudes between 1,500 and 2,000 m

[81] or in VO_2max and 3.2 km running performance after 4 weeks at 1,740 m [82]. It could seem that these altitudes were too low. Indeed (albeit in much less-controlled studies), no increases in performance have been reported in recent studies lasting 13 days–1 month at altitudes below 1,900 m [4, 83, 84] whereas sea level time trials (or similar) have been reported to be increased after approximately 3 weeks at altitudes between 2,100 and 2,650 m [85, 86]. Based on the present literature it is difficult to come to a clear-cut conclusion in regard to the effects of LHTH on potential gains in performance at sea level; however (1) LHTH may increase sea level performance in some but not all individuals (2) based on current knowledge it can be recommended to live at an altitude starting at 2,000 m, and finally (3) the duration of exposure should not be less than 3–4 weeks.

Live or Sleep High–Train Low

The general idea with live (or sleep) high–train low (LHTL) is to increase performance at sea level through an altitude-induced augmentation of red cell mass. As illustrated in Fig. 16.11b it indeed seems that the main reason for LHTL to increase performance is strictly related to an enhanced hemoglobin mass. The reason to train at sea level (and not at the resident altitude) is due to the fact that the absolute training intensity cannot be maintained at altitude, and that a reduction in training intensity may affect performance negatively. The use of LHTL was proven effective in increasing sea level performance in college runners by Levine and Stray-Gundersen [80]. As mentioned above, in this study VO_2max was increased following LHTH and LHTL, but running performance was only increased in the LHTL group [80]. Subsequently they confirmed LHTL to be effective also in elite male and female athletes (no control group in this study) [87]. In the latter study, subjects were enrolled in the training camps immediately after national championships when the athletes were supposed to be in peak physical condition. Before going into detail with the results from more recent LHTL studies, it should be highlighted that the studies undertaken by the group of Levine and coworkers seem more controlled and thoroughly conducted than many of the remaining studies within this area. Indeed, besides reducing confounding factors typically associated with training studies also an impressive test battery was applied. Of the tested variables, the only factor that was associated with improved sea level performance was an improvement in red cell mass when coupled with maintenance of training velocities and O_2 flux [88].

Confirming the above mentioned studies, it was recently demonstrated that red cell mass was increased following living at 2,500 m with concomitant training at lower altitudes for 24 days. At the same time also VO_2max was increased [89]. From the above studies is seems (1) living at 2,100–2,800 m for approximately 3 weeks may increase red cell mass, and (2) if at the same time training intensity can be maintained by descending to lower altitudes for training, then sea level endurance performance may be increased. It should be noted however that the term "train low" in these studies does not refer to training at sea level (this is also the case for the French studies described below). Albeit the training altitudes are low, i.e., between 1,200 and 1,800 m, it should be remembered that VO_2max (and hence also training intensity) may be reduced from as low as 580 m [2]. If training had been conducted at actual sea level, it could be speculated that more pronounced effects on subsequent sea level performance could have been observed. Finally, although performance is increased in these studies, the magnitude here most likely does not justify the costs for amateur athletes to engage in this type of activity; they should rather add an additional training session to their normal training routine.

For practical reasons it may not be convenient for athletes to spend time at altitude. To surpass this potential problem, studies have been conducted substituting altitude exposure with the use of "nitrogen housing" where indoor living areas are flushed with N_2 to decrease F_IO_2 and thus stimulate exposure to high altitude. From the many studies addressing the question whether

"Sleep high–train low" is effective in increasing sea level performance it has become clear that a certain cumulative hypoxia threshold has to be surpassed. For example, several studies from Australia found no increase in red cell mass or VO_2max after 8–10 h/day sleeping in hypoxia up to 3,000 m stimulated altitude for 3 weeks [90–92], except in one of their studies (approximately 46 nights at 2,860, 8–9 h/night) where an increase of 4.9 % was observed in Hb mass, but no changes in VO_2max were observed [93]. In a series of experiments from France, it was concluded that (1) in cross country skiers spending 11 h/day for a total of 18 nights (6 days 2,500 m + 6 days at 3,000 m + 6 days at 3,500 m) that red cell volume and performance were unaffected [94] (2) in swimmers spending 16 h/day for a total of 13 days (5 days 2,500 m + 8 days 3,000 m) red cell volume was increased by 8.5 % and VO_2max by 4.5 % [95] and (3) in runners that spend 14 h/day for 18 days (6 days 2,500 m + 12 days 3,000 m) red cell volume as well as VO_2max were increased [96]. Neya and coworkers studied long and middle distance runners before and after 29 nights (11 h/night) at 3,000 m and reported no change in total Hb mass or VO_2max [97]. Summarizing the above "sleep high–train low" studies it seems that Hb mass is increased if (1) the altitude exceeds 2,100 m altitude, (2) the duration of exposure is approximately 3 weeks or more, and (3) daily time in hypoxia exceed 14 h. The factors seem to depend on each other, i.e., inducing more severe hypoxia (for example 3,000 m) does not allow a concomitant reduction in daily exposure time. It may be advantageous to supplement with daily iron intake since ferritin levels of >35 ng mL^{-1} are necessary to obtain the optimal effect of hypoxia on erythropoiesis. It should also be acknowledged that after approximately 3 weeks after termination of the hypoxic stimulus, that the increase in Hb mass is usually reduced by 50 %, and that the time period from altitude training to competition should not exceed this period.

It should also be mentioned that data from two research groups support the idea that LHTL increases sea level performance not only by increasing VO_2max but also by increasing exercise economy [93, 97, 98] or an increase in muscle buffer capacity [22] (see also Chap. 9). Lundby et al. [99] analyzed the results of three independent studies but did not find changes of exercise economy after acclimatization to moderate or high altitude. Considering that most LHTL studies do not find changes in economy it seems doubtful that this is a major factor contributing to enhanced sea level performance.

The above conclusions are all based on non-blinded, non placebo-controlled studies. Such a study was however performed recently by Siebenmann [3] by making use of nitrogen flushed rooms. Following a 2 week lead-in in normal atmospheric conditions, elite athletes were exposed to either 3,000 m "altitude" or "sea level" for at least 16 h/day for a total of 4 weeks, and spend the remaining time of the day in normal conditions, i.e., outside the rooms. Subjects were tested before, after 3 and 4 weeks of treatment, and at 1 and 2 weeks following the termination of the treatment. In that study hemoglobin mass, VO_2max at sea level and at a stimulated altitude of 2,500 m, exercise economy and 26 km time trial performance did not differ between groups at any time point, and suggests that the conclusions from the previous studies may at least in part have been attributed to a placebo effect, and that the conclusions drawn previously should be taken with some caution. From the same study it was also concluded that cycling efficiency was not altered by LHTL, nor was mitochondrial efficiency [100] or anaerobic exercise performance or the expression of skeletal muscle proteins usually associated with exercise performance [101].

Live Low–Train High

The rationale for performing LLTH according to proponents of this strategy is that during exercise at sea level one of the main stimuli for training-induced adaptations is tissue hypoxia. By performing the training sessions in hypoxic conditions the oxygen partial pressure is lower [102, 103], and it may be speculated that training adaptations could be of greater magnitude.

Indeed, augmented transcription of HIF-1 is known to induce the expression of a variety of genes [104] of which many could also contribute to enhancement in performance if also increased at the protein level. Indeed LLTH-induced adaptations at the muscular level [105] seem to be the mayor consequence of this type of training and Hb concentration remains in general unchanged. Two studies have reported changes in hematocrit and Hb concentrations [106, 107] likely related to uncontrolled changes in plasma volume. One further argument [108] to consider in LLTH as compared to LHTL is that permanent exposure to severe hypoxia (5,000 m and higher) leads to a considerable deterioration of skeletal muscle tissue. This argument is however misleading because (1) LHTL is not recommended to be carried out at altitudes above 3,000 m (2) even at 4,559 m altitude muscle protein synthesis is not decreased [109], and (3) if nutritional intake and physical activity are maintained deterioration of skeletal muscles may occur at 5,250 m altitude [110], but not at 4,100 m altitude [111].

Since 2000 more than 15 studies have reported on the effects of LLTH. Despite this high number it is difficult to draw clear-cut conclusions for guidelines. Confounders include the variation of exposure duration (10 days–8 weeks), degree of hypoxic exposure (2,300–5,700 m), and the use of different subjects ranging from untrained to athletically active. Yet another reason for discrepancies between studies could be related to the great variation in exercise intensity (50–80 % of VO_2max). Since VO_2max is reduced at altitude, also the intensity that can be sustained during training is reduced, and this is the main criticism against LLTH studies. It appears that an effect of LLTH is mostly seen when the training is performed at the same absolute workload as at sea level, and hence the relative higher training intensity. This should also be considered as a confounding factor in those studies. It should also be noticed that almost all studies were conducted non-blinded, i.e., both the experimenter as well as the subjects were aware of the treatment which may obviously affect the outcome do to expectations. Levine and coworkers seem to have the most thoroughly controlled study in this sense, as subjects as well as investigators were blinded in the studies of Truijens et al. [112]. In this study no gains in performance in comparison to normoxic training were observed. Hoppeler et al. have conducted LLTH studies for a decade. In some of these studies no effect hereof on performance has been reported [113, 114] whereas in others an increase is reported [115]. Other recent studies have also found positive effects on sea level performance by using LLTH, and for a thorough description of all studies the reader is referred to a recent review on the topic [108]. The main outcome of this review however, is that no clear-cut conclusion can be drawn as to what extent LLTH improves sea level performance or not, and that the hypothesis of a distinct functional phenotype associated with successful LLTH results must be rejected. When strictly including double blinded and placebo-controlled studies, then LLTH does not increase performance.

Intermittent Hypoxia at Rest

Intermittent hypoxic exposure refers to either (1) repeated switching between breathing hypoxic and normoxic air during a session usually lasting between 60 and 90 min. Because the hypoxic exposures are brief (in some examples 5 min) the severity of hypoxia can be high (4,500–6,000 m) or (2) one session of daily hypoxic exposure, also of rather severe hypoxic magnitude (4,000–5,500 m). The precise mechanisms for such an approach to increase sea level performance are less clear than LHTL and LLTH, and at present the mechanisms remain obscure. Since the last edition of the book a high number of well-controlled studies all including a double blinded design have been performed. In one of the studies, 14 national-class distance runners completed a 4 week regimen (5:5 min hypoxia to normoxia ratio for 70 min, 5 times/week) of intermittent normobaric hypoxia or placebo control at rest. Following the experimental period there were no significant differences in VO_2max or 3,000-m time-trial performance [116]. Subsequently, the group performed a double-blind, randomized, placebo-controlled trial to examine the effects of 4 weeks of resting exposure to intermittent hypobaric hypoxia (3 h/day, 5 days/week at 4,000–5,500 m). No differences in VO_2max, performance [117] or exercise economy were reported [118]. Also others have reported similar results with similar protocols. During 15 consecutive days, 20 endurance-trained men were exposed each day to breathing either a gas mixture (11 % O_2 on days 1–7 and 10 % O_2 on days 8–15) or compressed air, six times for 6 min, followed by 4 min of breathing room air for a total of six consecutive cycles. The results of this study demonstrated that 1 h of intermittent hypoxic exposure for 15 consecutive days has no effect on aerobic or anaerobic performance [119]. In conclusion, the use of intermittent hypoxic exposure does not increase sea level performance, despite the advertised guarantees by manufacturers of such devices (see also the theme issue of Scand J Med Sci Sports. 2008 Aug;18 Suppl 1).

References

1. Buskirk ER, Kollias J, Akers RF, Prokop EK, Reategui EP. Maximal performance at altitude and on return from altitude in conditioned runners. J Appl Physiol. 1967;23:259–66.
2. Gore CJ, Hahn AG, Scroop GC, et al. Increased arterial desaturation in trained cyclists during maximal exercise at 580 m altitude. J Appl Physiol. 1996; 80:2204–10.
3. Siebenmann C, Robach P, Jacobs RA, et al. "Live high–train low" using normobaric hypoxia did not increase exercise performance in a double-blinded, placebo-controlled study. J Appl Physiol. 2012;112: 106–17.
4. Wehrlin JP, Hallen J. Linear decrease in. VO_2max and performance with increasing altitude in endurance athletes. Eur J Appl Physiol. 2006 Mar;96(4): 404–12.
5. Mollard P, Woorons X, Letournel M, et al. Role of maximal heart rate and arterial O_2 saturation on the decrement of VO_2max in moderate acute hypoxia in trained and untrained men. Int J Sports Med. 2007;28:186–92.
6. Wagner PD, Sutton JR, Reeves JT, Cymerman A, Groves BM, Malconian MK. Operation Everest II: pulmonary gas exchange during a simulated ascent of Mt Everest. J Appl Physiol. 1987;63:2348–59.
7. Amann M, Romer LM, Subudhi AW, Pegelow DF, Dempsey JA. Severity of arterial hypoxaemia affects the relative contributions of peripheral muscle fatigue to exercise performance in healthy humans. J Physiol. 2007;581:389–403.
8. Lundby C, Damsgaard R. Exercise performance in hypoxia after novel erythropoiesis stimulating protein treatment. Scand J Med Sci Sports. 2006;16:35–40.
9. Robach P, Calbet JAL, Thomsen JJ, et al. The ergogenic effect of recombinant human erythropoietin on VO_2max depends on the severity of arterial hypoxemia. PLoS One. 2008;3:e2996.
10. Saltin B. Aerobic and anaerobic work capacity at 2,300 meters. Schweiz Z Sportmed. 1966;14:81–7.
11. Schuler B, Thomsen JJ, Gassmann M, Lundby C. Timing the arrival at 2340 m altitude for aerobic performance. Scand J Med Sci Sports. 2007;17: 588–94.
12. Lundby C, Møller P, Kanstrup IL, Olsen NV. Heart rate response to hypoxic exercise: role of dopamine D2-receptors and effect of oxygen supplementation. Clin Sci. 2001;101:377–83.
13. Lundby C, Sander M, van Hall G, Saltin B, Calbet JAL. Maximal exercise and muscle oxygen extraction in acclimatizing lowlanders and high altitude natives. J Physiol. 2006;573:535–47.
14. Calbet JAL, Boushel R, Rådegran G, Søndergaard H, Wagner PD, Saltin B. Why is VO_2max after altitude acclimatization still reduced despite normalization of arterial O2 content? Am J Physiol Regul Integr Comp Physiol. 2003;284:R304–16.
15. Horstman D, Weiskopf R, Jackson RE. Work capacity during 3-wk sojourn at 4,300 m: effects of relative polycythemia. J Appl Physiol. 1980;49: 311–8.
16. Maher JT, Jones LG, Hartley LH. Effects of high-altitude exposure on submaximal endurance capacity of men. J Appl Physiol. 1974;37:895–8.
17. Beidleman BA, Muza SR, Rock PB, et al. Exercise responses after altitude acclimatization are retained during reintroduction to altitude. Med Sci Sports Exerc. 1997;29:1588–95.
18. Fulco CS, Rock PB, Cymerman C. Maximal and submaximal exercise performance at altitude. Aviat Space Environ Med. 1998;69:793–801.
19. Bassett DR, Kyle CR, Passfield L, Broker JP, Burke ER. Comparing cycling world hour records, 1967-1996: modeling with empirical data. Med Sci Sports Exerc. 1999;31:1665–76.
20. Dickinson ER, Piddington MJ, Brain T. Project Olympics. Schweiz Z Sportmed. 1966;14:305–8.
21. Peronnet F, Thibault G, Cousineau DL. A theoretical analysis of the effect of altitude on running performance. J Appl Physiol. 1991;70:399–404.
22. Juel C, Lundby C, Sander M, Calbet JA, Hall G. Human skeletal muscle and erythrocyte proteins involved in acid-base homeostasis: adaptations to chronic hypoxia. J Physiol. 2003;548:639–48.
23. Levine BD, Stray-Gundersen J, Mehta RD. Effect of altitude on football performance. Scand J Med Sci Sports. 2008;18:76–84.
24. Cunningham DJ, Petersen ES, Pickering TG, Sleight P. The effects of hypoxia, hypercapnia, and asphyxia on the baroreceptor-cardiac reflex at rest and during exercise in man. Acta Physiol Scand. 1972;86: 456–65.

25. Amann M, Pegelow DF, Jacques AJ, Dempsey JA. Inspiratory muscle work in acute hypoxia influences locomotor muscle fatigue and exercise performance of healthy humans. Am J Physiol. 2007;293:R2036–45.
26. Sutton JR, Reeves JT, Wagner PD, et al. Operation Everest II: oxygen transport during exercise at extreme simulated altitude. J Appl Physiol. 1988;64: 1309–21.
27. Alexander JK, Hartley LH, Modelski M, Grover RF. Reduction of stroke volume during exercise in man following ascent to 3,100 m altitude. J Appl Physiol. 1967;23:849–58.
28. Kahler RL, Gaffney TE, Braunwald E. The effects of autonomic nervous system inhibition on the circulatory response to muscular exercise*. J Clin Invest. 1962;41:1981–7.
29. Lundby C, Araoz M, van Hall G. Peak heart rate decreases with increasing severity of acute hypoxia. High Alt Med Biol. 2001;2:369–76.
30. Boushel R, Calbet J-AL, Radegran G, Sondergaard H, Wagner PD, Saltin B. Parasympathetic neural activity accounts for the lowering of exercise heart rate at high altitude. Circulation. 2001;104:1785–91.
31. Wagner PD. Reduced maximal cardiac output at altitude—mechanisms and significance. Respir Physiol. 2000;120:1–11.
32. Reeves JT, Groves BM, Sutton JR, et al. Operation Everest II: preservation of cardiac function at extreme altitude. J Appl Physiol. 1987;63:531–9.
33. Alexander JK, Grover JF. Mechanism of reduced cardiac stroke volume at high altitude. Clin Cardiol. 1983;6:301–3.
34. Naeije R. Pulmonary circulation in hypoxia. Int J Sports Med. 1992;13 Suppl 1:S27–30.
35. Butterfield GE, Gates J, Fleming S, Brooks GA, Sutton JR, Reeves JT. Increased energy intake minimizes weight loss in men at high altitude. J Appl Physiol. 1992;72:1741–8.
36. Brooks GA, Butterfield GE, Wolfe RR, et al. Increased dependence on blood glucose after acclimatization to 4,300 m. J Appl Physiol. 1991;70:919–27.
37. Lundby C, Van Hall G. Substrate utilization in sea level residents during exercise in acute hypoxia and after 4 weeks of acclimatization to 4100 m. Acta Physiol Scand. 2002;176:195–201.
38. Wagner PD. Origin of the lactate paradox: muscles or brain? J Appl Physiol. 2009;106:740–1.
39. Wagner PD, Lundby C. The lactate paradox: does acclimatization to high altitude affect blood lactate during exercise? Med Sci Sports Exerc. 2007; 39:749–55.
40. van Hall G, Lundby C, Araoz M, Calbet JAL, Sander M, Saltin B. The lactate paradox revisited in lowlanders during acclimatization to 4100 m and in high-altitude natives. J Physiol. 2009;587:1117–29.
41. Hansen J, Sander M. Sympathetic neural overactivity in healthy humans after prolonged exposure to hypobaric hypoxia. J Physiol. 2003;546:921–9.
42. Katayama K, Ishida K, Iwamoto E, Iemitsu M, Koike T, Saito M. Hypoxia augments muscle sympathetic neural response to leg cycling. Am J Physiol. 2011;301:R456–64.
43. Pronk M, Tiemessen I, Hupperets MDW, et al. Persistence of the lactate paradox over 8 weeks at 3800 m. High Alt Med Biol. 2003;4:431–43.
44. Amann M, Calbet JAL. Convective oxygen transport and fatigue. J Appl Physiol. 2008;104:861–70.
45. Subudhi AW, Miramon BR, Granger ME, Roach RC. Frontal and motor cortex oxygenation during maximal exercise in normoxia and hypoxia. J Appl Physiol. 2009;106:1153–8.
46. Subudhi AW, Olin JT, Dimmen AC, Polaner DM, Kayser B, Roach RC. Does cerebral oxygen delivery limit incremental exercise performance? J Appl Physiol. 2011;111:1727–34.
47. Imray CHE, Myers SD, Pattinson KTS, et al. Effect of exercise on cerebral perfusion in humans at high altitude. J Appl Physiol. 2005;99:699–706.
48. Rasmussen P, Stie H, Nielsen B, Nybo L. Enhanced cerebral CO_2 reactivity during strenuous exercise in man. Eur J Appl Physiol. 2006;96:299–304.
49. Siebenmann C, Sorensen H, Jacobs RA, Haider T, Rasmussen P, Lundby C. Hypocapnia during hypoxic exercise and its impact on cerebral oxygenation, ventilation and maximal whole body VO_2 uptake. Respir Physiol Neurobiol. 2013;185: 461–7.
50. Anholm JD, Foster GP. Con: hypoxic pulmonary vasoconstriction is not a limiting factor of exercise at high altitude. High Alt Med Biol. 2011;12:3313–7.
51. Naeije R. Pro: hypoxic pulmonary vasoconstriction is a limiting factor of exercise at high altitude. High Alt Med Biol. 2011;12:309–12.
52. Olfert IM, Loeckinger A, Treml B, et al. Sildenafil and bosentan improve arterial oxygenation during acute hypoxic exercise: a controlled laboratory trial. Wilderness Environ Med. 2011;22:211–21.
53. Faoro V, Lamotte M, Deboeck G, et al. Effects of sildenafil on exercise capacity in hypoxic normal subjects. High Alt Med Biol. 2007;8:155–63.
54. Ghofrani HA, Reichenberger F, Kohstall MG, et al. Sildenafil increased exercise capacity during hypoxia at low altitudes and at Mount Everest base camp: a randomized, double-blind, placebo-controlled crossover trial. Ann Intern Med. 2004;141: 169–77.
55. Richalet JP, Gratadour P, Robach P, et al. Sildenafil inhibits altitude-induced hypoxemia and pulmonary hypertension. Am J Respir Crit Care Med. 2005;171: 275–81.
56. Siebenmann C, Bloch KE, Lundby C, Nussbamer-Ochsner Y, Schoeb M, Maggiorini M. Dexamethasone improves maximal exercise capacity of individuals susceptible to high altitude pulmonary edema at 4559 m. High Alt Med Biol. 2011;12:169–77.
57. Fischler M, Maggiorini M, Dorschner L, et al. Dexamethasone but not tadalafil improves exercise capacity in adults prone to high-altitude pulmonary edema. Am J Respir Crit Care Med. 2009; 180:346–52.

58. Faoro V, Boldingh S, Moreels M, et al. Bosentan decreases pulmonary vascular resistance and improves exercise capacity in acute hypoxia. Chest. 2009;135:1215–22.
59. DeGraff AC, Grover RF, Johnson RL, Hammond JW, Miller JM. Diffusing capacity of the lung in Caucasians native to 3,100 m. J Appl Physiol. 1970;29:71–6.
60. Dempsey JA, Reddan WG, Birnbaum ML, et al. Effects of acute through life-long hypoxic exposure on exercise pulmonary gas exchange. Respir Physiol. 1971;13:62–89.
61. Lundby C, Calbet JAL, van Hall G, Saltin B, Sander M. Pulmonary gas exchange at maximal exercise in Danish lowlanders during 8 wk of acclimatization to 4,100 m and in high-altitude Aymara natives. Am J Physiol. 2004;287:R1202–8.
62. Schoene RB. Limits of human lung function at high altitude. J Exp Biol. 2001;204:3121–7.
63. Cerny FC, Dempsey JA, Reddan WG. Pulmonary gas exchange in nonnative residents of high altitude. J Clin Invest. 1973;52:2993–9.
64. Wagner PD, Araoz M, Boushel R, et al. Pulmonary gas exchange and acid-base state at 5,260 m in high-altitude Bolivians and acclimatized lowlanders. J Appl Physiol. 2002;92:1393–400.
65. Zhuang J, Droma T, Sutton JR, et al. Smaller alveolar-arterial O2 gradients in Tibetan than Han residents of Lhasa (3658 m). Respir Physiol. 1996; 103:75–82.
66. Hsia CCW, Carbayo JJP, Yan X, Bellotto DJ. Enhanced alveolar growth and remodeling in guinea pigs raised at high altitude. Respir Physiol Neurobiol. 2005;147:105–15.
67. McDonough P, Dane DM, Hsia CCW, Yilmaz C, Johnson RL. Long-term enhancement of pulmonary gas exchange after high-altitude residence during maturation. J Appl Physiol. 2006;100:474–81.
68. Johnson RL, Cassidy SS, Grover RF, Schutte JE, Epstein RH. Functional capacities of lungs and thorax in beagles after prolonged residence at 3,100 m. J Appl Physiol. 1985;59:1773–82.
69. Lundby C, Calbet J. Why are high altitude natives so strong at altitude? Adv Exp Med Biol. 2012.
70. Brutsaert TD. Do high-altitude natives have enhanced exercise performance at altitude? Appl Physiol Nutr Metab. 2008;33:582–92.
71. Brutsaert TD, Spielvogel H, Soria R, Caceres E, Buzenet G, Haas JD. Effect of developmental and ancestral high-altitude exposure on VO_2 peak of Andean and European/North American natives. Am J Phys Anthropol. 1999;110:435–55.
72. Brutsaert TD, Parra EJ, Shriver MD, et al. Spanish genetic admixture is associated with larger VO_2max decrement from sea level to 4,338 m in Peruvian Quechua. J Appl Physiol. 2003;95:519–28.
73. Marconi C, Marzorati M, Grassi B, et al. Second generation Tibetan lowlanders acclimatize to high altitude more quickly than Caucasians. J Physiol. 2004;556:661–71.
74. Niu W, Wu Y, Li B, Chen N, Song S. Effects of long-term acclimatization in lowlanders migrating to high altitude: comparison with high altitude residents. Eur J Appl Physiol. 1995;71:543–8.
75. Brutsaert TD, Haas JD, Spielvogel H. Absence of work efficiency differences during cycle ergometry exercise in Bolivian Aymara. High Alt Med Biol. 2004;5:41–59.
76. Marconi C, Marzorati M, Sciuto D, Ferri A, Cerretelli P. Economy of locomotion in high-altitude Tibetan migrants exposed to normoxia. J Physiol. 2005;569:667–75.
77. Lundby C, Millet GP, Calbet JA, Bärtsch P, Subudhi AW. Does altitude training increase performance in elite athletes? Br J Sports Med. 2012;46:792–5.
78. Mellerowicz H, Meller W, Woweries J, et al. Vergleichende untersuchungen über wirkungen von höhentraining auf die dauerleistung in meereshöhe. Sportarzt Sportmedizin. 1970;21:207–40.
79. Adams WC, Bernauer EM, Dill DB, Bomar JB. Effects of equivalent sea-level and altitude training on VO_2max and running performance. J Appl Physiol. 1975;39:262–6.
80. Levine BD, Stray-Gundersen J. "Living high-training low": effect of moderate-altitude acclimatization with low-altitude training on performance. J Appl Physiol. 1997;83:102–12.
81. Bailey DM, Davis B, Romer L, Castell L, Newsholme E, Gandy G. Implications of moderate altitude training for sea-level endurance in elite distance runners. Eur J Appl Physiol. 1998;78:360–8.
82. Gore CJ, Hahn AG, Burge CM, Telford RD. VO_2max and haemoglobin mass of trained athletes during high intensity training. Int J Sports Med. 1997;28: 477,82.
83. Svendenhag J, Piehl-Aulin K, Skog C, Saltin B. Increased left ventricular muscle mass after long-term altitude training in athletes. Acta Physiol Scand. 1997;161:63–70.
84. Friedmann B, Jost J, Rating T, et al. Effects of iron supplementation on total body hemoglobin during endurance training at moderate altitude. Int J Sports Med. 1999;20:78,85.
85. Gore CJ, Craig NP, Hahn AG, et al. Altitude training at 2690 m does not increase total haemoglobin mass or sea level VO_2max in world champion track cyclists. J Sci Med Sport. 1998;1:156–70.
86. Friedmann B, Frese F, Menold E, Kauper F, Jost J, Bärtsch P. Individual variation in the erythropoietic response to altitude training in elite junior swimmers. Br J Sports Med. 2005;39:148–53.
87. Stray-Gundersen J, Chapman RF, Levine BD. "Living high-training low" altitude training improves sea level performance in male and female elite runners. J Appl Physiol. 2001;91:1113–20.
88. Stray-Gundersen J, Levine BD. Live high, train low at natural altitude. Scand J Med Sci Sports. 2008;18:21–8.
89. Wehrlin JP, Zuest P, Hallen J, Marti B. Linear decrease in. VO_2max and performance with

increasing altitude in endurance athletes. Eur J Appl Physiol. 2006 Mar;96(4):404–12.
90. Roberts AD, Clark SA, Townsend NE, Anderson ME, Gore CJ, Hahn AG. Changes in performance, maximal oxygen uptake and maximal accumulated oxygen deficit after 5, 10 and 15 days of live high:train low altitude exposure. Eur J Appl Physiol. 2003;88:390–5.
91. Ashenden MJ, Gore CJ, Dobson GP, Hahn AG. "Live high, train low" does not change the total haemoglobin mass of male endurance athletes sleeping at a simulated altitude of 3000 m for 23 nights. Eur J Appl Physiol. 1999;80:479–84.
92. Ashenden MJ, Gore CJ, Martin DT, Dobson GP, Hahn AG. Effects of a 12-day "live high, train low" camp on reticulocyte production and haemoglobin mass in elite female road cyclists. Eur J Appl Physiol. 1999;80:472–8.
93. Saunders PU, Telford RD, Pyne DB, Hahn AG, Gore CJ. Improved running economy and increased hemoglobin mass in elite runners after extended moderate altitude exposure. J Sci Med Sport. 2009; 12:67–72.
94. Robach P, Schmitt L, Brugniaux J, et al. Living high–training low: effect on erythropoiesis and maximal aerobic performance in elite Nordic skiers. Eur J Appl Physiol. 2006;97:695–705.
95. Robach P, Schmitt L, Brugniaux J, et al. Living high–training low: effect on erythropoiesis and aerobic performance in highly-trained swimmers. Eur J Appl Physiol. 2006;96:423–33.
96. Brugniaux JV, Schmitt L, Robach P, et al. Eighteen days of "living high, training low" stimulate erythropoiesis and enhance aerobic performance in elite middle-distance runners. J Appl Physiol. 2006;100: 203–11.
97. Neya M, Enoki T, Kumai Y, Sugoh T, Kawahara T. The effects of nightly normobaric hypoxia and high intensity training under intermittent normobaric hypoxia on running economy and hemoglobin mass. J Appl Physiol. 2007;103:828–34.
98. Gore CJ, Hahn AG, Aughey RJ, et al. Live high:train low increases muscle buffer capacity and submaximal cycling efficiency. Acta Physiol Scand. 2001; 173:275–86.
99. Lundby C, Calbet JAL, Sander M, et al. Exercise economy does not change after acclimatization to moderate to very high altitude. Scand J Med Sci Sports. 2007;17:281–91.
100. Robach P, Siebenmann C, Jacobs RA, Rasmussen P, Nordsborg N, Pesta A, Gnaiger R, Diaz V, Christ A, Fiedler J, Crivelli N, Secher NH, Pichon A, Maggiorini M, Lundby C. The role of haemoglobin mass on VO_2max following normobaric 'live high-train low' in endurance-trained athletes. Br J Sports Med. 2012; 46:822–7.
101. Nordsborg NB, Siebenmann C, Jacobs RA, Rasmussen P, Diaz V, Robach P, Lundby C. Four weeks of normobaric "live high-train low" do not alter muscular or systemic capacity for maintaining pH and K^+ homeostasis during intense exercise. J Appl Physiol. 2012;112:2027–36.
102. Richardson RS, Duteil S, Wary C, Wray DW, Hoff J, Carlier PG. Human skeletal muscle intracellular oxygenation: the impact of ambient oxygen availability. J Physiol. 2006;571:415–24.
103. Richardson RS, Noyszewski EA, Kendrick KF, Leigh JS, Wagner PD. Myoglobin O2 desaturation during exercise. Evidence of limited O2 transport. J Clin Invest. 1995;96:1916–26.
104. Semenza GL, Shimoda LA, Prabhakar NR. Regulation of gene expression by HIF-1. Novartis Found Symp. 2006;272:2–8.
105. Vogt M, Puntschart A, Geiser J, Zuleger C, Billeter R, Hoppeler H. Molecular adaptations in human skeletal muscle to endurance training under simulated hypoxic conditions. J Appl Physiol. 2001;91:173–82.
106. Hendriksen IJ, Meeuwsen T. The effect of intermittent training in hypobaric hypoxia on sea-level exercise: a cross-over study in humans. Eur J Appl Physiol. 2003;88:396–403.
107. Meeuwsen T, Hendriksen IJ, Holewijn M. Training-induced increases in sea-level performance are enhanced by acute intermittent hypobaric hypoxia. Eur J Appl Physiol. 2001;84:283–90.
108. Hoppeler H, Klossner S, Vogt M. Training in hypoxia and its effects on skeletal muscle tissue. Scand J Med Sci Sports. 2008;18:38–49.
109. Holm L, Haslund ML, Robach P, et al. Skeletal muscle myofibrillar and sarcoplasmic protein synthesis rates are affected differently by altitude-induced hypoxia in native lowlanders. PLoS One. 2010;5:e15606.
110. Mizuno M, Savard GK, Areskog N-H, Lundby C, Saltin B. Skeletal muscle adaptations to prolonged exposure to extreme altitude: a role of physical activity? High Alt Med Biol. 2008;9:311–7.
111. Lundby C, Pilegaard H, Andersen JL, van Hall G, Sander M, Calbet JAL. Acclimatization to 4100 m does not change capillary density or mRNA expression of potential angiogenesis regulatory factors in human skeletal muscle. J Exp Biol. 2004;207:3865–71.
112. Truijens MJ, Toussaint HM, Dow J, Levine BD. Effect of high-intensity hypoxic training on sea-level swimming performances. J Appl Physiol. 2003;94: 733–43.
113. Geiser J, Vogt M, Billeter R, Zuleger C, Belforti F, Hoppeler H. Training high–living low: changes of aerobic performance and muscle structure with training at simulated altitude. Int J Sports Med. 2001;22:579,85.
114. Ventura N, Hoppeler H, Seiler R, Binggeli A, Mullis P, Vogt M. The response of trained athletes to six weeks of endurance training in hypoxia or normoxia. Int J Sports Med. 2003;24:166,72.
115. Dufour SP, Ponsot E, Zoll J, et al. Exercise training in normobaric hypoxia in endurance runners. I. Improvement in aerobic performance capacity. J Appl Physiol. 2006;100:1238–48.
116. Julian CG, Gore CJ, Wilber RL, et al. Intermittent normobaric hypoxia does not alter performance or

erythropoietic markers in highly trained distance runners. J Appl Physiol. 2004;96:1800–7.
117. Rodríguez FA, Truijens MJ, Townsend NE, Stray-Gundersen J, Gore CJ, Levine BD. Performance of runners and swimmers after four weeks of intermittent hypobaric hypoxic exposure plus sea level training. J Appl Physiol. 2007;103:1523–35.
118. Truijens MJ, Rodríguez FA, Townsend NE, Stray-Gundersen J, Gore CJ, Levine BD. The effect of intermittent hypobaric hypoxic exposure and sea level training on submaximal economy in well-trained swimmers and runners. J Appl Physiol. 2008;104:328–37.
119. Tadibi V, Dehnert C, Menold E, Bärtsch P. Unchanged anaerobic and aerobic performance after short-term intermittent hypoxia. Med Sci Sports Exerc. 2007;39:585–64.
120. Calbet JAL, Rådegran G, Boushel R, Saltin B. On the mechanisms that limit oxygen uptake during exercise in acute and chronic hypoxia: role of muscle mass. J Physiol. 2009;587:477–90.
121. Robach P, Lundby C. Is live high–train low altitude training relevant for elite athletes with already high total hemoglobin mass? Scand J Med Sci Sports. 2012;22:303–5.

Sleep

17

Yvonne Nussbaumer-Ochsner and Konrad E. Bloch

Abstract
Sleep at high altitude is perceived as fragmented and less restorative compared to lowland, and insomnia is one of the leading symptoms of acute mountain sickness. The physiological changes in sleep and in the breathing pattern during high altitude exposure consist mainly in a reduction of deep sleep and an increase in arousals which are related in part to periodic breathing with central apneas/hypopneas. This chapter provides an overview over the scientific literature on sleep and breathing in healthy newcomers at altitude, describes the changes that take place during acclimatization, and shows modalities to treat sleep disturbances at altitude. Recent studies performed in patients with the obstructive sleep apnea syndrome travelling to altitude are also summarized.

Introduction

Travelling to the mountains for recreational and professional reasons is quite popular although altitude exposure may have adverse health effects [1]. Of these, the high altitude insomnia and dyssomnia are quite common and characterized by difficulties falling asleep and frequent awakenings, sometimes associated with feelings of air hunger or suffocation. High altitude sleep disturbances are closely linked to periodic breathing and may occur with or without accompanying high altitude-related illnesses such as acute mountain sickness (AMS) or high altitude pulmonary edema (HAPE). The aim of this chapter is to review the literature on sleep of lowland residents at higher altitude, to analyze the interactions among sleep and breathing at altitude, and to review studies on treatment of high altitude insomnia and sleep-disordered breathing.

Y. Nussbaumer-Ochsner, M.D.
K.E. Bloch, M.D. (✉)
Pulmonary Division, Sleep Disorders Center, University Hospital of Zurich, Ramistrasse 100, 8091 Zurich, Switzerland
e-mail: yvonne.nussbaumer@hispeed.ch; konrad.bloch@usz.ch

Normal Sleep

It is estimated that humans spend approximately one third of their lives in sleep, a reversible state characterized behaviorally by reduced motor activity and response to environmental stimuli. Neurophysiologically sleep is defined by characteristics of the electro-encephalogram (EEG),

E.R. Swenson and P. Bärtsch (eds.), *High Altitude: Human Adaptation to Hypoxia*,
DOI 10.1007/978-1-4614-8772-2_17, © Springer Science+Business Media New York 2014

electro-oculogram (EOG), and electro-myogram (EMG). Two major types of sleep can be distinguished, i.e., non-rapid eye movement (NREM) sleep and rapid eye movement (REM) sleep. Over the course of a normal night sleep, NREM and REM sleep periods alternate in cycles lasting 60–90 min interrupted by brief periods of wakefulness. The EEG of stages 1 and 2 NREM sleep shows relatively low voltage and mixed frequency (3–7 Hz) while the EEG in stages 3 and 4 (slow wave sleep) shows a high amplitude and low frequency (≤2 Hz). REM sleep is characterized by phasic rapid eye movements, saw tooth waves in the EEG, and suppression of muscular tone. Sleep and wakefulness states have profound influences on the function and activity of many organ systems including on ventilation. During wakefulness the dual control of ventilation by behavioral influences and chemical feedback (chemoreceptor sensing of arterial PO_2, PCO_2, and pH) assures adequate and stable arterial oxygen and carbon dioxide levels. In contrast, during sleep, the wakefulness drive on ventilation is absent and breathing is mainly controlled by chemical feedback making it more susceptible to instability.

Sleep at Altitude

It is a common perception of lowlanders that sleep during a sojourn at altitude is often disturbed and not as refreshing as at home. In a study performed in a ski resort at 3,500 m in Iran, 46 of 100 unacclimatized hotel guests reported poor sleep quality in their first night after arrival based on a score ≥6 in the Groningen sleep quality questionnaire [2]. In the same study, 60 of 100 subjects indicated sleep disturbance based on the corresponding question within the Lake Louise AMS questionnaire. Subjects with high scores in the Groningen sleep quality questionnaire (≥6) were more likely to suffer from clinically relevant AMS (Lake Louise score ≥5) than those with lower scores on the sleep quality questionnaire underlining the association and potential interaction of sleep disturbance with other AMS symptoms such as headache, anorexia, nausea/vomiting, and dizziness. Since recent data indicate that sleep disturbances assessed subjectively correlate poorly with polysomnographic variables (Fig. 17.1) [3] studies on the effects of altitude on sleep should ideally comprise both subjective and objective evaluations.

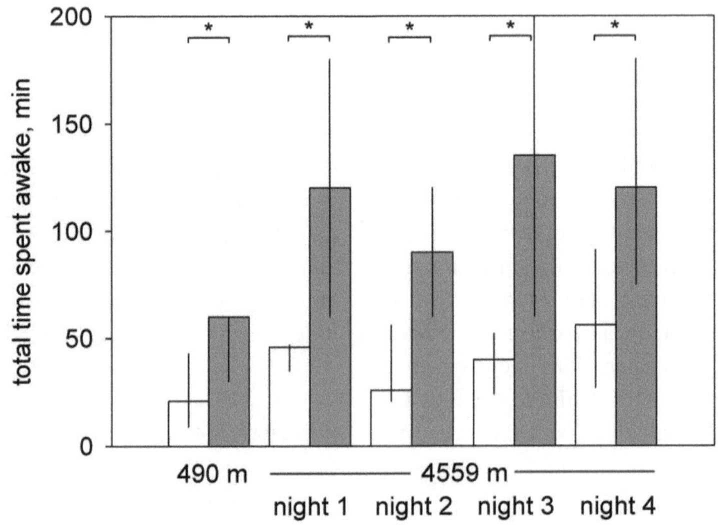

Fig. 17.1 Subjective and polysomnographic estimates of night-time spent awake. Fourteen mountaineers estimated the time spent awake during one night at 490 m and during four successive nights at 4,559 m. Comparisons to the corresponding wakefulness times measured by polysomnography reveal that subjective estimates (*grey bars*) consistently exceed objective measurements (*white bars*) by a variable degree. *Bars* and *vertical lines* represent medians and quartile ranges, $^*P<0.05$. Modified from Nussbaumer-Ochsner et al. [3]

Systematic Review of the Literature

Despite the high prevalence of subjective sleep disturbances at altitude a current review of the scientific literature reveals that sleep at altitude has been rarely studied with objective methods and most investigations in this field consist of uncontrolled case series observations. Studies evaluating the effect of altitude exposure on sleep and containing objective sleep data based either on neurophysiologic recordings (EEG, EOG, EMG) or on actimetry were systematically analyzed. Studies were classified according to the design as randomized, controlled studies ($n=3$ studies [4–6a], including a total of 76 subjects of which 51 were included in one of the studies [6]), case–control studies ($n=3$ studies, including a total of 82 subjects [7–9]), and observational studies ($n=13$ studies [10–23], including a total of 122 subjects). Field studies performed in the mountains [6–11, 15, 19, 20, 22, 23] were differentiated from studies simulating altitude by use of normobaric hypoxia (low inspired oxygen fraction [14, 21]) or hypobaric hypoxia (hypobaric chamber experiments [4, 5, 12, 13, 16–18]).

When assessing the available literature on sleep at altitude major limitations that prevent generalization of results become evident. Subjects were usually selected among healthy, highly trained mountaineers, athletes, or soldiers. The sample size is often inadequate and power calculation or confidence intervals are not presented. The study settings vary greatly even within the same investigation. For example, some studies at low altitude were carried out in a sleep laboratory while the altitude studies were performed in the much less comfortable and protected setting of a high altitude research facility. In high altitude field studies but also in hypobaric chamber studies, the effect of environmental influences (noise, temperature) cannot be neglected. Furthermore, the ascent rate, acclimatization time, physical activity during daytime, and the sleeping altitude vary between studies.

Summary of the Evidence on Sleep at Altitude

Within the limitations discussed above, and taking the more robust evidence from a few recent randomized or controlled studies into particular consideration some cautious conclusions can still be drawn. A trend of deteriorated sleep quality at altitude as illustrated in Fig. 17.2 appears to be consistent in most studies. Compared to sea level, the sleep efficiency was reduced and subjects revealed a decreased relative amount of slow wave sleep at high altitude, i.e., a more superficial sleep [23]. Further changes include a progressively increased frequency of arousals observed in three studies performed at altitudes ≥4,500 m [12, 13, 15, 23] while in three other studies performed at altitudes <3,200 m no trend of increasing arousals was observed [6, 16, 21]. In one study, no significant change in the arousal index was observed during repeated sleep studies performed over the course of ascent from sea level to 5,050 m [19]. Although some authors have claimed that total sleep time was maintained at higher compared to lower altitude there is not sufficient evidence to support this assumption. In fact, the opposite may be true since a reduced total sleep time at altitude was found in several studies [4, 8, 9, 13, 14, 24]. Whether acclimatization improves sleep quality at various altitudes has not been appropriately studied although preliminary reports suggest that this is indeed the case [25]. We have identified only two reports on sleep in children at altitude. In these studies actigraphic recordings suggested a more restless sleep in unacclimatized prepubertal children aged 11–12 years at 3,560 m [9] and in preverbal children aged 3–33 months at 3,109 m [26] compared to sea level. Research on sleep at altitude has been hampered in the past by the limited availability of polysomnography, the gold standard for objective assessment of sleep. Recently, actigraphy has been increasingly used as a simple tool suitable for altitude field studies and applicable over several weeks (Fig. 17.3).

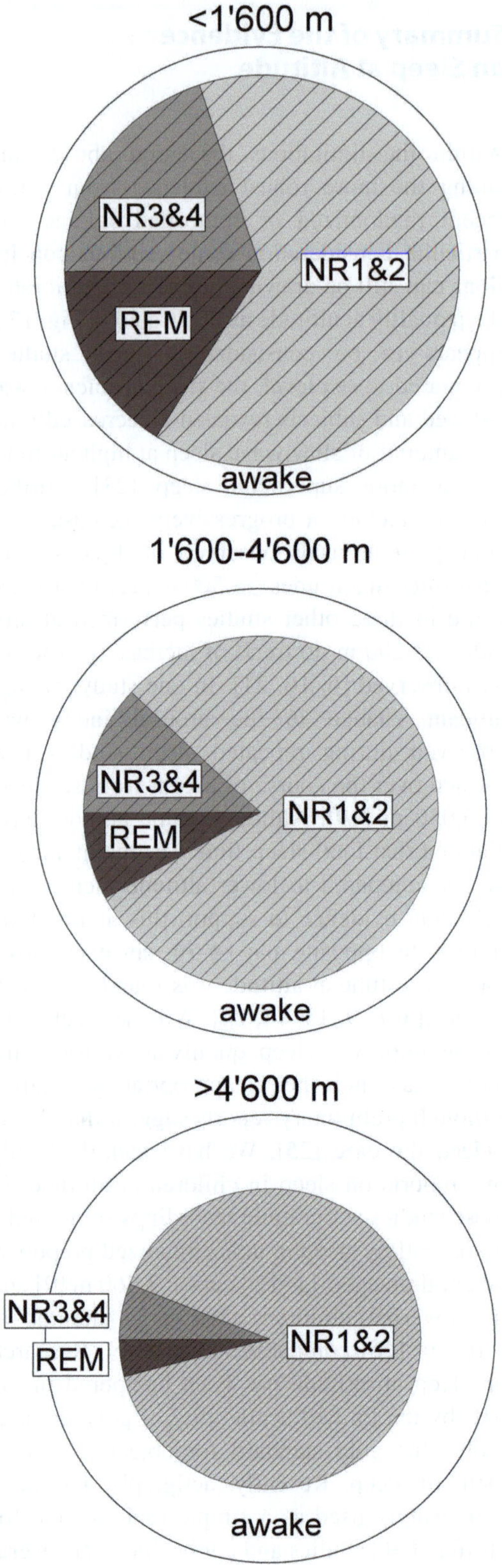

High Altitude Periodic Breathing

Periodic breathing, an oscillating pattern of waxing and waning of ventilation with periods of hyperventilation alternating with central apnea or hypopnea, is a prominent characteristic of sleep at high altitude and may even occur during wakefulness. Newcomers often wake up with a distressing sense of suffocation; take a few rapid, deep breaths before dozing off again. This sequence may repeat itself many times during a night at altitude and prevent refreshing sleep. The awakenings or brief arousals commonly occur at the transition from apnea to hyperpnea. In one study up to 52 % of nocturnal arousals were associated with periodic breathing cycles suggesting an important but not exclusive role of periodic breathing in causing sleep disruption at altitude [12]. In more recent field studies [3, 6b, 22, 25] the correlation of arousals and periodic breathing during exposure to hypoxia was less clear than in the cited hypobaric chamber study [12] or even completely absent. Of note, there may be a mutual interaction among arousals and periodic breathing since periodic breathing may induce arousals and they may promote in turn periodic breathing due to changes in ventilatory control during the sleep–wakefulness transition.

Fig. 17.2 Alterations in sleep structure at altitude. The trend of alterations in sleep structure in different altitude ranges is illustrated in a qualitative way based on various studies (see text). The *outer circles* of each panel represent the time in bed, the *white area* is the time spent awake, and the *pie chart* is total sleep time comprising NREM stages 1–4 (NR1&2 = superficial stages, NR3&4 = slow wave or deep sleep stages). The ratio of total sleep time (pie chart area) to time in bed (outer circle) is the sleep efficiency. The density of diagonal lines reflects the number of arousals. Little changes occur between sea level and 1,600 m. Starting at altitudes >1,600 m, sleep efficiency is increasingly reduced, and more time is spent in superficial sleep stages. At very high altitude sleep is fragmented by frequent arousals

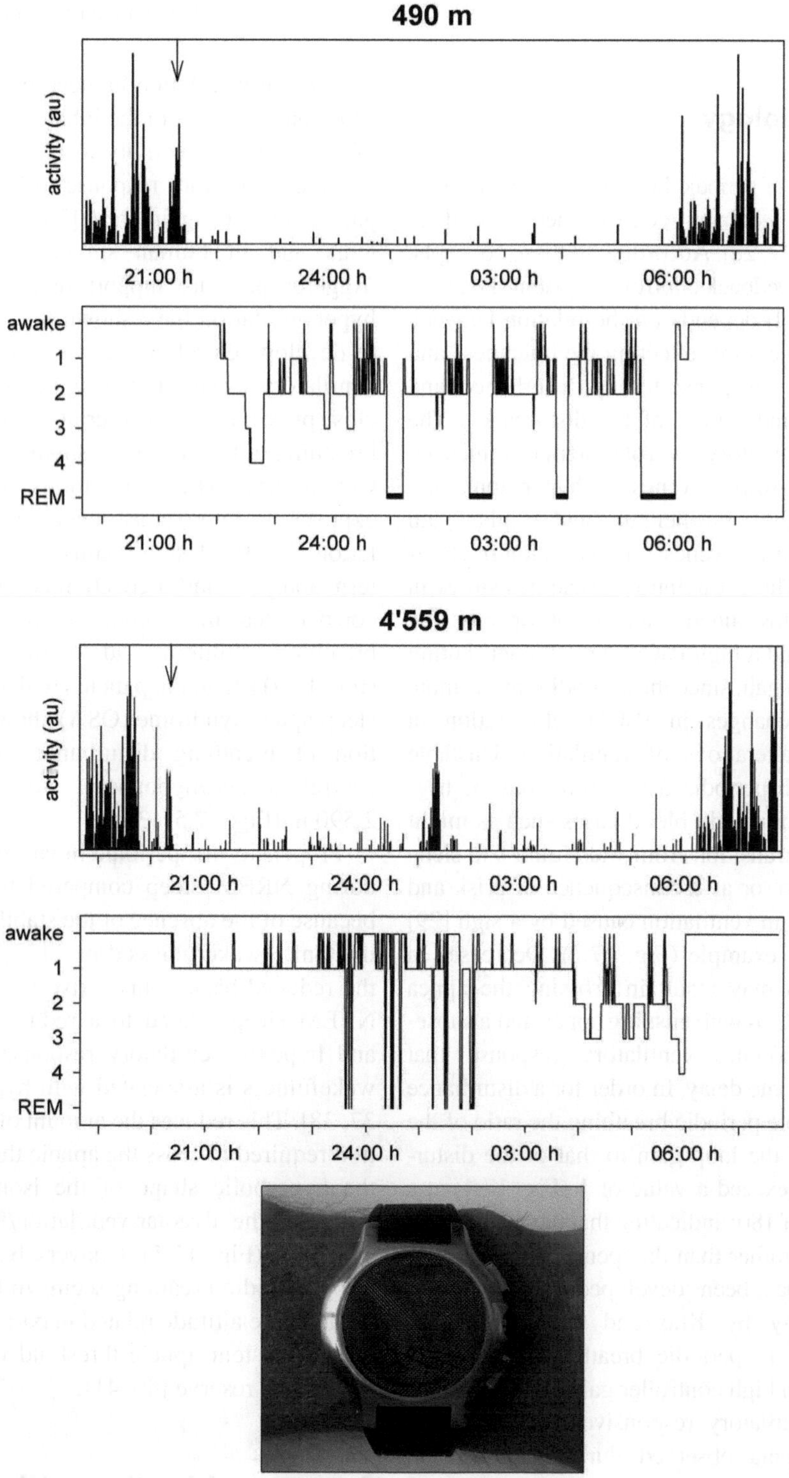

Fig. 17.3 Actigrams and hypnograms at 490 and 4,559 m. Actigram obtained in a subject during a night at 490 m (*top panel*) shows very rare spikes with low activity indicating nearly undisturbed sleep. The corresponding hypnogram shows a normal distribution of sleep stages and 4 NREM/REM cycles. In contrast, the actigram recorded during a night at 4,559 m (*third panel* from *top*) reveals frequent activity spikes reflecting movements during awakenings. The corresponding hypnogram (*lowest panel*) reveals predominantly superficial sleep stages with frequent awakenings, very rare deep sleep stages 3 and 4, and no REM sleep. *y*-Axis of actigrams = acceleration in arbitrary units. Lights off is marked by a *vertical arrow*. An actigraph specifically developed for use in mountaineers comprising a barometric pressure recorder is also shown. Modified from Nussbaumer-Ochsner et al. [3]

Pathophysiology

Many features of periodic breathing can be predicted by applying feedback control theory used in engineering [27, 28]. According to these concepts, stability of a feedback controlled system (i.e., ventilatory control) depends on the relation between loop gain, a measure reflecting the quickness and magnitude of a response to any disturbance, and the strength and timing of the disturbance. The gain of the ventilatory control system is influenced by the chemo-responsiveness to hypercapnia and hypoxia, i.e., the controller gain, and the plant gain reflecting the effectiveness of ventilation in eliminating CO_2. Thus, low intrathoracic gas stores in subjects with low functional residual capacity, low dead space, and a high $PaCO_2$ are all contributing to a high plant gain since these conditions promote exaggerated changes in $PaCO_2$ elimination in response to alterations of ventilation. Unstable breathing with periodic oscillations may be triggered by alterations in blood gases such as might occur during transition from wakefulness to sleep and vice versa, or as a consequence of brisk and large changes in ventilation caused by a sigh [29] or cough, for example (Fig. 17.3). Decreases in arterial PCO_2 may result in crossing the apnea threshold for CO_2 with ensuing apnea and a subsequent overshooting ventilatory response that occurs with some delay. In order for a disturbance to trigger stable periodic breathing the ratio of the magnitude of the loop gain to that of the disturbance has to exceed a value of 1 (Fig. 17.4) at a phase delay of 180° indicating that the disturbance is reinforced rather than dampened. The underlying theory has been developed and described mathematically by Kho and coworkers [27]. Susceptibility to periodic breathing is therefore enhanced by a high controller gain reflected in the increased ventilatory responsiveness to hypoxia and hypercapnia observed during exposure to hypobaric hypoxia at altitude [30, 31]. Consistently, the amount of periodic breathing in mountaineers at high altitude was correlated with their individual ventilatory sensitivity to hypoxia [32].

While the ventilatory response to CO_2 has been conventionally tested by recording the increase in ventilation induced by increasing the CO_2 concentration in the inhaled air, the concept of ventilatory sensitivity to CO_2 below eupnea, i.e., the ventilatory response to hypocapnia, has gained increasing interest [33]. Studies in animals and in human subjects using subject-triggered pressure support ventilation to induce hyperventilation have shown that the propensity to develop central apnea is enhanced by a high ventilatory sensitivity to CO_2 below eupnea and a close proximity of the arterial PCO_2 during stable breathing to the apnea threshold indicating a low CO_2 reserve. These conditions are met during exposure to hypoxia associated with hyperventilation [33–35]. Using noninvasive breathing pattern analysis and $PetCO_2$ recordings we have corroborated these concepts in spontaneously breathing children and adults at 3,650 m (Fig. 17.4) [9] and in patients with the obstructive sleep apnea syndrome (OSA) showing exacerbation of breathing disturbances with frequent central apneas/hypopneas when sleeping at 2,590 m (Fig. 17.5) [36].

Propensity for periodic breathing is increased during NREM sleep compared to wakefulness because of the absence of the stabilizing effect of the tonic "wakefulness drive" [28]. Furthermore, the reduced background drive to breathe during NREM sleep related to a reduced hypercapnic and hypoxic ventilatory response compared to wakefulness is associated with hypercapnia [30, 37, 38]. This reduces the amount of hyperventilation required to cross the apneic threshold due to the hyperbolic shape of the isometabolic line reflecting the alveolar ventilation/PCO_2 relationship [39] (Fig. 17.5). Conversely, during REM sleep periodic breathing seems to be uncommon at moderate altitude related in part to the absence of a consistent apneic threshold for CO_2 and a larger CO_2 reserve [40, 41].

Summary of Studies on High Altitude Periodic Breathing

Most of the studies on sleep at altitude contain data on ventilation as well. The studies reveal an increasing prevalence of periodic breathing with increasing

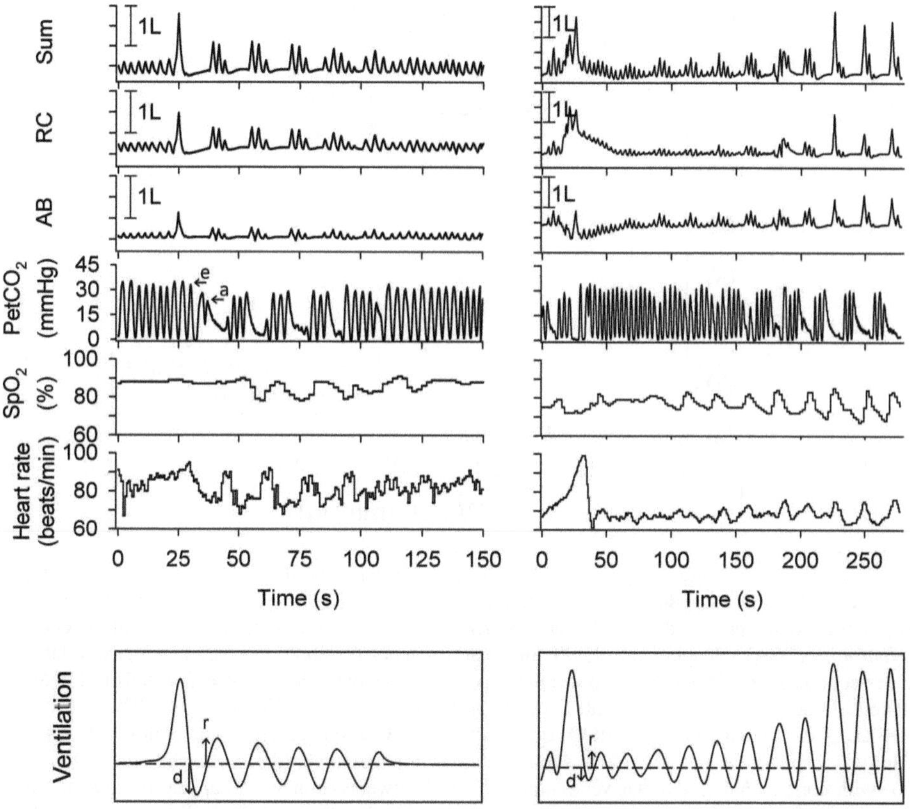

Fig. 17.4 Nocturnal breathing patterns in a 12-year-old child and her father at 3,450 m (Jungfrau Joch, Switzerland). Recordings were obtained with calibrated respiratory inductive plethysmography (RC, AB, Sum: rib cage and abdominal volume changes and their sum), capnography of exhaled air ($PetCO_2$), pulse oximetry (SpO_2), and ECG (heart rate). A short sequence of periodic breathing in the child (*left panels*) begins with a sigh as the initial disturbance. It drives the $PetCO_2$ (the surrogate of $PaCO_2$) from the eupneic level (*left $PetCO_2$ panel, arrow e*) to below the level where the drive to breathe stops (apnea threshold, *left $PetCO_2$ panel, arrow a*). A large difference between the eupneic PCO_2 and the apnea threshold (i.e., a large CO_2 reserve) prevents that apnea occurs with minor fluctuations in ventilation. In the child the breathing pattern stabilizes after only four cycles of periodic breathing indicating stable ventilatory control. The recording obtained in the adult (*right panels*) shows irregular, large breaths that trigger periodic breathing with progressively increasing fluctuations until indicating ventilatory control instability (modified from Kohler et al. [9]). The *lower panels* schematically display ventilation. In the child (*left*) the reduction in ventilation following the sigh represents a disturbance (*d*) that is counteracted by a moderate response (*r*); the ratio of response to disturbance (r/d = loop gain) is <1, ventilation is stable; if the ratio r/d is >1 as in the adult (right), the overshooting response results in control instability. Concepts are explained in [28]

altitude and hypoxemia. The reported number of periodic breathing cycles with central apnea/hypopnea ranges from 10 events/h at a simulated altitude of 2,650 m [42] to as much as 254 events/h at 7,620 m [13]. Correspondingly, with increasing altitude, the fraction of the night spent with periodic breathing increased from 34 to 58 % at 4,559 m [7, 8], to 68 % at 5,050 m [15, 43], and to 73–75 % at 7,620–8,050 m [13, 44]. Figure 17.6 illustrates the relation between the decrease in nocturnal oxygen saturation, the associated increase in minute ventilation, and in periodic breathing cycles measured in participants of an expedition ascending Mt. Muztagh Ata at 7,546 m [43]. The cycle length of periodic breathing has been found to decrease at increasing altitude (29–30 s at 3,560 m, 25–26 s at 4,559 m [8], and 19–23 s at 5,530–6,850 m [43]) consistent with ventilatory control theory predicting a higher controller gain with an accelerated response to alterations in blood gases.

Research on ventilatory adaptation to high altitude has been nearly exclusively focused on

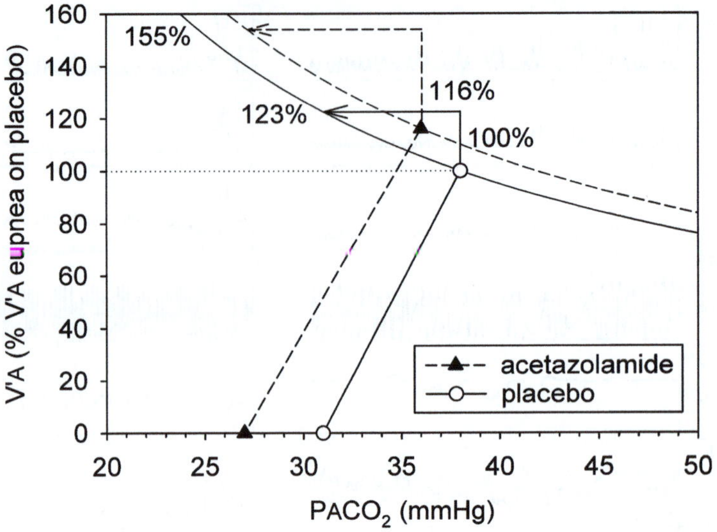

Fig. 17.5 Effect of acetazolamide on control of breathing at altitude. Mechanisms of the reduction in central apneas by acetazolamide in patients with the obstructive sleep apnea syndrome (OSA) sleeping at 2,590 m, from Nussbaumer et al. [36]. The diagram illustrates the hyperbolic relationship between alveolar ventilation (V'A) and alveolar PCO_2 ($PACO_2$). The eupneic $PetCO_2$ on placebo (37 mmHg) was taken as the surrogate of the eupneic alveolar PCO_2 with corresponding alveolar ventilation (V'A) defined as 100 %; $PetCO_2$ of 31 mmHg at V'A=0 is the apnea threshold. Assuming an increase in metabolic rate on acetazolamide (by an arbitrary amount of 10 %) related to hyperventilation, a second metabolic hyperbola is shown (*dashed line*). The corresponding eupneic $PetCO_2$ (35 mmHg) and apnea threshold $PetCO_2$ (27 mmHg) are also shown. Driving $PetCO_2$ from the eupneic level to the apnea threshold requires a greater ventilatory overshoot on acetazolamide (increase in V'A from 116 to 155 %) than on placebo (from 100 to 123 %). This is because the eupneic $PetCO_2$ on acetazolamide is moved to a steeper portion of the metabolic hyperbola while the CO_2 reserve (difference between eupneic and apneic $PetCO_2$) is similar to that on placebo. The slope of the lines connecting the apnea threshold with the corresponding eupneic $PetCO_2$ reflects the ventilatory sensitivity to CO_2 below eupnea which is similar on acetazolamide and on placebo

adults. Only one study investigated the nocturnal breathing pattern in prepubertal children aged 10–11 years in comparison to adults [9]. Pairs of children with their fathers underwent polygraphic sleep studies at the Jungfrau Joch (3,650 m) in Switzerland after rapid ascent by train. It was found that despite a similar degree of nocturnal hypoxemia (mean nocturnal oxygen saturation of 85 % vs. 84 %) and hypocapnia (mean of 32 vs. 32 mmHg) children had less periodic breathing than adults (32.5 vs. 54.1 periodic breathing cycles/h in the first night at 3,450 m, $P<0.05$). This was related to a lower apneic threshold for CO_2, a greater CO_2 reserve, and a shorter circulation time in children promoting a more stable control of breathing compared to adults (Fig. 17.4).

There is only little information on the effect of ventilatory acclimatization on periodic breathing at altitude. In a hypobaric chamber study simulating an altitude of 4,200 m in 7 healthy men, ventilation increased over the course of 4 days associated with progressive hypocapnia and gradually improved arterial oxygen saturation [31]. Changes were similar in wakefulness, NREM sleep, and REM sleep suggesting that supra-pontine influences were not essential for the acclimatization to chronic hypoxia despite the marked influence of the sleep/wakefulness state on the stability of ventilatory control. Periodic breathing was not reported in the cited study [31] but other investigations have revealed variable results in this regard during acclimatization. While some authors reported a decrease in the prevalence of periodic breathing over the course of 7 days at 4,300 m [30], others found no significant change during 6 days at 3,200 m [16] or even an increase during 28 days at 5,050 m [15]. In 34 mountaineers

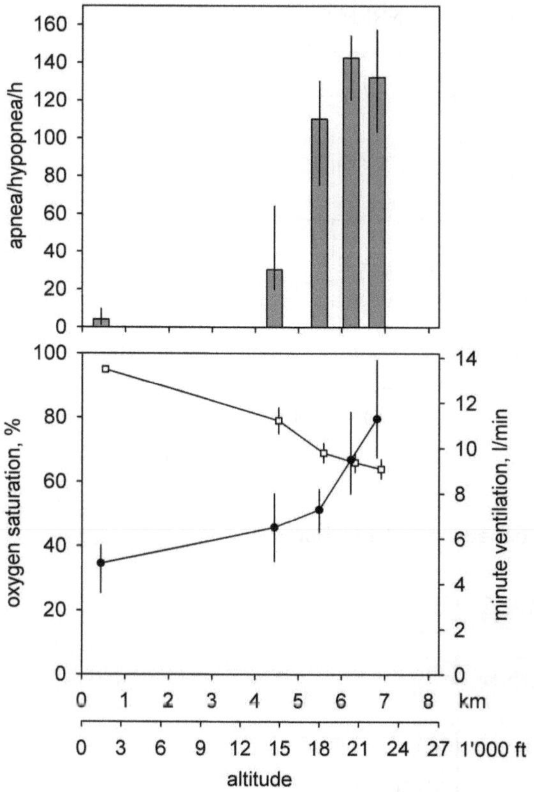

Fig. 17.6 Nocturnal oxygen saturation and minute ventilation in mountaineers during an expedition to extreme altitude. Nocturnal polygraphic monitoring was performed in 34 mountaineers during baseline examination in Zurich (490 m) and over the course of an expedition to Mt. Muztagh Ata at 4,497, 5,533, 6,265, and 6,865 m. With increasing altitude oxygen saturation dropped to lowest values of 64 % (*lower panel, open circles*) while minute ventilation (*closed circles*) and the apnea/hypopnea index increased to highest values of 11.3 L/min and 142.6/h. Values are medians (quartile ranges)

ascending to Muztagh Ata periodic breathing increased during acclimatization over the course of more than 2 weeks at altitudes between 3,730 and 6,850 m despite improving arterial oxygen saturation consistent with a progressive increase in the loop gain of the respiratory control system [43].

Sleep in Altitude-Related Illness

Insomnia/dyssomnia is a typical symptom of AMS [45, 46] and it is conceivable that other symptoms of AMS, in particular headache, nausea, and vomiting, may interfere with onset and continuity of sleep. In a study performed at 4,559 m at the Capanna Regina Margherita [8], the sleep quality and nocturnal breathing pattern were compared among subjects suffering from AMS and controls without AMS. Subjects with AMS had significantly reduced sleep efficiency (68 % in subjects with AMS vs. 85 % in controls) and lower mean nocturnal oxygen saturation (subjects with AMS 59 % vs. 73 % in controls) but a similar prevalence of periodic breathing (subjects with AMS 34 % of the night vs. 43 % in controls). In repeated sleep studies performed over the course of an ascent to the Pyramid research station at 5,050 m in Nepal, no significant changes in the arousal index, in slow wave sleep, or REM sleep were noted at increasing altitude although the Lake Louise score progressively increased up to a mean value of 8 at 5,050 m [19]. The lack of a significant association between measures of objective sleep quality and AMS in this study might have been related to the relatively small sample size of 12–14 subjects. Similarly, sleep efficiency at 4,559 m was not statistically different between 5 subjects with AMS, 8 subjects with HAPE, and 8 controls without AMS or HAPE [7]. However, in subjects developing HAPE after rapid ascent to 4,559 m, nocturnal periodic breathing and oxygen desaturation were more pronounced than in controls without HAPE [47]. Moreover, dexamethasone taken before ascent prevented severe hypoxemia and sleep disturbances while dexamethasone taken 24 h after arrival at 4,559 m increases oxygenation and deep sleep [25].

Sleep at Altitude in Patients with Preexisting Breathing Disorders

Until recently, data on sleep and breathing in patients with preexisting lung disease or disorders of control of breathing were virtually absent [48]. An exception is the OSA. Observations in small groups of patients with the OSA residing at moderate altitude in Colorado have revealed that altitude descent was associated with a decrease in the apnea/hypopnea index; apparently, continuous positive airway pressure (CPAP) requirements

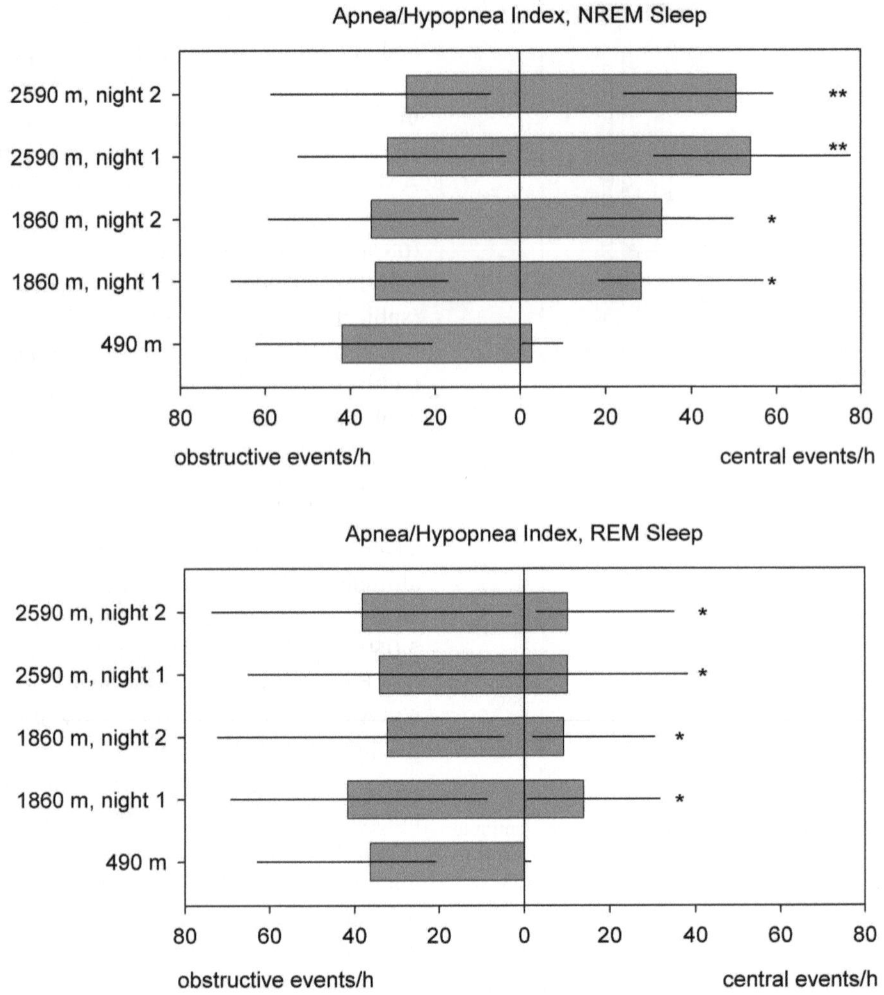

Fig. 17.7 Exacerbation of nocturnal breathing disturbances in patients with the OSA during an altitude sojourn. Sleep-related breathing disturbances at 490 m and in the first and second night at 1,860 and 2,590 m, respectively, in untreated patients with the OSA. The *bars* and *horizontal lines* represent median values and quartile ranges of obstructive and central apnea/hypopnea indices at the different locations. The *upper panel* represents events during NREM sleep and the *lower panel* corresponding values during REM sleep. At higher altitudes, the total apnea/hypopnea index increases significantly related to emerge of central apnea/hypopnea during NREM sleep. $^{*}P<0.01$ vs. 490 m, $^{**}P<0.01$ vs. 490 m and vs. 1,860 m. From Nussbaumer-Ochsner et al. [52]

were not significantly altered [49, 50]. Another observational study suggested that simulated altitude induced central apneas in OSA patients [51]. More recently, a randomized trial was performed in 34 OSA patients who were studied during a few days of CPAP withdrawal at 490, 1,860, and 2,590 m, respectively [52]. Analysis revealed that altitude aggravated hypoxemia, increased sleep-related breathing disturbances due to frequent central apneas/hypopneas (Fig. 17.7), and induced sleep fragmentation. Daytime evaluation at altitude demonstrated that the patients performed poorly in driving simulator tests and had an increase in blood pressure, cardiac arrhythmias, weight gain, and leg edema. These data emphasize the unfavorable health effects of altitude in OSA. Subsequent randomized, placebo-controlled trials revealed that OSA patients benefit from acetazolamide in combination with computer-autoadjusting CPAP during a stay at altitude [53] or from acetazolamide alone compared to no treatment at all (Fig. 17.5) [36]. While autoadjusting CPAP alone did not consistently eliminate the central sleep apneas emerging at altitude, the combined treatment with CPAP and acetazolamide provided optimal control of sleep-related breathing disturbances. During daytime patients did not suffer from symptoms of AMS or sleepiness and the treatment was well tolerated. According to manufacturer specifications, current CPAP devices incorporate altitude correction. However, treating patients at the same fixed CPAP at low and high altitude may be suboptimal since we observed that the mask pressure applied by autoadjusting CPAP devices increased at 2,590 m compared to 490 m by approximately 2 cm H_2O [53]. These results suggest that CPAP devices with autoadjusting may be suitable for use at altitude [54].

Treatment of Sleep Disturbances at Altitude

Sleep disturbances at altitude are caused by various factors including environmental conditions and physiologic changes such as periodic breathing and altitude-related illness. Sleep quality should therefore be improved by preventing high altitude-related illness through avoidance of physical over-exertion, appropriate ascent rate and acclimatization, and by optimizing sleeping conditions (e.g., through proper clothing). The use of pharmacological means should be reserved for those who suffer from severe sleep problems. Several drugs used to prevent and treat high altitude-related illness (AMS, cerebral edema, HAPE) including acetazolamide, dexamethasone, nifedipine, among others, may also improve sleep quality but are described elsewhere (see Chaps. 20 and 21). The following discussion will focus on studies that specifically evaluated the effect of oxygen or drugs on sleep and periodic breathing at altitude.

Oxygen

Since sleep disturbances and altitude-related illnesses are mainly caused by hypoxia providing supplemental oxygen is a rational way to treat these conditions. Raising the oxygen concentration at a given high altitude by 1 % increases the inspired oxygen partial pressure to a degree corresponding to an altitude reduction of approximately 300 m [55]. Raising F_IO_2 in the sleeping room at 3,800 m by 3 % by oxygen concentrators increased nocturnal oxygen saturation, reduced the amount of periodic breathing, and improved subjective sleep quality and some tests of cognitive daytime performance [56, 57].

Carbonic Anhydrase Inhibitors

The carbonic anhydrase inhibitor acetazolamide has been extensively tested and found to be an effective drug for prevention and treatment of AMS (see Chap. 20). Studies have also demonstrated improvement of nocturnal oxygen saturation and high altitude periodic breathing by acetazolamide and benzolamide [58–60]. Carbonic anhydrase inhibitors promote renal bicarbonate excretion thus counteracting the hypoxia-induced respiratory alkalosis [61] and stimulating ventilation through augmenting the hypercapnic ventilatory response [60]. Hyperventilation dampens periodic breathing by moving the $PaCO_2$ on the isometabolic hyperbola to a steeper portion so that a larger overshoot in ventilation is required to reduce the $PaCO_2$ by a given amount (Fig. 17.5). Hypocapnia mitigates an exaggerated increase in hypoxic ventilatory sensitivity through interaction at the chemoreceptor [62] and due to the reduced loop gain that results from the higher oxygen saturation associated with hyperventilation [63]. In a randomized comparison with theophylline, acetazolamide was equally effective in normalizing periodic breathing in 30 mountaineers sleeping at 3,454 m but acetazolamide provided a higher nocturnal oxygen saturation [64]. In otherwise untreated OSA patients we found that acetazolamide

(250 mg b.i.d.) improved oxygenation, nocturnal breathing disturbances, and sleep quality and it prevented excessive blood pressure rises at altitude [36]. Even though the amount of residual breathing disturbances was considerable, acetazolamide was superior to no therapy at all and may therefore be recommended for OSA patients at altitude if CPAP therapy is not feasible. An even better and nearly optimal control of breathing disturbances was achieved with the combined treatment by acetazolamide and autoadjusting CPAP [53].

Dexamethasone

Like acetazolamide, dexamethasone has also been found to effectively prevent and treat AMS and the drug prevents HAPE in susceptible subjects [65]. In a recent study in HAPE-susceptible subjects ascending from lowlands to 4,559 m within 24 h we found that dexamethasone (4 mg b.i.d.) taken before ascent improved nocturnal oxygen saturation and increased the amount of slow wave sleep at 4,559 m [25]. To our knowledge, no other studies specifically addressing effects of dexamethasone on sleep at altitude have been performed. The powerful action of the drug in the prevention and the treatment of AMS—and thereby also improving sleep—seems to outweigh the potential sleep disturbing effects which are not well established [66].

We refer the reader to a more extensive discussion on treatment of high altitude-related illness by acetazolamide and dexamethasone in Chaps. 20 and 21.

Theophylline

As mentioned above, theophylline (250 mg b.i.d.) has been evaluated as a drug for treatment of high altitude periodic breathing. At an altitude of 3,454 m breathing disturbances were nearly completely suppressed but oxygen saturation was lower than in subjects taking acetazolamide [64]. In another study performed at 4,559 m, theophylline (300 mg taken in the evening) was started 5 days before ascent and continued during a 5-day stay at altitude [67]. The drug alleviated symptoms of AMS slightly and reduced nocturnal periodic breathing by about 50 %. However, oxygen saturation and sleep structure were not improved. Theophylline may exert its action by an increase in ventilatory drive and in cardiac output but the mechanisms have not been studied in detail. Compared to acetazolamide theophylline has the disadvantage of not improving oxygen saturation, promoting cardiac arrhythmia but, on the other hand, does not alter taste or cause paresthesias. It may therefore be an alternative if acetazolamide is not tolerated.

Hypnotics

Hypnotics are commonly used to alleviate insomnia and dyssomnia at low altitude. Their application in the treatment of altitude insomnia has been met with reservation because of concerns that these drugs would aggravate hypoxemia and promote sleep apnea due to a ventilatory depressant effect. Several randomized, placebo-controlled trials have addressed this point by evaluating the effect of benzodiazepines and non-benzodiazepine hypnotics in subjects at simulated or real altitude in terms of sleep and ventilation. The results reveal that contrary to some negative expectations neither the benzodiazepine temazepam nor the non-benzodiazepine hypnotics zolpidem and zaleplon which act on the $GABA_A$ receptor had deleterious effects on oxygen saturation and ventilation but were effective in improving sleep quality at altitude and, in some studies, even next day performance. It can therefore be concluded that the cited hypnotics may be used in mountaineers at moderate altitude (up to 4,000 m) to treat insomnia in a safe setting that allows resting in a state of reduced arousability. Since zaleplon has a relatively short half-life (1 h) compared to zolpidem (2.4 h) it might be particularly suitable for use in the mountains.

Summary and Conclusions

Insomnia/dyssomnia and unrefreshing sleep are common complaints in newcomers at high altitude. Objective sleep recordings reveal frequent awakenings and arousals and reduced amounts of slow wave and REM sleep. Periodic breathing, a pattern of waxing and waning of ventilation with central apneas/hypopneas alternating with bouts of hyperventilation, is an important cause of sleep disruption at altitude. It occurs in healthy subjects at altitudes >2,000 m and is increasingly prevalent at higher altitude. It may be prevented or reduced by acetazolamide and oxygen administration. Temazepam and the non-benzodiazepine hypnotics zolpidem and zaleplone improve altitude insomnia without relevant adverse effects on ventilation. Because of the fact that the majority of published studies on sleep at altitude have been performed in highly selected, well trained healthy subjects, mostly men, and because of methodological flaws of many uncontrolled studies performed in specific settings at various altitudes, the results from the available literature on sleep at altitude cannot be generalized. Future studies should explore the effects of altitude on sleep and breathing accounting for potential confounders such as acclimatization, high altitude-related illness, and preexisting respiratory and other disorders and should evaluate treatments that might improve sleep quality at altitude in various settings.

References

1. Nussbaumer-Ochsner Y, Bloch KE. Air travel and altitude. In: Ayres JG, Harrison RM, Nichols GL, et al., editors. Environmental medicine. 1st ed. London: Hodder Arnold; 2010. p. 547–61.
2. Jafarian S, Gorouhi F, Taghva A, et al. High-altitude sleep disturbance: results of the Groningen Sleep Quality Questionnaire survey. Sleep Med. 2008;9(4):446–9.
3. Nussbaumer-Ochsner Y, Schuepfer N, Siebenmann C, et al. High altitude sleep disturbances monitored by actigraphy and polysomnography. High Alt Med Biol. 2011;12(3):229–36.
4. Mizuno K, Asano K, Okudaira N. Sleep and respiration under acute hypobaric hypoxia. Jpn J Physiol. 1993;43(2):161–75.
5. Muhm JM, Signal TL, Rock PB, et al. Sleep at simulated 2438 m: effects on oxygenation, sleep quality, and postsleep performance. Aviat Space Environ Med. 2009;80(8):691–7.
6. Latshang TD, Lo Cascio CM, Stadelmann K, et al. Are sleep, nocturnal breathing and daytime performance impaired at moderate altitude (1630-2590m)? Sleep. 2013, in press.
6b. Stadelmann K, Latshang TD, Lo Cascio CM. Quantitative changes in the sleep EEG at moderate altitude (1630 m and 2590 m). PLoS.One. 2013, in press.
7. Eichenberger U, Weiss E, Riemann D, et al. Nocturnal periodic breathing and the development of acute high altitude illness. Am J Respir Crit Care Med. 1996; 154(6 Pt 1):1748–54.
8. Erba P, Anastasi S, Senn O, et al. Acute mountain sickness is related to nocturnal hypoxemia but not to hypoventilation. Eur Respir J. 2004;24(2):303–8.
9. Kohler M, Kriemler S, Wilhelm EM, et al. Children at high altitude have less nocturnal periodic breathing than adults. Eur Respir J. 2008;32(1):189–97.
10. Reite M, Jackson D, Cahoon RL, et al. Sleep physiology at high altitude. Electroencephalogr Clin Neurophysiol. 1975;38(5):463–71.
11. Normand H, Barragan M, Benoit O, et al. Periodic breathing and O2 saturation in relation to sleep stages at high altitude. Aviat Space Environ Med. 1990; 61(3):229–35.
12. Khoo MC, Anholm JD, Ko SW, et al. Dynamics of periodic breathing and arousal during sleep at extreme altitude. Respir Physiol. 1996;103(1):33–43.
13. Anholm JD, Powles AC, Downey III R, et al. Operation Everest II: arterial oxygen saturation and sleep at extreme simulated altitude. Am Rev Respir Dis. 1992;145(4 Pt 1):817–26.
14. Matsuzawa Y, Kobayshi T, Fujimoto YS, et al. Nocturnal periodic breathing and arterial oxygen saturation in acute mountain sickness. J Wilderness Med. 1994;5:269–81.
15. Salvaggio A, Insalaco G, Marrone O, et al. Effects of high-altitude periodic breathing on sleep and arterial oxyhaemoglobin saturation. Eur Respir J. 1998; 12(2):408–13.
16. Zielinski J, Koziej M, Mankowski M, et al. The quality of sleep and periodic breathing in healthy subjects at an altitude of 3,200 m. High Alt Med Biol. 2000;1(4):331–6.
17. Beaumont M, Goldenberg F, Lejeune D, et al. Effect of zolpidem on sleep and ventilatory patterns at simulated altitude of 4,000 meters. Am J Respir Crit Care Med. 1996;153(6 Pt 1):1864–9.
18. Beaumont M, Batejat D, Coste O, et al. Effects of zolpidem and zaleplon on sleep, respiratory patterns and performance at a simulated altitude of 4,000 m. Neuropsychobiology. 2004;49(3):154–62.
19. Burgess KR, Johnson P, Edwards N, et al. Acute mountain sickness is associated with sleep desaturation at high altitude. Respirology. 2004;9(4):485–92.

20. Beaumont M, Batejat D, Pierard C, et al. Zaleplon and zolpidem objectively alleviate sleep disturbances in mountaineers at a 3,613 meter altitude. Sleep. 2007; 30(11):1527–33.
21. Hoshikawa M, Uchida S, Sugo T, et al. Changes in sleep quality of athletes under normobaric hypoxia equivalent to 2,000-m altitude: a polysomnographic study. J Appl Physiol. 2007;103(6):2005–11.
22. Johnson PL, Edwards N, Burgess KR, et al. Sleep architecture changes during a trek from 1400 to 5000 m in the Nepal Himalaya. J Sleep Res. 2010;19(1 Pt 2): 148–56.
23. Nussbaumer-Ochsner Y, Ursprung J, Siebenmann C, et al. Effect of short-term acclimatization to high altitude on sleep and nocturnal breathing. Sleep. 2012;35:419–423.
24. Nicholson AN, Smith PA, Stone BM, et al. Altitude insomnia: studies during an expedition to the Himalayas. Sleep. 1988;11(4):354–61.
25. Nussbaumer-Ochsner Y, Ursprung J, Siebenmann C, et al. Sleep and breathing in high altitude pulmonary edema susceptible subjects at 4559m. Sleep 2012;35: 1413–21.
26. Yaron M, Lindgren K, Halbower AC, et al. Sleep disturbance after rapid ascent to moderate altitude among infants and preverbal young children. High Alt Med Biol. 2004;5(3):314–20.
27. Khoo MC, Kronauer RE, Strohl KP, et al. Factors inducing periodic breathing in humans: a general model. J Appl Physiol. 1982;53(3):644–59.
28. White DP. Pathogenesis of obstructive and central sleep apnea. Am J Respir Crit Care Med. 2005; 172(11):1363–70.
29. Perez-Padilla R, West P, Kryger MH. Sighs during sleep in adult humans. Sleep. 1983;6(3):234–43.
30. White DP, Gleeson K, Pickett CK, et al. Altitude acclimatization: influence on periodic breathing and chemoresponsiveness during sleep. J Appl Physiol. 1987;63(1):401–12.
31. Berssenbrugge AD, Dempsey JA, Skatrud JB. Effects of sleep state on ventilatory acclimatization to hypoxia in humans. J Appl Physiol. 1984;57(4):1089–96.
32. Lahiri S, Maret K, Sherpa MG. Dependence of high altitude sleep apnea on ventilatory sensitivity to hypoxia. Respir Physiol. 1983;52(3):281–301.
33. Dempsey JA. Crossing the apnoeic threshold: causes and consequences. Exp Physiol. 2005;90(1):13–24.
34. Xie A, Skatrud JB, Dempsey JA. Effect of hypoxia on the hypopnoeic and apnoeic threshold for CO[2] in sleeping humans. J Physiol. 2001;535(Pt 1):269–78.
35. Xie A, Skatrud JB, Puleo DS, et al. Influence of arterial O2 on the susceptibility to posthyperventilation apnea during sleep. J Appl Physiol. 2006;100(1):171–7.
36. Nussbaumer-Ochsner Y, Latshang TD, Ulrich S, et al. Patients with obstructive sleep apnea syndrome benefit from acetazolamide during an altitude sojourn: a randomized, placebo-controlled, double-blind trial. Chest. 2012;141(1):131–8.
37. Douglas NJ, White DP, Weil JV, et al. Hypercapnic ventilatory response in sleeping adults. Am Rev Respir Dis. 1982;126(5):758–62.
38. Douglas NJ, White DP, Weil JV, et al. Hypoxic ventilatory response decreases during sleep in normal men. Am Rev Respir Dis. 1982;125(3):286–9.
39. Dempsey JA, Smith CA, Przybylowski T, et al. The ventilatory responsiveness to CO(2) below eupnoea as a determinant of ventilatory stability in sleep. J Physiol. 2004;560(Pt 1):1–11.
40. Berssenbrugge A, Dempsey J, Iber C, et al. Mechanisms of hypoxia-induced periodic breathing during sleep in humans. J Physiol. 1983;343:507–26.
41. Xi L, Smith CA, Saupe KW, et al. Effects of rapid-eye-movement sleep on the apneic threshold in dogs. J Appl Physiol. 1993;75(3):1129–39.
42. Kinsman TA, Hahn AG, Gore CJ, et al. Respiratory events and periodic breathing in cyclists sleeping at 2,650-m simulated altitude. J Appl Physiol. 2002; 92(5):2114–8.
43. Bloch KE, Latshang TD, Turk AJ, et al. Nocturnal periodic breathing during acclimatization at very high altitude at Mount Muztagh Ata (7,546 m). Am J Respir Crit Care Med. 2010;182(4):562–8.
44. West JB, Peters Jr RM, Aksnes G, et al. Nocturnal periodic breathing at altitudes of 6,300 and 8,050 m. J Appl Physiol. 1986;61(1):280–7.
45. Sampson JB, Cymerman A, Burse RL, et al. Procedures for the measurement of acute mountain sickness. Aviat Space Environ Med. 1983;54(12 Pt 1): 1063–73.
46. Roach RC, Bartsch P, Hackett PH et al. The Lake Louise acute mountain sickness scoring system. In: Sutton JR, Coates G, Huston CS. Hypoxia and molecular medicine. Proceedings of the 8th international hypoxia symposium. Queen City Printers, Burlington, Canada. 1993; p. 272–4.
47. Christ AL, Clarenbach CF, Senn O, et al. Pulmonary function and nocturnal ventilation in mountaineers developing high altitude pulmonary oedema. In: Roach R, Wagner PD, Hackett P, editors. Proceedings of the 14th International Hypoxia Symposium, held in Chateau Lake Louise, Alberta Canada, February 22–27, 2005. New York: Springer; 2005. p. 346. abstract.
48. Luks AM, Swenson ER. Travel to high altitude with pre-existing lung disease. Eur Respir J. 2007;29(4):770–92.
49. Patz D, Spoon M, Corbin R, et al. The effect of altitude descent on obstructive sleep apnea. Chest. 2006;130(6):1744–50.
50. Patz DS, Swihart B, White DP. CPAP pressure requirements for obstructive sleep apnea patients at varying altitudes. Sleep. 2010;33(5):715–8.
51. Burgess KR, Cooper J, Rice A, et al. Effect of simulated altitude during sleep on moderate-severity OSA. Respirology. 2006;11(1):62–9.
52. Nussbaumer-Ochsner Y, Schuepfer N, Ulrich S, et al. Exacerbation of sleep apnoea by frequent central events in patients with the obstructive sleep apnoea

syndrome at altitude: a randomised trial. Thorax. 2010;65(5):429–35.

53. Latshang TD, Nussbaumer-Ochsner Y, Ulrich-somaini S, et al. Effect of acetazolamide and autoCPAP therapy on breathing disturbances among patients with obstructive sleep apnea syndrome who travel to altitude: a randomized controlled trial. JAMA. 2012;308:2390–98.
54. Latshang TD, Bloch KE. How to treat patients with obstructive sleep apnea syndrome during an altitude sojourn. High Alt Med Biol. 2011;12(4):303–7.
55. West JB. Oxygen enrichment of room air to relieve the hypoxia of high altitude. Respir Physiol. 1995; 99(2):225–32.
56. Luks AM, van Melick H, Batarse RR, et al. Room oxygen enrichment improves sleep and subsequent day-time performance at high altitude. Respir Physiol. 1998;113(3):247–58.
57. McElroy MK, Gerard A, Powell FL, et al. Nocturnal O2 enrichment of room air at high altitude increases daytime O2 saturation without changing control of ventilation. High Alt Med Biol. 2000;1(3):197–206.
58. Hackett PH, Roach RC, Harrison GL, et al. Respiratory stimulants and sleep periodic breathing at high altitude. Almitrine versus acetazolamide Am Rev Respir Dis. 1987;135(4):896–8.
59. Sutton JR, Houston CS, Mansell AL, et al. Effect of acetazolamide on hypoxemia during sleep at high altitude. N Engl J Med. 1979;301(24):1329–31.
60. Swenson ER, Leatham KL, Roach RC, et al. Renal carbonic anhydrase inhibition reduces high altitude sleep periodic breathing. Respir Physiol. 1991; 86(3):333–43.
61. Swenson ER, Teppema LJ. Prevention of acute mountain sickness by acetazolamide: as yet an unfinished story. J Appl Physiol. 2007;102(4):1305–7.
62. Teppema LJ, van Dorp EL, Dahan A. Arterial [H+] and the ventilatory response to hypoxia in humans: influence of acetazolamide-induced metabolic acidosis. Am J Physiol Lung Cell Mol Physiol. 2010; 298(1):L89–95.
63. Wellman A, Malhotra A, Jordan AS, et al. Effect of oxygen in obstructive sleep apnea: role of loop gain. Respir Physiol Neurobiol. 2008;162(2):144–51.
64. Fischer R, Lang SM, Leitl M, et al. Theophylline and acetazolamide reduce sleep-disordered breathing at high altitude. Eur Respir J. 2004;23(1):47–52.
65. Maggiorini M, Brunner-La Rocca HP, Peth S, et al. Both tadalafil and dexamethasone may reduce the incidence of high-altitude pulmonary edema: a randomized trial. Ann Intern Med. 2006;145(7):497–506.
66. Born J, DeKloet ER, Wenz H, et al. Gluco- and antimineralocorticoid effects on human sleep: a role of central corticosteroid receptors. Am J Physiol. 1991;260(2 Pt 1):E183–8.
67. Kupper TE, Strohl KP, Hoefer M, et al. Low-dose theophylline reduces symptoms of acute mountain sickness. J Travel Med. 2008;15(5):307–14.

18 Reproduction and Growth

Susan Niermeyer

Abstract

Hypoxia, as well as other environmental and social factors, shapes reproduction and fetal/infant growth at high altitude. Although fertility and early fetal loss differ little from patterns observed at lower altitude, placental development does differ and the medical complications of pre-eclampsia and post-partum hemorrhage pose greater risks at altitude. Reduction in birth weight with increasing altitude of gestation occurs across high-altitude regions, resulting in an increased proportion of low-birth-weight infants (<2,500 g). However, the magnitude of reduction in fetal growth varies among population groups, depending on maternal ventilation, plasma volume expansion, blood flow to the uteroplacental circulation, and nutrient transport. Neonatal, infant, and child mortality are increased at high altitude. Respiratory problems account for the majority of neonatal and infant deaths. Exposure to hypoxia during the fetal and neonatal periods may have consequences that extend into adult life and span generations.

An Ecologic and Lifecyle Perspective on Reproduction and Growth

Reproduction and development during infancy and childhood present critical, vulnerable periods in the individual lifecycle and the health of a population. For humans living permanently at high altitude, the environment they inhabit adds risks for maternal and child health and survival. In addition to hypoxia, many high-altitude regions impose other constraints on the circumstances necessary for optimal fetal development and infant/child survival. These include impoverished living conditions, social inequalities, physically demanding livelihoods, nutritional deficiencies, and infectious diseases. Such determinants influence fertility, fetal loss and growth, neonatal survival, and child development. Populations resident at high altitude for millennia, for example the Tibetans in the Himalaya and the Aymara and Quechua of South America, attest to possibility of not only surviving but also thriving,

S. Niermeyer, M.D., Ph.D. (✉)
Pediatrics, Section of Neonatology, University of Colorado School of Medicine and Children's Hospital Colorado, 13121 E. 17th Avenue, Mail Stop 8402, Aurora, CO 80045, USA
e-mail: susan.niermeyer@ucdenver.edu

E.R. Swenson and P. Bärtsch (eds.), *High Altitude: Human Adaptation to Hypoxia*,
DOI 10.1007/978-1-4614-8772-2_18, © Springer Science+Business Media New York 2014

under such circumstances. However, consideration of the biological and behavioral adaptations necessary to do so aids understanding of the physiologic challenges. Addressing pregnancy complications such as pre-eclampsia and fetal growth restriction, neonatal mortality risks, and challenges to growth is essential to reduce mortality and improve the quality of survival of women and children at high altitude. Finally, the developmental origins of adult disease link the health of fetuses and newborn infants to their later health as adults in the current and successive generations.

Maternal Aspects of Reproduction

Hypoxia, as well as other environmental and social factors, shapes the reproductive pattern at high altitude. How these factors and the organism's response to them influence the ability of one generation to successfully reproduce the next in essence defines adaptation. From an evolutionary perspective, pregnancy represents the overlap between successive generations. Although the mortality risk is greater for the fetus, the death of a pregnant woman or mother impacts the evolutionary contribution of all; the fetus or neonate likely dies as well and thereby truncates the genetic contribution of the father. Fertility, maintenance of pregnancy to term, and the maternal medical complications during gestation all reflect the physical and socioecologic influences of the high-altitude environment. Although fertility and early fetal loss appear to follow the patterns observed at lower altitude, placental development differs and certain medical complications at high altitude place pregnant women at greater risk, most notably from pre-eclampsia and postpartum hemorrhage.

Fertility

Fertility describes the production of live offspring within a population. Although it has been proposed that hypoxia reduces fertility (number of live births), the complex interplay of behavioral and sociocultural factors that underlie fertility appear to exert the predominant effect at high altitude [1]. Historical observations in the Andean region raised the question whether hypoxia reduces fertility through impairing fecundity (the ability to conceive) and/or by increasing fetal loss, especially in newcomers to high altitude [2]. Early efforts to test this hypothesis using census data and community-level estimates from Andean countries proved inconclusive, but an increase in completed fertility was noted among Peruvian high-altitude natives who migrated to low altitude [3]. Other studies in the Andes and Himalaya do not show an altitude-related reduction in fertility, but rather suggest higher completed fertility at high than low altitudes [4–7]. Overall, fertility levels vary widely at any altitude. The completed fertility ratio expresses the number of children born per woman to a cohort of women by the end of their childbearing years. A review of Andean samples yielded estimates of completed fertility ratios from 5.8 to 9.1 among high-altitude populations and 4.6 to 8.3 among low-altitude comparison groups. Recent data from the Peruvian National Institute of Statistics and Informatics (INEI 2001) report a global fecundity rate of 3.7 (the number of expected births through a woman's reproductive period, having children at the prevailing rate for each age) for altitudes >2,000 m compared to 2.4–3.8 in sea level and jungle zones [5]. Estimates of completed fertility from Himalayan populations range from 3.2 to 7.4 [1].

In order to interpret and compare fertility rates at various altitudes, the set of proximate determinants of fertility must be evaluated, as these factors mediate environmental, behavioral, and sociocultural influences. In Andean and Himalayan women, the age of menarche may be delayed by as much as a year, consistent with delay in growth and maturation at high altitude [8–11]. Onset of menopause appears to be accelerated by a similar time [10, 12]. Shortening of the reproductive span is unlikely to have major demographic impact, however, due to predominant patterns of sexual behavior [1].

Data are accumulating on certain aspects of fecundability at high altitude. Ovarian cycle length in rural Aymara women averaged 29.1 days, which falls within the observed range for sea level [13]. The proportion of ovulatory cycles, as assessed by progesterone levels, was comparable in a sample of Chicago women and economically advantaged women in La Paz, Bolivia (91 % and 88 % respectively); however, it was far lower (45 %) in the poorer women of La Paz [14]. Mean progesterone levels in ovulatory cycles among poor women were lower than those in better off peers and substantially lower than in Chicago women. This continuum likely relates to differences in nutrition and physical demands during childhood development, as it correlates with body size [14, 15]. Despite having lower progesterone levels during the fertile period, rural Bolivian women conceived and carried pregnancies to full term [16]. Early pregnancy loss in rural Bolivian women was associated with higher follicular progesterone and low luteal/follicular progesterone ratios [17]. Sperm quality and quantity as well as testosterone levels appear equivalent in low and high-altitude men [18–21].

A higher probability of fetal loss under hypoxic conditions is logically attractive, but has not been substantiated. In any setting, accurate estimates of fetal loss are difficult to obtain, as the greatest risk for loss occurs in early pregnancy before a woman is aware of the conception. Using elevated human chorionic gonadotropin as a marker of implantation, the fetal loss rate in rural Aymara women was 30 %—very similar to the 31 % fetal loss rate reported among U.S. women using similar criteria [16]. Estimates of fetal loss based on reproductive histories among the Sherpa of Nepal and Tibetan women living at 3,000–4,000 m suggest approximately 10 % loss among recognized pregnancies [22, 23], although this figure likely reflects underreporting.

In addition to the strictly physiologic determinants of fertility, behavioral, sociocultural and economic factors reflect the ecology of the high-altitude environment. Age at marriage or entry into sexual union and the proportion of unmarried persons varies by geographic location and culture. In the Andes, lifelong domestic relationships are the norm, whereas 18–44 % of women of reproductive age in Nepal and 11–33 % in Ladakh are unmarried [8, 24–27]. Moreover, more flexible marriage patterns, including monogamy, polyandry, polygyny and polygynandry also impact fertility [24, 28]. Religious practices in the Himalaya can result in a high proportion of certain populations living as celibate monks or nuns [22]. Among those who do marry, at least one study found later age at marriage in higher altitude Himalayan populations [29]. Spousal absence due to seasonal work in agriculture or tourism may influence the frequency of sexual intercourse, but this effect did not appear to contribute to variation in fertility by altitude in Nepal [29].

The duration of lactational amenorrhea exerts a strong influence on fertility differences among populations without access to modern contraception. The duration of breast feeding varies by economic status, available weaning foods, and social forces that promote or delay weaning. Community studies in Peru and Bolivia suggest that rural, high-altitude women breast feed the longest when compared to their urban and low-altitude counterparts [1]. In the Himalaya, breastfeeding continues for 2–3 years, but introduction of complementary foods within the first months results in insufficient suckling stimulus to maintain anovulation and an earlier return to fecundity [28, 30]. The survival, contributions, and costs of children also influence childbearing patterns. In rural highland Peru, children traditionally generated more resources than they consumed, making high completed fertility desirable [31]; however, improving child survival and increasing urbanization may alter fertility trends in high-altitude environments now undergoing rapid social change.

Levels of fertility vary among populations resident across altitude ranges; however, there is no clear effect of hypoxia on any proximate determinant of fertility. Nonetheless, it is important to appreciate the complexity of the high-altitude environment in shaping behavioral and sociocultural practices may either produce fertility differentials or compensate for underlying physiologic differences caused by hypoxia.

Although fecundity and fetal loss are seemingly the most vulnerable elements of fertility at high altitude, well-controlled, minimally biased studies support no direct effect of hypoxia on these determinants. Rather, cultural practices appear to exert the greatest influence on fertility among high-altitude populations.

Pregnancy and Its Complications

Fetoplacental development is influenced by hypobaric hypoxia, as are the medical and obstetrical complications of pregnancy at high altitude. One of the best-documented effects of high altitude is intrauterine growth restriction (IUGR) resulting in reduction of birth weight. Hypoxia alters placental development from very early stages, and the adaptive physiologic responses to preserve oxygen delivery to the fetus give insight into the mechanisms in play. Altered placentation has implications for maternal health as well, with higher observed incidence of pre-eclampsia and hypertensive disorders of pregnancy at high altitude. Pre-eclampsia/eclampsia and postpartum hemorrhage are leading causes of maternal mortality everywhere, but they pose a special risk to women at high altitude.

Fetal Growth

Studies conducted over a half century in North America, South America, and Tibet have documented reduction in birth weight with increasing altitude of gestation. Birth weight decreases an average of 100 g per 1,000 m altitude gain [32]. Reduction in birth weight also increases the proportion of low-birth-weight infants (<2,500 g or 5.5 lb) approximately fourfold at high (>2,700 m) compared to low altitude in a North American population [33]. The separable influences of fetal growth restriction and prematurity on birth weight were first recognized at high altitude, giving rise to the concept of small for gestational age as a result of fetal growth restriction [34, 35]. At high altitude, lower birth weight results from restriction of intrauterine growth rather than shortened gestation. Fetal ultrasound studies in Denver (1,610 m) as compared to sea level demonstrated significant reduction in fetal subcutaneous fat tissue but not lean mass [36]. A large, population-based comparison of infants born in public hospitals near sea level (150 m) and at high altitude (3,000–4,400 m) in Peru showed reduction not only in percentiles for weight but also length and head circumference at high altitude [37].

The magnitude of the reduction in fetal growth with increasing altitude varies among population groups, with the longest resident populations experiencing the least effect and the shortest resident groups the greatest reduction in birth weight. The reduction in birth weight is greatest in North Americans (−352 g, $p<0.001$), intermediate in South Americans (−270 g in Peru, −282 g in Bolivia, $p<0.001$), and least in Tibetans (−72 g, $p>0.05$) [32]. Where woman of different ancestry reside at the same altitude, those from long-resident high-altitude populations give birth to heavier infants than women of low-altitude population ancestry. In Tibet, birth weights averaged 294–650 g more in Tibetan than Han women, and in Bolivia, the infants of Aymara women weighed 143 g more than infants born to women of European or mestizo ancestry [38–40]. Birth weight declined proportionately with the degree of European admixture among residents of La Paz, and decline in birth weight correlated most tightly with the degree of maternal admixture, suggesting possible involvement of epigenetic factors as well [41]. Protection from altitude-associated reduction in birth weight has also been observed for Nepalese Sherpa but not Ladakhis [42, 43].

Both genetic and developmental factors contribute to the observed birth weight differences between multigenerational high-altitude residents and newcomers. Genetic factors are recognized determinants of birth weight, suggesting the selection of adaptive variants among populations with long residence at high altitude [44, 45]. Developmental factors, such as nutrition, behavioral adaptations, and medical complications of pregnancy account for additional variability between geographic locations. In contrast to the effect of altitude on fertility, the reduction in birth weight is attributable to direct effects of high altitude rather than other variables such as maternal age, parity, body size, or prenatal care [46]. Socioeconomic factors appear to make relatively little contribution [47, 48]. Hypobaric hypoxia and maternal physiological mechanisms that mediate genetic and developmental factors together shape oxygen and nutrient transport during pregnancy at high altitude.

Maternal Oxygen and Nutrient Transport During Pregnancy

Pregnancy increases maternal ventilation, which in turn, increases arterial oxygen saturation at high altitude [32]. However, arterial oxygen content may actually *fall* during pregnancy as a result of the physiological anemia of pregnancy that arises from the expansion of circulating plasma volume in the second trimester [49]. In Colorado and Peru infants born to mothers with greater increases in hypoxic ventilatory sensitivity, higher levels of ventilation, and higher arterial oxygen content during pregnancy weighed more than those born to women with lesser increases in ventilation [50]. In Tibetan and Han women living at 3,600 m, higher ventilation and increased ventilatory sensitivity to hypoxia also correlated with infant birth weight, but differences in arterial oxygen saturation and content failed to account for the heavier birth weights of Tibetan infants [23, 51]. Lower hemoglobin levels among the Tibetans actually resulted in lower calculated levels of oxygen content near term than in the Han women. Analysis of birth weight and maternal hemoglobin at sea level and re-analysis of such data from high altitude demonstrated a significant negative relationship between maternal hemoglobin and birth weight, emphasizing the importance of appropriate plasma volume expansion [52].

Changes in the uteroplacental circulation under conditions of chronic hypoxia at high altitude more fully explain observed patterns of fetal growth restriction [32]. At sea level, in pregnancies near term nearly 15 % of maternal cardiac output is directed to the uteroplacental circulation [53]. Blood supply to this pregnancy-specific circuit is accomplished by a doubling of uterine artery diameter by mid-gestation and an increase in uterine artery flow velocity that continues throughout pregnancy. Colorado women at 3,100 m demonstrated changes in uterine artery diameter that were only about half of those at 1,600 m, resulting in one-third lower uterine artery blood flow. Pelvic blood flow distribution to the uterine artery was also diminished at high versus low altitude. Pre-eclamptic women at high altitude showed even less redistribution of pelvic blood flow to the uterine circulation and effectively no increase in uterine artery flow velocity near term. The reduction in uterine artery blood flow at high altitude in Colorado women was consistent with the observed decrease in birth weight [54]. In Tibetans, birth weight is relatively preserved at high altitude compared to Han pregnancies at the same elevation, yet Tibetan women had lower hemoglobin and lower arterial oxygen content likely due to greater plasma volume expansion. However, in Tibetans, more pelvic (common iliac) blood flow was directed toward the uterine artery, and the increase in uterine blood flow correlated with heavier birth weights [55]. A similar pattern of augmented uterine artery blood flow and oxygen delivery appears to protect Andeans from altitude-associated reductions in fetal growth [56, 57]. Uterine artery blood flow was twofold higher in Andean than European women at high altitude; these differences corresponded to greater fetal size among Andeans and a birth weight differential of 253 g after controlling for gestational age, maternal height, and parity. Study of angiogenic factors during pregnancy has shown that Andeans compared to Europeans at high altitude had lower sFlt-1 and sFlt-1/PlGF ratio during pregnancy, and these values corresponded to higher uterine artery blood flow and birth weight [58]. Among women residing at 1,600 and 3,100 m in Colorado, changes in uterine artery flow and the ratio of endothelin-1 to nitric oxide (ET-1/NOx) preceded changes in fetal growth and correlated with ultimate birth weight reduction [59]. Thus, an increase in maternal ventilation, plasma volume expansion, and increased blood flow to the uteroplacental circulation all play a role in maintaining oxygen delivery to support fetal growth.

Reduced nutrient transport to the fetus at high altitude may also contribute to growth restriction. Maternal plasma glucose levels during pregnancy, as well as in the nonpregnant state, averaged approximately 10 % lower in Peruvian women at 4,300 m as compared to those at 300 m [60]. Lower glucose levels may result from higher glucose utilization as suggested by an indirect measure of insulin sensitivity in near-term pregnant women at high and low altitude [61]. Small maternal body size and anthropometric indices of undernutrition are not strongly related to birth weights in a given population or between populations [23]. However, differences in nutrient transport and energy balance have been proposed as playing a role in fetal growth restriction at high altitude [32, 62]. Glucose transporters showed a 40 % reduction in the fetal syncytial basal membranes of placentas at high altitude compared to low [63]. With identification of ancestry using single nucleotide polymorphisms, admixture was not related directly to oxygen transport, but placental weight explained significant variation in birth weight. Decreased placental nutrient transport, through reduced placental blood flow, placental metabolism, or reduced transporter density, may influence fetal growth at high altitude [45, 64].

Placental Development

Impaired placentation and reduced uteroplacental blood flow represent a common pathway through which IUGR and pregnancy complications can occur at high altitude. Reduced uteroplacental blood flow has been documented in pregnancies affected by IUGR and pre-eclampsia at sea level [65, 66]. Early in pregnancy, normal embryonic and placental development occurs in an environment of relatively low oxygen tension. At the end of the first trimester, when the intervillous space enlarges to facilitate circulation of maternal blood, pO_2 increases. This increase correlates with maximal invasion of trophoblast cells, specialized epithelial cells of the placenta, into the maternal decidua to access and remodel the spiral arteries. Trophoblast differentiation and invasion are now recognized to be oxygen-regulated events mediated by HIF-1alpha and HIF-2alpha, in part mediated by TGFbeta(3), an inhibitor of trophoblast differentiation [67]. When trophoblast invasion is impaired, placentation is shallower, maternal spiral arteries retain more of their muscular walls, and they preserve their contractile sensitivity which can reduce uterine blood flow [68]. Under hypoxic conditions trophoblast stem cells fail to transition to an invasive phenotype and rapidly invade extracellular matrix [69]. Overexpression of HIF-1 has been shown in placentas from high altitude compared to sea level [70]. Culture of trophoblast cells under a gradient of oxygen concentrations demonstrated that low oxygen (3 %) inhibited trophoblast differentiation. Increased HIF-1alpha protein levels and activity correlated with the inhibition of trophoblast differentiation in low oxygen. In a murine model, constitutive expression of an oxygen-insensitive, active form of HIF-1alpha protein mimicked the effects of hypoxia and inhibited trophoblast differentiation [71]. HIF-1alpha mRNA and protein expression have been shown to be abnormally elevated in pre-eclamptic placental tissue compared to controls and TGFbeta(3) is overexpressed in pre-eclamptic pregnancy [67]. Mice lacking HIF-1alpha, HIF-2alpha, and the beta subunit of the HIF dimer (ARNT) showed abnormal placental morphogenesis, angiogenesis, and cell fate decisions, demonstrating that oxygen tension through HIF mediation serves as a critical regulator of trophoblast differentiation [72].

Pre-eclampsia

Pre-eclampsia is a leading cause of maternal and fetal mortality in both the industrialized and the developing world, and a contributing cause to fetal growth retardation at high altitude. Defined clinically as hypertension and proteinuria in an otherwise normotensive woman, its incidence is increased three- to fourfold at high compared to low altitudes (16 % and 3 % respectively) in Colorado [65, 66, 73, 74]. Where urine testing for protein is not routinely available, pre-eclampsia and hypertension are reported together. A Bolivian study reported a 17.6 % incidence of pre-eclampsia/gestational hypertension at 3,600 m as compared to 11.4 % at 300 m (OR

1.7, 95 % CI: 1.3–2.2) [73]. Studies of iliac and uterine artery blood flow in normotensive and pre-eclamptic/gestational hypertensive Andean women showed elevated end-arteriolar vascular resistance, rather than narrowed uterine artery diameter, was responsible for decreased uterine artery flow in those with hypertension/pre-eclampsia. The degree of uterine artery flow reduction correlated with disease severity and adverse fetal outcomes [75]. In a population-based study in Lhasa, Tibet at 3,560 m the incidence of pre-eclampsia/gestational hypertension was nearly double (5.9 % vs 10.3 %) among Han Chinese women compared to Tibetan women (OR 1.85, 95 % CI: 1.01, 3.33). The Han women who experienced pre-eclampsia/gestational hypertension had twice the odds of a poor neonatal outcome and more than four times the odds of having a low-birth-weight infant [76]. In three large hospitals in Lhasa, pre-eclampsia/gestational hypertension was the most common maternal complication, occurring in 18.9 % of births [77].

Placental morphologic and molecular changes in pre-eclampsia reflect the response to chronic high-altitude hypoxia. Placentas from pre-eclamptic women (at high and low altitude) are characterized by shallow trophoblast invasion and incomplete remodeling of maternal uterine vessels. Comparison of placental tissue from high-altitude pregnancies, in vitro organ culture models under 3 % oxygen, and pregnancies complicated by pre-eclampsia showed highly similar global gene expression profiles [78]. Both high-altitude and pre-eclamptic placental tissue showed increased expression of soluble vascular endothelial growth factor receptor-1 (sFlt-1), with mediation of the effect via HIF-1 [79].

Other Pregnancy Complications

A historical review of all deliveries in La Oroya, Peru (3,750 m) over a 15-year period documented that placental abruptions occurred three times more frequently than at sea level; the incidence increased in older women (6.8 % among women over 40 years) and with higher parity (3.4 % with parity greater than 4) [80]. Certain pregnancy complications, such as placental abruption and pre-eclampsia, may result in preterm delivery, either spontaneous or by medical intervention; however the marked disparities in mean birthweight observed between high and low altitudes or between population groups result from IUGR rather than prematurity. Recent studies confirm that mean gestational age at high altitude is reduced by only a few days at most; the mean gestational age of Tibetan and Han Chinese newborns at 3,650 m was 39.4 and 39.2 weeks respectively [76], consistent with observations in Colorado and South America [46, 48, 81].

Maternal Mortality

Maternal mortality is high in many countries where a substantial proportion of the population lives at high altitude [82]. While these statistics reflect availability, access, and quality of health care as much as underlying risk, there are threats to maternal survival that are potentiated by high altitude. The increased incidence of pre-eclampsia carries with it the potential for life-threatening complications of eclampsia and cerebral hemorrhage. Severe forms of pre-eclampsia, such as the HELLP syndrome (Hemolysis, Elevated Liver enzymes, Low Platelets) can lead to irreversible hepatic injury or postpartum hemorrhage [83]. In the Tibet Autonomous Region, postpartum hemorrhage is the leading cause of maternal mortality [84]. In that region, up to 85 % of deliveries occur at home without the possibility of prompt transfusion; large-volume blood loss paired with the dilutional anemia of pregnancy is likely made even more lethal by hypobaric hypoxia.

Reproduction is influenced by the high-altitude environment through sociocultural and physiological pathways. While there is no direct evidence for an effect of hypoxia on fertility, a variety of social, cultural, and religious practice contribute to the range of rates observed. Rates of fetal loss for populations resident at high altitude appear comparable to those at sea level. Fetal growth and pregnancy complications provide the strongest evidence for a direct effect of hypoxia at high altitude. Placental development appears to be regulated at least in part by oxygen-sensing mechanisms, and fetal growth reflects the interplay of genetics, oxygen delivery, and nutrient supply of the fetus. Pre-eclampsia and hemorrhage, major causes of maternal mortality in any setting, pose special hazards at high altitude.

Late Fetal, Neonatal, and Infant Health and Mortality

Maternal genetic, constitutive, obstetrical and medical factors continue to act as strong determinants of late fetal, neonatal, and to some extent, infant health and mortality. Estimates of mortality during the perinatal period, infancy, and childhood are elevated in many developing countries and countries in transition where a significant proportion of the population lives at high altitude. Table 18.1 summarizes mortality statistics for several countries with an estimated >20 % of the population residing at >2,500 m [85]. Although rates do vary regionally, and determinants such as poverty, women's status, armed conflict, and political commitment to health account for much of the variation, the fact remains that rates for the major indicators of child and maternal mortality are elevated. Within countries, mortality has been demonstrated to rise with increasing elevation. Data from Peru and Bolivia show this trend for neonatal, infant, and under-five mortality [86–89]. Respiratory problems account for the majority of neonatal and infant deaths in this region [90, 91]. Consistent with the theory of developmental origins of adult disease, exposure to hypoxia during the fetal and neonatal periods may have consequences that extend into adult life and span generations [92].

Stillbirths

Stillbirths, representing a combination of late fetal demises and intrapartum hypoxia-related deaths, likely result from a different set of factors than those mediating earlier fetal loss. In highly developed medical environments, intensive surveillance of fetal growth and indices of well-being with the availability of early delivery and neonatal intensive care have largely erased altitude differentials with respect to rates of stillbirth. In Bolivia, hypertensive complications of pregnancy increased the odds of stillbirth at high altitude (OR 6.0, 95 % CI: 2.2, 16.2) but not at

Table 18.1 Indicators of mortality in countries with estimated 20 % or more of population residing above 2,500 m and comparison indicators from predominantly low-altitude populations

Country	Under-5 mortality rate[a]	Percent of under-5 deaths due to pneumonia	Infant mortality rate[b]	Neonatal mortality rate[c]	Maternal mortality ratio[d]
Bolivia	65	17	52	27	420
Colombia	21	10	17	14	130
Peru	27	14	23	16	410
Bhutan	75	19	65	38	420
Kazakstan	73	17	63	32	210
Nepal	74	19	56	40	740
Tajikistan	71	20	59	38	100
Ethiopia	164	22	109	51	850
Singapore	3	9	3	1	30
United States	7	1	6	5	17
United Kingdom	6	2	5	4	13

Sources: Columns 1, 3–5 from UNICEF. State of the World's Children, 2007 and column 2 from UNICEF/WHO. Pneumonia: The Forgotten Killer of Children

[a]Probability of dying between birth and exactly 5 years of age expressed per 1,000 live births, for year 2005

[b]Probability of dying between birth and exactly 1 year of age expressed per 1,000 live births, for year 2005

[c]Probability of dying during the first 28 completed days of life expressed per 1,000 live births, for year 2000

[d]Probability of dying during or within 42 days of termination of pregnancy expressed per 100,000 live births, for year 2000

low altitude (OR 1.9, 95 % CI: 0.2, 17.5). The odds of fetal distress in labor were similarly increased (OR 7.3, 95 % CI: 3.9, 13.6). Populations living for multiple generations at high altitude in Peru had lower stillbirth rates than those with fewer generations at altitude [93]. In Lhasa, Tibet, the stillbirth rate was 0.8–1.4 % in Tibetan and Han Chinese delivering in large urban hospitals [77], while in Leh, Ladakh the stillbirth rate for facility deliveries was 4.5 % in a population with less antiquity at high altitude and less available medical care [30].

Neonatal Mortality

Neonatal mortality (deaths in the first 28 days) accounts for a very high proportion of total infant mortality (deaths in the first year) at high altitude. Globally, neonatal deaths account for approximately 50 % of infant deaths; a community-based ecological study in Ladakh documented an infant mortality rate near 20 % by retrospective structured interview methods, with 75–84 % of deaths in the first month [30]. Such a pattern of very high mortality, rather than reflecting prematurity—as in developed settings, relates to fetal growth restriction, low birth weight, and respiratory distress. Birth weight was a strong predictor of neonatal mortality. Low body mass left neonates more vulnerable to cold and decreased stores of fat and carbohydrate may have resulted in hypoglycemia, especially among stressed infants. Both of those pathways potentially led to respiratory distress. The high proportion of respiratory infections, diarrhea, and other infections contributing to infant mortality may have related to the relatively compromised immune defenses that accompany growth restriction or stunting at birth [30]. A prospective study of hospital deliveries in Lhasa, Tibet (1939 Tibetans, 511, Han, 84 Hui) demonstrated an overall neonatal mortality rate of 42/1,000 with 22.2 % classified as small for gestational age. Lower gestational age, vaginal delivery, fetal distress, and lack of prenatal care were associated with increased risk of death in multivariate logistic regression. Forty-five percent of newborns admitted to the pediatric ward required supplemental oxygen for respiratory problems related to fetal distress, prematurity, or low birth weight [94]. In Bolivia, newborn respiratory distress was increased at high compared to low altitude (OR 7.3, 95 % CI: 3.9, 13.6 $p<0.01$) [73]. Although respiratory causes play an important role in neonatal mortality at any altitude, the hypoxemia of high altitude potentiates mortality, especially where supplemental oxygen is not available for treatment. Household conditions in the Andes and other resource-limited highland regions often mirror those in the Himalaya, with household heating employing dung, coal, kerosene or wood, purposefully limited ventilation to conserve heat, and high levels of indoor air pollution with carbon monoxide and particulates [30]. Not only does postnatal exposure to indoor air pollution increase respiratory problems and potentially worsen hypoxemia through CO binding of hemoglobin, exposure during pregnancy also has been associated with lower birthweight [95, 96].

Beyond Infancy: Growth and Development in Childhood and Implications for Adult Health

Growth and Nutritional Status

Whether at sea level or at high altitude, growth serves as an overall indicator of child health. Although stunting or slow growth are often reported among high-altitude populations, recent studies from South America suggest that slower linear growth relates more to socioeconomic status and the availability of adequate nutrition than to hypoxia [97–99]. A positive secular trend in height, suggesting better growth with improved living conditions, supports this [100, 101]. Hypoxia may work indirectly to influence growth through blunting of appetite, symptomatic altitude-associated illness (link to cardiopulmonary chapter), more severe consequences of acute respiratory infection, and high energy costs of activity, as all these circumstances may decrease nutritional intake and increase energetic demands, resulting in poor growth [102, 103].

Deficiency of both macronutrients and micronutrients may result in malnutrition among children at high altitude. Where chronic food scarcity exists due to limitations on agricultural production or poverty, protein-energy malnutrition follows after fetal growth restriction, producing stunting from an early age. A community survey of Tibetan children demonstrated consistently better height-for-age among urban than rural children, but no direct association of stunting with altitude [98]. The high-altitude ecology uniquely predisposes to micronutrient deficiencies in vitamins A and D, iron, and iodine [104]. Vitamin A deficiency is most common where fruit and vegetable consumption is low; it predisposes young children to respiratory and gastrointestinal illness, blindness, and death. Iron deficiency affects children and women of childbearing age worldwide. At high altitude where erythropoietic demands are high and oxygen content of blood is reduced, anemia poses a risk for impaired neurodevelopment of children [105]. Evaluation of iron deficiency at high altitude requires correction of hemoglobin values or measurement of body iron stores. In Bolivia, body iron stores of mothers and their children less than 5 years correlated strongly, and women living above 3,000 m had significantly reduced iron stores [106]. Iodine deficiency is especially prevalent in the Himalaya, where iodine salts have washed away from soils and iodine-rich or supplemented foods have limited distribution. Iodine deficiency causes mental retardation and low intelligence quotient; severe deficiency can result in cretinism, birth defects, and stillbirth [104, 107]. Where cold climates require that infants and young children be kept inside or completely covered when outside, rickets (vitamin D deficiency) may be prevalent. Among Tibetan children in a community survey, 66 % had clinical signs of rickets, and vitamin D levels in a subsample confirmed deficiency [98]. While malnutrition in childhood produces well-recognized lifelong consequences, these effects may be compounded at high altitude by the late effects of fetal growth restriction.

Developmental Origins of Adult Disease

Under the hypothesis of developmental origins of adult disease, impaired fetal growth may predispose to coronary heart disease, hypertension, diabetes, and other illnesses later in life. David Barker and colleagues observed that districts in England with the highest mortality from cardiovascular disease in 55–74-year olds also recorded the highest neonatal mortality 55–74 years prior [108]. Low birth weight proved to be a mediating link, and multiple subsequent large-scale surveys conducted in England (Hertfordshire, Sheffield), Sweden, Finland, Wales, USA and India, showed mortality rates from coronary heart disease nearly doubled when comparing low (<5.5 lb or 2,500 g) to normal birth weights, with a slight upward trend reappearing at high birth weights (>9.5 lb or 4,300 g) [109, 110]. The relationships persisted after controlling for a range of lifestyle factors (smoking, exercise, employment, and alcohol consumption), suggesting that impaired growth in utero contributed to the mortality rise observed [92]. Subsequent studies using both human data and experimental animal preparations have shown that poor growth in utero not only affects the cardiovascular system but also the renal and endocrine systems, mediating the risk of hypertension and diabetes. Lower birth weight has been shown to predispose persons to other disorders including obesity, osteoporosis, schizophrenia, depression, cancers of the breast and ovary, and polycystic ovary syndrome [92]. While most studies have addressed the consequences of lower birth weight due to poor maternal nutrition, similar relationships have been observed when the birth weight reduction was due to other causes; however, no large-scale studies have tested the developmental origins hypothesis at high altitude.

The consequences of reduced birth weight at high altitude may be more important for the right side of the circulation than for the left side, which is where virtually all "fetal programming" studies to date have been performed [111]. Such "right-side" effects could influence the development of the lung and its circulation at the levels of the airways, vasculature and/or respiratory control. Experimental animal studies have shown that chronic hypoxia affects airway structure; rat pups whose mothers were exposed to 10 % F_IO_2 for 9 h on the last day of gestation and for 1–2 h after birth had a delayed increase in lung volume, impaired septation of gas exchange saccules, blunted expansion of gas exchange surface area, and accelerated thinning of the alveolar walls [112]. Consistent with chronic hypoxia affecting lung structure in humans are the observations that healthy newborns in La Paz (3,600 m) had 33–37 % greater pulmonary compliance, absolutely or on per kg, than babies born at low altitude (300 m) [113]. Restricted diet during pregnancy in mice has been shown to result in pulmonary vascular dysfunction in the offspring (impaired endothelium-dependent pulmonary artery vasodilation, exaggerated hypoxia-induced pulmonary hypertension, and right ventricular hypertrophy) related to epigenetic change [114]. Similarly, exposure of pregnant ewes to high-altitude chronic hypoxia during gestation resulted in lambs with basal pulmonary hypertension and an increased pulmonary vascular response to acute hypoxia at sea level, despite evidence of enhanced pulmonary NO function [115]. In relation to respiratory control, chronic perinatal hypoxia delays the onset and decreases the ventilatory sensitivity to hypoxia at maturity [116–118]. Maturation of the chemoreceptor pathway also differs by gender, with prepubertal female rats having higher HVRs than males [119]. Sex hormones may be involved since progesterone enhances HVR and reduces the occurrence of apneas in rat pups, and prenatal estradiol blockade blunts HVR during the neonatal period [120]. These latter observations clearly have implications for explaining the blunted HVR and the male preponderance seen in CMS (see Chap. 22).

Low birth weight at altitude continues to mediate mortality throughout the neonatal period and into infancy. Genetic adaptation and complications of pregnancy, as well as available medical facilities, strongly influence the rate of stillbirth. Again, cultural and socioeconomic factors influencing indoor air pollution and diet shape the trajectory of health and growth during infancy and childhood. Exposure to hypoxia during fetal life may result in specific alterations during critical periods in the development of the respiratory system that predispose to chronic mountain sickness in adult life. More generally, the link between low birth weight and other chronic diseases of adult life, such as coronary artery disease, hypertension, and diabetes, await further investigation.

The continuum of reproduction, perinatal health, and growth at high altitude presents opportunities for significant contributions from both clinical and basic science research. Increasingly accessible markers of conception and pregnancy health used with prospective methodology will permit greater understanding of early fetal loss, placental development, and complications of pregnancy. The high-altitude environment coupled with the tools of molecular biology provides an ideal opportunity to understand the pathogenesis of pre-eclampsia. There is a crucial need to address the management of pre-eclampsia and postpartum hemorrhage within medical systems at high altitude if high rates of maternal mortality are to be reduced. Similarly, very high rates of stillbirth and neonatal mortality reflect the need for greater emphasis on institutional delivery and availability of appropriate medical care, including neonatal resuscitation and supplemental oxygen at high altitude. Even if short-term morbidity and mortality associated with low birth weight can be modified, the long-term consequences for adult health warrant further research.

References

1. Vitzthum VJ, Wiley AS. The proximate determinants of fertility in populations exposed to chronic hypoxia. High Alt Med Biol. 2003;4:125–39.
2. Monge C. Acclimatization in the Andes. Baltimore: Johns Hopkins University Press; 1948.
3. Abelson AE. Altitude and fertility. Hum Biol. 1976; 48:83–92.
4. Cruz-Coke R, Crisoffanini A, Aspillaga M, Biancani F. Evolutionary forces in human populations in an environmental gradient in Arica, Chile. Hum Biol. 1966;4:421–38.
5. Gonzales GF. Peruvian contributions to the study on human reproduction at high altitude: from the chronicles of the Spanish conquest to the present. Respir Physiol Neurobiol. 2007;158:172–9.

6. Goldstein MC, Tsarong P, Beall CM. High altitude hypoxia, culture and human fecundity/fertility: a comparative study. Am Anthropol. 1983;85:28–49.
7. Gonzales GF. Determinantes biomedicos de la fertilidad humana en la altura. In: Gonzales GF, editor. Reproducíón Humana en la Altura. Lima, Peru: Consejo Nacional de Ciencia y Tecnología; 1993. p. 73–87.
8. Weitz CA, Pawson IG, Weitz MV, Lang SDR, Lang A. Cultural factors affecting the demographic structure of a high altitude Nepalese population. Soc Biol. 1978;25:179–95.
9. Vitzthum VJ. The home team advantage: reproduction in women indigenous to high altitude. J Exp Biol. 2001;204:3141–50.
10. Beall CM. Ages at menopause and menarche in a high-altitude Himalayan population. Ann Hum Biol. 1983;10:365–70.
11. Gonzales GF, Villena A, Ubilluz M. Age at menarche in Peruvian girls at sea level and at high altitude: effect of ethnic background and socioeconomic status. Am J Hum Biol. 1996;8:457–63.
12. Gonzales GF, Villena A. Age at menopause in central Andean Peruvian women. Menopause. 1997; 4:32–8.
13. Vitzthum VJ, Spielvogel H, Caceres E, Gaines J. Menstrual patterns and fecundity in non-lactating and lactating cycling women in rural highland Bolivia: implications for contraceptive choice. Contraception. 2000;62:181–7.
14. Vitzthum VJ, Bentley GR, Spielvogel H, et al. Salivary progesterone levels and rate of ovulation are significantly lower in poorer than in better-off urban-dwelling Bolivian women. Hum Reprod. 2002;17:1906–13.
15. Vitzthum VJ. Why not so great is still good enough: flexible responsiveness in human reproductive functioning. In: Ellison PT, editor. Reproductive ecology and human evolution. New York: Aldine de Gryter; 2001. p. 179–202.
16. Vitzthum VJ, Spielvogel H, Thornburg J. Interpopulational differences in progesterone levels during conception and implantation in humans. Proc Natl Acad Sci U S A. 2004;101:1443–8.
17. Vitzthum VJ, Spielvogel H, Thornburg J, West B. A prospective study of early pregnancy loss in humans. Fertil Steril. 2006;86:373–9.
18. Sobrevilla LA, Romero I, Moncloa F, Donayre J, Guerra-Garcia R. Endocrine studies at high altitude. III. Urinary gonadotrophins in subjects native to and living at a 14,000 feet and during acute exposure of men living at sea level to high altitudes. Acta Endocrinol. 1967;56:369–75.
19. Garcia-Hjarles MA. Espermatograma y bioquímica seminal de nativos de altura y pacientes con mal de montaña crónica. Arch Biol Med Exp. 1989;22:61–7.
20. Beall CM, Worthman CM, Stallings J, Strohl KP, Brittenham GM, Barragan M. Salivary testosterone concentration of Aymara men native to 3600 m. Ann Hum Biol. 1992;19:67–78.
21. Stallings JF, Vitzthum VJ, Worthman CM. Ecological correlates of diurnal variation in gonadal and adrenal activity in rural Bolivian Aymara men. J Phys Anthropol Suppl. 2000;30:289.
22. Lang SDR, Lang A. The Kunde Hospital and a demographic survey of the Upper Khumbu, Nepal. N Z Med J. 1971;74:1–8.
23. Moore LG, Young D, McCullough RE, Droma TS, Zamudio S. Tibetan protection from intrauterine growth restriction at high altitude. Am J Hum Biol. 2001;13:635–44.
24. Goldstein MC. Fraternal polyandry and fertility in a high Himalayan valley in northwest Nepal. Hum Ecol. 1976;4:223–33.
25. Goldstein MC. New perspectives on Tibetan fertility and population decline. Am Ethnol. 1981;8:721–38.
26. Ross JL. Culture and fertility in the Nepal Himalayas: a test of a hypothesis. Hum Ecol. 1984;12:163–81.
27. Wiley AS. The ecology of low natural fertility in Ladakh. J Biosoc Sci. 1998;30:457–80.
28. Levine NE. Differential child care in three Tibetan communities: beyond son preference. Popul Dev Rev. 1987;13:281–304.
29. Laurenson IF, Benton MA, Bishop AJ, Mascie-Taylor CGN. Fertility at low and high altitude in central Nepal. Soc Biol. 1985;32:65–70.
30. Wiley AS. An ecology of high-altitude infancy. Cambridge: Cambridge University Press; 2004.
31. Thomas RB. Energy flow at high altitude. In: Baker PT, Little MA, editors. Man in the Andes a multidisciplinary study of high altitude Quechua. Stroudsburg, PA: Dowden, Hutchinson and Ross, Inc.; 1976. p. 379–404.
32. Moore LG. Fetal growth restriction and maternal oxygen transport during high-altitude pregnancy. High Alt Med Biol. 2003;4:141–56.
33. Yip R. Altitude and birth weight. J Pediatr. 1987; 111:869–76.
34. Lichty JA, Ting RY, Bruns PD, Dyar E. Studies of babies born at high altitude. I. Relation of altitude to birth weight. AMA J Dis Child. 1957;93:666–9.
35. Lubchenco LO. Intrauterine growth as estimated from liveborn birth-weight data at 24 to 42 weeks of gestation. Pediatrics. 1963;32:793–800.
36. Galan HL, Rigano S, Radaelli T, et al. Reduction of subcutaneous mass, but not lean mass, in normal fetuses in Denver, Colorado. Am J Obstet Gynecol. 2001;185:839–44.
37. Gonzales GF, Tapia V. Birth weight charts for gestational age in 63,620 healthy infants born in Peruvian public hospitals at low and at high altitude. Acta Paediatr. 2009;98:454–8.
38. Zhoma ZX, Sun SF, Zhang JG, Huang SY, Moore LG. Fetal growth and maternal oxygen supply in Tibetan and Han residents of Lhasa (3658 m). FASEB J. 1989;3:A987.
39. Niermeyer S, Yang P, Shanmina, Drolkar, Zhuang J, Moore LG. Arterial oxygen saturation in Tibetan and Han infants born in Lhasa, Tibet. N Engl J Med. 1995;333:1248–52.

40. Haas LD, Frongillo EJ, Stepcik C, Beard J, Hurtado L. Altitude, ethnic and sex differences in birth weight and length in Bolivia. Hum Biol. 1980;52: 459–77.
41. Bennett A, Sain SR, Vargas E, Moore LG. Evidence that parent-of-origin affects birth-weight reductions at high altitude. Am J Hum Biol. 2008;20:592–7.
42. Smith CF. Comparative birthweight data for Sherpa women living at high and low altitudes in Nepal. Am J Phys Anthropol. 1993;16(Suppl):183.
43. Wiley AS. Neonatal size and infant mortality at high altitude in the western Himalaya. Am J Phys Anthropol. 1994;94:289–305.
44. Beall CM. Two routes to functional adaptation: Tibetan and Andean high-altitude natives. PNAS. 2007;104:8655–60.
45. Zamudio S, Postigo L, Illsley NP, et al. Maternal oxygen delivery is not related to altitude- and ancestry-associated differences in human fetal growth. J Physiol. 2007;582:883–95.
46. Jensen GM, Moore LG. The effect of high altitude and other risk factors on birthweight: independent or interactive effects? Am J Public Health. 1997;87: 1003–7.
47. Giussani DA, Phillips S, Anstee S, Barker DJP. Effects of altitude versus economic status on birth weight and body shape at birth. Pediatr Res. 2001;49:490–4.
48. Mortola JP, Frappell PB, Aguero L, Armstrong K. Birth weight and altitude: a study in Peruvian communities. J Pediatr. 2000;136:324–9.
49. McAuliffe F, Kametas N, Krampl E, Ernsting J, Nicolaides K. Blood gases in pregnancy at sea level and at high altitude. Br J Obstet Gynaecol. 2001;108: 980–5.
50. Moore LG. Human genetic adaptation to high altitude. High Alt Med Biol. 2001;2:257–79.
51. Moore LG, Zamudio S, Zhuang J, Sun S, Droma T. Oxygen transport in Tibetan women during pregnancy at 3658 m. Am J Phys Anthropol. 2001;114:42–53.
52. Nahum GG, Stanislaw H. Hemoglobin, altitude and birth weight: does maternal anemia during pregnancy influence fetal growth? J Reprod Med. 2004; 49:297–305.
53. Palmer SK, Zamudio S, Coffin C, Parker S, Stamm E, Moore LG. Quantitative estimation of human uterine artery blood flow and pelvic blood flow redistribution in pregnancy [review]. Obstet Gynecol. 1992;80:1000–6.
54. Zamudio S, Palmer SK, Droma T, Stamm E, Coffin C, Moore LG. Effect of altitude on uterine artery blood flow during normal pregnancy. J Appl Physiol. 1995;79:7–14.
55. Moore LG. Maternal O2 transport and fetal growth in Colorado, Peru and Tibet high-altitude residents. Am J Hum Biol. 1990;2:627–37.
56. Julian CG, Wilson MJ, Lopez M, et al. Augmented uterine artery blood flow and oxygen delivery protect Andeans from altitude-associated reductions in fetal growth. Am J Physiol Regul Integr Comp Physiol. 2009;296:R1564–75.
57. Vargas M, Vargas E, Julian CG, et al. Determinants of blood oxygenation during pregnancy in Andean and European residents of high altitude. Am J Physiol Regul Integr Comp Physiol. 2007;293: R1303–12.
58. Davila RD, Julian CG, Wilson MJ, et al. Do anti-angiogenic or angiogenic factors contribute to the protection of birth weight at high altitude afforded by Andean ancestry? Reprod Sci. 2010;17:861.
59. Julian CG, Galan HL, Wilson MJ, DeSilva W, Cioffi-Ragan D, Schwartz J, Moore LG. Lower uterine artery blood flow and higher endothelin relative to nitric oxide metabolite levels are associated with reductions in birth weight at high altitude. Am J Physiol Regul Integr Comp Physiol. 2008;295:R906–15.
60. Krampl E, Kametas NA, Cacho-Zegarra AM, Roden M, Nicolaides KH. Maternal plasma glucose at high altitude. Br J Obstet Gynaecol. 2001;108:254–7.
61. Krampl E, Kametas NA, McAuliffe F, Cacho-Zegarra AM, Nicolaides KH. Maternal serum insulin-like growth factor binding protein-1 in pregnancy at high altitude. Obstet Gynecol. 2002;99: 594–8.
62. Zamudio S. The placenta at high altitude. High Alt Med Biol. 2003;4:171–91.
63. Zamudio S, Baumann ME, Illsley NP. Effects of chronic hypoxia in vivo on the expression of human placental glucose transporters. Placenta. 2006;27: 49–55.
64. Zamudio S, Torricos T, Fik E, et al. Hypoglycemia and the origin of hypoxia-induced reduction in human fetal growth. PLoS One. 2010;5:e8551.
65. Lunell NO, Lewander R, Mamoun I, Nylund L, Sarby S, Thornstrom S. Uteroplacental blood flow in pregnancy induced hypertension. Scan J Clin Lab Invest Suppl. 1984;169:28–35.
66. Lunell NO, Sarby B, Lewander R, Nylund L. Comparison of uteroplacental blood flow in normal and in intrauterine growth-retarded pregnancy. Measurements with Indium-113m and a computer-linked gammacamera. Gynecol Obstet Invest. 1979;10: 106–18.
67. Caniggia I, Winter JL. Adriana and Luisa Castellucci Award lecture 2001. Hypoxia inducible factor-1: oxygen regulation of trophoblast differentiation in normal and pre-eclamptic pregnancies—a review. Placenta. 2002;23(Suppl A):S47–57.
68. Tissot van Patot M, Grilli A, Chapman P, et al. Remodelling of uteroplacental arteries is decreased in high altitude placentae. Placenta. 2003;24: 326–35.
69. Genbacev O, Joslin R, Damsky C, Poliotti B, Fisher S. Hypoxia alters early gestation human cytotrophoblast differentiation/invasion in vitro and models the placental defects that occur in preeclampsia. J Clin Invest. 1996;97:540–50.
70. Zamudio S, Wu Y, Ietta F, et al. Human placental hypoxia-inducible factor-1α expression correlates

with clinical outcomes in chronic hypoxia in vivo. Am J Pathol. 2007;170:2171–9.
71. Gultice AD, Kulkarni-Datar K, Brown TL. Hypoxia-inducible factor 1alpha (HIF1A) mediates distinct steps of rat trophoblast differentiation in gradient oxygen. Biol Reprod. 2009;80:184–93.
72. Cowden Dahl KD, Fryer BH, Mack FA, et al. Hypoxia-inducible factors 1alpha and 2alpha regulate trophoblast differentiation. Mol Cell Biol. 2005;25:10479–91.
73. Keyes LE, Armaza JF, Niermeyer S, Vargas E, Young DA, Moore LG. Intrauterine growth restriction, preeclampsia and intrauterine mortality at high altitude in Bolivia. Pediatr Res. 2003;54:20–5.
74. Palmer SK, Moore LG, Young DA, Cregger B, Berman JC, Zamudio S. Altered blood pressure course during normal pregnancy and increased preeclampsia at high altitude (3100 meters) in Colorado. Am J Obstet Gynecol. 1999;180:1161–8.
75. Browne VA, Toledo-Jaldin L, Davila RD, et al. High end-arteriolar resistance limits uterine artery blood flow and restricts fetal growth in preeclampsia and gestational hypertension at high altitude. Am J Physiol Regul Integr Comp Physiol. 2011;300: R1221–9.
76. Miller S, Tudor C, Thorsten V, et al. Comparison of maternal and newborn outcomes of Tibetan and Han Chinese delivering in Lhasa, Tibet. J Obstet Gynaecol Res. 2008;34:986–93.
77. Miller S, Tudor C, Nyima, et al. Maternal and neonatal outcomes of hospital vaginal deliveries in Tibet. Int J Gynaecol Obstet. 2007;98:217–21.
78. Soleymanlou N, Jurisica I, Nevo O, et al. Molecular evidence of placental hypoxia in preeclampsia. J Clin Endocrinol Metab. 2005;90:4299–308.
79. Nevo O, Soleymanlou N, Wu Y, et al. Increased expression of sFlt-1 in vivo and in vitro models of human placental hypoxia is mediated by HIF-1. Am J Physiol Regul Integr Comp Physiol. 2006;291: R1085–93.
80. Briceño G, Derezin Quintana T. Evaluación del desprendimiento prematuro de placenta en la altura. Rev Méd Inst Peru Segur Soc. 1994;3:27–32.
81. López Camelo JS, Campaña H, Santos R, Poletta FA. Effect of the interaction between high altitude and socioeconomic factors on birth weight in a large sample from South America. Am J Phys Anthropol. 2006;129:305–10.
82. Niermeyer S, Andrade Mollinedo P, Huicho L. Children's health and high altitude living Arch Dis Child. 2008. doi:10.1136/adc.2008.141838.
83. Moodley J. Maternal deaths due to hypertensive disorders in pregnancy. Best Pract Res Clin Obstet Gynaecol. 2008;22:559–67.
84. Tudor C, Miller S, Nyima, Sonam, Droyoung, Varner M. Preliminary progress report: randomized double-blind trial of Shi Byed 11, a Tibetan traditional medicine, versus misoprostol to prevent postpartum hemorrhage in Lhasa, Tibet. Int J Gynecol Obstet. 2006;94:S145–6.
85. The United Nations Children's Fund (UNICEF). The State of the World's Children 2007. New York; 2006.
86. Edmonston B, Andes N. Community variations in infant and child mortality in Peru. J Epidemiol Community Health. 1983;37:121–6.
87. Pan American Health Organization. Health conditions in the Americas. Scientific publication No. 549. Washington, DC: World Health Organization; 1994.
88. Mazess RB. Neonatal mortality and altitude in Peru. Am J Phys Anthropol. 1965;23:209–14.
89. Huicho L, Trelles M, Gonzales F. National and subnational under-five mortality profiles in Peru: a basis for informed policy decisions. BMC Publ Health. 2006;6:173.
90. Spector RM. Mortality characteristics of a high-altitude Peruvian population [Master's Thesis]. Pennsylvania State University; 1971.
91. Beall CM. The effects of high altitude on growth, morbidity, and mortality of Peruvian infants [PhD Dissertation]. Pennsylvania State University; 1976.
92. Barker D. Fetal origins of cardiovascular and lung disease. In: Lenfant C, editor. Lung biology in health and disease. New York: Marcel Dekker; 2001.
93. Gonzales GF, Tapia V, Carrillo CE. Stillbirth rates in Peruvian populations at high altitude. Int J Gynaecol Obstet. 2008;100:221–7.
94. Yangzom Y, Qian L, Shan M, et al. Outcome of hospital deliveries of women living at high altitude: a study from Lhasa in Tibet. Acta Paediatr. 2008;97: 317–21.
95. Boy E, Bruce N, Delgado H. Birth weight and exposure to kitchen wood smoke during pregnancy in rural Guatemala. Environ Health Perspect. 2002;110: 109–14.
96. Smith KR, Samet JM, Romieu I, Bruce N. Indoor air pollution in developing countries and acute lower respiratory infections in children. Thorax. 2000; 55:518–32.
97. Greksa LP. Effect of altitude on the physical growth of upper-class children of European ancestry. Ann Hum Biol. 1985;12:225–32.
98. Harris NS, Crawford PB, Yangzom Y, Pinzo L, Gyaltsen P, Hudes M. Nutritional and health status of Tibetan children living at high altitudes. N Engl J Med. 2001;344:341–7.
99. Leonard WR. Nutritional determinants of high-altitude growth in Nunoa, Peru. Am J Phys Anthropol. 1989;80:341–52.
100. Pawson IG, Huicho L, Muro M, Pacheco A. Growth of children in two economically diverse Peruvian high-altitude communities. Am J Hum Biol. 2001; 13:323–40.
101. Dittmar M. Secular growth changes in the stature and weight of Amerindian schoolchildren and adults in the Chilean Andes, 1972–1987. Am J Hum Biol. 1998;10:607–17.

102. Bozzini CE, Lezon CE, Norese MF, et al. Evidence from catch-up growth and hoarding behavior of rats that exposure to hypobaric air lowers the body-mass set point. Growth Dev Aging. 2005;69:81–8.
103. Smith TA, Lehmann D, Coakley C, Spooner V, Alpers MP. Relationships between growth and acute lower-respiratory infections in children aged <5y in a highland population of Papua, New Guinea. Am J Clin Nutr. 1991;53:963–70.
104. Huddleston B, Ataman E, de Salvo P, et al. Towards a GIS-based analysis of mountain environments and populations. Rome: Food and Agriculture Organization of the United Nations; 2003.
105. Walker SP, Wachs TD, Gardner JM, et al. Child development: risk factors for adverse outcomes in developing countries. Lancet. 2007;369:145–57.
106. Cook JD, Boy E, Flowers C, Daroca M. The influence of high-altitude living on body iron. Blood. 2005;106:1441–6.
107. Pretell EA, Delange F, Hostalek U, et al. Iodine nutrition improves in Latin America. Thyroid. 2004;14:590–9.
108. Barker DJP, Osmond C. Infant mortality, childhood nutrition, and ischaemic heart disease in England and Wales. Lancet. 1986;1:1077–81.
109. Burke J, Forsgren J, Palumbo P, et al. Association of birth weight and type 2 diabetes in Rochester, Minnesota. Diabetes Care. 2004;27:2512–3.
110. Gluckman P, Hanson M. Living with the past: evolution, development, and patterns of disease. Science. 2004;305:1733–6.
111. Moore LG, Niermeyer S, Vargas E. Does chronic mountain sickness (CMS) have perinatal origins? Respir Physiol Neurobiol. 2007;158:180–9.
112. Massaro GD, Olivier J, Massaro D. Short-term perinatal 10% O2 alters postnatal development of lung alveoli. Am J Physiol. 1989;257:L221–5.
113. Mortola JP, Rezzonico R, Fisher JT, et al. Compliance of the respiratory system in infants born at high altitude. Am Rev Respir Dis. 1990;142:43–8.
114. Rexhaj E, Bloch J, Jayet PY, et al. Fetal programming of pulmonary vascular dysfunction in mice: role of epigenetic mechanisms. Am J Physiol Heart Circ Physiol. 2011;301:H247–52.
115. Herrera EA, Riquelme RA, Ebensperger G, et al. Long-term exposure to high-altitude chronic hypoxia during gestation induces neonatal pulmonary hypertension at sea level. Am J Physiol Regul Integr Comp Physiol. 2010;299:R1676–84.
116. Eden GJ, Hanson MA. Effects of chronic hypoxia from birth on the ventilatory response to acute hypoxia in the newborn rat. J Physiol Lond. 1987; 392:11–9.
117. Okubo S, Mortola JP. Control of ventilation in adult rats hypoxic in the neonatal period. Am J Physiol Regul Integr Comp Physiol. 1990;259:R836–41.
118. Joseph V, Soliz J, Pequignot J, et al. Gender differentiation of the chemoreflex during growth at high altitude: functional and neurochemical studies. Am J Physiol Regul Integr Comp Biol. 2000;278: R806–16.
119. Mortola JP, Saiki C. Ventilatory response to hypoxia in rats: gender differences. Respir Physiol. 1996;106: 21–34.
120. Doan VD, Gagnon S, Joseph V. Prenatal blockade of estradiol synthesis impairs respiratory and metabolic responses to hypoxia in newborn and adult rats. Am J Physiol Regul Integr Comp Physiol. 2004;287:R612–8.

Human Evolution at High Altitude

19

Cynthia M. Beall

Abstract

This chapter reviews evidence that natural selection is acting or has acted on indigenous high-altitude populations of the Andean, Tibetan and East African plateaus and resulted in distinctive biological characteristics conferring vigor and health. It describes the results of classic era and genomic era approaches to detecting natural selection. Genomic era evidence of natural selection on high-altitude populations is accumulating rapidly and broadly supports that from the classic era. An important remaining step is to associate phenotypic with genomic variation and to associate them with survival and reproduction, the demographic currency of natural selection.

Now, can it be doubted, from the struggle each individual has to obtain subsistence, that any minute variation in structure, habits, or instincts, adapting that individual better to the new conditions, would tell upon its vigour and health? In the struggle it would have a better chance of surviving; and those of its offspring which inherited the variation, be it ever so slight, would also have a better chance. ….. Let this work of selection on the one hand, and death on the other, go on for a thousand generations, who will pretend to affirm that it would produce no effect…? ([1] p. 49)

C.M. Beall, Ph.D. (✉)
Department of Anthropology, Case Western Reserve University, Cleveland, OH 44106-7125, USA
e-mail: cmb2@case.edu

Introduction

Individuals moving to high altitude gradually encounter the "new condition" of hypobaric hypoxia and their responses vary widely. Darwin would have predicted that descendants from such a population resident at high altitude for a "thousand generations" would differ in "structure, habits, or instincts" from their colonizing ancestors. Consistent with such expectations, the "Andean man" – Quechua and Aymara populations residing on the altiplano – described by Carlos Monge C. and Alberto Hurtado starting in the 1930s had distinctive biological characteristics of [2–4]. The hypothesis that those distinctive characteristics result from evolution by natural selection has been considered formally since the 1960s [5].

E.R. Swenson and P. Bärtsch (eds.), *High Altitude: Human Adaptation to Hypoxia*,
DOI 10.1007/978-1-4614-8772-2_19,

This chapter reviews evidence that natural selection is acting or has acted on indigenous high-altitude populations of the Andean, Tibetan and East African plateaus and resulted in distinctive biological characteristics conferring vigor and health. It examines briefly the state of knowledge up to the end of the last century reviewed by Niermeyer et al. in the predecessor to this volume [6]. They provided an insightful analysis of phenotypic evidence leading to the inference that natural selection has acted or is acting on high altitude populations. Because natural selection works on phenotypes with genetic variance, the rapid growth of genetic data and investigation since then has enabled new advances in our understanding. This chapter presents first the classic strategies—relying largely on phenotypes and protein sequences—and then the genomic strategies relying on DNA variation and sequences alone or in combination with phenotypes.

Background

Some elements of the process of evolution by natural selection are straightforward at high altitude. The unavoidable stress of hypobaric hypoxia is clear and so is the presence of distinctive traits such as the very high hemoglobin concentrations of Andean or exhaled nitric oxide levels of Tibetan highlanders [7, 8]. Other elements of the process are more challenging to detect: identifying inherited variation in traits offsetting high-altitude hypoxia and associating them with survival and reproduction. As a result, the intriguing hypothesis that high altitude hypoxia has been or is an agent of natural selection on highlanders remains the subject of intense work.

The laboratories for testing the natural selection hypothesis are the highland areas of the world populated by the more than 83 million people in 35 countries who live at 2,500 m or above, an altitude commonly used as a threshold for physiological response to hypobaric hypoxia. The estimate is based on census data from 1990 to 2005 matched to a global elevation grid (see Appendix). The majority of those people live between 2,500 and 3,000 m (Figs. 19.1 and 19.2). The relatively few residents of the highest altitudes where the altitude difference from sea level is largest and the stress is most severe are particularly informative for studies of natural selection.

Another important consideration is the length of time, and number of generations, that natural selection has had the opportunity to act on resident populations. Long-term human occupation of high-altitude environments varies globally. People have been using the Tibetan Plateau for a very long time. A recent authoritative review stated that "Although the data are sparse, both archaeology and genetics suggest that the plateau was occupied in the Late Pleistocene, perhaps as early as 30,000 yr ago, and that these early peoples have left a genetic signature in modern Tibetans. ….. Three areas of the plateau…have evidence of permanent settlements dating from ca. 6500, 5900, and 3750 yr ago, respectively" ([9] p. 141). The Andean Plateau has been occupied for some 11,500 years [10]. Maternal and paternal DNA link Andean skeletal remains from 650 to 1100 AD to modern highlanders [11]. The large, deep chest morphology of pre-contact skeletons also link ancient inhabitants to modern highlanders [12]. Less is known about the East African plateau where there is evidence of occupation at 2,300–2,400 m in Ethiopia as long ago as 70,000 years ago although only as recently as 5,000 years ago at 2,500 m or more [10, 13]. East Africa could have the longest or the shortest period of human adaptation of the three plateaus. There is insufficient evidence to reliably link these sites with any of the many modern ethnic groups in Ethiopia.

Considering a conservative model of 11,000 years or 440 human generations of 25 years each, a mutation occurring at a frequency of 1/10,000 or 0.0001 could reach a frequency of 1/2 or 0.5 if it increased in frequency by ~2 % each generation. Alternatively, if a founding population arrived with a pre-existing allele having a frequency of 1/100 or 0.01 and if the allele were beneficial at high altitude, it could reach a frequency of 0.5 if it increased by ~1.5 % each generation. These increases are in the range reported for the lactase persistence allele that reached a frequency of 77/100 or 0.77 among northern Europeans in 5–10,000 years [14–16]. Therefore, there has probably been sufficient length of residence on all three plateaus for a new mutation or

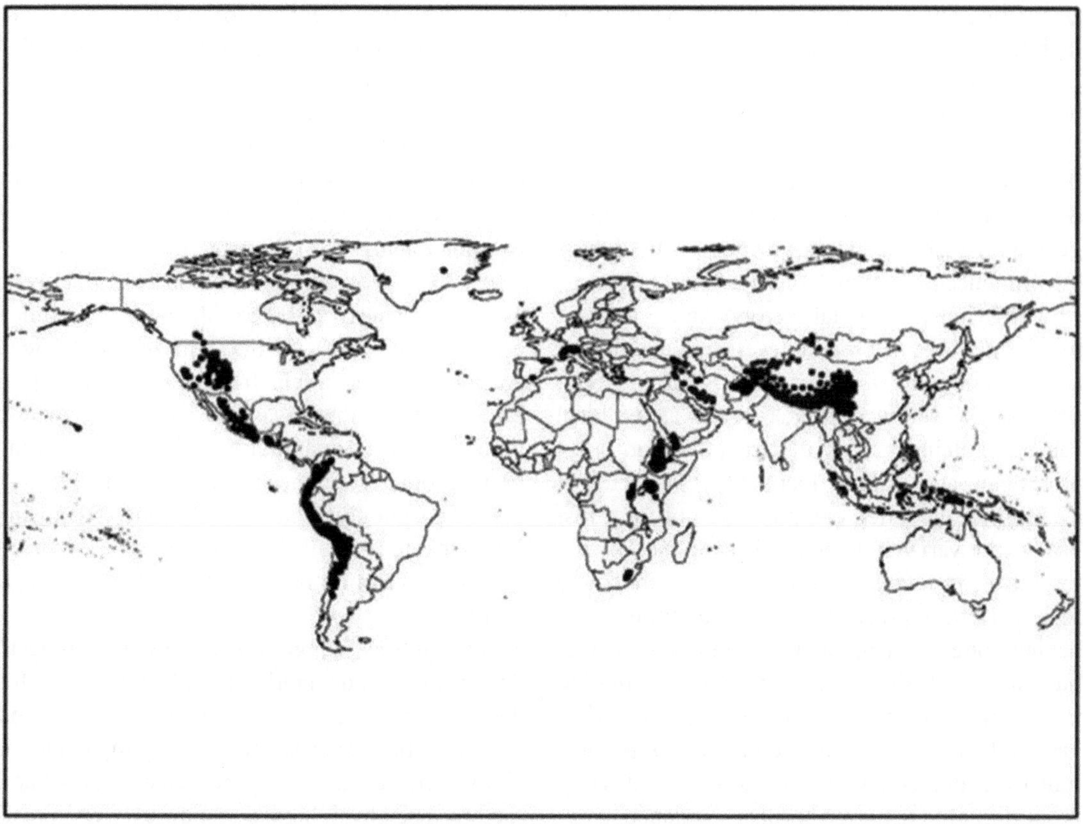

Fig. 19.1 Map of areas of the world where people live at 2,500 m or higher

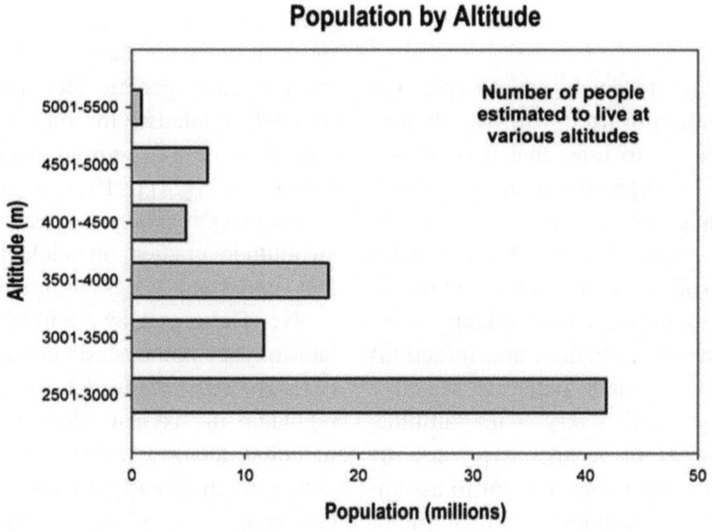

Fig. 19.2 The number of people estimated to live at various altitudes

an existing allele at a polymorphic locus to reach high frequencies in the indigenous populations.

Adaptive traits at high altitude, at least those we know about, are quantitative and continuously varying rather than present or absent as in the case of the lactase persistence phenotype. Some continuous traits are influenced by a single gene with a large quantitative effect, while others are influenced by many loci with small effects as well as by environmental factors. An example of the former is the protein level and activity of angiotensin-converting enzyme (ACE). About 45 % of the variance in ACE levels at sea level can be explained by an insertion/deletion polymorphism at a single locus [17, 18]. Height exemplifies the latter, complex, type of trait. Just 3 % of the variance in height is explained by the top 20 of 54 identified associated chromosomal regions [19]. In practice with natural populations having long lifespans and generations, such as humans, it is easier to detect natural selection on traits influenced by a single or a few loci with large effects because the potential change in a trait from one generation to the next is directly proportional to the large heritable variance.

Research Strategies

Classic Era

1. A classic strategy to detect natural selection correlates variation in traits with variation in the selective factor to infer that it is or was operating [20]. A large altitude range providing a wide range of the selective factor is desirable for such studies. Most studies implicitly or explicitly consider that hypobaric hypoxia is the selective factor although temperature, ultraviolet radiation, and infectious disease are among other potential selective factors that generally vary with altitude. Altitude gradients of a phenotype can be impractical for a single investigator to assemble. As a result, many studies contrast one or two high altitudes with a low altitude control or combine the results of studies at various altitudes to produce a composite gradient. Phenotypic variation across altitude gradients is well-documented for many traits. For example, birth weight declines linearly [21] while hemoglobin increases exponentially with altitude in the Andes [22, 23].

In contrast to the many phenotypic examples, decades of attempts to identify altitude gradients in allele frequencies were unsuccessful. For example, analyses of 22 protein coding loci and seven loci involved in red blood cell glycolysis assayed in nearly 2,100 people along an altitude gradient from 300 to 4,000 m in the Chilean Andes found no allele frequency gradient [24, 25]. Similarly, there was no gradient in genetic variance of four quantitative red blood cell traits (hemoglobin concentration, hematocrit, 2,3-diphosphoglycerate (DPG), and adenosine triphosphate (ATP) levels) as would be expected if natural selection had reduced variance at high altitude [26].

The authors' suggested reasons for failing to detect an altitude gradient remain relevant for designing and interpreting studies nowadays and bear repeating. They wrote "Not finding evidence for genetic selection may be because of several factors. The physiological responses seen in residents of high altitudes and assumed to be adaptive may simply be within the normal range of physiological response in human populations. The Aymara migration from the altiplano [Andean plateau] to lower altitudes may be too recent for genetic changes to be detected. Finally, the particular genetic loci studied may be too remotely related to the physiology of high altitude adaptation to be used in studying adaptation to hypoxia" ([24] p. 101).

Two recent successful instances of identifying an altitude gradient in allele frequency showed that important genes in crucial pathways can evolve. These genes were in a pathway that regulates oxygen homeostasis in multicellular animals [27]. The HIF1 pathway is named after the master regulator of oxygen homeostasis, the hypoxia inducible factor 1 (HIF) discovered a decade or more after those early studies of altitude gradients. One study reported that certain variants in a key oxygen sensor *EGLN1* in the oxygen homeostasis system were more frequent among high- than low-altitude Tibeto-Burman samples in India [28]. *EGLN1* is also known as *PHD2*. Table 19.1 lists

Table 19.1 Genetic loci mentioned in this chapter and identified by the HUGO gene nomenclature committee (HGNC) (available at http://www.genenames.org)

Approved symbol for gene	HGNC gene ID #	Name or alias	Chromosome location	General role in hypoxia sensing or responding to HIF
ACE	2707	Angiotensin I converting enzyme (peptidyl-dipeptidase A) 1	17q23	HIF induced
HIF1A	4910	Hypoxia inducible factor 1, alpha subunit (basic helix-loop-helix transcription factor)	14q23.2	Hypoxia stabilizes
EPAS1	3374	*HIF2A*, endothelial PAS domain protein 1	2p21-p16	Hypoxia stabilizes
EGLN1	1232	*PHD2*, egl nine homolog 1 (*C. elegans*)	1q42.1	Tags HIF-alpha for binding to VHL
VHL	2687	VHL1, von Hippel-Lindau tumor suppressor, E3 ubiquitin protein ligase	3p25.3	Targets HIF1A for degradation
PRKAA1	9376	*AMPKa1*, protein kinase, AMP-activated, alpha 1 catalytic subunit	5p12	Along with HIF1A regulates energy metabolism
NOS2A	7873	*iNOS*, nitric oxide synthase 2, inducible	17q11.2-q12	HIF1 induced
NOS3	7876	*eNOS*, nitric oxide synthase 3 (endothelial cell)	7q36	Hypoxia down-regulates
EPO	3415	Erythropoietin	7q21	HIF1 induced
EDN1	3176	Endothelin-1, *ET1*	6p24.1	HIF1 induced
CBARA1	1530	Mitochondrial calcium uptake 1	10q22.1	No known association with HIFs
VAV3	12659	vav 3 guanine nucleotide exchange factor	1p13.3	No known association with HIFs
ARNT2	16876	Aryl-hydrocarbon receptor nuclear translocator 2	15q25.1	Protein forms dimer with HIF1A
THRB	11799	Thyroid hormone receptor, beta	3p24.2	No known association with HIFs

the genes mentioned in this chapter along with their names and chromosomal locations. The second recent study confirmed variation in *EGLN1* along an altitude gradient among East Asian samples and added variation in *EPAS1* (also known as HIF2a or hypoxic inducible factor 2) [29]. *EPAS1* binds to another protein to form HIF2, a second transcription factor regulating the expression of hundreds of loci involved in the response to hypoxia [30] such as the erythropoietin (*EPO*) locus.

2. A second classic strategy involves perturbing natural populations in order to infer the past action of natural selection. This is implemented in human populations with the "migrant model" comparing residents and migrants from one altitude to another [31]. It interprets differences between high altitude residents and upward migrants in the mean values of a trait associated with good function or offsetting hypoxia as evidence for a genetic basis for the trait in the high-altitude population. This strategy is somewhat like comparing the phenotypes of contemporary high-altitude populations after thousands of years of opportunity for natural selection with those of the early colonists adapting to high altitude by acclimatization or development. An informative illustration is a comparison of two samples of highland Andean ancestry

Table 19.2 Differences in adaptive success between lifelong high-altitude natives and acclimatized newcomers (updated from ([6], p. 84)), types of supportive evidence for the past action of natural selection, and disease indicators of unsuccessful adaptation. The differences favor better function among natives and seem to differ from the normal range of homeostatic response to hypoxia. These patterns form a major body of evidence that natural selection has modified the biological characteristics of high-altitude natives

Altitude natives—acclimatized newcomer differences	Altitude gradient evidence	Altitude extremes evidence	Diseases indicating unsuccessful adaptation
Source: Niermeyer et al. [6]			
Less intrauterine growth retardation	Present	Present	Low birthweight, pre-eclampsia [131, 132]
Better neonatal oxygenation and involution of fetal cardiopulmonary characteristics	Not enough data	Present	Elevated pulmonary artery pressure, increased prevalence of patent ductus arteriosus [133, 134]
Enlarged lung volumes and deceased alveolar-arterial oxygen diffusion gradients	Not enough data	Present	–
Higher maximal exercise capacity	Present	Present	–
Better maintained increase in cerebral blood flow during exercise	Not enough data	Present	–
Lower hemoglobin concentration (Tibetans only)	Present	Present	Andean excessive erythrocytosis [135]
Less susceptibility to chronic mountain sickness (CMS) (Tibetans only)	Not enough data	Present	Chronic mountain sickness
Update since [6]			
Higher exhaled nitric oxide (Tibetans only)	Not enough data	Present	Excessive hypoxia pulmonary vasoconstriction [38]
Physiological responses to pregnancy favor good outcome	Not enough data	Present	Higher neonatal mortality [39, 136]

born and raised at high and at low altitude with two samples of lowland European ancestry born and raised at high and at low altitude. It revealed that both populations had larger forced vital capacities if they grew up at high altitude. However, a larger effect among those of Andean descent indicated that both developmental exposure and genetic ancestry contributed to the trait of large forced vital capacity at high altitude [32].

Relying largely on evidence obtained using these two strategies, Niermeyer et al. [6] identified seven "differences in adaptive success between natives and newcomers…." To the extent possible, those authors took care to consider the alternative hypothesis that developmental adaptation or acclimatization accounted for the differences. The differences in all traits were in the direction of better function for the indigenous high-altitude populations. The traits were less intrauterine growth retardation, better neonatal oxygenation and involution of fetal cardiopulmonary characteristics, enlarged lung volumes and decreased alveolar-arterial oxygen diffusion gradients, higher maximal work capacity, better maintained increase in cerebral blood flow during exercise, lower hemoglobin concentrations (Tibetans only), and less susceptibility to chronic mountain sickness (CMS) (Tibetans only) among natives than among newcomers (Table 19.2).

Subsequent work has added two more traits to the list: higher nitric oxide levels among highlanders (Tibetans only) and higher uterine artery blood flow and oxygen delivery during pregnancy. Acute exposure to hypoxia in lowlanders causes a decrease in exhaled nitric oxide followed by a return to baseline or slightly above [33–38], however they do not elevate levels to those observed among Tibetan highlanders [8, 37]. Similarly, women who migrated themselves or whose recent ancestors migrated, do not increase uterine artery blood flow to the same degree as highland women [39]. Table 19.2 also lists some diseases of the reproductive, cardiovascular, and hematological systems reflecting unsuccessful adaptation that may have beset early colonists such as pre-eclampsia, pulmonary hypertension, congenital heart anomalies, excessive erythrocytosis, and CMS. Genetic variants contributing to the differences in adaptive success would likely have increased in frequency along with those associated with decreased vulnerability to the diseases.

The "admixture model" adapts the migration strategy to take advantage of gene flow occurring in the Andes over the past 500 years of European and indigenous highlanders' co-residence and potential for intermarriage. Blending classic and genomic strategies, it quantifies individuals' proportion of Native American ancestry on a scale ranging from 0 to 100 % as assessed by a panel of ancestry informative single nucleotide polymorphisms (SNPs) [40]. For example, a sample of low-altitude natives with an average estimated "Native American Ancestry Proportion" (NAAP) of 85 % was perturbed by a trip to 4,338 m and tested after 10–12 h there. Higher NAAP correlated with lower hypoxic ventilatory response after 10 min of experimental hypoxia at high altitude and with lower ventilation during exercise, both characteristics of Andean highlanders that differ from those of Tibetans. The authors asserted that this was "the first direct evidence that ventilatory traits, probably unique to Andeans, have a population genetic basis. Our quantification of ancestry as an independent variable has led us to infer both a genetic mechanism and an evolutionary origin for these traits" ([40] p. R232). Continuing with that study design, lowlanders with a high proportion of Andean ancestry experienced a smaller fall in maximal oxygen consumption tested within 24 h at 4,338 m [41]. However, there was no similar evidence for forced vital capacity or maximal oxygen consumption [42]. Those results imply that natural selection produced a distinctive Andean gene pool at unknown loci influencing some although not all distinctive Andean traits.

3. A third classic strategy to detect natural selection is the cross-population strategy comparing multiple indigenous populations exposed to the same environment [43]. If phenotypes differ among populations exposed to the same stress, after taking into account potential confounding factors, it suggests that natural selection favored different responses. Table 19.3 presents the compilation by Niermeyer et al. [6] summarizing phenotypic differences between Tibetan and Andean highlanders. It includes less intrauterine growth retardation, greater reliance on redistribution of blood flow than elevated arterial oxygen content to increase uteroplacental oxygen delivery during pregnancy, higher resting ventilation and hypoxic ventilatory responsiveness, less hypoxic vasoconstriction as measured by lower pulmonary arterial pressure and resistance, less susceptibility to CMS among Tibetan than among Andean highlanders.

Additions to that list include evidence of higher nitric oxide levels among Tibetans than among Andean highlanders and the direct association of nitric oxide levels with pulmonary and systemic blood flow among Tibetans [38, 44, 45]. Another addition may be higher cerebral blood flow among Tibetan than among Andean highlanders [46]. The addition of lower percent of oxygen saturation of hemoglobin reflects findings of studies using the same protocol [47, 48]. A possible deletion is lower pulmonary artery pressure among Tibetans than Andean highlanders. Two recent studies of minimally elevated pulmonary artery pressure in samples of healthy, non-miners of Andean ancestry reported values very similar to the low values among Tibetans [49, 50]. The explanation for the difference between those and earlier studies of pulmonary hypertension among Andean highlanders could be that many, although not all, early studies of this trait were conducted with samples of miners in the Central Andes. There they may have been confounded by non-altitude-related environmental factors. Later studies found elevated cobalt levels in the water that exaggerated the hemoglobin concentration levels and have caused cardiomyopathy and elevated pulmonary artery

Table 19.3 Differences in adaptive success between Tibetan and Andean highlanders (updated from [6]), and relevant measures of genetic variance and findings from genomic studies. The measures inform about the possibility of ongoing selection for the traits because selection works on variation than can be inherited. This information guides study interpretation by directing attention to information needed to study ongoing natural selection

Tibetan and Andean highlanders—differences	Presence of genetic variance
Source: Niermeyer et al. [6]	
Tibetans have less intrauterine growth retardation	No data
Tibetans have greater reliance on redistribution of blood flow than elevated arterial oxygen content to increase uteroplacental oxygen delivery during pregnancy	No data
Tibetans have higher levels of resting ventilation and hypoxic ventilatory responsiveness	h^2 for resting ventilation is 0.32 and insignificant; for hypoxic ventilatory responsiveness it is 0.35 and 0.22, for Tibetans and Andean highlanders respectively; admixture studies identify Andean adaptations [40, 41, 91]
Tibetans have less hypoxic vasoconstriction as measured by lower pulmonary arterial pressure and resistance	No data
Tibetans have lower hemoglobin concentration	h^2 is 0.61 and 0.89 for Tibetans and Andean highlanders; respectively; Tibetans have high frequencies of alleles associated with lower hemoglobin concentration [89–91]
Tibetans are less susceptible to chronic mountain sickness	No data; variance in some loci associated with hemoglobin are excluded for Andean highlanders, genome-wide analyses found no associations [114]
Update since [6]	
Tibetans have lower percent of oxygen saturation of hemoglobin when measured using the same protocol	h^2 ranges from 0.33 to 0.47 for Tibetans and is insignificant for Andean highlanders; ACE and other genotypes contribute to variation among Andean highlanders [41, 118]
Tibetans have higher exhaled nitric oxide	No data

pressure at low altitude [51, 52]. Similarly, experimental evidence shows that iron status influences hemoglobin levels at high altitude [53] and that iron supplementation may reduce pulmonary artery pressure under hypoxia [54]. Those findings emphasize the need for considering a growing number of known confounding factors in order to have the most informative phenotypic data.

Information on East African highlanders, specifically of the Amhara ethnic group of Ethiopia, is sparse although accumulating. Consider the evidence for differences in adaptive success between high-altitude natives and acclimatized newcomers (Table 19.2), Amhara highlanders have large lung volumes, lower hemoglobin concentrations and higher oxygen saturations of hemoglobin [55, 56]. Considering the cross-population differences among highlanders (Table 19.3), the evidence remains sparse. With respect to intrauterine growth retardation as measured by birthweight, publications from Ethiopia are not very informative about altitude. They generally do not distinguish among ethnic groups, are usually put into the context of problems and interventions to raise birthweights and are available for the relatively low and narrow altitude range of 1,300–2,300 m. Mean birthweights for healthy singleton hospital deliveries were in the 3.1–3.3 kg range [57–63]. A focused study of altitude effects on intrauterine growth in East African would be valuable. The birthweights are 100–200 g lower than predicted for Tibetan, Andean or European populations at the corresponding altitudes although they are higher than predicted for Han Chinese [64].

The Amhara have the highest systolic pulmonary artery pressures among samples of the three populations along with high pulmonary blood flow and low pulmonary vascular resistance [65]. That pattern of high pressure accompanied by high flow and low resistance is different from the classic pattern based on the Andean model of high pressure and resistance owing to pulmonary vasoconstriction.

Continuing a trend of investigating blood flow, a study found that the cerebral circulation of Amhara was relatively insensitive to hypoxia as compared with Andean highlanders and concluded that such a response could contribute to high cerebral blood flow at altitude rather than the usual response of reducing cerebral blood flow [66].

With respect to hematological traits, early evidence suggested that Amhara adapted like the Tibetans in the sense of having little altitude-associated increase in hemoglobin concentration and uniquely in the sense of having little decrease in oxygen saturation of hemoglobin. Later work has shown that Amhara can increase hemoglobin concentration and decrease oxygen saturation somewhat at altitudes above 3,500 m [55, 65, 67]. That suggests that the Amhara threshold for hematological response lies in between Andean and Tibetan populations. However, excessive erythrocytosis measured in terms of very high hemoglobin concentration has not been observed.

Although measures of exhaled nitric oxide have not been made, urinary measures of total body nitric oxide synthesis show that Amhara at high and low altitude [65] have lower synthesis than Tibetans. Finally on the list evaluating possible population differences, a report on physiological responses to pregnancy found pre-eclampsia at about 5 % [68] in a hospital at 2,300 m. The rate in the US varies from roughly 1 to 5 % depending on ethnicity [69]. Because pre-eclampsia is generally more prevalent in developing countries [70], it is probably inappropriate to infer with these data any association with altitude. Overall, although data are sparse, they suggest the possibility that Amhara high-altitude natives in Ethiopia may represent a third pattern of population adaptation to high altitude, one with distinctive hematological and cardiovascular characteristics.

4. Quantitative and statistical genetics techniques are additional classic strategies. Quantitative genetics estimate the heritability (h^2), the proportion of total variance in a trait that is accounted for by biological kinship among individuals in a population. It may range from 0 to 1.0 with zero indicating no genetic contribution and one indicating no non-genetic contribution to variation in the trait. Because natural selection requires genetic variance, a significant h^2 indicates the potential for ongoing natural selection. A lack of genetic variance indicates no current potential for ongoing selection, perhaps because a past selective sweep removed it or perhaps another trait is preventing its expression. Tibetan samples generally have higher genetic variance, in the range of 0.3–0.7 commonly considered moderate to high levels, than Andean samples with the exception of hemoglobin concentration for which both have high variance (Table 19.3). Research using genomic techniques described below confirmed and extended that evidence of genetic variance.

Complex segregation analyses test the hypothesis that levels of a quantitative trait are inherited in Mendelian fashion. Such analyses detected a major gene at an unknown locus with an autosomal dominant mode of inheritance among Tibetans associated with 6–10 % higher percent of oxygen saturation of hemoglobin. Tibetan women estimated with high probability to have one or two copies of the inferred autosomal dominant allele had more than twice as many living children as compared with women estimated to be homozygous recessive for the low saturation allele [71]. That evidence linking inferred genotypes with offspring survival suggests that very strong natural selection is increasing the frequency of the inferred allele at the unknown locus.

The cross-species strategy compares multiple species exposed to the same environment [43]. Traits common to a wide range of organisms with long histories of high-altitude habitats may indicate expression of a common inherited response. The few data about our closest biological relatives, other primate species, come from Old World monkeys. Upon acute exposure, pig-tailed macaques increase hemoglobin concentration [72], suggesting an inherited response in common with human visitors to altitude and Andean highlanders. Mitochondrial DNA shows evidence of altitude differences in allele frequencies among snub-nosed macaques resident at 4,000 m [73].

Data are relatively abundant for more distantly related species. An illustration is deer mice populations resident along an altitude gradient in the Rocky Mountains of the U.S. that exhibit a

Table 19.4 Characteristics of genotypically and phenotypically adapted high-altitude animals (categories identified by [80]) compared with cross-population evidence about Tibetan and Andean high-altitude native phenotypes. Tibetans share several traits in the genotypically adapted category while Andean highlanders share all but one of the traits in the phenotypically adapted category. This was further evidence that past natural selection might be more evident among Tibetans

Trait	Genotypically adapted high-altitude animals	Tibetan high-altitude natives	Andean high-altitude natives	Phenotypically adapted high-altitude animals
High hemoglobin affinity (low p50)	Present	Absent	Absent	Absent
Moderate or absent polycythemia	Present	Present	Absent	Absent
Low venous pO_2	Present	No data	Present	Absent
Thin-walled pulmonary vascular tree that responds moderately to hypoxia	Present	Present	Absent	Absent
Absence of chronic mountain sickness (CMS)	Present	Present	Absent	Absent

parallel gradient of higher frequency at higher altitude of alleles in the beta chain of hemoglobin that increase oxygen affinity [74]. High oxygen affinity of hemoglobin has been reported for numerous high-altitude species and is accomplished by at least two different mechanisms, although people on the Andean and Tibetan Plateaus do not have the trait [75–78]. Instead, visitors, Tibetan and Andean highlanders achieve lower oxygen affinity as a result of high levels of the red blood cell enzyme 2, 3 DPG that lowers oxygen binding to hemoglobin [79]. High affinity enhances pulmonary loading while low affinity enhances tissue offloading of oxygen. That response, however, is offset by higher ventilation [76].

An influential cross-species analyses by Monge and Leon-Velarde proposed that high-altitude animals such as llamas and yak are "genotypically adapted" while others such as cows introduced in the past few 100 years are "phenotypically adapted" [80]. According to their analysis, genotypically adapted organisms have high hemoglobin affinity, moderate or absent polycythemia, low venous pO_2, thin-walled pulmonary vascular trees that respond moderately to hypoxia, and the absence of CMS. Table 19.4 presents the five traits characterizing the two categories of adaptations and indicates the phenotypic resemblance of Tibetan and Andean highlanders to one or the other category of adaptation. Tibetans exhibit three of the genotypically adapted phenotypic features—moderate or absent polycythemia, thin-walled pulmonary vasculature, little CMS, do not exhibit high hemoglobin affinity, and there are no data on venous pO_2. In contrast, Andean highlanders exhibit four of the phenotypically adapted phenotypic features but not a fifth, low venous pO_2 [81]. The concept of cryptic adaptive evolution has been proposed to explain the contrast between genotypically and phenotypically adapted species. It reasons that "In cases in which the acclimatization response to hypoxia is maladaptive, selection will favor an attenuation of the induced phenotypic change" ([79] p. 4125). Beneficial outcomes include maintaining adaptability to further stress and avoiding the costs of sustained acclimatization responses. Some of the Tibetan—Andean contrasts can be interpreted with this concept, for example the relatively dampened hemoglobin response and relatively low pulmonary artery pressures of the Tibetans. That is, those phenotypes may resemble those of lowlanders as a result of selection at high altitude.

Thus, a large body of mainly phenotypic data supports the hypothesis that natural selection has acted on indigenous highland populations of the Andean and Tibetan Plateaus. There are links with reproductive success. The situation on the East African Plateau is not clear because of lack of information. These informative studies have generally not dealt with specific genetic loci or variants. Thus it has not been possible to test formal population genetics "null models" of no selection. That situation has changed rapidly with the implementation of genomic strategies [82–85].

Genomic Era

Genomic-Wide Approach

Since the completion of the first survey of the human genome in 2000 and especially since the publication of the first haplotype map of the human genome in 2005 [86], the genomic era is providing an abundance of new strategies to detect "signals of natural selection" in the genome [83, 86–88]. The genomic era burst onto the high-altitude scene in 2010–2011 with the publication of seven articles reporting signals of selection detected in Tibetan samples, three of which also reported specific genotypes that associated with the distinctively low hemoglobin concentration of Tibetans [29, 89–95]. Extensive analysis and commentary followed as did a report on an East African sample [79, 95–103].

It is extraordinary that every one of the seven studies of Tibetans identified *EPAS1* (*HIF2A*) and four identified *EGLN1* (*PHD2*) loci as playing a role that population's genetic adaptation to high altitude. More than a dozen samples of Tibetans across a wide swathe of the plateau provided evidence. It is extraordinary for the degree of consensus and replication among studies with different designs and analyses as well as extraordinary because those two loci are central to the pathways regulating oxygen homeostasis in all vertebrates [104–106]. This ancient homeostatic biochemical system that participates in many biochemical pathways has apparently tolerated adaptive variants [107]. A change in HIF regulation by oxygen could be one route to efficiently cause many adaptive responses. Yet the results of that cascade and others it initiates could be too far-reaching and have maladaptive consequences. So far it appears that the Tibetan population benefits. The causal mutation(s) (and whether it is indeed new or unique to Tibetans) and the functional links between genotype and phenotype remained to be discovered. The extent to which hemoglobin levels are the target of selection or a pleiotropic manifestation of selection on another trait is another important point to address in future work.

Such genomic strategies differ fundamentally from the classic strategies seeking to discover and explain the distinctive phenotypes of high-altitude natives. Genomic strategies are potentially useful for identifying inductively areas of the genome and thus certain loci under selection that may underlie distinctive high-altitude phenotypes traits or even suggest new phenotypes to investigate. That is, they can be applied without prior knowledge of the relevant loci or phenotypes. This may be counterintuitive to many scientists studying high-altitude adaptations who are intensely interested in the phenotypes involved in maintaining homeostasis under severe and chronic stress. Yet, the two strategies are complementary and ask related questions using different data and outcome measures.

The basic concept of genomic signals of natural selection involves interpreting patterns of similarity or differences in variation emerging after genotyping SNPs at (usually) hundreds of thousands or even more than a million loci throughout the entire genome of individuals in samples of one or more populations. "This approach is based on the idea that natural selection introduced a local perturbation in the patterns of neutral genetic variation surrounding an advantageous allele relative to regions where variation is shaped only by genetic drift" ([83] p. 198).

Analyzing signals of natural selection can involve (a) one of several measures of allele frequency differences between two or more samples, ideally including the ancestral and descendent populations or (b) identifying haplotypes, measuring haplotype frequencies, and haplotype homozygosity within or between samples. Haplotypes are combinations of polymorphisms along a stretch of DNA; in the case of SNPs, many regions of the genome have just a few combinations. For the purposes of analyzing natural selection, haplotypes are generally identified as groups of alleles on a chromosome that are inherited together as a block because "they are descended from a single ancestral chromosome" [83]. A high frequency of a distinctive haplotype is evidence of past positive natural selection for a mutation in that group of SNPs inherited together. The causal mutation could be in a measured SNP or an unmeasured one inherited in the same block. If the haplotype is universal in a sample then that is evidence that a selective sweep occurred.

Genome-wide signals have different strengths to detect selection depending on what happened. A variety of signals of natural selection were measured in the studies of Tibetans because there was no prior knowledge of what had happened. Relevant features to consider when designing and interpreting studies include whether the current frequency of the selected allele or haplotype is moderate (50–80 %) or high (more than 90 %), whether selection occurred on standing or on new variants, and the length of time an allele existed before selection and how long since selection began [83, 108, 109]. For example, some alleles with frequencies in the range of 50–80 % in Tibetan samples had frequencies of just 10–30 % in the HapMap Han sample [94]. The moderately high frequency in the Tibetan samples means that an approach relying solely on cross-population analysis of haplotypes might have been unsuccessful because there is less statistical power when allele frequencies are moderate. However, *EPAS1* alleles associated with lowered hemoglobin concentration among Tibetans had larger differences with allele frequencies of 80 % or more among Tibetans and 20 % or lower among the HapMap Han [89]. That is in the range where cross-population extended haplotype homozygosity measures have the most statistical power [108, 109]. The SNP analyses in that study were based on a custom array designed to genotype *EPAS1* as a candidate locus. Many of those SNPs are not included on commercial arrays.

With respect to the age of the allele, some of the variants found in high frequency among Tibetans are found globally which implies that they are very old. If so, then selection had a long time to act before the ancestral population migrated to Tibet and for recombination to occur that would remove some types of selection signals. Similarly, because at least some Tibetans have been at altitude for tens of millennia, there would have been time for haplotypes to decay due to recombination. Both historical features could compromise approaches based on haplotype length [29, 109]. For such reasons, most studies report the results of multiple signals of selection.

The effect of the high-frequency-in-Tibetans alleles on hemoglobin concentration varied among samples. The largest was a 1.7 g/dL decrease in hemoglobin concentration with each additional copy of the Tibetan haplotype. That was reported for a sample that included people who would have been excluded from a study of normal variation because of low values in the range of anemia and high values in the range of polycythemia [90]. The smallest effect, yet still appreciable at around one-half a standard deviation, was 0.8 g/dL lower hemoglobin with each additional copy of the Tibetan SNP allele. That was reported for a sample that had been extensively screened to include only normal, healthy individuals outside the normal range of variation [89]. Both findings among Tibetans contrast with those from a genome-wide analysis among more than 24,000 low altitude Europeans. It detected tiny associations with hemoglobin concentration ranging only as high as 0.06 g/dL [110] and did not identify the loci that emerged as influential among Tibetans. The contrast illustrates again information based on loci identified or excluded on the basis of associations at low altitude may not apply to high-altitude populations. Interestingly, the loci associated with hemoglobin concentration among Tibetans variation was not associated with percent of oxygen saturation of hemoglobin [90, 91].

Similar genomic analyses of an Andean sample identified *EGLN1*, *PRKAA1*, and *NOS2A* as possible candidates for having undergone selection although no tests of association with high-altitude phenotypes have been reported [92, 93]. The Andean highlanders from two ethnic groups were compared with highland Tibetans, lowland Mesoamericans, Europeans, and East Asians. The SNPs identified in *EGLN1* in the Andean samples were not the same as those identified in the Tibetan samples described above. Those analyses are consistent with the known pattern of different phenotypic patterns among highland Tibetan and Andean natives (Table 19.3).

A highland Amhara population from Ethiopia was compared with two unrelated lowland populations and four completely different loci were identified as candidates for having undergone selection—*CBARA1*, *VAV3*, *ARNT2*, and *THR*. Those results support a hypothesis of a different history of natural selection. Tests of association with hemoglobin concentration were undertaken without finding significant results, perhaps because more than half of one of the low altitude samples had hemoglobin levels below the usual cutoff for anemia [96]. A different highland Amhara sample from Ethiopia was compared with an Amhara lowland sample. The samples were screened for potential confounding factors that could influence hemoglobin concentration. The study found no evidence that *EPAS1* or *EGLN1* variants associated with hemoglobin concentration or oxygen saturation of hemoglobin. Instead, using a genome-wide analysis, it detected a single variant at another locus not associated with a currently known gene that associated with hemoglobin concentration at both high and low altitudes [56]. These findings suggest the possibility of convergent evolution on the same phenotype of dampened hematological responses to high-altitude hypoxia using different genetic pathways.

In contrast to the genome-wide approaches, candidate gene approaches are hypothesis-driven and have been described succinctly as follows. "Such studies rely on intelligent reviews of all the available scientific literature and the proposal of a system thought to be of relevance (the *candidate system*)." From this, a key component is chosen (perhaps a rate-limiting enzyme) and its gene identified (the *candidate gene*). A polymorphic variant is then identified in the gene …. ([82] p. 125).

The HIFs and their target genes have been the focus of candidate analysis. The target genes are inducible by the family of transcription factors including HIF1 and HIF2 described above. Before publication of the evidence about *EPAS1*, attention focused on *HIF1A*.

A case–control study of Sherpas tested for an association of *HIF1A* and *VHL* variants with Acute Mountain Sickness symptoms at extremely high altitude. The VHL protein targets HIF1A and EPAS1 for ongoing degradation when oxygen levels are normal for low altitude. The study found no association, although it did acknowledge that it had low statistical power owing to the small sample size of 49 cases and 54 controls [111]. Another study reported a new sequence in the *HIF1A* gene among Sherpas as compared with Japanese and also reported other allele frequency differences at that locus [112]. In contrast, the *HIF1A* sequence of Andean highlanders did not differ from that of low-altitude controls [113]. A case–control study examining a number of genes involved in the response to hypoxia found no association with CMS in an Andean sample [114]. The Sherpa results suggested tentatively the possibility of a new adaptive mutation while the Andean did not. However, none of the studies of genome-wide variation identified *HIF1A* variants as likely to have undergone selection at altitude in either population. An explanatory hypothesis reasons that because HIF1 is expressed in all tissues, it may have little latitude for variation whereas HIF2 is expressed in fewer tissues and could have more focused effects [98].

With respect to variance in the response to HIF, the candidate gene first examined among high-altitude populations was *ACE*. It has a polymorphism associated with athletic ability and in some cases with successful adaptation to acute exposure to high altitude [115]. The variants are called the insertion (I, associated with better endurance and with metabolic efficiency) and deletion (D, associated with power and strength) alleles. At high altitude, I allele genotypes were associated with lower pulmonary artery pressure—better functional adaptation—in a sample of lowlanders acutely exposed [116]. However, they were associated with increased risk of pulmonary hypertension in a sample of Kyrgyz highlanders [117] and were not associated with pulmonary artery pressure in a sample of Amhara highlanders [65]. The contrasts may illustrate another example of the differences in adaptive success between lifelong high-altitude natives and acclimatized newcomers summarized in Table 19.2. Alternatively, they could indicate that ACE I/D polymorphism may be closely linked to another locus that is the true influence on phenotypes.

In contrast to the Kyrgyz and similar to the acutely exposed highlanders, Andean residents from high and low altitude with the ACE II genotype had some benefits. They had higher oxygen saturations of hemoglobin at rest and during exercise at 4,300 m [118]. ACE genotype accounted for ~4 % of the variance in oxygen saturation phenotype. That study quantified the proportion of Native American Ancestry for each participant in order to establish genetic homogeneity of the two samples and increase confidence that the I allele or a linked locus accounted for the difference in oxygen saturation. The *ACE* locus was not identified by the genome-wide studies as having been subject to natural selection at high altitude.

Other candidate genes such as some involved in pulmonary vascular tone have been analyzed including endothelin-1 (*EDN1*) and *NOS3* (also known as *eNOS*). Their proteins code for the vasoconstrictor endothelin and for an enzyme catalyzing the synthesis of the vasodilator nitric oxide, respectively. For example, two polymorphisms of *NOS3* were analyzed to test the hypothesis that alleles associated with higher rates of nitric oxide synthesis at low altitude would be associated with higher rates at high altitude and with less vasoconstriction among high-altitude Sherpas. One polymorphism at the *NOS3* locus was associated with higher levels of nitric oxide metabolites and one with lower levels while neither was associated with blood pressure as might be expected from a vasodilator [119]. This may be another illustration that candidate gene—phenotype associations identified in low-altitude Europeans samples may not be replicated in high-altitude, non-European samples. The *NOS2* locus rather than the *NOS3* locus was identified as a possible candidate for selection in the Andean population [92].

However, factors in addition to—or perhaps instead of—population differences can contribute to the outcome of genome-wide or candidate-gene analyses. One is the selection of appropriate controls that are likely to provide the clearest evidence of what happened. For example, the Yi population is more closely related to the Tibetans than the HapMap Han sample and may be a better low-altitude comparison sample [95]. Some studies address that issue with multiple control populations.

Another crucial factor influencing the detection of population differences is statistical power. Power to detect differences is lower with small sample sizes or small effects or when doing multiple statistical tests. The latter is particularly relevant for genomic strategies testing hundreds of thousands of hypotheses, one for each SNP, for example. The Bonferroni adjustment is a widely used method of multiple-testing correction that can illustrate this point. If a significance threshold of 0.05 is adopted and 20 separate tests of significance are performed, then the Bonferroni adjustment calls for dividing 0.05 by 20 for a p-value of 0.0025 as the appropriate significance threshold. If one million SNPS were tested for association with a single phenotype, then the 0.05 would be divided by 1,000,000 for a p-value of 0.000000005 as the appropriate significance threshold [120]. Going forward, it is very important to collect adequately sized samples.

Mitochondrial genotypes have been considered as well although with inconsistent results. One study concluded there was little evidence that mitochondrial mutations might contribute to successful adaptation among Tibetans, while another hypothesized a contribution by one haplotype over-represented among Tibetans as compared with Han Chinese could contribute [121, 122]. Neither reported phenotypes. A study of Andean women giving birth at >3,800 m collected information on mitochondrial DNA haplotypes and reproductive histories. Sixty-five percent of the sample had mt haplotype B (42–80 % of Andean highland populations have this haplotype [123]) and the rest had other New World haplotypes. Compared to the reproductive histories of women with mt haplotype B, the women with non-B haplotypes had more than three times the risk of losing offspring between conception and 1 month post-partum, partly due to ten times higher risk of neonatal mortality [124]. This could reflect very strong selection among Andean women.

Potential Confounding Factors

The classic and genomic strategies using phenotypes rely on accurate reflection of the response to hypoxia. Many studies carefully defined and controlled for a number of other influences such as smoking or health status, yet still there was uncertainty about the potential role of unknown confounding factors. Ironically some of the most relevant discoveries for discovering natural selection at high altitude may have come from recent gains in knowledge about potential confounding factors that influence expression of oxygen homeostasis-associated loci or high-altitude phenotypes. These could foil attempts to associate genotypes with phenotypes or even efforts to distinguish differences between high and low altitude. For example, a long-standing puzzle was the wide range of variation in mean hemoglobin concentration of samples of Andean highlanders. Samples from mining communities in the Central Peruvian Highlands were long known to have very high hemoglobin concentrations. High cobalt levels were detected in the ground water in one mining community. Cobalt stabilizes HIF1A; that could in turn induce transcription of target genes including *EPO*, perhaps partly explaining the very high mean values of some samples. Men with high cobalt levels in that community had higher EPO levels and higher hemoglobin concentration [52].

While cobalt can stabilize, iron is required to degrade HIF1A. Thus iron depletion or supplementation can exacerbate or dampen high-altitude pulmonary hypertension [54] although it is not known whether naturally occurring variation in iron levels influences variation in pulmonary blood pressure in natural populations.

Intergenerational influences on newborn phenotypes and perinatal origins of adult high-altitude phenotypes may further complicate efforts to link heritable variation to phenotype variation and survival to test the natural selection hypothesis. The relationship of birth weight to infant survival is appreciated as an example of balancing selection because newborns of very high or low birth weight have higher mortality [125]. Maternal effects on birth weight were demonstrated in a study of women of European ancestry delivering babies at 3,100 m in Colorado. Women born and raised at that altitude tended to give birth to infants of lower weight than migrant women [126], a suggestive sign of intergenerational epigenetic effects. Paternal effects on birth weight were demonstrated in a study of babies born to Andean, European or mixed parents that found that having an Andean father raised the birth weight slightly but significantly. The authors suggested that genomic imprinting—"modification of gene expression through the addition of molecules … to specific genes affecting intrauterine growth, based upon parental origin" ([127] p. 596) may be the mechanism.

Furthermore, perinatal events may influence adult phenotype. Retrospective data on a sample of young men with excessive erythrocytosis at 3,700 m in Bolivia indicated that they had had particularly low birth weights and all but one had experienced perinatal hypoxia [128]. Andean highlanders whose mothers had had pre-eclampsia had higher pulmonary artery pressures as teenagers than those who mothers had had normal pregnancies [50]. Similarly, Europeans who experienced hypoxia during the first week of life at low altitude had a larger pulmonary vascular response to acute high-altitude hypoxia as adults [129]. Unusual hypoxia perinatally may be over-represented among high-altitude residents with relatively unsuccessful adaptations. Thus, other environmental factors and individual life-history events may contribute to phenotype variance and confound the search for natural selection at high altitude.

Summary

In summary, indigenous high-altitude populations comprise an informative case study of evolution because they seem likely to have experienced natural selection favoring adaptations to the unavoidable, severe stress of hypobaric hypoxia. Beginning in the 1960s, scientists used classic strategies to detect natural selection in such populations [20]. A large body of evidence of distinctive phenotypes at high altitude, along with altitude gradients, cross-population and cross-species comparisons suggested the action of natural selection.

Recently, purely genomic strategies have been developed along with others that integrate genomic with phenotypic information. Recently findings from such studies confirmed that natural selection has occurred, thus supporting and expanding the large body of earlier evidence. So far, the genomic evidence is strong for one trait in one population—hemoglobin concentration among Tibetans. It is more indirect for Andean and East African populations and to a large extent rests on the failure to replicate the Tibetan findings. Detecting positive information on the genetic and phenotypic bases for the long-term success of these two populations is an important goal.

Future work integrating genomic and classical strategies will expand the phenotype–genome associations with the aim of building thoroughly integrated cases of natural selection linking phenotypic variation to heritable variation to survival at high altitude. That work will help explain the extent to which Andean, Tibetan, and East African differences in phenotypes as due to longer Tibetan residence [6], to chance difference in the natural experiments of colonization [130] or other reason to be discovered.

The key remaining gap in understanding the process of natural selection at high altitude is associating phenotypic with genomic variation and then determining that the "variation be it ever so slight, would also have a better chance....of survival" ([1] p. 49).

In summary, *Natural selection is not easy to detect* ([20] p. 97), however progress is under way.

Appendix: Global Population Estimates

The global population estimates were obtained by Ann Holstein, Manager of GIS Systems and Numeric Data Services at the Kelvin Smith Library, Case Western Reserve University using ESRI's ArcGIS Desktop 9.3 Geographic Information System (GIS) software. The following data sets were merged.

1. Global Digital Elevation Model (GTOPO30), developed by USGS EROS Data Center in 1996, represents gridded 30 arc seconds (+-1 km) elevation for the world (available at http://www1.gsi.go.jp/geowww/globalmap-gsi/gtopo30/gtopo30.html).
2. Gridded population of the world, version 3 (GPWV3) centroids, were acquired from SEDAC—Socioeconomic Data and Applications Center (available at http://sedac.ciesin.columbia.edu/). Each centroid includes a population value attribute for P05A, the UN-adjusted population estimates for 2005.

The spatial analysis methods were as follows. Contour lines were derived from the Global Digital Elevation Model at 500 m intervals. Line data were converted to polygons (areas), where the area located between two contour lines was defined and calculated by the software. Population centroids spatially located within polygons identified as over 2,500 m elevation were retained. The sum of the P05A field of centroids for each country having populations living above 2,500 m was calculated. These population values are listed in the table below.

Global population at Elevations above 2,500 m

		2005 Population Estimate
AFRICA		**12,849,539**
	Congo	740,595
	Ethiopia	10,874,518
	Kenya	616,229
	Lesotho	123,225
	Rwanda	182,515
	Tanzania	268,912
	Uganda	43,545
ASIA		**31,869,915**
	Afghanistan	3,234,170
	Bhutan	609,298
	China	16,018,246
	India	7,428,994
	Kyrgyz Republic	608,636
	Mongolia	96,718
	Myanmar	4,436
	Nepal	1,397,479
	Pakistan	1,408,409
	Russia	21,761
	Tajikistan	872,514
	Uzbekistan	169,254
NORTH AMERICA		**9,516,326**
	Greenland	537
	Guatemala	1,098,961
	Mexico	8,028,898
	United States	387,930
SOUTH AMERICA		**29,143,068**
	Argentina	417,503
	Bolivia	5,071,377
	Chile	289,843
	Colombia	10,426,067
	Ecuador	4,120,668
	Peru	8,456,051
	Venezuela	361,559
TOTAL		**83,378,848**

References

1. Darwin C, Wallace A. On the tendency of species to form varieties; and on the perpetuation of varieties and species by natural means of selection. In: Proceedings of the Linnaean society. 1858. p. 45–62.
2. Monge C. Acclimatization in the Andes. Reissued 1948 edition ed. Baltimore: The Johns Hopkins Press; 1978.
3. Hurtado A. Studies at high altitude: blood observations on the Indian natives of the Peruvian Andes. Am J Physiol. 1932;100:487–505.
4. Hurtado A. Respiratory adaptation in the Indian Natives of the Peruvian Andes. Studies at high altitude. Am J Phys Anthropol. 1932;17(2):137–65.
5. Baker PT. Human adaptation to high altitude. Science. 1969;163:1149–56.
6. Niermeyer S, Zamudio S, Moore LG. The people. In: Hornbein TF, Schoene RB, editors. High altitude an exploration of human adaptation. New York, NY: Marcel Dekker, Inc.; 2001. p. 43–100.
7. Hurtado A. Animals in high altitudes: resident man. In: Dill DB, editor. Handbook of physiology section 4: adaptation to the environment. Washington, DC: American Physiological Society; 1964. p. 843–59.
8. Beall CM, Laskowski D, Strohl KP, Soria R, Villena M, Vargas E, et al. Pulmonary nitric oxide in mountain dwellers. Nature. 2001;414:411–2.
9. Aldenderfer M. Peopling the Tibetan plateau: insights from archaeology. High Alt Med Biol. 2011; 12(2):141–7.
10. Aldenderfer MS. Moving Up in the World; archaeologists seek to understand how and when people came to occupy the Andean and Tibetan plateaus. Am Sci. 2003;91:542–9.
11. Fehren-Schmitz L, Warnberg O, Reindel M, Seidenberg V, Tomasto-Cagigao E, Isla-Cuadrado J, et al. Diachronic investigations of mitochondrial and Y-chromosomal genetic markers in pre-Columbian Andean highlanders from South Peru. Ann Hum Genet. 2011;75(2):266–83.
12. Weinstein KJ. Thoracic skeletal morphology and high-altitude hypoxia in Andean prehistory. Am J Phys Anthropol. 2007;134(1):36–49.
13. Pleurdeau D. Human technical behavior in the African middle stone age: the Lithic Assemblange of Porc-Epic Cave (Dire Dawa, Ethiopia). Afr Archaeol Rev. 2006;22(4):177–97.
14. Bersaglieri T, Sabeti PC, Patterson N, Vanderploeg T, Schaffner SF, Drake JA, et al. Genetic signatures of strong recent positive selection at the lactase gene. Am J Hum Genet. 2004;74(6):1111.
15. Itan Y, Powell A, Beaumont MA, Burger J, Thomas MG. The origins of lactase persistence in Europe. PLoS Comput Biol. 2009;5(8):e1000491.
16. Tishkoff SA, Reed FA, Ranciaro A, Voight BF, Babbitt CC, Silverman JS, et al. Convergent adaptations of human lactase persistence in Africa and Europe. Nat Genet. 2007;39(1):31–40.
17. Rigat B, Hubert C, Corvol P, Soubrier F. PCR detection of the insertion/deletion polymorphism of the human angiotensin converting enzyme gene (DCP1) (dipeptidyl carboxypeptidase 1). Nucleic Acids Res. 1992;20(6):1433.
18. Tiret L, Rigat B, Visvikis S, Breda C, Corvol P, Cambien F, et al. Evidence, from combined segregation and linkage analysis, that a variant of the angiotensin I-converting enzyme (ACE) gene controls plasma ACE levels. Am J Hum Genet. 1992; 51(1):197.
19. Weedon MN, Lango H, Lindgren CM, Wallace C, Evans DM, Mangino M, et al. Genome-wide association analysis identifies 20 loci that influence adult height. Nat Genet. 2008;40(5):575.
20. Endler JA. Natural selection in the wild. In: May RM, editor. Princeton, NJ: Princeton University Press; 1986.
21. Soria R, Julian CG, Vargas E, et al. Graduated effects of high-altitude hypoxia and highland ancestry on birth size. Pediatr Res. 2013 (in press).
22. Cosio G. Caracteristicas Hematicas y Cardiopulmonares Del Minero Andino. Bol Off Sanit Panama. 1972;72(June):547–57.
23. Villafuerte FC, Cardenas R, Monge CC. Optimal hemoglobin concentration and high altitude: a theoretical approach for Andean men at rest. J Appl Physiol. 2004;96(5):1581–8.
24. Ferrell RE, Bertin T, Barton SA, Rothhammer F, Schull WJ. The multinational Andean genetic and health program. IX. Gene frequencies and rate variants of 20 serum proteins and erythrocyte enzymes in the Aymara of Chile. Am J Hum Genet. 1980; 32:92–102.
25. Ferrell RE, Bertin T, Schull WJ. An electrophoretic study of glycolytic enzymes in a human population living at high altitude: the Aymara of Northern Chile and Western Bolivia. Hum Genet. 1981;56:397–9.
26. Chakraborty R, Clench J, Ferrell RE, Barton SA, Schull WJ. Genetic components of variation of red cell glycolytic intermediates at two altitudes among the South American Aymara. Ann Hum Biol. 1983;10(2):174–84.
27. Semenza GL. Hypoxia-inducible factors in physiology and medicine. Cell. 2012;148(3):399–408.
28. Aggarwal S, Negi S, Jha P, Singh PK, Stobdan T, Pasha MA, et al. EGLN1 involvement in high-altitude adaptation revealed through genetic analysis of extreme constitution types defined in Ayurveda. Proc Natl Acad Sci. 2010;107(44):18961–6.
29. Xu S, Li S, Yang Y, Tan J, Lou H, Jin W, et al. A genome-wide search for signals of high-altitude adaptation in Tibetans. Mol Biol Evol. 2011;28(2):1003–11.
30. Schödel J, Oikonomopoulos S, Ragoussis J, Pugh CW, Ratcliffe PJ, Mole DR. High-resolution genome-wide mapping of HIF-binding sites by ChIP-seq. Blood. 2011;117(23):e207–17.
31. Harrison GA. Human adaptability with reference to the IBP proposals for high altitude research. In:

Baker PT, Weiner JS, editors. The biology of human adaptability. Oxford: Clarendon; 1966. p. 509–19.
32. Brutsaert TD, Soria R, Caceres E, Spielvogel H, Haas JD. Effect of developmental and ancestral high altitude exposure on chest morphology and pulmonary function in Andean and European/North American Natives. Am J Hum Biol. 1999;11:385–95.
33. Duplain H, Sartori C, Lepori M, Egli M, Allemann Y, Nicod P, et al. Exhaled nitric oxide in high-altitude pulmonary edema. Am J Crit Care Med. 2000;162:221–4.
34. Busch T, Bärtsch P, Pappert D, Grunig E, Hildebrandt W, Elser H, et al. Hypoxia decreases exhaled nitric oxide in mountaineers susceptible to high-altitude pulmonary edema. Am J Respir Crit Care Med. 2001;163:368–73.
35. Brown DE, Beall CM, Strohl KP, Mills PS. Exhaled nitric oxide decreases upon acute exposure to high-altitude hypoxia. Am J Hum Biol. 2006;18(2):196–202.
36. Janocha AJ, Koch CD, Tiso M, Ponchia A, Doctor A, Gibbons L, et al. Nitric oxide during altitude acclimatization. N Engl J Med. 2011;365(20):1942–4.
37. Beall CM, Laskowski D, Erzurum SC. Nitric oxide in adaptation to altitude. Free Radic Biol Med. 2012;52(7):1123–34.
38. Donnelly J, Cowan DC, Yeoman DJ, et al. Exhaled nitric oxide and pulmonary artery pressures during graded ascent to high altitude. Respir Physiol Neurobiol. 2011;177:213–217.
39. Julian CG, Wilson MJ, Moore LG. Evolutionary adaptation to high altitude: a view from in utero. Am J Hum Biol. 2009;21(5):614–22.
40. Brutsaert T, Parra E, Shriver M, Gamboa A, Rivera-Chira M, Leon-Velarde F. Ancestry explains the blunted ventilatory response to sustained hypoxia and lower exercise ventilation of Quechua altitude natives. Am J Physiol Regul Integr Comp Physiol. 2005;289:R225–34.
41. Brutsaert TD, Parra EJ, Shriver MD, Gamboa A, Palacios JA, Rivera M, et al. Spanish genetic admixture is associated with larger V(O2) max decrement from sea level to 4338 m in Peruvian Quechua. J Appl Physiol. 2003;95(2):519–28.
42. Brutsaert TD, Parra E, Shriver M, Gamboa A, Palacios JA, Rivera M, et al. Effects of birthplace and individual genetic admixture on lung volume and exercise phenotypes of Peruvian Quechua. Am J Phys Anthropol. 2004;123(4):390–8.
43. Harvey PH, Purvis A. Comparative methods for explaining adaptations. Nature. 1991;351(6328): 619–24.
44. Erzurum SC, Ghosh S, Janocha AJ, Xu W, Bauer S, Bryan NS, et al. Higher blood flow and circulating NO products offset high-altitude hypoxia among Tibetans. Proc Natl Acad Sci. 2007;104(45):17593.
45. Hoit BD, Dalton ND, Erzurum SC, Laskowski D, Strohl KP, Beall CM. Nitric oxide and cardiopulmonary hemodynamics in Tibetan highlanders. J Appl Physiol. 2006;99:1796–801.
46. Jansen GF, Basnyat B. Brain blood flow in Andean and Himalayan high-altitude populations: evidence of different traits for the same environmental constraint. J Cereb Blood Flow Metab. 2011;31(2):706–14.
47. Beall CM, Strohl K, Blangero J, Williams-Blangero S, Brittenham GM, Goldstein MC. Quantitative genetic analysis of arterial oxygen saturation in Tibetan highlanders. Hum Biol. 1997;69(5):597–604.
48. Beall CM. Two routes to functional adaptation: Tibetan and Andean high-altitude natives. Proc Natl Acad Sci. 2007;104 Suppl 1:8655.
49. Schwab M, Jayet PY, Stuber T, Salinas CE, Bloch J, Spielvogel H, et al. Pulmonary-artery pressure and exhaled nitric oxide in Bolivian and Caucasian high altitude dwellers. High Alt Med Biol. 2008; 9(4):295.
50. Jayet PY, Rimoldi SF, Stuber T, Salmon CS, Hutter D, Rexhaj E, et al. Pulmonary and systemic vascular dysfunction in young offspring of mothers with preeclampsia. Circulation. 2010;122(5):488–94.
51. Morin Y, Têtu A, Mercier G. Cobalt cardiomyopathy: clinical aspects. Br Heart J. 1971;33(Suppl):175–8.
52. Jefferson JA, Escudero E, Hurtado ME, Pando J, Tapia R, Swenson ER, et al. Excessive erythrocytosis, chronic mountain sickness, and serum cobalt levels. Lancet. 2002;359:407–8.
53. Tufts DA, Haas JD, Beard JL, Spielvogel H. Distribution of hemoglobin and functional consequences of anemia in adult males at high altitude. Am J Clin Nutr. 1985;42:1–11.
54. Smith TG, Talbot NP, Privat C, Rivera-Ch M, Nickol AH, Ratcliffe PJ, et al. Effects of iron supplementation and depletion on hypoxic pulmonary hypertension: two randomized controlled trials. JAMA. 2009;302(13):1444.
55. Harrison GA, Kuchemann CF, Moore MAS, Boyce AJ, Baju T, Mourant AE, et al. The effects of altitudinal variation in Ethiopian populations. Philos Trans R Soc Lond B Biol Sci. 1969;256(805):147–882.
56. Alkorta-Aranburu G, Beall CM, Witonsky DB, Gebremedhin A, Pritchard JK, Di Rienzo A. The genetic architecture of adaptations to high altitude in Ethiopia. PLoS Genet. 2012;8(12):e1003110.
57. Teshome D, Telahun T, Solomon D, Abdulhamid I. A study on birth weight in a teaching-referral hospital, Gondar, Ethiopia. Cent Afr J Med. 2006;52(1–2): 8–11.
58. Nekatibeb G, G/Mariam A. Analysis of birth weight in Metu Karl hospital: South West Ethiopia. Ethiop Med J. 2007;45(2):195–202.
59. Feleke Y, Enquoselassie F. Maternal age, parity and gestational age on the size of the newborn in Addis Ababa. East Afr Med J. 1999;76(8):468–71.
60. Madebo T. A two year retrospective study of birth weight in Sidamo Regional Hospital. Ethiop Med J. 1994;32(4):255–60.
61. Sheferaw T. Some factors associated with birth weight in Jima, southwestern Ethiopia. Ethiop Med J. 1990;28(4):183–90.

62. Green-Abate C. Changes in birthweight distribution from 1973 to 1982 in Addis Ababa. Bull World Health Organ. 1986;64(5):711–4.
63. Andersen GS, Girma T, Wells JC, Kæstel P, Michaelsen KF, Friis H. Fat and fat-free mass at birth: air displacement plethysmography measurements on 350 Ethiopian newborns. Pediatr Res. 2011;70(5):501–6.
64. Moore LG, Charles SM, Julian CG. Humans at high altitude: hypoxia and fetal growth [Research Support, N.I.H., Extramural Research Support, Non--U.S. Gov't Research Support, U.S. Gov't, Non--P.H.S. Review]. Respir Physiol Neurobiol. 2011;178(1):181–90.
65. Hoit BD, Dalton ND, Gebremedhin A, Janocha A, Zimmerman PA, Zimmerman AM, et al. Elevated pulmonary artery pressure among Amhara highlanders in Ethiopia. Am J Hum Biol. 2011;23(2):168–76.
66. Claydon VE, Gulli G, Slessarev M, Appenzeller O, Zenebe G, Gebremedhin A, et al. Cerebrovascular responses to hypoxia and hypocapnia in Ethiopian high altitude dwellers. Stroke. 2008;39(2):336–42.
67. Beall CM, Decker MJ, Brittenham GM, Kushner I, Gebremedhin A, Strohl KP. An Ethiopian pattern of human adaptation to high-altitude hypoxia. Proc Natl Acad Sci U S A. 2002;99(26):17215–8.
68. Mekbeb T, Ketsela K. Pre-eclampsia/eclampsia at Yekatit 12 Hospital, Addis Ababa, Ethiopia (1987–1989). East Afr Med J. 1991;68(11):893–9.
69. Gong J, Savitz DA, Stein CR, Engel SM. Maternal ethnicity and pre-eclampsia in New York City, 1995–2003. Paediatr Perinat Epidemiol. 2012; 26(1):45–52.
70. López-Jaramillo P, García RG, López M. Preventing pregnancy-induced hypertension: are there regional differences for this global problem? J Hypertens. 2005;23(6):1121–9.
71. Beall CM, Song K, Elston RC, Goldstein MC. Higher offspring survival among Tibetan women with high oxygen saturation genotypes residing at 4,000 m. Proc Natl Acad Sci U S A. 2004;101(39):14300–4.
72. Buderer MC, Page N. Hemopoiesis in the pig-tailed monkey, *Macaca nemestrina* during chronic altitude exposure. Am J Physiol. 1972;223(2):346–52.
73. Yu L, Wang X, Ting N, Zhang Y. Mitogenomic analysis of Chinese snub-nosed monkeys: evidence of positive selection in NADH dehydrogenase genes in high-altitude adaptation. Mitochondrion. 2011; 11(3):497–503.
74. Storz JF, Sabatino SJ, Hoffmann FG, Gering EJ, Moriyama H, Ferrand N, et al. The molecular basis of high-altitude adaptation in deer mice. PLoS Genet. 2007;3(3):e45.
75. Ramirez JM, Folkow LP, Blix AS. Hypoxia tolerance in mammals and birds: from the wilderness to the clinic. Annu Rev Physiol. 2007;69:113–43.
76. Winslow RM, Monge CC, Statham NJ, Gibson CG, Charache S, Whittembury J, et al. Variability of oxygen affinity of blood: human subjects native to high altitude. J Appl Physiol. 1981;51(6):1411–6.
77. McCracken KG, Bulgarella M, Johnson KP, Kuhner MK, Trucco J, Valqui TH, et al. Gene flow in the face of countervailing selection: adaptation to high-altitude hypoxia in the betaA hemoglobin subunit of yellow-billed pintails in the Andes. Mol Biol Evol. 2009;26(4):815.
78. Storz JF, Moriyama H. Mechanisms of hemoglobin adaptation to high altitude hypoxia. High Alt Med Biol. 2008;9(2):148–57.
79. Storz JF, Scott GR, Cheviron ZA. Phenotypic plasticity and genetic adaptation to high-altitude hypoxia in vertebrates. J Exp Biol. 2010;213(Pt 24):4125–36.
80. Monge C, Leon-Velarde F. Physiological adaptation to high altitude: oxygen transport in mammals and birds. Physiol Rev. 1991;71(4):1135–71.
81. Beall CM. Detecting natural selection in high-altitude human populations. Respir Physiol Neurobiol. 2007;158(2–3):161–71.
82. Grocott M, Montgomery H. Genetophysiology: using genetic strategies to explore hypoxic adaptation. High Alt Med Biol. 2008;9(2):123–9.
83. Hancock AM, Di Rienzo A. Detecting the genetic signature of natural selection in human populations: models, methods, and data. Annu Rev Anthropol. 2008;37:197–217.
84. Sabeti PC, Schaffner SF, Fry B, Lohmueller J, Varilly P, Shamovsky O, et al. Positive natural selection in the human lineage. Science. 2006;312(5780): 1614–20.
85. Sabeti PC, Varilly P, Fry B, Lohmueller J, Hostetter E, Cotsapas C, et al. Genome-wide detection and characterization of positive selection in human populations. Nature. 2007;449(7164):913–8.
86. The International HapMap Consortium. A haplotype map of the human genome. Nature. 2005;437(7063): 1299–320.
87. Harris EE, Meyer D. The molecular signature of selection underlying human adaptations. Yearb Phys Anthropol. 2006;49:89–130.
88. Bamshad M, Wooding SP. Signatures of natural selection in the human genome. Nat Rev Genet. 2003;4(2):99.
89. Beall CM, Cavalleri GL, Deng L, Elston RC, Gao Y, Knight J, et al. Natural selection on EPAS1 (HIF2alpha) associated with low hemoglobin concentration in Tibetan highlanders. Proc Natl Acad Sci. 2010;107(25):11459–64.
90. Simonson TS, Yang Y, Huff CD, Yun H, Qin G, Witherspoon DJ, et al. Genetic evidence for high-altitude adaptation in Tibet. Science. 2010; 329(5987):72–5.
91. Yi X, Liang Y, Huerta-Sanchez E, Jin X, Cuo ZX, Pool JE, et al. Sequencing of 50 human exomes reveals adaptation to high altitude. Science. 2010; 329(5987):75–8.
92. Bigham A, Bauchet M, Pinto D, Mao X, Akey JM, Mei R, et al. Identifying signatures of natural selection

in Tibetan and Andean populations using dense genome scan data. PLoS Genet. 2010;6(9):e1001116.
93. Bigham AW, Mao X, Mei R, Brutsaert T, Wilson MJ, Julian CG, et al. Identifying positive selection candidate loci for high-altitude adaptation in Andean populations. Hum Genomics. 2009;4(2):79–90.
94. Peng Y, Yang Z, Zhang H, Cui C, Qi X, Luo X, Tao X, Wu T, et al. Genetic variations in Tibetan populations and high altitude adaptation at the Himalayas. Mol Biol Evol. 2011;28:1075–81.
95. Wang B, Zhang YB, Zhang F, Lin H, Wang X, Wan N, et al. On the origin of Tibetans and their genetic basis in adapting high-altitude environments. PLoS One. 2011;6(2):e17002.
96. Scheinfeldt LB, Soi S, Thompson S, Ranciaro A, Meskel DW, Beggs W, et al. Genetic adaptation to high altitude in the Ethiopian highlands. Genome Biol. 2012;13:R1.
97. Storz JF. Evolution. Genes for high altitudes. Science. 2010;329(5987):40–1.
98. MacInnis MJ, Rupert JL. 'ome on the Range: altitude adaptation, positive selection, and Himalayan genomics. High Alt Med Biol. 2011;12(2):133–9.
99. Rupert J. Will blood tell? Three recent articles demonstrate genetic selection in Tibetans. High Alt Med Biol. 2010;11(4):307–8.
100. Scheinfeldt LB, Tishkoff SA. Living the high life: high-altitude adaptation. Genome Biol. 2010;11(9):133–5.
101. Wilson MJ, Julian CG, Roach RC. Genomic analysis of high altitude adaptation: innovations and implications. Curr Sports Med Rep. 2011;10(2):59–61.
102. Cheviron ZA, Brumfield RT. Genomic insights into adaptation to high-altitude environments. Heredity. 2012;108:354–61.
103. van Patot MC, Gassmann M. Hypoxia: adapting to high altitude by mutating EPAS-1, the gene encoding HIF-2α. High Alt Med Biol. 2011;12(2):157–67.
104. Webster KA. Evolution of the coordinate regulation of glycolytic enzyme genes by hypoxia. J Exp Biol. 2003;206(Pt 17):2911–22.
105. Semenza GL. Hydroxylation of HIF-1: oxygen sensing at the molecular level. Physiology. 2004;19: 176–82.
106. Loenarz C, Coleman ML, Boleininger A, Schierwater B, Holland PW, Ratcliffe PJ, et al. The hypoxia-inducible transcription factor pathway regulates oxygen sensing in the simplest animal, Trichoplax adhaerens. EMBO Rep. 2011;12(1):63–70.
107. Brakefield PM. Evo-devo and accounting for Darwin's endless forms. Philos Trans Biol Sci. 2011;366(1574):2069–75.
108. Voight BF, Kudaravalli S, Wen X, Pritchard JK. A map of recent positive selection in the human genome. PLoS Biol. 2006;4(3):e72.
109. Pickrell JK, Coop G, Novembre J, Kudaravalli S, Li JZ, Absher D, et al. Signals of recent positive selection in a worldwide sample of human populations. Genome Res. 2009;19(5):826–37.
110. Ganesh SK, Zakai NA, van Rooij FJ, Soranzo N, Smith AV, Nalls MA, et al. Multiple loci influence erythrocyte phenotypes in the CHARGE Consortium. Nat Genet. 2009;41(11):1191–8.
111. Droma Y, Hanaoka M, Basnyat B, Arjyal A, Neupane P, Pandit A, et al. Adaptation to high altitude in Sherpas: association with the insertion/deletion polymorphism in the Angiotensin-converting enzyme gene. Wilderness Environ Med. 2008; 19(1):22–9.
112. Suzuki K, Kizaki T, Hitomi Y, Nukita M, Kimoto K, Miyazawa N, et al. Genetic variation in hypoxia-inducible factor 1alpha and its possible association with high altitude adaptation in Sherpas. Med Hypotheses. 2003;61(3):385–9.
113. Hochachka P, Rupert J. Fine tuning the HIF-1 'global' O2 sensor for hypobaric hypoxia in Andean high-altitude natives. BioEssays. 2003;25(5):515–9.
114. Mejía OM, Prchal JT, León-Velarde F, Hurtado A, Stockton DW. Genetic association analysis of chronic mountain sickness in an Andean high-altitude population. Haematologica. 2005;90(1):13.
115. Puthucheary Z, Skipworth JR, Rawal J, Loosemore M, Van Someren K, Montgomery HE. The ACE gene and human performance: 12 years on. Sports Med. 2011;41(6):433–48.
116. Kumar R, Qadar Pasha M, Khan A, Gupta V, Grover S, Norboo T, et al. Association of high-altitude systemic hypertension with the deletion allele of the angiotensin-converting enzyme (ACE) gene. Int J Biometeorol. 2003;48(1):10–4.
117. Morrell NW, Sarybaev AS, Alikhan A, Mirrakhimov MM, Aldashev AA. ACE genotype and risk of high altitude pulmonary hypertension in Kyrghyz highlanders. Lancet. 1999;353(March 6):814.
118. Bigham AW, Kiyamu M, León-Velarde F, Parra EJ, Rivera-Ch M, Shriver MD, et al. Angiotensin-converting enzyme genotype and arterial oxygen saturation at high altitude in Peruvian Quechua. High Alt Med Biol. 2008;9(2):167–78.
119. Droma Y, Hanaoka M, Basnyat B, Ariyal A, Neupane P, Pandit A, et al. Genetic contribution of the endothelial nitric oxide synthase gene to high altitude adaptation in Sherpas. High Alt Med Biol. 2006; 7(3):209–20.
120. Carroll KJ. Back to basics: explaining sample size in outcome trials, are statisticians doing a thorough job? Pharm Stat. 2009;8(4):333–45.
121. Luo Y, Gao W, Liu F, Gao Y. Mitochondrial nt3010G-nt3970C haplotype is implicated in high-altitude adaptation of Tibetans. Mitochondrial DNA. 2011;22(5–6):181–90.
122. Torroni A, Miller JA, Moore LG, Zamudio S, Zhuang J, Droma T, et al. Mitochondrial DNA analysis in Tibet: implications for the origin of the Tibetan population and its adaptation to high altitude. Am J Phys Anthropol. 1994;93:189–99.
123. Merriwether DA, Ferrell RE. The four founding lineage hypothesis for the new world: a critical reevaluation. Mol Phylogenet Evol. 1996;5(1):241–6.
124. Myres JE, Malan M, Shumway JB, Rowe MJ, Amon E, Woodward SR. Haplogroup-associated differences

in neonatal death and incidence of low birth weight at elevation: a preliminary assessment. Am J Obstet Gynecol. 2000;182(6):1599–605.
125. Haldane JB. Natural selection in man. Acta Genet Stat Med. 1956;6(3):321.
126. Weinstein RS, Haas JD. Early stress and later reproductive performance under conditions of malnutrition and high altitude hypoxia. Med Anthrop. 1977;1(1):25–54.
127. Bennett A, Sain SR, Vargas E, Moore LG. Evidence that parent-of-origin affects birth-weight reductions at high altitude. Am J Hum Biol. 2008;20(5):592.
128. Moore LG, Niermeyer S, Vargas E. Does chronic mountain sickness (CMS) have perinatal origins? Respir Physiol Neurobiol. 2007;158(2–3):180.
129. Sartori C, Allemann Y, Trueb L, Delabays A, Nicod P, Scherrer U. Augmented vasoreactivity in adult life associated with perinatal vascular insult. Lancet. 1999;353(June 26):2205–7.
130. Beall CM. Adaptations to altitude: a current assessment. Annu Rev Anthropol. 2001;30:423–46.
131. Keyes LE, Armaza JF, Niermeyer S, Vargas E, Young DA, Moore LG. Intrauterine growth restriction, preeclampsia, and intrauterine mortality at high altitude in Bolivia. Pediatr Res. 2003;54(1):20–5.
132. Zamudio S. High-altitude hypoxia and preeclampsia. Front Biosci. 2007;12:2967.
133. Penaloza D, Arias-Stella J. The heart and pulmonary circulation at high altitudes: healthy highlanders and chronic mountain sickness. Circulation. 2007; 115(9):1132.
134. Wu T. High altitude heart disease in children in Tibet. High Alt Med Biol. 2002;3(3):323–5.
135. Leon-Velarde F, Monge CC, Arregui A, Stanley C. Increased prevalence of excessive erythrocytosis with age in healthy high altitude miners. In: Sutton JR, Coates G, Remmers JE, editors. Hypoxia: the adaptations. Toronto: B.C. Decker Inc; 1990. p. 280.
136. Gonzales GF, Steenland K, Tapia V. Maternal hemoglobin level and fetal outcome at low and high altitudes. Am J Physiol Regul Integr Comp Physiol. 2009;297(5):R1477–85.

Acute Mountain Sickness and High Altitude Cerebral Oedema

20

Peter Bärtsch and Damian Miles Bailey

Abstract

This chapter summarises advances made over the last 12 years regarding our understanding of the pathophysiology and its clinical implications in acute mountain sickness (AMS) and high altitude cerebral oedema (HACE). Issues on the definition and diagnosis of AMS and HACE as well as determinants of incidence and susceptibility are discussed. Furthermore, new studies on prevention and treatment of AMS are critically evaluated. Findings on lung function, gas exchange, metabolism, hormonal response, markers of inflammation, changes in the autonomic nervous system, cerebral blood flow, and brain imaging are reviewed. The results of these examinations are incorporated into an overall concept relating to the underlying pathophysiology of acute mountain sickness and high altitude cerebral oedema.

Introduction

The goal of this chapter is (1) to summarise recent developments in our understanding of the cause, prevention and treatment of acute mountain sickness (AMS) and high altitude cerebral oedema (HACE) and (2) to discuss the pathophysiologic concepts that have evolved from this gain in knowledge. The chapter is based on 681 papers published between January 2000 and December 2011 that are found in Medline under the keywords "AMS" or "HACE". 90 % of these publications are written in English. Due to limited space we focused on original research papers that in our opinion have added significantly to the existing body of knowledge.

P. Bärtsch, M.D. (✉)
Division of Sports Medicine, Department of Internal Medicine, Medical University Clinic, University of Heidelberg, Heidelberg, Germany
e-mail: peter.bartsch@med.uni-heidelberg.de

D.M. Bailey, Ph.D. (✉)
Neurovascular Research Laboratory, Faculty of Life Sciences and Education, University of South Wales, Mid-Glamorgan, UK
e-mail: damian.bailey@southwales.ac.uk; o2radical@btinternet.com

Clinical Aspects of Acute Mountain Sickness

Diagnosis of AMS

Comparison between studies is difficult because of the different definitions. Many authors use the definition suggested by a consensus group [1] of

E.R. Swenson and P. Bärtsch (eds.), *High Altitude: Human Adaptation to Hypoxia*,
DOI 10.1007/978-1-4614-8772-2_20, © Springer Science+Business Media New York 2014

a Lake Louise (LL) score >2 with headache plus one additional symptom. This means that mild headache plus either mild dizziness or slight reduction of appetite or mild disturbance of sleep is considered AMS after a recent gain in altitude. At higher altitudes such symptoms occur almost universally and individuals with such symptoms do not usually consider themselves sick. The AMS-C score, which is also frequently used, has a cut-off score that is supposed to identify with reasonable sensitivity and specificity individuals who consider themselves sick. A direct comparison between these two scoring systems at 4,559 m indicates that an LL score >4 (independent of whether headache is present or not) [2] approximates to an AMS-C cut-off score ≥0.70 [3, 4]. Investigators that intend to unravel the pathophysiology of AMS are interested in comparing an abnormal vs. a normal response to hypoxia. Some investigators have therefore chosen to use more rigorous criteria such as the combination of an AMS-C score ≥0.70 and an LL score >4 [5]. These cut-off levels have been shown to identify individuals that consider themselves sick. Other investigators eliminate confounding effects of mild symptoms with questionable relevance by comparing groups with the highest vs. the lower scores [6, 7].

There are two further limitations of the Lake Louise score: (1) The inclusion of sleep quality precludes comparison of exposures of short duration, i.e. of 8 h during day time, with those that last over night and include sleeping in hypoxia. (2) The severity of symptoms has only 4 levels. It may be difficult to decide between no and mild headache. Since the LL score is often familiar to lay people, many study subjects may be aware that the answer to the question about headache results in their being classified as being sick or not sick. The assessment of AMS by the Lake Louise score may therefore be vulnerable to bias. Some investigators have suggested to use visual analogue scales for assessment of headache [8, 9]. The advantage of the AMS-C in these regards is obvious as it does not include sleep quality and as the subjects choosing between 6 levels of 11 symptoms will not be aware of the outcome regarding classification.

Exercise may be an additional confounder. Symptoms of AMS are non-specific and some can also be induced by exercise. Thus studies that measure AMS scores within a short time after the end of an ascent may assess a mixture of symptoms due to exhaustion and AMS.

In summary all these considerations regarding the definition of AMS point to the need to re-evaluate the diagnostic criteria for clinically relevant AMS. It may not be possible to use universal definitions since what is normal at 5,000 m may be abnormal at 3,000 m. A major problem is due to the fact that an abnormal response needs to be defined in a specific setting of altitude, ascent rate and degree of acclimatisation since these are the major independent predictors of AMS [10]. Considering all these limitations, differences in definitions of AMS limit comparison between studies and may explain controversial findings.

Determinants of Incidence

The threshold altitude for symptoms attributable to hypobaric hypoxia in resting individuals was shown to be between 7,000 and 8,000 ft [11]. Several studies add further evidence that history of AMS, level of residency or recent preceding altitude exposure and rate of ascent are the most important independent predictors of AMS for any given altitude (Fig. 20.1) [10, 12, 13]. Prolonging the ascent by a few days reduced severity and prevalence of AMS considerably when climbing to an altitude of 6,285 m in 12 vs. 16 days [14]. Gender, BMI and age were not associated with AMS in mountaineering populations [10, 12, 13] while severe obesity (average BMI of 35) is a risk factor for AMS as shown in a hypobaric chamber study [15]. Awareness and knowledge about AMS reduces the prevalence [16, 17], which is most likely due to the intake of drugs for prevention or slow ascent [16]. A small study suggests that individuals with systemic hypertension may be more prone to AMS [18].

Exercise was shown to exacerbate AMS at simulated altitude [19], but this finding has been challenged recently [20]. Aerobic capacity reduces AMS only in groups with identical ascent

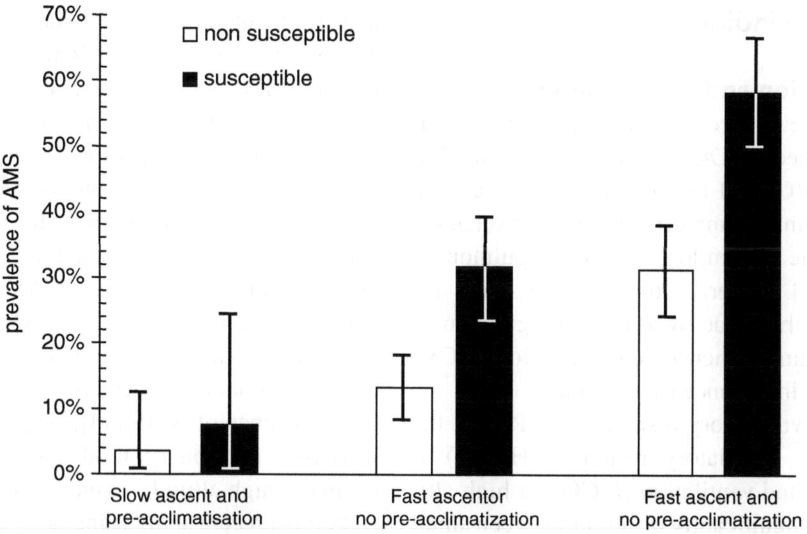

Fig. 20.1 Prevalence of AMS at an altitude of 4,559 m in relation to speed of ascent, pre-acclimatisation and susceptibility to AMS in 827 mountaineers. Slow ascent was defined as taking more than 3 days for the ascent above 2,000 m, and pre-acclimatisation was defined as having spent more than 4 days above 3,000 m within the last 2 months. Mountaineers who reported to suffer never or rarely from headache, and who had a low score on additional symptoms with exposures above 3,000 m were considered not to suffer from AMS (data from [10])

rates [14] or when ascent and descent occur in 1 or 2 days and increased aerobic capacity allows shorter exposure to high altitude [12, 13]. The former finding is compatible with the notion that a higher relative exercise intensity augments AMS.

Epidemiologic data based on mountaineers at extreme altitude (<5,500 m), such as Mt. Damavand (5,671 m) [13], Kilimanjaro (5,895 m) [21, 22] and Aconcagua (6,962 m) [23], has been added. The average overall ascent rate above 2,000 m on these mountains was 611 m/day over 8 days on Aconcagua, 780–980 m/day over 4 or 5 days on Kilimanjaro and 1,200–1,600 m/day in 2 or 3 days on Mt. Damavand. The comparison between studies regarding incidence of AMS is difficult since different definitions of AMS were used. Average values on these mountains range from 60 to 75 %.

Two studies indicate that the prevalence of AMS in children is similar to those obtained in accompanying adults, for preverbal children in whom AMS was assessed with a special children's Lake Louise score [24] at 3,109 m [25], and in prepubertal children at 3,450 m [26]. Furthermore, prevalence of AMS in adolescents at the same altitude was comparable with historic adult controls [26, 27]. These data indicate that up to an altitude of 3,500 m the prevalence of AMS is similar between children of all ages and adults. One group in South America reports an incidence of 90–100 % in children at 3,500 and 4,400 m in the Andes [28, 29]. Most likely differences in assessment, altitude, study setting and possibly cultural background account for these discrepancies.

Clinical Signs

Two studies have shown that mild subclinical ataxia that can only be demonstrated by posturography is present at altitudes of 4,300–4,559 m in all examined subjects independent of the presence or absence of AMS [30, 31]. Furthermore, it does not improve following 10 min of breathing oxygen [30]. In order to detect minor abnormalities in ataxia by clinical examination, a sharpened Romberg test has been evaluated [32] and also showed a lack of correlation with symptoms of AMS. In summary, it appears that clinically detectable ataxia is not present in AMS and is, as discussed below, an early sign of HACE [33].

Additional Findings

Lung Function and Gas Exchange

A previous review concluded that lower SaO_2 or PaO_2, widened $AaDO_2$, decreased forced vital capacity (FVC) and lower diffusion capacity in AMS all point to impairment of gas exchange, possibly due to mild interstitial pulmonary oedema [34]. Several studies have since re-investigated this issue by assessing regulation of ventilation, lung function, closing volume (CV) and thoracic impedance tomography.

Hypoxic ventilatory response (HVR) and the hypercapneic ventilatory response (HCVR) at low altitude and ventilation (P_ACO_2) at high altitude are not significantly different between small groups of mountaineers with and without AMS [35]. In larger groups poikilocapneic HVR at rest and during exercise are significantly lower in mountaineers who develop severe high altitude disease, with a big overlap between groups. A low HVR during exercise is, however, a minor, but significant, risk factor in a multivariate model of prediction [36]. The more severe hypoxaemia during sleep at 4,559 m cannot be attributed to hypoventilation [37]. Nevertheless, the increase of HVR at high altitude is delayed in AMS [35]. This and equal ventilation and P_ACO_2 between subjects with and without AMS despite significantly lower P_aO_2 in AMS subjects suggest that a lower ventilation may be a contributor to more severe hypoxaemia in AMS [35].

A study with 261 subjects showed no change in FVC at 4,559 m [38], while smaller studies reporting a reduction in FVC at comparable altitudes were accompanied by a reduction of FEV_1 or a peak flow [39] that was lower than predicted from the decreased gas density [40]. Thus, a reduced effort due to fatigue or illness may also contribute to the decrease in FVC in these studies [41]. In accordance with this hypothesis, reduced respiratory muscle strength was found after 12 h of exposure to hypobaric hypoxia (4,270 m) [42] and at altitudes of 3,600–4,559 m [43].

Closing volume (CV), an indicator of small airway closure, was increased in 77 % of 261 mountaineers and significantly more frequent in the 41 subjects with rales or radiographic evidence of interstitial pulmonary oedema [38]. There was, however, no significant association with AMS in this study. In another investigation at the same location, the increase of CV and the changes of CV at high altitude were not related to differences in lung volume or arterial oxygen saturation (S_PO_2). If we assume that increased CV indicates interstitial oedema, these data suggest that this is not related to AMS and not the only cause for lower S_PO_2 in AMS. Decreased thoracic impedance interpreted as indicating subclinical pulmonary oedema was found at 3,800 m in the absence of AMS [44], suggesting also that increased pulmonary fluid accumulation, if it occurs at high altitude, is independent of AMS.

A most recent study using body plethysmography at 4,559 m found no significant changes compared to baseline in subjects with and without AMS for CV, FVC, compliance, resistance and diffusion capacity [45]. Furthermore, minimal changes of these parameters in four patients with beginning, alveolar HAPE on chest radiographs suggest that pulmonary function testing and measurement of closing volume are not very sensitive for detecting mild interstitial pulmonary oedema.

From these investigations again we concluded that at present there are no reliable tools for field studies to prove or exclude the presence of mild pulmonary oedema at high altitude and to determine its potential role in the pathophysiology of AMS.

Echocardiography

Exposure to hypoxia, particularly in combination with exercise, may be associated with oedema formation in various tissues because of an increase in hormones causing sodium and water retention and because of increased capillary permeability as discussed in a previous review [34]. A small, clinically non-relevant pericardial effusion was detected by echocardiography in the majority of subjects during a stay of 7 days at 5,200 m. The effusion was not correlated with symptoms of AMS and not prevented by the intake of several antioxidants [46]. Interestingly, other studies using echocardiography by different experienced examiners at the altitude of 4,559 m failed to report these findings during exposures of 3–4 days.

Autonomic Nervous System

As discussed in a previous review exposure to hypoxia is associated with increased sympathetic activity measured by microneurography, heart rate variability, plasma levels and urinary excretion of catecholamines [34]. Higher norepinephrine and epinephrine in subjects with AMS suggest a higher sympathetic activity in this condition without establishing cause-and-effect relationship. Activation of the sympathetic nervous system may contribute to AMS by increasing metabolism and thus oxygen demand or by enhancing sodium retention supporting oedema formation, but the sympathetic nervous system could also be activated by stress due to illness. Furthermore, auto-oxidation of epinephrine to adrenochrome has the potential to yield the epinephrine semiquinone free radical which can further catalyse superoxide ($O_2^{\bullet -}$) formation [47]. This may prove relevant since increased neuro-oxidative stress might be involved in the pathophysiology of AMS (see later).

A most recent study demonstrates that a greater increase of arterial catecholamines in the first hours of exposure to hypoxia precedes symptoms of AMS [48]. Furthermore, AMS was associated with a heart rate and blood pressure variability pattern indicating greater sympathetic activity compared to non-AMS controls [49] which in some studies showed a decrease of autonomic activity in subjects without AMS particularly with passive [50] or slow ascent [50–52]. However, this is not a universal finding [53]. During acute exposure in low pressure chambers sympathetic activity may be triggered by a vasovagal reaction [52]. An acute 10-min exposure to normobaric hypoxia caused an increased low frequency component of the systolic blood pressure variability suggesting increased sympathetic activity that was confined to AMS-susceptible subjects [49]. In summary, AMS-prone individuals may have a higher sympathetic activity in hypoxia than non-AMS controls, and this could be relevant for the pathophysiology of AMS.

Cerebral Blood Flow

This topic is discussed in more detail in Chap. 7. In this paragraph we focus on the relevance of cerebral blood flow (CBF) for AMS. The increase of CBF in hypoxia was not consistently detectable within the first 6 h of hypoxia by transcranial Doppler and changes were not related to AMS [8, 54]. Single photon emission tomography (SPECT), however, demonstrated a clear increase in CBF after 40 min at a simulated exposure to 5,500 m without any major regional redistribution of blood [55]. Furthermore, it was shown by MRI using spin labelling that the increase of CBF is not different between grey and white matter, irrespective of susceptibility to AMS [56]. In addition, dynamic autoregulation was shown to be selectively impaired in subjects with AMS exposed to an altitude of 4,559 m [57] and during a comparable normobaric hypoxic exposure [8]. This suggests that the AMS-brain is less capable of buffering rapid surges in arterial pressure and thus potentially more vulnerable to mild over-perfusion and, thus by consequence, vasogenic oedema.

Analysis of cerebrospinal fluid (CSF) did, however, not provide evidence for any disruption of the blood–brain barrier (BBB) subsequent to an autoregulatory breakthrough in AMS [5]. Several studies have consistently failed to demonstrate any defining changes in the CSF [5], systemic [5, 8] and arterio-jugular venous concentration difference [58] of the astrocytic protein S100β considered a molecular surrogate of BBB leakage [59] in subjects when compared to those without AMS. Likewise, the CSF–blood concentration quotient of total protein and the immunoglobulins G, A and M also failed to distinguish AMS subjects from non-AMS controls [5]. While these findings exclude "major" disruption of the BBB, further studies focusing on the CSF–blood quotient of small molecules not synthesised by the brain combined with contrast-enhanced CT-MRI with gadolinium are warranted to determine if "minor" disruption of the BBB does indeed represent a significant event in AMS. It is conceivable that small molecules that exhibit toxic neurotransmitter-like properties such as products of digestion or gastrointestinal bacterial aminoacid adducts might cause minor disruption of the BBB and cause symptoms.

Estimates of Intracranial Pressure

Lumbar puncture after 18 h of exposure to 12 % F_IO_2 (corresponding to a PO_2 of an altitude of 4,559 m) was normal and not different between subjects with and without AMS [5]. Optical disc swelling (ODS, papilloedema) was found in 59 % of climbers at altitudes of 4,497–6,865 m [60]. Although ODS and AMS increased in parallel with altitude, there was no significant association between the two parameters, as 64 % of those with and 56 % of those without AMS had ODS. Similar findings were reported for optical nerve sheath diameter (ONSD) assessed by B-scan ultrasound during an active ascent of Mt. Everest [61]. The authors demonstrated an association between AMS and a physiologically insignificant increase in ONSD (+0.02 mm) at the intermediate altitudes of 2,000–3,700 m where symptoms of AMS typically present and at 6,400 m, ONSD increased as symptoms tended to resolve. This study only showed an association of AMS and ONSD with increasing altitude but not between AMS and ONSD at the highest altitude (6,400 m) [61]; the second study showed a correlation of ONSD and AMS scores in a large setting of 287 trekkers at 4,250 m [62]. These latter findings could point to an increased intracranial pressure (ICP) as suggested by the authors of these investigations. They could, however, merely be due to a general slightly increased leakiness of capillaries in hypoxia that also results in peripheral oedema, mild pericardial effusion, possibly mild interstitial pulmonary oedema and proteinuria, as discussed in part in a previous review [34]. The fact that several MRI studies found no indirect evidence of increased ICP, as discussed below, argues against the interpretation of these authors, as does a recent study using a confocal scanning laser ophthalmoscope [63]. At 4,559 m optical disc oedema was present in 79 % of the subjects and not related to AMS.

Direct measurement of ICP by telemetric monitoring through a burr hole in the skull in three individuals ascending to 5,730 m showed normal pressure at rest and an increase to 51 mmHg with exercise in the one subject with severe symptoms of AMS who had also a ventricular atrial shunt because of a post-traumatic hydrocephalus. It remained normal in one and rose about twofold to 36 mmHg in another asymptomatic subject during press-up exercise [64]. Because of the small number, the lack of an adequate control experiment in normoxia and the concomitant post-traumatic hydrocephalus, the interpretation of these data is difficult.

ODS and particularly increased ONSD were validated as an indicator of increased ICP by direct measurements at low altitude, particularly in the setting of trauma. Whether this also applies to altitude exposure is not known. Theoretically an increase in ODS or ONSD may merely reflect a general increase in permeability. Similar considerations apply to the significance of intra-ocular pressure as an indicator of ICP. Most studies reported, however, a decrease or no change of intra-ocular pressure up to altitudes of 5,200 m [65–67].

Brain Imaging

MRI studies after exposures of 6–16 h of normobaric or hypobaric hypoxia showed a minimal increase in total brain volume of 5–10 mL (<1 % of brain volume) [68], a small reduction of 1 mL of the inner cerebral spinal fluid volume [69] and an increased T2-signal intensity indicating that there is more water in the brain [68, 70]. A study performed in a hypobaric chamber over 32 h at 4,300 m showed a 2.8 % volume increase of predominantly grey matter [71]. While all these parameters were not significantly different between subjects with and without AMS, a decrease of the apparent diffusion coefficient indicating cell swelling (cytotoxic oedema) was significantly associated with AMS [68, 70]. The decrease of intracranial CSF volume and the increase in brain volume are already detectable after 40 min at simulated altitude (3,800 m) with a similar magnitude compared to measurements performed after 6–16 h and they were also not different between individuals considered resistant and susceptible to AMS [72].

Identification of Susceptible Individuals

Previous data suggested that susceptible individuals cannot be identified reliably by measuring HVR at low altitude [34]. Possible explanations were that HVR is not a major determinant of ventilation at high altitude and that ventilation at high altitude does not account for the variability in AMS. In addition HVR measured at low altitude does not correlate with measurements at high altitude, since the increase of HVR with acclimatisation is delayed in subjects when compared to those without AMS [35]. Accordingly, end-tidal PCO_2 did not correlate between measurements obtained during acute exposure at low altitude with the measurements performed at comparable altitudes during trekking [73].

In a retrospective study that selected susceptible and non-susceptible individuals from a cohort of 500 mountaineers, SaO_2 measured after 30 min of exposure to hypoxia or high altitude results in significantly lower SaO_2 values in those with compared to those without a history of AMS [74]. However, because of the large overlap of SaO_2 values between the groups this measurement has no clinical relevance for prediction or identification of AMS. This was shown in four prospective studies of large groups of unselected mountaineers [75–78].

Other tests for identification of susceptibility to AMS have been suggested, such as measuring SaO_2 during running uphill at high altitude [79], assessing hyperventilatory capacity [80] or respiratory rate after arrival at high altitude [81]. Positive findings may be due to particular circumstances of these studies. A large overlap of individual values between groups suggests again that statistical significance does not mean clinical relevance for prediction of susceptibility for AMS.

It is highly unlikely that any single test would be able to identify susceptible individuals with sufficient accuracy without taking several physiologic factors that are considered to be involved in the pathophysiology of AMS into account. For prediction of AMS in a specific setting, degree of acclimatisation, rate of ascent and final altitude should also be added to a multivariate approach.

Clinical Aspects of High Altitude Cerebral Oedema

The construction of the Golmud-Lhasa railway that runs for almost 800 km at altitudes above 4,000 m offered a great opportunity for gathering epidemiologic data on rare altitude-associated diseases such as HACE, since 74,735 workers were involved in the construction. The prevalence of HACE was found to be 0.26 %. The clinical data of 66 cases were presented recently [33]. The frequency of symptoms and signs of these HACE patients is shown in Table 20.1. Ataxia (unstable gait) was usually the first sign. Noteworthy was that headache was absent in 33 % of the cases. So far, there were only rare reports on HACE without preceding headache [82]. Twenty six patients developed coma on average within 18 h after the onset of symptoms. CT scans of 36 patients

Table 20.1 Symptoms and signs in HACE

Symptoms and signs	Number	Percent	Symptoms and signs	Number	Percent
Disturbance of consciousness	52	79	Apathy	24	36
Ataxia	48	73	Drowsiness	22	33
Headache	44	67	Abnormal reflexes	19	29
Anorexia	38	58	Psychological changes	18	27
Nausea	35	53	Vomiting	11	17
Papilloedema	24/52	46	Disorientation	8	14
Retinal haemorrhage	22/52	42	Hallucinations	2	3
Lassitude	27	41	Fingers and ankles clonus	1	2

Frequency of symptoms and signs in HACE in 66 cases of HACE that occurred during the construction of the Golmud-Lhasa railway according to Wu et al. [33]

showed oedema with compression of ventricles. MRI of the brain was performed in four subjects and confirmed the previously reported finding of oedema predominantly in the splenium of the corpus callosum [83]. Ataxia disappeared with treatment in the majority of patients within 3 days, but in 25 % of the patients it took 4–7 days. There was complete recovery except for one patient with persistent clonic convulsions. Another recently reported sequelae of HACE is a globus pallidus lesion in two cases [84].

Cerebral symptoms, such as global amnesia [85, 86], delirium [87], globus pallidus lesion [84] and sudden onset of ataxia [88], may in rare cases be caused by hypoxia in the absence of HACE. HACE may be excluded as an underlying cause when such symptoms occur in well-acclimatised and slowly ascending subjects without preceding symptoms of AMS and when they resolve rapidly with descent.

It is well known that microhaemorrhages can be found at autopsies in brains of HACE victims. Using MRI techniques highly sensitive to blood or blood remnants, haemosiderin deposits are detectable in the corpus callosum 2–31 months after recovery from HACE (Fig. 20.2) [89]. Such changes were not seen in subjects after severe AMS. These haemosiderin deposits indicate that microhaemorrhages occur in non-lethal HACE, and they may constitute a novel diagnostic MRI sign for HACE that can be detected many months to years after the event.

Pathophysiology of AMS and HACE

While it is well-established that patients with HACE exhibit extracellular (vasogenic) oedema subsequent to disruption of the BBB [83, 89], the situation with AMS is more complex, due in large part to the difficulties associated with clinical diagnosis. Traditionally, AMS has been considered a mild form of HACE and that both syndromes share a common pathophysiology linked by vasogenic oedematous brain swelling and intracranial hypertension at opposing ends of a clinical continuum [90]. However, recent studies employing diffusion-weighted (DW)-MRI have questioned this paradigm and since provided insight into alternative mechanisms. The subsequent discussion will critically appraise each of the traditional and newly emerging components currently implicated in the pathophysiology of the cerebral syndromes encountered at high altitude. These are summarised schematically in Fig. 20.3.

Phase I: The Stimulus

Hypoxia

Although hypoxia is not the immediate cause of AMS since symptoms typically take 6 h to evolve, it is the primary stimulus since symptoms typically become worse with increasing altitude [91] and relieved by normalising the inspiratory PO_2 [92]. Furthermore, it has been suggested that AMS-susceptible subjects are systemically more hypoxaemic or equally, more reactive to hypoxaemia for any given inspirate during wakefulness [6] and sleep [68] compared to their healthier counterparts. Tentative evidence suggests that a blunted HVR and sympathetic activation leading to activation of the renin–angiotensin system, fluid retention and subclinical interstitial pulmonary oedema may prove additional risk factors that further compound hypoxia subsequent to impaired gas exchange and thus account for the increased sensitivity-reactivity to arterial hypoxaemia. However, evidence of interstitial pulmonary oedema in AMS is at best indirect and inconsistent [38, 45] while fluid retention might prove the consequence rather than the direct cause of AMS [93].

Since a previous history has been identified as one of the most reliable predictors of illness during subsequent ascent, there may be some innate predisposition to AMS [10]. Subsequent investigations have focused on candidate gene polymorphisms in an attempt to identify potential genetic variants underlying susceptibility [94]. Recent interest has focused on the angiotensin-converting enzyme (ACE) gene since the insertion (I) allele has previously been associated with elite mountaineering status [95] and successful ascent to 8,000 m peaks [96]. However, studies have failed to support any association between ACE genotype and susceptibility to AMS [97, 98]. In addition, no biologically plausible association has been observed between polymorphisms of the heat shock protein (HSP) 70 family (A1 and A2) [99].

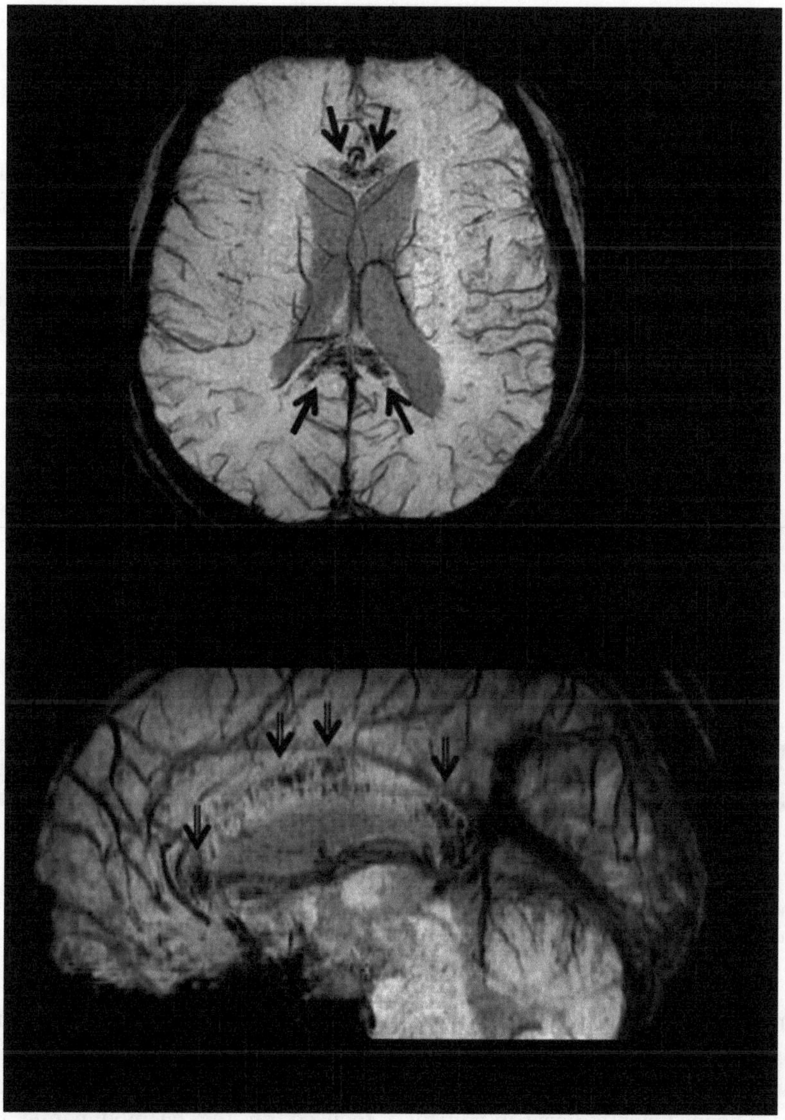

Fig. 20.2 T* MRI of a 65-year-old woman who had suffered from severe HACE 7 weeks before at 3,580 m shows microhaemorrhages in the corpus callosum (*arrows*) (courtesy of Michael Knauth, Professor of Neuroradiology, University Hospital Göttingen)

While the current evidence has failed to identify a genetic contribution to AMS, it is important to emphasise that the sample sizes of all the afore-mentioned studies are too small for genetic association studies. Furthermore, "genetophysiology" is very much in its infancy with only 20 out of ~25,000 genes in the human genome and only 50 out of ~ millions of polymorphisms currently assessed for roles in hypoxic (mal)-adaptation and altitude illness.

Phase II: The response

Impaired Focal Cerebral Oxygenation

While existing measurement techniques have failed to identify any evidence for a "global" cerebral O_2 deficit in the severely hypoxic–hypocapneic human ($SaO_2 < 80$ %, $PaCO_2 < 25$ mmHg) [100], positron-emission tomography studies have revealed striking spatio-temporal differences in the brain's "regional"

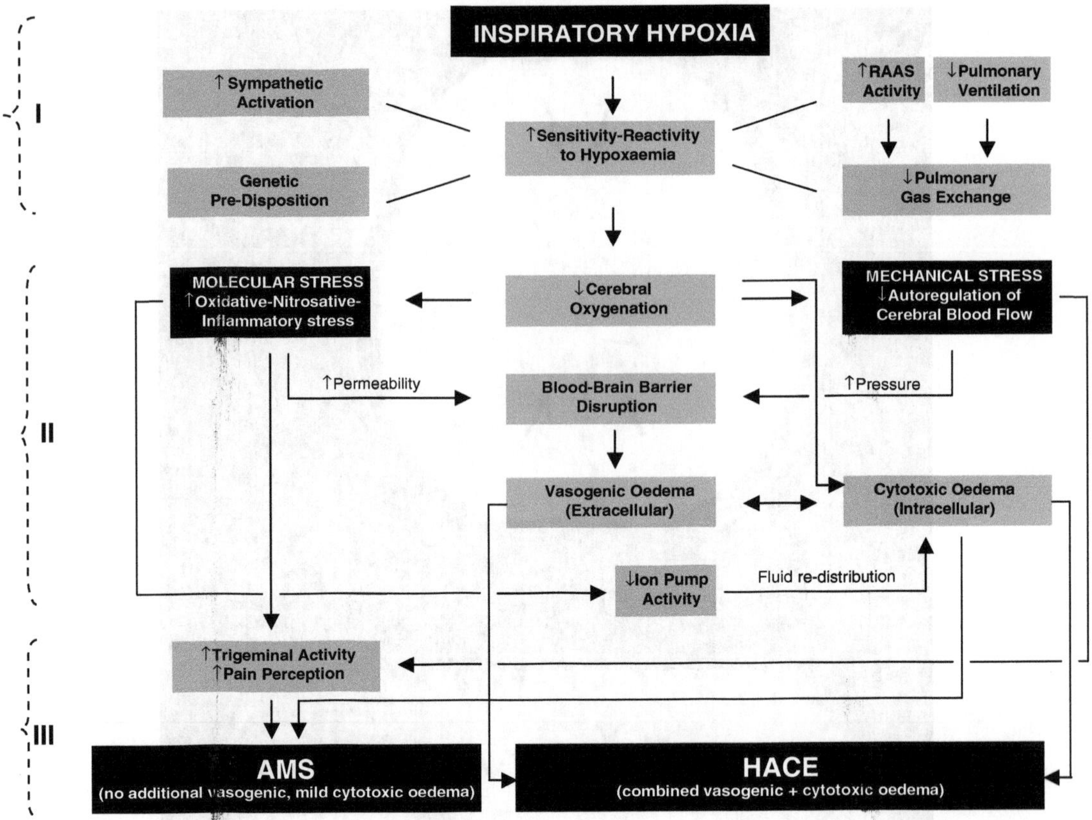

Fig. 20.3 Pathophysiology of acute mountain sickness (AMS) and high altitude cerebral oedema (HACE). Phase I and II describe the sequence of events common to both syndromes leading to varying degrees of blood–brain barrier disruption and vasogenic oedema (mild to severe). Unlike HACE, this is considered an incidental finding in AMS, with molecular activation of the trigeminovascular system (Phase III) the emerging mechanism of potential significance

energy consumption which is apparent even at rest during normoxia [101]. Focal prefrontal cortical deoxygenation determined via near infra-red spectroscopy has also been shown to occur in humans exposed to hypoxia and further compounded by exercise [102], though like all non-invasive approaches, there are inherent limitations associated with this technique [103]. Furthermore, neurotransmitters, the ubiquitous regulators of neuronal activity, are particularly vulnerable to hypoxia in light of their correspondingly high Michaelis constant for O_2 [104]. Thus, it is conceivable that even mild arterial hypoxaemia could result in focal cerebral deoxygenation and impaired neuronal communication.

Molecular-Mechanical Stress

Focal impairments in cerebral oxygenation have the potential to promote capillary leakage (molecular stress) and hyper-perfusion (haemodynamic stress), established risk factors that collectively conspire to disrupt the BBB. Much interest has focused on the haemodynamic pathway since increased CBF occurs in response to hypoxia [105–107] and impaired autoregulation has recently been documented in AMS [8, 57]. These findings suggest that the AMS-brain is less capable of "buffering" rapid surges in arterial pressure and thus potentially more vulnerable to hyperperfusion. A pressure-passive rise in regional CBF could translate into increased capillary hydrostatic pressure [108] though this is yet to be confirmed in humans. Elevated intravascular pressure could cause vasogenic oedema subsequent to hydrostatic disruption of the BBB.

Oxidative-nitrosative-inflammatory stress constitutes an alternative, albeit complementary, pathway with equal potential to further compound barrier disruption through its (molecular) impact on capillary permeability. A recent study provided direct electron paramagnetic resonance (EPR) spectroscopic evidence for an increased release of lipid-derived alkoxyl-alkyl ($LO^{\bullet}$-$LC^{\bullet}$) radicals (oxidative stress) and corresponding reduction in the vascular bioavailability of nitric oxide (nitrosative stress) across the hypoxic human brain in direct proportion to AMS symptom scores [58]. Furthermore, dietary antioxidant vitamin supplementation provided some, albeit mild, prophylactic benefit [109] though further research employing larger sample sizes with improved methods of delivering novel antioxidant vehicles across the BBB to the brain parenchyma is warranted.

Ascent to high altitude has consistently been associated with a mild, non-specific inflammatory response though its relationship to AMS remains equivocal. To date, only one study has identified a linear correlation between an increase in the systemic concentration of vasoactive eicosanoids and AMS symptom scores following ascent to the Vallot Observatory on Mt. Blanc (4,350 m) [110]. However, follow-up studies employing similar [111] and indeed alternative, complementary blood-borne biomarkers of inflammation [5, 112–115] have failed to confirm any such relationship. These findings do not exclude a pathogenic role for inflammation but merely highlight detection sensitivity issues associated with mostly cytokine detection using enzyme-linked immunosorbent assays (ELISA) and unavoidable limitations when attempting to correlate blood-borne biomarkers with different, often delayed, half-lives against subjective symptoms whose onset is more acute [8]. Furthermore, it is conceivable that the inflammatory response incurred during passive exposure to hypoxia is so mild that it is below the (as of yet undefined) physiological threshold required to trigger symptoms. Indeed, a recent study "primed" volunteers by infusing endotoxin intravenously during passive exposure to hypoxia in an attempt to further amplify the systemic inflammatory response and identified that 85 % of volunteers subsequently developed moderate-to-severe AMS compared to 9 % of the controls (saline infusion) [116]. Though this (pharmacological) approach is clearly supra-physiological, ascent to high altitude is none-the-less associated with strenuous physical exercise accompanied by damaging eccentric muscle contractions, established stimuli known to compound the inflammatory response to hypoxia [114, 117, 118]. It would be interesting to follow these studies up with separate infusions of recombinant TNF-α and IL-6 to determine the potential impact of individual cytokines on AMS susceptibility.

However, it is important to emphasise that these and related studies have consistently failed to provide molecular evidence for major BBB disruption in the form of an increase in the astrocytic protein S100β [8, 58] or CSF–blood protein concentration quotient [5]. However, in light of current structural DW-MRI evidence [5, 68], these findings do not discount the possibility of barrier disruption; they simply suggest that the breach is so subtle that it is likely beyond standard bio-molecular detection limits [119]. Follow-up studies employing transcerebral sampling techniques preceded by gadolinium-enhanced DW-MRI will allow this hypothesis to be tested more directly.

Animal models have demonstrated that hypoxia stimulates vascular endothelial growth factor (VEGF) resulting in vascular leakage and cerebral oedema tempting speculation that its expression may predispose to high altitude illness [120, 121]. Human studies have since failed to demonstrate any relationship between blood and/or CSF concentrations of "total" VEGF and AMS [5, 122]. This may be due to the fact that differences in the "free" concentration of VEGF mediated largely by its soluble receptor (sFlt-1) which serves to bind VEGF thus reducing vascular leak and angiogenesis [123] were not evaluated. This hypothesis remains equivocal, however, since studies measuring free VEGF and sFlt-1 have yielded conflicting results [124, 125].

BBB Integrity and Vasogenic Oedema

Figure 20.3 describes how this combination of molecular-mechanical stress "arrives" at the brain, ultimately compromising barrier integrity and encouraging the formation of extracellular (vasogenic) oedema. This was recently confirmed by diffusion-weighted MRI which established the "signature" rise in brain volume, T_2-relaxation time and apparent diffusion coefficient during hypoxia characteristic of mild vasogenic oedematous brain swelling [68, 70]. These changes were particularly pronounced in the splenium and genu of the corpus callosum, the same predilection site to that observed in HACE [83], likely the result of its unique vascular anatomy; densely packed horizontal fibres characterised by short arterioles that lack adrenergic tone may render it more susceptible to hyper-perfusion oedema in the setting of hypoxic cerebral vasodilatation [83].

It was originally suggested that AMS may be the direct consequence of elevated ICP caused by more pronounced vasogenic oedematous brain swelling [126]. This could prove one of the triggers underlying cephalalgia through stimulation of pain-sensitive fibres in the meninges/pial vessels and activation of the trigeminovascular system (TVS) [127]. This was an attractive hypothesis in light of an early study [128] that identified an increased T_2 signal in the white matter of subjects with moderate-to-severe AMS in whom clinical HACE had not yet developed (no ataxia or altered consciousness).

However, more recent DW-MRI studies have consistently failed to support this concept with no relationships observed between the hypoxia-induced increases in brain volume or T_2rt and cerebral AMS scores [68, 70]. Furthermore, the observation that intracranial volume (ICV) is not increased to any greater extent in the AMS-brain argues against intracranial hypertension as a significant event. As discussed earlier under "Estimates of Intracranial Pressure", increases in ODS and ONSD measured by ultrasound were often minor and not consistently related to AMS scores. Thus, there is no convincing human evidence to date to confirm that ICP is indeed raised in AMS [91]. Mild vasogenic oedematous brain swelling therefore appears to be an incidental finding that occurs in all brains exposed to the hypoxia of high altitude independent of AMS.

Intracranial–Intraspinal Buffering Capacity

Borne out of an original hypothesis developed over 20 years ago [129], the lack of correlation between brain swelling and AMS symptoms has since been ascribed to anatomical differences that determine how effectively the human skull can accommodate swelling through displacement of cranial CSF to extra-cranial compartments [90]. Thus, individuals characterised by a smaller ratio of cranial CSF to brain volume (popularised as the "tight-fit" brain) would be expected to be more prone to intracranial hypertension and thus by consequence AMS, since they are less capable of buffering the volume increase through changes in CSF dynamics due to an inadequate craniospinal axis CSF reserve.

To date, only one human study has provided indirect evidence to support the tight-fit hypothesis with AMS-susceptible subjects identified as having comparatively larger brain to ICV ratios that was apparent even at sea-level [68]. However, it would seem highly unlikely that the minor volumetric changes of 0.6–0.8 % of total brain volume would translate into any physiologically meaningful changes in the mechanical displacement of pain-sensitive structures capable of activating the TVS [91]. Indeed, it has been suggested that the longitudinal stretch of arteries or meninges is likely to be less than the elongations that would normally occur during pulsatile blood flow. The small increases in ICV documented in the literature likely occupy the flat-part of the ICP-volume curve and eminently well-buffered by semi-elastic membranes and changes in CSF dynamics [104]. Clearly, this remains a topic of ongoing debate and further studies are warranted to test this hypothesis.

Cerebral Venous Outflow Limitation

Attempts to explain the pathophysiology of AMS by haemodynamic mechanisms have traditionally focused on the consequences of increased "arterial inflow" to the brain. Many studies have demonstrated that CBF is increased with exposure to acute hypoxia due to the predominance of hypoxic vasodilatation over hypocapnic vasoconstriction (see chapter 7). Authors have consistently failed to demonstrate a relationship between increased blood flow in the internal carotid artery [130], middle cerebral artery [131] or brain tissue [131] and symptom severity. Recently, Wilson et al [132] hypothesised that a functional impairment in "venous outflow" at the level of the transverse venous sinus may prove a potential risk factor.

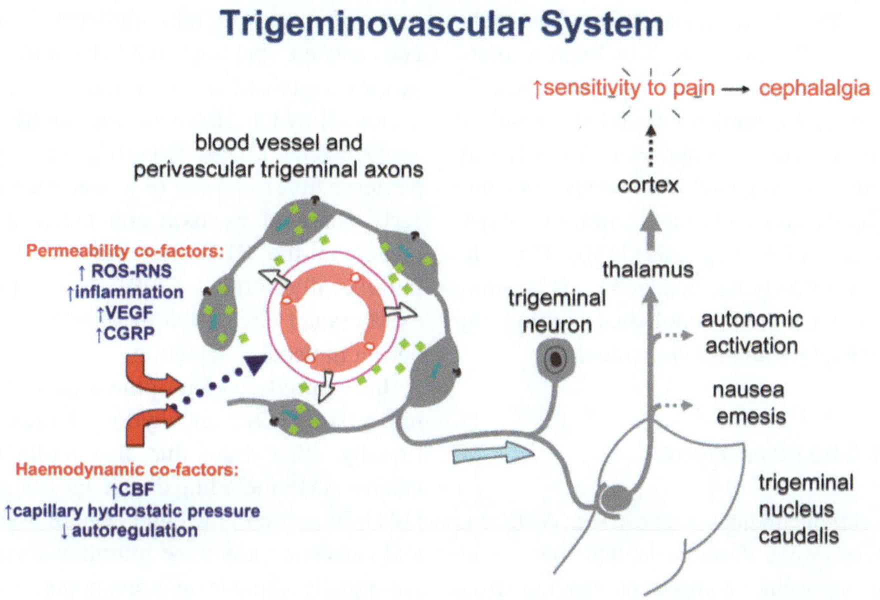

Fig. 20.4 Trigeminovascular system (TVS), trigeminal nerve fibres and ganglions surround meningeal vessels and those of other large vessels on the surface of the brain. When activated, they can lead to headache, nausea, emesis and autonomic activation. Activation of the TVS may occur through neurogenic inflammation triggered by afferent fibres from the brain stem or by free radicals and associated reactive oxygen-nitrogen species

In support, these authors identified that at low altitude, one or both transverse sinuses were narrower in subjects who subsequently developed AMS trekking to 5300 m. The functional significance of this venous anomaly was supported by significant correlations between retinal venous distension and headache during trekking as well as between retinal venous distension and cerebral venous engorgement during exposure to acute normobaric hypoxia [133]. Though an interesting finding, the underlying mechanisms that link venous distension to headache remain unclear. It is possible that distension of dural venous sinuses with dense trigeminal innervation [134] leads to headache in the absence of increased ICP since significant brain swelling [68–70] and elevated ICP [5] were not found in AMS. The latter findings are discussed in more detail in the sections on "Estimates of intracranial pressure" and "Brain imaging" in this chapter.

As discussed previously, the concept of the tight-fit hypothesis within the intracranial compartment as a whole is physiologically unlikely given the minor increase in brain volume (~7 mL in AMS) in the setting of an intracranial volume in excess of ~1300 mL [5]. The paper of Wilson et al [133], however, points to the possibility of more localized tight-fit constrained to the infratentorial space, since lower infratentorial CSF volumes were shown to correlate against headache scores.

Cytotoxic Oedema

The only defining morphological feature that distinguishes the AMS-prone from the healthy brain as revealed by more recent DW-MRI studies is a selective decrease in ADC scores [68, 70] taken to reflect intracellular (cytotoxic) oedema that likely coexists with extracellular vasogenic oedema. Since both studies failed to provide any evidence for additional oedema or swelling, the attenuation of the ADC likely reflects fluid redistribution from within the extracellular space (ECS) as intracellular (astrocytic) swelling proceeds without any additional increment in brain volume, oedema or ICP [135]. The temporal sequence and cause of this is unknown but may reflect ion pump suppression subsequent to down-regulation of Na^+/K^+-ATPase activity.

However, the critical question is whether the symptoms of AMS are caused by such a minor translocation of water from the ECS into the cells of the corpus callosum that entered as a result of mild vasogenic oedema in hypoxia. This is highly unlikely since oedema of the corpus callosum would typically give rise to a disconnection syndrome (e.g. associative agnosia) [136]. Thus, the mechanisms underlying headache, its leading symptom, remain to be established highlighting the need to explore alternative pathways.

Phase III: The Phenotype

Thus, the current findings suggest that AMS and HACE share similar features in that there is an underlying vasogenic component ranging from mild to severe. However, Phase III of the schema represents the point of "departure" from traditional theory [90] since the oedema cannot be held responsible for the symptoms of AMS. Furthermore, the clinical observation that HACE can develop rapidly in the absence of preceding AMS [137] and that in a large cohort of 66 cases with HACE 33 % had no headache (Table 20.1) [33] further questions the concept that the AMS-to-HACE symptom complex can be explained by this continuum of mild to severe cerebral oedema. Thus, there is a clear need to explore alternative pathways.

Pain and the Trigeminovascular System (TVS)

Though at present only a hypothesis, AMS may be the result of altered pain perception and trigeminovascular nociceptive input from the meningeal/pial vessels during hypoxia (Fig. 20.4). Direct inhibition of TVS activity with sumatriptan, a selective 5-hydroxytryptamine $(HT)_1$ receptor subtype agonist, points towards some, albeit preliminary neuroprotective potential. In the largest randomised trial to date, oral prophylaxis was shown to reduce the relative risk of AMS by 50 % [138] though design limitations including a general failure to control for headache history needs to be taken into account.

However, the recent observation that AMS did not influence the (steady-state) cerebral metabolism of the "migraine-molecule", calcitonin gene-related peptide (CGRP) [139], argues against sustained activation of the TVS as an important event. However, this finding does not exclude acute release from trigeminal perivascular nerve fibres, the site of nociception during the early phase of hypoxia and the lack of major breach in the BBB may have also prevented intrathecally formed CGRP from entering the extra-cranial circulation in sufficient amounts to permit molecular detection.

Interestingly, sumatriptan can also scavenge nitric oxide, $O_2^{\bullet -}$ and hydroxyl radicals [140]. Equally, other drugs that also confer protection against AMS including the analgesics gabapentin [141], ibuprofen [142] and acetaminophen [143] and carbonic anhydrase inhibitor acetazolamide are equally capable of scavenging free radicals [144–146]. Further research is required to test this alternative hypothesis and determine if activation of the TVS through oxidative-nitrosative-inflammatory stress does indeed represent the final common pathway given that it does not rely on any volumetric changes to the brain.

The Link to HACE

While HACE has traditionally been ascribed to vasogenic oedema [90], it is reasonable to assume that a cytotoxic component may also be present [83, 147] since hypoxaemia is likely more severe compared to AMS. Therefore, cell swelling should be even more pronounced in HACE. However, ADC mapping has not been performed immediately following evacuation from altitude in patients with HACE.

A recent study combined conventional T_2* with a novel, highly sensitive susceptibility-weighted MRI technique to reveal multiple "microbleeds" detected as haemosiderin deposits confined to the genu and splenium of the corpus callosum of patients with a history of HACE that had occurred up to 3 years ago [89]. Erythrocyte extravasation was taken to reflect cerebral capillary "stress failure" subsequent to cerebral hyperperfusion and severe BBB disruption. The failure to detect any haemosiderin deposits in subjects diagnosed with severe AMS reinforces the fundamental concept that vasogenic oedema is minor in AMS and is unlikely to account for symptoms and points to a novel diagnostic feature that further discriminates AMS from HACE.

Prevention and Treatment

Acetazolamide

A meta-analysis concluding that only 750 mg/day of acetazolamide is efficient for prevention of AMS [148] was questioned for doubtful interpretation of study results such as not considering different ascent rates and inappropriate exclusion of studies [149, 150]. This controversy triggered a series of investigations. Whereas a considerably underpowered study concluded that 250 mg bd was superior to 125 mg bd or placebo [151], three studies showed that 125 mg bd is effective. This dose of acetazolamide was more effective than a placebo in a low-risk setting ascending from 4,293 to 4,937 m in Himalayan trekkers [152], and in a study with a higher risk for AMS, in which subjects were transported from 1,600 to 4,300 m in a day [153]. The latter study yielded a number needed to treat of 3. Furthermore, acetazolamide 125 mg bd was equally effective as 375 mg bd during an ascent from 3,440 to 4,920 m in Himalayan trekkers and SaO_2 was not different between groups [154]. Paresthesia and disturbance of taste were frequent side effects while micturition (>3× at night) was not significantly different between groups. The results of these three studies also indicate that a dose of <5 mg/kg bodyweight, predominantly blocking the renal carbonic anhydrase [155], is sufficient for prevention of AMS.

In particularly susceptible individuals or when the risk for AMS is very high, such as in unacclimatised mountaineers ascending Kilimanjaro (5,859 m) within 5 days, a dose of 250 mg cannot reduce AMS incidence below 55–43 % [156]. The question remains, however, how much AMS is avoidable at all under such circumstances. Thus, it needs to be shown that in this setting, 750 mg/day is superior to 500 or 250 mg as suggested by the authors. In summary, in most settings, AMS incidence can be greatly reduced by 125 mg bd. Some particularly susceptible individuals or very rapid ascents of unacclimatised individuals to extreme altitude may justify 250 mg bd of acetazolamide.

Potentially Interesting Drugs Not Sufficiently Tested

Antioxidants

In a small trial in trekkers slowly ascending to Everest base camp (5,180 m), oral prophylaxis with a combination of dietary antioxidant vitamins (L-ascorbic acid, *dl*-α-tocopherol acetate and α-lipoic acid) for 3 weeks prior to ascent to high altitude and throughout the altitude sojourn was shown to improve systemic oxygenation and reduce the severity of AMS [109]. In contrast, an identical antioxidant cocktail failed to provide prophylaxis in a follow-up study [157]. However, the assimilation of lipid-soluble antioxidants into biological cell membranes was clearly suboptimal given the lack of a priming phase since supplementation commenced upon arrival to high altitude. Furthermore, unlike the study by Bailey and Davies [109], subjects refrained from strenuous physical activity which is an established pro-oxidant stimulus [158] known to compound the oxidative stress response to high altitude and increase corresponding susceptibility to AMS [159]. Clearly, further studies employing larger sample sizes and in settings where the risk of AMS is considerably higher are warranted. Furthermore, researchers need to consider alternative "ideal" antioxidants and methods of delivery across the BBB [160] in order to maximise their (potential) prophylactic benefits. Cellular hypoxia has been shown to increase the mitochondrial formation of ubisemiquinone and $O_2^{\bullet-}$ radicals [161, 162] which likely give rise to the blood-borne formation of lipid-derived $LO^{\bullet}$-$LC^{\bullet}$ species that have consistently been shown to be elevated in AMS [58]. Thus, the strategic delivery of cerebral mitochondria-specific antioxidants to quench these upstream species may prove a more effective approach [163]. Indeed, the flavonoid quercetin was recently shown to reduce oxidative-inflammatory stress and subsequent (cerebral) oedema formation more effectively than dexamethasone in an animal model of HACE [164].

Sufficiently Tested Drugs Not to Be Recommended

Gingko Biloba

Gingko biloba extract (GBE) has received much attention because it is of natural herbal origin and its availability without prescription. GBE is composed of many components with different biological actions that may be beneficial for the prevention of altitude illness, such as increasing NO bioavailability and endogenous antioxidants or reducing free radical production and lung leak in animal experiments [165]. However, a large, well controlled 3-arm-trial showed that GBE was no better than placebo for preventing AMS in a low-risk setting in contrast to the threefold reduction of AMS with acetazolamide 250 mg bd in this trial [166]. Several small trials have yielded controversial results [167–170]. The discrepancies can in part be attributed to insufficient statistical power of these small trials and variations in the composition of preparations [165].

Theophylline

Several pharmacological effects of theophylline, a phosphodiesterase inhibitor, are of potential benefit for improving sleep at high altitude and avoiding AMS, such as increase of central respiratory drive with reduction in periodic breathing, suppression of microvascular permeability in the brain and lung, a decrease of pulmonary artery pressure and a decrease of CBF. Theophylline 250 mg bd at 3,454 m [171] and 300 mg/day at 4,559 m reduces periodic breathing [172], without significantly affecting oxygen saturation. There was a small decrease of AMS scores without reduction of the overall incidence of AMS at 4,559 m [172]. At 3,454 m, theophylline 250 mg bd caused only a transient minor decrease of AMS scores [173], and the dose had to be reduced in 20 % of the subjects because of side effects, such as tachycardia and tremor. During a 10-h exposure at a simulated altitude of 4,500 m, theophylline (250 mg bd) was not better than placebo, while acetazolamide (250 mg bd) reduced AMS significantly [69].

Because of the minor effects on symptom scores of AMS, the higher probability for unpleasant side effects and the lack of improvement of oxygen saturation during sleep, theophylline is clearly inferior to acetazolamide and should therefore not be recommended for prevention of AMS or treatment of periodic breathing at high altitude.

Drugs Without Effect

Leukotriene Modifying Drugs

Previous investigations had shown that leukotriene concentrations were increased in urine and plasma in individuals acutely exposed to simulated altitudes of 4,300 m [34]. Therefore, it was hypothesised that these mediators of inflammation could be involved in the pathophysiology of AMS. Three studies [174–175] showed no effects of lipoxygenase inhibitors or leukotriene receptor blockers. Furthermore, there was no difference in urinary leukotriene E_4 excretion between treatment groups. These studies demonstrate that leukotrienes do not play a causal role in the pathophysiology of AMS, and that leukotrienes are therefore not a target for preventing AMS.

Magnesium

Because of its neuroprotective effect oral magnesium was unsuccessfully tested for prevention of AMS after rapid ascent to 4,559 m. In subjects with AMS intravenous application of magnesium reduced symptoms somewhat but not to a clinically relevant extent [177].

Suggestions for Future Research

Clinical Practice

Prediction of susceptibility to AMS is difficult and almost impossible in those who have never been to high altitude before. Models for prediction of susceptibility to AMS need to be evaluated in prospective studies.

Pathophysiology

Unlike HACE, the definitive mechanisms underlying AMS remain largely unresolved justifying the need for further research. Given the complexity associated with what are mostly subjective sensations of pain, it is recommended that future studies adopt a functionally integrated translational approach with a specific focus on the three following areas.

Genetics

Within the migraine field, studies have identified "susceptibility genes" that encode for altered synaptic (notably glutamatergic) transmission and/or neuronal excitability [178, 179]. Given that AMS shares similar phenotypical characteristics and is thus likely to be equally "polygenic", it would be interesting to explore these and additional polymorphic oxidative-nitrosative-inflammatory stress pathways, to address the concept that AMS represents an "ionopathy" characterised by free radical-mediated ion channel dysfunction [119].

Brain Structure–Function

The rapidly advancing neuroimaging field is ideally placed to confirm if cerebral haemodynamic function is indeed selectively impaired in AMS and more fundamentally, what structures are involved during transmission of the cephalic pain signal. Positron-emission tomography (PET) and functional MRI can provide unique insight into regional (as opposed to global) alterations in CBF though complementary modalities including magnetoencephalography, voxel-based morphometry, diffusion tensor imaging, transcranial magnetic stimulation, proton magnetic resonance spectroscopy (MRS), fluorescence-mediated molecular tomography, multiphoton microscopy and nanoparticle imaging are constantly evolving to enable simultaneous structural–functional assessment [180–182]. Indeed, an earlier $H_2{}^{15}O$-labelled PET study identified an exaggerated increase in hypothalamic blood flow during acute hypoxia [183] which, if confirmed in AMS, would provide a physiological basis for functional impairment. From a clinical perspective, the processing of neuropathic pain in patients involves the complex interplay between functional networks collectively referred to as the "pain matrix" which includes the primary and secondary somatosensory cortices, anterior cingulate cortex, insular cortex, prefrontal cortex, thalamus, basal ganglia and cerebellum [184]. Furthermore, acute migraine attacks have been associated with activation of the locus coeruleus within the dorsolateral brainstem [185, 186]. However, to what extent this matrix becomes activated during AMS remains unknown. Unfortunately, trigeminal fibres are not within the resolution of currently available imaging techniques [187] demanding alternative approaches to assess the pathogenic role of TVS activity in AMS (*vide infra*). Using ^{1}H-MRS, differences in cerebral energy metabolism can be addressed via targeted detection of lactate [188], γ-aminobutyric acid [189], *N*-acetylaspartylglutamate [190], the redox-reactive metabolites glutathione [191] and ascorbate [192] and the neuronal–glial integrity biomarkers, *N*-acetylaspartate and myoinositol [193]. Follow-up studies combining transfer function analysis of focal dynamic CA applied to discrete regions of interest (including the corpus callosum) with gadolinium-enhanced DW-MRI are required to more accurately determine to what extent hypoxia impairs BBB integrity and whether minor disruption is indeed a risk factor for AMS. This needs to be complemented by recent pharmacogenomic approaches employing the targeted delivery of nootropic radical scavengers to the BBB and brain parenchyma [160] and thus facilitate a more comprehensive examination of the possible role, if indeed any, that the BBB plays in the aetiology of AMS. Acute exercise may provide a novel priming stimulus to optimise antioxidant delivery and uptake to the AMS-prone brain since it has recently been shown to cause a subtle increase in BBB permeability subsequent to a free radical-mediated impairment in dCA [194]. Finally, MRI has the potential to measure ICP non-invasively [195] and thus resolve whether intracranial hypertension is a risk factor for AMS.

Molecular-Metabolic

The source, mechanisms and consequences of free radical formation in humans remain to be established and invasive approaches that involve arterio-jugular venous sampling and simultaneous measurement of CBF has the potential to provide unique insight into cerebral exchange kinetics [58]. Since the $O_2^{\bullet -}$ has been identified as the primary species responsible for impaired dCA subsequent to activation of K_{Ca} channels in rodent cerebral vascular smooth muscle cells [196], it would be of interest to measure this directly in humans via EPR spectroscopic detection of the novel *β*-phosphorylated nitrone spin trap, 5-(diethoxy-phosphoryl)-5-methyl-1-pyrroline *N*-oxide [194]. This would also facilitate interrogation of the reaction that drives the oxidative inactivation of NO ($O_2^{\bullet -} + NO \xrightarrow{k=16-20\times10^9 Ms^{-1}} ONOO^-$) [39] that has been held responsible for vascular endothelial dysfunction and impaired cerebral O_2 transport in AMS [8, 58]. Proteomic–metabolonomic analyses of oxidative post-translational protein modifications and associated reactants would provide additional complementary insight into AMS redox homeostasis [197, 198]. Further studies also need to confirm if catalytic "free iron" serves as the upstream trigger for hydroxyl radical formation and the elevated oxidative-nitrosative-inflammatory stress response in AMS, thus raising the possibility that iron chelation via desferrioxamine may provide neuroprophylactic benefit [58, 198]. Finally, additional pharmacological intervention with CGRP-receptor antagonists including the non-peptide drug BIBN 4096 BS, recently shown to be effective in the treatment of migraine [199], will help clarify the potential role of TVS activation in AMS, the next frontier in its complex pathophysiology.

References

1. Roach RC, Bärtsch P, Hackett PH, et al. The Lake Louise acute mountain sickness scoring system. In: Sutton JR, Houston CS, Coates G, editors. Hypoxia and molecular medicine. Burlington: Queen City Printers Inc.; 1993. p. 272–4.
2. West JB. Con: headache should not be a required symptom for the diagnosis of acute mountain sickness. High Alt Med Biol. 2011;12:23–5.
3. Maggiorini M, Müller A, Hofstetter D, Bärtsch P, Oelz O. Assessment of acute mountain sickness by different score protocols in the Swiss Alps. Aviat Space Environ Med. 1998;69(12):1186–92.
4. Bärtsch P, Müller A, Hofstetter D, et al. AMS and HAPE scoring in the Alps. In: Sutton JR, Houston CS, Coates G, editors. Hypoxia and molecular medicine. Burlington: Queen City Press; 1993. p. 265–71.
5. Bailey DM, Roukens R, Knauth M, Kallenberg K, Christ S, Mohr A, et al. Free radical-mediated damage to barrier function is not associated with altered brain morphology in high-altitude headache. J Cereb Blood Flow Metab. 2006;26(1):99–111.
6. Loeppky JA, Icenogle MV, Charlton G, Conn CA, Maes D, Riboni K, et al. Hypoxemia and acute mountain sickness: which comes first? High Alt Med Biol. 2008;9:271–9.
7. Hampson NB, Camporesi EM, Stolp BW, Moon RE, Shook JE, Griebel JA, et al. Cerebral oxygen availability by NIR spectroscopy during transient hypoxia in humans. J Appl Physiol. 1990;69:907–13.
8. Bailey DM, Evans KA, James PE, McEneny J, Young IS, Fall L, et al. Altered free radical metabolism in acute mountain sickness: implications for dynamic cerebral autoregulation and blood-brain barrier function. J Physiol. 2009;587:73–85.
9. Iversen HK, Olesen J, Tfelt-Hansen P. Intravenous nitroglycerin as an experimental model of vascular headache. Basic characteristics. Pain. 1989;38: 17–24.
10. Schneider M, Bernasch D, Weymann J, Holle R, Bärtsch P. Acute mountain sickness: influence of susceptibility, pre-exposure and ascent rate. Med Sci Sports Exerc. 2002;34(12):1886–91.
11. Muhm JM, Rock PB, McMullin DL, Jones SP, Eilers KD, Space DR, et al. Effect of aircraft-cabin altitude on passenger discomfort. N Engl J Med. 2007; 357:18–27.
12. Wagner DR, D'Zatko K, Tatsugawa K, Murray K, Parker D, Streeper T, et al. Determinants of summit success and acute mountain sickness. Med Sci Sports Exerc. 2008;40:1820–7.
13. Ziaee V, Yunesian M, Ahmadinejad Z, Halabchi F, Kordi R, Alizadeh R, et al. Acute mountain sickness in Iranian trekkers around Mount Damavand (5761 m) in Iran. Wilderness Environ Med. 2003; 14:214–9.
14. Bloch KE, Turk AJ, Maggiorini M, Hess T, Merz T, Bosch MM, et al. Effect of ascent protocol on acute mountain sickness and success at Muztagh Ata, 7546 m. High Alt Med Biol. 2009;10:25–32.
15. Ge RL, Chase PJ, Witkowski S, Wyrick BL, Stone JA, Levine BD, et al. Obesity: associations with acute mountain sickness. Ann Intern Med. 2003;139(4):253–7.
16. Gaillard S, Dellasanta P, Loutan L, Kayser B. Awareness, prevalence, medication use, and risk factors of acute mountain sickness in tourists trekking around the Annapurnas in Nepal: a 12-year follow-up. High Alt Med Biol. 2004;5:410–9.
17. Vardy J, Vardy J, Judge K. Can knowledge protect against acute mountain sickness? J Public Health (Oxf). 2005;27:366–70.
18. Ledderhos C, Pongratz H, Exner J, Gens A, Roloff D, Honig A. Reduced tolerance of simulated altitude

(4200 m) in young men with borderline hypertension. Aviat Space Environ Med. 2002;73:1063–6.
19. Roach RC, Loeppky JA, Icenogle MV. Acute mountain sickness: increased severity during simulated altitude compared with normobaric hypoxia. J Appl Physiol. 1996;81(5):1908–10.
20. Schommer K, Hammer M, Hotz L, Menold E, Bärtsch P, Berger MM. Exercise intensity typical of mountain climbing does not exacerbate acute mountain sickness in normobaric hypoxia. J Appl Physiol. 2012;113:1068–74.
21. Kalson NS, Thompson J, Davies AJ, Stokes S, Earl MD, Whitehead A, et al. The effect of angiotensin-converting enzyme genotype on acute mountain sickness and summit success in trekkers attempting the summit of Mt. Kilimanjaro (5,895 m). Eur J Appl Physiol. 2009;105:373–9.
22. Karinen H, Peltonen J, Tikkanen H. Prevalence of acute mountain sickness among Finnish trekkers on Mount Kilimanjaro, Tanzania: an observational study. High Alt Med Biol. 2008;9:301–6.
23. Pesce C, Leal C, Pinto H, Gonzalez G, Maggiorini M, Schneider M, et al. Determinants of acute mountain sickness and success on Mount Aconcagua (6962 m). High Alt Med Biol. 2005;6(2):158–66.
24. Yaron M, Waldman N, Niermeyer S, Nicholas R, Honigman B. The diagnosis of acute mountain sickness in preverbal children. Arch Pediatr Adolesc Med. 1998;152:683–7.
25. Yaron M, Niermeyer S, Lindgren KN, Honigman B. Evaluation of diagnostic criteria and incidence of acute mountain sickness in preverbal children. Wilderness Environ Med. 2002;13:21–6.
26. Bloch J, Duplain H, Rimoldi SF, Stuber T, Kriemler S, Allemann Y, et al. Prevalence and time course of acute mountain sickness in older children and adolescents after rapid ascent to 3450 meters. Pediatrics. 2009;123:1–5.
27. Maggiorini M, Bühler B, Walter M, Oelz O. Prevalence of acute mountain sickness in the Swiss Alps. Br Med J. 1990;301:853–5.
28. Moraga FA, Pedreros CP, Rodríguez CE. Acute mountain sickness in children and their parents after rapid ascent to 3500 m (Putre, Chile). Wilderness Environ Med. 2008;19:287–92.
29. Moraga FA, Osorio JD, Vargas ME. Acute mountain sickness in tourists with children at Lake Chungará (4400 m) in Northern Chile. Wilderness Environ Med. 2002;13:31–5.
30. Baumgartner RW, Bärtsch P. Ataxia in acute mountain sickness does not improve with short-term oxygen inhalation. High Alt Med Biol. 2002;3:283–92.
31. Cymerman A, Muza SR, Beidleman BA, Ditzler DT, Fulco CS. Postural instability and acute mountain sickness during exposure to 24 hours of simulated altitude (4300 m). High Alt Med Biol. 2001;2:509–14.
32. Johnson BG, Wright AD, Beazley MF, Harvey TC, Hillenbrand P, Imray CHE, the Birmingham Medical Research Expeditionary Society. The sharpened Romberg test for assessing ataxia in mild acute mountain sickness. Wilderness Environ Med. 2005;16:62–6.
33. Wu T, Ding S, Liu J, Jia J, Dai R, Liang B, et al. Ataxia: an early indicator in high altitude cerebral edema. High Alt Med Biol. 2006;7:275–80.
34. Bärtsch P, Roach R. Acute mountain sickness and high-altitude cerebral edema. In: Hornbein TF, Schoene R, editors. High altitude—an exploration of human adaptation. New York: Marcel Dekker; 2001. p. 731–76.
35. Bärtsch P, Swenson ER, Paul A, Jülg B, Hohenhaus E. Hypoxic ventilatory response, ventilation, gas exchange, and fluid balance in acute mountain sickness. High Alt Med Biol. 2002;3:361–76.
36. Richalet JP, Larmignat P, Poitrine E, Letournel M, Canoui-Poitrine F. Physiological risk factors of severe high altitude illness: a prospective cohort study. Am J Respir Crit Care Med. 2012;185:1092–198.
37. Erba P, Anastasi S, Senn O, Maggiorini M, Bloch KE. Acute mountain sickness is related to nocturnal hypoxemia but not to hypoventilation. Eur Respir J. 2004;24(2):303–8.
38. Cremona G, Asnaghi R, Baderna P, Brunetto A, Brutsaert T, Cavallaro C, et al. Pulmonary extravascular fluid accumulation in recreational climbers: a prospective study. Lancet. 2002;359:303–9.
39. Compte-Torrero L, Botella de Maglia J, de Diego-Damiá AG-PL, Ramírez-Galleymore P, Perpiña-Tordera M. Changes in spirometric parameters and arterial oxygen saturation during a mountain ascent to over 3000 meters. Arch Bronconeumol. 2005; 41:547–52.
40. Senn O, Clarenbach CF, Fischler M, Thalmann R, Brunner-La-Rocca H, Egger P, et al. Do changes in lung function predict high-altitude pulmonary edema at an early stage? Med Sci Sports Exerc. 2006;38(9):1565–70.
41. Mason NP, Barry PW, Pollard AJ, Collier DJ, Taub NA, Miller MR, et al. Serial changes in spirometry during an ascent to 5300 m in the Nepalese Himalayas. High Alt Med Biol. 2000;1:185–95.
42. Deboeck G, Moraine JJ, Naeije R. Respiratory muscle strength may explain hypoxia-induced decrease in vital capacity. Med Sci Sports Exerc. 2005; 37(5):754–8.
43. Fasano V, Paolucci E, Pomidori L, Cogo A. High-altitude exposure reduces inspiratory muscle strength. Int J Sports Med. 2007;28:426–30.
44. Mason NP, Petersen M, Mélot C, Imanov B, Matveykine O, Gautier MT, et al. Serial changes in nasal potential difference and lung electrical impedance tomography at high altitude. J Appl Physiol. 2003;94:2043–50.
45. Dehnert C, Luks AM, Schendler G, Menold E, Berger MM, Mairbäurl H, et al. No evidence for interstitial lung oedema by extensive pulmonary function testing at 4,559 m (Author Correction). Eur Respir J. 2010;36:699.
46. Thompson AAR, Baillie JK, Toshner M, Maxwell SRJ, Webb DJ, Irving JB. Pericardial effusions in

healthy lowlanders after acute ascent to high altitude. Heart. 2006;92(4):539–40.
47. Bors W, Michel C, Saran M, Lengfelder E. The involvement of oxygen radicals during the autoxidation of adrenalin. Biochim Biophys Acta. 1978; 540:162–72.
48. Kamimori GH, Ryan EJ, Otterstetter R, Barkley JE, Glickman EL, Davis HQ. Catecholamine levels in hypoxia-induced acute mountain sickness. Aviat Space Environ Med. 2009;80:376–80.
49. Lanfranchi PA, Colombo R, Cremona G, Baderna P, Spagnolatti L, Mazzuero G, et al. Autonomic cardiovascular regulation in subjects with acute mountain sickness. Am J Physiol Heart Circ Physiol. 2005;289:H2364–72.
50. Chen Y-C, Lin F-C, Shiao G-M, Chang S-C. Effect of rapid ascent to high altitude on autonomic cardiovascular modulation. Am J Med Sci. 2008; 336:248–53.
51. Sevre K, Bendz B, Hankø E, Nakstad AR, Hauge A, Kåsin JI, et al. Reduced autonomic activity during stepwise exposure to high altitude. Acta Physiol Scand. 2001;173:409–17.
52. Loeppky JA, Icenogle MV, Maes D, Riboni K, Scotto P, Roach RC. Body temperature, autonomic responses, and acute mountain sickness. High Alt Med Biol. 2003;4:367–73.
53. Koehle MS, Guenette JA, Warburton DER. Oximetry, heart rate variability, and the diagnosis of mild-to-moderate acute mountain sickness. Eur J Emerg Med. 2010;17:119–22.
54. Baumgartner R, Spyridopoulos I, Bärtsch P, Maggiorini M, Oelz O. Acute mountain sickness is not related to cerebral blood flow. A decompression chamber study. J Appl Physiol. 1999;86:1578–82.
55. Pagani M, Ansjön R, Lind F, Uusijärvi J, Sumen G, Jonsson C, et al. Effects of acute hypobaric hypoxia on regional cerebral blood flow distribution: a single photon emission computed tomography study in humans. Acta Physiol Scand. 2000;168:377–83.
56. Dyer EA, Hopkins SR, Perthen JE, Buxton RB, Dubowitz DJ. Regional cerebral blood flow during acute hypoxia in individuals susceptible to acute mountain sickness. Respir Physiol Neurobiol. 2008; 160:267–76.
57. Van Osta A, Moraine JJ, Melot C, Mairbaurl H, Maggiorini M, Naeije R. Effects of high altitude exposure on cerebral hemodynamics in normal subjects. Stroke. 2005;36(3):557–60.
58. Bailey DM, Taudorf S, Berg RMG, Lundby C, McEneny J, Young IS, et al. Increased cerebral output of free radicals during hypoxia; implications for acute mountain sickness? Am J Physiol Regul Integr Comp Physiol. 2009;297:R1283–93.
59. Kanner AA, Marchi N, Fazio V, Mayberg MR, Koltz MT, Siomin V, et al. Serum S100B: a noninvasive marker of blood-brain barrier function and brain lesions. Cancer. 2003;97:2806–13.
60. Bosch MM, Barthelmes D, Merz TM, Bloch KE, Turk AJ, Hefti U, et al. High incidence of optic disc swelling at very high altitudes. Arch Ophthalmol. 2008;126:644–50.
61. Sutherland AI, Morris DS, Owen CG, Bron AJ, Roach RC. Optic nerve sheath diameter, intracranial pressure and acute mountain sickness on Mount Everest: a longitudinal cohort study. Br J Sports Med. 2008;42:183–8.
62. Fagenholz PJ, Gutman JA, Murray AF, Noble VE, Camargo Jr CA, Harris NS. Optic nerve sheath diameter correlates with the presence and severity of acute mountain sickness: evidence for increased intracranial pressure. J Appl Physiol. 2009;106:1207–11.
63. Willmann G, Fischer MD, Schatz A, Schommer K, Messias K, Zrenner E, et al. Quantification of optic disc edema during exposure to high altitude shows no correlation to acute mountain sickness. PLos One. 2011;6(11):e27022.
64. Wilson MH, Milledge J. Direct measurement of intracranial pressure at high altitude and correlation of ventricular size with acute mountain sickness: Brian Cummins' results from the 1985 Kishtwar Expedition. Neurosurgery. 2008;63:970–5.
65. Somner JEA, Morris DS, Scott KM, MacCormick JC. What happens to intraocular pressure at high altitude? Invest Ophthalmol Vis Sci. 2007; 48:1622–6.
66. Pavlidis M, Stupp T, Georgalas I, Georgiadou E, Moschos M, Thanos S. Intraocular pressure changes during high-altitude acclimatization. Graefes Arch Clin Exp Ophthalmol. 2006;244:298–304.
67. Cymerman A, Rock PB, Muza SR, Lyons TP, Fulco CS, Mazzeo RS, et al. Intraocular pressure and acclimatization to 4300 m altitude. Aviat Space Environ Med. 2000;71:1045–50.
68. Kallenberg K, Bailey DM, Christ S, Mohr A, Roukens R, Menold E, Steiner T, Bärtsch P, et al. Magnetic resonance imaging evidence of cytotoxic cerebral edema in acute mountain sickness. J Cereb Blood Flow Metab. 2007;27:1064–71.
69. Fischer R, Vollmar C, Thiere M, Born C, Leitl M, Pfluger T, et al. No evidence of cerebral oedema in severe acute mountain sickness. Cephalalgia. 2004; 24:66–71.
70. Schoonman G, Sándor P, Nirkko A, Lange T, Jaermann T, Dydak U, Kremer C, Ferrari M, Boesiger P, Baumgartner R. Hypoxia-induced acute mountain sickness is associated with intracellular cerebral edema: a 3 T magnetic resonance imaging study. J Cereb Blood Flow Metab. 2008;28:198–206.
71. Mórocz IA, Zientara GP, Gudbjartsson H, Muza S, Lyons T, Rock PB, et al. Volumetric quantification of brain swelling after hypobaric hypoxia exposure. Exp Neurol. 2001;168:96–104.
72. Dubowitz DJ, Dyer EAW, Theilmann RJ, Buxton RB, Hopkins SR. Early brain swelling in acute hypoxia. J Appl Physiol. 2009;107:244–52.
73. Grant S, MacLeod N, Kay JW, Watt M, Patel S, Paterson A, et al. Sea level and acute responses to hypoxia: do they predict physiological responses and acute mountain sickness at altitude? Br J Sports Med. 2002;36:141–6.

74. Burtscher M, Flatz M, Faulhaber M. Prediction of susceptibility to acute mountain sickness by SaO2 values during short-term exposure to hypoxia. High Alt Med Biol. 2004;5(3):335–40.
75. O'Connor T, Dubowitz G, Bickler PE. Pulse oximetry in the diagnosis of acute mountain sickness. High Alt Med Biol. 2004;5(3):341–8.
76. Roach RC, Greene ER, Schoene RB, Hackett PH. Arterial oxygen saturation for prediction of acute mountain sickness. Aviat Space Environ Med. 1998;69:1182–5.
77. Chen HC, Lin WL, Wu JY, Wang SH, Chiu TF, Weng YM, et al. Change in oxygen saturation does not predict acute mountain sickness on Jade Mountain. Wilderness Environ Med. 2012; 23:122–7.
78. Wagner DR, Knott JR, Fry JP. Oximetry fails to predict acute mountain sickness or summit success during a rapid ascent to 5640 meters. Wilderness Environ Med. 2012;23:114–21.
79. Tannheimer M, Albertini N, Ulmer HV, Thomas A, Engelhardt M, Schmidt R. Testing individual risk of acute mountain sickness at greater altitudes. Mil Med. 2009;174:363–9.
80. Hayat A, Hussain MM, Aziz S, Siddiqui AH, Hussain T. Hyperventilation capacity—a predictor of altitude sickness. J Ayub Med Coll Abbottabad. 2006;18:17–20.
81. Jafarian S, Gorouhi F, Ghergherechi M, Lotfi J. Respiratory rate within the first hour of ascent predicts subsequent acute mountain sickness severity. Arch Iran Med. 2008;11:152–6.
82. Thomassen O, Skaiaa SC. High-altitude cerebral edema with absence of headache. Wilderness Environ Med. 2007;18:45–7.
83. Hackett PH, Yarnell PR, Hill R, Reynard K, Heit J, McCormick J. High-altitude cerebral edema evaluated with magnetic resonance imaging. JAMA. 1998;280(22):1920–5.
84. Jeong JH, Kwon JC, Chin J, Yoon SJ, Na DL. Globus pallidus lesions associated with high mountain climbing. J Korean Med Sci. 2002;17:861–3.
85. Litch JA, Bishop RA. High-altitude global amnesia. Wilderness Environ Med. 2000;11:25–8.
86. Litch JA, Bishop RA. Transient global amnesia at high altitude. N Engl J Med. 1999;340:1444.
87. Basnyat B. Case report: delirium at high altitude. High Alt Med Biol. 2002;3:69–71.
88. Firth PG, Bolay H. Transient high altitude neurological dysfunction: an origin in the temporoparietal cortex. High Alt Med Biol. 2004;5:71–5.
89. Kallenberg K, Dehnert C, Dörfler A, Schellinger PD, Bailey DM, Knauth M, et al. Microhemorrhages in nonfatal high-altitude cerebral edema. J Cereb Blood Flow Metab. 2008;28:1635–42.
90. Hackett PH, Roach RC. High-altitude illness. N Engl J Med. 2001;345:107–14.
91. Bärtsch P, Bailey D, Berger M, Knauth M, Baumgartner M. Acute mountain sickness: controversies and advances. High Alt Med Biol. 2004; 5:110–24.
92. Bärtsch P, Baumgartner RW, Waber U, Maggiorini M, Oelz O. Comparison of carbon-dioxide-enriched, oxygen-enriched, and normal air in treatment of acute mountain sickness. Lancet. 1990;336:772–5.
93. Bärtsch P, Shaw S, Weidmann P, Hildebrandt W, Bucler T, Biollaz J. Fluid retention in acute mountain sickness: cause or consequence? In: Ohno H, Kobayashi T, Masuyama S, Nakashima M, editors. Progress in mountain medicine and high altitude physiology. Matsumoto: Press Committee of the 3rd World Congress on Mountain Medicine and High Altitude Physiology; 1998. p. 234–9.
94. Rupert JL, Koehle MS. Evidence for a genetic basis for altitude-related illnesses. High Alt Med Biol. 2006;7:150–67.
95. Montgomery HE, Marshall R, Hemingway H, Myerson S, Clarkson P, Dollery C, Hayward M, Holliman DE, Jubb M, World M, et al. Human gene for physical performance. Nature. 1998;393:221–2.
96. Thompson J, Raitt J, Hutchings L, Drenos F, Bjargo E, Loset A, Grocott M, Montgomery H. Angiotensin-converting enzyme genotype and successful ascent to extreme altitude. High Alt Med Biol. 2007; 8:278–85.
97. Dehnert C, Weymann JMHE, Woods D, Maggiorini M, Scherrer U, Gibbs JSR, Bärtsch P. No association between high-altitude tolerance and the ACE I/D gene polymorphism. Med Sci Sports Exerc. 2002; 34:1928–33.
98. Koehle MS, Wang P, Guenette JA, Rupert JL. No association between variants in the ACE and angiotensin II receptor 1 genes and acute mountain sickness in Nepalese pilgrims to the Janai Purnima Festival at 4380 m. High Alt Med Biol. 2006;7:281–9.
99. Li FZ, Zhou CZ, Jiang CZ, Sun SY, He M, Zhang SY, et al. [Relationship between heat stress protein 70 gene polymorphisms and the risk of acute mountain sickness] [Article in Chinese]. Zhonghua Lao Dong Wei Sheng Zhi Ye Bing Za Zhi. 2004;22:413–5.
100. Møller K, Paulson OB, Hornbein TF, Colier WNJM, Paulson AS, Roach RC, et al. Unchanged cerebral blood flow and oxidative metabolism after acclimatization to high altitude. J Cereb Blood Flow Metab. 2002;22:118–26.
101. Kennedy C, Sakurada O, Shinohara M, Jehle J, Solokoff L. Local cerebral glucose utilization in the normal conscious macaque monkey. Ann Neurol. 1978;4:293–301.
102. Subudhi AW, Dimmen AC, Roach RC. Effects of acute hypoxia on cerebral and muscle oxygenation during incremental exercise. J Appl Physiol. 2007;103:177–83.
103. Hoshi Y. Towards the next generation of near-infrared spectroscopy. Philos Trans A Math Phys Eng Sci. 2011;369:4425–39.
104. Hackett PH. The cerebral etiology of high-altitude cerebral edema and acute mountain sickness. Wilderness Environ Med. 1999;10:97–109.

105. Baumgartner RW, Bärtsch P, Maggiorini M, Waber U, Oelz O. Enhanced cerebral blood flow in acute mountain sickness. Aviat Space Environ Med. 1994;65:726–9.
106. Jensen JB, Wright AD, Lassen NA, Harvey TC, Winterborn MH, Raichle ME, et al. Cerebral blood flow in acute mountain sickness. J Appl Physiol. 1990;69:430–3.
107. Sørensen SC, Lassen NA, Severinghaus JW, Coudert J, Zamora MP. Cerebral glucose metabolism and cerebral blood flow in high altitude residents. J Appl Physiol. 1974;37:305–10.
108. Lassen NA, Harper AM. High-altitude cerebral oedema. Lancet. 1975;2:1154.
109. Bailey DM, Davies B. Acute mountain sickness; prophylactic benefits of antioxidant vitamin supplementation at high altitude. High Alt Med Biol. 2001;2:21–9.
110. Richalet J-P, Hornych A, Rathat C, Aumont J, Larmignat P, Rémy P. Plasma prostaglandins, leukotrienes and thromboxane in acute high altitude hypoxia. Respir Physiol. 1991;85:205–15.
111. Roach JM, Muza SR, Rock PB, Lyons TP, Lilly CM, Drazen JM, et al. Urinary leukotriene E (4) levels increase upon exposure to hypobaric hypoxia. Chest. 1996;110:946–51.
112. Kleger G-R, Bärtsch P, Vock P, Heilig B, Roberts LJI, Ballmer PE. Evidence against an increase of capillary permeability in subjects exposed to high altitude. J Appl Physiol. 1996;81:1917–23.
113. Hartmann G, Tschöp M, Fischer R, Bidlingmaier C, Riepl R, Tschöp K, et al. High altitude increases circulating interleukin-6, interleukin-1 receptor antagonist and C-reactive protein. Cytokine. 2000;12:246–52.
114. Bailey DM, Ainslie PN, Jackson SK, Richardson RS, Ghatei M. Evidence against redox regulation of energy homoeostasis in humans at high altitude. Clin Sci. 2004;107(6):589–600.
115. Julian CG, Subudhi AW, Wilson MJ, Dimmen AC, Pecha T, Roach R. Acute mountain sickness, inflammation and permeability: new insights from a blood biomarker study. J Appl Physiol. 2011;111:392–9.
116. Bailey DM, Taudorf S, Berg RM, Lundby C, Pedersen BK, Moller K. Inflammatory "priming" predisposes to acute mountain sickness. In: Roach RC, Wagner PD, Hackett PH, editors. Hypoxia and the circulation. Banff: Springer; 2007. p. 300.
117. Bailey DM, Davies B, Young IS, Hullin DA, Seddon PS. A potential role for free radical-mediated skeletal muscle soreness in the pathophysiology of acute mountain sickness. Aviat Space Environ Med. 2001;72:513–21.
118. Bailey DM, Davies B, Castell LM, Collier DJ, Milledge JS, Hullin DA, et al. Symptoms of infection and acute mountain sickness; associated metabolic sequelae and problems in differential diagnosis. High Alt Med Biol. 2003;4:319–31.
119. Bailey DM, Bärtsch P, Knauth M, Baumgartner RW. Emerging concepts in acute mountain sickness and high-altitude cerebral edema: from the molecular to the morphological. Cell Mol Life Sci. 2009;66: 3583–94.
120. Schoch HJ, Fischer S, Marti HH. Hypoxia-induced vascular endothelial growth factor expression causes vascular leakage in the brain. Brain. 2002; 125:2549–57.
121. Xu F, Severinghaus JW. Rat brain VEGF expression in alveolar hypoxia: possible role in high-altitude cerebral edema. J Appl Physiol. 1998;85(1):53–7.
122. Maloney J, Wang D, Duncan T, Voelkel N, Ruoss S. Plasma vascular endothelial growth factor in acute mountain sickness. Chest. 2000;118:47–52.
123. Kendall RL, Wang G, Thomas KA. Identification of a natural soluble form of the vascular endothelial growth factor receptor, FLT-1, and its heterodimerization with KDR. Biochem Biophys Res Commun. 1996;226:324–8.
124. Tissot van Patot MC, Leadbetter G, Keyes LE, Bendrick-Peart J, Bekckey VE, Christians U, et al. Greater free plasma VEGF and lower soluble VEGF receptor-1 in acute mountain sickness. J Appl Physiol. 2005;98:1626–9.
125. Schommer K, Wiesegart N, Dehnert C, Mairbäurl H, Bärtsch P. No correlation between plasma levels of vascular endothelial growth factor or its soluble receptor and acute mountain sickness. High Alt Med Biol. 2011;12:323–7.
126. Roach RC, Hackett PH. Frontiers of hypoxia research: acute mountain sickness. J Exp Biol. 2001;204:3161–70.
127. Sanchez del Rio M, Moskowitz MA. High altitude headache—lessons form headaches at sea level. In: Roach RC, Wagner PD, Hackett PH, editors. Hypoxia: into the next millenium. New York: Kluwer Academic/Plenum Publishers; 1999. p. 145–53.
128. Matsuzawa Y, Kobayashi T, Fujimoto K, Shinozaki S, Yoshikawa S, Yamaguchi S, et al. Cerebral edema in acute mountain sickness. In: Ueda G, Reeves JT, Sekiguchi M, editors. High-altitude medicine. Matsumoto: Shinshu University Press; 1992. p. 300–4.
129. Ross RT. The random nature of cerebral mountain sickness. Lancet. 1985;1:990–1.
130. Reeves JT, Moore LG, McCullough RE, McCullough RG, Harrison G, Tranmer BI, Micco AJ, Tucker A, Weil JV. Headache at high altitude is not related to internal carotid arterial blood velocity. J Appl Physiol. 1985;59:909–15.
131. Baumgartner R, Spyridopoulos I, Bärtsch P, Maggiorini M, Oelz O. Acute mountain sickness is not related to cerebral blood flow. A decompression chamber study. J Appl Physiol. 1999;86: 1578–82.
132. Wilson MH, Imray CHE, Hargens AR. The headache of high altitude and microgravity—similarities with clinical syndromes of cerebral venous hypertension. High Alt Med Biol. 2011;12:379–86.
133. Wilson MH, Davagnanam I, Holland G, Dattani RS, Tamm A, Hirani SP et al. Cerebral venous system and anatomical predisposition to high-altitude headache. Ann Neurol, i2013;73:381-89.

134. Strassman AM, Levy D. Response properties of dural nociceptors in relation to headache. J Neurophysiol. 2006;95:1298–306.
135. Rosenblum WI. Cytotoxic edema: monitoring its magnitude and contribution to brain swelling. J Neuropathol Exp Neurol. 2007;66:771–8.
136. Catani M, ffytche DH. The rises and falls of disconnection syndromes. Brain. 2005;128:2224–39.
137. Clarke C. High altitude cerebral oedema. Int J Sports Med. 1988;9:170–4.
138. Jafarian S, Gorouhi F, Salimi S, Lotfi J. Sumatriptan for prevention of acute mountain sickness: randomized clinical trial. Ann Neurol. 2007;62:273–7.
139. Bailey DM, Taudorf S, Berg RMG, Jensen LT, Lundby C, Evans KA, et al. Transcerebral exchange kinetics of nitrite and calcitonin gene-related peptide in acute mountain sickness—evidence against trigeminovascular activation? Stroke. 2009;40:2205–8.
140. Ikeda Y, Jimbo H, Shimazu M, Satoh K. Sumatriptan scavenges superoxide, hydroxyl, and nitric oxide radicals: in vitro electron spin resonance study. Headache. 2002;42:888–92.
141. Jafarian S, Gorouhi F, Salimi S, Lotfi J. Low-dose gabapentin in treatment of high-altitude headache. Cephalalgia. 2007;27:1274–7.
142. Gertsch JH, Lipman GS, Holck PS, Merritt A, Mulcahy A, Fisher RS, et al. Prospective, double-blind, randomized, placebo-controlled comparison of Acetazolamide versus ibuprofen for prophylaxis against high altitude headache: the headache evaluation at altitude trial (HEAT). Wilderness Environ Med. 2010;21:236–43.
143. Harris NS, Wenzel RP, Thomas SH. High altitude headache: efficacy of acetaminophen vs. ibuprofen in a randomized, controlled trial. J Emerg Med. 2003;24:383–7.
144. Asanuma M, Nishibayashi-Asanuma S, Miyazaki I, Kohno M, Ogawa N. Neuroprotective effects of non-steroidal anti-inflammatory drugs by direct scavenging of nitric oxide radicals. J Neurochem. 2001;76:1895–904.
145. Bailey DM, Brugniaux JV, Pietri S, Culcasi M, Swenson ER. Redox-regulation of neurovascular function by acetazolamide: complementary insight into mechanisms underlying high-altitude acclimatisation. J Physiol. 2012;590(Pt 15):3627–8.
146. Prouillac C, Vicendo P, Garrigues JC, Poteau R, Rima G. Evaluation of new thiadiazoles and benzothiazoles as potential radioprotectors: free radical scavenging activity in vitro and theoretical studies (QSAR, DFT). Free Rad Biol Med. 2009;46:1139–48.
147. Hackett PH, Roach RC. High altitude cerebral edema. High Alt Med Biol. 2004;5(2):136–46.
148. Dumont L, Mardirosoff C, Tramèr MR. Efficacy and harm of pharmacological prevention of acute mountain sickness: quantitative systematic review. BMJ. 2000;321:267–72.
149. Bärtsch P, Schneider M. Pharmacological prevention of acute mountain sickness—same ascent rates must be used to assess effectiveness of different doses of acetazolamide. BMJ. 2001;322:48.
150. Severinghaus JW. Sightings: diamox debate. High Alt Med Biol. 2001;2:9.
151. Carlsten C, Swenson ER, Ruoss S. A dose-response study of acetazolamide for acute mountain sickness prophylaxis in vacationing tourists at 12,000 feet (3630 m). High Alt Med Biol. 2004;5(1):33–9.
152. Basnyat B, Gertsch JH, Johnson EW, Castro-Marin F, Inoue Y, Yeh C. Efficacy of low-dose Acetazolamide (125 mg BID) for the prophylaxis of acute mountain sickness: a prospective, double-blind, randomized, placebo-controlled trial. High Alt Med Biol. 2003;4(1):45–52.
153. van Patot MC, Leadbetter III G, Keyes LE, Maakestad KM, Olson S, Hackett PH. Prophylactic low-dose acetazolamide reduces the incidence and severity of acute mountain sickness. High Alt Med Biol. 2008;9:289–93.
154. Basnyat B, Gertsch JH, Holck PS, Johnson EW, Luks AM, Donham BP, et al. Acetazolamide 125 mg BD is not significantly different from 375 mg BD in the prevention of acute mountain sickness: the prophylactic acetazolamide dosage comparison for efficacy (PACE) trial. High Alt Med Biol. 2006;7(1): 17–27.
155. Swenson ER. Carbonic anhydrase inhibitors and ventilation: a complex interplay of stimulation and suppression. Eur Respir J. 1998;12:1242–7.
156. Kayser B, Hulsebosch R, Bosch F. Low-dose acetylsalicylic acid analog and acetazolamide for prevention of acute mountain sickness. High Alt Med Biol. 2008;9:15–23.
157. Baillie JK, Thompson AAR, Irving JB, Bates MGD, Sutherland AI, MacNee W, et al. Oral antioxidant supplementation does not prevent acute mountain sickness: double-blind, randomized placebo-controlled trial. Q J Med. 2009;102:341–8.
158. Bailey DM, Young IS, McEneny J, Lawrenson L, Kim J, Barden J, Richardson RS. Regulation of free radical outflow from an isolated muscle bed in exercising humans. Am J Physiol Heart Circ Physiol. 2004;287(4):H1689–99.
159. Roach RC, Maes D, Sandoval D, Robergs RA, Icenogle M, Hinghofer-Szalkay H, et al. Exercise exacerbates acute mountain sickness at simulated high altitude. J Appl Physiol. 2000;88:581–5.
160. Scherrmann J-M. Drug delivery to brain via the blood-brain barrier. Vascul Pharmacol. 2002; 38:349–54.
161. Guzy RD, Schumacker PT. Oxygen sensing by mitochondria at complex III: the paradox of increased reactive oxygen species during hypoxia. Exp Physiol. 2006;91:807–19.
162. Bailey DM, Lawrenson L, McEneny J, Young IS, James PE, Jackson SK, et al. Electron paramagnetic spectroscopic evidence of exercise-induced free radical accumulation in human skeletal muscle. Free Radic Res. 2007;41:182–90.

163. Kagan VE, Wipf P, Stoyanovsky D, Greenberger JS, Borisenko G, Belikova NA, et al. Mitochondrial targeting of electron scavenging antioxidants: regulation of selective oxidation vs random chain reactions. Adv Drug Deliv Rev. 2009;61:1375–85.
164. Patir H, Sarada SK, Singh S, Mathew T, Singh B, Bansal A. Quercetin as a prophylactic measure against high altitude cerebral edema. Free Radic Biol Med. 2012;53:659–68.
165. van Patot MC, Keyes LE, Leadbetter GI, Hackett PH. Gingko biloba for prevention of acute mountain sickness: does it work? High Alt Med Biol. 2009;10: 33–43.
166. Gertsch JH, Basnyat B, Johnson EW, Onopa J, Holck PS. Randomised, controlled trial of ginkgo biloba and acetazolamide for prevention of acute mountain sickness: the prevention of high altitude illness trial (PHAIT). BMJ. 2004;328:797–9.
167. Gertsch JH, Seto TB, Mor J, Onopa J. Gingko biloba for the prevention of severe acute mountain sickness (AMS) starting one day before rapid ascent. High Alt Med Biol. 2002;3:29–37.
168. Chow T, Browne V, Heileson HL, Wallace D, Anholm J. Ginkgo biloba and acetazolamide prophylaxis for acute mountain sickness—a randomized, placebo-controlled trial. Arch Intern Med. 2005;165(3):296–301.
169. Moraga FA, Flores A, Serra J, Esnaola C, Barriento C. Ginkgo biloba decreases acute mountain sickness in people descending to high altitude at Ollagüe (3696 m) in Northern Chile. Wilderness Environ Med. 2007;18:251–7.
170. Leadbetter G, Keyes LE, Maakestad KM, Olson S, Tissot van Patot MC, Hackett PH. Gingko biloba does—and does not—prevent acute mountain sickness. Wilderness Environ Med. 2009;20: 66–71.
171. Fischer R, Lang SM, Leitl M, Thiere M, Steiner U, Huber RM. Theophylline and acetazolamide reduce sleep-disordered breathing at high altitude. Eur Respir J. 2004;23:47–52.
172. Küpper TE, Strohl KP, Hoefer M, Gieseler U, Netzer CM, Netzer NC. Low-dose theophylline reduces symptoms of acute mountain sickness. J Travel Med. 2008;15:307–14.
173. Fischer R, Lang SM, Steiner U, Toepfer M, Hautmann H, Pongratz H, Huber RM. Theophylline improves acute mountain sickness. Eur Respir J. 2000;15:123–7.
174. Muza SR, Kaminsky D, Fulco CS, Banderet LE, Cymerman A. Cysteinyl leukotriene blockade does not prevent acute mountain sickness. Aviat Space Environ Med. 2004;75:413–9.
175. Grissom CK, Richer LD, Elstad MR. The effects of a 5-lipoxygenase inhibitor on acute mountain sickness and urinary leukotriene E-4 after ascent to high altitude. Chest. 2005;127:565–70.
176. Luks A, Henderson WR, Swenson ER. Leukotriene receptor blockade does not prevent acute mountain sickness induced by normobaric hypoxia. High Alt Med Biol. 2007;8:131–8.
177. Dumont L, Lysakowski C, Tramer MR, Junod JD, Mardirosoff C, Tassonyi E, et al. Magnesium for the prevention and treatment of acute mountain sickness. Clin Sci. 2004;106(3):269–77.
178. Vecchia D, Pietrobon D. Migraine: a disorder of brain excitatory-inhibitory balance? Trends Neurosci. 2012;35:507–20.
179. Freilinger T, Anttila V, de Vries B, Malik R, Kallela M, Terwindt GM, et al. Genome-wide association analysis identifies susceptibility loci for migraine without aura. Nat Genet. 2012;44(7):777–82.
180. Kupers R, Kehlet H. Brain imaging of clinical pain states: a critical review and strategies for the future. Lancet Neurol. 2006;5:1033–44.
181. Stephenson DT, Arneric SP. Neuroimaging of pain: advances and future prospects. J Pain. 2008; 9:567–79.
182. Lee JH, Huh YM, Jun YW, Seo JW, Jang JT, Song HT, et al. Artificially engineered magnetic nanoparticles for ultra-sensitive molecular imaging. Nat Med. 2007;13:95–9.
183. Buck A, Schirlo C, Jasinsky V, Weber B, Burger C, von Schulthess GK, et al. Changes of cerebral blood flow during short-term exposure to normobaric hypoxia. J Cereb Blood Flow Metab. 1998;18:906–10.
184. Tracey I, Johns E. The pain matrix: reloaded or reborn as we image tonic pain using arterial spin labelling. Pain. 2010;148:359–60.
185. Weiller C, May A, Limmroth V, Jüptner M, Kaube H, von Schayck R, et al. Brain stem activation in spontaneous human migraine attacks. Nat Med. 1995;1:658–60.
186. Afridi SK, Matharu MS, Lee L, Kaube H, Friston KJ, Frackowiak RSJ. A PET study exploring the laterality of brainstem activation in migraine using glyceryl trinitrate. Brain. 2005;128:932–9.
187. Sánchez del Rio M, Linera JA. Functional neuroimaging of headaches. Lancet Neurol. 2004;3:645–51.
188. Edden RAE, Harris AD, Murphy K, Evans CJ, Saxena N, Hall JE, et al. Edited MRS is sensitive to changes in lactate concentration during inspiratory hypoxia. J Magn Reson Imaging. 2010;32:320–5.
189. Terpstra M, Ugurbil K, Gruetter R. Direct in vivo measurement of human cerebral GABA concentration using MEGA-editing at 7 Tesla. Magn Reson Med. 2002;47:1009–12.
190. Edden RAE, Pomper MG, Barker PB. In vivo differentiation of N-acetyl aspartyl glutamate from N-acetyl aspartate at 3 Tesla. Magn Reson Med. 2007;57:977–82.
191. Terpstra M, Henry PG, Gruetter R. Measurement of reduced glutathione (GSH) in human brain using LCmodel analysis of difference-edited spectra. Magn Reson Med. 2003;50:19–23.
192. Terpstra M, Gruetter R. H NMR detection of vitamin C in human brain in vivo. Magn Reson Med. 2004;51:225–9.
193. Fayed N, Modrego PJ, Morales H. Evidence of brain damage after high-altitude climbing by means of magnetic resonance imaging. Am J Med. 2006; 119:168.e1–6.

194. Bailey DM, Evans KA, McEneny J, Young IS, Hullin DA, James PE, et al. Exercise-induced oxidative-nitrosative stress is associated with impaired dynamic cerebral autoregulation and blood-brain barrier leakage. Exp Physiol. 2011; 96:1196–207.
195. Alperin NJ, Lee SH, Loth F, Raksin PB, Lichtor T. MR-intracranial pressure (ICP): a method to measure intracranial elastance and pressure noninvasively by means of MR imaging: baboon and human study. Radiology. 2000;217:877–85.
196. Zagorac D, Yamaura K, Zhang C, Roman RJ, Harder DR. The effect of superoxide anion on autoregulation of cerebral blood flow. Stroke. 2005; 36:2589–94.
197. Gelfi C, De Palma S, Ripamonti M, Eberini I, Wait R, Bajracharya A, et al. New aspects of altitude adaptation in Tibetans: a proteomic approach. FASEB J. 2004;18(3):612–4.
198. Kumar V, Calamaras TD, Haeussler DJ, Colucci W, Cohen RA, McComb ME, et al. Cardiovascular redox and ox stress proteomics. Antioxid Redox Signal. 2012;17:1528–59.
199. Olesen J, Diener HC, Husstedt IW, Goadsby PJ, Hall D, Meier U, et al. Calcitonin gene-related peptide receptor antagonist BIBN 4096 BS for the acute treatment of migraine. N Engl J Med. 2004; 350:1104–10.

High-Altitude Pulmonary Edema (HAPE)

21

Robert B. Schoene and Erik R. Swenson

Abstract

Much of the clinical impact of acute altitude illnesses stems from fluid accumulation in interstitial spaces and nowhere is this more apparent than in the lungs as the edema escapes into the alveoli to cause life-threatening hypoxemia. This chapter will update our knowledge of HAPE over the past decade about the vasculature, alveolar epithelium, innervation, immune response, and genetics of the lung in hypoxia, as well as prophylactic and therapeutic strategies to reduce the toll of this most common alpine life-threatening illness.

Introduction

Much of the clinical impact of acute altitude illnesses stems from fluid accumulation in interstitial spaces and nowhere is this more apparent than in the lungs as the edema escapes into the alveoli to cause life-threatening hypoxemia. This chapter will update our knowledge of HAPE over the past decade about the vasculature, alveolar epithelium, innervation, immune response, and genetics of the lung in hypoxia, as well as prophylactic and therapeutic strategies to reduce the toll of this most common alpine life-threatening illness.

Clinical Presentation

HAPE has an incidence of 0.2–15 % depending on altitude, ascent rate, exertion, gender, age, infection, individual susceptibility, and underlying health problems associated with pulmonary hypertension [1]. With past radiographically proven HAPE, the incidence may approach 60% [2]. Men are more susceptible than women [3], reflecting possibly the advantages of ventilatory stimulation from progesterone [4] and lesser hypoxic pulmonary vasoconstriction (HPV) from estrogen [5]. HAPE has no age dependence, although aging-related increases in pulmonary vascular resistance (PVR), which might also

R.B. Schoene
Clinical Professor of Medicine, Division of Pulmonary and Critical Care Medicine, Department of Medicine, University of Washington, Seattle, WA, USA
Bay Area Pulmonary/Critical Care Medical Associates, Berkeley, CA, USA
e-mail: rbschoene@gmail.com

E.R. Swenson, M.D. (✉)
VA Puget Sound Health Care System, University of Washington, Seattle, WA, USA
e-mail: eswenson@u.washington.edu

E.R. Swenson and P. Bärtsch (eds.), *High Altitude: Human Adaptation to Hypoxia*,
DOI 10.1007/978-1-4614-8772-2_21, © Springer Science+Business Media New York 2014

extend to HPV, would predict greater HAPE susceptibility. Although HAPE can develop in sedentary persons, exercise [6] and its pulmonary hemodynamic consequences (discussed below) are important precipitating factors.

Symptoms, signs, and physiologic changes in lung function typical of pulmonary edema [1–3, 7] evolve in 2–4 days after ascent, often preceded or accompanying AMS (see Chap. 20), but can occur later with further ascent. Arterial saturations can fall as low as 40 % and PaO_2s in the low 20 mmHg range. HAPE in its severest stage with profound hypoxemia can lead to high-altitude cerebral edema [8]. Repeat occurrences of HAPE do not always involve infiltrates in the same areas, which suggests that fixed structural aspects of lung parenchyma or vessels do not account for the timing of edema or its locale [9]. A special exception is unilateral absence of a pulmonary artery, in which edema always occurs in the contralateral lung receiving the entire cardiac output [10].

It has been suggested that many persons (50–75 %) may have subclinical HAPE that resolves spontaneously despite remaining at altitude [11–13]. This incidence equals that of AMS, which itself can cause mild gas exchange impairment [14] by unknown mechanisms perhaps related to altered autonomic influences on the pulmonary circulation and/or airways leading to ventilation-perfusion mismatching. Subclinical HAPE may be considerably overestimated without radiography [15] because indirect measures of interstitial edema such as spirometry, closing volume, and/or transthoracic impedance can vary for other reasons related to mountaineering including intense exercise and increased cardiac output, cold/dry air-induced bronchoconstriction, and hypocapnia [16]. With radiographically mild HAPE, only modest abnormalities were detectable [15] suggesting many lung function parameters are not sensitive enough to detect small changes in interstitial fluid and may require high-resolution tissue density measurements by CT or MR imaging. Reentry HAPE occurs when long-term high-altitude residents return to high altitude following a brief low-altitude sojourn. It has a strong familial basis and afflicts children more than adults [17], perhaps due to a twofold greater magnitude of HPV in preteen children (age 6–9) compared to teenagers (age 14–16) when tested 40 h after ascent to high altitude [18].

Pathophysiology

From its first modern descriptions, pulmonary hypertension and HAPE have been inextricably linked suggesting a primary hemodynamic basis. However, an inflammatory reaction and differences in hypoxia-sensitive, active alveolar fluid reabsorption may at times be contributory.

Hemodynamics

Mean PA pressures by catheterization in untreated HAPE range from 35 to 95 mmHg with normal PA wedge pressures. Noninvasive echocardiographic estimations of systolic PA pressure substantiate these data [2, 19] and show no evidence for left ventricular systolic or diastolic dysfunction [20, 21]. Excessive PA pressure precedes HAPE [2] and any interventions (descent, oxygen, or drugs) which lower PA pressure improve gas exchange and outcomes [19, 22, 23].

HAPE susceptibles have several hypoxic responses putting them at risk; the most important is strong HPV. Although their resting PA pressures are at the high end of normal at low altitude, exaggerated responses with normoxic exercise and sleep [24, 25] point to a constitutional hyperreactivity of the pulmonary circulation (Fig. 21.1). Their relatives have not been studied, but children and their fathers at 3,450 m show similar PA pressure increases [26], suggesting HPV is, in part, genetically determined as is the hypoxic ventilatory response (HVR) [27]. The prevalence of heightened pulmonary vascular responsiveness in the general population may be as high as 10 % [28] and may contribute to the out of proportion pulmonary hypertension that develops later in life in patients with sleep apnea, heart failure, and chronic lung disease.

PA catheterization studies at 4,559 m [29] found the exaggerated rise in PA pressure in HAPE susceptibles at 4,559 m led to an increased microvascular pressure above 20 mmHg in those

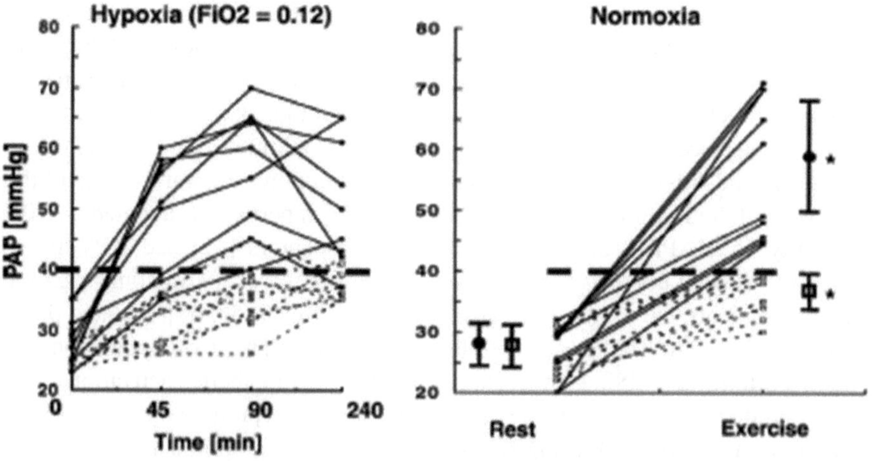

Fig. 21.1 PAP in HAPE-susceptible individuals (*continuous lines* and *filled symbols*) and in non-susceptible controls (*dashed lines* and *open symbols*) during exposure to normobaric hypoxia (*left*) and before and during exercise on a bicycle ergometer (*right*). The highest PAP recordings during exercise (75–150 W) are shown (From ref [62])

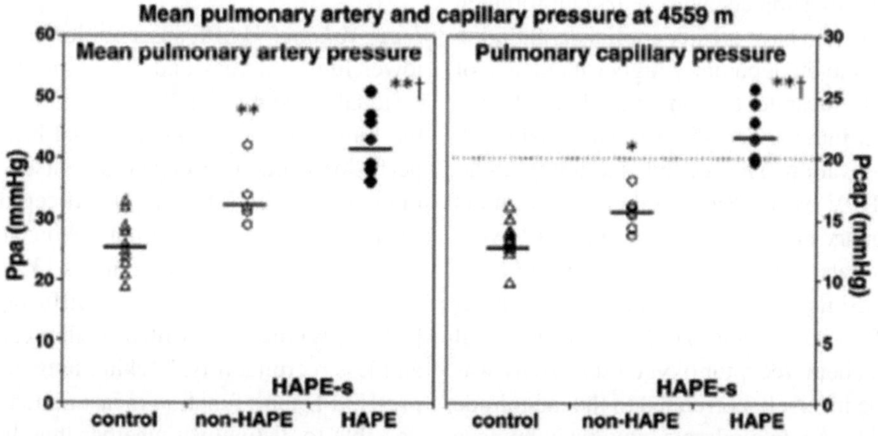

Fig. 21.2 Mean pulmonary artery pressure (P_{pa}) and pulmonary capillary pressure (P_{cap}) in 14 controls and in 16 high-altitude edema susceptible (HAPE-s) subjects at high altitude. HAPE-s is further divided in those who developed HAPE (HAPE) and those who did not develop HAPE (non-HAPE). *Bars* indicate the mean values in each group. *$p<0.05$, **$p<0.01$ vs. control, †$p<0.01$ vs. non-HAPE

developing HAPE (Fig. 21.2). This threshold for edema is similar to animal work in showing a PO_2-independent microvascular pressure of 17–24 mmHg [30]. It is interesting that this same pressure range in normoxic rats imposed by left atrial pressure elevation also reduces active alveolar epithelial sodium reabsorption [31]; which is discussed below in the "Alveolar Fluid Clearance" section. It should be emphasized that microvascular pressure rather than upstream PA pressure elevation is more crucial because strong HPV alone need not necessarily lead to HAPE, as shown in a study of adults with a history of perinatal hypoxia [32, 33].

High hypoxic PA pressures and PVR in HAPE-susceptibles is the sum of many influences including those intrinsic to vascular smooth muscle, but also to neuro-humoral responses, lung volume, and vascular endothelium.

Neuro-humoral Responses: HAPE susceptibles have a lower isocapnic HVR set largely by the peripheral chemoreceptors [34, 35], which results in lower alveolar and arterial PO_2s at the same P_IO_2, and thus a stronger HPV stimulus. Similar, but only indirect measures of lesser ventilatory response to hypoxia (lower arterial saturation and higher end-tidal CO_2) have been shown in a group of high-altitude adolescents with susceptibility to HAPE (reentry HAPE) when compared to resistant controls [36]. These data from awake subjects have now been also shown to apply to sleep in HAPE-susceptible persons at high altitude [37, 38]. Lower HVR also leads to a smaller fall in P_ACO_2 and less hypocapnic inhibition of HPV [39]. HVR and HPV may be linked in two ways. Lower arterial PO_2 itself may increase PA pressure because isolated hypoxic perfusion of the bronchial artery (supplying the vasa vasorum of pulmonary arteries) increases PA pressure [40]. This finding may be relevant to subjects with a patent foramen ovale (PFO) who have greater arterial desaturation at altitude and an apparently higher incidence of HAPE [41], despite the potential benefit of a right to left intracardiac shunt in minimizing PA pressure elevation. The second is a direct modulation of HPV by the peripheral chemoreceptors via pulmonary innervation. In mechanically ventilated animals, vagotomy [42] and carotid body ablation [43] increase HPV. In humans, the magnitude of poikilocapnic HVR as measure of peripheral chemoreceptor oxygen sensitivity was found to be inversely correlated to the magnitude of HPV [44]. Acute moderate hypoxia in humans causes diuresis and natriuresis which are correlated to higher HVR [45]. HAPE susceptibles have lower HVR and may be disadvantaged by a limited diuretic response [46], in part by greater activation of the renin-angiotensin system [47] and sympathetic nervous system [48], but also less chemoreceptor-mediated natriuresis.

Increased Sympathetic Tone: Increased sympathetic tone with hypoxia, especially in the absence of strong opposing peripheral chemoreceptor input, may also contribute to stronger HPV, in an analogous fashion to those with low HVR who are more susceptible to cerebral-mediated hypoxic ventilatory depression [49]. HAPE susceptibles have increased skeletal muscle sympathetic tone during hypoxia at low and high altitude [47], although this study, similar to the HPV studies, suffers from lack of control on the strength of the hypoxic stimulus due to HVR differences. Stimulation by cerebral hypoxia of the lung sympathetic innervation augments HPV [50] via alpha receptors [51] and HPV is reduced with autonomic blockade in some [52] but not in all studies [46, 47, 53, 54]. Isolated perfusion of the dog brain with hypoxic-hypercapnic blood causes intense sympathetic activation, increased PA pressures, and pulmonary edema [55]. The impact of this study is diminished somewhat by the use of hypoxic-hypercapnic blood, because cerebral hypercapnic hypoxia is a potent stimulus for sympathetic activation, which would be less with the arterial alkalemia and hypoxemia typical at high altitude.

Lung Volume: HAPE susceptibles have 10–15 % lower lung volumes and 30 % lower functional residual capacity (FRC) [34, 35, 56, 57]. FRC is the lung volume at which normal breathing and perfusion occur and lung volume itself is a determinant of PVR [58]. HAPE susceptibles show greater arterial desaturation while supine that resolves with high tidal volumes [35] indicative of a lower FRC. Their lower diffusing capacity [57, 59] is consistent with a smaller capillary bed and less recruitability. Lacking lung biopsies and pre-HAPE lung function measurements, it is not possible to distinguish whether they have intrinsically different lung structure or that an episode of HAPE heals with a small loss in volume and capillary bed.

Vascular Endothelium: There are differences of the vasculature itself in HAPE susceptibility, although it has not been studied whether differences in PA vascular smooth muscle responsiveness to hypoxia exist. Nitric oxide (NO) and endothelin 1 (ET-1) are key endothelial-derived vasoactive mediators. Lung NO production in hypoxia is reduced in HAPE susceptibles as measured in exhaled gas (Fig. 21.3), alveolar lavage fluid, and blood [60–63]. Systemic vascular endothelial NO generation as a surrogate for the pulmonary circu-

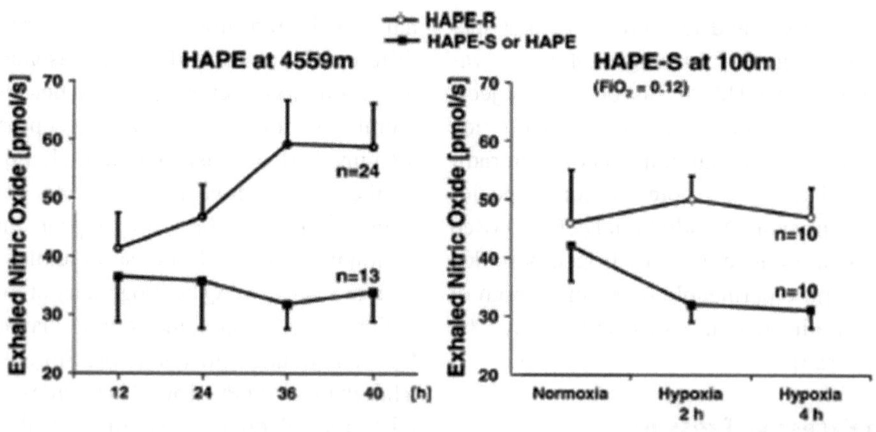

Fig. 21.3 (**a**) Exhaled NO after 40 h at 4,559 m in individuals developing HAPE (*left*) and in individuals not developing HAPE (HAPE-R) despite identical exposure to high altitude. (**b**) Exhaled NO in individuals with (HAPE-S) and without susceptibility (HAPE-R) to HAPE after 4 h of exposure to hypoxia ($FIO_2 = 0.12$) at low altitude (elevation 100 m) (From refs [61, 62])

lation is reduced more in hypoxic HAPE susceptibles [64, 65], possibly as a result of greater circulating concentrations of asymmetric dimethyl arginine (ADMA), an endogenous metabolite of arginine and inhibitor of endothelial NO synthase [65]. Circulating ET-1 is elevated almost threefold at high altitude and to a greater degree in HAPE susceptibles and correlates with the rise in PA pressure [64–66]. Other vascular mediators studied in over 400 subjects at 3,500 m with likely hypertensive effect on the pulmonary vasculature are greater in HAPE susceptibles, including plasma concentration of serotonin, 8-iso prostaglandin F, renin, and aldosterone [65].

The study of HPV continues to identify new sensing, signalling and effector mechanisms and pathways, of which several warrant mention. In addition to the critical role of NO, two other endogenously produced gases, *carbon monoxide and hydrogen sulfide,* may be potentially important HPV modulators [67, 68], but have not been studied at high altitude or in HAPE susceptibles. Iron supplementation and iron chelation reduce and increase HPV respectively [69, 70], possibly via altered *HIF metabolism* [71]. In two rat strains with differing pulmonary hypoxic responses, HIF-1 activity and HIF-mediated protein expression were higher in the strain with lesser pulmonary hypertension [72]. In contrast, mice with heterozygous HIF 1-alpha deficiency have weaker acute and chronic hypoxic responses in isolated pulmonary vascular smooth myocytes and pulmonary vessels than wild type mice [73, 74]. Carotid body sensitivity to hypoxia in these same HIF 1-alpha deficient heterozygote mice is depressed [75], although this does not appear to diminish the HVR. Interestingly a recent report finds that low-altitude HAPE-susceptible individuals compared to HAPE-resistant persons and high-altitude natives in India have distinct polymorphisms in the EGLN-1 gene (HIF-prolyl hydroxylase-20) that acts to regulate HIF-1 alpha activity [76]. These polymorphisms are associated with differences in SaO_2 and pulmonary artery pressures at high altitude as would be predicted for HAPE susceptibility. Further supporting pharmacological evidence for HIF-1alpha mediation of HPV was demonstrated in mice by reduction in hypoxic pulmonary hypertension with digoxin, a known inhibitor of HIF-1alpha transcriptional activity [77]. At present it is not clear how HIF-dependent gene transcription affects HPV, but it likely involves alterations in pulmonary vascular smooth muscle calcium signalling [77].

A compelling case is emerging that hypoxia increases *reactive oxygen species* (*ROS*) generation (see Chap. 1), which is an upstream signal for HPV [78, 79]. While it is clear that altitude increases stable circulating markers of ROS production [80, 81] and may play a role in AMS, it also appears that persons with higher HPV generate more ROS and less bioactive NO species across the lung [82]. In support of this, it was recently shown that HAPE-susceptible subjects have lower plasma concentrations of superoxide dismutase, an enzyme that converts oxygen radical (O_2^-) to H_2O_2, a less potent oxidant species [65]. Isolated human PA endothelial cells exposed to 3 % oxygen produce more hydrogen peroxide and become more permeable to albumin, both of which are diminished in vitro and in vivo by antioxidants [83].

Site(s) of Excessive Pressure and Leak in HAPE

Three theories have been proposed and none are mutually exclusive: (1) trans-arteriolar leakage, (2) hypoxic venoconstriction leading to capillary pressure elevation, and (3) uneven regional HPV with overperfusion in certain areas. Small arterioles are a site of leakage with markedly increased PA pressure [84, 85] possibly because their endothelial cells in vitro have a 20-fold greater permeability than more downstream microvascular arterial endothelial cells [86]. Pulmonary veins also constrict with hypoxia [87], thus increasing resistance and pressure downstream of the fluid filtration region. Uniform arteriolar and venoconstriction, alone or in combination, may contribute to edema formation; however, they cannot explain the patchy radiographic appearance of early HAPE unless there is regional HPV heterogeneity leading to uneven distribution of perfusion with high flow in those areas of lesser vasoconstriction. In these regions pressures might rise to the threshold of 17–24 mmHg [88], possibly aggravated further by increased venous resistance [89]. Recent investigations in non-exercising animals with microspheres [90] and humans with magnetic resonance imaging [91, 92] demonstrate that HPV is indeed uneven. Although exercise in hypoxia does not significantly further increase regional perfusion heterogeneity above that with hypoxia alone [93, 94], the much higher pressures [24, 25] should greatly increase the risk of injury in the more highly perfused areas (overperfusion edema).

The basis of uneven regional HPV is unknown but may involve inhomogeneous localization of smooth muscle, both in thickness and longitudinal distribution along the arterial tree. Intrinsic differences in local endothelial vasoactive mediator production or expression and heterogeneity of membrane ion channels and receptors may be also invoked, as exist for endothelial derived NO in the horse between dorsal and ventral lung regions [95]. It does not appear that uneven regional HPV in HAPE susceptibles is due to greater resting V_A/Q heterogeneity at low altitude, which would lead some regions having a lower P_AO_2 and to undergo a more precipitous drop with a fall in inspired PO_2 [56]. Unevenness of regional HPV may decrease with time at altitude since slow ascent prevents HAPE even in susceptible individuals and HAPE rarely occurs after the first 5 days at a given altitude. There may be rapid remodeling and generalized homogenous muscular hypertrophy of all arterioles or greater NO production [61], both of which would lead to a more even blood flow distribution and microvascular protection. Another protective factor may be upregulated gene transcription and protein expression for collagen and other extracellular matrix constituents that strengthen the alveolar capillary barrier [96].

Although hypoxia even in the absence of pressure changes can also increase alveolar–capillary permeability [97, 98], the evidence best supports overperfusion edema occurring in some regions as a result of high blood flow under large driving gradients with resultant increased microvascular pressures exceeding the capacity of the lungs to maintain a fluid-free air space. What is lacking is definitive in vivo evidence that edema develops in areas of high flow. This will require studies using a single imaging modality capable of simultaneously resolving blood flow and changes in extravascular fluid content.

Inflammation

Alveolar lavage fluid in some mountaineers with HAPE [99, 100] revealing significant neutrophilia and elevations of several proinflammatory cytokines and neutrophil chemotactic factors engendered the idea that inflammation might be causal in HAPE. Work in hypoxic animals with viral [101] or endotoxin [102] administration showing

Table 21.1 Bronchoalveolar lavage characteristics in early HAPE

	550 m		4,559 m		
	CONT (n=8)	HAPE-S (n=9)	CONT (n=6)	HAPE-s (pre-HAPE) (n=6)	HAPE-s (ill) (n=3)
Leukocyte count (×10³/mL)	8.1	6.3	9.5	8.1	9.8
Macrophages (%)	94	95	83	82	85
Neutrophils (%)	1	0	0	1	1
Red cell count (×10³/mL)	3	4	10	26*	623*
Total protein (mg/dL)	1.8	1.5	14*	34*	163*
PAP systolic(mmHg)	22	26	37*	61*	81*

Bronchoalveolar lavage (BAL) at 550 m and on the second day at 4,559 m in 8 control subjects (CONT) and in 9 HAPE-susceptible subjects (HAPE-S) of whom 3 had pulmonary edema at the time of BAL. Of those 6 HAPE-S without pulmonary edema at the time of BAL (pre-HAPE), 4 developed HAPE within 18 h after BAL. *$p<0.05$ vs. 550 m (From ref [62])

more edema and prevention of hypoxic edema with corticosteroid pretreatment, a classic anti-inflammatory therapy [103], further supported the concept. Ten percent oxygen given to rats causes leukocyte adhesion to systemic capillaries, ROS formation, depletion of endogenous NO, and increased permeability [104] and this can be blocked by dexamethasone [105]. The systemic capillary changes were due to local mast cell degranulation triggered by monocyte chemoattractant protein-1 generated by hypoxic alveolar macrophages [106]. Likewise, similar findings of increased ROS production, inflammation, and upregulation of the proinflammatory gene transcription factor, nuclear factor kappa beta (NFkB), were found in the lungs of rats and mice exposed altitudes >18,000 ft [83, 107–109] and were attenuated by antioxidant treatment [106]. More recently [110] it has been shown in mice exposed to 10 % hypoxia for 1 day that lung cells uniquely express hypoxia-induced mitogenic factor (MIHF), which is proangiogenic and vasoconstricting, but also stimulates release of the proinflammatory cytokine, monocyte chemotactic protein (MCP-1). However, the relevance of these studies to human HAPE is questioned because evidence for systemic capillary inflammation and leakage has not been found [111] and HAPE in humans develops at much lower altitudes and lesser hypoxia than studied in these rodent models.

The noninflammatory characteristics of alveolar fluid in some cases of HAPE [99] and no in vivo thrombin and fibrin formation except in very advanced HAPE [112] or differences in platelet activation with hypoxia between HAPE susceptibles and HAPE-resistant subjects [113] are more consistent with any inflammation occurring as a secondary response to alveolar–capillary barrier disruption and edema. When lavage was performed in climbers within a day of ascent to 4,559 m only mild alveolar hemorrhage and increased protein concentrations in the airspace (Table 21.1) (Fig. 21.4) were found both in those ill with HAPE (HAPE-ill) and in those who developed HAPE (pre-HAPE) within the next 24 h [62]. There was a strong correlation between the magnitude of pulmonary hypertension by echocardiography and hemorrhage and protein elevation in the alveolar space (Fig. 21.5). In contrast, there were no increases in alveolar neutrophils and proinflammatory mediators (tumor necrosis factor-α, interleukins 1 and 8) early in course of HAPE. Alveolar macrophages harvested at sea level and at high altitude showed no differences in TNF, IL-8, IL-6, and IL-1 production between the HAPE-resistant and HAPE-susceptible subjects when stimulated in vitro under normoxic or hypoxic conditions, before or after endotoxin stimulation [114], and in rat alveolar epithelial cells, macrophages, and pulmonary artery smooth muscle cells, mild hypoxia (5 % O_2) in fact led to an attenuation of proinflammatory gene and protein expression [115]. Lack of inflammation in early HAPE was recently shown in rats that were made to continuously walk slowly in hypobaric hypoxia (4,800 m) for up to 48 h. Despite greater hemorrhagic lung edema and histological evidence of capillary stress failure, there were no

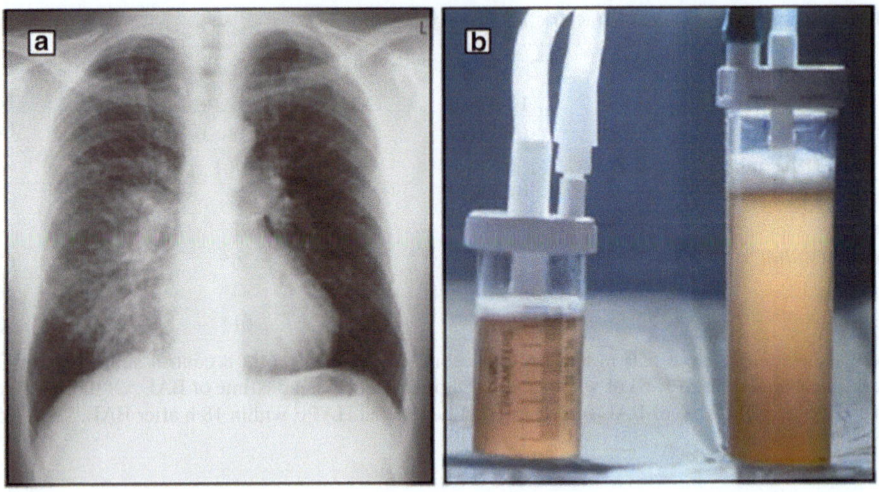

Fig. 21.4 Chest radiograph with bronchoalveolar lavage fluid aliquots (first and fifth) from a representative subject with early high-altitude pulmonary edema. The radiograph shows interstitial and alveolar infiltrates and the lavage performed after the X-ray was taken shows mild alveolar hemorrhage (From ref [62])

significant increases in leukotriene B4 and no correlations of edema or arterial hypoxemia with neutrophils in the lavage fluid [116]. In a model of greater hypoxic stress (10 % O_2) lasting 168 h but without exercise, inflammation as determined by increased lung mRNA for IL-1, IL-6, and TNF-alpha was not increased [117]. Increases in circulating interleukin-6 (IL-6) in humans at altitude [111, 118–120] continue to be taken as evidence for an inflammatory effect of hypoxia that might lead to HAPE [118]. This idea is problematic because exercising muscle releases IL-6 in proportion to intensity and duration of work both in normoxia and hypoxia [120, 121]. With passive ascent to high altitude, there is little to no increase in IL-6, even in HAPE-susceptible persons [122]. Although IL-6 is considered a classic proinflammatory cytokine, it may have equally important anti-inflammatory and endothelial permeability protective actions [121, 123]. Studies in IL-6 knockout and overexpressing mice should be instructive in determining the role of IL-6 at high altitude.

What initiates secondary inflammation is not clear? Sustained high pressures in untreated HAPE of sufficient duration may trigger inflammation [124] or the inflammation represents healing of a disrupted alveolar–capillary barrier that occurs in the most severe cases of HAPE. Alveolar hemorrhage, whose breakdown products including free hemoglobin and its subsequent degradation to heme metabolites are neutrophil chemoattractants [125]. These bind to danger and pathogen-associated molecular pattern cell surface proteins of the innate immune system to engage inflammatory signalling pathways [126].

Despite evidence, particularly in humans given above, against primary inflammatory alteration of the alveolar–capillary barrier in HAPE, animal models at much higher altitude or with greater normobaric hypoxia [105, 107, 108, 127] do support an element of initial co-contributing risk. Emerging evidence shows that severe hypoxia can induce inflammation via HIF-1 alpha and NFkB-linked gene regulation [127]. Thus, it may be possible that HAPE is primarily a pressure-related pathology, but if the hypoxia is severe enough, increased capillary permeability from activation of inflammatory cascades will also contribute to alveolar edema. Drugs such as nifedipine, the best studied pulmonary vasodilator for HAPE prevention (see below), could conceivably also act to limit inflammation and edema in the lung with severe hypoxia by suppressing NFkB, although in a rat study used to investigate this hypothesis the dosing was 50–100 times higher than that used in humans [128] and lower doses were not tested. Furthermore, it is likely

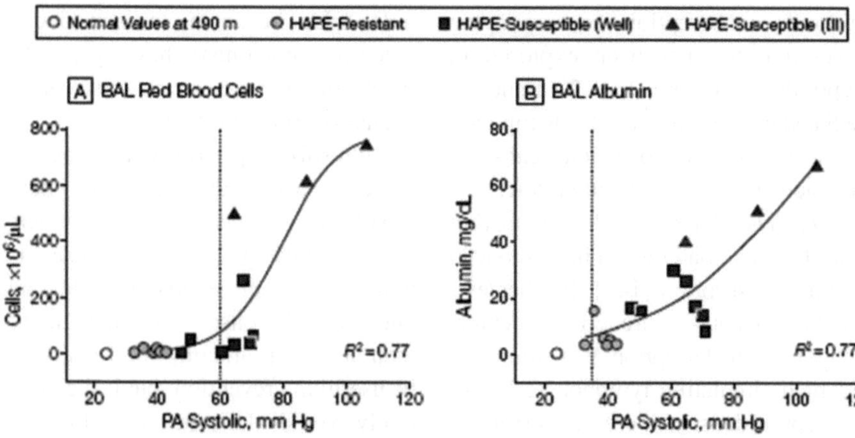

Fig. 21.5 Individual BAL red blood cell count and albumin concentrations plotted against pulmonary artery systolic pressures at high altitude (4,559 m). *BAL* indicates bronchoalveolar lavage; *HAPE* high-altitude pulmonary edema. The *vertical lines* denote a threshold systolic PA pressure (>60 mmHg) above which red blood cell (**a**) appear in the BAL fluid in contrast to the lower pressure (35 mmHg) at which albumin leakage occurs (**b**). The *open circles* in the *lower left* of both panels show the normal values for these at low altitude. The correlation coefficients are given for the best-fit curves of the values at high-altitude ($P<0.05$ for both curves) (From ref [62])

that any concurrent process altering alveolar–capillary barrier permeability, such as preceding respiratory viral infections [129, 130], will lower the edema threshold and also explain why HAPE in humans can occur in some at a modestly low altitude [131].

Alveolar Fluid Clearance

Active sodium transport from alveolar space into the interstitium is important in normal lung fluid balance and a strong argument has been made for this process at high altitude. Hypoxia in vitro decreases transepithelial sodium transport by reducing expression and activity of the apical epithelial sodium channel (ENaC) and basolateral Na^+/K^+ ATPase [132] in cultured alveolar epithelial cells possibly by an impairment of beta-2 adrenergic receptor signalling [133, 134]. These and many other in vitro studies have used 1–3 % O_2 (reviewed in [135]), but recently it was found that 5 % O_2 had much less effect [115]. Whether or not this translates into reduced alveolar fluid clearance (AFC) with hypoxia in vivo is also conflicting. Some studies find depression of AFC [136, 137] acutely and mice partially deficient in ENaC develop greater accumulation of lung water in hypoxia [138]. However, in another study of rats exposed to 10 % O_2 for 5 days, there was no AFC depression in the first 3 days, after which it rose by 25–30 % [139]. Interestingly, a link between inflammation and HAPE may arise from viral infection-related downregulation of ENaC activity and diminished AFC [140]. In addition to hypoxia itself, raised intracapillary pressure around the level at which alveolar and interstitial fluid begin to accumulate also reduces active alveolar epithelial sodium reabsorption [31], but the mechanism is unknown.

Whether alveolar sodium transport differences underlie HAPE susceptibility has been addressed with transepithelial nasal potential (NP) differences as a surrogate for alveolar epithelial ion transport. Reduced NP differences indicative of decreased sodium reabsorption by ENaC have been reported in HAPE-susceptibles [141–143]. This has been questioned because NP differences between HAPE-susceptibles and controls at high altitude could not be attributed to differences in ENaC activity [143–145], but rather to differences in chloride secretion, which contributes a large fraction of the NP difference in the nasal mucosa, but not in the alveolar epithelium. The role of the cystic fibrosis transmembrane regulator (CFTR), a chloride channel and possible pathway for chloride reabsorption accompanying reabsorbed sodium, has not been explored in HAPE susceptibility, but in rats beta-2 adrenergic receptor-mediated upregulation of CFTR function is necessary for increased alveolar fluid reabsorption [146] and newborn pigs with either heterozygous or homozygous CFTR gene knockout (but before the onset of lung pathology) have reduced alveolar sodium reabsorption [147]. Further evidence of the role of beta-2 adrenergic receptor-mediated lung fluid reabsorption was found in humans, in which circulating lymphocyte beta-2 adrenergic receptor density increase correlated with greater decrease in lung water during a 17-h normobaric hypoxic exposure [148].

To assess the relevance of active alveolar epithelial fluid reabsorption, inhaled salmeterol and oral dexamethasone, both of which are known to upregulate membrane ENaC and Na^+/K^+ ATPase (reviewed in [149, 150]), were studied at 4,559 m. Both drugs reduced HAPE in susceptible climbers [141, 151] when the drugs were begun one day before ascent. Owing to multiple actions of beta-2 adrenergic agonists (HPV inhibition, stimulation of HVR and ventilation, tightening of cell-to-cell contacts, and upregulation of NO production [152–155]), the contribution of enhanced active AFC to the positive outcome of the salmeterol study remains uncertain. Protection by dexamethasone [151] was not corroborated with indirect measures of enhanced active alveolar sodium and fluid reabsorption (NP difference and expression of leukocyte mRNA for sodium transporting proteins), but rather to a surprising reduction of PA pressure discussed below. It still remains the case that we need more selective and specific drugs to evaluate the role of active AFC in HAPE. In addition to deleterious effects of reduced NO in HAPE-susceptibles on PVR, NO may have a permissive and stimulatory effect on alveolar Na^+ reabsorption as shown in cell culture studies [156]. Two studies [157, 158] of endothelin add another face to this vasoconstrictor, that of inhibiting AFC by activation of endothelial cell ET-B receptors. The clinical importance of this possible effect of ET-1 on AFC and HAPE awaits human studies with selective ET-B receptor antagonists, because three studies at high altitude with bosentan (a nonselective ET-A/B receptor antagonist) have yielded conflicting results on PA pressure, exercise capacity, and gas exchange [66, 159, 160].

Unresolved questions surrounding the importance of active alveolar fluid reabsorption in HAPE are whether a reduced capacity is central in the earliest stages of the disease by failing to maintain sufficient airspace clearance or only later after the onset of edema. In the first case, then agents stimulating fluid reabsorption would be useful in prevention, but if the latter then these likely would alone be only effective treatment. Another is whether interstitial and alveolar edema, either alone or in combination with reduced alveolar PO_2 occurring with consequent lower local ventilation in these areas, link reduced fluid reabsorption to increased pulmonary vascular tone.

Mechanisms of Increased Alveolar–Capillary Barrier Permeability

Traditionally, pulmonary edema has been categorized as non-cardiogenic (increased permeability with exudative characteristics: high protein concentrations and inflammatory mediators with normal or only modestly elevated intravascular pressures) or cardiogenic (elevated hydrostatic pressures leading to a noninflammatory protein-poor transudative leak). Lavage and catheterization findings in nascent HAPE [29, 62] reveal characteristics of a hydrostatic but non-cardiogenic noninflammatory edema suggesting pressure-induced alterations to the normal permeability of the alveolar–capillary barrier or frank traumatic injury.

Stress Failure

Earlier work [161, 162] established that cardiogenic edema does lead to alveolar protein accumulation. The concept was further advanced for HAPE [163], when it was shown that discrete ultrastructural disruptions in the alveolar capillary barrier develop in rabbit lungs with very high transmural pressures typical of severe HAPE. These included disruptions in cell membranes,

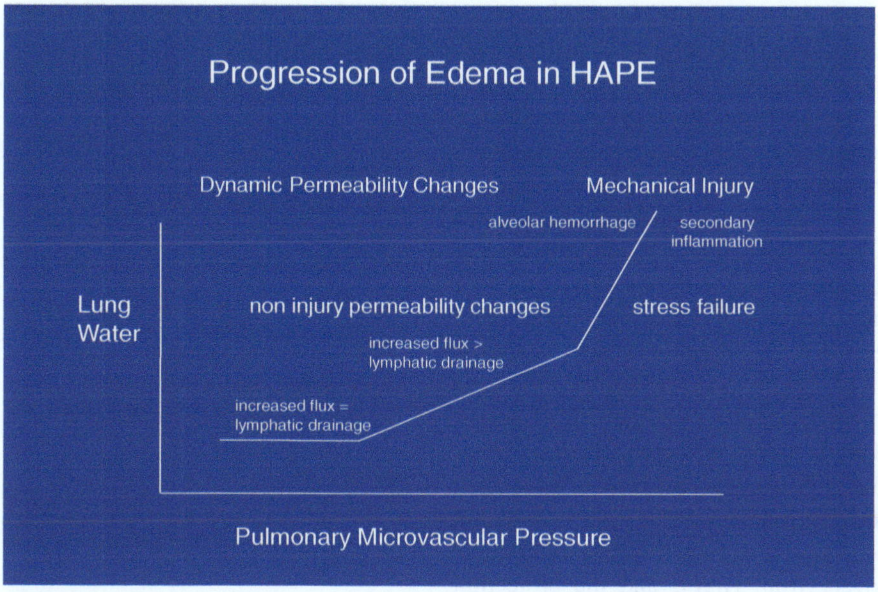

Fig. 21.6 Schematic sequence of events in the progression of edema with pulmonary artery pressure rise in HAPE from dynamic changes in alveolar capillary barrier permeability to mechanical injury

between cells, and in the basement membrane. These changes, also shown in rats exposed to rapid simulated hypobaric "ascents" to 8,800 m [164] for 1 day and to 4,700 m for 2 days with exercise as an additional stress [116], were termed capillary "stress failure" and ascribed to stretch and deformation of the extracellular collagen matrix in excess of their load-bearing capacity. Despite their traumatic-like appearance, even allowing red cell egress, these discontinuities can quickly close with pressure reduction [163]. It has been proposed that these hydrostatic disruptions of the alveolar–capillary barrier permit the leak of vascular endothelial growth factor (VEGF) from the alveolar air space (where it is in high concentration) to the capillary endothelium whose VEGF receptors are activated to promote vascular leak [165]. Elevated serum VEGF concentrations, however, have not been detected in patients with HAPE [166] or in HAPE susceptibles at high altitude [122].

Dynamic Alterations in Permeability

Before the onset of the obvious injurious changes in the alveolar capillary barrier noted above, there is an earlier phase of less severe hydrostatic pressure-induced permeability changes in the intact barrier (Fig. 21.6). This is evidenced by the fact that beta-adrenergic agonists [167] and gadolinium [168] reduce normoxic hydrostatic edema at constant vascular pressure. However, in a rat model of hypoxic edema at constant vascular pressure, terbutaline did not prevent the hypoxic permeability increase [97] suggesting hypoxia may limit beta-adrenergic signaling [133, 134]. Other factors such as hypoxic degradation of the glycocalyx lining the luminal surface of the vascular endothelium, which serves as a barrier to fluid extravasation [169, 170], may be involved. Vascular permeability changes with hypoxia are partially opposed by endothelial cell expression of adrenomedullin-2, a peptide that is upregulated by HIF-1 and is protective in other forms of lung injury [171], and by sphingosine-1-phosphate, an endogenous lipid that promote barrier enhancement via actin and junctional protein rearrangement [172]. Dynamic changes in transcellular leakage via vesicle formation and fusion to create pathways that traverse the cell [173, 174] and paracellular pathways via alterations in gap junction assembly [175] arise from signals initiated when cells are deformed by pressure or stretch. These responses may represent a preemptive attempt of the alveolar–capillary barrier to lower stress forces temporarily and prevent damage to the basement membranes [176] and precede the more profound "stress failure" changes. There was no correlation between ultrastructural lesions and leakiness [177] suggesting that much of the permeability changes are not due to histologically evident stress failure disruptions. Lavage data at high altitude in climbers and in subjects before and after hypoxic exercise [62, 178] support this in showing a very mild protein leak even in the HAPE-resistant subjects (Table 21.1).

If hydrostatic forces persist, then gene upregulation and production of collagen and other extracellular matrix proteins are initiated to strengthen and remodel the alveolar–capillary barrier within hours [96]. These observations may explain the rapid recovery from HAPE and the protection from recurrence when reascending only several days later [179]. In reentry HAPE, there may be a reverse de-modeling of the vasculature when these people descend which then puts them at risk upon returning to high altitude.

Prevention and Treatment

Ascent Rates and Activity Level

Slow ascent is the most effective form of prevention even in susceptible individuals [1]. Persons should not ascend with any symptoms of altitude illness and should descend when mild symptoms do not improve after a day of rest. Vigorous exercise should be avoided during the first days by individuals with HAPE susceptibility and by those with symptoms of altitude illness or after a rapid ascent to altitudes above 3,500–4,000 m. Those with unilateral absence of a pulmonary artery [10], pneumonectomy [180], upper airway obstruction [181], cardiopulmonary conditions predisposing to pulmonary hypertension [1], or those with hypoventilation syndromes [182, 183] should be very circumspect about heavy exertion. As pointed out earlier, susceptibility to HAPE may be increased during and shortly after any infection.

Table 21.2 HAPE-susceptibility characteristics

Hemodynamic
• Exaggerated hypoxic pulmonary vasoconstriction (HPV)
• Greater normoxic exercise-induced PA pressure elevation
• Augmented sympathetic tone with hypoxia
• Reduced vascular endothelial nitric oxide production
• Increased vascular endothelial endothelin production
Pulmonary
• Smaller lung volumes
• Reduced recruitment of diffusing capacity with hypoxia and exercise
• Possibly reduced alveolar epithelial Na^+/H_2O reabsorptive capacity
Ventilatory and renal
• Lower hypoxic ventilatory responsiveness (HVR)
• Possibly reduced natriuretic response to acute hypoxia

Prediction of Susceptibility: Phenotypic and Genotypic Characteristics

Although as a group, HAPE susceptibles have physiological characteristics and responses to hypoxia that arguably set them at risk (Table 21.2), these responses are not easily tested except in specialized laboratories. Systolic pulmonary artery pressure in hypoxia (2 h at an F_IO_2 of 0.12) identified HAPE susceptibles with a specificity of 93 % and a sensitivity 77 % [184]. Lung volumes and HVR did not improve the identification. Recently it was proposed that a brief 8 min bout of hypoxic exercise ($F_IO_2=0.12$ at 30 % VO_2 max) could identify HAPE susceptibility by a greater than 19 mmHg rise in PA systolic pressure by echocardiography [185]. This study, however, should be considered preliminary and needs validation in a larger group, due to the small numbers of subjects studied, the difficulty in assessing PA pressures by echocardiography during exercise, and the fact that increases rather than absolute pressures were measured. Interestingly, systemic hypertension seems to be associated with stronger HPV [186], but this has not been explored as a risk factor in HAPE. Given the low prevalence of HAPE, such testing described above may cause an overestimation of HAPE susceptibility. Further studies are needed to determine if a detectable PFO is a true risk factor for HAPE. In the last analysis, general screening of trekkers or mountaineers for HAPE susceptibility is not necessary, since illness can be avoided by slower ascent rates that permit adaptation of the pulmonary microvasculature to increasing pressures by remodeling [96].

There has been considerable interest in identifying genetic markers that might more readily predict HAPE susceptibility. Numerous candidates based upon reported and hypothesized differences in vasoactive and inflammatory pathways have been sought. The most studied thus far is vascular endothelial nitric oxide synthase (eNOS) and the data are far from compelling. Others include angiotensin converting enzyme (ACE), angiotensin receptor, surfactant proteins, coagulation factors, endothelin, tyrosine hydroxylase, HLA major histocompatibility loci, cytochrome P450, VEGF, bone morphogenic protein receptor-2, heat shock protein, beta-2 adrenergic receptor, aldosterone synthase, and EGLN-1 or HIF-prolyl hydroxylase 2 (for reviews see [187, 188]). Although statistically significant differences have been found, the differences are often not large and could not be confirmed in other ethnic groups. Furthermore, in some cases we do not know whether the detected allele has any functional impact on the biology of the transcribed protein. The major difficulties for these genome studies are that HAPE susceptibility will likely not be limited to a single gene, studies to date are considerably underpowered, ethnic differences may exist, and non-genomic determinants may be equally important, such epigenetic modification, microRNA control of gene expression, and post-translational protein modification.

Pharmacological Prophylaxis

Drug prophylaxis decisions should focus on any previous HAPE occurrence and a risk-benefit discussion. Nifedipine which inhibits HPV is the drug of choice for a history of unquestionable HAPE when slow ascent is not possible [2]. High-dose inhaled salmeterol [141] is an alternative choice, although the dose shown to be effective may cause tremor and tachycardia and is only 50 % as effective as nifedipine. Efficacy in general in stimulating fluid reabsorption to limit pulmonary edema remains unproven as demonstrated in the failure of inhaled high-dose albuterol to hasten AFC in ARDS [189].

Acetazolamide, long used for AMS prevention, blunts or abolishes HPV in animals and man [190, 191]. It was successful in an animal model of mild HAPE [192] and in children in Colorado with reentry HAPE (Peter Hackett, personal communication). Acetazolamide did not lower PA pressure in Himalayan trekkers and because no control subjects became ill, it could not be ascertained whether it might prevent HAPE [193]. The failure to lower PA pressures, similar to another study of subjects also at altitude for 10 days before testing [194], suggests that acetazolamide is not effective in already partially or fully acclimatized subjects. Acetazolamide prevents hypoxia-mediated increases in PA smooth muscle cytosolic calcium by a mechanism not involving carbonic anhydrase inhibition [195]. In vivo, acetazolamide may act by stimulating AFC [196] and by its classic stimulant action to increase ventilation and alveolar PO_2, thus reducing the main HPV stimulus. Recently it was reported that thiadiazoles, such as acetazolamide, are ROS scavenging [197]. This may be relevant not only to how acetazolamide reduces HPV but another action of acetazolamide in reducing HAPE if it is proven that increased ROS production is causal. A controlled trial in HAPE susceptibles needs to be undertaken.

Tadalafil (a long acting phosphodiesterase-5 inhibitor) was equally effective [151] in HAPE susceptibles as nifedipine in reducing the recurrence rate to <10 % vs. the 50 % effectiveness of salmeterol. The effectiveness of tadalafil was unsurprising given it blocks cGMP breakdown (the mechanism of action of PDE-5 inhibitors in amplifying NO effects to diminish HPV), as shown with sildenafil [198]. In addition to the vasodilatory effect of NO, increased NO or cGMP reduces pulmonary vascular endothelial permeability [199], a finding that helps to explain why inhibition of NO synthesis in the isolated perfused lung increases edema even when the perfusion pressure was held constant [200]. Despite the effectiveness of tadalafil in reducing HAPE occurrence in a susceptible population, sildenafil, another PDE-5 inhibitor, did not significantly reduce PA systolic pressure at 5,200 in a group of climbers without known HAPE susceptibility [201].

In the same study examining tadalafil, dexamethasone (a glucocorticoid) was 100 % effective in preventing HAPE [151]. By its reduction of HPV and improvement in arterial oxygenation, it increased hypoxic exercise capacity [202]. Dexamethasone was chosen because it was thought that it might be more selective and specific in upregulating alveolar fluid reabsorption than salmeterol. Its potent prophylaxis, however, was mediated by a striking reduction in PA pressures as in the tadalafil group, but with even better arterial oxygenation both awake [151] and during sleep [37]. A leading explanation for its efficacy is its ability to upregulate pulmonary vascular eNOS and NO production [203] as indirectly suggested by higher urinary cGMP excretion [151] and thus reduce PA pressure [19] and strengthen the alveolar capillary barrier [199, 200]. A sympatholytic effect (lower heart rates) may be another contributing factor [151]. Another relevant effect is increased surfactant production and secretion even in adult lungs [204]. Surfactant reduces alveolar lining surface tension, which in turn reduces negative forces at the air–liquid interface and thus lowers the alveolar–capillary transmural pressure difference [205]. This mechanism may explain the reduction in vascular permeability in hypoxic mice treated with dexamethasone in which there was no change in PA pressure [103]. Dexamethasone-mediated effects may require gene transcription rather than more rapid non-genomic actions, because its efficacy was reduced [206] if given 1 day after arrival (late prophylaxis) rather than 1 day before as early prophylaxis.

Several other available drugs that inhibit or might inhibit HPV but have not been tested for efficacy in HAPE prevention include the other PDE-5 inhibitors, vardenafil and sildenafil; minoxidil, an activator of ATP-gated K^+ channels that reduces HPV in HAPE susceptibles [207]; statins [208], which upregulate endothelial cell NO production; ACE inhibitors [209, 210]; angiotensin II receptor blockers [211]; iron [70]; Rho kinase inhibitors [212]; and stimulators of cyclic GMP [213]. Although endothelin receptor blockers, such as bosentan, reduce HPV acutely [214], they are not advised presently due to their significant fluid-retaining activity [66, 159] as well as lack of efficacy with more chronic hypoxic exposure [159].

Lastly, biologic agents that reduce transendothelial permeability, such as keratinocyte growth factor-2 (KGF-2) via inhibition of apoptosis and upregulation of active salt and water transport as recently reported in a rat model of HAPE [215] may offer both prophylactic and therapeutic possibilities.

Non-pharmacologic strategies have received much less study, but two recent studies are worth noting. In the first study [216], staged ascent of subjects to a lower altitude for 7 days in a hypobaric chamber (Pb = 548 mmHg) before then ascending to 4,300 m (460 mmHg) resulted in a lower estimated mean PA pressure than that occurring without staging (25 vs. 37 mmHg). Ischemic preconditioning of one leg (arterial occlusion by cuff application for four cycles of 5 min occlusion followed by 5 min release), a technique that improves hypoxic and ischemic tolerance, reduced HPV by roughly 35 % 3 h later [217]. Whether either of these means of reducing HPV will be effective for HAPE prevention remains to be determined.

Treatment

The treatment of HAPE requires a proper diagnosis utilizing available history, physical exam, laboratory, and imaging data. Other possibilities including pulmonary embolism, bronchitis, pneumonia, congestive heart failure, and myocardial infarction should be entertained if symptom onset is too early or there are atypical features. Of these pulmonary embolism may be the most relevant alternate diagnosis. D-dimer appears to provide good discrimination because in a large series at high altitude, only 1 out of 31 patients with HAPE had a diagnostic D-dimer elevation [218]. Recently for the diagnosis and monitoring of HAPE in the field, chest ultrasonography using a handheld portable unit was useful [219, 220] in detecting typical "comet tail" echogenic reflections arising from engorged septal lymphatics abutting the pleural surface indicative of pulmonary edema. The technique has promise but further validation in those with other underlying lung and cardiac diseases is needed.

The treatment of HAPE, unlike prophylaxis, includes a variety of strategies for which no controlled trials exist (reviewed in [221]). Immediate improvement of oxygenation is paramount. In a remote area without medical care, descent has first priority. The tourist with HAPE in an alpine resort may remain avoiding exercise if the arterial oxygen saturation can be kept above 90 % by low-flow oxygen (2–4 L/min) with monitoring by family or friends and easy access to clinical care if needed. A recent randomized controlled trial of nifedipine and oxygen vs. oxygen alone for soldiers with HAPE brought down to 1,370 m found no difference in outcomes [222]. Relief of symptoms is achieved within hours and complete clinical recovery usually occurs within 2–3 days. Severe and advanced cases need to be hospitalized or evacuated to low altitude.

Mortality is around 50 % when either descent or other treatment is not possible [223]. Without either oxygen or descent, portable hyperbaric chambers [224] or a continuous positive airway pressure "helmet" [225] can be initiated. Treatment with slow release nifedipine should be started until descent is underway [23]. Potent loop diuretics, such as furosemide, are not recommended in the field because victims maybe already volume depleted and any further ensuing volume contraction will cause greater renin-angiotensin system activation and increase PVR [226]. Inhaled nitric oxide, although technically difficult to provide, is effective [19] suggesting that other inhaled NO donors such as nitroglycerin, nitroprusside and nitrite, or inhaled prostacyclin analogues may be more practical. Whether multiple drugs should be administered has not been formally tested, but they are nonetheless often employed.

Summary and Directions for Future Research

HAPE is well established as a consequence of exaggerated hypoxic HPV and sufficient transmission of high PA pressure and blood flow to portions of the pulmonary capillary bed, most likely due to regional unevenness in HPV with a possible contribution by venoconstriction. Although strong HPV is a characteristic shared by most individuals who develop HAPE, there probably can be no absolute resistance to HAPE, even in "non-susceptible" individuals if altitude gained and ascent rate are high enough or if other factors such as a concurrent respiratory infection transiently arise. The fluid leak in humans with HAPE (and in animal models) affirms the concept that increased pulmonary capillary pressure can lead to a permeability type edema in the absence of inflammation and challenges the classical paradigm that hydrostatic stress can only lead to ultrafiltration of protein-poor fluid.

Further developments in HAPE pathophysiology and clinical management will be greatly abetted by work on several fronts. The first is development of a large animal model that better mimics the entire time course, physical activity, and extent and injury characteristics that occur in humans. Presently models of HAPE in smaller animals such as the rat show only slight increases in lavage indices of permeability (e.g., 2–3 fold elevation in alveolar protein vs. the 20–50 fold increase in humans), less alveolar hemorrhage, greater systemic permeability, and in many cases lack any pulmonary vascular measurements to better assess the sequence of events that leads to a permeability leak. Even a recent pig model [227] of 48 h of 10 % oxygen (but without exercise) showed only modest increases in permeability and gas exchange abnormalities. Second, in order to assess definitively whether active alveolar sodium reabsorption plays a role in the development and resolution of HAPE, drugs and/or genetic engineering in animals that only affect alveolar epithelial ion transport are sorely needed. Third, although many mediators of inflammation and HPV have been studied, their roles are judged largely on the basis of blood values, not changes in lung tissue or airspace concentrations, whose measurement may be more illuminating, especially in conjunction with known inhibitors or agonists. Fourth, given the astonishing results with dexamethasone in HAPE prophylaxis and PA pressure and exercise at high altitude, some of which may be attributed to blunting of the sympathetic nervous system response to hypoxia, the role of lung innervation in HAPE deserves more study. Lastly, many more persons with a history of HAPE are needed for genetic assessment of susceptibility. This may be advanced by the recent creation of a HAPE registry (http://iharc.partners.org) that hopefully will serve also as a database for genetic studies and analysis of preventative/treatment practices and as a source of subjects for controlled trials.

References

1. Luks AM, Swenson ER. Travel to high altitude with pre-existing lung disease. Eur Respir J. 2007;29: 770–92.
2. Bärtsch P, Maggiorini M, Ritter M, et al. Prevention of high altitude pulmonary edema by nifedipine. N Engl J Med. 1991;325:1284–9.
3. Hultgren HN, Honigman B, Theis K, et al. High altitude pulmonary edema at a ski resort. West J Med. 1996;164:222–7.
4. Regensteiner J, Woodward W, Hagerman D, et al. Combined effects of female hormones and metabolic rate on ventilatory drives in women. J Appl Physiol. 1989;66:808–13.
5. Lahm T, Crisostomo PR, Markel TA, et al. Selective estrogen receptor-alpha and estrogen receptor-beta agonists rapidly decrease pulmonary vasoconstriction by a nitric oxide-dependent mechanism. Am J Physiol. 2008;295:R1486–93.
6. Rashid H, Hashmi SN, Hussain T. Risk factors in high altitude pulmonary oedema. J Coll Physicians Surg Pak. 2005;15:96–9.
7. Clarenbach C, Senn O, Christ AL, Fischler M, Maggiorini M, Bloch KE. Lung function and breathing pattern in subjects developing high altitude pulmonary edema. PLoS One. 2012;7:e41188.
8. Fagenholz PJ, Gutman JA, Murray AF, et al. Evidence for increased intracranial pressure in high altitude pulmonary edema. High Alt Med Biol. 2007;4:331–6.
9. Vock P, Brutsche MH, Nanzer A, et al. Variable radiomorphologic data of high altitude pulmonary edema—features from 60 patients. Chest. 1991;100: 1306–11.
10. Schoene RB. Fatal high altitude pulmonary edema associated with absence of the left pulmonary artery. High Alt Med Biol. 2001;2:405–6.
11. Cremona G, Asnaghi R, Baderna P. Pulmonary extravascular fluid accumulation in recreational climbers: a prospective study. Lancet. 2002;359:303–9.
12. Mason NP, Petersen M, Mélot C, et al. Serial changes in nasal potential difference and lung electrical impedance tomography at high altitude. J Appl Physiol. 2003;94:2043–50.
13. Senn O, Clarenbach CF, Fischler M, et al. Do changes in lung function predict high-altitude pulmonary edema at an early stage? Med Sci Sports Exerc. 2006;38:1565–70.
14. Loeppky JA, Icenogle MV, Charlton GA, et al. Hypoxemia and acute mountain sickness: which comes first? High Alt Med Biol. 2008;9:271–9.
15. Dehnert C, Luks AM, Schendler G, et al. No evidence for interstitial lung oedema by extensive pulmonary function testing at 4,559 m. Eur Respir J. 2010;35:812–20.
16. Swenson ER. Con: most climbers do not develop subclinical interstitial edema. High Alt Med Biol. 2011;12:125–8.
17. Penaloza D, Sime F, Ruiz L. Pulmonary hemodynamics in children living at high altitudes. High Alt Med Biol. 2008;9:199–207.
18. Alleman Y, Stuber T, de Marchi SF, Rexhaj E, Sartori C, Scherrer U, Rimoldi SF. Pulmonary artery pressure and cardiac function in children and adolescents after rapid ascent to 3,450 m. Am J Physiol. 2012;302:H2646–53.
19. Scherrer U, Vollenweider L, Delabays A, et al. Inhaled nitric oxide for high-altitude pulmonary edema. N Engl J Med. 1996;334:624–9.
20. Bernheim AM, Kiencke S, Fischler M, et al. Acute changes in pulmonary artery pressures due to exercise and exposure to high altitude do not cause left ventricular diastolic dysfunction. Chest. 2007; 132:380–7.
21. Hanaoka M, Kogashi K, Droma Y, et al. Myocardial performance index in subjects susceptible to high altitude pulmonary edema. Intern Med. 2011;50: 2967–73.
22. Hackett PH, Roach RC, Hartig GS, et al. Effect of vasodilators on pulmonary hemodynamics in high altitude pulmonary edema: a comparison. Int J Sports Med. 1992;13:S68–71.
23. Oelz O, Ritter M, Jenni R, et al. Nifedipine for high altitude pulmonary edema. Lancet. 1989;2(8674): 1241–4.
24. Eldridge MW, Podolsky A, Richardson RS, et al. Pulmonary hemodynamic response to exercise in subjects with prior high-altitude pulmonary edema. J Appl Physiol. 1996;81:911–21.

25. Grünig E, Mereles D, Hildebrandt W, et al. Stress Doppler echocardiography for identification of susceptibility to high altitude pulmonary edema. J Am Coll Cardiol. 2000;35:980–7.
26. Kriemler S, Jansen C, Linka A, et al. Higher pulmonary artery pressure in children than in adults upon fast ascent to high altitude. Eur Respir J. 2008;32: 664–9.
27. Weil JV. Variation in human ventilatory control-genetic influence on the hypoxic ventilatory response. Respir Physiol Neurobiol. 2003;135:239–46.
28. Grunig E, Weismann S, Ehlken N, et al. Stress Doppler echocardiography in relatives of patients with idiopathic and familial pulmonary arterial hypertension. Circulation. 2009;119:1747–57.
29. Maggiorini M, Mélot C, Pierre S, et al. High-altitude pulmonary edema is initially caused by an increase in capillary pressure. Circulation. 2001;103:2078–83.
30. Homik LA, Bshouty RB, Light RB, et al. Effect of alveolar hypoxia on pulmonary fluid filtration in in situ dog lungs. J Appl Physiol. 1988;65:46–52.
31. Saldias FJ, Azzam ZS, Ridge KM, et al. Alveolar fluid reabsorption is impaired by increased left atrial pressure in rats. Am J Physiol. 2001;281:L591–7.
32. Sartori C, Allemann Y, Trueb L, et al. Augmented vasoreactivity in adult life associated with perinatal vascular insult. Lancet. 1999;353:2205–7.
33. Dehnert C, Greiner S, Albers D, Scheurlen F, Maggiorini M, Mereles D, Grünig E, Bärtsch P. Eine abnormel hypoxische pulmonale Vasokonstriktion ist nicht gleichbedeutend mit der Anfälligkeit zum Höhenlungenödem. Dtsch Z Sportmed. 2011;62:180.
34. Hohenhaus E, Paul A, McCullough RE, et al. Ventilatory and pulmonary vascular response to hypoxia and susceptibility to high altitude pulmonary oedema. Eur Respir J. 1995;8:1825–33.
35. Schirlo C, Pavlicek V, Jacomet A, et al. Characteristics of the ventilatory responses in subjects susceptible to high altitude pulmonary edema during acute and prolonged hypoxia. High Alt Med Biol. 2002;3:267–76.
36. Hyers TM, Scoggin CH, Will DH, et al. Accentuated hypoxemia at high altitude in subjects susceptible to high altitude pulmonary edema. J Appl Physiol. 1979;46:41–6.
37. Eichenburger U, Weiss E, Riemann D, Oelz O, Bartsch P. Nocturnal periodic breathing and the development of acute high altitude illness. Am J Respir Crit Care Med. 1996;154:1748–54.
38. Nussbaumer-Ochsner Y, Schuepfer N, Ursprung J, Siebenmann C, Maggiorini M, Bloch KE. Sleep and breathing in high altitude pulmonary edema susceptible subjects at 4559 meters. Sleep. 2012;10:1413–21.
39. Balanos GM, Talbot NP, Dorrington KL, et al. Human pulmonary vascular response to 4 h of hypercapnia and hypocapnia measured using Doppler echocardiography. J Appl Physiol. 2003;94:1543–51.
40. Marshall BE, Marshall C, Magno M, et al. Influence of bronchial arterial PO_2 on pulmonary vascular resistance. J Appl Physiol. 1991;70:405–15.
41. Allemann Y, Hutter D, Lipp E, et al. Patent foramen ovale and high-altitude pulmonary edema. J Am Med Assoc. 2006;296:2954–8.
42. Wilson LB, Levitzsky MC. Chemoreflex blunting of hypoxic pulmonary vasoconstriction is vagally mediated. J Appl Physiol. 1989;66:782–91.
43. Chapleau M, Wilson T, Gregory J, et al. Chemoreceptor stimulation interferes with hypoxic pulmonary vasoconstriction. Respir Physiol. 1988;71:185–200.
44. Albert TJ, Swenson ER. Peripheral chemoreceptor responsiveness and hypoxic pulmonary vasoconstriction in humans. High Alt Med Biol. 2013 (in press).
45. Swenson ER, Duncan TB, Goldberg SV, et al. Diuretic effect of acute hypoxia in humans: relationship to hypoxic ventilatory responsiveness and renal hormones. J Appl Physiol. 1995;78:377–783.
46. Bärtsch P, Pfluger N, Audetat M, et al. Effects of slow ascent to 4559 m on fluid homeostasis. Aviat Space Environ Med. 1991;62:105–10.
47. Duplain H, Vollenweider L, Delabays A, et al. Augmented sympathetic activation during short-term hypoxia and high-altitude exposure in subjects susceptible to high-altitude pulmonary edema. Circulation. 1999;99:1713–8.
48. Bärtsch P, Shaw S, Franciolli M, et al. Atrial natriuretic peptide in acute mountain sickness. J Appl Physiol. 1988;65:1929–37.
49. Weiskopf RB, Gabel RA. Depression of ventilation during hypoxia in man. J Appl Physiol. 1975;39: 911–5.
50. Ming Z, Wang D. Sympathetic innervation of pulmonary circulation and its role in hypoxic pulmonary vasoconstriction. J Tongji Med Univ. 1989;9:153–9.
51. Brimioulle S, Vachiery J-L, Brichant J-F, et al. Sympathetic modulation of hypoxic pulmonary vasoconstriction in intact dogs. Cardiovasc Res. 1997;34:384–92.
52. van Heerden PV, Cameron PD, Karanovic A, et al. Orthodeoxia—an uncommon presentation with bilateral sympathectomy. Anaesth Intensive Care. 2003;31:581–3.
53. Liu C, Smith TG, Balanos GM, et al. Lack of involvement of the autonomic nervous system in early ventilatory and pulmonary vascular acclimatization to hypoxia in humans. J Physiol. 2007;579:215–25.
54. Lodato RF, Michael JR, Murray PA. Absence of neural modulation of hypoxic pulmonary vasoconstriction in conscious dogs. J Appl Physiol. 1988;65: 1481–7.
55. Irwin DC, Subudhi AW, Klopp L, et al. Pulmonary edema induced by cerebral hypoxic insult in a canine model. Aviat Space Environ Med. 2008;79:472–8.
56. Podolsky A, Eldridge MW, Richardson RS, et al. Exercise-induced V_A/Q inequality in subjects with prior high-altitude pulmonary edema. J Appl Physiol. 1996;81:922–32.
57. Steinacker J, Tobias P, Menold E, et al. Diffusing capacity and exercise in subjects with previous high altitude pulmonary edema. Eur Respir J. 1998;11: 643–50.

58. Thomas L, Griffo Z, Roos A. Effect of lung negative pressure inflation on pulmonary vascular resistance. J Appl Physiol. 1961;16:451–6.
59. Guleria JS, Pande JN, Khanna PK. Pulmonary function in convalescents of high altitude pulmonary edema. Dis Chest. 1969;55:434–7.
60. Busch T, Bärtsch P, Pappert D, et al. Hypoxia decreases exhaled nitric oxide in mountaineers susceptible to high altitude pulmonary edema. Am J Respir Crit Care Med. 2001;163:368–73.
61. Duplain H, Sartori C, Lepori M, et al. Exhaled nitric oxide in high-altitude pulmonary edema: role in regulation of pulmonary vascular tone and evidence for a role against inflammation. Am J Respir Crit Care Med. 2000;162:221–4.
62. Swenson ER, Maggiorini M, Mongovin S, et al. Pathogenesis of HAPE: inflammation is not an etiologic factor. J Am Med Assoc. 2002;287:2228–35.
63. Ahsan A, Mohd G, Norboo T, et al. Heterozygotes of NOS3 polymorphisms contribute to reduced nitrogen oxides in high altitude pulmonary edema. Chest. 2006;130:1511–9.
64. Berger M, Hesse C, Dehnert C, et al. Hypoxia impairs systemic endothelial function in individuals prone to high-altitude pulmonary edema. Am J Respir Crit Care Med. 2005;172:763–7.
65. Ali Z, Mishra A, Kumar R, et al. Interactions among vascular tone modulators contribute to high altitude pulmonary edema and augmented vasoreactivity in highlanders. PLoS One. 2012;7:244049.
66. Modesti PA, Vanni S, Morabito M, et al. Role of endothelin-1 in exposure to high altitude: acute mountain sickness and endothelin-1 (ACME-1) study. Circulation. 2006;114:1410–6.
67. Zhang F, Kaide JI, Yang L, et al. CO modulates pulmonary vascular response to acute hypoxia: relation to endothelin. Am J Physiol. 2004;286:H137–44.
68. Madden JA, Ahlf SB, Dantuma MW, et al. Precursors and inhibitors of hydrogen sulfide synthesis affect acute hypoxic pulmonary vasoconstriction in the intact lung. J Appl Physiol. 2012;112:411–8.
69. Smith TG, Balanos GM, Croft OP, et al. Increase in pulmonary arterial pressure caused by hypoxia depends on iron status. J Physiol. 2008;586:5999–6005.
70. Smith TG, Talbot NP, Privat C, et al. Effects of iron supplementation and depletion on hypoxic pulmonary vasoconstriction. J Am Med Assoc. 2009;302:1444–50.
71. Knowles HJ, Raval RR, Harris AL, et al. Effect of ascorbate on the activity of hypoxia-inducible factor in cancer cells. Cancer Res. 2003;63:1764–9.
72. Engebretsen BA, Irwin D, Valdez ME, et al. Acute hypobaric hypoxia (5486 M) induces greater pulmonary HIF-1 activation in Hilltop compared to Madison rats. High Alt Med Biol. 2007;8:812–21.
73. Shimoda LA, Manalo DJ, Sham JS, et al. Partial HIF-1alpha deficiency impairs pulmonary arterial myocyte electrophysiological responses to hypoxia. Am J Physiol. 2001;281:L202–8.
74. Yu AY, Shimoda LA, Iyer NV, et al. Impaired physiological responses to chronic hypoxia in mice partially deficient for hypoxia-inducible factor 1alpha. J Clin Invest. 1999;103:691–6.
75. Kline DD, Peng YJ, Manalo DJ, et al. Defective carotid body function and impaired ventilatory responses to chronic hypoxia in mice partially deficient for hypoxia-inducible factor 1 alpha. Proc Natl Acad Sci. 2002;99:821–6.
76. Mishra A, Mohammad G, Thinlas T, Pasha MAQ. EGLN-1 variants influence its expression and SaO_2 levels to associate with high altitude pulmonary edema. Clin Sci. 2012;124:479–89.
77. Abud EM, Maylor J, Undem C, et al. Digoxin inhibits development of hypoxic pulmonary hypertension in mice. Proc Natl Acad Sci. 2012;109:1239–44.
78. Ward JPT, McMurtry IF. Mechanisms of hypoxic pulmonary vasoconstriction and their roles in pulmonary hypertension: new findings for an old problem. Curr Opin Pharmacol. 2009;9:287–96.
79. Frazziano G, Moreno L, Moral-Sanz J, et al. Neutral sphingomyelinase, NADPH oxidase and reactive oxygen species. Role in acute hypoxic pulmonary hypertension. J Cell Physiol. 2011;226:2633–40.
80. Behn C, Araneda OF, Llanos AJ, et al. Hypoxia-related lipid peroxidation: data, implications and approaches. Respir Physiol Neurobiol. 2007;158:143–50.
81. Heinicke I, Boehler A, Rechsteiner T, et al. Moderate altitude but not additional endurance training increases markers of oxidative stress in exhaled breath condensate. Eur J Appl Physiol. 2009;106:599–604.
82. Bailey DM, Dehnert C, Luks AM, et al. High-altitude pulmonary hypertension is associated with a free radical-mediated reduction in pulmonary nitric oxide bioavailability. J Physiol. 2010;588:4837–47.
83. Irwin DC, McCord JM, Nozik-Grayek E, et al. A potential for reactive oxygen species and the HIF-1alpha-VEGF pathway in hypoxia-induced pulmonary vascular leak. Free Radic Biol Med. 2009;47:55–61.
84. Hakim TS, Kelly S. Occlusion pressures vs. micropipette pressures in the pulmonary circulation. J Appl Physiol. 1989;67:1277–85.
85. Whayne TF, Severinghaus JW. Experimental hypoxic pulmonary edema in the rat. J Appl Physiol. 1968;25:729–32.
86. Parker JC, Stevens T, Randall J, et al. Hydraulic conductance of pulmonary microvascular and macrovascular endothelial cell monolayers. Am J Physiol. 2006;291:L30–7.
87. Gao Y, Raj JU. Role of veins in regulation of pulmonary circulation. Am J Physiol. 2005;288:L213–26.
88. Younes M, Bshouty Z, Ali J. Longitudinal distribution of pulmonary vascular resistance with very high pulmonary flow. J Appl Physiol. 1987;62:344–58.
89. Hillier SC, Graham JA, Godbey PS, et al. Hypoxic vasoconstriction in pulmonary arterioles and venules. J Appl Physiol. 1997;82:1084–90.
90. Lamm WJ, Starr IR, Neradilek B, et al. Hypoxic pulmonary vasoconstriction is heterogeneously distrib-

uted in the prone dog. Respir Physiol Neurobiol. 2004;144:281–94.
91. Hopkins SR, Garg J, Bolar DS, et al. Pulmonary blood flow heterogeneity during hypoxia and high-altitude pulmonary edema. Am J Respir Crit Care Med. 2005;171:83–7.
92. Dehnert C, Risse F, Ley S, et al. Magnetic resonance imaging of uneven pulmonary perfusion in hypoxia in humans. Am J Respir Crit Care Med. 2006;174: 1132–8.
93. Hopkins SR, Kleinsasser A, Bernard S, et al. Hypoxia has a greater effect than exercise on the redistribution of pulmonary blood flow in swine. J Appl Physiol. 2007;103(6):2112–9.
94. Kuwahira I, Moue Y, Urano T, et al. Redistribution of pulmonary blood flow during hypoxic exercise. Int J Sports Med. 2001;22:393–9.
95. Pelletier N, Robinson NE, Kaiser L, et al. Regional differences in endothelial function in horse lungs: possible role in blood flow distribution? J Appl Physiol. 1998;85:537–42.
96. Berg JT, Breen EC, Fu Z, et al. Alveolar hypoxia increases gene expression of extracellular matrix proteins and platelet-derived growth factor-B in lung parenchyma. Am J Respir Crit Care Med. 1998; 158:1920–8.
97. Dehler M, Zessin E, Bartsch P, Mairbaurl H. Hypoxia causes permeability oedema in constant-pressure perfused rat lungs. Eur Respir J. 2006;27:600–6.
98. Ogawa S, Koga S, et al. Hypoxia-induced increased permeability of endothelial monolayers occurs through lowering of cellular cAMP levels. Am J Physiol. 1992;262:C546–54.
99. Schoene RB, Swenson ER, Pizzo CJ, et al. The lung at high altitude: bronchoalveolar lavage in acute mountain sickness and pulmonary edema. J Appl Physiol. 1988;64:2605–13.
100. Kubo K, Hanaoka M, Hayano T, et al. Inflammatory cytokines in BAL fluid and pulmonary hemodynamics in high-altitude pulmonary edema. Respir Physiol. 1998;111:301–10.
101. Carpenter TD, Reeves JT, Durmowicz AG. Viral respiratory infection increases susceptibility of young rats to hypoxia-induced pulmonary edema. J Appl Physiol. 1998;84:1048–54.
102. Ono S, Wescott JY, Chang SW, et al. Endotoxin priming followed by high altitude causes pulmonary edema in rats. J Appl Physiol. 1993;74:1534–42.
103. Stelzner TJ, O'Brien RF, Sato K, et al. Hypoxia-induced increases in pulmonary trans-vascular protein escape in rats. Modulation by glucocorticoids. J Clin Invest. 1988;82:1840–7.
104. Gonzalez NC, Wood JG. Leukocyte-endothelial interactions in environmental hypoxia. Adv Exp Med Biol. 2001;502:39–60.
105. Chao J, Viets Z, Donham P, Wood JG, Gonzalez NC. Dexamethasone blocks the systemic inflammation of alveolar hypoxia at several sites in the inflammatory cascade. Am J Physiol. 2012;303:H168–77.
106. Chao J, Wood JG, Gonzalez NC. Alveolar macrophages initiate the systemic microvascular inflammatory response to alveolar hypoxia. Respir Physiol Neurobiol. 2011;178:439–48.
107. Shukla D, Saxena S, Purushothaman J, et al. Hypoxic preconditioning with cobalt ameliorates hypobaric hypoxia induced pulmonary edema in rat. Eur J Pharmacol. 2011;656:101–9.
108. Shukla D, Saxena S, Jayamurthy P, et al. Hypoxic preconditioning with cobalt attenuates hypobaric hypoxia-induced oxidative damage in rat lungs. High Alt Med Biol. 2009;10:57–69.
109. Sarada S, Hindari P, Mishra C, et al. Role of oxidative stress and NFkB in hypoxia-induced pulmonary edema. Exp Biol Med. 2008;233:1088–98.
110. Yamaji-Kegan K, Su Q, Angelini DJ, Meyers AC, Cheadle C, Johns RA. Hypoxia-induced mitogenic factor (HIMF/FIZZ1/RELMalpha) increases lung inflammation and activates pulmonary microvascular endothelial cells via an IL-4-dependent mechanism. J Immunol. 2010;185:5539–48.
111. Kleger G-R, Bärtsch P, Vock P, et al. Evidence against an increase in capillary permeability in subjects exposed to high altitude. J Appl Physiol. 1996;81:1917–23.
112. Bärtsch P, Lammie B, Huber I, et al. Contact phase of blood coagulation is not activated in edema of high altitude. J Appl Physiol. 1989;67:1336–40.
113. Lehmann T, Mairbaurl H, Pleisch B, et al. Platelet count and function at high altitude and high-altitude pulmonary edema. J Appl Physiol. 2006;100:690–4.
114. Swenson ER, Mongovin S, Maggiorini M, et al. Alveolar macrophage interleukin-6 response to hypoxia and lipopolysaccharide in HAPE-susceptible and -resistant mountaineers. Am J Respir Crit Care Med. 2001;163:A618.
115. Urner UM, Hermann IK, Booy C, Roth-Z Graggen B, Maggiorini M, Beck-Schimmer B. Effect of hypoxia and dexamethasone on inflammation and ion transporter function in pulmonary cells. Clin Exp Immunol. 2012;169:119–228.
116. Bai C, She J, Goolaerts A, et al. Stress failure plays a major role in the development of high-altitude pulmonary oedema in rats. Eur Respir J. 2010;35: 584–91.
117. Rassler B, Marx G, Reissig C, et al. Time course of hypoxia-induced lung injury in rats. Respir Physiol Neurobiol. 2007;159:45–54.
118. Hagobian TA, Jacobs KA, Subudhi AW, et al. Cytokine responses at high altitude: effects of exercise and antioxidants at 4300 m. Med Sci Sports Exerc. 2006;38:276–85.
119. Imoberdorf R, Garlick PJ, McNurlan MA, et al. Enhanced synthesis of albumin and fibrinogen at high altitude. J Appl Physiol. 2001;90:528–37.
120. Mazzeo RS, Donovan D, Fleshner M, et al. Interleukin-6 response to exercise and high-altitude exposure: influence of alpha-adrenergic blockade. J Appl Physiol. 2001;91:2143–9.

121. Pederson BK, Fabriao M. Muscle as an endocrine organ: focus on muscle-derived interleukin 6. Physiol Rev. 2008;88:1379–406.
122. Pavlicek V, Marti HH, Grad S, et al. Effects of hypobaric hypoxia on vascular endothelial growth factor and the acute phase response in subjects who are susceptible to HAPE. Eur J Appl Physiol. 2000;81: 497–503.
123. Kolliput N, Waxman AB. IL-6 cytoprotection in hyperoxic acute lung injury occurs via suppressor of cytokine signaling-1-induced apoptosis signal-regulating kinase-1 degradation. Am J Respir Mol Biol. 2009;40:313–24.
124. Kuebler WM, Ying X, Singh B, et al. Pressure is pro-inflammatory in lung venular capillaries. J Clin Invest. 1999;104:495–502.
125. Graca-Souza AV, Arruda MA, de Freitas MS, et al. Neutrophil activation by heme: implications for inflammatory processes. Blood. 2002;99:4160–5.
126. Lee SK, Ding JL. A perspective on the role of extracellular hemoglobin on the innate immune system. DNA Cell Biol. 2013;32:36–40.
127. Voelkel NF, Mizuno S, Bogaard HJ. The role of hypoxia in pulmonary vascular diseases: a perspective. Am J Physiol. 2013;L457–65.
128. Sarada S, Veeramohan, Himadri P, et al. Nifedipine inhibits hypoxia-induced transvascular leakage through down regulation of NFkB. Respir Physiol Neurobiol. 2012;183:26–44.
129. Das BB, Wolfe RR, Chan KC, et al. High-altitude pulmonary edema in children with underlying cardiopulmonary disorders and pulmonary hypertension living at altitude. Arch Pediatr Adolesc Med. 2004;158:1170–6.
130. Durmowicz AG, Nooordeweir E, Nicholas R, et al. Inflammatory processes may predispose children to develop HAPE. J Pediatr. 1997;130:838–40.
131. Gabry AL, Ledoux X, Mozziconacci M, et al. High-altitude pulmonary edema at moderate altitude (<2,400 m). Chest. 2003;123:49–53.
132. Wodopia R, Ko HS, Billian J, et al. Hypoxia decreases proteins involved in epithelial electrolyte transport in A549 cells and rat lung. Am J Physiol. 2000;279:L1110–9.
133. Baloglu E, Ke A, Abu-Taha IH, et al. In vitro hypoxia impairs beta 2 adrenergic receptor signaling in primary rat alveolar epithelial cells. Am J Physiol. 2009;296:L500–9.
134. Baloglu E, Reingruber T, Bärtsch P, et al. β2-Adrenergics in hypoxia desensitize receptors but blunt inhibition of reabsorption in rat lungs. Am J Respir Cell Mol Biol. 2011;45:1059–68.
135. Zhou G, Dada LA, Sznajder JI. Regulation of alveolar epithelial function by hypoxia. Eur Respir J. 2008;31:1107–13.
136. Vivona ML, Matthay M, Chabaud MB, et al. Hypoxia reduces alveolar epithelial sodium and fluid transport in rats: reversal by beta-adrenergic agonist treatment. Am J Respir Cell Mol Biol. 2001;25:554–61.
137. Nagyova B, O'Neill M, Dorrington KL. Inhibition of active sodium absorption leads to a net liquid secretion into in vivo rabbit lung at two levels of alveolar hypoxia. Br J Anaesth. 2001;87:897–904.
138. Egli M, Duplain H, Lepori M, et al. Defective respiratory amiloride-sensitive sodium transport predisposes to pulmonary oedema and delays its resolution in mice. J Physiol. 2004;560:857–65.
139. Sakuma T, Hida M, Nambu Y, Osanai K, Toga H, Takahashi K, Ohya N, Inour M, Watanabe Y. Effects of hypoxia on alveolar fluid transport capacity in rat lungs. J Appl Physiol. 2001;91:1766–74.
140. Chen XJ, Seth S, Yue G, et al. Influenza virus inhibits ENaC and lung fluid clearance. Am J Physiol. 2004;287:L366–73.
141. Sartori C, Allemann Y, Duplain H, et al. Salmeterol for the prevention of high-altitude pulmonary edema. N Engl J Med. 2002;346:1631–6.
142. Sartori C, Duplain H, Lepori M, et al. High altitude impairs nasal trans-epithelial sodium transport in HAPE-prone subjects. Eur Respir J. 2004;23:916–20.
143. Mairbäurl H, Weymann J, Möhrlein A, et al. Nasal potential difference at high altitude: evidence for secretion. Am J Respir Crit Care Med. 2003;167: 862–7.
144. Mairbäurl H, Schwobel F, Hoschele S, et al. Altered ion transporter expression in bronchial epithelium in mountaineers with high-altitude pulmonary edema. J Appl Physiol. 2003;95:1843–50.
145. Betz T, Mairbaurl H, Dehnert C, et al. Can combined effects of decreased nasal potential and exaggerated hypoxic pulmonary vasoconstriction explain HAPE susceptibility. In: International hypoxia symposium, Lake Louise, Canada; 2013 (in press).
146. Mutlu GM, Adir Y, Jameel M, et al. Interdependency of beta-adrenergic receptors and CFTR in regulation of alveolar active Na+ transport. Circ Res. 2005;96:999–1005.
147. Li X, Comellas AP, Karp PH, Ernst SE, Moniger TO, Gansemer ND, Taft PJ, Pezzulo AA, Rector MV, Stoltz DA, McCray PB, Welsh MJ, Zabner J. CFTR is required for maximal transepithelial liquid transport in pig alveolar epithelia. Am J Physiol. 2012;303:L152–8.
148. Johnson MW, Taylor BJ, Hulsebus ML, Johnson BD, Synder EM. Hypoxia induced changes in lung fluid balance in humans is associated with beta-2 adrenergic receptor density on lymphocytes. Respir Physiol Neurobiol. 2012;183:157–65.
149. Clerici C, Planes C. Gene regulation in the adaptive process to hypoxia in lung epithelial cells. Am J Physiol. 2009;296:L267–74.
150. Mairbäurl H. Role of alveolar epithelial sodium transport in high altitude pulmonary edema (HAPE). Respir Physiol Neurobiol. 2006;151:178–91.
151. Maggiorini M, Brunner-La Rocca HP, Peth S, et al. Both tadalafil and dexamethasone may reduce the incidence of high-altitude pulmonary edema: a randomized trial. Ann Intern Med. 2006;145:497–506.

152. Bärtsch P, Mairbäurl H. Salmeterol for the prevention of high-altitude pulmonary edema. N Engl J Med. 2002;347:1283.
153. Easton PA, Katagiri M, Johnson MW, et al. Salbutamol on respiratory muscle function and ventilation in awake canines. Respir Physiol Neurobiol. 2008;161:253–60.
154. Adding LC, Agvald P, Artlich A, et al. Beta-adrenoceptor agonist stimulation of pulmonary nitric oxide production in rabbit. Br J Pharmacol. 1999; 126:833–9.
155. Kolluru GK, Tamilarasan KP, Rajkumar AS, et al. Nitric oxide/cGMP protects endothelial cells from hypoxia-mediated leakiness. Eur J Cell Biol. 2008; 87:147–61.
156. Hardiman KM, McNicholas-Bevensee CM, Fortenberry J, et al. Regulation of amiloride-sensitive Na^+ transport by basal nitric oxide. Am J Respir Cell Mol Biol. 2004; 30:720–8.
157. Berger MM, Rozendal CS, Schieber C, et al. The effect of endothelin-1 on alveolar fluid clearance and pulmonary edema formation in the rat. Anesth Analg. 2009;108:225–31.
158. Comellas A, Briva A, Dada L, et al. Endothelin impairs alveolar epithelial function via ETB receptor. Am J Respir Crit Care Med. 2009;179:113–22.
159. Scheult RD, Ruh K, Foster GP, et al. Prophylactic bosentan does not improve exercise capacity or lower pulmonary artery systolic pressure at high altitude. Respir Physiol Neurobiol. 2009;165:123–30.
160. Faoro V, Boldingh S, Moreels M, et al. Bosentan decreases pulmonary vascular resistance and improves exercise capacity in acute hypoxia. Chest. 2009;135:1215–22.
161. Szidon JP, Pietra GG, Fishman AP. The alveolar-capillary membrane and pulmonary edema. N Engl J Med. 1972;286:1200–4.
162. Bachofen H, Schürch S, Weibel ER. Experimental hydrostatic pulmonary edema in rabbit lungs: barrier lesions. Am Rev Respir Dis. 1993;147:997–1004.
163. West JB. Thoughts on the pulmonary blood-gas barrier. Am J Physiol. 2003;285:L501–13.
164. West JB, Colice GL, Lee YJ, et al. Pathogenesis of high-altitude pulmonary oedema: direct evidence of stress failure of pulmonary capillaries. Eur Respir J. 1995;8:523–9.
165. Kaner RJ, Crystal RG. Pathogenesis of high altitude pulmonary edema: does alveolar epithelial lining fluid vascular endothelial growth factor exacerbate capillary leak? High Alt Med Biol. 2004;5: 399–409.
166. Hanaoka M, Droma Y, Naramoto A, et al. Vascular endothelial growth factor high altitude pulmonary edema. J Appl Physiol. 2003;94:1836–40.
167. Parker JC, Ivey CL. Isoproterenol attenuates high vascular pressure-induced permeability increases in isolated rat lungs. J Appl Physiol. 1997;8:1962–7.
168. Parker JC, Cl I, Tucker JA. Gadolinium prevents high airway pressure-induced permeability increases in isolated rat lungs. J Appl Physiol. 1998;84: 1113–8.
169. Ward BJ, Donnelly JL. Hypoxia induced disruption of the cardiac endothelial glycocalyx: implications for capillary permeability. Cardiovasc Res. 1993; 27:384–9.
170. Annecke T, Fischer J, Hartmann H, Tscheop J, Rehm M, Conzen P, Sommerhof CP, Becker BF. Shedding of the coronary endothelial glycocalyx: effects of hypoxia-reoxygenation vs ischaemia/reperfusion. Br J Anaesth. 2011;107:679–86.
171. Pfeil U, Aslam M, Paddenberg R, et al. Intermedin/adrenomedullin-2 is a hypoxia-induced endothelial peptide that stabilizes pulmonary microvascular permeability. Am J Physiol. 2009;297:L837–45.
172. Belvitch P, Dudek SM. Role of FAK in S1P-regulated endothelial permeability. Microvasc Res. 2012;83:22–30.
173. Dvorak A, Feng D. Vesiculo-vacuolar organelle: a new endothelial cell permeability organelle. J Histochem Cytochem. 2001;49:419–31.
174. DeFouw DO. Ultrastructural features of alveolar epithelial transport. Am Rev Respir Dis. 1983;127: S9–13.
175. Cavanaugh KJ, Oswari J, Margulies SS. Role of stretch on tight junction structure in alveolar epithelial cells. Am J Respir Cell Mol Biol. 2001;25: 584–91.
176. Curry FE, Clough GH. Flow-dependent changes in microvascular permeability—an important adaptive phenomenon. J Physiol. 2002;543:729–32.
177. Maron MB, Fu Z, Mathieu-Costello O, et al. Effect of high transcapillary pressures on capillary ultrastructure and permeability coefficients in dog lung. J Appl Physiol. 2001;90:638–48.
178. Eldridge MW, Braun RK, Yoneda KY, et al. Effects of altitude and exercise on pulmonary capillary integrity: evidence for subclinical high-altitude pulmonary edema. J Appl Physiol. 2006;100:972–80.
179. Litch JA, Bishop RA. Reascent following resolution of high altitude pulmonary edema (HAPE). High Alt Med Biol. 2001;2:53–5.
180. Chou YT, Wang CL, Kao KC, et al. High altitude pulmonary edema in a patient with previous pneumonectomy. J Formos Med Assoc. 2007;106:320–2.
181. McDonald SE, Peacock AJ, Harvey JE. High altitude pulmonary edema triggered by vocal cord stenosis. High Alt Med Biol. 2004;5:450–2.
182. Richalet JP, Letournel M, Salama J. Holmes-Adie syndrome associated with high altitude pulmonary edema and low chemo-responsiveness to hypoxia. Clin Auton Res. 2011;21:55–6.
183. Richalet JP, Chenivesse C, Larmignat P, et al. High altitude pulmonary edema, down syndrome, and obstructive sleep apneas. High Alt Med Biol. 2008;9:179–81.
184. Dehnert C, Grunig E, Mereles D, et al. Identification of individuals susceptible to high-altitude pulmonary oedema at low altitude. Eur Respir J. 2005;25:1–7.

185. Mounier R, Amonchot A, Caillot N, et al. Pulmonary arterial systolic pressure and susceptibility to high altitude pulmonary edema. Respir Physiol Neurobiol. 2011;179:294–9.
186. Guazzi MD, Alimento M, Berti M, et al. Enhanced hypoxic pulmonary vasoconstriction in hypertension. Circulation. 1989;79:337–43.
187. MacInnis MJ, Koehle MS, Rupert JL. Evidence for a genetic basis for altitude-related illness: 2010 update. High Alt Med Biol. 2010;11:349–68.
188. Luo Y, Zou Y, Gao Y. Gene polymorphisms and high altitude pulmonary edema: a 2011 update. Respiration. 2012;84:155–62.
189. Matthay MA, Brower RG, Carson S, et al. Randomized, placebo-controlled clinical trial of an aerosolized β_2-agonist for treatment of acute lung injury. Am J Respir Crit Care Med. 2011;184:561–8.
190. Höhne C, Pickerodt PA, Francis RC, et al. Pulmonary vasodilation by acetazolamide during hypoxia is unrelated to carbonic anhydrase inhibition. Am J Physiol. 2007;292:L178–84.
191. Teppema LJ, Balanos GM, Steinback CD, et al. Effects of acetazolamide on ventilatory, cerebrovascular, and pulmonary vascular responses to hypoxia. Am J Respir Crit Care Med. 2007;175:277–81.
192. Berg JT, Ramanathan S, Swenson ER. Inhibitors of hypoxic pulmonary vasoconstriction prevent high-altitude pulmonary edema in rats. Wilderness Environ Med. 2004;15:32–7.
193. Basnyat B, Hargrove J, Holck PS, et al. Acetazolamide fails to decrease pulmonary artery pressure at high altitude in partially acclimatized humans. High Alt Med Biol. 2008;9:209–16.
194. Faoro V, Huez S, Giltaire S, et al. Effects of acetazolamide on aerobic exercise capacity and pulmonary hemodynamics at high altitudes. J Appl Physiol. 2007;103:1161–5.
195. Shimoda LA, Luke T, Sylvester JT, et al. Inhibition of hypoxia-induced calcium responses in pulmonary arterial smooth muscle by acetazolamide is independent of carbonic anhydrase inhibition. Am J Physiol. 2007;292:L1002–12.
196. Runyon M, Bhargava M, Wangensteen D, et al. Acetazolamide stimulates alveolar fluid clearance in ventilated adult rats. Am J Respir Crit Care Med. 2005;171:A561.
197. Prouillac C, Vicendo P, Garrigues JC, et al. Evaluation of new thiadiazoles and benzothiazoles as potential radioprotectors: free radical scavenging activity in vitro and theoretical studies. Free Radic Biol Med. 2009;46:1139–48.
198. Ghofrani HA, Reichenberger F, Kohstall MG, et al. Sildenafil increased exercise capacity during hypoxia at low altitudes and at Mount Everest Base Camp. Ann Intern Med. 2004;141:169–77.
199. Kuebler WM, Yang Y, Samapati R, et al. Vascular barrier regulation by PAF, ceramide, caveolae and NO- an intricate signalling network with discrepant effects in the pulmonary and systemic circulation. Cell Physiol Biochem. 2010;26:29–40.
200. Mundy AL, Dorrington KL. Inhibition of nitric oxide synthesis augments pulmonary oedema in isolated perfused rabbit lung. Br J Anaesth. 2000;85:570–6.
201. Bates MG, Thompson AA, Baillie JK, et al. Sildenafil citrate for the prevention of high altitude pulmonary hypertension: double blind, randomized, placebo-controlled trial. High Alt Med Biol. 2011;12:207–14.
202. Fischler M, Maggiorini M, Dorschner L, et al. Dexamethasone but not tadalafil improves exercise capacity in adults prone to high-altitude pulmonary edema. Am J Respir Crit Care Med. 2009;180: 346–52.
203. Murata T, Hori M, Sakamoto K, et al. Dexamethasone blocks hypoxia-induced endothelial dysfunction in organ-cultured pulmonary arteries. Am J Respir Crit Care Med. 2004;170:647–55.
204. Young SL, Ho YS, Silbajoris RA. Surfactant apoprotein in adult rat lung compartments is increased by dexamethasone. Am J Physiol. 1991;260:L161–7.
205. Albert RK, Lakshminarayan S, Hildebrandt J, et al. Increased surface tension favors pulmonary edema formation in anesthetized dog lungs. J Clin Invest. 1979;63:1015–8.
206. Maggiorini M, Streit M, Siebenmann C, et al. Dexamethasone decreases systemic inflammatory and stress response and favors vasodilation in HAPE susceptibles at 4559 m. In: International hypoxia symposium, Lake Louise, Canada; 2009. p. 58.
207. Peth S, Karle C, Dehnert C, et al. K+ channel activation with minoxidil stimulates nasal-epithelial ion transport and blunts exaggerated hypoxic pulmonary hypertension. High Alt Med Biol. 2006;7:54–63.
208. Laufs U, Fata VL, Liao JK. Inhibition of 3-hydroxy-3-methylglutaryl (HMG)-CoA reductase blocks hypoxia-mediated down-regulation of endothelial nitric oxide synthase. J Biol Chem. 1997;272: 31725–9.
209. Cargill RI, Lipworth BJ. Lisinopril attenuates hypoxic pulmonary vasoconstriction in humans. Chest. 1996;109:424–9.
210. Niazova ZA, Batyraliev TA, Aikimbaev KS, et al. High-altitude pulmonary hypertension: effects of captopril on pulmonary and systemic arterial pressures. J Hum Hypertens. 1996;10 Suppl 3:S141–2.
211. Camelo JS, Hehre D, Devia C, et al. The role of angiotensin II receptor-1 blockade in the hypoxic pulmonary vasoconstriction response in newborn piglets. Neonatology. 2008;93:263–8.
212. Kojonazarov B, Myrzaakhmatova A, Sooronbaev T, Ishizaki T, Aldashev A. Effects of fasudil in patients with high-altitude pulmonary hypertension. Eur Respir J. 2012;39:496–8.
213. Lang M, Kojonazarov B, Tian X, Kalymbetov A, Weissmann N, Grimminger F, Kretschmer A, Stasch JP, Seeger W, Ghofrani HA, Schermuly RT. The soluble guanylate cyclase stimulator riociguat ameliorates pulmonary hypertension induced by hypoxia and SU5416 in rats. PLoS One. 2012;7:e43433.

214. Pham I, Wuerzner G, Richalet JP, Peyrard S, Azizi M. Bosentan effects in hypoxic pulmonary vasoconstriction: preliminary study in subjects with or without high altitude pulmonary edema-history. Pulm Circ. 2012;2:28–33.
215. She J, Goolaerts A, Sehen J, Bi J, Tong L, Goa L, Song Y, Bai C. KGF-2 targets alveolar epithelia and capillary endothelia to reduce high altitude pulmonary edema in rats. J Cell Mol Med. 2012;16: 3074–84.
216. Baggish AL, Fulco CS, Muza S, et al. The impact of moderate-altitude staging on pulmonary arterial hemodynamics after ascent to high altitude. High Alt Med Biol. 2010;11:139–45.
217. Foster GP, Westerdahl DE, Foster LA, et al. Ischemic preconditioning of the lower extremity attenuates the normal hypoxic increase in pulmonary artery systolic pressure. Respir Physiol Neurobiol. 2011;179:248–53.
218. Khan DA, Hashim R, Mizra TM, et al. Differentiation of pulmonary embolism form high altitude pulmonary edema. J Coll Physicians Surg Pak. 2003;13:267–70.
219. Fagenholz PJ, Gutman JA, Murray AF, et al. Chest ultrasonography for the diagnosis and monitoring of high altitude pulmonary edema. Chest. 2007; 131:1013–8.
220. Pratali L, Cavana M, Sicari R, et al. Frequent subclinical high-altitude pulmonary edema detected by chest sonography as ultrasound lung comets in recreational climbers. Crit Care Med. 2010;38:1818–23.
221. Luks AM. Do we have a 'best practice' for treating high altitude pulmonary edema? High Alt Med Biol. 2008;9:111–4.
222. Deshwal R, Iqbal M, Basnet S. Nifedipine for the treatment of high altitude pulmonary edema. Wilderness Environ Med. 2012;23:7–10.
223. Taber RL. Protocols for use of portable hyperbaric chambers for treatment of high altitude disorders. J Wild Med. 1990;1:181–92.
224. Koch RO, Hinterhuber L, Faulhaver M, et al. A successful therapy of high altitude pulmonary edema with a CPAP helmut on Lenin Peak. Clin J Sport Med. 2009;19:72–3.
225. Fagenholz PJ, Gutman JA, Murray AF, Harris NS. Treatment of high altitude pulmonary edema at 4240 m in Nepal. High Alt Med Biol. 2007;8:139–46.
226. Kiely DG, Cargill RI, Lipworth BJ. Effects of furosemide and hypoxia on the pulmonary vascular bed in man. Br J Clin Pharmacol. 1997;43:309–13.
227. Kleinsasser A, Lervin DL, Loeckinger A, et al. A pig model of high altitude pulmonary edema. High Alt Med Biol. 2003;4:465–74.

Chronic Mountain Sickness

22

Fabiola León-Velarde, María Rivera-Ch, Luis Huicho, and Francisco C. Villafuerte

Abstract

More than 140 million people live above 2,500 m worldwide, about 80 million in Asia, and approximately 35 million in the Andean mountains. The greatest population density is located above 3,500 m. Chronic mountain sickness (CMS) is one of the most important high-altitude pathologies in the majority of mountainous regions of the world. Its hallmark sign is excessive erythrocytosis (EE). In more advanced and severe stages, high-altitude pulmonary hypertension (HAPH) appears frequently, with related remodeling of pulmonary arterioles and right ventricular hypertrophy.

This chapter summarizes CMS clinical features, physiology, pathology, pathogenesis, epidemiology, and genetics. It is based on a systematic review of worldwide literature, with emphasis in the Andes, including the literature from pioneering work conducted several decades ago. The role of the evolution of erythrocytosis and of ventilatory function in the development of hypoxemia is highlighted. Hematologic and pulmonary systems are affected by several risk factors including age, obesity, sleep disorders, menopause, air, and metal pollution, and therefore, these aspects are analyzed as the basis of secondary CMS. We also examine how hypoxia and/or EE affect plasma volume, pulmonary hemodynamics, autonomic nervous system, kidneys, and endocrine function. A section on prevention and treatment discusses different available treatments and future therapeutic and prevention prospects.

F. León-Velarde, D.Sc. (✉) • M. Rivera-Ch, D.Sc.
F.C. Villafuerte, D.Phil.
Facultad de Ciencias y Filosofía, Departamento Académico de Ciencias Biológicas y Fisiológicas, Universidad Peruana Cayetano Heredia, Av. Honorio Delgado 430, Lima 31, Peru
e-mail: fabiola.leon-velarde@upch.pe

L. Huicho, D.Sc.
Departamento Académico de Pediatría, Universidad Peruana Cayetano Heredia, Lima, Peru

Departamento Académico de Pediatría, Universidad Nacional Mayor de San Marcos, Lima, Peru

Departamento Académico de Pediatría, Instituto de Salud del Niño, Lima, Peru

Introduction

Chronic mountain sickness (CMS) affects people who are native or long-time residents of high-altitude (HA). It is a unique manifestation of chronic hypoxia characterized by excessive erythrocytosis (EE) for the altitude of residence, severe hypoxemia, and in many cases, particularly in severe CMS, pulmonary hypertension. At present, there is no data on the mortality and expected life span of patients with CMS.

E.R. Swenson and P. Bärtsch (eds.), *High Altitude: Human Adaptation to Hypoxia*,
DOI 10.1007/978-1-4614-8772-2_22, © Springer Science+Business Media New York 2014

CMS usually begins insidiously in adult life, but much of the pathophysiology resolves when the patient moves to lower altitudes.

Carlos Monge-M offered the first description of CMS in 1925 [1]. The patient was a native of Cerro de Pasco, a mining town at 4,300 m in the Peruvian Andes. His symptoms disappeared with descent to Lima (sea level) but recurred upon returning home. In 1928 Monge-M published an extensive article in Spanish [2] on cases of CMS from Cerro de Pasco (4,300 m) and Puno (3,800 m), an agricultural town, and thenceforth, the condition has borne his name: Monge's disease. He considered CMS to be a "loss of acclimatization" because it developed only after prolonged exposure to altitude in previously well-acclimatized subjects.

This chapter presents the findings on CMS that have been reported mainly in the Andes. However, allusions to research in other mountainous areas of the world extend and illustrate the concepts and factors involved in the development of CMS. The chapter will cover the clinical and epidemiological expressions of CMS, its physiological mechanisms and pathophysiology, and its prevention and treatment.

The clinical and epidemiological aspects of CMS include the clinical description of the disease, symptoms, signs, laboratory findings, and the geographical distribution and altitude relationships. This section also examines the physiological mechanisms by which other pathologies, as well as age and gender-specific factors, are likely to influence the development of CMS. We also present an integrative approach encompassing the clinical, pathophysiological, and epidemiological aspects of the disease. The section on mechanisms and pathophysiology describes the evolution of erythrocytosis and the role of ventilatory function in the development of hypoxemia. In the section on organ effects of hypoxia, we examine how hypoxia and/or EE affect plasma volume, pulmonary hemodynamics, autonomic nervous system, kidney, and endocrine function. The section on prevention and treatment discusses the different treatments available and future therapeutic and prevention possibilities.

Clinical and Epidemiological Aspects

Symptoms and Signs of CMS

In the late 1990s, an International Working Group was established by the International Society of Mountain Medicine in order to generate a consensus and unify information into a consistent nomenclature and diagnostic criteria of signs and symptoms for CMS.

The consensus defines CMS as "a clinical syndrome that occurs to natives or long-life residents above 2,500 m. It is characterized by EE (females [Hb] $\geq$19 g/dL; males [Hb] $\geq$21 g/dL), severe hypoxemia, and in some cases moderate or severe pulmonary hypertension, which may evolve to cor pulmonale, leading to congestive heart failure. The clinical picture of CMS gradually disappears after descending to low altitude and reappears after returning to high altitude" [3].

The consensus also agreed that the most common symptoms of CMS are headache, dizziness, dyspnea, sleep disturbance (insomnia, hypersomnia), tinnitus, physical and mental fatigue, alterations of memory, loss of appetite, and bone and muscle pain. Patients with CMS usually have intermittent or permanent cyanosis, venous dilation in hands and feet, and clubbing of the fingers and toes. In advanced stages of the disease, right heart failure secondary to excessive pulmonary hypertension usually appears [2, 4, 5].

The laboratory findings show hemoglobin concentration ([Hb]) and hematocrit (Hct) above normal values for the altitude of residence. Arterial O_2 partial pressure (PaO_2) and arterial O_2 saturation (SaO_2) are lower, while arterial CO_2 partial pressure ($PaCO_2$) is higher than normal for a given altitude. Bicarbonate concentration is also higher than normal, indicating lesser

Table 22.1 Blood parameters in sea-level subjects (SL), high-altitude normal Andeans (HA), and Andeans with chronic mountain sickness (CMS) living at 4,300 m

	SL (*N*=15) Mean±SD	HA (*N*=15) Mean±SD	CMS (*N*=55) Mean±SD
Age (years)	42±7	44±9	45±10
Hb (g/dL)	13.1±0.9	16.5±0.9[a]	23.1±1.6[b,c]
Hct (%)	43.7±2.7*	49±3[a]	69±5[b,c,**]
SaO_2 (%)	97±1	89±2[a]	81±4[b,c]
PaO_2 (Torr)	93±14	49±2[a]	42±4[b,c]
$PaCO_2$ (Torr)	38±4	24±1[a]	29±3[b,c]
HCO_3^- (mmol/l)	25.7±2.5	18.4±1.1[a]	20.3±1.9[b,c]
pHa	7.45±0.02	7.5±0.03[a]	7.47±0.02[b,c]

Source: Adapted from references [38, 129]
**N*=20; ***N*=47. Values are significant when $P<0.001$
[a] HA significantly different from SL
[b] CMS significantly different from SL
[c] CMS significantly different from HA

respiratory alkalosis with partial renal compensation [2, 4–6]. All the above laboratory aspects are shown in Table 22.1.

Geographical Distribution

Epidemiological data reveals considerable variation in the prevalence of CMS in the different high-altitude populations. The prevalence of CMS in natives of Qinghai, Tibetan plateau, is 1.21 % as compared with 5.59 % in Han immigrants [7]. Using the same criteria as Leon-Velarde et al. [8] and at comparable altitude, a CMS prevalence of 0.91 % was reported in Tibetans [9], while in Andeans the prevalence was of 5 %. As CMS has not been reported in Ethiopians so far, these data suggest that Tibetans and Ethiopians might be protected from CMS due to genetic factors. It is believed that several physiological differences reflect a successful adaptation to a high-altitude environment in Tibetans and Ethiopians [10]. These two groups, contrary to Andeans and Han immigrants, have a preserved hypoxic ventilatory responsiveness [10–12], minimal hypoxic pulmonary hypertension [13, 14], and lower [Hb] values [10, 15, 16].

Several studies have shown that Tibetans maintain higher alveolar ventilation than Andeans or Han immigrants, reflected in lower $P_{ET}CO_2$ values [12, 17, 18]. Lower [Hb] values are also observed in Tibetans, normally 1–4 g/dL lower than in Andean highlanders at similar altitudes [15, 16]. This lower [Hb] would be associated to the reduced prevalence of EE and CMS in Tibetans compared to Andeans when the same cut-off values are employed.

Recently, Beall and coworkers [19] used genomic and candidate gene approaches to search for evidence if the severe reduction in O_2 availability associated with life at high altitudes was likely to produce a genetic selection. A genome-wide allelic differentiation scan (GWADS) comparing native highlanders of the Tibetan Plateau (3,200–3,500 m) with closely related lowland Han individuals revealed a genome-wide significant divergence across eight SNPs located near the *EPAS1* gene. This gene encodes the transcription factor HIF2α, which stimulates production of red blood cells and thus increases [Hb] in blood. In a separate cohort of Tibetans residing at 4,200 m, they identified 31 *EPAS1* SNPs in high linkage disequilibrium that correlated significantly with hemoglobin concentration. [Hb] was, on average, 0.8 g/dL lower in the major allele homozygotes compared with the heterozygotes. The alleles associated with lower [Hb] were correlated with the signal from the GWADS study and were observed at greatly elevated frequencies in the Tibetan compared with the Han. High [Hb] is the outstanding sign of CMS; thus, the authors hypothesized that a low [Hb] might offer a plausible mechanism for selection of Tibetans for life at high altitude.

Mechanisms of CMS

The pathophysiological sequence of CMS traditionally involves hypoventilation, aggravated hypoxemia, and EE. The relationship between these variables, however, is complex, and the sequence not always straightforward.

This section will discuss the findings on these variables in CMS patients and their possible interrelation.

Ventilatory Function

Andean highland populations living above 3,000 m have a blunted ventilatory response to increasing hypoxia and breathe less compared to acclimatized newcomers. For a review of control of breathing at HA, the reader is referred to León-Velarde and Richalet [20]. CMS patients are relatively hypoxic and hypercapnic when compared to healthy HA natives.

The acute HVR for both HA natives and subjects with CMS is around one-third of the values for SL natives [21]. HA natives and subjects with CMS do not differ significantly from each other in any of the variables associated with the HVR, such as the hypoxic ventilatory decline (see Chap. 3 for details). In addition, when the peripheral chemoreflex contribution at $P_{ET}O_2 = 52.5$ Torr is calculated from the value for the acute HVR multiplied by the reduction in saturation associated with reducing $P_{ET}O_2$ from 200 Torr to 52.5, the contribution for the sea-level (SL) group is clearly much larger than for the HA and CMS groups, but there is no difference between the HA and CMS group. Overall, recent studies provide evidence that lower levels of ventilation in CMS subjects, i.e., the loss of the drive to breathe, may arise from mechanisms other than reductions in the peripheral chemoreflex sensitivity to hypoxia or that its contribution is quite marginal.

The central respiratory chemoreceptor drive is set in response to PCO_2. Air breathing end-tidal PCO_2 ($P_{ET}CO_2$) or arterial PCO_2 is used to evaluate the level of resting ventilation among different subjects, because ventilation is an inadequate indicator to compare individuals, due to varying metabolic rate, dead space fraction, and respiratory rate. As early as 1942, Hurtado [22] found that SaO_2 of patients with CMS is lower than that corresponding to the physiological altitude level. He therefore postulated that hypoventilation, secondary to a reduced sensitivity of the respiratory center to CO_2, is an important factor in the etiology of this disease [23].

Table 22.2 Results from fast and slow ventilatory responses to CO_2 at 52.5 Torr in sea-level subjects (SL), HA natives (HA), and CMS subjects (CMS)

	SL ($N=25$) Mean ± SD	HA ($N=25$) Mean ± SD	CMS ($N=14$) Mean ± SD
$P_{ET}CO_2$ (Torr)	39.2 ± 1.2	26.8 ± 2.1[a]	29.5 ± 1.8[b,c]
Gp (l/min/Torr)	1.9 ± 1.2	2.1 ± 1.4	1.8 ± 1.1
Tp (s)	10.9 ± 7.2	12.0 ± 6.9	12.7 ± 6.8
dp (s)	4.5 ± 3.4	5.0 ± 2.3	6.1 ± 3.0
Gc (l/min/Torr)	2.3 ± 1.0	3.2 ± 1.9[a]	3.3 ± 1.2[b]
Tc (s)	107 ± 107	175 ± 112[a]	231 ± 93[b]
dc (s)	9.0 ± 5.8	12.0 ± 5.3[a]	11.2 ± 5.4
B (Torr)	33.3 ± 3.8	25.5 ± 2.6[a]	28.6 ± 2.3[b,c]
No of breaths	544 ± 88	514 ± 82	481 ± 85[b]

Values are significant when $P < 0.05$

Source: Modified from [6, 26]

Tp peripheral chemoreflex time constant, *dp* peripheral chemoreflex delay, *dc* central chemoreflex delay, *B* bias term equivalent to $P_{ET}CO_2$ for which ventilation in the absence of hypoxia is reflected

[a] HA significantly different from SL

[b] CMS significantly different from SL

[c] HA significantly different from CMS

There have been a number of studies regarding the ventilatory response to CO_2 in patients with CMS [24, 25] but only one that attempted to separate the peripheral (fast) and central (slow) components of the ventilatory response to CO_2 [26]. A protocol employing bursts of 8 Torr increases in $P_{ET}CO_2$ was used to assess the difference in the speed of response of the peripheral and central chemoreflexes and separate their relative contributions to the overall respiratory response [27]. CMS subjects show a lower ventilatory sensitivity to CO_2, compared to euoxic HA and SL natives, but a higher response compared to SL natives in hypoxia. Altogether these findings suggest that the $P_{ET}CO_2$ in the absence of hypoxia is significantly higher for CMS when compared with the normal HA natives (Table 22.2).

Overall, our interpretation is that in CMS the central CO_2 chemoreceptors have been reset to operate around a resting $P_{ET}CO_2$ closer to SL values, i.e., around a higher $P_{ET}CO_2$ value than that normal for the altitude of residence. The mechanisms responsible for this change in central chemosensitivity to CO_2 are not entirely clear. However, changes in the activity of certain neurotransmitters known to act as modulators in the control of breathing such as glutamate, and/or GABA, might be involved and merit investigation.

Although the effect of blunted ventilation on Hct is logical in the sense of a cause–effect relationship, a possible effect of Hct on ventilation has been suggested by two studies in which CMS patients were acutely hemodiluted and ventilation rose [28, 29] similar to the results in many other studies in normals and patients with lung disease, which show greater ventilation after isovolemic hemodilution. The mechanism by which Hct per se could modulate ventilation is not entirely clear, but it may involve the greater efficiency of blood carbon dioxide transport afforded by a higher CO_2-carrying capacity and larger Haldane effect of polycythemic blood, so that less ventilation is needed to maintain satisfactory tissue and arterial PCO_2. Also, it is interesting to consider the possibility that erythrocytosis per se does not modulate ventilation. In light of new findings on the role of Epo in ventilatory control, it is feasible that changes in circulating Epo, consequence of hemodilution, may have caused an increase in ventilation.

Whether hypoventilation precedes and causes erythrocytosis or vice versa (or both pathways coexist and reinforce each other) will be a difficult issue to settle and may require longitudinal studies in persons begun before the onset of CMS.

Erythrocytosis, Erythropoietin, and CMS

Erythrocytosis, and the consequent rise in hemoglobin concentration [Hb], increases O_2 content of blood but not necessarily O_2 delivery to tissues. Increased [Hb] and Hct have long been considered beneficial at high altitude because the increased oxygen-carrying capacity of blood would compensate for decreased arterial oxygen saturation. The increase in [Hb] should act to maintain tissue oxygen transport, thereby reducing the need to increase cardiac output. This concept, however, has been challenged because with a low HbO_2 saturation (low PO_2), increasing [Hb] would only help up to a certain point, and thus, only a moderate increase in [Hb] would benefit O_2 transport. It has been suggested that above a certain [Hb] value, compensatory erythrocytosis is not longer beneficial because of elevated blood viscosity and increased blood volume, both of which would lead to congestive symptoms [30, 31]. These also might affect pulmonary artery pressure, distribution of pulmonary blood flow, and pulmonary ventilation–perfusion relationships. The resulting impairment in pulmonary gas exchange and greater hypoxemia will further stimulate erythropoiesis [29, 32]. Interestingly, some studies have shown that when these subjects undergo hemodilution, blood oxygenation improves and the symptoms are reduced dramatically [28, 29, 33].

In support of the idea that EE can become counterproductive, Monge-C [34, 35] and Villafuerte and coworkers [31] used mathematical modeling and found that venous O_2 partial pressure (PvO_2) reaches a maximum at an [Hb] value of 15 g/dL and varies little until values exceed 17 g/dL (corresponding to an altitude of 3,200 m and below the maximum CaO_2). Above this value, PvO_2 declines despite an increasing [Hb]. Also, based on an analysis of arterial–venous O_2 content differences at different [Hb] and SaO_2 values, Reeves and Leon-Velarde [30] have suggested that up to a certain value (SaO_2 of 87 % and a [Hb] of 17.5 g/100 mL), increasing [Hb] has a cardiac output sparing effect because of greater oxygen extraction. However, when hypoxemia worsens, an elevated or excessive [Hb] no longer supports efficient O_2 transport, and cardiac output would have to increase, yet the literature shows no evidence of this in CMS [36–38].

EE is seldom accompanied by an elevated blood volume [22, 36, 39], but cardiac output, peripheral resistance, and systolic blood pressure in CMS patients are similar to those of healthy individuals at the same altitude [36, 40]. Some recent studies, however, have shown that systemic pressure in CMS individuals is slightly elevated when compared to healthy highlanders [41, 42]. Although elevated blood viscosity would tend to decrease venous return, the increased blood volume would counteract this effect, and therefore, cardiac output would be maintained. The rise in peripheral resistance due to increased blood viscosity is most probably counteracted by the increased aortic elasticity and vasodilation in high-altitude residents [43–45]. Contrary to what is observed for systolic pressure, some studies have found consistently increased diastolic blood pressure in persons suffering severe CMS [37, 40] which is probably related to increased blood volume.

Studies of the physiology of transgenic mice overexpressing erythropoietin (Epo) have shed insightful information on the hypoxia-independent effects of EE. Under normoxic conditions, this interesting mouse model (tg6) exhibits a 12-fold elevation in plasma Epo concentration, Hct values in the range of 85–90 %, and 75 % larger blood volume compared to its wild-type counterparts [46]. Although CMS patients are rarely this polycythemic, Jefferson and colleagues [47] reported one subject with CMS at Cerro de Pasco with a Hct of 91 %. Despite EE, increased viscosity, and elevated blood volume, tg6 mice have normal cardiac output and blood pressure values. The excessive Hct would result in elevated peripheral resistance; however, the increased shear stress on vascular endothelium and Epo itself induces the expression of endothelial nitric oxide synthase (eNOS) and an enhanced production of NO, which in turn causes vasodilation counteracting the effect of Hct on peripheral resistance [48]. Interestingly to date, it has not been investigated whether this compensation in mice is present in CMS patients. On the venous side, due to increased blood volume, tg6 mice show significantly elevated central venous pressure [46] which could result in elevated diastolic pressure. It is possible then that some of the mechanisms operating in tg6 mice are also present in CMS individuals. An enhanced expression of eNOS in CMS might have a role, for example, in the normal peripheral resistance found in CMS patients. The role of Epo in the development of EE in CMS is controversial. The physiological or genetic findings to date do not support the hypothesis of a primary abnormality of the Epo response as a cause of CMS. However, differences in downstream signaling pathways of the Epo receptor and posttranslational changes of the Epo receptor, for example, need to be explored.

Epo concentration in the blood of healthy HA natives is elevated when compared with SL residents [16, 49–52]. However, the average serum Epo concentration in CMS patients is not different from healthy HA residents, and there is a poor correlation between serum Epo concentration and SaO_2. Although an increased Epo concentration is not necessarily the sole cause for EE in CMS, it has been also suggested [16, 49, 51] that two subgroups of CMS may exist, those with elevated Epo and those with "normal" Epo levels, or that the over-responders might represent the extreme of the reversed feedback mechanisms relating hypoxia, Epo production, and EE. For further elaboration see also Spivak [52], Bozzini et al. [53], and Haase et al. [54].

It is also possible that alterations in the circadian rhythm of Epo secretion contribute to EE. Bernardi and coworkers [55] have shown that in healthy Andeans, the circadian rhythm of Epo is almost identical to that reported at SL [56], with highest values during late evening and their lowest at 8:00 a.m. Epo values were similar to those reported at SL, and similarly, they had ample circadian variation, reaching on average difference of 40 % from day to night. The study showed that in Andeans with EE the day–night variation was abolished, and therefore, the circadian rhythm of Epo (with nocturnal reduction) was disrupted.

Secondary CMS and Risk Factors

CMS and Pulmonary Dysfunction

CMS is defined by EE in natives who are assumed to have normal respiratory function. When there is an obvious respiratory disease (or another underlying condition such as heavy smoking and carboxyhemoglobinemia), which triggers EE, the disease can be called secondary CMS [5]. However, in practice it may be difficult to separate cases of CMS with purely diminished ventilation from those having additional pulmonary dysfunction aggravating the condition. It has been shown that even common chronic lower respiratory tract disorders may predispose to CMS [8].

CMS and Air Pollution

Worldwide, the greatest impact on health caused by air pollution occurs among the poorest and most vulnerable populations. In the developing world, approximately 76 % of all global air pollution is caused by indoor particulate matter. Use of biomass fuels, such as firewood, charcoal, dung, or crop residues for cooking, heating, and lighting, is common in several high-altitude populations because of their low cost. Burning on open fires or traditional stoves generates large amounts of particulate matter as well as carbon monoxide (CO), hydrocarbons, and other compounds [57]. There is evidence that the use of biomass fuels is associated with COPD and exacerbation of inflammatory lung conditions [58, 59]. Although, these conditions could lead to secondary CMS, pollution by CO could aggravate hypoxia at the tissue level and therefore contribute to increase the signs and symptoms of primary CMS. This hypothesis is supported by a recent study carried out in the Peruvian Andes at 4,100 m that showed that the CMS score was slightly but significantly increased in persons using biomass fuels for cooking [60]. Additional research is required to confirm whether air pollution by CO or particulates represent a trigger or a significant contributor to CMS.

CMS, Age, and Other Underlying Conditions

With normal aging at or near SL, PaO_2 decreases approximately linearly [61, 62], from about 95 Torr at the age of 20 to about 75 Torr at the age of 75 as a result of changes in lung compliance and greater ventilation–perfusion mismatch [62]. This decrease in ventilation is also reflected in slight increasing values of $PaCO_2$ with age [63]. Furthermore, vital capacity, together with oxygen saturation, in HA men decreases more with age than at SL [64].

Whittembury and Monge-C [65] have also shown an inverse relationship between age and hematocrit at three different altitudes. They showed that as altitude increases the correlation is stronger. At moderate altitudes, however, the correlation gets weaker because of the small increase in Hct and the significant variability of Hb concentration at high altitude. Later, Sime and coworkers [66] at Morococha at 4,540 m showed a very good correlation between ventilation, Hct, and age. They showed clearly at this altitude that as age increases, ventilation falls and Hct rises. However, this correlation does not imply that all people that age at HA will develop CMS but reveals that age is an important risk factor. This picture becomes clear when the effect of age on the prevalence of EE and CMS is investigated. Monge and coworkers [67] and Leon-Velarde and Arregui [68] have shown that the prevalence of EE and CMS clearly increases with age. For example, in the age range of 20 29 years, the prevalence of EE is about 7 %, while in the age group of 50–59 years and over 60 years, the prevalence is about 15 % and 27 %, respectively.

Vargas and Spielvogel [69], by studying young (mean age 22.3 years) and old (mean age 46.7 years) CMS patients and healthy young and old controls (mean age 22 years and 43.5 years, respectively) in La Paz, Bolivia, suggested that lower PaO_2 and SaO_2 in the young CMS group compared to control groups is the result of an excessively reduced hypoxic ventilatory response, whereas hypoventilation and diminished lung function are responsible of lower blood saturation in the older CMS group.

In addition to the role played by age in the etiology of CMS, other additional factors such as ethnicity [70, 71], ventilation–perfusion mismatch [28, 72], and obesity [64, 68] might be involved in the development of CMS.

A fall in blood O_2 saturation produced during sleep at HA has also been considered a risk factor that contributes to EE [73–75]. It has been suggested that sleep disturbances of various types, including sleep apneas and hypopneas, might contribute to nighttime desaturation.

However, studies of sleep at HA in the Andes have shown that there are no major sleep abnormalities in CMS patients [74, 75]. Simple and occasional hypopneas were the most frequent abnormalities found, but these were equally frequent in CMS and healthy highlanders. Interestingly, SaO_2 values in the CMS group were consistently around or below 80 % during the entire night period, whereas in the healthy highlander group, O_2 saturation was above 80 % most of the time [75]. The O_2 saturation difference between groups was moderate, in the range of 3 %. However, nighttime saturation showed a significant correlation with Epo levels during the morning, indicating that these relatively small changes could have contributed to EE.

CMS in Women: Gender Differences

Some years ago, CMS was considered an almost exclusively male condition, since little was then known about its prevalence and associated risk factors in women. Premenstrual women have been thought to be protected because female hormones, progestin and estrogen, increase alveolar ventilation and the hypoxic ventilatory response [76]. In addition low concentrations of androgenic hormones may reduce this stimulus to erythropoiesis [77]. Menstruation like phlebotomy may protect premenopausal women from developing EE, and any associated iron deficiency would act similarly. Unfortunately, no data are available to shed light on these possibilities.

León-Velarde et al. [78] measured Hct, SaO_2, and peak expiratory flow rates (PEFR) in pre- and postmenopausal high-altitude women. After menopause women have higher Hct, lower SaO_2 and lower PEFR, and, significantly, a higher frequency of symptoms associated with CMS. Premenopausal women resident at high altitude have higher oxygenation and lower [Hb] levels, likely from the ventilatory stimulation of higher levels of progesterone in the luteal phase of the menstrual cycle [79] as well as in pregnancy, in combination with rising estrogen [80, 81]. In rats, female sex hormones suppress the erythropoietic and cardiopulmonary responses during chronic exposure to hypoxia [77]. The postmenopausal decrease of female hormones could depress alveolar ventilation and PaO_2, stimulate erythropoiesis, increase viscosity, decrease tissue perfusion, and lead to further erythrocytosis and, ultimately, CMS. Santolaya and coworkers [82] have also shown in women of 55 years of age or older that PaO_2 values were below those of age-matched men living at the same altitude. Arterial PCO_2 showed a sharp rise after age 40, reaching values similar to or above those of men. These observations support the notion that ventilatory drive in women has an important hormonal component and that this stimulus decreases after menopause.

CMS and Metals

Cellular studies have shown a possible effect of certain divalent metals, such as cobalt, zinc, and nickel to enhance erythropoiesis [83] by stimulation of erythropoietin (Epo) expression via inhibition of enzymes that enhance degradation of the hypoxia-inducible transcription factors, HIF prolyl, and asparaginyl hydroxylases [84]. There is a concern about the possible role of contamination by metals in CMS, particularly in cities with high mining activity.

A study carried out in Cerro de Pasco in the central Andes of Peru found that elevated serum levels of cobalt were associated with EE [47]. 52 % of the subjects with an average Hct of 76 % had serum cobalt concentrations at least ten times higher than healthy subjects. Given that cobalt induces the expression of Epo and promote erythropoiesis, the authors suggested a cause–effect relationship between cobalt and EE with drinking water as a possible source of contamination. However, cobalt was not detected in water samples taken from the city reservoirs. Unfortunately, this study did not report serum Epo values. A later study [55] carried out in the same population showed that serum cobalt values were normal in healthy and CMS individuals and that there was a poor correlation with hematologic variables. Nevertheless, it should be noted that the burden and toxicity of metals is better assessed by tissue samples or urinary excretion. Unfortunately, none of the studies mentioned above analyzed tissue or urine samples.

Although evidence at the cellular level, and in vivo, in animals and humans, is compelling [85–89], direct evidence is still needed for the role of cobalt on the development of EE and CMS in high-altitude populations.

Recently, Villafuerte and coworkers [90] found that levels of copper and lead in the blood of healthy highlanders and CMS patients were within the normal range and there was no difference between the CMS and the control group. They found that serum levels of zinc have a significant positive correlation with Hct and are higher in residents with CMS. This correlation, however, occurs within the physiological zinc concentration range, with only few participants having values above normal. Although, cellular studies have shown that zinc can stimulate the HIF system and cause increased Epo production, serum Epo concentration values in healthy and CMS highlanders are similar. Thus, at present, we cannot infer a cause–effect relationship between increasing levels of zinc within the physiological range and Hct.

It would be logical that the association between increasing serum zinc levels within the physiological range and increasing Hct could be a consequence of Hct itself on circulating zinc concentration. Bernal and coworkers [91] have shown that hypoxia mediates release of intracellular zinc stores in the pulmonary circulation due to the action of hypoxia-induced production of nitric oxide (NO) on metallothionein to cause the release of bound zinc. By binding and releasing zinc, metallothionein may regulate zinc levels within the body. Not only does hypoxia directly cause an increase in the biosynthesis of NO in pulmonary vessels [91], but the shear stress caused by high Hct induces expression of eNOS and production of NO [48].

Organ Effects of CMS

Kidney Function

HA erythrocytosis results in elevated Hct and plasma volume contraction [33, 92, 93], yielding a negative correlation between these parameters in HA natives [94, 95]. Lozano and Monge [94] showed a reduced effective renal plasma flow (ERPF) and glomerular filtration rate (GFR) with increased filtration fraction (FF) in healthy highlanders. Differences in total renal blood flow were not significant. The reduction in GFR and ERPF and the increase in FF were greater in CMS patients. The study also showed a positive correlation between Hct and FF.

Given that HA erythrocytosis is accompanied by hypoxemia, Gonzales and coworkers [96] studied renal hemodynamics in a group of highlanders at high altitude (Morococha, 4,500 m), and 24 h and 30 days after their arrival to Lima (sea level) and confirmed decreased GFR and ERPF at HA, but also showed that after 24 h at SL values remained unchanged. This finding suggested that hypoxemia was not responsible for the hemodynamic changes found at high altitude and that Hct was a major determinant. In contrast, after 30 days in Lima, hemodynamic parameters decreased to almost SL values and so did Hct, showing that the alterations of renal hemodynamics observed at HA are totally reversible when Hct reaches SL values.

Renal tubular function at HA was studied by Monge and coworkers [97] in healthy lowlanders, healthy highlanders, and CMS patients. The study showed that healthy highlanders are in a new steady state of acid–base equilibrium with low $PaCO_2$ and normal bicarbonate tubular reabsorption. CMS patients have a higher maximal transport rate of bicarbonate because of their higher $PaCO_2$, which is known to act as a stimulus for bicarbonate reabsorption.

Early observations by Monge-M and Monge-C found proteinuria in some CMS patients [5]. Rennie and coworkers [98] studied SL natives (Lima) and in high-altitude natives at two different altitudes (4,650 and 4,710 m). Urinary protein excretion rate, although within normal limits, was higher in HA natives. Creatinine clearance was lower at the highest altitude, and this population also had the highest Hct values. Proteinemia and the excretion of strong electrolytes were similar in the three groups.

Jefferson and colleagues [41] found significant proteinuria in CMS patients and elevated serum urate strongly correlated with Hct values. Uricemia is most probably the consequence of increased urate production and a normal fractional excretion. The study also showed significant higher blood pressure in the CMS group associated with decreased renin levels and increased proximal sodium reabsorption.

Excessive Erythrocytosis and Pulmonary Artery Pressure

Hypoxic vascular remodeling and pulmonary vasoconstriction lead to HAPH. In CMS, HAHP may be aggravated by EE due to increased blood viscosity and increased nitric oxide scavenging due to excessive [Hb] [99]. In some cases however, excessive HAPH can occur without EE, suggestive of particularly strong vascular remodeling and excessive vasoconstrictor response to hypoxemia. The possible contribution of excessively elevated Hct on pulmonary artery pressure (PAP) in CMS was studied using isovolemic hemodilution. Winslow and coworkers [29, 33] have shown that decreasing Hct from 62 % to 42 % reduces symptomatology and lowers PAP by 40 % at 4,300 m. PAP continued to decrease days after the procedure. Hct reduction was also accompanied by an increase in ventilation, and besides the positive effect of this on alveolar PO_2, hypocapnia also developed. Given that hypocapnia decreases PAP, it is possible that greater hypocapnia contributed to the sustained fall in PAP. Manier and collaborators [28] observed a 20 % decrease in pulmonary vascular resistance and a smaller decrease in PAP (8 %, but not significant) after performing less aggressive isovolemic hemodilution than Winslow et al. in reducing Hct from 66 % to 55 % in 7 CMS patients in La Paz, Bolivia, at 3,600 m.

It has been suggested that iron status might also play a role in pulmonary hypertension in CMS. Recently, Smith et al. [100] showed that isovolemic phlebotomy (2 L within 4 days to reduce Hct from 73 % to 59 %) and the consequent reduction of iron availability resulted in an increase (11 % on day 5 and a maximum of 29 % over the following 7 days) rather than a decrease in PAP in CMS patients at 4,300 m. This effect, which was evident over several weeks, is consistent with the pulmonary sequelae of acute iron chelation [101, 102]. However, post-phlebotomy, infusion of iron did not lower the elevated PAP, at least in the time frame of the study. The rise of PAP after isovolemic hemodilution is contrary to what should be expected from the reduction of blood viscosity and pulmonary vascular resistance. However, it is possible that iron deficiency effects in the Winslow et al. and Manier et al. studies were masked by the effects of hypocapnia on PAP [103] or a rise in cardiac output. Unfortunately, no measurements of iron status were made in these studies in order to make a proper comparison. Iron status in CMS patients in the study by Smith et al. was within the normal range, implying that the greater PAP of the CMS group would not be a consequence of iron deficiency due to EE. Further studies are required to elucidate the role of iron in pulmonary hypertension in CMS patients.

Endocrine Function and CMS

Thyroid status could be important in CMS because thyroxine increases ventilation and Hct. Pretell [104] showed that thyroid function (as assessed by T4) in healthy young HA males was the same as that at SL but declined with age with an inverse relationship with Hct. In CMS, T4 was low and TSH and the T3/T4 ratio were high. Although iodine deficiency was found in a large percentage of the population, these findings were also seen in HA natives without iodine deficiency. They concluded that thyroid function is lower at HA and more severely depressed in cases of CMS. These changes were reversible upon descent to sea level [105]. However, Gonzales and coworkers [50] did not find an association between CMS and thyroid function. No additional studies have been carried out in order to have a better explanation for these contradictory findings.

Patients with CMS at 4,300 m have a lower urinary excretion of androgenic hormones after the administration of human chorionic gonadotropin, which was not found in normal HA natives [106]. Aldosterone has a normal increase with orthostatic challenge in high-altitude natives, but the response is impaired in patients with CMS [107]. These patients also have a decreased plasma cortisol response to ACTH when compared to normal natives [108]. It has been shown that plasma aldosterone correlates with changes in arterial pH in hypoxic COPD patients and appears to be related to changes in hypoxia-induced rise in pH, as could be the case in CMS patients [109]. The decreased plasma cortisol response to ACTH when compared to normal natives still awaits a plausible explanation.

Sex hormones seem to play a role in the development of EE in women (see CMS and women), and recently, elevated serum levels of 17-alpha-hydroxyprogesterone and testosterone and lower concentration of dehydroepiandrosterone sulfate (DHEAS) have been reported in HA Peruvian males with EE [50], indicating a high androgenic activity in CMS.

Autonomic Nervous System and Cerebral Hemodynamics

Altitude exposure is associated with an increase in sympathetic activity, the extent of which depends on the genetic background, the altitude, the rate of ascent, and the time at altitude [110]. CMS subjects show a reduced response to stimulation of peripheral chemoreceptors, as well as a reduction in the baroreflex control of heart rate and blood pressure [55, 110]. The decreased baroreflex control is directly correlated with CMS score and [Hb]. These findings are interpreted as suggestive of a functional, reversible central depression rather than of the presence of an organic dysautonomia in CMS. At the level of peripheral vasculature and blood pressure regulation, responses of forearm vascular resistance to carotid baroreceptor stimulation in HA residents with and without CMS, both at their altitude of residence and shortly after descent to SL, showed that the baroreflex "set point" was higher in CMS at altitude, whereas at SL values were similar [111, 112].

Cerebral circulation normally shows efficient autoregulation to minimize the effect on flow changes in cerebral perfusion pressure. Autoregulation as assessed from the correlation between flow and pressure during orthostatic stress shows impairment in CMS patients [111]. These findings accord with another study that found reduced cerebrovascular sensitivity to CO_2 in the presence of hypobaric hypoxia in subjects with CMS [113].

As for cerebrovascular responses to hypoxia and hypercapnia, CMS subjects respond similarly to controls, although in both groups at altitude the sensitivity of the cerebral circulation to hypoxia is less than that in SL residents. Shortly after descent to SL, however, sensitivity increases, and sensitivity to hypercapnia during hypoxia decreases [114]. Unlike Andean natives, Ethiopians, a population where CMS has not been reported, show efficient cerebral circulation autoregulation. This, together with their increased $P_{ET}CO_2$, may lead to higher cerebral blood flow and may be advantageous for life at HA [115].

Genetics of CMS

If there is a genetic component in CMS, the obvious candidate genes seem to be those involved in key components in the cardiopulmonary system and metabolic efficiency that are sensitive to hypoxia, i.e., genes known to be regulated by HIF and/or by hypoxia. Some alleles of HIF-responsive genes are more prevalent (G allele of eNOS polymorphism Glu298Asp in Sherpas and ACE I allele in HA Kyrgyz) or less prevalent (ACE D allele in HA Andeans) in the different HA populations, but no different prevalence of these alleles has been found in CMS patients [116].

The discovery of HIF has been a great step forward in our understanding of responses to hypoxia (see Chaps. 1 and 2). HIF is a transcription factor which is a key regulator of oxygen homeostasis, and in addition to Epo,

HIF-responsive genes include regulators of cellular energy metabolism, iron metabolism, catecholamine metabolism, vasomotor control, and angiogenesis, suggesting an important role in the coordination of oxygen supply and cellular metabolism [84]. Modifications within the HIF pathway may give rise to the differing responses of normal humans to hypoxia. Appenzeller and coworkers [117] found a correlation between the high levels of expression of several genes including HIF, symptoms at HA, and CMS. The failure of these values to normalize at low altitudes may indicate the alterations in regulators of HIF-α subunit, resulting in higher levels in the absence of hypoxia stimulus. However, Mejia et al. [118] found no evidence of association with genetic markers located in close proximity to HIF-1α gene and to genes that regulate its stabilization and degradation, such as von Hippel–Lindau (VHL), "prolyl hydroxylase domain containing" 1, 2, 3 (*PHD1*, *PHD2*, PHD3), and phosphatase and tensin homologue deleted on chromosome ten (*PTEN*) in CMS and high-altitude controls.

GWADS revealed eight single-nucleotide polymorphisms (SNPs) located near *EPAS1*, a member of the HIF family of transcription factors that encodes HIF2α [19]. Moreover, a candidate gene analysis of *EPAS1* conducted on two independent cohorts of highlanders determined that genotype had a large effect on phenotype; i.e., the difference in [Hb] between the genotypes was 0.8 g/dL, which was related with being a Tibetan, having a lower [Hb] and being better adapted to HA [19]. CMS should wait for these type of studies, in order to be able to determine if polymorphic variations in HIF2α (EPAS1) might contribute to the high [Hb] characteristic of CMS and, thus, to the poor adaption to high altitude of these subjects. These studies raise indeed the possibility that these polymorphic variations in HIF2α might contribute to the marked differential susceptibility to erythrocytosis, reduced plasma volume, and pulmonary hypertension in humans at HA.

In CMS, serum Epo levels are variable, but in patients with very high Hct values, Epo is usually high [51]. Mejía et al. [118] performed a genetic association analysis including Epo and Epo-receptor (EpoR) genes as functional candidate genes from cell lines derived from the high-altitude natives. There were no variations in the coding sequences, intron/exon boundaries, or in any regulatory domain. The association between the polymorphisms linked to Epo and EpoR and erythrocytosis failed to show evidence of a major monogenic contribution of the loci tested to EE in CMS. This seems to indicate that the excessive erythropoiesis might be caused by factors other than changes in the Epo gene, perhaps in post-receptor Epo signaling pathways or alterations in the sensitivity to Epo caused by changes in the abundance of soluble EpoR.

VEGF, another HIF-responsive gene product, with effects on vascular and other tissues was studied by Appenzeller et al. [117]. They found that VEGF165 subunits (but not VEGF121) were significantly more expressed in native CMS patients at 4,300 m compared to controls. VEGF165 was negatively correlated with arterial saturation in CMS patients. The authors suggested that these findings could be related to VEGF angiogenic effects or to a possible neuroprotective role, as shown in several neurodegenerative diseases [119].

Hypoxia-related gene expression was significantly higher in Ethiopian highlanders at their native altitude compared to Andean and Himalayan highlanders at their resident altitudes and immediately after descent to low altitude [120]. Of note, normoxic (794 m) Ethiopians and normoxic Himalayan controls showed no effect on expression levels except for upregulation of PDP1 (phosphatase that dephosphorylates the E1 alpha subunit of pyruvate dehydrogenase). Conversely, in Peruvians, in Lima (normoxia), significant increases were found in many genes, except for HIF prolyl hydroxylases 1 and 2 and Epo, which were downregulated. Upregulation of pyruvate dehydrogenase kinases (PDK1, PDK2) related to "aerobic glycolysis," which favors lactate accumulation in the presence of hypoxia and reduces free radical formation, was significantly higher in Ethiopians in hypoxia than in Peruvians at their native altitude, suggesting that Andean tissues were less adapted to hypoxia. PDP2 was highly predictive of CMS scores in the Andes and Himalayas, suggesting that pyruvate metabolism, heavily dependent on PDP2, is disturbed in this disease. This reinforces the proposition that decreased expression of some mitochondrial genes in response to hypoxia might be important in the pathogenesis of CMS.

Expression of hypoxia-responsive genes was compared in children of Andean highlanders from Perú, with and without CMS and in SL controls, seeking markers that might predict the development of CMS later in adulthood [121]. Gene expression was assessed at HA and then immediately after arrival at SL. HA children showed higher expression of genes regulated by HIF and lower levels of those involved in glycolysis and in the TCA cycle. PDK1 and HPH3 mRNA expression were lowest in children of CMS fathers at altitude. At SL the pattern of gene expression in all groups was similar. The molecular signatures of children of CMS patients showed a pattern suggestive of impaired adaptation to hypoxia. At altitude children of CMS had defective coupling between glycolysis and mitochondrial respiration, which may be a key mechanism/biomarker for adult CMS. If confirmed by further studies, these findings may pave the way for early screening of CMS susceptibility in Andeans, with obvious public health impact.

Treatment and Prevention

Treatment

There are several potential targets in the pathophysiological sequence of CMS that are potentially amenable to modification through non-pharmacological and pharmacological interventions (Fig. 22.1). In brief, these aim to lower Hct by descent to lower altitudes or phlebotomy, by inhibiting directly the erythropoietic response to reduce the red cell mass and thus blood hyperviscosity or by stimulating the hypoxic ventilatory response with pharmacologic agents.

Descent to low altitudes is the ideal treatment and leads to prompt improvement of signs and symptoms and reduction of EE. However, very often patients cannot descend for economic, personal, and family reasons.

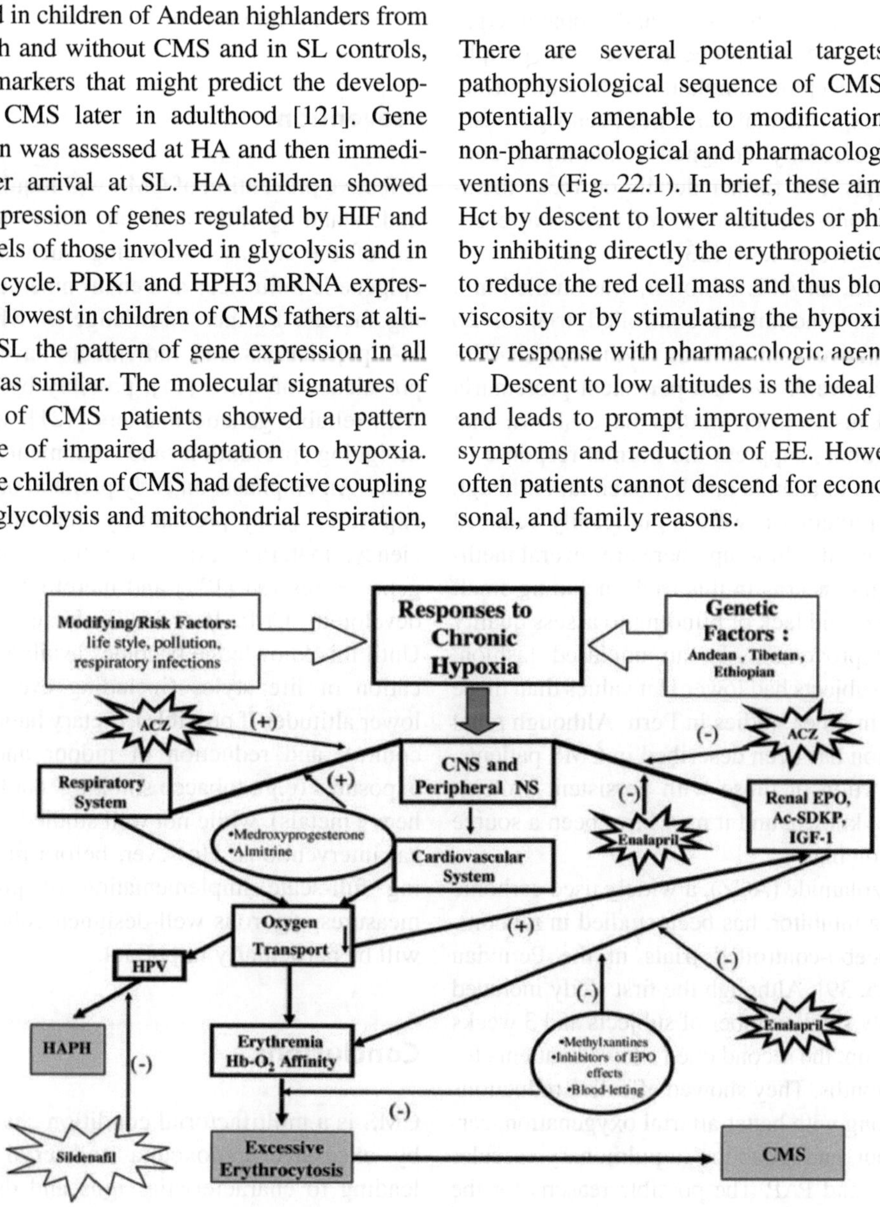

Fig. 22.1 Pathophysiology of CMS and the effects of acetazolamide and other pharmacological agents. Reproduced from reference [130] with Elsevier permission

Although used for a long time, there are no clinical trials on the efficacy and safety of phlebotomy with or without isovolemic hemodilution. Difficulties that render it problematic include invasiveness, practicality, acceptance, and transient effects. Isovolemic hemodilution is said to be much better tolerated, as it may avoid hypovolemia and cardiovascular collapse [122].

A variety of agents that include angiotensin-converting enzyme (ACE) inhibitors, methylxanthines, adrenergic blockers and dopaminergic antagonists, medroxyprogesterone, and peripheral chemoreceptor stimulants such as almitrine, basically experimental, have not been rigorously tested in carefully designed randomized controlled trials. For further details on these therapeutic options, the reader is referred to a recent comprehensive review [123].

Enalapril, an ACE inhibitor, was studied in a small open randomized controlled trial in 26 Bolivian natives with altitude erythrocytosis (Hct equal or above 55 %) and persistent proteinuria [124] because earlier studies have shown that ACE inhibitors suppress the normal response of the kidneys to anemia and hypoxemia. Although Hct and proteinuria were significantly reduced over 2 years of follow-up, there are several methodological concerns in this trial, including small sample size and lack of blinding to assess quality of life improvements in an unbiased fashion. Enrolled subjects had lower Hct values than those included in other studies in Peru. Although renal dysfunction has been described in CMS patients, the proportion of those with persistent proteinuria is not known, and it may have been a source of selection bias.

Acetazolamide (ACZ), a widely used carbonic anhydrase inhibitor, has been studied in randomized, placebo-controlled trials in the Peruvian Andes [35, 39]. Although the first study included a relatively small number of subjects and 3 weeks of follow-up, the second one followed patients for up to 6 months. They showed effective reductions in Hct along with better arterial oxygenation, cardiac output, and reduction in pulmonary vascular resistance and PAP. The possible reasons for the success of ACZ include its well-known ventilatory stimulant effects to improve oxygenation, particularly during sleep at HA, and so reduce the stimulus for Epo production [125], as well as recent studies showing that ACZ reduces hypoxic pulmonary vasoconstriction [126].

Due to the chronic relapsing nature of CMS, it is clear that both enalapril and ACZ need further large non-inferiority clinical trials that assess their comparative efficacy on physiological parameters, quality of life, and ideally on mortality with several years of follow-up.

Prevention

Effective prevention of CMS will require a better understanding of the pathophysiology and genetics of the disease, including such aspects as epigenetic influences and environmental modifying factors. Further knowledge of why certain HA populations and individuals have different prevalence of CMS [11], possibly by screening with reliable genetic markers [121], will aid in designing prevention and treatment interventions. An emphasis on early prenatal or postnatal exposures (e.g., prematurity, placental insufficiency, maternal exposures) that may modify gene expression [127] and therefore the risk of developing CMS later in life is surely needed. Until this knowledge becomes available, modification of life styles (including even living at lower altitudes if possible), dietary habits, weight control, and reduction of indoor and outdoor exposures (e.g., tobacco smoking, cooking fuels, heavy metals), while not well studied, seem logical interventions. However, before recommending full-scale implementation of preventative measures, rigorous well-designed cohort studies will be particularly important.

Conclusions

CMS is a multifactorial condition characterized by excessive hypoxemia and erythrocytosis, leading to characteristic signs and debilitating symptoms. Studies in HA Andean natives show an age-related decline in ventilation. Therefore, a failure of the hyperventilatory capacity and/or inefficient oxygenation of the blood may be the underlying pathophysiological cause of excessive hypoxemia. When the hypoxemic condition is severe, it is accompanied by pulmonary arterial hypertension, leading in advanced cases to right heart failure. Epidemiological studies, and the search for risk factors, confirm the presence of a number of underlying pathophysiological factors in CMS. Most of them are present in lesser degrees in apparently normal HA natives and are more prominent in those with the clinical picture of CMS.

This condition, and the fact that other HA adapted mammalian species do not develop similar features of CMS [128], shows the lack of adaptation of some human populations to life at high altitudes. The absence of CMS in animal species native to HA is related to several adaptations, not only of the oxygen transport chain but also of the pulmonary circulation and cellular mechanisms. Species adapted to HA do not have attenuated respiratory sensitivity to acute hypoxia, hypertrophied peripheral chemoreceptors, pulmonary vasoconstriction, right ventricular hypertrophy, or increased Hct. High hemoglobin–oxygen affinity is the most typical genetic adaptation in high-altitude mammals, birds, and amphibians [33, 128], perhaps by allowing for greater O_2 uptake by the lungs at an equivalent Hct in the face of lowered ambient oxygen availability.

In regard to genetic studies, they have given us a clue about the contribution of gene expression to the emergence and progress of CMS, but published results to date are not entirely consistent. The number of subjects used has been too small to allow definitive conclusions, most likely, because CMS is polygenic and thus several genes account for small changes in risk. More powerful approaches such as genome-wide association studies are needed to tackle these problems. Gene–gene or gene–environment interactions should also be taken into account in the search for a genetic cause of these high-altitude diseases. The genetic background of the populations is so vast that the findings could represent also genetic markers for other susceptibilities, irrespective of its effects on the downstream effectors in the suspected systems affected by CMS and/or HAPH.

Future Directions

Future studies should address some of the questions discussed in this chapter. There is great variability in susceptibility to CMS; thus, one of the challenges is to explain not only the causes for its signs and symptoms but also why at any given altitude some individuals are more likely to develop CMS than others. In addition to the physiological variables that could lead to the pathophysiological sequence of CMS, other etiologic factors, of genetic origin, must also be considered. Studies at the molecular level looking for the genetic and molecular basis of the disease, i.e., variability in HVR, HCVR, and HPV and sensitivity to Epo, should be pursued. Furthermore, studies should be undertaken at the microvascular level to understand O_2 distribution in vessels of various sizes and the distribution of Hct and viscosity in these vessels, as possible contributors to the symptoms of the disease and its progression. Future clinical trials must include adequate randomization, inclusion of large representative patient cohorts, assessment of washout and withdrawal effects, selection of relevant clinical and nonclinical outcomes, long-term follow-up, compliance, concomitant therapy, intention-to-treat-analysis, and safe.

References

1. Monge-M C, Sobre un caso de Enfermedad de Vaquez, In: Comunicacion presentada a la Academia Nacional de Medicina. 1925: Lima. Reimpresa en: Carlos Monge, Obras Vol 2. D. Bigio, Editor. 1988, CONCYTEC: Lima. p. 571–577.
2. Monge-M C, La enfermedad de los Andes, in Anales de la Facultad de Medicina. 1928, Universidad de Lima. Carlos Monge, editor: Lima. p. 1–309.
3. Leon-Velarde F, Maggiorini M, Reeves JT, et al. Consensus statement on chronic and subacute high altitude diseases. High Alt Med Biol. 2005;6: 147–57.

4. Monge CC, Leon-Velarde F, Arregui A. Chronic mountain sickness. In: Lenfant C, editor. High altitude. An exploration of human adaptation. New York: Marcel Dekker; 2001. p. 815–38.
5. Monge-M C, Monge-C C. High altitude disease: mechanism and management. Springfield, IL: Charles C. Thomas; 1966.
6. Monge CC, León-Velarde F. El Reto Fisiológico de Vivir en los Andes. Lima: IFEA, UPCH; 2003.
7. Wu TY, Li W, Li Y, et al. Epidemiology of chronic mountain sickness: ten years study in Quingai-Tibet. In: Ohno H, Kobayashi K, Masuyama S, Nakashima M, Matsumoto M, editors. Progress in mountain medicine and high altitude physiology. Matsumoto: Press Committee of the Third World Congress; 1998.
8. Leon-Velarde F, Arregui A, Vargas M, et al. Chronic mountain sickness and chronic lower respiratory tract disorders. Chest. 1994;106:151–5.
9. Wu TY, Zhang Q, Jin B, et al. Chronic mountain sickness (Monge's disease). An observation in Quingai-Tibet plateau. In: Ueda G, editor. High altitude medicine. Matsumoto, Japan: Shinshu University Press; 1992. p. 314–24.
10. Beall CM, Decker MJ, Brittenham GM, et al. An Ethiopian pattern of human adaptation to high-altitude hypoxia. Proc Natl Acad Sci U S A. 2002; 99:17215–8.
11. Beall CM. Andean, Tibetan and Ethiopian patterns of adaptation to high-altitude hypoxia. Integr Comp Biol. 2006;46:18–24.
12. Beall CM, Strohl KP, Blangero J, et al. Ventilation and hypoxic ventilatory response of Tibetan and Aymara high altitude natives. Am J Phys Anthropol. 1997;104:427–47.
13. Groves BM, Droma T, Sutton JR, et al. Minimal hypoxic pulmonary hypertension in normal Tibetans at 3,658 m. J Appl Physiol. 1993;74:312–8.
14. Gupta ML, Rao KS, Anand IS, et al. Lack of smooth muscle in the small pulmonary arteries of the native Ladakhi. Is the Himalayan highlander adapted? Am Rev Respir Dis. 1992;145:1201–4.
15. Beall CM, Brittenham GM, Strohl KP, et al. Hemoglobin concentration of high-altitude Tibetans and Bolivian Aymara. Am J Phys Anthropol. 1998;106:385–400.
16. Winslow RM, Chapman KW, Gibson CC, et al. Different hematologic responses to hypoxia in Sherpas and Quechua Indians. J Appl Physiol. 1989; 66:1561–9.
17. Moore LG. Comparative human ventilatory adaptation to high altitude. Respir Physiol. 2000;121: 257–76.
18. Zhuang J, Droma T, Sun S, et al. Hypoxic ventilatory responsiveness in Tibetan compared with Han residents of 3,658 m. J Appl Physiol. 1993;74:303–11.
19. Beall CM, Cavalleri GL, Deng L, et al. Natural selection on EPAS1 (HIF2alpha) associated with low hemoglobin concentration in Tibetan highlanders. Proc Natl Acad Sci U S A. 2010;107:11459–64.
20. Leon-Velarde F, Richalet JP. Respiratory control in residents at high altitude: physiology and pathophysiology. High Alt Med Biol. 2006;7:125–37.
21. Leon-Velarde F, Gamboa A, Rivera-Ch M, et al. Selected contribution: Peripheral chemoreflex function in high-altitude natives and patients with chronic mountain sickness. J Appl Physiol. 2003; 94:1269–78. discussion 1253–4.
22. Hurtado A. Chronic mountain sickness. JAMA. 1942; 120:1278–83.
23. Hurtado A. Some clinical aspects of life at high altitudes. Ann Intern Med. 1960;53:247–58.
24. Lahiri S, DeLaney RG, Brody JS, et al. Relative role of environmental and genetic factors in respiratory adaptation to high altitude. Nature. 1976;261: 133–5.
25. Severinghaus JW, Bainton CR, Carcelen A. Respiratory insensitivity to hypoxia in chronically hypoxic man. Respir Physiol. 1966;1:308–34.
26. Fatemian M, Gamboa A, Leon-Velarde F, et al. Selected contribution: ventilatory response to CO_2 in high-altitude natives and patients with chronic mountain sickness. J Appl Physiol. 2003;94:1279–87. discussion 1253–4.
27. Pedersen ME, Fatemian M, Robbins PA. Identification of fast and slow ventilatory responses to carbon dioxide under hypoxic and hyperoxic conditions in humans. J Physiol. 1999;521(Pt 1): 273–87.
28. Manier G, Guenard H, Castaing Y, et al. Pulmonary gas exchange in Andean natives with excessive polycythemia–effect of hemodilution. J Appl Physiol. 1988;65:2107–17.
29. Winslow RM, Monge CC, Brown EG, et al. Effects of hemodilution on O_2 transport in high-altitude polycythemia. J Appl Physiol. 1985;59:1495–502.
30. Reeves JT, Leon-Velarde F. Chronic mountain sickness: recent studies of the relationship between hemoglobin concentration and oxygen transport. High Alt Med Biol. 2004;5:147–55.
31. Villafuerte FC, Cardenas R, Monge CC. Optimal hemoglobin concentration and high altitude: a theoretical approach for Andean men at rest. J Appl Physiol. 2004;96:1581–8.
32. Cruz JC, Diaz C, Marticorena E, et al. Phlebotomy improves pulmonary gas exchange in chronic mountain polycythemia. Respiration. 1979;38: 305–13.
33. Winslow RM, Monge CC. Hypoxia, polycythemia and chronic mountain sickness. Baltimore, MD: John Hopkins University Press; 1986.
34. Monge CC. Hemoglobin regulation in hypoxemic polycythemia. Adjustments to high altitude. In International symposium on acclimatization, adaptation, and tolerance to High Altitude. 1983: NIH, editor, Baltimore, MD.
35. Monge CC. Regulacion de la concentracion de hemoglobina en la policitemia de altura: modelo matematico. Bull Inst Fr Etud Andines. 1990;19: 455–67.

36. Ergueta J, Spielvogel H, Cudkowicz L. Cardiorespiratory studies in chronic mountain sickness (Monge's syndrome). Respiration. 1971;28: 485–517.
37. Penaloza D, Sime F. Chronic cor pulmonale due to loss of altitude acclimatization (chronic mountain sickness). Am J Med. 1971;50:728–43.
38. Richalet JP, Rivera-Ch M, Maignan M, et al. Acetazolamide for Monge's disease: efficiency and tolerance of 6-month treatment. Am J Respir Crit Care Med. 2008;177:1370–6.
39. Claydon VE, Norcliffe LJ, Moore JP, et al. Orthostatic tolerance and blood volumes in Andean high altitude dwellers. Exp Physiol. 2004;89:565–71.
40. Peñaloza D, Sime F, Ruiz L. Cor pulmonale in chronic mountain sickness: present concept of Monge's disease. In: Porter R, Knight J, editors. High altitude physiology: cardiac and respiratory aspects. Edinburgh: Churchill Livingstone; 1971. p. 41–60.
41. Jefferson JA, Escudero E, Hurtado ME, et al. Hyperuricemia, hypertension, and proteinuria associated with high-altitude polycythemia. Am J Kidney Dis. 2002;39:1135–42.
42. Richalet JP, Rivera M, Bouchet P, et al. Acetazolamide: a treatment for chronic mountain sickness. Am J Respir Crit Care Med. 2005;172: 1427–33.
43. Saldaña M, Arias-Stella J. Studies on the structure of the pulmonary trunk. III. The thickness of the media of the pulmonary trunk and ascending aorta in high altitude natives. Circulation. 1963;27:1101–4.
44. Saldaña M, Arias-Stella J. Studies on the structure of the pulmonary trunk. II. The evolution of the elastic configuration of the pulmonary trunk in people native to high altitudes. Circulation. 1963;27: 1094–100.
45. Saldaña M, Arias-Stella J. Studies on the structure of the pulmonary trunk. I. Normal changes in the elastic configuration of the human pulmonary trunk at different ages. Circulation. 1963;27:1086–93.
46. Wagner KF, Katschinski DM, Hasegawa J, et al. Chronic inborn erythrocytosis leads to cardiac dysfunction and premature death in mice overexpressing erythropoietin. Blood. 2001;97:536–42.
47. Jefferson JA, Escudero E, Hurtado ME, et al. Excessive erythrocytosis, chronic mountain sickness, and serum cobalt levels. Lancet. 2002;359:407–8.
48. Ruschitzka FT, Wenger RH, Stallmach T, et al. Nitric oxide prevents cardiovascular disease and determines survival in polyglobulic mice overexpressing erythropoietin. Proc Natl Acad Sci U S A. 2000;97:11609–13.
49. Dainiak N, Spielvogel H, Sorba S, et al. Erythropoietin and the polycythemia of high-altitude dwellers. Adv Exp Med Biol. 1989;271:17–21.
50. Gonzales GF, Gasco M, Tapia V, et al. High serum testosterone levels are associated to excessive erythrocytosis of Chronic Mountain Sickness in men. Am J Physiol Endocrinol Metab. 2009;296(6):E319–25.
51. Leon-Velarde F, Monge CC, Vidal A, et al. Serum immunoreactive erythropoietin in high altitude natives with and without excessive erythrocytosis. Exp Hematol. 1991;19:257–60.
52. Spivak JL. Erythropoietin: a brief review. Nephron. 1989;52:289–94.
53. Bozzini CE, Alippi RM, Barcelo AC, et al. The biology of stress erythropoiesis and erythropoietin production. Ann N Y Acad Sci. 1994;718:83–92. discussion 92–3.
54. Haase VH. Hypoxic regulation of erythropoiesis and iron metabolism. Am J Physiol Renal Physiol. 2010;299:F1–13.
55. Bernardi L, Roach RC, Keyl C, et al. Ventilation, autonomic function, sleep and erythropoietin. Chronic mountain sickness of Andean natives. Adv Exp Med Biol. 2003;543:161–75.
56. Wide L, Bengtsson C, Birgegard G. Circadian rhythm of erythropoietin in human serum. Br J Haematol. 1989;72:85–90.
57. Naeher LP, Brauer M, Lipsett M, et al. Woodsmoke health effects: a review. Inhal Toxicol. 2007;19: 67–106.
58. Fullerton DG, Bruce N, Gordon SB. Indoor air pollution from biomass fuel smoke is a major health concern in the developing world. Trans R Soc Trop Med Hyg. 2008;102:843–51.
59. Fullerton DG, Semple S. Air pollution and health: indoor air pollution in the developing world is the real key to reducing the burden of ill health. Thorax. 2008;63:288. author reply 288.
60. Chirinos A, Malpartida N, Matos C, et al. Exposure to indoor biomass fuel and to high altitude (4100m) on health status and Chronic Mountain Sickness: effect of consumption of maca. High Alt Med Biol. 2010;11:A24.
61. Loew PG, Thews G. The dependency of the arterial oxygen pressure on age in the working population. Klin Wochenschr. 1962;40:1093–8.
62. Sorbini CA, Grassi V, Solinas E, et al. Arterial oxygen tension in relation to age in healthy subjects. Respiration. 1968;25:3–13.
63. Arai Y, Sherpa NK, Horie Y, et al., Arterial blood gas change with aging in Sherpa, in Progress in Mountain Medicine and High Altitude Physiology, Hideki Ohno, Toshio Kobayashi, Shigeru Nakashima, and Michiro Matsumoto, Editors. 1992, Press Committee of the 3rd World Congress in Mountain Medicine and High Altitude Physiology. Matsumoto, Japan.
64. Leon-Velarde F, Arregui A, Monge C, et al. Aging at high altitudes and the risk of Chronic Mountain Sickness. J Wild Med. 1993;4:183–8.
65. Whittembury J, Monge CC. High altitude, haematocrit and age. Nature. 1972;238:278–9.
66. Sime F, Monge C, Whittembury J. Age as a cause of chronic mountain sickness (Monge's disease). Int J Biometeorol. 1975;19:93–8.
67. Monge CC, Leon-Velarde F, Arregui A. Increasing prevalence of excessive erythrocytosis with age among healthy high-altitude miners. N Engl J Med. 1989;321:1271.

68. Leon-Velarde F, Arregui A. Desadaptacion a la vida en las grandes alturas. Lima: Institut français d'études andines (IFEA); 1994.
69. Vargas E, Spielvogel H. Chronic mountain sickness, optimal hemoglobin, and heart disease. High Alt Med Biol. 2006;7:138–49.
70. Moore LG. Human genetic adaptation to high altitude. High Alt Med Biol. 2001;2:257–79.
71. Moore LG, Niermeyer S, Zamudio S. Human adaptation to high altitude: regional and life-cycle perspectives. Am J Phys Anthropol. 1998;Suppl 27: 25–64.
72. Kreuzer F, Tenney SM, Mithoefer JC, et al. Alveolar-arterial oxygen gradient in Andean natives at high altitude. J Appl Physiol. 1964;19:13–6.
73. Kryger M, Weil J, Grover R. Chronic mountain polycythemia: a disorder of the regulation of breathing during sleep? Chest. 1978;73:303–4.
74. Normand H, Vargas E, Bordachar J, et al. Sleep apneas in high altitude residents (3,800 m). Int J Sports Med. 1992;13 Suppl 1:S40–2.
75. Spicuzza L, Casiraghi N, Gamboa A, et al. Sleep-related hypoxaemia and excessive erythrocytosis in Andean high-altitude natives. Eur Respir J. 2004; 23:41–6.
76. Tatsumi K, Hannhart B, Moore LG. Hormonal influences on ventilatory control. In: Dempsey JA, Pack AI, editors. Regulation of breathing. New York: Marcel Dekker, Inc.; 1995. p. 829–64.
77. Ou LC, Sardella GL, Leiter JC, et al. Role of sex hormones in development of chronic mountain sickness in rats. J Appl Physiol. 1994;77:427–33.
78. Leon-Velarde F, Ramos MA, Hernandez JA, et al. The role of menopause in the development of chronic mountain sickness. Am J Physiol. 1997;272:R90–4.
79. Leon-Velarde F, Rivera-Chira M, Tapia R, et al. Relationship of ovarian hormones to hypoxemia in women residents of 4,300 m. Am J Physiol Regul Integr Comp Physiol. 2001;280:R488–93.
80. Goodland RL, Reynolds JG, Pommerenke WT. Alveolar carbon dioxide tension levels during pregnancy and early puerperium. J Clin Endocrinol Metab. 1954;14:522–30.
81. Takano N, Sakai A, Iida Y. Analysis of alveolar PCO_2 control during the menstrual cycle. Pflugers Arch. 1981;390:56–62.
82. Santolaya BR, Arraya CJ, Vecchiola DA, et al. Gases y pH en sangre arterial en 176 hombres y 162 mujeres sanas trabajadores no mineros residentes a 2899 mts de altura. Rev Hosp Roy H Glover. 1982;2:7–18.
83. Hirsila M, Koivunen P, Xu L, et al. Effect of desferrioxamine and metals on the hydroxylases in the oxygen sensing pathway. FASEB J. 2005;19: 1308–10.
84. Semenza GL. Hydroxylation of HIF-1: oxygen sensing at the molecular level. Physiology (Bethesda). 2004;19:176–82.
85. The mystery of the Quebec beer-drinkers' cardiomyopathy. Can Med Assoc J, 1967. 97: p. 930–1.
86. Goldberg MA, Dunning SP, Bunn HF. Regulation of the erythropoietin gene: evidence that the oxygen sensor is a heme protein. Science. 1988;242: 1412–5.
87. Saxena S, Shukla D, Saxena S, et al. Hypoxia preconditioning by cobalt chloride enhances endurance performance and protects skeletal muscles from exercise-induced oxidative damage in rats. Acta Physiol (Oxf). 2010;200:249–63.
88. Shrivastava K, Ram MS, Bansal A, et al. Cobalt supplementation promotes hypoxic tolerance and facilitates acclimatization to hypobaric hypoxia in rat brain. High Alt Med Biol. 2008;9:63–75.
89. Shrivastava K, Shukla D, Bansal A, et al. Neuroprotective effect of cobalt chloride on hypobaric hypoxia-induced oxidative stress. Neurochem Int. 2008;52:368–75.
90. Miranda LF, Macarlupu JL, Leon-Velarde F, et al. Elevated serum zinc levels and excessive erythrocytosis. High Alt Med Biol. 2010;11:A91.
91. Bernal PJ, Leelavanichkul K, Bauer E, et al. Nitric-oxide-mediated zinc release contributes to hypoxic regulation of pulmonary vascular tone. Circ Res. 2008;102:1575–83.
92. Monge CC, Cazorla A, Whittembury G, et al. A description of the circulatory dynamics in the heart and lungs of people at sea level and at high altitude by means of the dye dilution technique. Acta Physiol Lat Am. 1955;5:198–210.
93. Sanchez C, Merino C, Figallo M. Simultaneous measurement of plasma volume and cell mass in polycythemia of high altitude. J Appl Physiol. 1970;28:775–8.
94. Lozano R, Monge C. Renal function in high-altitude natives and in natives with chronic mountain sickness. J Appl Physiol. 1965;20:1026–7.
95. Monge CC, Lozano R, Marchena C, et al. Kidney function in the high-altitude native. Fed Proc. 1969;28:1199–203.
96. Gonzales E. Hemodinamica Renal en el Nativo de Altura estudiado a nivel del mar. Lima: Universidad Peruana Cayetano Heredia; 1971.
97. Monge C, Lozano R, Carcelen A. Renal excretion of bicarbonate in high altitude natives and in natives with Chronic Mountain Sickness. J Clin Invest. 1964;43:2303–9.
98. Rennie D, Marticorena E, Monge C, et al. Urinary protein excretion in high-altitude residents. J Appl Physiol. 1971;31:257–9.
99. Deem S, Swenson ER, Alberts MK, et al. Red-blood-cell augmentation of hypoxic pulmonary vasoconstriction: hematocrit dependence and the importance of nitric oxide. Am J Respir Crit Care Med. 1998;157:1181–6.
100. Smith TG, Talbot NP, Privat C, et al. Effects of iron supplementation and depletion on hypoxic pulmonary hypertension: two randomized controlled trials. JAMA. 2009;302:1444–50.
101. Balanos GM, Dorrington KL, Robbins PA. Desferrioxamine elevates pulmonary vascular resis-

tance in humans: potential for involvement of HIF-1. J Appl Physiol. 2002;92:2501–7.
102. Smith TG, Balanos GM, Croft QP, et al. The increase in pulmonary arterial pressure caused by hypoxia depends on iron status. J Physiol. 2008;586: 5999–6005.
103. Balanos GM, Talbot NP, Dorrington KL, et al. Human pulmonary vascular response to 4 h of hypercapnia and hypocapnia measured using Doppler echocardiography. J Appl Physiol. 2003;94: 1543–51.
104. Pretell EA. Cambios en la función tiroidea en nativos de altura. In: IV Congreso Nacional de Medicina. Lima: Asociacion Medica Peruana "Daniel Alcides Carrión"; 1989.
105. Guerra-Garcia R, Llaque WR, Crandall ED. Observaciones sobre la funcion endocrina de pacientes con mal de montana cronico (MMC) estudiados a nivel del mar. Ica: Sociedad Peruana de Endocrinología; 1977. p. 80.
106. Guerra-Garcia R, Llerena LA, Garayar D, et al. Función endocrina hipófiso testicular en nativos de altura y en pacientes con mal de montaña crónico. Cusco: Sociedad Peruana de Endocrinología; 1973. p. 40.
107. Villena A, Zorrilla R, Guerra-Garcia R. Respuesta ortostática de aldosterona sérica en nativos normales y residentes de la altura y en pacientes con "mal de montaña crónico". In: Congreso Peruano de Endocrinología. Lima: Sociedad Peruana de Endocrinología; 1987.
108. Guerra-García R, Gonez C, Zubiate M, et al. Función suprerrenal en nativos de altura y en pacientes con mal de montaña crónico. Cusco: Sociedad Peruana de Endocrinología; 1973. p. 42.
109. Adnot S, Andrivet P, Chabrier PE, et al. Plasma levels of atrial natriuretic factor, renin activity, and aldosterone in patients with chronic obstructive pulmonary disease. Response to O_2 removal and to hyperoxia. Am Rev Respir Dis. 1990;141:1178–84.
110. Hainsworth R, Drinkhill MJ, Rivera-Chira M. The autonomic nervous system at high altitude. Clin Auton Res. 2007;17:13–9.
111. Claydon VE, Norcliffe LJ, Moore JP, et al. Cardiovascular responses to orthostatic stress in healthy altitude dwellers, and altitude residents with chronic mountain sickness. Exp Physiol. 2005;90: 103–10.
112. Moore JP, Claydon VE, Norcliffe LJ, et al. Carotid baroreflex regulation of vascular resistance in high-altitude Andean natives with and without chronic mountain sickness. Exp Physiol. 2006;91:907–13.
113. Roach R, Passino C, Bernardi L, et al. Cerebrovascular reactivity to CO_2 at high altitude and sea level in Andean Natives. Clin Auton Res. 2001;11:183.
114. Norcliffe LJ, Rivera-Ch M, Claydon VE, et al. Cerebrovascular responses to hypoxia and hypocapnia in high-altitude dwellers. J Physiol. 2005;566:287–94.
115. Claydon VE, Gulli G, Slessarev M, et al. Cerebrovascular responses to hypoxia and hypocapnia in Ethiopian high altitude dwellers. Stroke. 2008;39:336–42.
116. Leon-Velarde F, Mejia O. Gene expression in chronic high altitude diseases. High Alt Med Biol. 2008;9:130–9.
117. Appenzeller O, Minko T, Qualls C, et al. Gene expression, autonomic function and chronic hypoxia: lessons from the Andes. Clin Auton Res. 2006;16:217–22.
118. Mejia OM, Prchal JT, Leon-Velarde F, et al. Genetic association analysis of chronic mountain sickness in an Andean high-altitude population. Haematologica. 2005;90:13–9.
119. Storkebaum E, Lambrechts D, Carmeliet P. VEGF: once regarded as a specific angiogenic factor, now implicated in neuroprotection. Bioessays. 2004;26:943–54.
120. Xing G, Qualls C, Huicho L, et al. Adaptation and mal-adaptation to ambient hypoxia; Andean, Ethiopian and Himalayan patterns. PLoS One. 2008; 3:e2342.
121. Huicho L, Xing G, Qualls C, et al. Abnormal energy regulation in early life: childhood gene expression may predict subsequent chronic mountain sickness. BMC Pediatr. 2008;8:47.
122. Klein HG. Isovolemic hemodilution in high altitude polycythemia. In: Adjustments to High Altitude - Proceedings of the International Symposium on Acclimatization, Adaptation and Tolerance to High Altitude. 1983: NIH, editor, Baltimore, MD.
123. Rivera-Ch M, Leon-Velarde F, Huicho L. Treatment of chronic mountain sickness: critical reappraisal of an old problem. Respir Physiol Neurobiol. 2007;158: 251–65.
124. Plata R, Cornejo A, Arratia C, et al. Angiotensin-converting-enzyme inhibition therapy in altitude polycythaemia: a prospective randomised trial. Lancet. 2002;359:663–6.
125. Swenson ER. Carbonic anhydrase inhibitors and ventilation: a complex interplay of stimulation and suppression. Eur Respir J. 1998;12:1242–7.
126. Swenson ER. Carbonic anhydrase inhibitors and hypoxic pulmonary vasoconstriction. Respir Physiol Neurobiol. 2006;151:209–16.
127. Moore LG, Niermeyer S, Vargas E. Does chronic mountain sickness (CMS) have perinatal origins? Respir Physiol Neurobiol. 2007;158:180–9.
128. Monge CC, Leon-Velarde F. Physiological adaptation to high altitude: oxygen transport in mammals and birds. Physiol Rev. 1991;71:1135–72.
129. Maignan M, Rivera-Ch M, Privat C, et al. Pulmonary pressure and cardiac function in chronic mountain sickness patients. Chest. 2009;135:499–504.
130. Rivera-Ch M, Huicho L, Bouchet P, et al. Effect of acetazolamide on ventilatory response in subjects with chronic mountain sickness. Respir Physiol Neurobiol. 2008;162:184–9.

High Altitude and Common Medical Conditions

23

Andrew M. Luks and Peter H. Hackett

Abstract

Increasing numbers of people are traveling to high altitude for work or pleasure. Given the prevalence of medical conditions in the general population, it is likely that many of these travelers will have one or more underlying medical problems. Unsure of how they will tolerate high altitude, these patients often seek input from their primary care physicians or travel clinical providers to determine if it is safe for them to make such a sojourn and, if so, what precautions should be taken during their trip to avoid problems that might lead to unplanned interruption of their trip. Clinicians faced with these concerns must address whether the underlying medical condition could be adversely affected by the hypoxic environment or alters the traveler's risk for developing high-altitude illness. This chapter provides information to guide clinicians in answering these questions as they pertain to patients with a wide variety of medical problems including pulmonary diseases such as chronic obstructive pulmonary disease, asthma, and obstructive sleep apnea; cardiac problems including coronary artery diseases, cardiomyopathy, and adult congenital heart diseases; as well as gastrointestinal, endocrine, hematologic, neurologic, and renal disorders. For each disorder we consider the primary challenges faced by those patients at altitude and provide recommendations for pretravel assessment as well as risk mitigation during the trip. The chapter concludes by considering medication use at high altitude and, in particular, whether medications used for treatment of underlying disorders have the potential for adverse interactions with medications used in the prophylaxis and treatment of acute altitude illness and whether the dose and choice of altitude illness medication needs to be altered depending on the patients underlying health issues.

A.M. Luks, M.D. (✉)
Division of Pulmonary and Critical Care Medicine, Harborview Medical Center, University of Washington, 325 Ninth Avenue, Box 359762, Seattle, WA 98104, USA
e-mail: aluks@u.washington.edu

P.H. Hackett, M.D.
Clinical Professor of Surgery, Division of Emergency Medicine, Department of Surgery, Altitude Research Center, University of Colorado Denver School of Medicine, Denver, CO, USA

Institute for Altitude Medicine, 500 W. Pacific Avenue, Telluride, CO, USA
e-mail: hackett@hypoxia.net

E.R. Swenson and P. Bärtsch (eds.), *High Altitude: Human Adaptation to Hypoxia*,
DOI 10.1007/978-1-4614-8772-2_23, © Springer Science+Business Media New York 2014

Introduction

Increasing numbers of people are traveling to high altitude for work or pleasure. Given the prevalence of medical conditions in the general population, it is likely that many sojourners will have some underlying medical problem. Unsure of how they will tolerate high altitude, these patients seek input from their medical providers who, in turn, must address two questions: (1) will the underlying medical condition be adversely affected by the hypoxic environment, and (2) will the underlying condition alter the traveler's risk for developing high-altitude illness, i.e., acute mountain sickness (AMS), high-altitude cerebral edema (HACE), or high-altitude pulmonary edema (HAPE)? The purpose of this chapter is to provide information to guide clinicians in answering these questions.

The topic has been covered in depth in an earlier review on the topic [1]. This chapter will focus on updating the current state of knowledge on the conditions discussed in that review and present information regarding several other conditions not addressed previously. For each of the topics, we focus on the effects of exposure lasting days to weeks and will not consider the issue of relocation to high-altitude or lifelong residence.

At What Altitude Are Patients at Risk for Problems?

There are no firm definitions in the literature as to what constitutes "high altitude," as the thresholds used for this designation vary between studies or review articles. A common approach is to make a distinction between differing degrees of altitude exposure such as that put forth by Bartsch and Saltin [2] in which they designate 2,000–3,000 m as moderate altitude, 3,000–5,500 m as high altitude, and >5,500 m as extreme altitude.

In general, the risk of acute altitude illness begins above 2,000 m and increases at higher elevations. Several physiologic responses that may impact those with underlying medical problems begin at roughly the same altitude. The hypoxic ventilatory response and hypoxic pulmonary vasoconstriction, for example, begin when alveolar PO_2 falls below 70 mmHg [3], a value reached in most individuals at or near 2,000 m. Of course, 2,000 m should not be viewed as a strict threshold below which patients are fine and above which they are at risk for problems. Rather, the risks of altitude travel likely occur along a spectrum of elevation gain and represent a complex interplay between the severity of underlying disease, interactions between its pathophysiology and acute hypoxia, and ultimate altitude attained and rate of ascent (degree and rate of onset of hypoxia).

Pulmonary Disorders

Of all the patients traveling to high altitude, patients with pulmonary disorders face some of the greatest challenges in this environment due to the potential for worsening hypoxemia and the ventilatory demands that may be placed on patients with altered lung mechanics. Several of the most common disorders are considered in the space that follows. Patients with pulmonary disease who are planning travel to high altitude should give consideration to monitoring oxygenation with pulse oximetry using a small handheld device. Further details about pulse oximetry monitoring at high altitude are available in a separate review on this issue [4].

Chronic Obstructive Pulmonary Disease

The effect of high altitude on patients with chronic obstructive pulmonary disease (COPD) has been reviewed extensively elsewhere [5] and will be summarized here. To date, only a single study has examined COPD patients in an actual high-altitude environment. Graham and Houston [6] studied 8 patients with COPD (mean FEV_1 1.27 L) and found that resting P_aO_2 fell from a sea level baseline of 66 to 54 mmHg after 3 h at 1,920 m. Further data on COPD patients during commercial flight consistently demonstrates that

P_aO_2 often falls below 50 mmHg when patients with FEV_1 between 1 and 1.5 L are exposed to hypobaric or normobaric hypoxia simulating altitudes between 2,348 and 3,050 m [7, 8]. Mild exercise leads to further declines in arterial oxygen tension [9], while supplemental oxygen administration rapidly reverses hypoxemia [10].

The degree of hypoxemia in these patients is relevant to planning travel to high altitude. Several leading guidelines, for example, recommend initiation of supplemental oxygen for commercial flight when the arterial PO_2 is predicted to be below 50 [11] or 55 mmHg [12] at 2,440 m. Recognizing that these thresholds are arbitrary and lack supporting data, it is reasonable to ask whether COPD patients require supplemental oxygen with travel to similar altitudes in the mountains.

The key issue becomes identifying those patients who might require supplemental oxygen. In the commercial flight literature, Gong et al. [13] proposed that sea level $P_aO_2 < 72$ mmHg correctly identified such patients, but subsequent studies have questioned this and suggest that prediction equations that take into account FEV_1 [14] or baseline maximum exercise capacity [8] might have better predictive capability. Sea level pulse oximetry does not accurately identify those individuals in whom the P_aO_2 will fall below 50 mmHg [15].

The utility of this literature is limited in several respects. First, the studies generally include only patients with baseline FEV_1 of 1.0–1.5 L and exclude patients with carbon dioxide retention. Second, and more importantly, the studies from the commercial flight literature involve short exposures (several hours) and, as a result, do not account for ventilatory acclimatization that occurs over the first several days at high altitude, leading to improvements in arterial oxygenation [6]. Finally, it is noteworthy that in the study by Graham and Houston, as well as studies from the commercial flight literature, there were no reports of adverse events, and patients only developed mild symptoms including dyspnea, fatigue, and headache, suggesting that chronically hypoxemic patients may tolerate further hypoxemia at high altitude to some extent because they are already partially "acclimatized" to impaired oxygenation. Given the latter two issues, it is possible that strict adherence to prediction equations for supplemental oxygen may lead to unnecessary use in some patients and that a more flexible approach is warranted [16].

Beyond the issue of oxygenation, the literature provides little conclusive evidence regarding the effects of high altitude on the degree of airflow obstruction as the studies on pulmonary mechanics in hypoxia have largely shown conflicting data [17–20]. There is currently no evidence that patients with bullous disease are at increased risk for pneumothorax due to the drop in barometric pressure at high altitude [21, 22]. There is also no information regarding COPD patients with carbon dioxide retention at terrestrial or simulated high altitude, but from a theoretical standpoint, there is strong reason to suspect these patients may experience problems. Because carbon dioxide retention is a sign of severe ventilatory impairment, these patients may be unable to raise ventilation sufficiently at rest or with exertion at high altitude and, as a result, may develop worsening symptoms, hypoxemia, and exercise limitation. In addition, many patients with carbon dioxide retention develop pulmonary hypertension, which, as discussed below, may predispose to HAPE and other complications at high altitude.

Recommendations for pretravel evaluation and other aspects of management of COPD patients at high altitude are provided in Table 23.1.

Interstitial Lung Disease

Only two studies have evaluated the effect of simulated high altitude on patients with interstitial lung diseases such as pulmonary fibrosis. Seccombe et al. [9] studied 15 patients with unspecified forms of interstitial lung disease (mean FEV_1 1.83 L, 55 % predicted) at a simulated altitude of 2,440 m (8,000 ft) and found that resting P_aO_2 fell from 84 to 51 mmHg after 20 min at rest and to 41 mmHg after walking 50 m. The decreased oxygenation was associated with increased Borg dyspnea scores.

Table 23.1 Recommendations for patients with underlying lung disease

General recommendations
• SpO_2 monitoring can be considered for any patient with lung disease
• "Home" oxygen is available in many high-altitude locations and can be arranged before or after arrival at high altitude
Chronic obstructive pulmonary disease
• Avoid travel above 2,000 m without supplemental oxygen in patients with:
– $FEV_1 < 1$ L
– CO_2 retention
– Pulmonary hypertension
– Active or recent exacerbation. Symptoms should be back at baseline prior to any planned travel
• Patients with FEV_1 1.0–1.5 L: assess need for supplemental oxygen using prediction equations with baseline FEV_1 [14] or the hypoxia inhalation test [188]:
– If the P_aO_2 is predicted to be <50 mmHg: travel with portable oxygen or a prescription to fill at destination
– If already on supplemental oxygen, increase flow rate by the ratio of higher to lower barometric pressure
• Patients with $FEV_1 > 1.0$ L can monitor pulse oximetry upon arrival and seek attention for severe hypoxemia or increasing dyspnea and fatigue
• Patients with $FEV_1 < 25$ % predicted should use dexamethasone rather than acetazolamide for AMS prophylaxis or treatment
• Travel with supply of rescue inhalers and oral steroids in event of exacerbation
Interstitial lung disease
• Avoid high-altitude travel with TLC < 40 % predicted or supplemental oxygen requirement at home elevation
• Patients with less severe disease: assess for supplemental oxygen [23] and travel with portable oxygen or prescription to fill at destination if predicted $P_aO_2 < 50$ mmHg
• Patients not requiring supplemental oxygen should monitor pulse oximetry upon arrival and seek attention for severe hypoxemia or increasing dyspnea and fatigue
Asthma
• Avoid high-altitude travel in patients with moderate persistent or severe asthma or in patients with recent or active exacerbations or recent increase in rescue inhaler use. Following an exacerbation, symptoms should be back at baseline prior to the planned sojourn
• Patients with mild-intermittent or mild-persistent asthma can ascend to high altitude and should remain on preexisting medications and carry a supply of rescue inhalers and steroids for exacerbations
• Patients traveling in cold environments must keep inhalers warm and be aware that the number of puffs per inhaler may be decreased at elevations above 3,000 m [189]

Cystic fibrosis
• High-altitude travel should be avoided with baseline $FEV_1 < 30\ \%$ predicted
• Patients with an FEV_1 30–50 % of predicted should undergo pretravel assessment to determine need for supplemental oxygen [44] and should travel with portable oxygen or prescription to fill at destination if predicted $P_aO_2 < 50$ mmHg
• Patients not requiring supplemental oxygen should monitor pulse oximetry upon arrival and seek attention for severe hypoxemia or increasing dyspnea and fatigue
• Continue preexisting airway clearance techniques, prophylactic antibiotics, and mucolytic therapy during sojourn
Pulmonary hypertension
• Patients with moderate to severe pulmonary hypertension (mean PA pressure ≥35 mmHg or systolic pressure ≥50 mmHg) should avoid high-altitude travel without supplementary oxygen. Patients should:
– Be counseled regarding recognition and management of HAPE
– Monitor oxygen saturation during their high-altitude stay
– Use nifedipine 30 mg SR BID as HAPE prophylaxis if not already on vasodilator therapy
• Patients with mild pulmonary hypertension (mean PA pressure ≤35 mmHg or systolic pressure <50 mmHg) may travel to high altitude but might consider prophylaxis with nifedipine SR 30 mg BID and consider monitoring oxygen saturation and symptoms following arrival at high altitude
Obstructive sleep apnea
• Patients with moderate to severe obstructive sleep apnea should travel to high altitude with their CPAP machine if reliable power access can be assured
• Acetazolamide should be used in addition or as an alternative to CPAP to decrease the incidence and severity of central sleep apnea
• Patients using older generation machines (>10 years old) that lack pressure-compensating features will need to adjust the intended CPAP pressure at high altitude [72]
Neuromuscular disorders
• Patients with hypoventilation at sea level should travel with supplemental oxygen and bi-level positive airway pressure for use at night
• Patients with blunted hypoxic ventilatory responses should monitor pulse oximetry and travel with a prescription for supplemental oxygen to fill if severe hypoxemia develops
• Patients with kyphoscoliosis and other disorders associated with pulmonary hypertension should be screened with echocardiography and managed according to guidelines for pulmonary hypertension
• Patients with bilateral diaphragmatic paralysis should consider travel with supplemental oxygen for use at night

Christensen et al. [23] exposed 17 patients with worse restrictive lung diseases (mean FEV_1 1.4 L, 49 % predicted; mean TLC 3.2 L, 52 % predicted; mean P_aO_2 78 mmHg at sea level) to the same simulated altitude and found that the P_aO_2 fell to 49 mmHg at rest and 38 mmHg during 20 W exercise. Administration of supplemental oxygen at rest (2 LPM) or with exercise (4 LPM) raised the P_aO_2 above 50 mmHg. Caution must be applied in interpreting these data, however, as interstitial lung disease is heterogeneous, a factor not addressed in the studies noted above. In addition, these data apply to those with severe impairment; patients with less severe disease are likely to do better.

The clinical implications of worsening hypoxemia at high altitude in interstitial lung disease patients are not clear. Increased hypoxemia increases the risk of AMS in patients without lung disease [24, 25], but studies that specifically examined patients with lung disease have reported varying results. Honigman et al. [26], for example, reported an increased risk of AMS in patients with underlying lung disease traveling to 1,900–2,900 m, while Roach et al. [27] found no relationship between AMS risk and decreased pulmonary function in older individuals (mean age 70 years) traveling to similar altitudes. These studies, however, did not focus specifically on patients with interstitial lung disease and, instead, looked at patients with a variety of unspecified lung diseases. As with severe COPD patients, those with severe interstitial lung disease develop secondary pulmonary hypertension, a risk factor for HAPE or other complications at high altitude.

Recommendations for pretravel evaluation and management in interstitial lung disease patients at high altitude are provided in Table 23.1.

Asthma

The impacts of high altitude on asthma are multiple. On the one hand, lower air density and decreased dust mites [28–31] benefit asthma patients, while hypoxia [32], hypocapnia [33], and decreased air temperature and humidity [34] may trigger increased airway reactivity. Epidemiologic studies have documented increased incidence of asthma or exercise-induced bronchoconstriction (EIB) among cross-country skiers [35, 36] and ski mountaineers [37], individuals whose activity requires very high ventilation in cold environments; many had no history of EIB at lower altitude.

Overall, the number of studies of asthma patients during actual acute high-altitude exposure is limited, making it difficult to draw firm conclusions. Golan et al. [38] reported that 20 % of adventure travelers with asthma, a large percentage of whom were participating in high-altitude trekking, reported an exacerbation during their trip while 16 % reported their "worst ever" asthma attack during the trip. Frequent rescue inhaler use (≥3 times per week) prior to the trip and intense physical activity during the trip were risk factors for worsening symptom control. The study, however, did not control for the altitudes attained or for transit through cities with very poor air quality. Louie and Pare [39] studied 5 asthmatic patients during a trek in the Nepal Himalaya and demonstrated that peak expiratory flow fell by an average of 76 LPM (~13 % change) between sea level and 4,100 m, but subjects were taking dexamethasone and acetazolamide which might have skewed the results. In more controlled studies, Cogo et al. [40] and Allegra et al. [41] assessed mild asthmatic patients at sea level and altitudes as high as 5,050 m and documented decreased bronchial hyperreactivity to hypoosmolar aerosol or methacholine at high altitude. Stokes et al. have also documented that well-controlled mild asthmatics can ascend to 5,895 m without exacerbations of their disease and with no increase in rate of AMS compared to non-asthmatic climbers [42], while Huismans et al. [43] reported that asthma patients traveling to 6,410 m in Tibet did not experience increasing asthma symptoms with increasing altitude and did not see significant increases in their need for asthma medications.

In addition to these few studies that suggest that high-altitude travel is likely safe in patients with mild, well-controlled disease, it is noteworthy that, aside from the poorly controlled study by Golan et al. [38], the literature contains no reports of unexpected asthma exacerbations at high altitude or an association between asthma and altitude illness [1].

Recommendations for the management of asthma patients traveling to high altitude are provided in Table 23.1.

Cystic Fibrosis

As the quality of care and life expectancy of cystic fibrosis (CF) patients improves, it is likely that more of these patients will travel to high altitude. Inconsistent effects of high altitude on pulmonary function have been reported, with one study [44] reporting an increase in FEV_1 and FVC at high altitude and others reporting either no change [45] or a small decrease [46] in these values. What is consistent across studies, however, is the fact that travel to moderate altitudes (2,000–3,000 m) is associated with impaired oxygenation; P_aO_2 falls to near or below 50 mmHg at these altitudes, with the largest decrease in those with more severe disease [44, 46], and exercise further exacerbates hypoxemia [44, 47]. The hypoxia inhalation test commonly used to predict the need for supplemental oxygen may not be as useful in cystic fibrosis patients [44, 48], while pretravel spirometry [44] or prediction equations different than those used for COPD patients [49] may have greater utility.

The consequences of impaired oxygenation, however, are unclear, since in the studies noted above subjects had either none or minor symptoms. This may be related to the fact that the studies examined younger patients with fewer comorbidities than COPD patients. A limitation to all of these studies is that they involved short exposure (hours) and, as a result, may not reliably capture some problems that occur with longer exposures. Reports of CF patients with low baseline FEV_1 (~1 L) developing pulmonary hypertension and cor pulmonale during longer time at high altitude [50] serve as a warning that severe complications are still possible. No data assess the frequency or severity of CF exacerbations at high altitude compared to sea level.

Recommendations for the management of cystic fibrosis patients planning high-altitude travel are provided in Table 23.1.

Pulmonary Hypertension

An important factor in HAPE pathogenesis is exaggerated pulmonary vascular response to hypoxia (see Chaps. 5 and 22), which leads to elevated pulmonary arterial and capillary pressures that promote the transit of red blood cells, protein, and fluid from the vascular space into the pulmonary interstitium and alveolar space [51, 52]. Numerous case reports demonstrate that preexisting pulmonary hypertension exacerbates this pathophysiology and increases risk of HAPE. These include patients with both anatomic [53–55] and nonanatomic causes [56–58] for pulmonary hypertension, although the exact level of pulmonary hypertension necessary to increase the risk of HAPE is unclear. The patient's underlying pulmonary vascular resistance, hypoxic pulmonary vasoconstriction responsiveness, and rate and height of ascent likely determine a continuum of risk. Interestingly, persons with chronic pulmonary hypertension from high-altitude residence are also susceptible to HAPE following reascent after a period at lower elevation [59, 60].

No studies or case reports examine patients with idiopathic pulmonary arterial hypertension. Because extensive vascular remodeling occurs in these patients, it is possible that their altered pulmonary vasculature provides a measure of protection against HAPE. In the absence of data regarding these patients at altitude, it is difficult to offer advice; the more prudent course is to consider them at increased risk for either HAPE or exacerbation of their baseline symptoms. Recent data suggest that relatives of patients with idiopathic pulmonary arterial hypertension also demonstrate abnormal pulmonary vascular responses to exercise and hypoxia [61], but whether they should be also considered at high risk for problems at high altitude is unknown.

One factor that may warrant attention when assessing the risk for HAPE in these patients is the adequacy of their iron stores. Recent work, for example, has suggested that iron deficiency may play an important pathophysiological role in the idiopathic and heritable forms of pulmonary arterial hypertension [62]. Given data suggesting that iron administration prior to high-altitude travel blunts pulmonary vascular responses in normal individuals traveling to 4,340 m [63], it is worth considering whether iron supplementation in pulmonary hypertension patients prior to high-altitude travel would be of benefit in preventing HAPE. No studies to date have specifically examined this question.

HAPE is not the only potential complication in patients with pulmonary hypertension traveling to high altitude, as patients may be at risk for worsening right ventricular (RV) function, ischemia, and impaired oxygenation. Although deterioration has not been reported at high altitude, case reports have documented this phenomenon during commercial flight in patients with morbid obesity [64] and severe kyphoscoliosis [65], two entities associated with pulmonary hypertension. Increased RV afterload could also cause RV dilation, leading to decreased subendocardial blood flow, ischemia, and chest pain, particularly during exertion.

Finally, exercise desaturation, a common feature at sea level in both adult [66] and pediatric [67] pulmonary hypertension patients, would likely worsen at high altitude and cause even greater exercise limitation.

Recommendations regarding the management of pulmonary hypertension patients at high altitude are summarized in Table 23.1.

Obstructive Sleep Apnea

The lower air density at high altitude may lead one to expect a decrease in the severity of obstructive sleep apnea with high-altitude exposure. This hypothesis was investigated by Burgess et al. [68] who performed sleep studies at sea level and high altitude on patients with baseline moderate OSA and found that the obstructive respiratory distress index (RDI) fell from 25.5 ± 14.4 events/h at sea level to 0.5 ± 0.7 events/h at 2,750 m. As might be expected given the lower ambient oxygen tensions, average and nadir nocturnal oxygen saturations were also lower at 2,750 when compared to sea level. A similar decrease in the obstructive apnea-hypopnea index with increasing altitude was seen in an earlier study on normal individuals with low baseline obstructive sleep apnea indices [69]. In both studies, however, the decrease in obstructive events was offset by a marked increase in central apnea. In moderate OSA patients, for example, the RDI for central events rose from 0.4 events/h at 60 m to 78.8 events/h at 2,750 m, much higher than the sea level obstructive RDI [68].

Subsequent field studies have demonstrated similar increases in the number of central apneas but, interestingly, no change in the number of obstructive events despite the fall in barometric pressure and air density [70, 71]. Use of acetazolamide in these patients decreased the number of central events but had no significant effect on the number of obstructive events in these patients. Importantly, acetazolamide also reduced nocturnal transcutaneous PCO_2, improved sleep efficiency and subjective impressions of insomnia, and prevented blood pressure elevations in OSA patients at altitude, findings which suggest that acetazolamide may be beneficial for these patients when they are unable to use their CPAP devices at altitude [70].

Patients using older CPAP machines (>10 years) should be aware that some of these machines lack the ability to compensate for changes in ambient pressure. Such machines may not deliver the intended pressure in the hypobaric environment at high altitude and may require adjustment [72].

Recommendations for the management of obstructive sleep apnea patients at high altitude are provided in Table 23.1.

Neuromuscular Diseases

Patients with various forms of neuromuscular disease, such as the muscular dystrophies or diaphragmatic paralysis, may be at risk for problems at high altitude. Those with impaired ventilation, for example, might be unable to mount an adequate ventilatory response to hypoxia and therefore suffer significant hypoxemia, difficulty with exertion, and higher risk of altitude illness. Although this issue has not been studied systematically, clinical experience suggests that muscular dystrophy patients do indeed experience such problems. Patients with Parkinson's disease [73] and myotonic dystrophy [74] may have blunted hypoxic ventilatory responses at sea level that could predispose to significant hypoxemia at high altitude, while patients with bilateral diaphragmatic paralysis can become more hypoxemic in the supine position and, as a result, may experience significant desaturation during sleep at high altitude [75]. Finally, many neuromuscular disease patients have associated disorders such as obstructive sleep apnea, nocturnal hypoventilation, or pulmonary hypertension, which could predispose to complications during high-altitude travel. These issues have been reviewed in greater detail elsewhere [5]. Recommendations for managing these patients during high-altitude travel are provided in Table 23.1.

Pediatric Chronic Lung Disease

As medical care improves, the life expectancy of children with respiratory complications during the neonatal period, referred to as chronic lung disease or bronchopulmonary dysplasia, is increasing. Aside from a report of a single individual who developed HAPE at 2,927 m at age 13 [76], little is known about how these children fare at high altitude. Clinicians should be aware, however, that they often develop long-term complications such as pulmonary hypertension or airflow limitation. Chronic lung disease patients whose families are planning high-altitude travel should be screened for such abnormalities and managed with caution including monitoring of pulse oximetry and possible use of supplemental oxygen. Patients with baseline pulmonary hypertension should be managed according to guidelines provided above for adult pulmonary hypertension patients. Identifying the onset of AMS in preverbal children can be difficult; alternative diagnostic criteria for AMS in infants and preverbal children have been proposed [77]. This and other topics regarding children at high altitude are discussed in a greater detail in a recent review of this issue [78].

Cardiovascular Disorders

Coronary Artery Disease

In response to both exercise and hypoxemia, coronary arteries dilate in an effort to preserve myocardial blood flow and oxygen delivery. Individuals with coronary artery disease (CAD) may have impaired coronary vasodilation and thus may be at risk for adverse cardiac events at high altitude. A summary of these risks is provided below and the reader is referred to Chap. 6 and several prior reviews on the topic [1, 79, 80] for further details.

Resting hypoxemia alone does not appear to unmask CAD [81, 82], while exercise in acute hypoxia may provoke ischemic changes on electrocardiography in persons with CAD [83]. For patients with stable CAD at sea level, the ischemia threshold is lower during the initial few days at moderate altitudes (~2,500 m) [84] but subsequently recovers to pre-ascent levels as acclimatization occurs over 5 days. The ability to raise coronary blood flow in response to exercise is also decreased in patients with CAD during exposure to a simulated altitude of 2,500 m [85]. Patients who have undergone revascularization following a myocardial infarction tolerate exertion at high altitude; a recent study [86] demonstrated that 6 months after revascularization for acute coronary syndrome, patients with normal left ventricular function and no evidence of ischemia on exercise testing at low altitude could perform maximal, symptom-limited exercise tests after rapid ascent to 3,454 m without electrocardiographic evidence of myocardial ischemia or significant arrhythmia. Cold temperature can aggravate ischemia at sea level due to increased sympathetic activity [87]; this issue has never been systematically studied at high altitude but is a concern, particularly with travel in the winter months, given that air temperature decreases as altitude increases. There is no evidence of life-threatening arrhythmias in patients with CAD at high altitude. Several studies [88, 89] have documented cases of sudden cardiac death at high altitude in a general population, but whether high altitude increases the risk of this problem specifically in CAD patients is unknown.

Recommendations for the management of patients with CAD planning to travel to high altitude are provided in Table 23.2.

Hypertension

In hypertensive patients ascending to high altitude, systolic blood pressure rises to a modest extent (10–15 mmHg on average) but then declines over a period of days to weeks [27, 90–92]. Marked variability is found in these responses, however [27, 90], and the lack of a way to predict the response makes advising these patients difficult. The literature is limited by the modest elevations (<3,500 m) and the fact that poorly controlled hypertensive patients were not included in these studies. There is no evidence that hypertension increases risk of AMS. Although there are theoretical reasons to support the use of alpha-blockers and calcium channel blockers for control of elevated blood pressure at high altitude, no systematic studies have been conducted to evaluate their utility in this regard.

Recommendations for the management of hypertensive patients traveling to high altitude are provided in Table 23.2. A more complete discussion of hypertension at high altitude is also available in a separate review [93].

Arrhythmia

There is some evidence that healthy people without preexisting cardiac arrhythmia may occasionally experience sinus tachycardia, sinus bradycardia, premature atrial or ventricular contractions, incomplete right bundle branch block, atrial flutter, and non-conducted p-waves following ascent to high altitude [84, 92, 94], but very little information exists regarding people with known preexisting arrhythmias, such as atrial fibrillation, traveling to this environment. In one of the only studies of this issue, Wu et al. [92] performed electrocardiograms (ECG) and 24-h ambulatory ECG on 42 people with a variety of preexisting arrhythmias including sinus arrhythmia, premature atrial or ventricular beats, first-degree AV block, and incomplete right bundle branch block working on the Qinghai-Tibet Railroad. Aside from one case of asymptomatic Wolff-Parkinson-White syndrome, no life-threatening arrhythmias were noted in these patients when assessments were made 1 week and 3 months after arrival between 4,500 and 5,056 m and there were no reports of exacerbations of their underlying arrhythmia.

With regard to patients with implanted pacemakers or defibrillators, Weilenmann et al. [95] exposed 13 patients with implanted single-chamber pacemakers to a simulated altitude of 4,000 m and noted no changes in ventricular stimulation thresholds. While this result suggests certain pacemaker patients may tolerate high altitude, the 30-min duration of exposure was considerably shorter than many people would experience and did not include exercise. Anecdotal reports suggest patients with implanted cardiac defibrillators (ICDs) may experience shocks at high altitude, but only a single study has systematically investigated this question. Kobza et al. surveyed 217 patients with ICDs who traveled to altitudes above 2,000 m and found that 4 % experienced an ICD shock during their sojourn [96]. It was not clear from the study whether the shocks were due to the patient's underlying cardiac condition or a change in the device's defibrillation threshold at high altitude.

Recommendations regarding the management of patients with pacemakers, defibrillators, or underlying arrhythmia are provided in Table 23.2.

Heart Failure

In one of the few studies to assess heart failure patients at terrestrial altitude, Erdman et al. [97] took patients with CAD with documented ejection fraction <45 % to 2,500 m by cable car for a few hours and found no evidence of ischemia, pulmonary edema, or other complications; in

Table 23.2 Recommendations for patients with underlying cardiac disease

Coronary artery disease
• Patients with coronary artery disease should be risk stratified using exercise treadmill testing or nuclear imaging studies in normoxia
• Avoid high-altitude travel with:
– Unstable or severe angina
– Objective evidence of myocardial ischemia at low levels of exertion
– Recent acute coronary syndrome (<3–6 months, no revascularization)
• Patients with stable CAD that permits exercise at sea level can ascend to high altitude and engage in physical activity but should:
– Avoid exertion during the first several days after arrival
– Reduce exercise levels to slightly lower than at sea level
– Avoid heavy exertion in cold temperatures
• Patients should remain on regular medications while at high altitude
• Patients on warfarin or clopidogrel should limit activities in remote areas to reduce the risk of trauma and severe bleeding
• Patients should travel with copy of recent electrocardiogram
• Supplemental oxygen can be prescribed for the first several days at altitude if concerns persist following sea level testing
Hypertension
• Patients with poorly controlled or labile hypertension at sea level should monitor blood pressure following ascent
• Medication adjustments and/or addition of medications should be made according to a prearranged plan if:
– Systolic pressure >180 mmHg or diastolic pressure >120 mmHg and symptoms including chest pain, vision changes, headache, or dyspnea
– Systolic pressure >220 mmHg or diastolic pressure >140 mmHg without symptoms
– Patients on longer sojourns at altitude may desire better control
• After adjustments, patients should continue monitoring and alter medications accordingly
• Patients should return to their original regimen upon descent to lower elevation
• Nocturnal oxygen can be considered to help stabilize blood pressure at high altitude
Arrhythmia
• Ensure pacemakers and implantable defibrillators are functioning properly prior to high-altitude travel
• Avoid high-altitude travel if preexisting arrhythmias are not under adequate control or, if such travel is necessary, consider supplemental oxygen
Heart failure
• Avoid high-altitude travel with poorly compensated heart failure
• Patients with stable, compensated heart failure can ascend to moderate altitudes (<3,000 m) and engage in physical activity; slightly below the level they can perform at sea level
• Monitor body weight and blood pressure regularly and travel with prearranged plan for adjusting diuretic or antihypertensive dosing if increasing weight or severely elevated blood pressure develops
• Consider traveling with supplemental oxygen
Adult congenital heart disease
• Patients should be evaluated prior to planned high-altitude travel using:
– Echocardiography to document pulmonary artery pressure
– Cardiopulmonary exercise testing
• High-altitude travel should be avoided with:
– Moderate to severe pulmonary hypertension
– Low maximum exercise capacity (<20 mL/kg/min)
• Right-to-left shunting *and* anemia or impaired cardiac output

addition, they found a decrement in exercise capacity similar to healthy controls. Agostoni et al. [98] performed cardiopulmonary exercise tests on 38 patients with heart failure and various levels of exercise impairment, at sea level and at simulated altitudes of 1,000, 1,500, 2,000, and 3,000 m. As expected, work rate decreased with increasing altitude with the greatest decrements in those patients with the lowest exercise capacity at baseline. None of the tests were interrupted due to arrhythmia, angina, or ischemia; each hypoxic exposure lasted for only the duration of the exercise study.

While these studies suggest patients with stable heart failure can tolerate short duration exposures to high altitude, clinicians working in high-altitude areas report that decompensation of heart failure is a common occurrence in tourists [99]. In addition, some theoretical concerns warrant attention. Given their propensity towards fluid overload, these patients may have higher pulmonary hydrostatic pressures that increase the risk for HAPE. Similarly, AMS is often associated with fluid retention and, to the extent this occurs in a heart failure patient, it could lead to volume overload, worse peripheral edema, and even pulmonary edema. From a theoretical standpoint, acetazolamide, with its diuretic effect, would be an effective means to prevent AMS or worsening heart failure in these patients, but this has not yet been studied. Marked blood pressure elevations could also increase left ventricular afterload and impair systolic function.

Questions also arise as to how well patients will tolerate use of beta-blockers at altitude, and Agostoni et al. [100], for example, have shown that carvedilol, a commonly used medication in heart failure patients, may reduce ventilatory responses in acute hypoxia which could predispose to hypoxemia and possibly altitude illness. More recently, Valentini et al. [101] compared the beta-blockers nebivolol and carvedilol on exercise following ascent to 4,559 m and found that peak oxygen consumption decreased less at high altitude following treatment with nebivolol compared to carvedilol. The decrease in maximum heart rate was also less with nebivolol, while peak minute ventilation actually increased on this agent. The study only examined healthy subjects, however, and how well this data translates to cardiomyopathy patients using beta-blockers at high altitude remains unclear.

See Table 23.2 for management recommendations for heart failure patients traveling to high altitude.

Congenital Heart Disease

As with heart failure, there is little information regarding congenital heart disease patients at high altitude. Case series and case reports demonstrate that a rare abnormality, unilateral absence of a pulmonary artery, is associated with an increased risk for developing HAPE [53, 54], while limited evidence suggests patients with Down's syndrome, a disorder associated with cardiac abnormalities, may also be predisposed to this problem [58]. Allemann et al. [102] argue that patent foramen ovale (PFO) may predispose to HAPE, but whether PFO is causal in HAPE or rather a sequela of the marked rise in pulmonary artery pressure that occurs during hypoxia or exercise in HAPE-susceptible individuals [103, 104] could not be established. Finally, Garcia et al. [105] performed cardiopulmonary exercise tests on 11 patients with history of Fontan procedure for correction of tricuspid atresia and noted that submaximal exercise was well tolerated at simulated altitude of 3,050 m, despite these patients lacking functional right ventricles. It was only with maximal exercise that the lack of a functional right ventricle impaired their ability to generate sufficient cardiac output.

In addition to the limited literature, drawing conclusions about congenital heart disease patients is difficult because of the wide variety of congenital disorders and their associated hemodynamic consequences and effect on oxygenation. From a theoretical standpoint, two groups of patients might be at risk for problems at high altitude. The first group is those in whom the defect, or its repair, is accompanied by pulmonary hypertension. As noted earlier, such patients might be at increased risk for HAPE or worsening right ventricular function. The second group is those patients with right-to-left shunts, who may already be hypoxemic at sea level.

With ascent to altitude, hypoxic pulmonary vasoconstriction could lead to further rises in pulmonary vascular resistance and greater right-to-left shunting. It should be noted, however, that many of these patients are polycythemic [106, 107] and thus may defend oxygen delivery despite very low arterial oxygen tension. In addition, since cardiac index is a more important determinant of oxygen delivery than arterial oxygen content in these patients [108], those with preserved cardiac function may better tolerate high altitude.

Management recommendations for adult congenital heart disease patients traveling to high altitude are provided in Table 23.2.

Hematologic Disorders

Thromboembolic Disease

Given the numerous reports documenting episodes of venous or arterial thrombosis at high altitude including lower extremity deep venous thrombosis [109], pulmonary embolism [56], cerebral venous thrombosis [110], portal system thrombosis [111], ischemic stroke [112], and popliteal artery thrombosis [113], the question arises as to whether high-altitude exposure provokes thromboembolic events.

To date, however, there is no compelling evidence that risk of thromboembolic events is increased in normal individuals. In one of the few systematic studies using venous thromboembolism as an end point, Anand et al. [114] retrospectively reviewed records from 20,257 military hospital admissions in India over a 3-year period and found a higher incidence of venous thromboembolism among residents of high-altitude areas. Jha et al. [112] used a similar retrospective approach and reported a higher incidence of stroke in soldiers stationed at high altitude compared to those at lower elevation, while Khalil and Saeed report that high-altitude exposure was the only risk factor for pulmonary embolism in 50 % of Indian soldiers who developed pulmonary embolism [115]. Each of these studies, unfortunately, has important methodological limitations that make it difficult to draw conclusions about the risk of thromboembolism in short-term travelers to high altitude.

Other studies have tried to assess the risk of thromboembolic events at high altitude by examining changes in various coagulation parameters during hypoxic exposure, but they have yielded conflicting results regarding platelet count and function [116–118], bleeding time [119, 120], fibrinolytic activity [121, 122], and thrombin formation [123, 124].

Rather than hypoxia increasing the risk of thromboembolism in *all* individuals, it may be that hypoxia increases risk in those with underlying coagulopathy. Schreijer et al. [125] exposed healthy individuals to hypobaric hypoxia equivalent to 1,800–2,100 m in an aircraft and noted increased levels of thrombin-antithrombin complexes only in those individuals with Factor V Leiden mutation and oral contraceptive use. Supporting a role for interaction between hypoxia and thrombophilia, many of the patients in the case reports noted above as well as some others were subsequently found to have protein C deficiency [110], antiphospholipid antibody syndrome [126], hyperhomocysteinemia [127], or S-C hemoglobinopathy [128]. However, no further studies have examined the relationship between hypoxia and thrombophilia, and, as a result, no definitive statements can be made about the risk of thromboembolic events in these patients at high altitude.

See Table 23.3 for recommendations regarding patients with underlying coagulopathy during high-altitude travel.

Hemoglobinopathy

High-altitude travel should be avoided in patients with sickle-cell anemia, as hypoxia predisposes to sickling and vaso-occlusive crises [129]. This problem has also been described with pressurized airplane flights or driving over mountain passes as low as 2,500 m [130]. The incidence of problems in patients with sickle-cell trait has not been well established, but two recent reports suggest these patients may also be at risk for splenic crises. Tiernan [131] described two Caucasian men with sickle-cell trait who developed splenic crises

Table 23.3 Recommendations for patients with hematologic disorders

Thromboembolic disease
• There is no need to screen for thrombophilia in asymptomatic persons prior to high-altitude travel
• There is no indication for people with underlying coagulopathy who are not already on anticoagulation to start warfarin or clopidogrel for their high-altitude sojourn. Patients may consider starting an aspirin regimen
• Individuals on preexisting anticoagulation should continue their preexisting regimen at altitude
• Patients on warfarin or dabigatran should limit activities in remote areas to reduce the risk of trauma and severe bleeding
• Individuals on warfarin taking trips of greater than 1 week in duration should monitor prothrombin times during or immediately after their stay [190]
• Make efforts to maintain hydration and adequate mobility
• Patients who develop arterial or venous thrombosis at high altitude should be screened for thrombophilia upon return to their home residence
Hemoglobinopathy
• Patients with sickle-cell disease should avoid travel to high altitude without supplemental oxygen
• Patients with sickle-cell trait may travel to high altitude but should:
– Maintain adequate hydration
– Seek medical attention or descend with the onset of left-upper quadrant abdominal pain or left-sided pleuritic chest pain
Anemia
• Thalassemia seems to cause no complications at high altitude
• Consider packed red blood cell transfusion prior to high-altitude travel for patients with hematocrit <20 %
• Anemic patients planning prolonged sojourns to high altitude (>2 weeks) should consider oral iron supplementation
• Patients on EPO therapy staying at altitude >2 weeks should have close follow-up of hematocrit and EPO dosing
Polycythemia vera
• Patients should maintain adequate hydration and mobility and consider low-dose aspirin to decrease risk of thrombotic events
• PV patients with history of gastrointestinal complications should avoid high-dose aspirin or dexamethasone for management of acute altitude illness
• Patients should monitor hematocrit with high-altitude sojourns lasting more than 2 weeks

at elevations between 2,700 and 2,900 m, while Franklin and Compeggie describe four cases of splenic syndrome that occurred during moderate exercise at altitudes between 1,675 and 3,660 m [132]. Sickle-cell disease patients are also at risk for developing pulmonary hypertension [133], which, as noted above, can predispose to HAPE.

Recommendations for patients with underlying hemoglobinopathy are provided in Table 23.3.

Anemia

Although anemia impairs oxygen delivery and would likely impair exercise performance, there are still no data in the literature to suggest that anemia increases the risk of high-altitude illnesses and no data indicating which hematocrit levels are safe for high-altitude travel. Thalassemia, a common form of anemia worldwide, has not been reported to cause problems at altitude.

Patients using exogenous erythropoietin (EPO) for chronic anemia may need reduced doses at high altitude. Brookhart et al. [134] studied data from over 300,000 dialysis-dependent patients in the USA and found that patients living above 1,800 m received 19 % less EPO and had higher hematocrits than patients living at sea level. These results agreed with those of an older, smaller study documenting decreased EPO requirements, but a higher incidence of thrombosis and hypertension, in patients living at high altitude [135].

Management recommendations for anemic patients are provided in Table 23.3.

Polycythemia Vera

Little is known about the response of individuals with polycythemia vera (PV) to high altitude. While a higher hematocrit could theoretically improve oxygen delivery in acute hypoxia, dehydration due to either inadequate fluid intake or altitude-induced diuresis could lead to increased blood viscosity and worsening symptoms of their disease. While PV is associated with increased risk of thrombosis at sea level [136], it is unknown whether acute hypoxia increases the risk of thromboembolism further. PV patients are also at risk for gastrointestinal complications such as gastroduodenal erosions and peptic ulcer disease [137], which, as discussed further below, could increase the risk of gastrointestinal bleeding at high altitude.

A final question for these individuals is whether acute hypoxia will lead to a rise in serum erythropoietin levels and, therefore, hematocrit, in the days following ascent. These individuals typically have been shown to have low serum erythropoietin levels at baseline [138], but no studies have examined whether these levels change in response to acute hypoxia.

Recommendations for PV patients at high altitude are provided in Table 23.3.

Diabetes Mellitus

Most of the available information on diabetic patients at high altitude derives from studies performed during climbing expeditions to Kilimanjaro [139, 140], Cho Oyu [141], and Aconcagua [142]. These focus exclusively on patients with well-controlled Type I diabetes, and the literature contains little to no information about patients with poorly controlled Type I disease or Type II disease of any severity. As a result, caution must be applied in extrapolating the data to these patient groups.

Based on limited available information, it does not appear that Type I diabetes mellitus impairs well-being, adaptation, or exercise performance at high altitude. None of the reports from the climbing expeditions, for example, reported a difference in AMS incidence or Lake Louise AMS scores between diabetic and nondiabetic climbers [139–141]. A study of a general tourist population at lower elevations between 1,900 and 3,000 m also found no difference in AMS incidence between those with and without either type of diabetes [26]. Healthy Type I diabetic patients also appear to have the capacity for normal ventilatory and hematologic adaptation. Pavan et al. [143], for example, measured arterial blood gases and hematologic parameters at 3,700 or 5,800 m during an ascent of Cho Oyu and noted no differences in P_aO_2, PCO_2, bicarbonate, and hematocrit between the healthy controls and Type I diabetic patients at either elevation. Finally, patients with Type I diabetes do have the capacity to ascend to extreme elevations as evidenced by the fact that in all but one of the expeditions noted above, diabetic patients successfully summited the peak in question with success rates similar to those of the healthy controls [140–142]. In one of the Kilimanjaro expeditions [139], none of the diabetic climbers summited, but it is difficult to attribute this to their diabetes as there was a very low success rate among the healthy controls on this expedition as well.

Beyond adaptation and exercise performance is the question of what happens to insulin requirements and glycemic control with ascent to high altitude. Data on insulin requirements are mixed, as Moore et al. [139] reported reduced insulin requirements during an ascent of Kilimanjaro, while Pavan et al. [141] and Admetlla et al. [142] noted increased requirements during significantly longer expeditions with different ascent rates on Cho Oyu and Aconcagua, respectively. Moore et al. [139] reported that about 50 % of the measured glucose values during their Kilimanjaro expedition fell outside the target range of 6–14 mmol/L (108–252 mg/dL), while 4 of the 15 climbers developed ketonuria. Pavan et al. [141] also reported an increase in hemoglobin A1C levels from 7 to 7.9 % over the course of a 5-week expedition on Cho Oyu. Whether hemoglobin A1C is elevated on high-altitude sojourns of shorter durations is not known. In addition, while hyperglycemia is a major concern, episodes of hypoglycemia have also been noted in several of these reports [139, 142].

Avoiding problems with hyper- and hypoglycemia at high altitude is made more difficult by the fact that it is unclear how well glucometers function this environment. Many early studies reported problems with glucometer accuracy in hypoxia [139, 144–147] with only a single report noting good reliability of glucometer measurements without adequate documentation to support this claim [142]. More recent studies suggest that later generation monitors may perform better than earlier models. In particular, glucose dehydrogenase-based systems may perform better than glucose oxidase systems because the reaction pathway in the former does not involve oxygen [148, 149].

Increasing numbers of diabetic patients are using insulin pumps rather than intermittent subcutaneous injections in order to regulate their blood glucose levels. Only a single study has examined how well these devices function in hypobaric hypoxia. King et al. [150] examined ten insulin pumps during airplane flight with a 200 mmHg decrease in barometric pressure as well as in a hypobaric chamber and noted that bubbles formed and expanded in the pumps tested in the hypobaric chamber. Excess insulin administration was noted to occur during the airplane flight, a finding the authors attributed to bubble formation and expansion which suggests that diabetic patients using pumps at high altitude may need to reduce their basal insulin rates or rely on intermittent injections when traveling at high altitude.

Finally, it is reasonable to question if diabetic patients are at increased risk for retinal hemorrhage at high altitude, which commonly affect normal individuals on trips to extreme altitude. Moore et al. [139] used ophthalmoscopy and noted asymptomatic retinal hemorrhages in 2 of 15 diabetic climbers, one of whom had preexisting retinopathy, during an ascent of Kilimanjaro, while Leal et al. [151] performed retinography on 7 diabetic climbers ascending a 7,100-m peak and noted the development of asymptomatic hemorrhages in only one climber, who also had preexisting diabetic retinopathy. The latter study may have underestimated the incidence of small hemorrhages, however, because 5 of the post-ascent studies were performed 2 weeks after the expedition. Overall, given the small number of patients in these studies, particularly those with preexisting retinopathy, it is difficult to make broad claims about the risk of retinal hemorrhage in all diabetic patients. Exposure to high altitude should be avoided, however, in any patient with severe diabetic retinopathy [152].

For those with less severe retinopathy who travel to high altitude, the question may arise as to whether use of aspirin or other nonsteroidal anti-inflammatory medications for headache or AMS symptoms increases the risk of retinal hemorrhage. The risk does not appear to be increased at sea level [153, 154], but this question has not been studied at high altitude and it likely best to rely on acetaminophen for those with history of retinopathy.

Recommendations for managing diabetic patients before and during high-altitude travel are provided in Table 23.4.

Table 23.4 Recommendations for patients with diabetes mellitus

- Increase frequency of glucose monitoring following ascent to high altitude
  - Due to concerns about glucometer performance, patients should not seek overly strict glucose control in early stages of their sojourn
- Pretravel assessment is warranted to identify comorbid conditions (e.g., coronary artery disease) that could predispose to problems at high altitude
- Patients should use more recent generation monitors that assess blood glucose levels based on the glucose dehydrogenase reaction
- Patients using insulin pumps should consider decreasing their basal insulin rate or switching to intermittent subcutaneous injections during their high-altitude stay until they learn more about how their pump functions at high altitude
- Screening retinal exams should be considered before travel to sleeping elevations >4,000 m and travel avoided in patients with ischemic or proliferative retinopathy
- Patients with diabetic retinopathy should avoid aspirin for treatment of headache or acute mountain sickness
- Only well-controlled diabetics experienced with exercise at sea level should engage in vigorous exercise following ascent
- Patients who opt for pharmacologic prophylaxis for AMS should avoid dexamethasone [184]

Obesity

With the rising incidence of obesity in the general population, it is worthwhile to consider their possible risks at high altitude. Two studies showed that obese individuals may be at greater risk for developing AMS. Ri-Li et al. [155] exposed 9 obese and non-obese men to a simulated altitude of 3,658 m for 24 h and noted that obese men had a higher incidence of AMS and lower oxygen saturation values during sleep than the non-obese men. Although this was a short duration chamber study, these results are in agreement with those of Honigman et al. [26] who found a higher incidence of AMS among obese members of a general tourist population traveling to altitudes between 1,920 and 2,950 m. While these studies are highly suggestive of a link between obesity and AMS, no prospective studies have been conducted to further investigate this issue. It is also not clear from the existing studies the degree of obesity at which risk might increase.

In addition to concerns about AMS, there are also theoretical reasons why very obese individuals may be at increased risk for developing HAPE or other complications. These individuals often develop obesity hypoventilation syndrome, a disorder of ventilatory control that can result in pulmonary hypertension and right heart failure [156]. Given the large number of case reports documenting the occurrence of HAPE in individuals with various forms of pulmonary hypertension [5], there is concern that patients with obesity hypoventilation and underlying pulmonary hypertension would be at risk for this as well. The greater degree of alveolar hypoxia at high altitude may also trigger a further rise in pulmonary artery pressures and induce right heart dysfunction, worsening dyspnea or ischemia [64].

Table 23.5 summarizes recommendations for the management of obese patients before and during high-altitude travel.

Table 23.5 Recommendations for obese patients

• Obese patients should be counseled about the importance of slow ascent, acclimatization rules, and recognition of altitude illness
• Individuals with obesity hypoventilation syndrome and/or pulmonary hypertension should avoid travel to high altitude. If such travel is necessary:
– Individuals should travel with supplemental oxygen or a prescription that can be filled at high altitude
– Individuals should monitor their oxygen saturation periodically during their stay at high altitude
• Individuals on CPAP therapy for management of obstructive sleep apnea should continue such therapy at high altitude if access to power can be assured

Gastrointestinal Diseases

Little information is available to provide guidance for those with underlying gastrointestinal disorders such as prior gastrointestinal (GI) bleeding or liver disease.

Gastrointestinal Bleeding

Reports from construction of the Qinghai-Tibet Railroad suggest an increased risk of gastrointestinal bleeding at altitude. Wu et al. [157] found a 0.49 % incidence of hematemesis, melena, or hematochezia among 13,502 railroad workers between 3,500 and 4,900 m. Eighty-four percent of cases occurred within 3 weeks of arrival at high altitude, and the greatest incidence was seen at the highest elevations. Endoscopy performed on all affected individuals revealed gastric and duodenal ulcers, gastric erosions, and hemorrhagic gastritis. The authors found that concurrent use of alcohol, aspirin, and dexamethasone were risk factors for gastrointestinal bleeding, but this was not based on formal statistical analysis. A few other reports have also documented GI bleeding at high altitude. Saito [158], for example, noted upper gastrointestinal bleeding in 5 of 52 Mt. Everest climbers, while Liu [159] reported that among Chinese soldiers stationed between 3,700 m and 5,380 m for 1 year, the incidence of GI bleeding was 0.8 % among all patients and 1.5 % among those patients with AMS. How this incidence rate compares to a similar population at sea level is not clear.

More recently, Fruehauf et al. [160] performed endoscopy on 26 asymptomatic climbers before and after ascent to 4,559 m over 22 h and noted gastric or duodenal erosions/ulcers, hemorrhagic gastritis or duodenitis, and reflux esophagitis in 28 % of individuals on day 2 and 61 % of individuals on day 4 at high altitude. None of these climbers had abnormalities on their pre-ascent endoscopy, and none experienced active gastrointestinal bleeding while at 4,559 m. This study and the others noted above do not establish a definitive link between hypoxia and gastrointestinal bleeding, but do suggest that patients with poorly controlled esophagitis, gastritis, or peptic ulcer disease or those taking long courses aspirin or other nonsteroidal anti-inflammatory medications may be at risk for problems at high altitude.

Table 23.6 Recommendations for patients with gastrointestinal disease

Gastrointestinal bleeding
• No data indicate that travelers with history of upper GI bleeding should use ulcer prophylaxis during a high-altitude sojourn
• Travelers with history of upper GI bleeding should avoid nonsteroidal anti-inflammatory medications and aspirin and rely on acetaminophen to treat headaches or musculoskeletal pain
• Travelers with history of upper GI bleeding who opt to use pharmacologic prophylaxis for AMS should avoid dexamethasone [191, 192]
Chronic liver disease
• Chronic liver disease patients should be screened for portopulmonary hypertension and hepatopulmonary syndrome, and high-altitude travel should be avoided in people with those conditions
• Portopulmonary hypertension patients who must travel to high altitude should travel with supplemental oxygen and use nifedipine or a phosphodiesterase inhibitor
• Hepatopulmonary syndrome patients who must travel to high altitude should monitor oxygen saturation following ascent [184]
• Acetazolamide is contraindicated in chronic liver disease due to increased risk of hepatic encephalopathy [184]

Chronic Liver Disease

Although no formal studies have addressed the risk of high-altitude travel in patients with chronic liver disease, strong theoretical reasons why two groups of liver disease patients might have problems should be considered. Hepatopulmonary syndrome is marked by intrapulmonary shunts and affects up to 47 % of patients with cirrhosis [161]. These individuals have increased alveolar-arterial oxygen difference at baseline, which worsens with upright positioning. Given that intrapulmonary shunting is the reason for baseline hypoxemia, one would predict exaggerated hypoxemia at high altitude, which would not improve with supplemental oxygen. Second, portopulmonary hypertension is a form of pulmonary arterial hypertension that occurs in up to16% of patients with cirrhosis and portal hypertension [162]. Given the numerous case reports and case series suggesting that patients with underlying pulmonary hypertension are at risk for HAPE at high altitude, there is reason to suspect portopulmonary hypertension patients might be at risk as well. No cases have been documented in the literature thus far, but Bogaard et al. [163] have reported an individual who developed hemoptysis and dyspnea at 2,700 m and was found upon further evaluation at sea level to have pulmonary hypertension (PA pressure 80/30 on right heart catheterization) and Abernethy malformation, a form of persistent portosystemic shunt. Chest imaging was not obtained at high altitude but the history is highly suggestive of HAPE.

Recommendations for the management of patients with gastrointestinal bleeding or chronic liver disease are provided in Table 23.6. Recommendations regarding the choice and dose of medication for prophylaxis or treatment of acute altitude illness are provided in Table 23.10.

Chronic Kidney Disease

A number of theoretical concerns should be considered in patients with chronic kidney disease who want to travel to high altitude. Renal insufficiency, for example, impairs urinary concentration and dilution capacity, and thus, these patients may be at risk for volume depletion or volume overload. In one of the few studies of kidney patients at high altitude, Mairbaurl et al. [164] demonstrated that dialysis-dependent patients had greater weight gain between dialysis sessions at 2,000 m compared to 576 m. Volume overload may increase risk for pulmonary edema or AMS.

Chronic kidney disease patients may have blunted erythropoietic responses to high altitude. Several studies have, in fact, shown little to no change in hemoglobin concentration, EPO production, and reticulocyte count over 2 weeks at altitudes between 2,000 and 4,600 m [165–167]. Interestingly, despite the blunted erythropoietic responses at high altitude, chronic kidney disease patients on exogenous EPO therapy require lower doses than at sea level and may be at risk for more thrombotic events and hypertension [134, 135].

While low hematocrit might be tolerated at low elevation, the blunted hematologic response at high altitude could decrease oxygen delivery and limit exercise capacity. At the same time, many patients with chronic kidney disease have a metabolic acidosis, which may stimulate minute ventilation, increase alveolar and arterial PO_2, and theoretically protect against AMS. This issue has not been systematically studied.

Finally, 40 % of patients with chronic kidney disease have mild to moderate pulmonary hypertension [168], which, as noted above, may predispose to HAPE, particularly if individuals also have impaired ability to make urine and regulate their volume status.

Identifying which patients are at risk for these problems based solely on creatinine clearance or glomerular filtration rate (GFR) is difficult as the degree of impaired concentrating or diluting capacity, metabolic acidosis, or anemia will vary based on the type of disease (glomerular versus interstitial), the patient's medication regimen, and other factors. Nevertheless, as patients approach Stage IV (GFR 15–30 mL/min) or Stage V (GFR < 15 mL/min) chronic kidney disease, these problems are likely to increase and caution should be advised for high-altitude travel.

Table 23.7 Recommendations for patients with chronic kidney disease

- No clear indications for routine pharmacologic prophylaxis against high-altitude illness in patients with chronic kidney disease
- When pharmacologic prophylaxis is indicated, selection and dose of medications should be adjusted based on renal function (Table 23.10) [184]
- Patients should monitor body weight closely and travel with a prearranged plan for altering diuretic regimen for weight gain or fluid retention
- Patients on chronic EPO therapy should have close follow-up of their hematocrit and EPO dosing with prolonged high-altitude sojourns (>2–4 weeks)

Recommendations for the management of chronic kidney disease patients at high altitude are summarized in Table 23.7 and have also been reviewed elsewhere [169]. Recommendations for medication choices and dose adjustments in the management of acute altitude illness are provided in Table 23.10.

Ophthalmological Conditions

Surgery to Correct Refractive Errors

The risk of high-altitude travel in patients who have undergone surgery to correct refractive errors will vary based on the particular procedure. Patients with a history of radial keratotomy (RK) may be at particular risk. Hypoxia of high altitude causes uneven swelling of the cornea in post-RK eyes, which, in turn, leads to hyperopic (far-sighted) shifts that can markedly impair vision. These changes may occur at altitudes as low as 3,000 m and generally develop after more than 24 h at a given altitude [152]. Further studies suggest that an alternative correction procedure, photorefractive keratectomy (PRK), is much better for patients going to high altitude.

Table 23.8 Recommendations for patients with or planning refractive surgery

Individuals planning refractive surgery prior to high-altitude travel should opt for LASIK or PRK rather than RK
Individuals with history of RK traveling to altitudes >5,000 m should travel with glasses with increasing plus power to compensate for hyperopic shifts

Hypoxia also causes corneal swelling in PRK-treated eyes, but the swelling is uniform throughout the cornea and, as a result, no refractive changes occur [170].

RK and PRK have largely been supplanted by another refractive procedure, laser-assisted in situ keratomileusis (LASIK), and the available data suggest that although some blurring of vision may occur in hypoxia, ascent to extreme elevations may still be tolerated following this procedure. In a controlled study using goggles to simulate a hypoxic environment around the eyes, Nelson et al. [171] demonstrated small myopic shifts in LASIK corneas. Several case reports [172, 173] have documented similar myopic shifts at altitudes above 5,500 m (~18,000 ft), although the climbers in each of these reports still reached their summit objectives. Dimmig and Tabin [174] reported subjective visual experiences on 12 post-LASIK eyes in 6 Mount Everest climbers who all ascended above 7,900 m, with 4 of them reaching the 8,850 m summit. Five of the six climbers had no visual changes up to 8,000 m. Three summiteers reported perfect vision on the summit, while two reported blurred vision at 8,200 and 8,700 m, respectively, which improved with descent. The time between LASIK surgery and the climb does not appear to have played a role in these cases, as the climbs were from 6 weeks to 3 years after surgery. Climbers with history of LASIK or PRK should be aware that both procedures may cause or exacerbate dry eye problems. Since the lower humidity could lead to contact lens intolerance [152], this issue should be addressed prior to travel to high altitude.

Recommendations for patients with a history of or planning to have refractive surgery are provided in Table 23.8. Information about travel with other ocular disorders at high altitude has been reviewed elsewhere [152].

Intraocular Pressure and Glaucoma

Variable results have been reported regarding the effect of high altitude on intraocular pressure with studies reporting an increase, decrease, or no change [175–177]. The reasons for these discrepant results are unclear but may result from failure of most studies to correct for changes in corneal thickness or account for the effects of cold and exercise, differences in measurement techniques, and the altitudes at which the studies were conducted. Somner et al. [178] corrected for many of these issues in their study of 76 individuals who ascended to 5,200 m without exertion and documented a statistically significant, but clinically insignificant increase in IOP from 11.4 mmHg at sea level to 12.4 mmHg at 5,200 m. IOP eventually decreased below sea level values by day 7, and there was no relationship noted between IOP and development of AMS or high-altitude retinopathy.

There is no evidence that high-altitude exposure provokes episodes of acute narrow angle glaucoma or worsens open angle glaucoma. However, hypoxia at high altitude could increase optic nerve damage in patients whose IOP is elevated at the time of their sojourn. Commonly used to prevent acute altitude illness, acetazolamide may have the added benefit in these patients of decreasing intraocular fluid production and thereby helping limit any rise in IOP [152].

Neurologic Conditions

A variety of common neurologic conditions may affect travelers to high altitude. A complete discussion of all the potential disorders is available elsewhere [1, 179], while some of the more common underlying disorders, headaches and seizures, are considered below.

Headaches

Individuals who experience frequent headaches at sea level are not predisposed to headaches following ascent to high altitude but may be at risk for increased severity of headaches when they do, in fact, occur [180]. In addition to the fact that hypoxia may trigger migraine headaches in people with a known history of the disorder [181], both anecdotal reports [1] and more systematic studies [182] suggest that a history of migraine headaches may predispose to acute altitude illness following ascent. Migraine attacks may also be more severe at high altitude and may be accompanied by focal neurologic deficits [183]. It may be difficult to distinguish a migraine headache from the headache associated with AMS, but the presence of aura, resemblance of symptoms to those of migraines at sea level, and the presence of focal neurologic deficits would suggest the individual is experiencing a migraine headache and may, as a result, require different treatment than would be used for AMS.

Seizures

People taking medications for known seizure disorder are not at risk for increased frequency or severity of seizures following ascent to high altitude, although there are several unpublished reports of seizures occurring in individuals with a remote history of seizures or who were subsequently diagnosed with an underlying seizure disorder following their return to sea level [179]. Patients with well-controlled seizure disorders can trek or do other activities at high altitude but should avoid technical climbing due to the risk posed to the climber and his/her partner should a seizure occur while on the climbing route. Patients using topiramate for seizure prophylaxis should avoid concurrent use of acetazolamide as topiramate also has carbonic anhydrase activity and combined use of the medications can result in nephrolithiasis [184].

Recommendations for the management of various forms of neurologic disease at high altitude are summarized in Table 23.9.

Table 23.9 Recommendations for patients with neurologic disorders

Migraine headaches
• Individuals should continue their normal migraine prophylaxis
• Headaches that are different in character than normal migraines or do not respond to standard treatment should be treated as AMS
• A trial of oxygen breathing may help distinguish migraine from high-altitude headache
Cerebrovascular diseases
• Patients with recent transient ischemic attack or cerebrovascular accident (<90 days) should avoid high altitude
• Patients on warfarin, dabigatran, or clopidogrel should limit activities in remote areas to reduce the risk of trauma and severe bleeding
• Patients on warfarin therapy planning a prolonged sojourn (>1 weeks) should obtain follow-up prothrombin times during or immediately following their stay at altitude
• Patients with known, unsecured intracranial aneurysms or AVM should avoid exertion at high altitude
Seizures
• Patients on preexisting antiseizure medications should continue those medications at high altitude
• Patients not currently on antiseizure medications should consider restarting prior medications during a prolonged high-altitude sojourn (>2 weeks) due to theoretical concerns about unmasking seizure disorders
• Patients should limit concurrent use of acetazolamide and topiramate to 3–5 days [184]

Raynaud's Phenomenon and Collagen Vascular Diseases

Raynaud's phenomenon is a disorder of vasomotor control associated with episodic pallor or cyanosis of distal extremities due to vasospasm and limitations in arterial blood flow which occurs either in isolation (primary Raynaud's phenomenon) or in association with collagen vascular diseases such as scleroderma. Given the pathophysiology of the disorder, concern has been raised about the safety of high-altitude travel in Raynaud's phenomenon patients, particularly when such travel is associated with cold exposure. Some sources [185] warn about the possibility of increased frequency and severity of attacks in this environment, but data supporting this claim are limited.

A recent study [186] examined this question in greater detail by surveying 142 people with Raynaud's phenomenon who travel to high altitude during the winter and summer months. Respondents reported spending between 5 and 7 days per month above 2,440 m (8,000 ft) and engaging in a wide variety of activities. Eighty-nine percent of respondents, in fact, reported participation in winter sports and only 22 % reported changing their pattern of activities at high altitude due to their Raynaud's phenomenon. There was considerable heterogeneity in the respondents' perceptions of the frequency, duration, and severity of attacks at high altitude compared to their home elevation, but the study suggests that by using a variety of preventive and treatment strategies, these individuals are able to tolerate travel to and activity in this environment. Ninety-eight percent of the respondents in this survey had primary Raynaud's phenomenon, however, and the conclusions cannot be extended to those with secondary disease. In the study, 15 % of respondents also reported a history of frostbite at high altitude. There was no control group to determine if this rate is increased relative to the general population exposed to similar conditions. In addition, the data are limited by the fact that it was based on self-report; nonetheless, this study suggests that Raynaud's patients must be vigilant about preventing this problem in high and cold environments.

There are no data regarding the risk of high-altitude travel in other forms of collagen vascular diseases such as scleroderma or systemic lupus erythematosus.

Pregnancy

The safety of travel to altitude during pregnancy has been reviewed elsewhere [1, 187]. This question has not been extensively studied but the available evidence and clinical experience suggest that travel to modest elevations of 3,000 m is safe and well tolerated at all stages of uncomplicated pregnancy. Travel into remote areas and exertion at levels greater than those done at home should be avoided, as should travel to high altitude with high-risk pregnancies, such as hypertension or placental insufficiency. Care should be taken to avoid altitude illness as worsening hypoxemia during HAPE, for example, could lead to more clinically significant fetal hypoxemia. Slow ascent is preferred to pharmacologic prophylaxis for this purpose, but altitude illness medications can be used if truly necessary. Recommendations for travel to high altitude during pregnancy are provided in Table 23.10, while the safety of altitude illness medications in pregnant or lactating women is described in Table 23.13.

Table 23.10 Recommendations for pregnant patients

Patients should have a checkup, including possible ultrasound, to ensure that the pregnancy is low risk
High-altitude travel should be avoided in any complicated or high-risk pregnancy
Sleeping altitudes >3,000 m should probably be avoided
Avoidance of altitude illness through proper acclimatization is important
Patients should avoid trauma during skiing, cycling, climbing, or other activities
Patients should exercise at levels lower than at home and avoid overexertion
Patients should maintain adequate hydration
Patients should avoid travel into remote areas

Preexisting Medical Conditions and Altitude Illness Medications

A variety of medications including acetazolamide, dexamethasone, nifedipine, phosphodiesterase inhibitors, and salmeterol are used in the prevention and treatment of high-altitude illness. It is noteworthy that the efficacy and safety of these medications has largely been established based on studies with only healthy individuals. Many underlying medical conditions, however, affect medication choices or cause potential adverse drug interactions. For example, patients with liver disease should avoid acetazolamide, while patients with chronic kidney disease may need a lower dose of the medication or may need to use dexamethasone for AMS prevention, depending on their creatinine clearance. A full discussion of these medication issues is provided elsewhere [184], and a summary of these recommendations is provided in Tables 23.11 and 23.12. The safety of these medications in pregnant women is described in Table 23.13.

Table 23.11 Dose adjustments for altitude illness medications in patients with underlying renal and liver disease

Medication	Renal insufficiency	Hepatic insufficiency
Acetazolamide	Avoid use in patients with GFR < 10 mL/min, metabolic acidosis, hypokalemia, hypercalcemia, and hyperphosphatemia or recurrent nephrolithiasis	Acetazolamide use is contraindicated
Dexamethasone	No contraindication and no dose adjustments necessary	No contraindication and no dose adjustments necessary
Nifedipine	No contraindication and no dose adjustments necessary	Best to avoid. If use is necessary, give at reduced dose (10 mg BID)
Tadalafil	Dose adjustments necessary if GFR < 50 mL/min; If GFR 30–50 mL/min, use 5 mg dose, maximum 10 mg in 48 h; If GFR < 30 mL/min, no more than 5 mg	Child's Class A and B: maximum 10 mg daily
		Child's Class C: Do not use tadalafil
Sildenafil	Dose adjustments necessary if GFR < 30 mL/min	Dose reductions recommended. Starting dose 25 mg TID
		Avoid use in patients with known esophageal or gastric varices
Salmeterol	No contraindication and no dose adjustments necessary	Insufficient data. Best to avoid the medication in these patients

Table 23.12 Important considerations in the selection of altitude illness medications for patients with underlying medical issues

Medication	Other major dosing issues
Acetazolamide	Avoid in patients on chronically high doses of aspirin
	Avoid in patients with ventilatory limitation (FEV_1 < 25 % predicted)
	Caution in patients with documented severe sulfa allergy
	Limit concurrent use with topiramate and ophthalmic carbonic anhydrase inhibitors to 3–5 days
Dexamethasone	Expect elevated blood glucose values when used in diabetic patients
	Avoid in patients at risk for peptic ulcer disease or upper gastrointestinal bleeding
	Caution in patients at risk for amoebiasis or strongyloidiasis
Nifedipine	Caution in patients taking medications metabolized by CytP450 3A4 and 1A2 pathways
	Caution during concurrent use with other antihypertensive medications
Tadalafil and sildenafil	Increased risk of gastroesophageal reflux
	Caution in patients taking medications metabolized by CytP450 3A4 pathway
	Avoid concurrent use of nitrates or alpha-blockers
Salmeterol	Potential for adverse effects in patients with coronary artery disease prone to arrhythmia
	Avoid concurrent use of beta-blockers
	Avoid concurrent use of monoamine oxidase inhibitors or tricyclic antidepressants

Table 23.13 Use of high-altitude medications in pregnant or lactating women

Medication	Safety during pregnancy	Safety during breast-feeding
Acetazolamide	Category C	Not established
Dexamethasone	Category C	Debate about safety[a]
Nifedipine	Category C	Compatible
Tadalafil	Category B	Not established
Sildenafil	Category B	Not established
Salmeterol	Category C	Not established

Pregnancy Category B: presumed safe based on studies in animals but no data from humans
Pregnancy Category C: no human studies demonstrating harm but animal studies show evidence of teratogenicity
[a]Some sources claiming compatibility with breast-feeding [193], while others recommend against use in this situation [194]

Conclusions

With the large number of people traveling to high altitude, clinicians are increasingly being asked to evaluate the safety of such travel in patients with underlying medical conditions. A slowly growing body of literature provides some insight into how to approach patients with underlying conditions but significant gaps in our knowledge remain, and considerably more research is warranted to clarify the risks and appropriate management in these travelers. Until such information is available, a careful approach emphasizing the current data and an understanding of the pathophysiology of the underlying disease and its interactions with the hypoxia-induced physiologic changes at high altitude will be vital to ensure a safe sojourn for the aspiring traveler.

References

1. Hackett P. High altitude and common medical conditions. In: Hornbein TF, Schoene RB, editors. High altitude: an exploration of human adaptation. New York: Marcel Dekker; 2001. p. 839–85.
2. Bartsch P, Saltin B. General introduction to altitude adaptation and mountain sickness. Scand J Med Sci Sports. 2008;18 Suppl 1:1–10. Epub 2008/09/09.
3. Barer GR, et al. Stimulus–response curves for the pulmonary vascular bed to hypoxia and hypercapnia. J Physiol. 1970;211(1):139–55.
4. Luks AM, Swenson ER. Pulse oximetry at high altitude. High Alt Med Biol. 2011;12(2):109–19. Epub 2011/07/02.
5. Luks AM, Swenson ER. Travel to high altitude with pre-existing lung disease. Eur Respir J. 2007;29(4): 770–92.
6. Graham WG, Houston CS. Short-term adaptation to moderate altitude. Patients with chronic obstructive pulmonary disease. JAMA. 1978;240(14):1491–4.
7. Akero A, et al. Hypoxaemia in chronic obstructive pulmonary disease patients during a commercial flight. Eur Respir J. 2005;25(4):725–30.
8. Christensen CC, et al. Development of severe hypoxaemia in chronic obstructive pulmonary disease patients at 2,438 m (8,000 ft) altitude. Eur Respir J. 2000;15(4):635–9.
9. Seccombe LM, et al. Effect of simulated commercial flight on oxygenation in patients with interstitial lung disease and chronic obstructive pulmonary disease. Thorax. 2004;59(11):966–70.
10. Berg BW, et al. Oxygen supplementation during air travel in patients with chronic obstructive lung disease. Chest. 1992;101(3):638–41.
11. Standards for the diagnosis and care of patients with chronic obstructive pulmonary disease. Am J Respir Crit Care Med. 1995;152:S112–S3.
12. Aerospace Medical Association Medical Guidelines Task Force. Medical Guidelines for Airline Travel, 2nd ed. Aviat Space Environ Med. 2003;74(5 Suppl): A1–19.
13. Gong Jr H, et al. Hypoxia-altitude simulation test. Evaluation of patients with chronic airway obstruction. Am Rev Respir Dis. 1984;130(6):980–6.
14. Dillard TA, et al. Hypoxemia during air travel in patients with chronic obstructive pulmonary disease. Ann Intern Med. 1989;111(5):362–7.
15. Akero A, et al. Pulse oximetry in the preflight evaluation of patients with chronic obstructive pulmonary disease. Aviat Space Environ Med. 2008;79(5): 518–24.
16. Luks AM. Do lung disease patients need supplemental oxygen at high altitude? High Alt Med Biol. 2009;10(4):321–7.
17. Finkelstein S, et al. Pulmonary mechanics at altitude in normal and obstructive lung disease patients. Aerosp Med. 1965;36:880–4.
18. Dillard TA, et al. Lung function during moderate hypobaric hypoxia in normal subjects and patients with chronic obstructive pulmonary disease. Aviat Space Environ Med. 1998;69(10):979–85.
19. Astin TW, Penman RW. Airway obstruction due to hypoxemia in patients with chronic lung disease. Am Rev Respir Dis. 1967;95(4):567–75.
20. Libby DM, et al. Relief of hypoxia-related bronchoconstriction by breathing 30 per cent oxygen. Am Rev Respir Dis. 1981;123(2):171–5.
21. Tomashefski JF, et al. Effects of altitude on emphysematous blebs and bullae. Aerosp Med. 1966; 37(11):1158–62.
22. Yanda RL, Herschensohn HL. Changes in lung volumes of emphysema patients upon short exposures to simulated altitude of 18,000 feet. Aerosp Med. 1964;35:1201–3.
23. Christensen CC, et al. Effect of hypobaric hypoxia on blood gases in patients with restrictive lung disease. Eur Respir J. 2002;20(2):300–5.
24. Burtscher M, et al. Prediction of susceptibility to acute mountain sickness by SaO2 values during short-term exposure to hypoxia. High Alt Med Biol. 2004;5(3):335–40.
25. Roach RC, et al. Arterial oxygen saturation for prediction of acute mountain sickness. Aviat Space Environ Med. 1998;69(12):1182–5.
26. Honigman B, et al. Acute mountain sickness in a general tourist population at moderate altitudes. Ann Intern Med. 1993;118(8):587–92.
27. Roach RC, et al. How well do older persons tolerate moderate altitude? West J Med. 1995;162(1):32–6.
28. Spieksma FT, et al. High altitude and house-dust mites. Br Med J. 1971;1(740):82–4.

29. Boner AL, et al. Influence of allergen avoidance at high altitude on serum markers of eosinophil activation in children with allergic asthma. Clin Exp Allergy. 1993;23(12):1021–6.
30. van Velzen E, et al. Effect of allergen avoidance at high altitude on direct and indirect bronchial hyperresponsiveness and markers of inflammation in children with allergic asthma. Thorax. 1996;51(6):582–4.
31. Valletta EA, et al. Peak expiratory flow variation and bronchial hyperresponsiveness in asthmatic children during periods of antigen avoidance and reexposure. Allergy. 1995;50(4):366–9.
32. Denjean A, et al. Mild isocapnic hypoxia enhances the bronchial response to methacholine in asthmatic subjects. Am Rev Respir Dis. 1988; 138(4):789–93.
33. van den Elshout FJ, et al. Effects of hypercapnia and hypocapnia on respiratory resistance in normal and asthmatic subjects. Thorax. 1991;46(1):28–32.
34. Kaminsky DA, et al. Peripheral airways responsiveness to cool, dry air in normal and asthmatic individuals. Am J Respir Crit Care Med. 1995;152(6 Pt 1):1784–90.
35. Larsson K, et al. High prevalence of asthma in cross country skiers. BMJ. 1993;307(6915):1326–9.
36. Pohjantahti H, et al. Exercise-induced bronchospasm among healthy elite cross country skiers and non-athletic students. Scand J Med Sci Sports. 2005;15(5):324–8.
37. Durand F, et al. Undiagnosed exercise-induced bronchoconstriction in ski-mountaineers. Int J Sports Med. 2005;26(3):233–7.
38. Golan Y, et al. Asthma in adventure travelers: a prospective study evaluating the occurrence and risk factors for acute exacerbations. Arch Intern Med. 2002;162(21):2421–6.
39. Louie D, Pare PD. Physiological changes at altitude in nonasthmatic and asthmatic subjects. Can Respir J. 2004;11(3):197–9.
40. Cogo A, et al. Bronchial asthma and airway hyperresponsiveness at high altitude. Respiration. 1997; 64(6):444–9.
41. Allegra L, et al. High altitude exposure reduces bronchial responsiveness to hypo-osmolar aerosol in lowland asthmatics. Eur Respir J. 1995;8(11):1842–6.
42. Stokes S, et al. Bronchial asthma on Mount Kilimanjaro is not a disadvantage. Thorax. 2008;63(10):936–7.
43. Huismans HK, et al. Asthma in patients climbing to high and extreme altitudes in the Tibetan Everest region. J Asthma. 2010;47(6):614–9. Epub 2010/07/17.
44. Fischer R, et al. Lung function in adults with cystic fibrosis at altitude: impact on air travel. Eur Respir J. 2005;25(4):718–24.
45. Thews O, et al. Respiratory function and blood gas variables in cystic fibrosis patients during reduced environmental pressure. Eur J Appl Physiol. 2004; 92(4–5):493–7.
46. Rose DM, et al. Blood gas-analyses in patients with cystic fibrosis to estimate hypoxemia during exposure to high altitudes in a hypobaric-chamber. Eur J Med Res. 2000;5(1):9–12.
47. Ryujin DT, et al. Oxygen saturation in adult cystic fibrosis patients during exercise at high altitude. Pediatr Pulmonol. 2001;32(6):437–41.
48. Oades PJ, et al. Prediction of hypoxaemia at high altitude in children with cystic fibrosis. BMJ. 1994;308(6920):15–8.
49. Kamin W, et al. Predicting hypoxia in cystic fibrosis patients during exposure to high altitudes. J Cyst Fibros. 2006;5:223–8.
50. Speechly-Dick ME, et al. Exacerbations of cystic fibrosis after holidays at high altitude–a cautionary tale. Respir Med. 1992;86(1):55–6.
51. Swenson ER, et al. Pathogenesis of high-altitude pulmonary edema: inflammation is not an etiologic factor. JAMA. 2002;287(17):2228–35.
52. Maggiorini M, et al. High-altitude pulmonary edema is initially caused by an increase in capillary pressure. Circulation. 2001;103(16):2078–83.
53. Hackett PH, et al. High-altitude pulmonary edema in persons without the right pulmonary artery. N Engl J Med. 1980;302(19):1070–3.
54. Rios B, et al. High-altitude pulmonary edema with absent right pulmonary artery. Pediatrics. 1985; 75(2):314–7.
55. Torrington KG. Recurrent high-altitude illness associated with right pulmonary artery occlusion from granulomatous mediastinitis. Chest. 1989;96(6):1422–4.
56. Nakagawa S, et al. High-altitude pulmonary edema with pulmonary thromboembolism. Chest. 1993; 103(3):948–50.
57. Naeije R, et al. High-altitude pulmonary edema with primary pulmonary hypertension. Chest. 1996; 110(1):286–9.
58. Durmowicz AG. Pulmonary edema in 6 children with Down syndrome during travel to moderate altitudes. Pediatrics. 2001;108(2):443–7.
59. Das BB, et al. High-altitude pulmonary edema in children with underlying cardiopulmonary disorders and pulmonary hypertension living at altitude. Arch Pediatr Adolesc Med. 2004;158(12):1170–6.
60. Wu T. A Tibetan with chronic mountain sickness followed by high altitude pulmonary edema on reentry. High Alt Med Biol. 2004;5(2):190–4.
61. Grunig E, et al. Stress Doppler echocardiography in relatives of patients with idiopathic and familial pulmonary arterial hypertension: results of a multicenter European analysis of pulmonary artery pressure response to exercise and hypoxia. Circulation. 2009;119(13):1747–57. Epub 2009/03/25.
62. Rhodes CJ, et al. Iron deficiency in pulmonary arterial hypertension: a potential therapeutic target. Eur Respir J. 2011;38(6):1453–60. Epub 2011/04/12.
63. Smith TG, et al. Effects of iron supplementation and depletion on hypoxic pulmonary hypertension: two randomized controlled trials. JAMA. 2009;302(13): 1444–50.
64. Toff NJ. Hazards of air travel for the obese: Miss Pickwick and the Boeing 747. J R Coll Physicians Lond. 1993;27(4):375–6.
65. Noble JS, Davidson JA. Cor pulmonale presenting in a patient with congenital kyphoscoliosis following

intercontinental air travel. Anaesthesia. 1999; 54(4):361–3.
66. American Thoracic Society; American College of Chest Physicians. ATS/ACCP Statement on cardiopulmonary exercise testing. Am J Respir Crit Care Med. 2003;167(2):211–77.
67. Smith G, et al. Safety of maximal cardiopulmonary exercise testing in pediatric patients with pulmonary hypertension. Chest. 2009;135(5):1209–14.
68. Burgess KR, et al. Effect of simulated altitude during sleep on moderate-severity OSA. Respirology. 2006;11(1):62–9.
69. Burgess KR, et al. Central and obstructive sleep apnoea during ascent to high altitude. Respirology. 2004;9(2):222–9.
70. Nussbaumer-Ochsner Y, et al. Patients with obstructive sleep apnea syndrome benefit from acetazolamide during an altitude sojourn: a randomized, placebo-controlled, double-blind trial. Chest. 2012;141(1):131–8. Epub 2011/06/11.
71. Nussbaumer-Ochsner Y, et al. Exacerbation of sleep apnoea by frequent central events in patients with the obstructive sleep apnoea syndrome at altitude: a randomised trial. Thorax. 2010;65(5):429–35.
72. Fromm Jr RE, et al. CPAP machine performance and altitude. Chest. 1995;108(6):1577–80.
73. Serebrovskaya T, et al. Hypoxic ventilatory responses and gas exchange in patient with Parkinson's disease. Respiration. 1998;65(1):28–33.
74. Carroll JE, et al. Ventilatory response in myotonic dystrophy. Neurology. 1977;27(12):1125–8.
75. Sandham JD, et al. Acute supine respiratory failure due to bilateral diaphragmatic paralysis. Chest. 1977;72(1):96–8.
76. Lovering AT, et al. Excessive gas exchange impairment during exercise in a subject with a history of bronchopulmonary dysplasia and high altitude pulmonary edema. High Alt Med Biol. 2007; 8(1):62–7.
77. Yaron M, et al. Evaluation of diagnostic criteria and incidence of acute mountain sickness in preverbal children. Wilderness Environ Med. 2002;13(1):21–6.
78. Yaron M, Niermeyer S. Travel to high altitude with young children: an approach for clinicians. High Alt Med Biol. 2008;9(4):265–9.
79. Bartsch P, Gibbs JS. The effect of altitude on the heart and lungs. Circulation. 2007;116:2131–202.
80. Dehnert C, Bartsch P. Can patients with coronary heart disease go to high altitude? High Alt Med Biol. 2010;11(3):183–8.
81. Burchell HB, et al. The stress and the electrocardiogram in the induced hypoxemia test for coronary insufficiency. Am Heart J. 1948;36(3):373–89.
82. Alexander J. Coronary heart disease at altitude. Tex Heart Inst J. 1994;21:261–6.
83. Khanna PK, et al. Exercise in an hypoxic environment as a screening test for ischaemic heart disease. Aviat Space Environ Med. 1976;47(10):1114–7.
84. Levine BD, et al. Effect of high-altitude exposure in the elderly: the Tenth Mountain Division study. Circulation. 1997;96(4):1224–32.
85. Wyss CA, et al. Influence of altitude exposure on coronary flow reserve. Circulation. 2003;108(10):1202–7.
86. Schmid JP, et al. Safety and exercise tolerance of acute high altitude exposure (3454 m) among patients with coronary artery disease. Heart. 2006;92(7):921–5.
87. Lassvik C, Areskog NH. Angina pectoris during inhalation of cold air. Reactions to exercise. Br Heart J. 1980;43(6):661–7.
88. Burtscher M. Risk of cardiovascular events during mountain activities. Adv Exp Med Biol. 2007;618:1–11.
89. Burtscher M, et al. Sudden cardiac death during mountain hiking and downhill skiing. N Engl J Med. 1993;329(23):1738–9.
90. Palatini P, et al. Effects of low altitude exposure on 24-hour blood pressure and adrenergic activity. Am J Cardiol. 1989;64(19):1379–82.
91. Savonitto S, et al. Effects of acute exposure to altitude (3,460 m) on blood pressure response to dynamic and isometric exercise in men with systemic hypertension. Am J Cardiol. 1992;70(18):1493–7.
92. Wu TY, et al. Who should not go high: chronic disease and work at altitude during construction of the Qinghai-Tibet railroad. High Alt Med Biol. 2007;8(2):88–107.
93. Luks AM. Should travelers with hypertension adjust their medications when traveling to high altitude? High Alt Med Biol. 2009;10(1):11–5.
94. Woods DR, et al. High altitude arrhythmias. Cardiology. 2008;111(4):239–46.
95. Weilenmann D, et al. Influence of acute exposure to high altitude and hypoxemia on ventricular stimulation thresholds in pacemaker patients. Pacing Clin Electrophysiol. 2000;23(4 Pt 1):512–5.
96. Kobza R, et al. Leisure-time activities of patients with ICDs: findings of a survey with respect to sports activity, high altitude stays, and driving patterns. Pacing Clin Electrophysiol. 2008;31(7):845–9.
97. Erdmann J, et al. Effects of exposure to altitude on men with coronary artery disease and impaired left ventricular function. Am J Cardiol. 1998;81(3):266–70.
98. Agostoni P, et al. Effects of simulated altitude-induced hypoxia on exercise capacity in patients with chronic heart failure. Am J Med. 2000; 109(6):450–5.
99. Hackett P. High altitude medicine. In: Auerbach PS, editor. Wilderness medicine. 5th ed. Philadelphia, PA: Mosby Elsevier; 2007.
100. Agostoni P, et al. Carvedilol reduces exercise-induced hyperventilation: a benefit in normoxia and a problem with hypoxia. Eur J Heart Fail. 2006; 8(7):729–35.
101. Valentini M, et al. Effects of beta-blockade on exercise performance at high altitude: a randomized, placebo-controlled trial comparing the efficacy of nebivolol versus carvedilol in healthy subjects. Cardiovasc Ther. 2012;30(4):240–8. Epub 2011/09/03.
102. Allemann Y, et al. Patent foramen ovale and high-altitude pulmonary edema. JAMA. 2006;296(24): 2954–8.

103. Grunig E, et al. Stress Doppler echocardiography for identification of susceptibility to high altitude pulmonary edema. J Am Coll Cardiol. 2000; 35(4):980–7.
104. Dehnert C, et al. Identification of individuals susceptible to high-altitude pulmonary oedema at low altitude. Eur Respir J. 2005;25(3):545–51.
105. Garcia JA, et al. The role of the right ventricle during hypobaric hypoxic exercise: insights from patients after the Fontan operation. Med Sci Sports Exerc. 1999;31(2):269–76.
106. Broberg CS, et al. Adult patients with Eisenmenger syndrome report flying safely on commercial airlines. Heart. 2007;93(12):1599–603.
107. Perloff JK, et al. Risk of stroke in adults with cyanotic congenital heart disease. Circulation. 1993;87(6):1954–9.
108. Berman Jr W, et al. Systemic oxygen transport in patients with congenital heart disease. Circulation. 1987;75(2):360–8.
109. Shlim DR, Papenfus K. Pulmonary embolism presenting as high-altitude pulmonary edema. Wilderness Environ Med. 1995;6(2):220–4.
110. Boulos P, et al. Superior sagittal sinus thrombosis occurring at high altitude associated with protein C deficiency. Acta Haematol. 1999;102(2):104–6.
111. Anand AC, et al. Symptomatic portal system thrombosis in soldiers due to extended stay at extreme altitude. J Gastroenterol Hepatol. 2005;20(5):777–83.
112. Jha SK, et al. Stroke at high altitude: Indian experience. High Alt Med Biol. 2002;3(1):21–7.
113. Fagenholz PJ, et al. Arterial thrombosis at high altitude resulting in loss of limb. High Alt Med Biol. 2007;8(4):340–7.
114. Anand AC, et al. Thrombosis as a complication of extended stay at high altitude. Natl Med J India. 2001;14(4):197–201.
115. Khalil KF, Saeed W. Pulmonary embolism in soldiers serving at high altitude. J Coll Physicians Surg Pak. 2010;20(7):468–71. Epub 2010/07/21.
116. Sharma SC. Platelet count in temporary residents of high altitude. J Appl Physiol. 1980;49(6):1047–8.
117. Maher JT, et al. Human coagulation abnormalities during acute exposure to hypobaric hypoxia. J Appl Physiol. 1976;41(5 Pt 1):702–7.
118. Hudson JG, et al. The effect of high altitude on platelet counts, thrombopoietin and erythropoietin levels in young Bolivian airmen visiting the Andes. Int J Biometeorol. 1999;43(2):85–90. Epub 1999/11/07.
119. Doughty HA, Beardmore C. Bleeding time at altitude. J R Soc Med. 1994;87(6):317–9.
120. Bartsch P, et al. Coagulation and fibrinolysis in acute mountain sickness and beginning pulmonary edema. J Appl Physiol. 1989;66(5):2136–44.
121. Bartsch P, et al. Fibrinogenolysis in the absence of fibrin formation in severe hypobaric hypoxia. Aviat Space Environ Med. 1988;59(5):428–32.
122. Mannucci PM, et al. Short-term exposure to high altitude cause coagulation activation and inhibits fibrinolysis. Thromb Haemost. 2002;87(2):342–3.
123. Bendz B, et al. Association between acute hypobaric hypoxia and activation of coagulation in human beings. Lancet. 2000;356(9242):1657–8.
124. Bartsch P, et al. Hypobaric hypoxia. Lancet. 2001;357(9260):955–6.
125. Schreijer AJ, et al. Activation of coagulation system during air travel: a crossover study. Lancet. 2006;367(9513):832–8.
126. Basnyat B, et al. A language barrier, abdominal pain, and double vision. Lancet. 2001;357(9273):2022.
127. Ashraf H, et al. Pulmonary embolism at high altitude and hyperhomocysteinemia. J Coll Physicians Surg Pak. 2006;16(1):71–3.
128. Heffner JE, Sahn SA. High-altitude pulmonary infarction. Arch Intern Med. 1981;141(12):1721.
129. Green RL, et al. The sickle-cell and altitude. Br Med J. 1971;4(5787):593–5. Epub 1971/12/04.
130. Mahony BS, Githens JH. Sickling crises and altitude. Occurrence in the Colorado patient population. Clin Pediatr. 1979;18(7):431–8. Epub 1979/07/01.
131. Tiernan CJ. Splenic crisis at high altitude in 2 white men with sickle cell trait. Ann Emerg Med. 1999;33(2):230–3.
132. Franklin QJ, Compeggie M. Splenic syndrome in sickle cell trait: four case presentations and a review of the literature. Mil Med. 1999;164(3):230–3.
133. Lee MT, et al. Pulmonary hypertension in sickle cell disease. Clin Adv Hematol Oncol. 2007;5(8):645–53; 585.
134. Brookhart MA, et al. The effect of altitude on dosing and response to erythropoietin in ESRD. J Am Soc Nephrol. 2008;19(7):1389–95.
135. Hussein MM, et al. Low-dose recombinant human erythropoietin in dialysis patients living at high altitude. Nephrol Dial Transplant. 1992;7(2):173–4.
136. Spivak JL. Polycythemia vera: myths, mechanisms, and management. Blood. 2002;100(13):4272–90.
137. Torgano G, et al. Gastroduodenal lesions in polycythaemia vera: frequency and role of Helicobacter pylori. Br J Haematol. 2002;117(1):198–202.
138. Carneskog J, et al. Plasma erythropoietin by high-detectability immunoradiometric assay in untreated and treated patients with polycythaemia vera and essential thrombocythaemia. Eur J Haematol. 1998; 60(5):278–82.
139. Moore K, et al. Extreme altitude mountaineering and Type 1 diabetes; the Diabetes Federation of Ireland Kilimanjaro Expedition. Diabet Med. 2001;18(9): 749–55.
140. Kalson NS, et al. Climbers with diabetes do well on Mount Kilimanjaro. Diabet Med. 2007; 24(12):1496.
141. Pavan P, et al. Extreme altitude mountaineering and type 1 diabetes: the Cho Oyu alpinisti in Alta Quota expedition. Diabetes Care. 2003;26(11):3196–7.
142. Admetlla J, et al. Management of diabetes at high altitude. Br J Sports Med. 2001;35(4):282–3.
143. Pavan P, et al. Metabolic and cardiovascular parameters in type 1 diabetes at extreme altitude. Med Sci Sports Exerc. 2004;36(8):1283–9.

144. Gautier JF, et al. Influence of simulated altitude on the performance of five blood glucose meters. Diabetes Care. 1996;19(12):1430–3.
145. Giordano BP, et al. Performance of seven blood glucose testing systems at high altitude. Diabetes Educ. 1989;15(5):444–8.
146. Pecchio O, et al. Effects of exposure at an altitude of 3,000 m on performance of glucose meters. Diabetes Care. 2000;23(1):129–31.
147. Fink KS, et al. Effect of high altitude on blood glucose meter performance. Diabetes Technol Ther. 2002;4(5):627–35.
148. Oberg D, Ostenson CG. Performance of glucose dehydrogenase-and glucose oxidase-based blood glucose meters at high altitude and low temperature. Diabetes Care. 2005;28(5):1261. Epub 2005/04/28.
149. de Mol P, et al. Accuracy of handheld blood glucose meters at high altitude. PLoS One. 2010;5(11):e15485. Epub 2010/11/26.
150. King BR, et al. Changes in altitude cause unintended insulin delivery from insulin pumps: mechanisms and implications. Diabetes Care. 2011;34(9):1932–3. Epub 2011/08/06.
151. Leal C, et al. Diabetic retinopathy at high altitude. High Alt Med Biol. 2008;9(1):24–7.
152. Mader TH, Tabin G. Going to high altitude with preexisting ocular conditions. High Alt Med Biol. 2003;4(4):419–30.
153. Kohner EM. Aspirin for diabetic retinopathy. BMJ. 2003;327(7423):1060–1.
154. Chew EY, et al. Effects of aspirin on vitreous/preretinal hemorrhage in patients with diabetes mellitus. Early Treatment Diabetic Retinopathy Study report no. 20. Arch Ophthalmol. 1995;113(1):52–5.
155. Ri-Li G, et al. Obesity: associations with acute mountain sickness. Ann Intern Med. 2003; 139(4):253–7.
156. Alpert MA. Obesity cardiomyopathy: pathophysiology and evolution of the clinical syndrome. Am J Med Sci. 2001;321(4):225–36.
157. Wu TY, et al. High-altitude gastrointestinal bleeding: an observation in Qinghai-Tibetan railroad construction workers on Mountain Tanggula. World J Gastroenterol. 2007;13(5):774–80.
158. Saito A. The medical reports of the China-Japan-Nepal Friendship Expedition to Mt. Qomolungma/ Sagamartha (Everest). Jpn J Mount Med. 1989; 9:83–7.
159. Liu MF. Upper alimentary bleeding at high altitude. In: Lu YD, Li KX, Yin ZY, editors. High altitude medicine and physiology. Tianjing: Tianjing Science and Technology Press; 1995. p. 586.
160. Fruehauf H, et al. Unsedated transnasal esophagogastroduodenoscopy at 4559 M (14,957 ft)—endoscopic findings in healthy mountaineers after rapid ascent to high altitude. Gastroenterology. 2010;138 (5 Suppl 1):S483–4.
161. Hopkins WE, et al. Frequency and significance of intrapulmonary right-to-left shunting in end-stage hepatic disease. Am J Cardiol. 1992;70(4):516–9.
162. Benjaminov FS, et al. Portopulmonary hypertension in decompensated cirrhosis with refractory ascites. Gut. 2003;52(9):1355–62.
163. Bogaard HJ, et al. A 31-year-old man with hemoptysis at high altitude and abnormal hepatic biochemistry tests. Chest. 2007;132(3):1088–92.
164. Mairbaurl H, et al. Exercise performance of hemodialysis patients during short-term and prolonged exposure to altitude. Clin Nephrol. 1989;32(1):31–9.
165. Blumberg A, et al. Effect of altitude on erythropoiesis and oxygen affinity in anaemic patients on maintenance dialysis. Eur J Clin Invest. 1973; 3(2):93–7.
166. Mairbaurl H, et al. Increase in Hb-O2-affinity at moderate altitude (2000 m) in patients on maintenance hemodialysis. Clin Nephrol. 1989;31(4):198–203.
167. Quick J, et al. Stimulation of erythropoietin in renal insufficiency by hypobaric hypoxia. Nephrol Dial Transplant. 1992;7(10):1002–6.
168. Abassi Z, et al. Pulmonary hypertension in chronic dialysis patients with arteriovenous fistula: pathogenesis and therapeutic prospective. Curr Opin Nephrol Hypertens. 2006;15(4):353–60.
169. Luks AM, et al. Chronic kidney disease at high altitude. J Am Soc Nephrol. 2008;19(12):2262–71.
170. Mader TH, et al. Refractive changes during 72-hour exposure to high altitude after refractive surgery. Ophthalmology. 1996;103(8):1188–95.
171. Nelson ML, et al. Refractive changes caused by hypoxia after laser in situ keratomileusis surgery. Ophthalmology. 2001;108(3):542–4.
172. Boes DA, et al. Effect of high-altitude exposure on myopic laser in situ keratomileusis. J Cataract Refract Surg. 2001;27(12):1937–41.
173. White LJ, Mader TH. Refractive changes at high altitude after LASIK. Ophthalmology. 2000;107(12):2118.
174. Dimmig JW, Tabin G. The ascent of Mount Everest following laser in situ keratomileusis. J Refract Surg. 2003;19(1):48–51.
175. Brinchmann-Hansen O, Myhre K. Blood pressure, intraocular pressure, and retinal vessels after high altitude mountain exposure. Aviat Space Environ Med. 1989;60(10 Pt 1):970–6.
176. Bayer A, et al. Intraocular pressure measured at ground level and 10,000 feet. Aviat Space Environ Med. 2004;75(6):543–5.
177. Ersanli D, et al. Intraocular pressure at a simulated altitude of 9000 m with and without 100% oxygen. Aviat Space Environ Med. 2006;77(7):704–6.
178. Somner JE, et al. What happens to intraocular pressure at high altitude? Invest Ophthalmol Vis Sci. 2007;48(4):1622–6.
179. Baumgartner RW, et al. Going high with preexisting neurological conditions. High Alt Med Biol. 2007; 8(2):108–16.
180. Silber E, et al. Clinical features of headache at altitude: a prospective study. Neurology. 2003;60(7):1167–71.
181. Schoonman GG, et al. Normobaric hypoxia and nitroglycerin as trigger factors for migraine. Cephalalgia. 2006;26(7):816–9.

182. Richalet JP, et al. Physiological risk factors for severe high-altitude illness: a prospective cohort study. Am J Respir Crit Care Med. 2012;185(2):192–8. Epub 2011/11/11.
183. Murdoch DR. Focal neurological deficits and migraine at high altitude. J Neurol Neurosurg Psychiatry. 1995;58(5):637.
184. Luks AM, Swenson ER. Medication and dosage considerations in the prophylaxis and treatment of high-altitude illness. Chest. 2008;133(3):744–55.
185. Grissom CK, DeLoughery TG. Chronic diseases and wilderness activities. In: Auerbach PS, editor. Wilderness medicine. 5th ed. Philadelphia, PA: Mosby Elsevier; 2007.
186. Luks AM, et al. Can people with Raynaud's phenomenon travel to high altitude? Wilderness Environ Med. 2009;20(2):129–38.
187. Jean D, Moore LG. Travel to high altitude during pregnancy: frequently asked questions and recommendations for clinicians. High Alt Med Biol. 2012;13(2):73–81.
188. Kelly PT, et al. Air travel hypoxemia vs. the hypoxia inhalation test in passengers with COPD. Chest. 2008;133(4):920–6.
189. Roggla G, Moser B. The function of metered dose inhalers at moderate altitude. J Travel Med. 2006;13(4):248; author reply 248–9.
190. Van Patot MC, et al. Risk of impaired coagulation in warfarin patients ascending to altitude (>2400 m). High Alt Med Biol. 2006;7(1):39–46.
191. Messer J, et al. Association of adrenocorticosteroid therapy and peptic-ulcer disease. N Engl J Med. 1983;309(1):21–4.
192. Nielsen GL, et al. Risk of hospitalization resulting from upper gastrointestinal bleeding among patients taking corticosteroids: a register-based cohort study. Am J Med. 2001;111(7):541–5.
193. World Health Organization. Breastfeeding and maternal medication. Geneva: World Health Organization; 2002.
194. Product Information: Decadron®, dexamethasone. West Point, PA: Merck & Co.; 1997.

Index

E.R. Swenson and P. Bärtsch (eds.), *High Altitude: Human Adaptation to Hypoxia*,
DOI 10.1007/978-1-4614-8772-2, © Springer Science+Business Media New York 2014

MIX
Papier aus verantwortungsvollen Quellen
Paper from responsible sources
FSC® C105338

If you have any concerns about our products,
you can contact us on
ProductSafety@springernature.com

In case Publisher is established outside the EU,
the EU authorized representative is:
Springer Nature Customer Service Center GmbH
Europaplatz 3, 69115 Heidelberg, Germany

Printed by Libri Plureos GmbH
in Hamburg, Germany